Routledge Revivals

I0885470

A HISTORY OF THE ART OF WAR IN THE SIXTEENTH CENTURY

A HISTORY OF
THE ART OF WAR
IN THE SIXTEENTH CENTURY

by

SIR CHARLES OMAN

K.B.E., Hon. D.C.L. Oxon. and Hon. LL.D. Cambridge and Edinburgh, F.B.A.

FELLOW OF ALL SOULS, OXFORD, CHICHELE PROFESSOR OF MODERN HISTORY IN THE UNIVERSITY OF OXFORD

ROUTLEDGE REVIVALS

Routledge
Taylor & Francis Group

First published in 1937 by Methuen & Co. Ltd.

This edition first published in 2018 by Routledge
2 Park Square, Milton Park, Abingdon, Oxon, OX14 4RN
and by Routledge
711 Third Avenue, New York, NY 10017

Routledge is an imprint of the Taylor & Francis Group, an informa business

© 1937 Taylor & Francis

All rights reserved. No part of this book may be reprinted or reproduced or utilised in any form or by any electronic, mechanical, or other means, now known or hereafter invented, including photocopying and recording, or in any information storage or retrieval system, without permission in writing from the publishers.

Publisher's Note
The publisher has gone to great lengths to ensure the quality of this reprint but points out that some imperfections in the original copies may be apparent.

Disclaimer
The publisher has made every effort to trace copyright holders and welcomes correspondence from those they have been unable to contact.
A Library of Congress record exists under ISBN: 38001590

ISBN 13: 978-1-138-55764-2 (hbk)
ISBN 13: 978-1-138-56347-6 (pbk)
ISBN 13: 978-1-315-12171-0 (ebk)

A HISTORY OF
THE ART OF WAR
IN THE SIXTEENTH CENTURY

BY THE SAME AUTHOR

THINGS I HAVE SEEN

A HISTORY OF THE ART OF WAR IN THE MIDDLE AGES, A.D. 378–1485

ENGLAND BEFORE THE NORMAN CONQUEST

STUDIES IN THE NAPOLEONIC WARS

THE SIXTEENTH CENTURY

A HISTORY OF THE ART OF WAR IN THE SIXTEENTH CENTURY

by

SIR CHARLES OMAN

K.B.E., HON. D.C.L. OXON. AND HON. LL.D. CAMBRIDGE AND EDINBURGH, F.B.A.

FELLOW OF ALL SOULS, OXFORD, CHICHELE PROFESSOR OF MODERN HISTORY IN THE UNIVERSITY OF OXFORD

WITH 33 MAPS AND 12 PLATES

METHUEN & CO. LTD. LONDON

36 Essex Street, Strand, W.C.2

First published in 1937

PRINTED IN GREAT BRITAIN

PREFACE

SINCE the publication in 1924 of the second edition of my *Art of War in the Middle Ages*, I have been asked by many friends to continue the survey of military science through that most interesting period the sixteenth century. The subject has received very imperfect treatment by technical historians, who generally hurry on to the days of Gustavus Adolphus and Cromwell, Turenne and Marlborough—concerning whom there is much narrative and comment. But there is little or none on Gonsalvo de Cordova, Alexander Farnese, or Henry of Navarre. As long as my time was to a great extent occupied by Parliamentary duties, I was unable to find leisure to put together into a logical thesis the results of much varied reading in the authors of a period that extends from Fornovo to Nieuport, starting with Comines and Machiavelli and going down to La Noue and Francis Vere. In the beginning we are still in the Middle Ages—Flodden Field or Novara might almost have been fought in the fifteenth century. At the end a formal battle like Nieuport might almost have been fought in the Thirty Years War. This volume is the result of an attempt to sum up the fundamental alterations in the Art of War between 1494 and 1600, and is intended to serve as an outline of military theory and practice between those dates.

It does not, of course, purport to give a complete annalistic record of all campaigns, nor to detail all the changes in military architecture and armament which were characteristic of the century, though I have endeavoured to furnish some account of them wherever they influence the course of events. I have rather tried to explain the strategy, tactics, and organization of armies, and especially (if I may use the phrase) the military psychology of the period. The Great Wars of Italy, the French 'Wars of Religion,' and the War of Independence in the Netherlands all require to be studied from the point of view of contemporary thought. So Machiavelli's misguided forecasts, and La Noue's *Paradoxes* and *Commentaries* are of much more importance than drill books, treatises on artillery, or records of personal adventure. But I am far from underrating the interest of the mental outlook on the times of such personages as Bayard's *Loyal Serviteur*, or that self-centred old swashbuckler Blaise de Montluc. Indeed, one of the gaps in this survey is caused by the lack of any autobiographical record by English military men, till we have arrived at the very end of the Tudor period, when material begins to become abundant from Francis Vere and his less-remembered contemporaries. Patten's narrative of the Pinkie campaign is essentially that of a civilian though an eye-witness. For the earlier Tudor times we have to work from scattered letters and dispatches, and chaotic statistics in *State Papers Domestic*.

The battles, sieges, and campaigns which I have selected for detailed treatment are all typical examples of the military development of the period, each emphasizing some important phenomenon. For example, Fornovo illustrates the disastrous end of the old Italian theory of manœuvres unaccompanied by serious fighting. Ravenna is the first example of a battle won by a completely dominant artillery. The Garigliano and Pavia are 'victories by surprise,' where an incautious enemy is caught before he can get his army into proper array. Marignano and Pinkie are demonstrations of the fact that an old-fashioned infantry-army is helpless against the combination of all arms. Coutras settled the much-debated question as to the relative value of shallow or deep formation for cavalry. Arques is a fine example of the defence of a defile by very inferior numbers. The military moral of each of the other battles or sieges dealt with in detail is emphasized in the course of the narrative dealing with them.

I visited the sites of many of the battles and leaguers described in their volume, e.g. Fornovo, Ravenna, Flodden, Bicocca, Pavia, Boulogne, Calais, Vienna, Pinkie, St. Denis, and others. But in most cases three centuries have obliterated the old topography. The ground at Vienna, and St. Denis is built over by featureless suburbs. At Ravenna and Calais modern drainage has changed the whole outline. Flodden and Fornovo were about the only battlefields which seemed essentially unchanged. The old enceinte of Rhodes is intact, but I regret that I was not able to visit the island. In most instances I have been driven back on to more-or-less contemporary maps and plans—some of them a full century too late for certain correctness.

In my earlier book I could give no absolutely contemporary battle-pictures: this disability comes to an end in the middle of our period—from Pavia onward there are many contemporary representations of battles and sieges, some of which I have utilized. The Pinkie and Nieuport pictures were designed by eye-witnesses. The Huguenot battles of the great series by Perissot and Tortorelle are at least drawn from the narratives of combatants by artists who engraved them a very few years after the event. I must express my high satisfaction at the way in which Mr. Richard Cribb has dealt with my battle-plans, which have been made admirably clear. My thanks are due to Messrs. Longmans, Green & Co. Ltd. for their courtesy in allowing me to reproduce the German woodcut of the Siege of Rhodes from Major Porter's *Knights of Malta*. Bodley's Librarian and the Curator of the Ashmolean Museum have also been kind in allowing me to utilize the drawing by John Ramsay of Pinkie, and the panoramic picture of Pavia from their respective collections.

My gratitude, so often expressed with regard to former books, must once more be rendered to the kindly hand and eyes which revised my proofs.

C. OMAN

OXFORD, *April* 1937

CONTENTS

BOOK I

BOOK II

THE GREAT BATTLES OF THE ITALIAN WARS, AND THEIR TACTICAL MEANING

BOOK III

THE LATER YEARS OF THE GREAT WARS (1527–59)

BOOK IV

MILITARY HISTORY OF ENGLAND UNDER THE TUDORS

BOOK V

THE WARS OF RELIGION IN FRANCE (1562–98)

BOOK VI

THE REVOLT OF THE NETHERLANDS AND THE DUTCH WAR OF INDEPENDENCE (1568–1609)

BOOK VII

THE TURKISH ATTACK ON CHRISTENDOM (1520–1606)

PLATES

BOOK I

CHAPTER I

PRELIMINARY DATA

THE military history of the fifteenth century may be described, not inaccurately, as being shut up in many water-tight compartments—(1) The wars of the English and the French; (2) the wars of the Castilians with their neighbours of Granada, Portugal, and Aragon; (3) the wars of the Italian States with each other, all waged by proxy with hired bands of condottieri; (4) the wars of the Hussites with the Germans and Hungarians; (5) the wars of the Kings of Hungary with the Ottoman Sultans, or with their Austrian neighbours; (6) the wars of the Russians with their former masters, the Tartars of the Steppes; (7) the continual bickering of the Danes and Swedes; (8) the aggressive wars of the Swiss Confederates with the Hapsburgs, the Lords of Milan, and later with the Duke of Burgundy. These can all be treated as separate stories, having few and infrequent cross-relations with each other.

Very rarely the stories get tangled together by the appearance of a belated Crusade, such as the attempt to crush the indomitable Hussites by the joint attack of the Germans and the Hungarians—which was possible because Sigismund, the Holy Roman Emperor, was also king of Hungary, and titular king of the revolted Bohemian sectaries. A second case was the Crusade of Ladislas of Poland, which ended so unhappily at Varna in 1444. He was king of his own country, elective king of Hungary, and captain of the inadequate crusading army collected by the propaganda of the Council of Basle, and headed by Cardinal Julian Cesarini. Hence not only Poles and Magyars but a miscellaneous collection of Western contingents marched under the banner of the Cross. Varna saw the end of a real Crusade—the last of the long series—but later attempts at an international movement of this sort failed entirely, from inter-state jealousy. The abortive end of one killed Pope Pius II, who broke his heart sitting at Ancona waiting for contingents which never turned up (1464). The project of another inspired Charles VIII in 1494, but while he was really set on the reconquest of Constantinople—perhaps of Jerusalem—by a league of Christian states, his neighbours could only see that his first stage was the conquest of Naples, on a most flimsy claim, from another Christian king. Very naturally they fell upon him, with the Pope egging them on! The Crusade-theory lingered on into the sixteenth century, had a grip on the mind of Charles V—though he was hampered by too many other problems—and is set forth at length in the interesting 'Commentaries' of the old Huguenot captain La Noue. This admirable idealist dreamed of a Pan-Christian movement against Islam, and vainly conceived of a series of campaigns in which the Duke of Alva should lead one of two co-operating armies, and Henry of Navarre another![1] But the advocate of toleration and common Christian zeal against the encroaching Turk preached to a generation which was inspired by one or other of two incompatible ideals—the crushing of 'Heresy,' or, alternatively, the smashing up of the Papacy, and the establishment of national churches of the 'Reformed' sort.

But in the fifteenth century, if we put aside the international military aspirations which had the Crusade-theory at their base, there were very few other crises in which the wars of one of the eight categories quoted above in our first paragraph became connected with those of another. Alfonso of Aragon's occupation of Naples was not a Spanish attempt to conquer Italy, but a large specimen of condottiere warfare, practically Italian and not international. Charles the Rash was a more universally disturbing element, since, in the building up of his visionary 'middle-kingdom' of Burgundy, he contrived to clash with France, with the Empire, and with the Swiss. But he perished, leaving behind him little more than an unexpected advertisement for the military power of the Confederates, who (despite of Morgarten, Sempach, and Arbedo) had hitherto been a practically unknown factor in European politics.

It is barely needful to mention the claims of the Orleans dukes to get a footing in Italy, on their female descent from the Visconti lords of Milan, or the designs of Louis XI on the northern borders of the kings of Aragon—which led to no

[1] See the most interesting 21st and 22nd chapters of his *Quatre Paradoxes Militaires*, of which there will be much mention in later pages of this volume.

serious clash. The wars of the Middle Ages ended with something more ominous—the Flemish campaigns which had as their base the competition between Hapsburg and Valois for the Burgundian inheritance—typified by the rivalry for the hand of the heiress, Mary of Burgundy. But Maximilian of Hapsburg did not represent Germany, but only his own house, and his decisive battle of Guinegate (1479) was won with a purely Burgundian, *i.e.* Netherlandish, army. This war was not a strife of France and Germany, but a continuation of old contests running back to Roosebeke, Mons-en-Pevèle, and Courtray, which turned on the ambition of French kings to get control of what we should now call Belgium.

Everything is changed when we shift on from the fifteenth to the sixteenth century, and all the old local groups of war gradually grow into one single complex, in which Spain, Italy, France, Germany, the Netherlands, Hungary, and the Ottoman Empire are all involved together. Even England, cut off for sixty years from continental campaigns, joins in, intermittently and rather ineffectively, while at the other end of Europe the Sultan's main contest with Christendom on the Danube and the Mediterranean is slightly affected by the Russian expansion down the Volga and the Don, which impinged upon the Tartar vassals of the Porte, and once brought a Turkish army so far east as Astrachan, where it died of Russian frosts. About the only region of Europe in which the wars of the first half of the sixteenth century do not fit into the general scheme is Scandinavia—where the successful attempt of Sweden to cut itself out of the Union of Kalmar can hardly be brought into touch with other politics—though Charles V once lent his brother-in-law, Christian of Denmark, some landsknechts and some money.

The Pan-European complications which set nine-tenths of Europe ablaze started with the great Italian ventures of Charles VIII and Louis XII—the scheme of a French domination in Italy by the conquest of Naples and Milan. This brought in not only all the other Italian powers of any importance —the Sforzas of Milan, the Venetians, the Pope, the Florentines, the Dukes of Savoy and Ferrara, and the Marquises of Mantua and Montferrat—but Ferdinand of Aragon, interested for his kingdom of Sicily, the Emperor Maximilian, who was, after all, still the suzerain in theory of northern Italy, and not least the Swiss, who were at this moment fascinated by the idea of getting hold of any convenient scraps on their borders—whether Savoyard, Milanese, or German—which could be won by lending military aid, wherever it could be best put on the market.

It resulted that all the various forms of warlike efficiency which prevailed in any corner of Western Europe were drawn into the general strife—the gendarmerie of the *compagnies d'ordonnance* which was the pride of France—the trained mercenary bands of the condottieri who had fought the wars of Italy for the last two centuries—the Swiss phalanx of pikemen, which had achieved such startling success in the recent strife with Burgundy—the German landsknechts whom the Emperor Maximilian had raised and trained in strict imitation of the Swiss—and the light horse 'genitors' of Spain, who had learned their tactics in the long Moorish wars. We may add that the problem of the use of firearms, small and great, became a prominent feature quite early in the struggle. The Spaniards, Swiss, and Germans had already arquebusiers in their ranks, while the French and Italians were still sticking to the old crossbow. But, on the other hand, the French had, all through the commencement of the wars, a decided superiority in artillery; the Italians owned cannon, but were not yet skilled in their use; nor had the Swiss any proper provision of them; the Germans had progressed further in the theory than in the practice of the employment of field artillery, and the same was the case with the Spaniards, who had, it is true, brought forward guns for siege work in their Moorish wars, but were not accustomed to their skilful utilization in the field.

All these methods of warfare and of military equipment came into conflict before the fifteenth century was quite finished, starting from the invasion of Italy by Charles VIII in 1494. We may add that right down to the middle of the sixteenth century the archaic, but still not ineffective, English long-bow was occasionally seen in continental wars, when Henry VIII indulged in one or other of his misguided campaigns in northern France, in hope of playing a part of some importance in the world-conflict. It had no new triumphs, but was still considered by English soldiers to be decidedly superior for practical work to the new-fangled arquebus. It remained in practical use down to the days of Elizabeth, and even in the latest years of her reign there was a lively controversy between the admirers of the long-bow and those of the arquebus and musket—which is dealt with in its proper place.

During the first twenty years of the great Italian wars, Spain and the Holy Roman Empire were separate powers, whose policy was swayed by the individual interests of the Emperor Maximilian and of Ferdinand of Aragon—for it was Ferdinand who was primarily interested in Italy, though as long as she lived, his consort Isabella allowed herself to be guided by him in this sphere of politics. Spain and the Empire were by no means always in co-operation. And for several years there was no prospect of their being united under one dynasty, for Ferdinand and Isabella had a son named John, the natural heir of both Castile and Aragon, who attained marriageable years. It was not till his unexpected and much-lamented death in 1497, and that of his eldest sister in 1498, that the prospect arose of the possible dynastic union of Spain and the Hapsburg dominions, owing to the marriage of Juana, the second daughter of Ferdinand and Isabella, to the short-lived Philip the Handsome, son of Maximilian. Their son Charles, born in 1500, became natural heir to both his grandfathers. Ferdinand hated the idea. The moment that Isabella died (1504) he deliberately married a second time, in the hope of getting issue on whom the kingdoms of Aragon and Sicily (if not that of Castile) would devolve (1505), as well as all his conquests in Italy.

Only the failure of his young queen, Germaine of Foix, to bear him a son and successor, brought about, at his decease in 1516, the union of the two great houses. The election of Charles as Emperor in 1519, in succession to his grandfather Maximilian, placed him in a position which no prince since Charlemagne had occupied, since he was not only the sole owner of the heritages of Hapsburg, Burgundy, Castile, and Aragon, and the possessor of half Italy, but also the titular head of the great hierarchy of the Holy Roman Empire. This not only gave him the precedence over all other rulers in Europe, but a chance of making imperial power in Germany a reality, after two centuries of impotence. For no emperor before had ever possessed such resources, territorial, financial, and military. Nothing looked more likely than the establishment of a Hapsburg domination over all Central and Southern Europe. This was made all the more possible by the fact that Charles possessed a share of the diplomatic ability of his grandfather Ferdinand, no small portion of the magnanimous and adventurous temperament of his other grandfather, the Emperor Maximilian, a strong sense of duty, and an untiring power of work. Though not a genius, he was a prince of far more than average capacity, and his character was not stained by any of the outstanding vices which have sometimes ruined the careers of great monarchs.

This being so, it may appear extraordinary that Charles V did not rise to what seemed to be his obvious destiny, and after many wars and many victories, retired in his old age to a monastery, as a broken old man who had failed to achieve his purpose.

The reasons for the frustration of his career were three: one psychological, and two military. With the first we have nothing to do, as we are concerned with the art of war, not with the general history of culture and religion in Europe. This was, of course, the outbreak of the Protestant Reformation —a religious movement far more serious in its results than any preceding phenomenon of a more or less similar character—such as the wave of indignation that followed on the Great Schism, or the unrest that resulted in certain regions from the teaching of Wyclif in England or John Huss in Bohemia. The disruption of Germany that started with the rise of the Lutheran protest was undoubtedly one of the main causes of the failure of Charles V in his life's work.

But there were two military causes to be considered. The first was the irruption of the Ottoman Turks into Central Europe: they had been checked on the Danube for a hundred years, and it was a wholly new danger that threatened Christendom when they overran Hungary (1526), thundered at the gates of Vienna (1529), and gave the Emperor a suddenly developed Eastern war-front, which distracted his attention, and drew his armies away from France and Italy. Nor was it only on the Danube that the new peril was obvious. A few years before the disaster of Mohács had ruined Hungary and laid bare not merely the Hapsburg dominions but the whole of Germany, another danger had already begun to develop. It was not only on land that the Turkish Empire, under Soliman the Magnificent (1520–66), became in the time of Charles V a public terror to all Europe. When the pirate chiefs who had conquered Algiers and Tunis from decadent dynasties (1516–17) offered their homage to the Sultan, and placed their fleets at his disposal, the western Mediterranean ceased to be the undisturbed *mare nostrum* of the Italians, where for five hundred years no non-Christian flag had been seen. A new and serious naval peril grew up for Italy and southern Spain, and the Emperor had twice, during his short intervals of peace on land, to direct his efforts against the pirate-scourge which was distracting his attention. His two great African campaigns against Tunis (1535) and Algiers (1541) were the one a transient success, the other a dreadful disaster, which destroyed an army which would have been invaluable in Italy a year later. It was the misfortune of Charles V that he had to deal with the most capable and ambitious of all the Sultans who ever reigned at Constantinople, a monarch as steadfast and obstinate as himself, and even more vigorous, for he died in his tent, still campaigning, at seventy-three, while Charles had retired, worn out, to his monastery at fifty-six.

If he had been engaged with the Turk alone, Charles might probably have restored the bounds of Christendom in the East, and swept the Mediterranean free from Barbary pirates. But the Turk was only his secondary enemy, against whom he marched or sailed in moments of special crisis or opportunity. His real foe was the King of France, still set beyond all reason on continuing the vain venture which had been commenced by Charles VIII, the conquest of Italy. It passes all reason to discover why this wild dream, after it had led to a dozen lost campaigns, and drained the best blood of France in a dozen disastrous battles, was still pursued with such obstinacy—not only by Francis I, whose military career had started with his one great victory of Marignano, but by Henry II, who never had any such an intoxicating glimpse of personal glory. It was not only the kings who were besotted, but the whole nation. As La Noue complained, the young men of his early days read too much of Amadis de Gaul, and their elders too much of Machiavelli. He doubted which was the more dangerous literary pabulum. For romances of chivalry, in which glory, wealth, and love are gained by mere knight-errantry, were demoralizing the young nobles and squires, who had lands and vassals at home, whose welfare they neglected while running abroad in search of 'adventure.' And also the study of Italian statecraft removed all traces of morality from the minds of princes and ministers, and taught them to laugh at the binding force of oaths, or the common Christian virtues of justice and mercy. Most of the romances are full of *amours deshonnêtes* and of exploits plainly impossible by any mortal man. What is the credit of winning a combat if you have magic arms and armour, and are backed by a fairy or a magician ?[1] So the romances are fatal to the conceptions of duty and common sense, just as a study of Machiavelli saps all sense of righteousness in dealing with one's subjects or one's neighbours.

That there is a good deal of truth in La Noue's analysis of the mental condition of the ruling classes in France in the first half of the sixteenth century is obvious. Otherwise they would have refused to back their kings in their wild Italian schemes. This never was the case—an appeal to the charms of 'glory' and adventure seldom failed in its effect. Criticism, if heard, was only on the lines that profit should be got rather by a solid attempt to win Flanders and Brabant, than by distant trans-alpine campaigns in Naples or Venetia. Chivalrous personal ambition combined with national pride to make the King's summons always effectual in raising a new army, even after the most disastrous failures. It is this that forms the most surprising feature in the mentality of the epoch ; it seemed impossible to teach the French that the dream of domination in Italy was a hopeless delusion ; and the strife was renewed again and again, the moment that Charles V was distracted by some urgent problem in Germany, Hungary, or on the High Seas.

The most immoral policy of all was alliance with the Turk, the common enemy of Christendom, which both Francis I and Henry II adopted without shame. There was some protest against it even among the French nobility : but the feeling of the man in the street was undoubtedly expressed by the old swashbuckler Montluc when he wrote :[2] 'Christian princes on the other side made much ado about our master calling in the Turk to his aid. But against one's enemy one may "make an arrow of any sort of wood." For my own part, I would call up all the devils of hell to help me break my enemy's head, when he is trying to break mine. I would do it cheerfully, and then may God pardon me.' This outburst refers to the time when Barbarossa's corsairs were driving off thousands of Christian men and women, to be sold in the slave market of Constantinople, while their French allies looked on, content to

[1] See the most interesting '*septième discours*' in vol. i. p. 189–289 of the *Discours Politiques*.

[2] This is of the time when Barbarossa's Algerine fleet was sacking Nice—with French help. *Montluc* ii. p. 137.

have handed over to them the depopulated walls (1543).[1] One does not envy the task of the verbose French bishop who had to explain to the Venetian Senate that his master's policy was all for the good of Christendom, since Charles V was a much more sinister figure than the Sultan. Was he not the oppressor of Popes, and the sworn ally of that pestilent schismatic and heretic Henry Tudor, King of England? The Venetians, to whom Turkish naval predominance would mean the inevitable loss of Cyprus, Crete, and Corfu, were not favourably impressed by these arguments.

As it turned out, the repeated attacks of the French kings on the Emperor, at any moment when he was engaged with the Turk, or worried by princely rebellions in Germany, just sufficed to keep him from making a success of his reign. He had to have armies everywhere—on the Pyrenees, in Italy, in Flanders, on the Meuse and Rhine, and in his precarious African garrisons. And money to pay them was not always forthcoming—for the treasures of Mexico and Peru, on which his son could depend, were only just beginning to drift across the Atlantic to his treasury. When his mercenary troops were not paid, they indulged in mutinies—especially the Landsknechts—and fell upon the civil population, or deserted. The princes of the Empire grudged the money and contingents which they ought to have provided under the rule of the *Matricula*. Some of them, like Charles of Guelders and Robert of Bouillon, openly sided with the French in the earlier years of the reign, and let the enemy into frontier fortresses. Later on, the Protestants of the Schmalkaldic League were far more dangerous, and took most of northern Germany into a rebellion that came most inopportunely (1547), when a new king, more unscrupulous and quite as obstinate as his father, Francis I, had just mounted the throne of France.[2] Though victorious for a moment over

[1] The Bishop of Valence made a long exculpatory speech to the Venetian Senate, which may be found in the Appendix to vol. i. of Montluc's *Memoires*. It is a shameless production, trying to prove that the Emperor was as guilty as Francis, since he had offered to pay the Sultan a tribute for the restoration of the kingdom of Hungary. France could accept without scandal the naval help sent by the Grand Signor, because the checking of Charles V was the thing most necessary to Christendom. Had not the Emperor enlisted the aid of Henry of England, a notorious heretic and schismatic? And had not France, from the time of Charlemagne to that of St. Louis, done so much in the way of crusading for Christendom that now she could be trusted not to betray it? (pp. 414–25).

[2] The battle of Mühlberg was won by the Emperor in April 1547, a few days after the accession of Henry II on March 31.

the German rebels at Mühlberg, and vainly in hopes that he had settled the religious troubles of Germany by the *Interim*, Charles was forced to face the last and most unhappy of all his wars, attacked at once by Henry II, by Sultan Soliman, and by the German rebellion that flared up again under the leadership of the unscrupulous Maurice of Saxony. After several years of indecisive and sometimes disastrous campaigning—the French had taken Metz, and had once more got a lodgment in central Italy at Siena, the Turk was threatening Naples, while he had failed to get his son Philip elected as Emperor in his stead—the favourite scheme of his old age—Charles abdicated, being now no more than a gout-ridden invalid, and retired to the remote Estremaduran Monastery of Juste. The legend that he spent his last months in the vain endeavour to make a number of clocks keep exactly the same time, is probably an allegory, hinting that he had so many complicated pieces of political machinery in so many regions under his charge at the same time, that the happy moment when all should be right simultaneously never arrived.

So from the invasion of Italy by Charles VIII of France in 1494, down to the peace of Cateau-Cambrésis (April 3, 1559), there elapsed sixty-six years of almost continuous warfare between France and the Hapsburg dynasty, with little profit to either. Of all the Italian lands which the French kings had held for shorter or longer terms, there remained to Henry II in 1559 only the Alpine marquisate of Saluzzo. On the other hand, France made some small but valuable acquisitions on her northern frontier—Elizabeth of England resigned herself for the moment to the loss of Calais, which her sister Mary had let slip by careless mismanagement.[1] The Holy Roman Empire lost the three fortress-bishoprics in Lorraine—Metz, Toul, and Verdun. All other conquests on other frontiers were mutually restored—which meant that the Spanish crown kept Milan and Naples, so often contested, and the strong places on the Belgian frontier, which the French kings had so often coveted. The Duke of Savoy returned to Turin, and recovered those parts of his dominion which the French had been holding in 1559. The claim of the d'Albrets on the kingdom of Navarre dropped through—save for the small patch of it on the north side of the Pyrenees, which gave to the gallant Henry IV a

[1] But for years she tried to get it back by plots with the Huguenots, or threats to the French court.

royal title and a single mountain city as his sole inheritance from the old kings.

On the whole, then, the Hapsburgs had been the winners in the long game. But as they split into two branches—Philip II holding Spain and the Indies, the old Burgundian lands in the Netherlands, and the dominating position in Italy, while his uncle Ferdinand got the Imperial title and the old German lands of his ancestors—there was no individual of the house who held the position or the power of Charles V. Nevertheless the resources left to Philip II were tremendous, and but for his unhappy personality he might have had almost as good a chance of controlling all Europe as his father had enjoyed. For France fell into the civil wars of Huguenot and Catholic, and the Ottoman Empire passed from the hands of the great Soliman into those of a sot—who was followed on the divan by a miser. Philip's procrastination and indecision, his distrust of his best servants and his nearest kin, his leaden hand and his faithless pen, his secret treacheries and his open persecutions, saved Europe from falling under the black tyranny of a fanatic.

But his armies fought many a good fight, as we shall show in a later chapter, and their organization and efficiency were the admiration of all his contemporaries. He himself was neither a general nor an organizer, and the strength which he wielded was due to his father, and to the line of great generals bred in the Italian wars. He himself, shut up in the Escurial—an ascetic Tiberius in a very bleak Capreae—contributed nothing to the wars but orders difficult, and sometimes impossible, to execute, received with dismay by the reluctant commanders of his formidable hosts.

CHAPTER II

MILITARY GEOGRAPHY IN THE GREAT WARS (1494–1559)

THE most striking feature in the general aspect of the great struggle between the Hapsburgs and the Valois, which went on with certain short breaks for a period of sixty years, is that most of the more important and interesting campaigns of its earlier time were fought in the Italian peninsula, though the rivals had long continuous frontiers on the Pyrenees, in the Belgian Netherlands, and along the borders of western Germany. On the Spanish-French frontier there was much bickering, but no great invasions on either side. The Spaniards never got farther than Bayonne—which they once besieged but failed to take (1523). The French had for one short period possession of Navarre, into which they had come ostensibly to restore an expelled Francophil king, but they were driven out after a few months. In all the wars we find tedious operations around Fonterabia and Perpignan, which were obviously regarded by both sides as side-shows of no primary importance. The Pyrenees are a formidable obstacle, with no more than three difficult points of passage at their western end,[1] and two at their eastern end on the Mediterranean,[2] Their central length was impracticable for a hundred and fifty miles. It is easier to understand the hesitation of the French to commit themselves to an invasion of Spain on a large scale —the occupation of Navarre in 1521 was made for local political reasons and with a moderate force—than to comprehend the reasons for which Ferdinand of Aragon and his grandson Charles the Emperor never made a serious attempt to overrun the fertile lands of Aquitaine. Ferdinand once lured an English auxiliary army to Fonterabia, and spoke of a great enterprise to be carried out by its aid (1512), but made no use of it—it

[1] The coast road by Irun, and the passes of Mayya and Roncesvalles.

[2] The coast road Perpignan-Barcelona and the inland pass through Cerdagne on to Puycerda.

merely attracted the attention of the French while he was conquering Navarre, and never crossed the Bidassoa. Charles did once dispatch an army across that river, which forced its way as far as Bayonne, and was blocked there by that not very distinguished general, Lautrec. But this was obviously a diversion, as is shown by the moderate strength of the army of invasion—the Emperor's thoughts were all in Italy.

In the later years of the struggle there were no operations of any importance along the Pyrenean front, save the fruitless siege of Perpignan by the Dauphin Francis in 1543. It remains a source of wonder that, during the not infrequent periods of the long contest when the Emperor was in complete possession of the disputed lands in Italy, he made no attempt to break into south-western France, and apply pressure from that side. Provence was twice invaded (1523 and 1536), Guienne and Languedoc never, though there were bases both in Biscay and in Roussillon from which an attack on a large scale could easily have been planned. Spanish soldiers by the ten thousand were poured into Italy, but in Spain itself there was never more than a moderate covering force provided, along the two practicable fronts. This remains a subject of wonder. The entry into Languedoc or Gascony is far easier than the entry into Provence.

The only explanation that can be provided for part of this puzzle is that, at any rate in the early years of the wars, an irruption into Guienne would have brought England into the problem. Henry VIII had never forgotten the South-French lands which Henry VI had lost, and the army which he sent to Fonterabia in 1512 was intended to occupy Bayonne and Bordeaux. The treaty which he made with King Ferdinand stipulated that any conquests made in Aquitaine should be held by the English. But Ferdinand had no desire to see the old Plantagenet holding in English rather than in French hands; hence his deliberate wrecking of the plan. The danger was not quite so imaginary as might be thought—there were discontents in Guienne: the mildness of the old English administration had left pleasant memories behind it, and malcontents at Bordeaux, even as late as 1548, were heard to murmur the name of England at the time of the great 'Gabelle Rising.' Still it must be conceded that the possibility of a re-establishment of the old duchy of Aquitaine was a very chimerical idea, and can hardly have had any great effect on the general schemes of policy of the Emperor Charles—he pretended to consent to it in 1523, but his promised attack on the south for Henry's benefit never came off. It would not have profited him.

Much more obvious than the notion of invading France from across the Pyrenees was that of bringing pressure to bear against her northern frontiers on the side of Picardy and Champagne, and here there was constant and sometimes rather important fighting in progress, though never till the later years of the struggle did events in the north have the same decisive effect as events in Italy. Considering the shortness of the distance between Paris and the frontier of the Hapsburg possessions in the Netherlands, which then ran so much farther south than in later centuries, it is surprising that the main contest was not fought out on this front. Montmédy, Maubeuge, Valenciennes, Cambray, Landrécies, Douai, were imperialist fortresses, and the French line of defence lay from Mezières by Guise, St. Quentin, and Peronne, to Abbeville, places all far inside the boundary of modern France. North of them again were Montreuil and Boulogne, dangerous advanced positions. There were no Alps or Pyrenees along this border, where the whole countryside was good fighting-ground. When France was in the ascendant, why was not Flanders overrun, and when the Hapsburg eagle was triumphant, why was not the short push of a hundred miles to Paris ever carried through? Charles in his farthest advance once got to Reims and Soissons—his English allies on one occasion made Montdidier, well south of Amiens—their stopping-point (1522), though on another occasion (1544) they got no farther than Montreuil. From Soissons or Montdidier the distance to Paris was about fifty miles—three days' march—but the last push was never made. In each case the excuse which contemporary historians proffer is that the invading army was short of supplies, and had in front of it considerable forces gathered for the defence of the French capital. Charles contented himself with dictating a moderately advantageous peace at Crépy in 1544. The Duke of Suffolk at Montdidier in 1523 pleaded the non-appearance of the imperialist reinforcements that had been promised him, early frosts, and the fact that La Tremouille's army in front of him was intact, when he made an ignominous retreat without a battle. Philibert of Savoy in 1557 was the only imperialist commander who resolved to make a desperate rush on Paris, after his victory at St. Quentin : he was absolutely prohibited from doing so by his master, Philip II, and the French had time to rally.

It may be observed that, from the other point of view, when the French cause was in the ascendant, it might have been expected that a serious invasion would have been launched against the Belgian Netherlands, where France had old ambitions going back for centuries, and where Flanders and Artois had been in theory fiefs of the French crown. Louis XI had made a vigorous attempt to rake in those regions, and Louis XIV in the seventeenth century was destined to carry out in a great measure his predecessor's designs. But all the four kings who reigned between 1494 and 1559 treated the Netherlands as a secondary theatre of war, when they had their hands comparatively free, and poured their main armies into Italy rather than into the lands of the Scheldt and Sambre. Yet Flanders, Artois, and Namur were prizes quite as attractive as Milan, and much more easily to be seized. If in the later wars there were large concentrations of force on the northern front, it was because the Emperor was at last putting in pressure from that direction, and it was necessary to hold him back. It was with the greatest reluctance that Francis I in 1543 and 1544, and Henry II in 1557, called back reinforcements from the army of Italy, because the armies of the Hapsburgs were striking deep into Champagne or Picardy.

Only at long-spaced intervals during the larger part of the struggle did the war develop any importance on the north-eastern, as opposed to the north-western, frontier of France. The Dukes of Guelders and Bouillon, those persistent rebels against the Holy Roman Empire, were never able to let their French patron into the Rhineland—at most they gave trouble on the Meuse. Bayard's famous defence of Mezières in 1521 was not against an enemy of any overpowering strength, and the German force which entered Champagne in 1523, and was defeated by Claude of Guise, was rated at the modest total of 12,000 men of all arms. With such numbers nothing of importance could be accomplished; the days of small armies were over.

In the year 1543 it looked for a moment as if the war might spread into Germany, when the Duke of Cleves rose in rebellion and asked for aid from France, for Francis sent a considerable force under his son, the Duke of Orleans, to attack the duchy of Luxemburg, and to open up communication with the rebel.

But though the bands of Cleves raided as far as the gates of Antwerp, and though Orleans conquered all Luxemburg save the single fortress of Thionville, they never succeeded in getting into communication. And the only result of the campaign was to draw the Emperor with his main army into western Germany, where he crushed the rebel duke, and came down through the Netherlands for the first serious attack on the northern frontier of France that had been seen since Suffolk hurried back from Montdidier twenty years before. To face the Emperor was urgent; the French were driven out of Luxemburg with ease, and their King, abandoning all other enterprises, came up to defend Picardy and Champagne.

One of the military puzzles of the time is to discover why there was no decisive battle about Landrécies in September 1543—both Charles and Francis were in full force, with no great advantage of position or number on either side. There were continuous deployments and manœuvres, and plenty of opportunities for a general action—but Francis, after having thrown supplies into the beleagued Landrécies, drew back, and Charles did not resume the siege but retired to St. Omer—pleading the approach of winter. Clearly it would have been to his advantage to risk a battle, since he not only had the better army, but also the more to gain by victory. Francis had the last levy of his realm with him, and Paris close at his back; his defeat would have been fatal. Charles could have endured misfortune with much less ruinous results. But he was no Napoleon—as is sufficiently shown again by the inadequate result of his advance into Champagne in the following year.

An even greater reluctance to push things to a decisive result, at the risk of a possible but improbable check, was shown by Philip II after the victory of St. Quentin, when he laid a positive prohibition on Philibert of Savoy's proposal to march straight on Paris, disregarding all fortresses still untaken, while the enemy's field army was in a state of rout and dispersion. But, as we shall see in a later chapter, a reluctance to put the fate of war to the arbitrament of battle, to push an advantage to extremes, was a common fault of kings and generals during all the final stages of the great struggle of Hapsburg and Valois. After half a century strategy had drifted back into the same state in which it is found in Italy during the age of the old condottieri—much manœuvring with few general actions was again the order of the day, as in the times of Colleone and Carmagnola.

By far the most interesting campaigns from first to last are these fought on the Italian front, where one would have supposed that the formidable barrier of the Alps would have made invasion far more difficult, whether from west to east, or from east to west, than in the plains of Picardy and Champagne, or of Flanders and Hainault. But as a matter of fact, there was never during the whole series of wars a competent attempt made by either side to defend the entire line of the Alps from the Swiss border to the sea. The nearest approach to such a scheme was when the Italo-Swiss confederates in 1515 made disposition to cover the passes, with a central reserve at Susa in Piedmont, ready to reinforce the point on which the enemy might make his thrust. But the enemy crossed the mountains at unguarded gaps, and descended on the rear of the defenders, while their general asked plaintively whether the French were birds, to fly over such obstacles.

The real explanation of the ease with which the French launched successive expeditions into Lombardy is that during the first half of the wars they had the rulers of Savoy as their allies, and when 'the porter of the Alps' opened the doors, passage was simple, if occasionally laborious from mere physical hindrances. The houses of France and Savoy were linked by marriage, and the Duke Philip II, the regent-Duchess Blanche, Philibert II, and Charles III all through his early years, were successively committed to the French cause, and always gave an invading army a welcome at Turin. It was not till 1536 that this long subjection came to an end. But when Charles III, now a middle-aged man, broke with his uncle Francis I, and refused him a passage over his lands, a French army took him unawares, and drove him from well-nigh all his heritage before the Emperor could succour him. From 1536 to the end of the wars in 1559, the French never lost control of Turin and all the central passes, so that there could not be, and never was, any possibility for the ruler of Lombardy to prevent an invading army from debouching into Piedmont. When the Constable Bourbon in 1523, and the Emperor Charles in 1536, launched retaliatory invasions westward into Provence, they had to use the southern passes, because the northern ones were still in hostile hands.

It may be added that the minor princes of the Western Alps were generally on the side of the Valois kings. Ludovico, Marquis of Saluzzo, was for many years a commander in the armies of Charles VIII and Louis XII, and the important fortress of Casale, the stronghold of the Marquises of Montferrat, was often a French base of operations. Gianludovico of Saluzzo broke with France in 1536, but the greater part of his Marquisate and its useful pass remained in French hands till the end of the wars, when it was left to Henry II by the Treaty of Cateau-Cambrésis—the only relic of all the transalpine conquests which four successive kings of France had won in fifty years of fighting.

The French invasions of Italy, therefore, were seldom complicated by any necessity for forcing any of the possible debouches into the plains of the Po. But when an army of invasion had descended into Lombardy the problems began. They differed greatly at various periods of the war. At the first start in 1494, Charles VIII had as his objective the kingdom of Naples, at the other end of the Italian peninsula. He was able to reach its capital, in February 1495, because he not only enjoyed the friendship of Savoy, but had, at that moment, the alliance of the Duke of Milan, across whose front he was about to march, and could scare the Florentines and the Pope into granting him free passage. His enemy Ferdinand II of Naples fled to Sicily, after seeing his armies retreat and disintegrate, while his fleet had proved unable to dispute the command of the Tyrrhenian Sea. After no fighting worthy of mention, Charles VIII found himself master of the whole kingdom of Naples, save a few outlying fortresses of the south. The house of Alfonso the Magnanimous had been profoundly unpopular among its subjects: the military strength of his descendants had been wholly dependent on hired bands of mercenaries, and the native nobility were mostly hostile. Their greatest families had belonged to the old 'Angevin' faction, and welcomed the arrival of its representative—even if his title was not quite clear.[1] Charles had himself crowned in great state, but soon began to rouse the jealousy of his local supporters by lavish grants of land and offices to his French followers, and to win unpopularity by the plundering habits of his mercenary army. Never-

[1] This claim to the Angevin inheritance rested on a cession of his rights to Naples by Charles of Anjou, Count of Maine, to Louis XI. This was to the detriment of his cousin Réné of Lorraine, son of the daughter of 'Good King Réné,' the last king of his line. But Naples was a kingdom where no 'Salic Law' prevailed, indeed the record of its two Queens regnant, the two Joannas, is sufficiently notorious. Charles of Maine could not cede the rights of his uncle's descendants in the female line.

theless he was in military possession of practically the whole kingdom of Naples, when the storms burst upon him which clear-sighted politicians had seen to be inevitable.[1]

All the Italian powers of the greater sort, Milan, Venice, and the Pope, had been terrified by the sudden and unexpected collapse of the King of Naples, and had banded themselves together to resist the danger of a complete French domination in Italy. They had drawn into their league both the Emperor Maximilian and the two sovereigns of Spain. The treaty of alliance was signed at Venice on March 31, 1495, only a month after Charles had made his triumphant entry into Naples.

The strategical position then produced was that the French king was in possession of Naples, but was completely cut off from his base. Between him and the Alps there was nothing friendly save the faction-ridden city of Florence, and a strong French garrison at Asti, in Piedmont, under the Duke of Orleans. The situation would have been much less dangerous if Charles had been in secure command of the sea. But though he had a considerable fleet in hand, it could not hope to compete with the united naval forces of Venice and Spain. Communication between Naples and Marseilles was intermittent, and often intercepted, all through these wars.

The problem thus developed was not to be settled in one campaign, but its solution was inevitable from the first. No foreign power can hold Naples, when central Italy is not in its hands, unless it has complete command of the sea. Subsequent history showed that a foreign power, even when it was beginning to show signs of decay, might retain its grip under that one essential condition. For a century and a half the heirs of Charles V, three Philips, and Charles II, ruled the Neapolitan kingdom as a viceroyalty harshly held down by alien governors. But Spain was mistress of the western Mediterranean in the sixteenth and seventeenth centuries. France never had such a predominance in the period with which we have to deal. It was therefore a wild endeavour on the part of Charles VIII, Louis XII, and Francis I to hold on to the old and doubtful Angevin claim. Lautrec's invasion of Naples in 1528, ruined by the intervention of Andrea Doria's Genoese fleet, proves the case.[2] If the sea were not always secure, Naples could only

[1] As did Commines, vii. cap. 19.

[2] When Doria drove off the French fleet and occupied Gaeta, the fate of Lautrec's army was sealed.

be held by a power dominant also at Rome, Florence, Milan, and Turin—if not also at Venice. This France would never be: the game was a hopeless one.

But there was a secondary French claim on an important part of Italy. If it were insane from a military point of view to persist in attempts to hold Naples, the design on Milan, which came to the point a little later, does not appear quite so reckless. For the duchy of Milan is not, like Naples, four hundred miles distant from the French border. The Orleans claim to the duchy came from the descent of the family from Valentina Visconti, the wife of that Louis, first Duke of Orleans, who had been murdered by John of Burgundy nearly a hundred years back. The last of the Visconti lords of Milan had been slain in a republican outbreak in 1447, and after a three-year interval the Sforzas had made themselves masters of the duchy. Ludovico Sforza, 'il Moro,' was the fourth of his house to hold possession of it, and his position had been fully legalized by the Emperor, for Milan was an Imperial fief. To raise claims founded on a remote connection with the Visconti 'tyrants' was as out of date as it was absurd. But successive heads of the house of Orleans had advertised their pretensions, and endeavoured to stir up trouble. This did not much matter [1] while the elder line of Valois reigned in France. But when, on the death of Charles VIII, last of that race, his distant cousin, Louis of Orleans, came to the French throne, the claim was urged by a much more formidable pretender, who declared his intentions by assuming the title of Duke of Milan immediately after his accession. In the following year he invaded the duchy and conquered it with ease, when the unhappy Ludovico Sforza had been betrayed by his Swiss mercenaries. He was handed over to the French King, who kept him a close prisoner at Loches till he died in 1510.

For thirteen years [1500–12] Louis of Orleans was in possession of the Milanese duchy, but this was used by him only as a step towards the reconquest of Naples, and the gaining of that general predominance in Italy on which his predecessor had been set. By the infamous Treaty of Granada (1500) King Ferdinand of Aragon betrayed the cause of his cousins of Naples, who had been restored by Spanish arms to their lost

[1] Though one of the reasons for Ludovico Sforza's desertion of the French cause in 1495 was his suspicion of Louis of Orleans, who lay at Asti with a large French force.

throne after the expulsion of the armies of Charles VIII in 1496, and agreed to partition their kingdom with Louis XII. King Frederic, last of the house of Alfonso the Magnanimous, could make no resistance against such allies, and vanished from history as a pensioned captive in France. Hence came the second French occupation of Naples, and the northern half of the Neapolitan kingdom, while the Spaniards took the southern half. The partners in such an iniquitous enterprise were bound to quarrel, and the result was a precise repetition of the original French disaster of 1496. Even though Milan was in the hands of King Louis, his base was too far away, and his sea communications were too uncertain. The rash venture ended in the disaster of the Garigliano (December 29, 1503) and the surrender of Gaeta (January 1, 1504).

Louis had to give up Neapolitan ambitions, and by the Treaty of Lyons (February 25, 1504) resigned his half of the kingdom to the Spaniards, who were to maintain their hold upon the ' Two Sicilies ' without a break till the days of Marlborough and Eugène, when the Austrian Hapsburgs made an end of the long and blighting Spanish domination of Italy, and replaced it by their own equally alien and unpopular régime. But this was, in essence, a demonstration of the fact that Spain was no longer mistress of the western Mediterranean. English fleets had appeared for the first time in those waters—and the capture of Gibraltar in 1702 was the first evidence of the fact that a new period of naval history had begun. The second was the destruction of Cardinal Alberoni's forlorn fleet off Cape Passaro by Admiral Byng in 1719, when Spain made an attempt to reconquer her old Italian viceroyalty, landed an army in Sicily, but saw its communication with its base at once severed by a superior naval power.

But dismissing the fate of Naples as a settled matter after 1504—despite of vain after-adventures in 1525, 1528, and 1557, all doomed to disaster—there remained the very important problem as to whether the French might not at least become masters of northern Italy. Milan was in the hands of Louis XII, and the Duke of Savoy (not to speak of the Marquis of Saluzzo) gave him free passage through the Alps. The only independent power surviving in the north was the Venetian Republic. If Venice were stripped of her lands in the Lombard plain, all Italy north of the Po would be French. When the intrigues which ended in the formation of the League of

Cambray (December 1508) are considered, it is hard to understand how the other members of that iniquitous conspiracy were induced to join in a plan of robbery by which the best spoils of Venice were to go to France. The gains of Ferdinand of Spain, the Emperor Maximilian, and the Pope, though appreciable, would be trifling compared with those of King Louis.[1] In April a French army crossed the Alps, established its base at Milan, and took in hand the conquest of the Venetian provinces. There was but one attempt to meet the invader in the open—the battle of Agnadello (May 14, 1509)—in which the great condottieri the Count of Pitigliano and Bartolo de Alviano, with one of those mercenary armies which were all that Venice ever put in the field, were easily discomfited after much manœuvring. They tried to hold their own in a chosen position, but were edged out of it by flank manœuvres, and in the plain their heterogeneous bands could not resist the French fury. Louis took all the western cities of the Venetian *terra firma*, Bergamo, Brescia, Peschiera, Crema, and Cremona, while his ally the Emperor occupied Vicenza, Verona, and Bassano. It seemed as if the last day of Venice as a land power had come—nothing remained to her outside the lagoons save Padua, which made a most creditable resistance against the Emperor, and some outlying territory on the side of Friuli.

That Venice contrived to protract the war for two years, till her oppressors fell out, is surprising, and only comprehensible when it is remembered that the Pope and the King of Spain had got all that they wanted and had ceased to push on, while the Emperor was beginning to be seized with doubts as to the intention of his French ally—'he was the most suspicious man in the world,' [2] and in this case not without reason. The result in 1510 much resembled the conjunction of affairs in 1495, after Charles VIII's first conquest of Naples. The French had conquered all western Lombardy and were trespassing into those regions of Venetia which the Emperor had earmarked for himself. Once more it looked as if a French domination of all northern Italy, if not this time of all Italy,

[1] Ferdinand got various towns in the heel of Italy, Brindisi, Otranto, etc., which had been pledged to Venice for war expenses some years back. The Pope was aiming at Forli Faënza and Ravenna, and to assert his much-disputed suzerainty over the duchy of Ferrara.

[2] 'N'estoit possible trouver un meilleur prince, mais ung mal avoit en luy, quil ne se fioit en personne,' says Bayard's *Loyal Servitor*, cap. xxxviii., and again: 'Le plus supsonneux homme du Monde,' cap. lxi.

was in process of establishing itself. Hence the great project of Pope Julius II, to 'expel the barbarians'—though to get rid of the alien Frenchman he had to buy the support of the equally alien Spaniard.

It is surprising to find that King Louis, when he found that his allies had leagued with the Venetians, and that he had no friend left in north Italy save the Duke of Ferrara—whose duchy the Pope coveted—did not draw in his armies to defend the Milanese and his recent conquests in western Lombardy, but launched an offensive against the Papal States, which led to the fatal victory of Ravenna (April 11, 1512). Here he lost his brilliant general, Gaston de Foix, and his army was so shaken by its hard-won success, that the successors of Gaston gave up the offensive, and recrossed the Po. While they were at Ravenna, a Swiss and Venetian army had slipped behind their backs, and was threatening Milan, while the Emperor had not only thrown up his French alliance, and ordered his troops who had been serving under Gaston to return to Germany, but had openly made a pact with Venice. By the end of 1512 the French had been manœuvred out of the Milanese, and had retired as far as Alessandria, having lost all their gains in northern Italy save a few isolated fortresses, which were bound to fall sooner or later from starvation. King Louis refused to give up the game, and in the next year heavy reinforcements crossed the Alps to pick up the shattered army of Italy.

Once more the Milanese was in danger of being overrun. The endeavour came to a disastrous end at Novara (June 6, 1513) where the Swiss phalanx won its last decisive victory. Though almost unaided by their allies, for there were no Spanish and only a few hundred Italian horse with them, the Swiss made complete havoc of the army of La Tremouille. Fighting in the traditional order of the three columns in échelon, they drove off the French gendarmerie, and cut the French infantry—mostly mercenary landsknechts—to pieces. Once more the French were thrust completely out of northern Italy, and back to the line of the Alps. Maximilian Sforza, the heir of the imprisoned Ludovico il Moro, was restored to the duchy of Milan. But he had a very unhappy existence; for his Swiss allies occupied his fortresses, annexed his northern Alpine borders—what is now the canton of Ticino—raised his taxes, and left him only a show of authority.

After Novara the end of the French attempt to establish

a transalpine empire in Italy might reasonably have come to an end—all the more easily because Louis XII, who was so devoted to the adventure, and who had actually reigned in Milan for thirteen years, died at midwinter 1514–15. Unfortunately for Italy, and for the rest of Christendom, his young and ambitious successor, Francis of Angoulême, took up the unhappy old game—intoxicated, as it seems, by the ambition to show himself a mighty man-at-arms, and a conquering hero in the style of Charlemagne. The idea of re-establishing a French empire in Italy was as insane as ever, but his first campaign (for his future unhappiness) was preposterously successful. He crossed the Alps unopposed, by a hazardous march, and beat at Marignano (August 15, 1515) the Swiss army, which maintained Duke Maximilian on his uncomfortable throne at Milan, and made him intolerable to his subjects by perpetual exactions and oppressions. Marignano was a very unscientific fight, but it was a turning-point in the history of the Art of War—the 'steam-roller' attack of the Swiss array of columns in échelon was brought to a stand, and repelled with fearful loss by an effective if haphazard combination of cavalry charges and salvos of field artillery. There were to be no more Gransons or Morats; and Novara was to be the last triumph of the old Swiss style.

Any chance that this brilliant campaign of King Francis was to lead to a permanent establishment of French power in north Italy was made impossible not merely by geographical hindrances, but by the oncoming dynastic complications. Only seven months after Marignano, died Ferdinand of Aragon (February 22, 1516), and by his death his grandson Charles, already master of all the 'Burgundian' lands, from Flanders to Franche Comté, received possession of Spain, and of all Spain's outlying dominions—Naples, Sicily, and Sardinia—not to speak of the as-yet-unrealized resources of the newly discovered America.[1] This accumulation of kingdoms would have sufficed by itself to make French designs for an Italian empire impracticable. But when in 1519 Kaiser Maximilian, the only survivor of the elder generation, passed away and left all the Austrian lands to his grandson, the idea of a France predominant in Europe ought to have been dropped by any French king or statesman of ordinary sense. The odds were too great.

[1] Technically his mother, the mad Queen Juana, who still lived, was Queen of Castile; but he acted as her representative and co-regent

Nevertheless Francis and his son Henry II persisted in the project for nearly forty disastrous years, though France was now surrounded by hostile territory on every one of her land frontiers. Why the struggle was able to continue so long has already been demonstrated. Without the Turk and the convulsions of the Reformation in Germany the contest must have come to an end much earlier. The French kings made five successive attempts between 1521 and 1559 to renew the struggle for domination in Italy. In each case the attack was beaten off, yet the vanquished but obstinate head of the house of Valois renewed the war, whenever another promising opportunity of assailing the Emperor seemed to have arrived. The most astonishing part of the contest is that, scorning all geographical hindrance, both Francis and Henry tried again and again to win back not only Milan but Naples. Preposterous attempts to get a footing once more in the south of Italy, while the Emperor was in command of the sea, occurred in 1525, 1527–28, 1553–55, and 1557. The southern enterprises of Albany in 1525, of Lautrec in 1527–28, of Strozzi and Montluc in 1553–55, and of Guise in 1557 each resulted in the ruin of an army sent on a wildly impossible task. For to establish a holding in southern Italy required either that the invader should have permanent possession of a supremacy at sea, or else be in occupation of the whole of the northern and central region of the great peninsula.

The French twice, once under Louis XII and once under Francis, got possession of the duchy of Milan for a term of years, more by reason of the divisions and selfishness of their enemies than by the real practicability of the enterprise. But both these occupations occurred before Charles V had united the forces of Spain, the Empire, and the Low Countries. After his accession the enterprise became hopeless, even while the Duke of Savoy, whose dominion lay between France and Milan, was dominated by French influence, or in later years the victim of an armed French occupation. Francis and Henry could never spend their whole strength on an Italian war—they might send armies over the Alps, but never their whole striking force, when defence was also required on the Pyrenees, on the borders of Champagne and Burgundy, and in the far north beyond the Somme. The nature of the Lombard plain, with no real defensive position upon it between the Adige and the Alps—for the line of the Ticino (or any other of the rivers that flow into the Po) is untenable—is such that it may be lost or won by a single great battle. There is no natural front of defence between Verona and Mantua on one side and the foot-hills of Piedmont on the other. Hence the way in which, after a decisive defeat in the neighbourhood of Milan, French armies always had to give back to the foot of the Alps. The same thing was to be seen in later centuries, whether the push was French from the west, or Austrian from the east. When once there has been a general action, which ruined the main army of one of the combatants, he cannot find any line in the plains of Lombardy on which to rally, but the Frenchman retires to the Alps, or the Austrian to the Mincio-Adige positions, which were guarded in later centuries by the famous 'Quadrilateral' of Verona, Mantua, Peschiera, and Legnago.

The only military fact which, in the days before artillery reached its later developments, prevented the results of one of the decisive battles from being complete and instantaneous in northern Italy, was the immense amount of fortified places. Obstinate garrisons, which refused to acknowledge the logic of a lost campaign, often held out for an unconscionable time. And just at this period the resources of fortification were beginning to outrun those of the artillerist. In the fifteenth century, as has been shown in an earlier volume, the improved gun had demonstrated the impotence of mediaeval walls and castles of the old style—as witness the fall of Constantinople and the theatrical speed with which the brothers Bureau battered the English out of Normandy and Guienne. In the sixteenth century the engineer had his revenge on the gunner, and scientific fortification developed so fast that the predominance of artillery came to an end for a time. Hence the whole military annals of the century tended gradually to become a record of sieges—a fact most evident in the story of the Dutch-Spanish struggle in its later decades, but progressively visible in the Italian wars that filled its central years. After Pavia pitched battles became rare, and sieges innumerable and tedious. Even before Pavia they were tending to grow longer and often fruitless, as the engineer—generally an accomplished Italian mathematician—replaced the old high-lying walls of towns with new, low, bastioned enceintes, shielded with ditch and glacis and thickly gunned on points that gave cross-fire. The fortification of places like Milan or Verona was an immense task—but well repayed the trouble and expense.

Victorious sixteenth-century generals disliked leaving unsubdued fortresses behind them, and the 'mopping up' of a region full of strong points held by resolute garrisons took time. This fact it is which explains in part the comparatively disappointing sequel of several successful campaigns. The same phenomenon was to be seen two centuries later in the wars of Louis XIV and Marlborough.

CHAPTER III

STRATEGY AND TACTICS IN THE GREAT ITALIAN WARS

WE have already shown that in the history of the Art of War, the break between the Middle Ages and modern times can be fixed very definitely at the start of Charles VIII of France on his great expedition of 1494, which was intended by the feckless young King to be the commencement of a Crusade, but which developed into an attempt to build up a French Empire in Italy—the first of a series of such attempts which were to last for a period of over sixty years. It seems certain that Charles was honestly convinced that he was predestined to be the champion of Christendom against the Turk, and that his occupation of Naples on the way eastward, as heir to the Angevin claim on that realm, was in his estimation to be an incident of secondary importance in his victorious march to Constantinople and Jerusalem.

He was under the impression that he had cleared his path by the one necessary preliminary precaution—the securing of the neutrality and alliance of Ferdinand and Isabella of Spain by the very surprising Treaty of Barcelona [1493]. By this he had restored to them his father's acquisitions in Rousillon and Cerdagne, without any counterbalancing advantage save their supposed friendship. If the Spanish danger were removed, he cared little for the views of the various states of Italy —involved in their usual rivalries and intrigues. Nor was he, perhaps, very wrong in his estimate of their importance. Without the unexpected Spanish intervention, he might not only have overthrown the shaky throne of Alfonso of Naples, but have viewed with contempt the machinations of Pope Alexander, Ludovico of Milan, and the Venetians. That he marched all down Italy almost without opposition, 'marking his billets day by day with chalk' [1] as the Pope sarcastically

[1] Comines, iii. cap. 14.

observed, is a sufficient proof of the terror which his appearance inspired. But the Spanish intervention changed the whole face of affairs, and when once Gonsalvo de Cordova had landed in Calabria, there was to be no further thought of Constantinople or Jerusalem. The affair that had begun was a struggle for domination in Italy between Spain and France, and it was to endure for a couple of centuries—with a break in the middle during the sixty years of the French Wars of Religion, which put France out of the game for two generations, and left Spain temporarily triumphant.

When King Charles crossed the Alps, with his head full of reminiscences of Charlemagne and Godfrey of Bouillon, and his chancellor's wallet duly stuffed with documents ceding to him all the rights of the last of the Paleologi on the vanished Byzantine Empire, and of the younger branch of Anjou on the Neapolitan kingdom, it can hardly be supposed that he realized that he was opening out a series of interesting technical experiments in the Art of War. He marched in great state, with his glittering *gendarmerie*, his train of field artillery, which had no match in Europe, and his blocks of Swiss pikeman—all the best material of the kind, as contemporary Europe supposed. How these weapons were to be utilized, and against what sort of hostile tactics, was the problem. Probably the King conceived of his army as a sort of steam-roller, which would crush down everything that stood in its way. His captains were well acquainted with the ridiculous details of fifteenth-century Italian campaigning—those battles of manœuvre, at which Guicciardini and Machiavelli laughed, where general actions were fought with a casualty list of one man killed and two wounded, and armies surrendered when they were technically outflanked or cut off from their base.[1] They may even have been looking up Turkish tactics, with a view to operations beyond the Adriatic—there is evidence that the name and moral of Nicopolis—the last French disaster in the East—was known to some[2] French soldiers of the sixteenth century. But what was actually before them were complicated problems of the combined employment of heavy and light cavalry, of pikemen as confronted or assisted by men bearing firearms, and of the best manner of making use of the smaller and larger field artillery. There were to be many disastrous experiments, and

[1] See *Art of War in the Middle Ages*, vol. ii. pp. 306–309, for examples.
[2] De La Noue, *Discourses*, p. 613.

many inconclusive results, as all manners of combination of arms were tried by one side or the other.

When dealing with the last fifteenth-century battles of continental Europe—Montl'héry, Granson, Morat, etc., we had arrived at the period when the two distinctive tactical novelties were the revindication of heavy cavalry, and the triumph of the Swiss technique of masses of pikemen moving in échelon. The English system of dismounting the men-at-arms and flanking them with archery, which had started at Dupplin and gone on successfully down to Agincourt and Verneuil, had passed away, except in England itself, where it was to continue not only till Bosworth (1485), but till Flodden (1514). But on the Continent it had been recognized that the system was essentially a defensive one, only effective against an enemy who was obliging enough to attack an English army placed in a good position. And the two last English fights in France had demonstrated its weakness. At Formigny (1450) the little army of Gough and Kyriel had been bombarded by artillery in its position, till it was forced to come out and counter-attack the pestilent guns—with disastrous results. At Castillon (1453) Talbot had failed hopelessly in an assault upon an intrenched line well garnished with artillery. Oddly enough we shall find in the first section of our discussion of Renaissance tactics two battles that remind us of Formigny and Castillon. At Ravenna the Spanish-Papal army was nicely placed in a defensive position, but forced to take the offensive by persistent pounding by gunfire, which was becoming unendurable. At Bicocca (1522) the Swiss columns tried, like Talbot at Castillon, to run over a line of trenches by sheer impact, but were so shattered by the fire of greater and smaller firearms that it was turned back in complete rout.

These two battles bear testimony to the fact that we are coming to the times when firearms were beginning to play an important part in tactics. But in the early years of the Great Wars they could hardly be called normal or typical exhibitions of the military art of the period, and there were still battles to come where artillery and arquebuses had little to do with the fate of the day. Sometimes the victory was won by the charge of heavy cavalry, as at Fornovo and the first Seminara (1495), sometimes by the wild impact of masses of pikemen, as at Novara (1513). Some of the acutest thinkers of the time still maintained that artillery was too slow-firing and slow-moving to be effective against a general who knew how to manœuvre, and was not intending to run in headlong frontally against guns in line. Machiavelli, writing as late as 1515, in his *Arte della Guerra*, was of opinion that a commander, with lightly moving and well-trained troops, could count on so making his attack that the enemy would not be able to get more than two or three volleys at him.

But on the whole we find two tendencies gradually growing—the progressive importance of firearms, and (partly in consequence of that progress) the utilization of field entrenchments, which would make cavalry charges less and less practicable. With both these developments the name of Gonsalvo de Cordova may be associated, as the first general who made decisive use both of the smaller firearms and of field fortifications. His triumph at Cerignola (April 1503) shows the combination of the two in very clear purpose, though his 'crowning mercy' of the Garigliano (December 1503) was not a battle won by any such devices, but was a simple 'victory by surprise,' a plunge into the scattered cantonments of an enemy who was not expecting to be attacked. At Ravenna, as we have already mentioned, the day was settled by a superiority of artillery, though the enemy had entrenched himself. At the very confused fight of Marignano (1515), the Swiss pike-phalanx was broken by a combination of the old-fashioned cavalry charge with overpowering salvos of artillery—but each would have been insufficient for victory if not helped by the other. Bicocca (1522) was won in the classic style of Gonsalvo de Cordova, by arquebus fire from impregnable entrenchments. At Pavia (1525) there was no question of entrenchments in the actual fighting, though both sides had been 'digging themselves in' for three weeks before the day of battle. Quitting their elaborate lines, the Imperialists turned one end of the French position by a night march, and forced King Francis to attack them on unprepared ground with hastily moved troops, coming up at irregular intervals into an array that was never properly formed. But the decisive blow in the actual fighting was given by the Spanish arquebusiers, who gradually shot down the French cavalry, which had broken into the centre of the Imperialist army, but was not properly supported by its late-arriving infantry.

After Pavia pitched battles of primary importance became rare. The only one, indeed, which was fought in the good old style, by two armies deliberately marching against each other with intent to come to close quarters, was Ceresole (1544). St. Quentin (1557) was not of this sort, for the French had no desire for a general action, but were caught while manœuvring by an enemy who suddenly struck home. In many ways it was comparable to Henry VIII's solitary victory of Guinegate (1513), *alias* the Battle of the Spurs, where the French, bent on demonstrations only, were suddenly attacked by a rapidly moving force of all arms while their infantry was miles to the rear, and gave way in disorder, without being able to form up for a regular fight. Gravelines (1558), the last battle of the war, was fought by two small forces, each detached from its main army which was lying far south in Picardy. Its character reverted somewhat to the ancient type, as the horse on each side fought sharply before the infantry became engaged. But its termination can be paralleled only by that of Pinkie—fought ten years before (1547)—among all battles; the decision was given by gunfire from a fleet belonging to one of the parties, when the other had incautiously taken up a position of which a wing rested on the sea, without any thought of possible naval intervention.

For many years after the Great Wars began, the mediaeval system of dividing an army into 'vaward,' 'main-battle,' and 'rearward' continued to prevail, in name at least. Each of these sections should have been composed of cavalry and infantry combined, and normally the wings would be more or less equal, while the 'main-battle' would be much larger than either. But this system gradually became one of mere administrative units, with little or no reference to position in the actual line of battle. And one wing might be very much smaller than the other, or occasionally was dispensed with altogether. At Novara and Bicocca the Swiss fought with only two divisions, not the normal three, as did the English at Flodden. At Pavia the Imperialist 'rearward' contained only 2000 men out of an army of 20,000, and all the heavy cavalry were in the centre corps, while the light cavalry went on with the 'vaward.' It is of this battle that du Bellay humorously observes that as a matter of fact the French main-battle became *their* 'vaward,' and the Imperialist 'rearward' became *their* 'vaward'—meaning that, as things went, the names implied nothing. As armies grew larger—and by the end of the war there were sometimes 50,000 men in line, corps of both arms got distributed just as they came upon the field, with no reference to the old idea that a 'rearward' should always be upon the left and a 'vaward' upon the right of a central 'main-battle.' Indeed, an army sometimes deployed with a second line in reserve, and that reserve might be formed of elements of any arm and from any division of the three. The term 'wing' [*esle*, *flügel*] comes into use in describing battle-array, replacing the old 'vaward' and 'rearward,' though the centre is still often called the 'battle.'

And so a line, as the years went on, might be formed by troops of either arm in any juxtaposition, sometimes with all the cavalry on one wing, sometimes with a concentration of it in the centre. But as a rule there were always flanking bodies of light horse, kept some way out on the flanks, not so much for utilization in the general clash of two armies, but for services of information and scouting, or—in the case of a victory—for the pursuit of broken hostile troops.

The use of the heavy cavalry for desperate charges to break the hostile line became less frequent as firearms grew more efficient, and the combination of pike and arquebus gave the infantry not only the resisting power which the old Swiss phalanx had possessed, but also the capacity for thinning the ranks of charging cavalry while it was striving with the pikes. Pavia, of course, was the battle where this form of tactics won its supreme vindication. In later fights when the heavy horse were let loose, it was generally to fall on infantry already engaged with other infantry from the flank, or to sweep away the hostile cavalry as a preliminary measure before attacking the enemy's main central block of foot, as at Ceresole and at several fights in the French 'Wars of Religion.'

But, as we shall see when investigating the later campaigns of the war, general engagements became very rare after Pavia, the will to fight for the pleasure of victory, which was predominant in the earlier years—especially among French generals—seems to have died out, and the risk of thinning away an army by misplaced pugnacity seems to have been felt in a fashion that was unknown in the earlier years of the contest. If success could be achieved by manœuvre, by cutting the enemy's line of communication, or starving him out in the face of impregnable lines, or distracting him by a sudden transference of troops to an unguarded front, it was no longer considered less creditable than success won in a pitched battle.

We shall note occasions in these later campaigns when armies of first-rate strength lay opposite each other so close that a general action seemed inevitable, yet drew off because each was loth to try the final arbitrament of war. This was especially notable in the years 1553–54, when Charles V and Henry II faced each other on the borders of Picardy, each with his main army at his back, yet parted without any decisive action—as Edward III and Philip of Valois had done at La Flamengerie (1339) many generations before, on fields not far distant. The risk to the attacking party in assailing an enemy in a good position seemed too great, when the consequences might be complete disaster. Before one of the rare general actions of the later wars a French commander—the Count of Enghien—actually sent to Paris for leave to fight a general action (Ceresole, 1544), a thing which would have seemed incredible to Gaston de Foix and his contemporaries. Hence many campaigns which seem most disappointing to the student of the Art of War, when a battle seemed to be the inevitable consequence of the situation, yet never took place—e.g. during the invasion of Provence by the Emperor in 1536, or the deadlock in Champagne in 1544.

This phenomenon has many side-causes: it was not always mere reluctance to take a risk—which might mean the destruction of an army or the loss of a province—which led to inconclusive manœuvring or retreat. When we read the decidedly tedious records of the later Italian campaigns, we are astonished to find how often cautious generals refused to put matters to the hazard of battle, and moved off. Sometimes the risk was starvation—an army which had exhausted its supplies, and harried the neighbouring region bare, and had lost secure communication with its base, almost always drew back, instead of delivering an attack. Sometimes, and this was a very characteristic feature of these wars, the reasons for a retreat were as much financial as strategical. The mercenaries whom both sides were using on the largest scale, were very badly disciplined troops. When pay was many months in arrears, the individual soldier deserted the colours without scruple. This was true on both sides, but the French, in the later years of the war, were under the special disadvantage that the Swiss, who formed the core of their infantry force, did not abscond individually, but went off in whole corps, after making representations through their captains to the effect that their contract had been broken, and that they considered themselves no longer bound by it. And they were such an important element in the French army that we never find an attempt made to treat them as mutineers and apply force to them.

On several occasions a body of several thousand men went off home in the midst of a strategical deadlock, or a crisis of manœuvring. At Bicocca (1522), as we shall presently see, the Swiss captains offered their French commander Lautrec the choice between delivering a battle immediately, or seeing them march away home. A few days before Pavia 6000 Swiss went off, even though no pay was owing! In these cosmopolitan and heterogeneous armies desertion often meant not mere absconding, but going over to the enemy. Du Bellay complains that in 1521, at the siege of Parma, of 5000 Italian foot in the French garrison 3000 deserted, and the majority of them enlisted in the enemy's army.[1] This sort of thing continued right down to the Thirty Years' War in the next century. Gustavus and Wallenstein enlisted deserters wholesale.

Landsknechts were quite as bad as Italians in this way. Every French army had several thousands of them in its ranks. They were all Germans, technically, therefore, rebels fighting against their sovereign the Emperor. But there never seems to have been any difficulty in recruiting them on a large scale—it was done as a rule through the German princes who were opposed to the Hapsburg domination, and had no scruple in helping the French against their suzerain. In the first half of the war the chief leaders of sedition were Charles of Egmond, Duke of Guelders, and Robert de la Marck, Duke of Bouillon, each of whom is found raising without any trouble corps of 5000 or 6000 landsknechts for the French service. When expelled from their own dominions by the Imperial armies, they became mere condottiere-generals in the French pay, but the astounding thing is that they never seem to have had any difficulty in keeping their corps afoot by finding new recruits. Apparently the landsknecht had so little national feeling that it was rather indifferent to him whether he served for or against the Emperor. And he would enlist under any competent and popular leader who offered good pay. The only general sentiment of a non-material kind that seems to have affected him was a dislike for the Swiss. When they met in battle they never gave quarter to each other: one was

[1] Du Bellay, i. p. 177.

as bad as the other. This was called 'bad war,' as opposed to 'good war,' in which the normal practice of taking prisoners and getting ransom for them was prevalent. On one occasion we find a French general, after a series of massacres and counter-massacres, giving formal permission to his Swiss to kill even prisoners in camp, who had been brought in after a surrender ! [1]

As far as one can make out, there must have been thousands of these German mercenaries who served indifferently on either side. On one occasion in 1512 we read of a special summons by the Emperor Maximilian for Germans in the French service to throw up their employment and come home. But a considerable proportion disobeyed, headed by a captain who explained that he had nothing to lose in Germany, and had a good job in the service of King Louis. This was evidently a prevalent conception in the landsknecht mind.[2]

It was this absolutely conscienceless mentality on the part of the mercenary troops which made the position of a general so difficult. If pay or provisions failed, he might see his army melt away without remedy. Hence the curious fact that enemies, conscious of the distress in the hostile camp, while their own was comparatively in good order, deliberately waited to see the balance of force changing—the historians of the war make no secret of the fact that tactical advantages were sometimes neglected by both sides, because there was a general impression that a few weeks more of privations and bankruptcy would ruin the opponent. The starving-match was not only physical but financial—because the Swiss or the landsknecht would go on strike or march off, not only because he was being kept on short commons, but because he was weeks or months in arrears of his pay. We shall see the phenomenon arising not only before Bicocca and before Pavia, but much later. It was common during the French and Dutch 'Wars of Religion' in the later half of the century.

These armies, on both sides, were singularly leaky, if we may use the term. Inspired neither by loyalty to a sovereign, nor devotion to a cause (such as was often to be found later in the Huguenot and the Dutch wars) they were at the best held together by *esprit de corps*. Some of the bands undoubtedly followed a favourite captain, whatever side he

[1] The general was Bonnivet. See Du Bellay, ii. 314–15.

[2] Bayard, lv. p. 213.

might choose, as did apparently the Italian 'Black Bands' of Giovanni dei Medici,[1] and the landsknechts of Georg von Fründsberg, but this was exceptional rather than normal. The national French and Spanish units were less given to desertion to the other side, but we must not forget that when the Constable Bourbon defied his master, Francis I, and joined the Emperor, he was followed by a very considerable number of gentlemen of his personal clientèle. And that even Spaniards would desert, when they considered themselves wronged, was shown by the case of the great engineer Pedro Navarro, who made no scruple of transferring his admirable talents of fortification and road-making to the side of the old enemy.

It was hazardous to seek battle with an army whose morale was low, since some unit might fail in the moment of the clash. Even war-tried troops were known to belie their reputation—like the Swiss of King Francis on the day of Pavia, or the French horse at the Battle of the Spurs. Hence much cautious manœuvring and avoidance of a decision. A general who met no disaster was sometimes as much esteemed as one who secured a positive success—such was the case with La Tremouille in 1523, in his campaign against the English along the Somme. The invaders marched much where they pleased, burning right and left as far as Roye and Montdidier, but as they never could force him to fight, and turned back from a harried land when winter drew near, he was regarded as a master of Fabian tactics. Picardy had been ravaged from end to end, but the army was intact. If it had fought and been beaten, there would have been nothing between the English and the walls of Paris. But La Tremouille's reputation for cleverness in avoiding battle was nothing to that of Prosper Colonna, on whom his Italian admirers bestowed the nickname of 'Fabius Cunctator,' a classical reminiscence from the fame of the ancient Roman general who foiled Hannibal by refusing to fight him.

[1] When Giovanni was wounded at the siege of Pavia his corps melted away at once: *Ses soldats, étants sans chef, se débandèrent, de sorte qu'ils devienrent à rien.*—Du Bellay, ii. p. 388.

CHAPTER IV

THE FRENCH ARMY IN THE GREAT ITALIAN WARS

IN the later years of Louis XI the core of the French armies had consisted of the *compagnies d'ordonnance* which his father had organized, each consisting of a hundred 'lances,' which meant not only the hundred men-at-arms but their retinue of 'archers,' *gros valets*, *couteilleurs*, etc. The archers, despite of their name, were mounted men and armoured.[1] Many of them were drawn from the noblesse. The famous Montluc, though of the best blood of Gascony (as he never ceased to boast), started in 1521 as a simple archer in the company of Thomas de Lescun, which, as he observed, was considered quite an honourable post in those days, 'though since then everything has got degraded.' He was only promoted to be a man-at-arms in the company of the Marshal de Foix two years after.[2]

Besides the *compagnies d'ordonnance* there served as heavy cavalry the King's personal retinue, his archers of the guard, and his 'pensioners' or 'gentlemen of the *vingt escus*'[3]—also many volunteers, who were under no proper organization, and sometimes caused trouble by turning up when they were not expected. If France was invaded—which happened less frequently than might have been expected, it was possible to call out the 'arrière ban' of all those who owed fiefs held under military tenure. Professional soldiers did not think much of the men-at-arms who came out on this obligation.[4] The best of the noblesse had already gone to the front in the

[1] As, of course, were the famous Scottish archers of the royal bodyguard. 'Le archers de la Garde avoient des coignées, qui étaient pendues à l'arson de la selle des chevaux, ils les mirent en besogne et donnoient de rudes coups sur l'armets des Espagnolz,' says Bayard's 'Loyal Serviteur,' describing the battle of Ravenna, liv. p. 208. Archers are mentioned as taking part in the general cavalry charge, also in Comines' account of the battle of Fornovo, and much later at Ceresole.

[2] Montluc, i. pp. 20 and 24.

[3] So called from the 20 crowns a month of their pay.

[4] See the very interesting pages of La Noue, 320–27.

companies or as volunteers. 'Fort peu de gentilhommes se trouvent dans l'arrière ban : ils courent quasi tous à la solde, les honneurs, et les récompenses militaires, et n'y trouvent que gens de petite expérience. Ainsi ne les employoit-on qu'a garder les provinces esloignées des dangers de la guerre.' Many military fiefs had been given to the Church, others by royal leave had been sold to courtiers, financiers, or merchants. The sort of deputies that these clerics or civilians sent to the levy were very worthless. However, in times of need and crisis—*e.g.* during Charles V's invasion of Provence, the 'arrière ban' was called out, and apparently did not distinguish itself.

There were no light cavalry in the French armies at the commencement of the Great Wars, save some mounted cross-bowmen, who were intended to be used for skirmishing alone —the cross-bow is a very bad weapon for use on horseback. We hear of them as early as the day of Fornovo. But experience of the trouble caused by the Venetian 'Stradiots,' and later by the Spanish 'genetaires,' led French kings to raise more 'chevaux légers,' who were to carry out the same functions of exploration, raiding, and light expeditions that the Stradiots and genetaires discharged for the enemy. The Stradiots were especially admired, and Louis XII succeeded in raising a considerable number of these raiders, trained in the practice of war as it went on in the Balkan peninsula. How far they were really Albanian or other Eastern folks it is not easy to judge—but they were trained in the fashion of the old Venetian mercenaries. We hear of French stradiots in the campaign of Novara (1513) and in many later affairs for thirty years.

When the cross-bow went out, we often read for a time of *arquebusiers à cheval*, but the arquebus was little better for a horseman than the cross-bow, and by the end of the Great Wars many light horse had developed into 'pistoleers'—a sort of cavalry whose efficiency was first learnt from the Germans. The *chevaux légers* wore breastplate and open helmet, with mail sleeves, but no leg armour. They were often used in line of battle, but were not expected to have the same weight and impetus as the gendarmerie. By the middle of the century the 'pistoleers' had developed the very unsound habit of working on the 'caracole,' a system of riding up by successive ranks to fire, and then wheeling off to let the next rank make its discharge. La Noue, as we shall see later on, rightly condemned these tactics. Bad 'pistoleers' let fly too early, knowing that they have to retreat and reload as fast as they can. Rear-rank men fire in the air. When the first rank has gone back to rearrange itself, cowards take the opportunity of riding off to the rear instead of halting. Pistol-fire is no good except at very close range, and only very tough and experienced troopers have the resolution to continue a 'caracole' manœuvre, without getting clubbed or scattered.[1]

Unlike the *chevaux légers* the heavy gendarmes of the old *compagnies d'ordonnance* charged in line, *en haye*, invariably, and this habit was continued all through the Italian wars, down to the time of the French civil 'Wars of Religion' which filled the last dismal forty years of the sixteenth century. The tactical problem as to whether the line or the squadron-in-column was preferable, exercised all French military writers for a generation. In the end, as we shall see, the deep order—borrowed from the Germans—prevailed, and its adoption synchronized with the rejection of the lance, which disappeared from the French army for two hundred years. From Henry IV to Napoleon there were no lancers in the army list,[2] and they only reappeared when the Emperor, after 1806, first enlisted Poles armed with their national weapon, and afterwards turned many dragoon regiments into *chevaux légers lanciers*. This reminds us of an eighteenth-century experiment, when the later Bourbons adopted from the Hungarians the name and equipment of the 'hussar.' But the *houssards* of Louis XV had sabre and carbine, never the lance.

So much for the cavalry arm: it remains to speak of the foot-soldiery. Louis XI in the middle of his reign had intended to constitute a national infantry force, by the institution of the *Francs Archers*, of whom he raised at least 16,000 men. But the experiment proved a failure, and after the disaster of the battle of Guinegate (1479), where the indiscipline and weakness of the *Francs Archers* lost the day,[3] the King took no further interest in them, and turned to the Swiss. He had witnessed their efficiency in his own battle of St Jacques-by-Basle (1444),

[1] See La Noue, *Discourses*, pp. 444–45. Yet he held what he called the 'paradox' that the deep formation of the German 'reiters' was preferable to the *en haye* formation of the French heavy cavalry.

[2] Except a 'freak' regiment of Uhlans raised by Maurice de Saxe at his own expense, and disbanded at his death.

[3] At Novara and Ravenna the French pike-battalions were all landsknechts, the Swiss being at the moment hostile.

and had noted their three victorious campaigns against his rival Charles the Bold. He took 6000 of these pikemen into his service—their pay was 4½ Rhenish gold florins per month—and retained them as a permanent force for the remaining five years of his life. This, as Machiavelli remarks, was an unwise policy for a very wise King ; to trust foreign mercenaries rather than one's own born subjects is a fundamental error.

Though some of Louis's 6000 Swiss were paid off after his death, a large nucleus was retained by the regents who followed him, and Charles VIII brought it up to a considerable corps before starting on his Italian campaign in 1494—8000 men in all. The French commanders of the succeeding age regarded the Swiss pikemen as the dominating power in war, and on the rare occasions when the Swiss were hostile, and could not be enlisted, hired German landsknechts, trained on the same principle, as the best substitute that could be found.[1] Such French infantry as did appear in the earlier years of the Italian war were mainly cross-bowmen from Gascony, the district of France where military spirit was highest, according to that very typical Gascon Montluc. We occasionally hear of Picards and Bretons, but they appear with much less regularity. The French infantry, raised by captains who received commissions from the King, but were not taken into permanent service, seem to have been rather a haphazard assembly, always liable to be disbanded, and looking forward to plunder rather than to regular pay. They are often called *aventuriers* as opposed to the gendarmerie and the regularly hired Swiss and landsknechts.

Arquebusiers were known, but there were very few of them in the early years of the war : it was only in the second generation that the arquebus superseded the cross-bow. Montluc remarks that in 1523, when he was ensign in the company of Monsieur de la Clotte, he had only six arquebusiers with him, and they were all deserters from the Spanish army. ' Encore en ce temps la il n'y avait point d'arquebusiers parmi notre nation.' He then proceeds to remark that he wishes that the arquebus had never been invented. ' Would to God that this unhappy weapon had never been devised, and that so many brave and valiant men had never died by the hands of those who are often cowards and shirkers, who would never dare to look in the face those whom they lay low with their wretched bullets. They are tools invented by the devil to make it easier for us to

[1] Comines, vi. chap. iii.

kill each other.' [1] The day had gone by when a certain commander used to order that quarter should never be given to men carrying firearms,[2] but they were still hated and despised, and it took some time to teach French generals that they must rather be encouraged, and introduced on the largest scale possible.[3]

Apparently the date at which the systematic introduction of firearms appears in the French army may be fixed after the battle of Pavia (1525), at which every soldier realized that it was the Spanish arquebusiers who had a decisive part in determining the event of the day. But it was recognized that they were helpless against cavalry, if they had no covering force of pikes attached to them. As La Noue observes, a body of 5000 arquebusiers marching across open country and intercepted by 1500—or even 600 or 700—horse would probably come to grief. But if they had a few thousand pikes with them they would not have the least difficulty in driving off the cavalry.[4] He quotes as the clearest example of the efficiency of a mixed force of pikemen and arquebusiers, assailed by swarms of horsemen, a retreat made by the Spanish captain Alvaro de Saude in one of the African campaigns of Charles V. With only 4000 men he marched through a mob of 18,000 or 20,000 Moorish horse, received and beat off dozens of charges, and came through with a loss of only 80 men, while 'the barbarians' lost at least 700. But of course the enemy had neither infantry nor guns—which made all the difference. Incidentally we may remark that Saude's exploit foreshadows precisely the achievements of Kleber's small squares at the battle of Mount Tabor in 1799.

For a long time the French commanders, after they had recognized the necessity of the combination of pike and arquebus, persisted in using foreign pikemen to serve their purpose, and so we find Swiss or landsknechts used to support the squads of arquebusiers, and not native bands. Even when compelled by necessity to arm Frenchmen with the pike, they seem to have looked upon them as less reliable than the foreigners. These troops were called 'corselets' from the breastplate with which (like the best of the landsknechts) they were

[1] Montluc, i. p. 2.

[2] This amiable person was the condottiere Vitelli.

[3] The first French-made arquebuses were reckoned poor stuff, and officers tried at all costs to get Italian weapons of the manufacture of Milan, says Brantôme.

[4] Cf. La Noue's *Discours*, pp. 387, 450, 455.

provided. La Noue says that several causes might be alleged for their comparative inefficiency—the first was that the best recruits were by now wishing to serve with firearms, preferring to shoot, rather than to be shot at without power to retaliate, as pikemen often had to do. The second, that captains only looked out for men with stout shoulders, not for men with stout hearts, and would enlist any big men that they could lay hands upon, without regard to their *morale*, whereas among the Swiss and even among the landsknechts the very best material served with the pike, and the 'doppelsöldner,' or soldiers with extra pay, who fought in the front ranks, were formidable veterans. Montluc often mentions that he dismounted to fight with his pikemen to keep them in heart, and made his officers do the same, in order to be sure of steadiness at the first clash.

A very clear cause of the recognized inferiority of the French infantry was that, outside the standing force of Swiss, the units were perpetually being raised in a hurry, and disbanded when a crisis was over. Montluc mentions that during the unexpected Spanish invasion of southern France in 1523, which ended in the abortive siege of Bayonne, King Francis raised fourteen or fifteen companies (*enseignes*, to be accurate) of foot in Gascony or Guienne,[1] But the moment that the Spaniards had retreated into Navarre, half of these new corps were disbanded, and finally only two were kept under arms. The officers and men were thrown upon the world, and had to seek service where they could get it—mainly in the army of Italy, to which Montluc himself repaired, when he found himself thus stranded. At each of the four treaties of peace which diversify the Italian wars, all the foot companions were disbanded, save a very few kept for garrison duty.

There were as yet no permanent units larger than the band or company (*enseigne*) ; several of these might be thrown together for some time under a single chief, but the organization was not permanent, and the word 'regiment' has not yet appeared. The first attempt to create something like an organized force of native infantry was made by Francis I in 1531, when he raised the so-called 'legions,' a name very typical of the Renaissance, when not only students but soldiers were interested in classical antiquity. The 'legions' were quite large bodies, 6000 strong, which is an obvious echo from the Roman original. 'Quite a good invention,' says Montluc, if it had only been well

[1] Montluc, i. pp. 25 and 43.

carried through, for though ordinances and our laws were kept up for some time, after a bit everything went to pieces. For the right way to get a good permanent army is to do like the Romans, and keep one's people accustomed to war. Whether this move was for the best I can not be sure, but I would always rather trust my own countrymen rather than foreigners.' [1]

There were to be four legions, named Picardy, Champagne, Normandy, and Languedoc, each to be composed of a mixture of pikemen and arquebusiers, with a small additional proportion of halberdiers—obviously a copy of the organization already in use by the Spaniards. Originally Francis intended to raise seven legions, but his financial straits forced him to reduce the number to four. Each was composed of six 'bands' of 1000 men, a figure much resembling the later battalion with which we shall grow familiar. The very scanty corps of officers only included for each 1000 men a captain, two lieutenants, two ensigns, and ten centurions (centenniers). The arquebusiers were to be only 300, compared with 600 pikemen (' corselets ') and 100 halberdiers: they were to be the special charge of one of the two lieutenants of the band. The commander of the whole legion was called a colonel; this is the first time the title appears formally in the French military annals.[2] But, as we shall see, it had been already in use among the Spaniards.

The system appears to have been a bad one—an infantry division of 6000 men without any attached cavalry, guns, or engineers, is an unwieldy body, and the idea seems to have prevailed of considering it as a mediaeval 'battle,' or great mass, such as were the Swiss columns which had been the terror of Europe during the preceding generation. The record of the legions seems to be unsatisfactory—in 1543, 10,000 of them, set to defend Luxemburg against a German army of Charles V, gave way very weakly, and many deserted,[3] so that the fortress was lost. There was also little good heard of them in their campaign against the English of Henry VIII, at the time of the siege of Boulogne in 1544. The Marshal de Vielleville declares in his memoirs that in 1557 he found his legionaries

[1] Montluc, i. 91–2.

[2] But that it was used a little earlier in common parlance is shown by the fact that in 1528 Montluc says that his company and some others had Pedro Navarro as their 'colonel.'

[3] Du Bellay, book x. p. 60, says that of many companies there were not thirty men left with the colours, and that he did not believe that either Champagne or Normandie had much over 300 men under arms—all the rest had shirked back to France.

little better than peasants, and most of them had only four or five months' service. The legions, we are somewhat surprised to find, did not disappear when the peace of Cateau Cambréis brought the Great Wars to an end. For even during the last twenty years of the great struggle between Valois and Hapsburg all French commanders had preferred to have Swiss or landsknechts under their hand rather than their native infantry.

The legions of Normandy, Picardy, and Champagne were still in existence, though with a reduced effective, and no longer on the vast 6000-man scale, during the Wars of Religion—they formed a nucleus of trained troops in the hand of the government, always ready to be opposed to the hastily raised bands of the Huguenots. The legion of Languedoc broke up in 1562, when its colonel, the Baron des Adrets, took off the greater part of it to join the rebels. The legion of Normandy was disbanded in 1593. Picardy and Champagne, after many reformations and reorganizations, went on into the seventeenth century, and may be considered ancestors of the 'regiments' of the same name, which were famous in many a war of the later Louis. The name 'legion' died out with the sixteenth century—as had already the unwieldy number of rank and file attributed to it: the 'regiment' which was to replace it was a much less ponderous formation.

The non-legionary French infantry was, as we have already said, a very casual and improvised force, always being raised and disbanded, as the long stretches of war and the short intervals of peace succeeded each other during the struggle of France and the Hapsburg power. After Cateau Cambrésis it naturally disappeared altogether. By the end of one of the longer periods of war there were, of course, corps which had been under arms for a good many continuous years, and had come to be considered veterans. Such were the 'vielles bandes de Piémont,' of which we read during the latter period of the reign of Henry II. They had obviously much greater value than the hastily raised corps which were to be found at the beginning of a war-period. The normal practice was that at the start of a new war, or after a great disaster or crisis, when an army had been destroyed, the King commissioned well-known captains to raise 'enseignes' or 'compagnies' or 'bandes' from the drifting mass of old soldiers out of employment, who were always to be found, supplemented by volunteer recruits. In the earliest period of the wars these corps must have been predominantly cross-bowmen; but after 1525 arquebusiers only were sought after. French pikemen seem hardly to be mentioned before Ravenna (1512), but are forthcoming in all the later fights.

Montluc, among the many disappointments of his life which he is always so careful to chronicle, gives one of 1527. He was commissioned to raise a new company in Gascony, and in a few days had collected no less than 800 men, of whom quite 400 were arquebusiers, though firearms were still rather a novelty in the French army. When he reached Alessandria, in Piedmont, on his way to the front, half of his arquebusiers were requisitioned from him, to give some 'shot' to the company of M. d'Ausun—which had none.[1] When it was that a 'band' or 'company' came to be considered incomplete without a certain allowance of 'corselets' (pikemen), in spite of the general prejudice against French pikemen, is hard to discover. Certainly it was long before the end of the reign of Francis I: we have distinct mention of bands with pikemen in their ranks, as early as the battle of Ravenna (1512). Yet thirty years after, at Ceresole (1544), they were still so distrusted that Montluc takes no mean credit to himself for having beaten a column of landsknechts in frontal pike-fighting with his own bands of Gascons.[2]

It is rather surprising to find in the later years of the war a considerable number of Italian units serving in the French army. With the officers this is not difficult to understand—there was a very important faction of Neapolitan nobles of the old Angevin party, who had lost their estates for siding with Charles VIII at the first invasion. The most notable of these were the San Severino family, the house of Carraccioli, a much ramified and once wealthy race. There were also exiled Florentines. Piero dei Medici died in the French service at the battle of the Garigliano; his illegitimate cousin Giovanni was the founder of the 'Black Bands' which had considerable repute in the army of Francis I. We find officers like Trivulzio, Strozzi, Birague, Visconti, Malaspina, Sforza, in high command. Some were experienced condottieri, who had no national prejudice, and took service wherever they could get it; they were welcomed by the French kings for their technical

[1] Montluc, i. p. 56. Just after this spoliation occurred another of Montluc's usual pieces of bad luck—he was wounded in the trenches before Vigevano, and was out of the service till the spring of the following year.

[2] Montluc, ii. p. 283.

knowledge and military skill. But it is more strange to note that numerous 'bands' of the rank and file of Italian mercenaries were normally employed, though every French writer is wont to allude to them with some disparagement, and to quote cases where they were the first to give way at a critical moment, when Swiss or landsknechts would have kept their ranks. Frequently Italians were found useful as light horse and arquebusiers, elements in which the French army was notoriously deficient for many years. But there were certainly men-at-arms and pikemen among them, though the French considered Spanish and German horse, and Swiss and German pikemen superior to anything that came out of Italy.[1] But to keep up numbers, in the time when the force of France was failing, any military material had to be utilized. It is curious to find Italian foot-companies employed not only in their own country, but in the northern campaigns against the English. Montluc had two with him in the attempt to storm the lower town of Boulogne (1545) under captains named Cesare de Porto and Geronimo Migrano, and he was not at all satisfied with their conduct.[2] Probably these units were mainly arquebusiers—the French were still short of foot-soldiers skilled in the use of firearms.

Through all the early stages of the war, the French had a decided preponderance over all their enemies in the matter of artillery. The great royal train which had been organized under Charles VII by the brothers Bureau, and which had so successfully blasted the English out of the castles of Normandy and Guienne in 1450–53, had never been allowed to fall behind the times. Jacques de Genouillac, the successor of Gaspard Bureau as Grand Master of the Ordnance, is said to have been adding continually new technical improvements. When Charles VIII crossed the Alps in 1494, his long train of cannon, small and great, provoked the wonder of the Italians. They were mostly of bronze, drawn by horses, and able to keep up with the marching speed of an army. The condottieri in their indecisive wars had only employed a few big iron guns, painfully drawn by oxen at the rear of the host, and slow to get into action. Such was the artillery which Machiavelli had

[1] Comines' account of the behaviour of the Italian men-at-arms at the battle of Fornovo shows a deep contempt for them. Montluc shows the same feeling on many occasions.

[2] Montluc, ii. 310.

known in his youth, and which left on him such an impression of inefficiency that he regarded it as practically useless when he wrote the *Arte della Guerra*, and only allowed that it might get off one or two discharges before battle was joined. The French guns moved much more rapidly, and the gunners had reached a skill in rapid reloading, and of change of direction of fire, which had previously been unknown. We shall note that their action was of decisive importance at Ravenna (1512) and Marignano (1525). Still, as has been already observed, the cases of battles won by artillery fire were rare compared with the number of those in which the matter was settled by the infantry or the cavalry. And in campaigns of manœuvring, ending in a fight which developed unexpectedly, there was generally little time to form up the line of artillery, which was mainly useful in set battles, where the defensive party had settled down into a position, and sometimes even entrenched itself more or less, with its guns placed in commanding positions. At Ceresole, however, as late as 1544, both sides indulged in a three-hour cannonade to little effect.

CHAPTER V

THE SPANISH ARMY, GONSALVO DE CORDOVA, AND THE ARQUEBUSIERS

ON the occasions where we have had to deal with Spanish armies in the earlier centuries, we have seen that they were originally a cavalry organization, with two sorts of horsemen: the heavy man-at-arms in the knightly equipment prevalent all over the rest of Western Europe, and the 'genitors' or 'genetaires,' who were peculiar to Spain. Of infantry, much despised in comparison with the horse, there were many cross-bowmen, either mercenaries or town levies (ballasteros). The Basques had once been famous for javelin-men, and the sword-and-buckler infantry of the Aragonese are occasionally mentioned. But the cavalry was considered all-important, and the national and most impressive part of it was the horde of 'genitors' whose tactics had been developed in centuries of war with the Moorish light horse. Armed with steel-cap and mail-shirt only, for their essential weapon was the javelin not the lance, and casting the javelin would have been impossible if the rider was cumbered with cuirass and brassards, they rode upon completely unarmoured horses. Down to the fifteenth century they seem to have retained the round Moorish shield for protection against the darts of the enemy—at any rate the capering 'genitors' of the great picture of the battle of Fuente-higuerola, illustrated in the preceding volume, have shields (sometimes bearing armorial devices) just like the Granadan horse with whom they are contending.[1] The tactics of the 'genitors' were to swarm round the enemy, to overwhelm him with darts, to draw off if he charged in mass, but to hang upon his flanks and charge him when he grew tired, or fell into disorder. They carried swords for close combat, when their store of javelins should have been exhausted.

In support of the 'genitors' there marched squadrons of heavy men-at-arms, the King's bodyguard, the contingents of

[1] See Illustration in vol. ii. p. 180.

the great nobles, and of the four Military Orders. Their number was not very great compared with the extent of the realms of Castile and Aragon, since the national preference lay in the development of the light horse. When Spain was drawn into the great Italian wars, the proportion of men-at-arms was always low in her armies. The new development was in the matter of infantry, which became all important. When first Gonsalvo de Cordova crossed into Calabria in 1495 with 600 horse and 1500 foot, it was on his 500 'genitors' that he relied, not on his 100 men-at-arms or his rather miscellaneous infantry, of whom many were cross-bowmen, the majority Aragonese sword-and-buckler men, and a few arquebusiers.

From causes which it is impossible to discover, the Spaniards had taken to the smaller firearms much earlier than the French or the English or the Italians, to all of whom they were unfamiliar in 1495—though the Burgundian dukes, the Hussites, and even the Swiss, had already begun to employ them a generation earlier. Gonsalvo's first—and only disastrous—battle at Seminara, in the very toe of the Calabrian peninsula, seems to have set him searching for new tactics. His 'genitors' were completely driven off by the charge of the French gendarmes, and the Swiss pikemen ran over his miscellaneous infantry in one rush. Henceforth he not only set himself for a time to avoiding battles, and adopting *guerillero* methods of surprising detachments and cutting off convoys, but took to providing himself with pikemen to cover his cross-bowmen and arquebusiers. The sword-and-buckler men were placed in rear of the pikemen, with directions to run in with their short weapons, when the pikes should have got locked in a frontal crash. This was the same function which the halberdiers were supposed to discharge in the Swiss service, but the short stabbing sword was much more effective than the halberd in a jammed formation, since the latter required not only strong arms but room to swing the ponderous weapon. Moreover, the arquebusiers were multiplied as fast as possible, and the cross-bowmen cut down in numbers.

Gonsalvo also adopted the system of 'digging himself in' whenever possible, and receiving charges rather than delivering them. His first great triumph, the investing and staving out of Montpensier's army at Atella (July 1496), was achieved, without any general action, by a combination of partial attacks and careful entrenchment. His second, the battle of Cerignola

(April 1503) was a case of inducing an enemy to charge in upon a prepared position. The Duke of Nemours had taken the offensive, with the usual combination of gendarmerie and Swiss pikes, hoping to run down the enemy by a determined rush. The Spaniard had taken up his position on the lower slopes of a vine-clad hill, and had set himself to throw up a hasty field-fortification at its foot. He dug a ditch along his whole front, made a bank behind it of the excavated earth, and planted it with stakes grubbed up from the vineyards. The whole line was held by infantry; the comparatively few Spanish men-at-arms were in reserve, the 'genitors' were sent out to worry and delay the advancing French. Nemours, arriving in front of the enemy late in the afternoon, and prevented from reconnoitring Gonsalvo's position by the 'genitors' who were hanging about his vanguard, was persuaded by his captains to attack headlong in échelon—the men-at-arms leading as the right échelon, the Swiss pikes and Gascon crossbowmen formed as a 'main-battle' and 'rereward.' He had no knowledge of the existence of the ditch, and his leading horsemen were precipitated into it; the mass had to halt, and while the general was riding along the front, looking in vain for a gap, he was shot by an arquebusier. The cavalry came to a stand in disorder. Meanwhile the French and Swiss infantry arrived on the ground and charged the centre of the Spanish position, but entirely failed to cross the trenches, owing to a hail of arquebus-fire, which brought down all their leaders while they were trying to scramble up the bank of loose earth, after having jumped down into the ditch. When it was clear that the attack had failed, and that the enemy was morally beaten, Gonsalvo ordered his whole line to advance, his cavalry from the reserve coming round the two flanks of the French, while his infantry passed the ditch, and fell on the Swiss and Gascons frontally. The enemy, having no leader to issue orders, since Nemours had been shot, gave way in all directions; most of the gendarmerie got off, but the infantry, retreating in disorder, were dreadfully cut up by the 'genitors,' who pursued them for many miles. The French train of artillery, which had never got into effective action, was captured on the road behind the battlefield; so was the whole of the baggage of Nemours' army.

Fabrizio Colonna, the celebrated Italian condottiere, who had been with Gonsalvo's cavalry-reserve this day, made the very sarcastic remark that it was neither the courage of the troops nor the steadfastness of the general that won the day, but a little ditch, and a parapet of earth, and the arquebus.[1] But surely Gonsalvo deserves all the credit for having made the combination of entrenchment and small-arms fire. Of course, the enemy had been most obliging!

The Spanish cavalry was never the decisive factor in any of the Italian campaigns. The 'genitors' were always useful, and superior to any other light horse that the enemy put into the field, particularly to the 'stradiots' whom the Venetians also, and the French occasionally, employed. But the heavy men-at-arms, though fairly good in quality—whatever Machiavelli may say against them—were always too few to cope with the French [2]—hence the way in which we find Gonsalvo and his successors habitually enlisting Italian condottiere men-at-arms, though aware of their doubtful quality. But such Spanish heavy cavalry as there was was reckoned useful and reliable, and French narratives such as that of Bayard's 'Loyal Servitor' repeatedly mention knights of distinction, such as Pedro de la Paz and Alonzo Sotomayor, who were well worth fighting. One nasty habit, however, is laid to their charge—that of aiming at the horse rather than the horseman in close combat, which was reckoned unknightly. The 'Loyal Servitor' says that they had a proverb, 'Muerto el caballo perdido el hombre,' [3] and regularly practised horse-killing as good tactics. This came out very strongly in the celebrated combat of thirteen French with thirteen Spanish knights in front of Barletta in 1503, during a truce which had been agreed upon by Gonsalvo and the Duke of Nemours. Though this was supposed to be a chivalrous contest, 'lesditz Espaignoles ne taschérent pas aux hommes, mais à tuer les chevaulx, ce qu-ils firent jusqu' au nombre de onze, et ne resta à cheval que le seigneur d'Oroze et le bon chevalier Bayard. Mais cette tromperie ne servit guères aux Espaignolz : et lesditz seigneur d'Oroze et le bon chevalier leur livroient aspres assaultz, et quand la grosse troupe les vouloit charger, se retirent derrière les chevaulz morts de leur compagnons, ou ils éstoient comme contre un rempart. Pour conclusion les Espaignolz, quoique

[1] See Paulus Jovius, *Vitae Illustrorum Virorum*, pp. 253–55.

[2] When the campaign of 1536 was starting, the emperor's muster-rolls show us that there were only 580 Spanish men-at-arms in Italy. Clonard, iii. 326.

[3] Bayard, liv. p. 47.

ilz fussent treize à cheval contre deux, ne sœurent obtenir le champ jusqu' a la nuyt feust survenu.' [1] Evil habits (such as the use of poison gas) easily spread from one enemy to another, and much later in the wars we find a French knight confessing that, he in his turn, once practised the ungentlemanly trick when in a desperate strait.

At Gonsalvo's 'crowning mercy,' the battle of the Garigliano (December 29, 1503), he completely destroyed the French army of Italy—both the wrecks which had escaped from Cerignola, and the heavy reinforcements which Louis XII had sent down to join them under the Marquis of Mantua, the condottiere-prince who plays a considerable part in all these wars. At Fornovo he had commanded against the French—seven years later he is in charge of a French army, and apparently saw nothing very odd in the change. National feeling in Italy was still very far in the future! When this Gonzaga of Mantua retired, really or officially sick from November chills, his place was taken by Ludovico, Marquis of Saluzzo—just such another princely professional soldier. The campaign to which the battle, or rather rout, gave its conclusion, reads much more like a modern than a mediaeval series of operations, and must be dealt with in detail, when we are considering the great general actions of the war. There was no set and orderly clash of armies in array, but a series of manœuvres, depending on the maintenance by Gonsalvo's army of the line of the lower Garigliano—then swollen by autumn rains and ending near the sea in the celebrated 'Marshes of Minturnae,' where Marius of old sought refuge in the mud. The game was won this time by 'major tactics,' not by armament—the Spaniard having attacked, at mid-winter and in wild weather, an enemy who had left his lines along the river undermanned. A sudden and unexpected passage of the river by a concealed bridge brought the Spaniards into the middle of the French cantonments, and the Marquis of Saluzzo, failing to concentrate in time his much dispersed detachments, was chased into Gaeta after a series of rear-guard skirmishes.[2] A few days later he capitulated, on much the same terms that Junot got from Sir Hew Dalrymple at the Convention of Cintra—leave to depart by sea

[1] Bayard, xxiii. p. 77.

[2] It is amusing to find that the 'Loyal Servitor' of Bayard has no more to tell of this campaign than a rather incredible story of how his master held a bridge alone against *two hundred* Spanish men-at-arms till succour came. Bayard, chap. xxxv. There is nothing about the disaster in which the whole army was routed.

on evacuating Gaeta and the other forts which he was holding, and abandoning all their guns and military stores.

If armament seems to count for little at the Garigliano, it is otherwise with the next battle in which the Spaniards took a leading part. Ramon de Cardona, the successor of Gonsalvo as viceroy of Naples, was endeavouring to relieve Ravenna, then closely pressed by Gaston de Foix and likely to fall. Cardona, trying the same game of fortification which his predecessor had carried to such a successful end at Cerignola, entrenched himself, and asked to be attacked. The experiment failed disastrously, because the French, who had a superior artillery, bombarded the entrenchments, and after a while forced the Spanish horse to take an unintended and hopeless offensive. But, as will be told elsewhere, the battle was largely remembered not so much for the Spanish defeat, as for the frightful slaughter which the sword-and-buckler men and the arquebusiers made of the French infantry—landsknechts and Gascons—whom they beat most effectively in the centre of the battle. Though the day ended in disaster, owing to the complete rout of the Spanish cavalry, the good effect of the sword-and-buckler fighting was the main thing which struck contemporaries; it is one of the main arguments used by Machiavelli in his *Arte della Guerra* for his scheme for arming infantry with short weapons, and not with the unwieldy pike which the Swiss and landsknechts had made famous. He was scholar enough to quote, as parallel instances, the way in which the short sword of the Roman legionary in days of old had hewed its way through the Macedonian phalanx—a formation much like the Swiss pike-column—which had formed the best part of the army of Pyrrhus of Epirus, at the battles of Heraclea and Beneventum seventeen centuries before.[1]

Of Ravenna much must be said in detail elsewhere, since it was a fight of high tactical interest. One of the notable features in it is the fact that the Spaniards were making tentative experiments in the way of forming units larger than the company, though much smaller than the ' tercio ' which was to form the normal basis of organization twenty years later. As early as 1505, King Ferdinand had made a list of twenty superior officers, who were to be entrusted with charge of several companies, apparently some of pikemen and some of arquebusiers, with a proportion of sword-and-buckler men. It is not clear

[1] Cf. Machiavelli, *Principe*, § 26, and Guicciardini, book x. 285.

that the companies were permanently attached to each other, or that the pike and arquebus elements were always of the same strength. These officers are called 'coloneles'—which the French and English shortened into 'colonels' by the middle of the century; originally it would seem that the proper title was 'cabo de colunela'—head of a column. But the perversion into 'coronel' is found as early as 1508.[1] It is interesting to find that of the twenty original colonels of 1505, five were killed at the battle of Ravenna—each at the head of his 'colonelcy.' Of the twelve actual colonels of 1512 present at Ravenna, no less than eleven fell. The strength of these bodies must be gauged from the fact that eleven were in the field that day, but the whole showed only 9000 present. Apparently, then, a 'colonelcy' was probably in theory 1000 strong, and composed of about four companies of 250 men—or conceivably five, allowing for one of 'sword-and-buckler men.' No doubt they were all under strength on the battle morning.

The old corps were practically destroyed at Ravenna—only 3000 men at most got away. But when the army was reformed, the organization seems to have remained the same, perhaps with a strengthening of the proportion provided with firearms. In the battles of the middle section of the Great Wars the Spanish generals practised intrenching more carefully then ever. The object was to induce the enemy to take the offensive against a prepared position well garnished with arquebusiers, so that even if they succeeded in closing, it would be in a thinned and disordered array, which the pikemen could easily cast back. In the next Spanish battle after Ravenna, that against the Venetians at La Motta (October 7, 1513), the Viceroy Cardona saw his horse badly cut up, but the hostile infantry advance came to a disastrous stop, along a trench thickly lined with arquebusiers. The credit of the day fell entirely to the Spanish infantry, commanded on that occasion by the Marquis of Pescara—an unlucky combatant at Ravenna, but fated twelve years later to be the most prominent actor in the great victory of Pavia, where the arquebus settled everything.

There does not seem much to be discovered of either tactical interest or details of changing armament by a study of the Spanish conquest of Navarre, which was conducted in the same year during which Cardona was fighting his battles in Italy, We read of manœuvring, of a long siege of Pampeluna by the

[1] For all this, see Clonard, vol. ii. pp. 414-16—careful documentary evidence.

French, and of their final retreat across the Pyrenees, abandoning the kingdom of the d'Albrets, which came to an end in 1513, though its vain title remained to their descendants, along with the single town of St. Jean Pied-du-Port. This was the only scrap of real Navarrese territory that was in the possession of the prince who made his name famous as ' Henry of Navarre ' seventy years after.

But military history on its Spanish side commences again to be of note when we reach the period of the central campaigns in Lombardy which started with the second occupation of Milan by the French after the victory of Marignano. Bicocca (1522) was somewhat of a repetition of Gonsalvo's victory at Cerignola—a complete and disastrous repulse of an attempt to storm by heavy infantry columns a well-garnished position lined with arquebusiers supported by pikes. The frontal hindrance on this occasion was not, however, an intrenchment, but a deep-sunk country lane, ten feet deep—which came to much the same thing. Lautrec, the French commander, had not wished to make a frontal attack, but the Swiss mercenaries who formed the bulk of his infantry, persisted in bringing on a battle despite of his remonstrances, and got a most bloody repulse. The heads of their columns were shot to pieces by arquebus fire helped by artillery fire, and such disorderly remnants as crossed the lane were driven back by a general charge of Spanish and German pikemen.

At Pavia (1525) the Spanish arquebusiers got more credit than at any of their earlier successes, for this was not a battle in which they were attacked while under good cover of an entrenched position, but a case where they were on the offensive and in open ground. It will be necessary to describe this complicated battle at length elsewhere. Here it must suffice to say that the army in which they were serving had turned the flank of the French position, and forced Francis I to come out and form ' front to flank ' on unexpected ground. The desperate charges of the French gendarmerie are recounted to have failed not so much from the frontal resistance which they met, as from the perpetual rolling fire of arquebuses which beat upon them from the side. The Spaniards had moved forward and turned the King's flank, and could not be driven away by any small or partial charge.

It will be remembered that it was in 1533 that Francis I took in hand the reorganization of the French army into 'legions.' It is interesting to find that it was in the following year 1534 that we find the first mention of the creation of large infantry units in the Spanish host, thrice the size of the 'colonelcies' of the previous generation. These were the 'Tercios' whose name was famous for many a day. Apparently the style came from the old military tradition that an army in array was divided into three divisions—van, main-battle, and rear—as we have so often seen. A 'tercio,' or third, was therefore a body sufficient to make up one of these normal divisions.

The most striking fact in the 'tercio' organization is that we find only pikemen and arquebusiers. There is no mention of the once-celebrated 'sword-and-buckler men' whom Machiavelli so much admired, and we discover only the modest number of eight halberdiers, and they apparently attached to the person of the commander of the 'tercio' as bodyguard, or perhaps rather as orderlies. The companies or bands, which used to be separate units, though often employed in blocks of any number from two to ten, now form one large regiment —almost we would say brigade. This curious and interesting organization of the three original great 'tercios' of the old army of Italy—named Lombardy, Naples, and Sicily—as approved by Charles V in 1534, is worth giving, for with slight modifications it continued in use for the core of the Spanish infantry for two centuries. These heavy corps of over 3000 men of all ranks were organized as follows:

(1) Staff—

1 maestre de campo [colonel]	at 40 escudos *per mensem*.
1 serjeanto mayor [major]	at 20 ,, ,,
1 furriel mayor [adjutant]	at 20 ,, ,,
1 municionario [quartermaster and armament officer]	at 10 ,, ,,
1 staff captain [capitan barrichel]	at 12 ,, ,,
1 lieutenant of the above captain	at 6 ,, ,,
1 physician	at 12 ,, ,,
1 surgeon and 1 apothecary	at 10 ,, ,,
1 chief chaplain	at 12 ,, ,,
1 drum-major	at 10 ,, ,,
8 halberdiers, attached to the colonel, each	at 4 ,, ,,

Total of regimental staff, 19 persons, costing 194 escudos *per mensem*.

(2) The 12 [sometimes only 10] companies, 6 of arquebusiers, 6 of pikemen [or 5 and 5], each consisted of—

1 captain	at 15 escudos *per mensem*.
1 alferez [ensign]	at 12 ,, ,,

1 serjeant	. . . .	at 10 escudos *per mensem*.
10 corporals	. . . .	at 4 " "
1 'furriel' [quartermaster]	. . . . .	at 3 " "
240 privates; the pikeman had all 3 escudos *per mensem*, but among the arquebusiers there were extra allowances for distinguished soldiers, which brought some of them up to 4 escudos *per mensem*.		
1 captain's page [batman]	. . . . .	at 4 escudos *per mensem*.
1 drummer and 1 fifer	. . . . . .	at 3 escudos each.
1 chaplain	. . . .	at 10 escudos.

Total in the companies, 3096 of all ranks for a 12-company 'tercio.' Cost for a pike-company 815 escudos *per mensem*, for the arquebusier companies something more, owing to the 'extra testoon' given to first-class privates, which brought the total to sums varying up from 900 to 950 escudos.

There are many points to notice in this table, the first of its kind available.

(1) The arquebusier is reckoned more valuable than the pikeman, the premium to the best shots in the company raising their pay to 4 escudos a month, while no pikeman draws more than 3.

(2) The rank of lieutenant in a company has not yet come into being; it is clear that the single sergeant in each company must have been practically treated as a commissioned officer. But, even so, the allowance of officers—only 3 to a company of 258, is very low. The only lieutenant mentioned is the assistant of the staff captain. Later, under Philip II every company had a lieutenant, who ranked above the alferez, and also four sergeants (now clearly non-commissioned officers) instead of one.

(3) The allowance of chaplains is enormous, one chief and 12 company clergy—13 in all to the equivalent of a brigade. In this, as might have been expected, the Spanish army was exceptional!

(4) On the other hand, the medical staff is very moderate—only 3 to 3096 persons!

(5) The brigade music runs to a chief and 24 other ranks; the drummers and fifers draw the same pay as a pikeman. The enormous salary of the drum-major (10 escudos) is explained by the fact that he was not only a director of music but a sort of specialist: 'he is expected not only to

teach marches, calls to arms, and calls to retreat, but to be acquainted with the military music of other armies, even the Turkish, so that he can judge the meaning of the sounds heard in the enemy's camp, or line of march, or battle array, and report to the colonel.' [1]

While the 'tercios' remained at their very high establishment, the units which Charles V and Philip II raised during the wars of the middle and later years of the century were usually much smaller in size. They were habitually called 'regiments,' and their commanders were 'colonels'; but their strength varied in the most irregular fashion. Four or five companies, and no more, put together formed a regiment, sometimes named after its colonel, sometimes after the province in which it had been originally raised. The number of 'tercios viejos' of the original army of Charles V and Philip II was limited, the units being so large; they all had local designations, like the first three which have been mentioned above.[2] 'Sicilia' and 'Lombardia'[3] survived for a very long time, but 'Naples' was disbanded in disgrace for a series of mutinies by Philip II in 1566.[4]

The muster-roll of the Spanish army in July 1536, when Charles V was commencing his third war with Francis I, happens to have been preserved, and is interesting as showing the exact force of the Spanish and non-Spanish elements in the Emperor's army of Italy. There are found--

Heavy cavalry [gente d'armas]	580
Spanish Light Cavalry	1,730
'Tercios' of Naples and Sicily	5,000
28 unregimented companies	4,850

At the same time the Emperor had in Italy of foreign troops:

German cavalry	2,060
Italian cavalry	950
German infantry	24,080
Italian infantry	25,903

[1] For an example of the way in which a knowledge of military music might prove profitable, see Bayard's escape from Rebecco, during a night surprise, when he got off because he knew that the drums heard behind him were Spanish infantry drums 'beating the alert,' and could be nothing else. Bayard, cap. lxiii.

[2] The Tercios of Portugal, 'Armada,' *i.e.* Marines, and Brabant were early corps. All this from the invaluable Conde de Clonard's *History of the Spanish Army*, vol. iii.

[3] Lombardia changed its name to Principe after Italy was lost in the eighteenth century.

[4] But a 'Tercio Nuevo de Napoles' was raised to replace it, which changed its name in 1718 to 'Corona' now that Naples was no longer Spanish.

The Italian contingents included those of the duchy of Milan, the Marquises of Saluzzo and Mantua, of the newly made Duke of Florence, and the Duke of Savoy. The rest were mercenaries.

A muster-roll of the national Spanish cavalry existing in the Iberian Peninsula two years later, at a time when the French were active on the Pyrenean front (1538), gives the very modest total of 961 men-at-arms and 655 'Ginetes.' It is clear from these figures, and from those quoted above, that Spanish armies had always a sad deficiency in the mounted arm, according to sixteenth-century ideas. In the army of Italy in 1536 there were only 5320 horse in an army of 67,155 of all arms and all nations. We find at Ceresole in 1544 under 1000 to 18,000 foot.

Of the numbers, organization, and tactics of the Spanish armies in the second half of the century, and more particularly during the Great Wars that followed the Insurrection of the Netherlands, much will be gathered from a later chapter. The tactics are particularly noteworthy, and were the admiration of all professional soldiers. The main points to be noted were the supersession of the old light horse by squadrons furnished with firearms, both mounted arquebusiers, and *herreruelos* who carried a shorter weapon which could be more easily used from the saddle, and the multiplication of small foot regiments as opposed to the immense 'tercios' of the earlier wars. The combination of pike and firearms continued, as in other European armies, far into the seventeenth century. The heavy cavalry, never proportionately numerous, continued to use the lance long after it had been abandoned in other armies. It was only discontinued by an edict of Philip III at the very end of the Netherland 'Wars of Religion.' The least admirable trait of the very formidable Spanish troops of this later generation was their proneness to mutinies or strikes on account of the non-appearance of their pay—a foible in which they rivalled the landsknechts and Swiss of the earlier years of the century. A regiment would put its commander under arrest, appoint an *eletto* or substitute, generally a veteran non-commissioned officer, and settle down in its cantonments, living on the country-side by organized requisitions, till the Viceroy of the Netherlands produced the missing dollars. How maddening this habit could be to commanders-in-chief hardly requires explanation. Only men like Alexander of Parma or Spinola could cope with it. But of this more in its proper place.

CHAPTER VI

THE SWISS AND THE PIKE-PHALANX

IN the narrative of the wars of the fifteenth century no small space had to be devoted to the growth and perfection of Swiss military tactics. And after the disasters which ended the dream of Charles the Bold for the establishment of his 'middle kingdom' of Burgundy, the eyes of soldiers all over Europe were turned on to the new form of warlike efficiency which had been proved to exist in the dense columns of the pikemen, who swept everything before them at Granson, Morat, and Nancy. Imitation is the best form of flattery, and before Charles the Bold had been twenty years in his grave, many professional soldiers had begun to think out the problem of how the Swiss tactics could best be reproduced. On the other hand, there were others who were pondering how those tactics could be met and frustrated. For some forty years the steam-roller-like advance of the three great masses of pikes in échelon struck dismay into enemies, and flattened out everything that dared to stand in its way. Its last typical triumph was at Novara (1513), its first notable check at Marignano (1515); after the bloody battle of Bicocca (1522) hostile generals thought that they had discovered the secret of how to deal with the Swiss.[1] And it is said that they themselves never again displayed the same magnificent confidence in victory which they had hitherto shown after that very depressing failure.[2] Certainly three years later, at Pavia, they showed that they had lost some of their old fury, and indeed disappointed the King of France in the most unexpected fashion.

The psychology of the Swiss during the period of their military predominance is very curious. They seem to have had little desire to build up a 'greater Switzerland' by foreign

[1] But on a very small scale Gonsalvo de Cordova had tried the same trick of getting them to attack intrenchments well manned. See p. 53.

[2] Guicciardini, xix. p. 128. Cf. Montluc, i. p. 24, and Du Bellay, ii. pp. 220–21.

conquest, contenting themselves with small territorial annexation near at hand; Vaud, taken from the dukes of Savoy, was an obvious acquisition: the nibblings off the northern edge of the duchy of Milan—Bellinzona, Lugano, and the Valtelline were mere trifles. They never annexed the lands of the minor ecclesiastical states which bordered on them—such as the bishoprics of Basle and Sion, and the Abbacy of St. Gall, but contented themselves with treating them as 'allies'—subject-allies of course, and leaving them alone. Only once does it appear that they displayed any broader views of conquest—this was during the three years 1512–15, when, having restored Maximilian Sforza to his duchy of Milan, they took over all authority from him, garrisoned his towns, raised his taxes, and allowed him to issue no political orders. But they did not actually annex the Milanese, and apparently were more set on getting money than on extending their borders.

Money, it must be confessed, was at the root of all the actions of these formidable and rather enigmatic mountaineers. 'Point d'argent, point de Suisse,' as was so often remarked. Indeed, the policy of the Confederates, speaking generally, was to exploit the foreigner to the utmost possible limit—but this renaissance 'fremdenindustrie' was carried out by the simple expedient of selling military aid, but withdrawing it with the sharpest accuracy when payment ran short. At one time or another the dukes of Milan, the Venetians, and the Pope bought their pikes, but the King of France was the most usual purchaser, and indeed after the so-called 'perpetual peace' of 1516 he enjoyed almost a monopoly of their costly service.

These Swiss troops were most tiresome material: if pay was a little in arrears they went on a passive strike, or simply marched back to their mountains. They had the strangest ideas of their privileges. When the Marshal de Montmorency once told them off for siege operations, they answered 'that they were always ready to fight in the open field, but that it was not their business to assault breaches.'[1] Four days before the battle of Pavia, while the French and the Imperialists were already facing each other in their trenches, a corps of 6000 Grisons suddenly marched off from the French camp. They had received news that the enemy had captured Chiavenna, their chief fortress in the Valtelline, and was threatening their

[1] Du Bellay, ii. p. 210; *ibid.* iii. p. 385.

own borders, and on a call from home simply departed, though they were *not* in arrears of pay. At Bicocca, the defeat already named, in 1522, when they *could* complain that there was something owing to them, 'their captains came to M. de Lautrec and offered him one of three choices—to pay their wages on the nail, or to give them leave to go home at once, or to attack the enemy without a moment's delay.' The unfortunate general pointed out to them that Imperialists were starving, and would probably have to retreat in a very few days, and to abandon their fortified position. The Swiss would listen to no arguments—if the attack were not made that afternoon they would start home next morning. As they formed the best part of Lautrec's infantry, he groaned and gave way. It has been already mentioned they were foiled by a sunk-road and a continued blast of Spanish arquebus fire, and so the duchy of Milan was lost. 'I have seen,' says Montluc, 'the perversity of these people cause us the loss of many a town, and wreck the King's campaigns. It is true that they are veritable soldiers, and form the backbone of an army, but you must never be short of money if you want them—and they will never take promises in lieu of cash.' [1]

The cases just quoted may pass as what Montluc called mere 'perversity.' But there are more sordid incidents to mention. The Swiss captains who had restored Ludovico Sforza to his duchy of Milan in February 1500, and who for two months had governed it for him—or at least in his name—found themselves faced in April by a new and formidable French invasion led by the great condottiere Trivulzio, and aided by other Swiss bands. Offered subsidies on a large scale by Louis XII, the captains threw up their alliance with Sforza, and ordered their troops to return home. 'Dog does not eat dog.' All that the Duke could obtain was leave to retire in disguise among the marching pikemen. But he was recognized, and the French were allowed by the Swiss captains to carry him off to a lifelong captivity.[2] Little better was the record of 1515: once more the Swiss had restored a Sforza—Maximilian, son of Ludovico—to the Milanese duchy. Francis I set to work upon them not merely with an army, but with wholesale bribery. The Duke was in arrears with the pay for his pikemen—Francis offered 400,000 gold crowns down,

[1] Montluc, i. p. 24.

[2] See Guicciardini, iii. p. 278, for details of this black business.

300,000 more for the possession of Bellinzona, Locarno, and Lugano, then in Swiss hands, and an annual subsidy to each canton. There was an animated debate in the camp of the Confederates—some held that the King's terms were too good to be refused; a written agreement was drawn up, and the contingents of Bern, Fribourg, Soleure, and Valais actually sent their message of approval to Francis, and went off home.

The remaining contingents, moved by the appeals of their countryman, Matthias Schinner, cardinal bishop of Sion, who demonstrated that Francis had been excommunicated by the Pope, resolved to fight. They made a surprise-attack on the King's cantonments, which ended by their bloody repulse at Marignano (September 13). This did not prevent the conclusion of the 'Perpetual Peace' with France in the following November, and the making of Switzerland into what was practically a French recruiting-ground. For the next thirty years no French army went forth without a large Swiss contingent. It had been determined that the landsknechts, who had been used as a 'colourable substitute' when the Confederates were alienated, were not equal to the genuine article—as witness certain happenings at Ravenna (1512) and Novara (1515) mentioned elsewhere. The fact was to be verified again at Ceresole, thirty years later, when the Swiss swept the Germans away once more.

The result reminds one of the cynical dictum of the Greek philosopher about the ladies—'it is impossible to get on without them—or with them.' They were far more formidable in battle than landsknechts, native French, or Italian infantry. In short, they earned their pay while they were getting it. But they were tiresome to a degree—they violated terms of surrender granted to a hostile garrison, and pillaged towns that had capitulated. Their own officers often failed to control them—a Swiss corps seems to have regarded itself as a sort of trades union, which could overrule its presidents and committees. They were unnecessarily cruel in battle—and took no prisoners—especially from among the landsknechts, whom they hated as a sort of impostors who were trying to turn their own patented methods against them. But their 'lightning strikes,' to use a modern term, were the most exasperating thing to the French generals who had to utilize them. These might occur not only when the monthly pay had not turned up, but when something had happened to offend the personal, corps, or national pride of the contingent. And the commander-in-chief could not apply military discipline, for the Swiss were a law unto themselves. They had, indeed, a strict camp-code of their own, and would punish by death or otherwise individuals who had offended or defrauded their comrades :[1] but it could not be used by any one but themselves.

The Swiss had practically no cavalry of their own ; when any men-at-arms are found with them they are almost exclusively French or Italian. But they were not wholly insensible to the importance of the arquebus—especially after their experiences at Cerignola, and still more at Bicocca, and the three regulation columns of pikes were generally garnished with a certain amount of firearms on the flanks. In exceptional cases the proportion was known to go up to one in five or one in six. But the ' shot ' were not considered part of the striking force, which lay entirely in the pikes : they were only intended to cover the flanks of the columns, or to go out in front to skirmish before the charge began. One hears of no battle in which the Swiss arquebusiers played any decisive part—while in the Spanish army the firearms are cited frequently as the important element—especially at Pavia. Similarly the Swiss had a small proportion of cannon—less than any other national force : they are sometimes mentioned, but never accomplished anything particularly effective. There was always a '*büchsenmeister*' appointed in later campaigns. But it was not indeed to be expected that the Helvetian mountains would breed either cavalry or artillery.

A clear distinction must be drawn between the organization of a Swiss force in the cases where the Diet of Cantons had agreed on a national war, and fitted out a federal army of its own, and the cases where, as was usual in the later wars, the Swiss contingent was only a hired unit in a French army composed of all manner of cosmopolitan troops. Examples of the former organization may be found not only in all the earlier wars with Charles of Burgundy and the Emperor Maximilian, but in the cases when the Confederates interfered as an independent political power, with an army which was entirely, or mainly, their own. In the campaigns of Novara and Marignano they furnished nearly the whole of the force opposed to the French, being only accompanied by small contingents of the

[1] See, for example, in Giovio, xi. p. 226, the awful fate of a Swiss captain who had secret intrigues with the French—he was hung, drawn, and quartered.

Duke of Milan's men, mainly cavalry. But in the vast majority of the engagements of the second half of the wars, as at Bicocca, Pavia, or Ceresole, they formed but one corps (though that a very important one) in a long line of battle, and were supposed to be under the orders of the French commander-in-chief—which they did not always obey, as will be seen.

In the case of a national war and a common expedition, each canton, and each of the 'allies' also, had to supply a contingent proportional to its population, which served under the cantonal banner. But there were many districts which from their importance were allowed to have a minor flag (*fähnlein*) of their own—these being sometimes rural fractions of a large canton, sometimes small towns within the canton, which had an entity of their own, *e.g.* under the big banner of Bern, there were small flags for Thun, the Haslithal, Nidau, Aarau, Bienne, etc.; under Zürich for Winterthur, Rapperschwyll, and Regensberg; under Lucerne for Willisau and the Entlibuch, etc. In the central column at the battle of Granson the Burgundians counted no less than forty standards—some cantonal, some local. There would appear to have been a rough rule that if a district contributed less than 200 men to a levy, it must not bring out its sub-flag, but be contented to serve under the great cantonal banner.

The cantons being so widely different in size—Bern must have sent out six or eight times the contingent of Uri, Zug, or Glarus—there was a serious problem to be solved when the three normal columns of attack were being formed up, since there was no common unit of size, and one of the three obligatory masses of pikes might be made of the troops of three large cantons or of ten small ones, as was decided before the battle. There is unfortunately little to be gathered about the system of brigading[1]—but apparently the three original cantons, the Waldstädte, generally marched together, and Bern's very large block of men was sometimes made up with the modest contingent of Valais. At Bicocca, where the Swiss did not form the whole of Lautrec's French army, we are told that they arranged themselves in two columns only, each 4000 strong, of which the one was composed of all the troops of the rural cantons and allies (Uri, Schwytz, Unterwalden, Glarus, Appenzell, Grisons, Valais, etc.), and the other of those of the

[1] At Marignano, where Bern was absent, Zürich and the eastern cantons had the right; the old central cantons the 'battle'; Lucerne, Basle, Fribourg the left.

city cantons (Bern, Zürich, Basle, Schaffhausen, etc.). But this may have been a very abnormal arrangement.

But in the later periods of the war, in which there was no general levy of the whole Confederacy, but only a contingent granted to the King of France—to be raised as might be determined—it would seem that contractors, old condottieri of experience, were authorized to raise units. It appears that inter-cantonal jealousy was always sufficient to make these captains keep their bands homogeneous, and confine their recruiting to men of their own particular district. Hence we find in one case a corps of 6000 men from the three leagues of the Grisons—a very large number for that mountain land—and in another no less than 3000 from the Gruyère district of Canton Fribourg alone.[1]

The most extraordinary thing about the Swiss contingents is that they never seem to have been under the proper control of a commander-in-chief, but worked under a committee of captains, who used to meet in council and discuss tactics before a battle, often with grave dissension and consequent mismanagement. No single Swiss officer all through these wars seems to have risen to the position of a capable general—we find instead a war-council of old soldiers of the mental calibre of an intelligent sergeant-major, or at best of a battalion commander. Something of this may be explained by the fact that in the democratic rural cantons the men elected the officers. They did not, it is true, elect demagogues, but they did choose to be commanded by war-scarred veterans of no education or inventive talent. Things were a little better in the 'aristocratic' districts, where the contingent-commander was often a magistrate or ex-magistrate from one of the predominant families of the canton. But even the old Bernese aristocracy did not produce generals, but at best colonels: and inter-cantonal jealousy at a war-council tended to result in the majority voting down any officer who showed a dangerous preference for new-fangled ideas, and for abandoning the time-hallowed practices of the Confederate army. But in their own limited sphere of tactics the old captains showed much skill, in surprises, ambuscades, forced marches, and 'demonstrations.'

A cantonal contingent, or in later times a band raised for

[1] These 'Gruyères' did not do well at Cerisoles, and incur much angry talk from Montluc, who says that they may live in Switzerland, but are no more like Swiss than a donkey is to a Spanish jennet (ii. p. 285).

service under the French, seems to have had a very moderate staff. There was the *Hauptman*, or commander, a *Venner*, or banner-bearer, who was second in command, and took over charge if the hauptman was killed or wounded, an *untervenner*, or under-ensign, a captain of the arquebusiers (*schützenhauptman*), a clerk (*schreiber*) a quartermaster (*proviantmeister*), a chaplain, a surgeon, two horn-blowers, and an executioner (*scharfrichter*). Each of the allied districts (*zugewandete orte*) had a flag (*fähnlein*) whose bearer was the commander of its contingent. There is an extraordinary lack of under-officers of every sort—no signs of sergeants or corporals. Apparently discipline was supposed to work itself by means of public opinion, and fidelity to the cantonal oath, or (later) the oath to the captain commanding, if the unit was one personally raised for foreign service. The company was somewhat like a trades-union, exercising a sharp restraint on ' blacklegs,' such as the man who went away from the banner in pursuit of private booty, or (far worse) the first pikeman who bolted to the rear in an unlucky crisis. The feeling of corps-pride and corps-responsibility was so keen that (as Machiavelli remarks) the first man who made a clear display of panic was certain to be handed over to the hangman, without any pity being shown by his comrades. There are countless records of stubborn resistance to the last man in time of defeat, and when the column had finally to recognize the impossibility of victory, it generally went off with closed ranks, if diminished numbers, and never dispersed. One of the Swiss habits which went for efficiency, but was naturally hateful to the enemy, was the refusal to make prisoners, even if the prisoners looked capable of paying a good ransom. In their great battles, unlike the French and the Germans, they knocked on the head every enemy who went down, without thinking of his value in *écus* or *gulden*. Hence they got similar treatment from the enemy, who would always have old scores to pay off. When Swiss and landsknechts met, it was always a very murderous business.[1]

With an open field before them, not cumbered with water-courses or intrenchments, the Swiss columns were always very formidable in their assault, and could run down any other infantry, and beat off cavalry charges with their hedgehog-front of eighteen-foot pikes. Hence the generals in command

[1] All this comes from Von Elgger's excellent *Kriegswesen und Kriegskünst der Eidgenossen im XIV, XV, und XVI Jahrhunderte*, which is invaluable for detail.

against them used to take up a position which was as unlike an open field as possible—a hillside tangled with walls and vineyards, or (if no hill was available), a camp entrenched with fosse and palisade, or ground cut up with steep or rocky patches of irregular ups and downs. For a column of the heavy deep sort, such as the Swiss employed, had to hang together at all costs. Gaps in its formation, caused by stray obstacles, would allow the enemy to slip in between the divided ranks, with horse, or with close-combat troops, like the Spanish sword-and-buckler men. The oddest device for breaking the order of the front of an advancing column was that devised by Pedro Navarro, the Spanish engineer, at Ravenna. He had evidently been studying the military pictures of late-Roman type, such as those shown in the appendices to Vegetius or in Vulturius. For he produced two-wheeled carts, with a long spear projecting from the front, and two scythe-like sharp blades on each side, which were to be trundled against the front of a hostile mass. In addition they had two or three fixed arquebuses strapped upon them, which were intended to be discharged simultaneously, when the cart was shoved in to the enemy's front line.[1] A big irregularly-shaped opening, caused by the impact of such an engine, would (as it was thought) make a nasty disarrangement of the orderly hedge of pikes. Unfortunately, we have no indication whether these machines were of any effect, or indeed whether they were used at all in the manner intended.

When once the advance of a Swiss force in its normal order of three columns in échelon, the right leading, had started, it could be foiled by a capable opponent in several ways. The simplest was to lure it on to attack an intrenched position garnished with artillery and arquebusiers and with a solid backing of pikemen, in case the Swiss should cross the ditch in spite of all difficulties—as at Bicocca and on a smaller scale at the second Seminara. A more scientific measure was to make a flank attack on the leading column with cavalry or light troops, which would compel it to halt in order to beat off the side-assault. For if the pikes on the flanks had to come down to resist a charge, it is clear that the pikes in front must stop—otherwise the column would break in two. If a frontal

[1] See Guicciardini, x. p. 281. He compares them to the scythed-chariots of the ancients—but they were not drawn by horses, but intended to be shoved forward by men.

hindrance were provided—*e.g.* artillery as at Marignano—the halted column would be shot to pieces. Or it might be blocked by other pike-bearing troops, bringing about the 'push of pike,' in which the front ranks on both sides would go down wholesale, and a mass of corpses or wounded would grow up checking all forward movement, and breaking the impetus of the assault. At Ceresole, toward the end of the wars (1544), Montluc tells us that he used an invention of his own, the placing of one rank of arquebusiers in the column of pikes, between the first and second ranks, with orders to hold their fire till the clash was just coming, and then to blaze into the front of the enemy, which would bring down his whole leading rank, where would be the officers and the picked men. 'I told M. de Tais, three or four days before the battle, that before one of our men had been piked, I would have the whole of the captains in the enemy's front rank killed off.' And so it was—'la se fit une grande tuerie—il n'y avait coup qui ne portait.' [1]

This may have been so—but one reflects that when the volley had been delivered, the arquebusiers (who would take two or three minutes to reload) would be quite helpless if the second rank of the enemy came on steadily. And they could not escape—being jammed between two ranks of their own pikes, whom they must have incommoded seriously. Moreover, if the trick was so ingenious, might not the enemy also adopt it? This actually happened at Ceresole, as is told in the proper place.

A very serious problem for the gross infantry column of many thousand men began to develop in the second half of the sixteenth century, when cavalry proceeded to furnish themselves with firearms, and the 'pistoleer' came to the front. The idea that a squadron, about to charge in, might deliver a volley of pistol-shots among the pikemen, and so produce practicable gaps in their serried front at the moment of contact, fascinated many tacticians, and was put into practice first by the German reiters who carried pistols and sword but no lance. It had its successes, but was largely vitiated by the introduction of the manœuvre called the 'caracole' of which some mention has already been made. When horsemen were taught to fire by ranks, but not to charge in after the first fire, but to wheel about and allow the second rank to fire in turn, and so on for six or eight ranks of pistoleers, the impetus was

[1] Montluc, ii. pp. 286–87.

lost, and there was the gravest danger of the whole squadron getting into disorder, unless it was composed of very steady veteran troopers. But this has been dealt with elsewhere.

The whole tendency of tactics in the middle and later years of the century was in the direction of abandoning the use of the vast column of pikes, such as prevailed in 1500, and of substituting smaller corps, in which pike and arquebus were regularly combined. Instead of the arquebusier being attached in smaller or greater numbers to the pike-column, he became a member of a smaller unit, in which the two weapons acted together on certain fixed principles. The Spaniards stopped raising 'tercios' of 3000 men, and changed over to a system of small regiments. So did the French and the Italians.

When the 'Wars of Religion' began to prevail in Switzerland, as in France and Germany, it is curious to find that the Protestant cantons—Bern, Zürich, Basle, etc.—still kept true to the French subsidiary treaty, and never failed to allow their contingent to be recruited for the armies of Charles IX and Henry III, though they were often employed against a Huguenot enemy. 'Business was Business' in the sixteenth century no less than in the twentieth ! [1]

[1] See Von Elgger's *History of the Swiss Confederation*, vol. ii. p. 150.

CHAPTER VII

THE GERMANS—LANDSKNECHTS AND REITERS

IN the not particularly interesting annals of fifteenth-century Germany, when decisive battles are rare, and wars seem to settle nothing, there is little that requires notice. The numerous defeats of the German knighthood before the *Wagenburg* fortification of the Hussites, the clash before Neuss with the Burgundian army of Charles the Bold, the Emperor Maximilian's early campaign against the French (Guinegate, 1479) and his short conflict with the Swiss (Dornach, 1499, etc.) have been mentioned in another place.[1]

The consequences, however, of Maximilian's disastrous acquaintance with the Swiss phalanx are of real importance, though the combats themselves seem to have been one-sided affairs. The Emperor, albeit a devoted admirer of the old chivalry, and an expert on armour-improvements and the technique of tilting, found himself obliged for practical purposes to turn his attention from the romance of the Past to the unpleasant realities of the Present. The mountaineers had been tiresome enemies of the house of Austria for the last two hundred years, and had gradually stripped his ancestors of their original feudal holdings in the valley of the Aar—including the time-honoured castle of Hapsburg itself. But they had not yet disclaimed allegiance to the Empire—though they ceased sending deputies to its Diets. Hitherto they had not encroached very far; till now the Rhine had been their farthest limit of raiding in the north, the Alps in the south. But now they had come far afield; their armies had run all over Lorraine and Franche Comté; they had passed the Gothard and begun to pick up scraps of the duchy of Milan. Old cities of the Empire like Basle and Schaffhausen linked themselves to the Confederates [1501]. When would the encroachment stop? Mühlhausen in Alsace [1466],[2] Rottweil in Swabia [1491] got

[1] See *Art of War in the Middle Ages*, vol. ii. pp. 272–74.

[2] The federation was renewed in 1513.

into the league as 'allies'—perhaps all the imperial cities of southern Germany might have followed their example.[1] The reason for the sudden expansion of the sphere of influence of the Swiss was a purely military one—certainly it was neither their constitution nor their culture which would attract their neighbours into amalgamation with them. It was simply the fact that they had become the most formidable fighting force in Central Europe, able to defend their independence against all enemies whatsoever, and not merely to hold the mountain lands, but to march far and wide over the plains, defying all the chivalry and cross-bowmen of Burgundy or Milan.

Maximilian had obviously to stop further Swiss expansion. The formation of the so-called Swabian League, by which all the princes and towns of south-western Germany bound themselves together for mutual protection, had proved a failure. Indeed Basle and Schaffhausen, very famous Swabian towns, deliberately got out of it in 1501. A more effective measure appeared to be the formation of a body of national troops, which should copy Swiss tactics, and meet the pike-column by its own methods and weapons. Hence came Maximilian's creation of the 'landsknechts' in—so far as can be made out—the year 1486. The word was not quite new: it can be found used as far back as 1417, but not in its later technical sense of disciplined infantry armed with the pike, and drilled to operate in heavy columns in the Swiss fashion.[2]

The word, as Dr. Delbrück very clearly proved, does *not* mean men armed with a lance—and in fact no landsknecht ever was so armed, for the lance was the horseman's weapon, and a pike is *not* a lance. Nor does it mean 'men of their own land'—as opposed to alien Swiss; nor 'men of the open lands as against men of the hills'; nor 'men for the defence of the land,' nor 'men not raised by the towns but by the countryside.' It was a very vague word originally, which Maximilian chose to specialize into a particular meaning—perhaps to disguise the fact that he was raising a permanent standing army, for whose appearance the Diet of the Empire had no particular enthusiasm.[3] All mediaeval parliaments disliked standing armies.

The first use of the word in its new sense comes, oddly enough, from a document drawn up at Zürich in October

[1] The very important Constance *nearly* came in.

[2] See Delbrück's *Kriegskunst*, iv. p. 9. [3] *Ibid.* 10.

1486, and laid before the Swiss Diet, complaining that a Swabian knight named Conrad Gäschuff, who was training troops for the Emperor, had been boasting in public that he was so arming and drilling Swabian and other landsknechts that in a short time one of them would be better than any two Swiss. This was a foolish outburst, as was to be proved in many a battle: but the interesting fact in it is that Conrad is using 'landsknecht' in the sense of a soldier armed and trained in the Swiss fashion.

Apparently the use of the word and the form of armament was spreading very quickly all round Germany, for in the very next year, when the famous condottiere Martin Swartz took a German mercenary band of foot-soldiers to serve in England in the fruitless rebellion of Lambert Simnel, they were armed with long pikes; and chroniclers abroad call them landsknechts, though the English do not know the name. He and they perished by the cloth-yard shaft of the English archery at Stoke Field [1] (June 16, 1487).

From 1496 onward the word 'landsknecht' is used frequently by all writers, some taking the precaution of noting, when they first employ it, that they mean the Emperor's new pikemen, trained in the Swiss method. There had, of course, been many German foot-soldiers all through the Middle Ages, but they had been cross-bowmen, hand-gun men,[2] or spearmen of the old sort, not drilled in large masses like the Swiss, nor armed with a pike of the enormous length which the Swiss had made popular. There is a curious note in Montluc's account of the battle of Ceresole (1544) in which he says that the difference between the landsknechts and the Swiss is that the former hold the pike very low down, and always keep the point slanted somewhat upwards, while the Swiss grasp the weapon nearer its middle, and keep the point down, which is much better in 'push of pikes,' when the staves are crossed.[3] It is curious to note that in the large panoramic picture of the battle of Pavia in the Ashmolean Museum at Oxford, the few surviving men of the French king's Swiss who are still fighting are represented with precisely this downward slant of their points. I presume that the advantage was that, when two pike-columns

[1] Molinet, iii. c. 158, says that the arrow was so effective because the German pikemen had little armour. This was also the case with the Swiss pikemen.

[2] There are said to have been German hand-gun men at 2nd Kossovo (1456); probably they had learnt the use of the weapon from the Hussites.

[3] Montluc, ii. p. 279.

had clashed, men whose points were held high were likely to find them shoved up well over the heads of the enemy, while those whose points were down would get a better chance of thrusting at the body of an opponent.

Maximilian's first experiences when using his landsknechts against the Swiss were all unfortunate: at Hardt, Frastenz, and Dornach (1497) they were always beaten—either because they were less well trained than the enemy, or because the *morale* of troops accustomed to victory was stiffer than that of troops trying a new experiment, even if well trained in it. But at any rate they made a much better show in these battles than any other foot-soldiery which the Swiss had ever encountered. The Emperor was not discouraged, and continued all his days to persevere with the pike. He encouraged knights and officers to dismount and fight among their men, and recalled the fact that he himself at Guinegate had sent his horse away, and stood on foot among his Flemings. Every one conceded that the landsknechts were better than any French or Italian infantry, and as good as the Spaniards, if they seldom could cope with their old enemies the Swiss, whose most ruthless massacre of them was at Novara (1514), when 4000 of them were fighting under the French flag.

The most extraordinary fact about the landsknechts is that, unlike the Swiss, they seem to have been singularly indifferent about their nationality. Except in their normal civil wars between canton and canton, the Confederates would not fight each other. The disgraceful end of the reign of Ludovico Sforza of Milan came about because there were Swiss both with him and with his enemies, and the two bodies refused to cross pikes—the smaller party betrayed the duke who was paying them, and surrendered him to the larger party.[1] Landsknechts seem to have had no such scruples: there were always thousands of them in an army of Francis I or Henry II, which was fighting against the Holy Roman-German Emperor. Originally these bands were raised by German rebel princes, like Charles of Guelders or Robert of Bouillon,[2] who were in arms against their sovereign. And sometimes in later years we find a German noble, like Philip of Cleves, or the 'Rhinegrave' Philip of Solms, or a count of

[1] See above, p. 65.

[2] Who brought to Novara a corps of 4000 Germans, led by himself and his two sons. Du Bellay, i. pp. 17–18.

Saarbrücken, in command of the 'lanquesnets' as the French called them. But more frequently the numerous German companies in an army of Francis I were led by soldiers of fortune, of humble birth and obviously of no sense of nationality. It is curious, however, to find that the large body of 'lanquesnets' who served in the Navarre campaign of 1521 were commanded by Richard de la Pole, rightful heir in Yorkist eyes to the crown of England, 'the White Rose,' as he was called. He was a great friend of the Chevalier Bayard and apparently a person of some humour.[1] He fell at Pavia in the French ranks in 1525: at that battle the numerous companies of Germans were commanded by Francis, brother of the Duke of Lorraine, another rebel against the Emperor.

An anecdote illustrative of the feelings of the landsknecht leader may be quoted from the time of the battle of Ravenna. The Emperor and the King of France were not at war at the moment, but were on the verge of it. Maximilian sent secret letters to Jacob Empser, who commanded the 5000 Germans in the French army, bidding him take off his troops at once and return to Germany. Jacob showed the letters to the French commander, and said that he was going to keep them dark, as a battle was at hand. His contingent was dreadfully mauled at Ravenna, and he himself was killed.[2] More angry letters coming from Vienna reached the hands of the surviving German captains: most of them agreed that they must go: but one, very much of the professional soldier, said that he had little to lose in Germany, and prevailed on about 800 of his landsknechts to stop in the French service when Maximilian declared war.

Another humorous anecdote, from the Navarrese campaign mentioned above, may be worth quoting as an example of landsknecht mentality. The French were besieging the small fortress of Puente la Reyna, where the wall had been breached. A detachment of the Germans was ordered to form the storming party: to the surprise of Bayard, who was carrying the order, the interpreter attached to the company came forward to say that it was the custom that at assaults of breaches the troops received double pay for the month. Unless the donation were promised them, they would not move. Bayard, greatly enraged, said that time was pressing, and that they should have the money—though he had never heard of such a custom—when

[1] Bayard, p. 225.

[2] *Ibid.* p. 195.

they had taken the place. But they tarried so long that he turned to a company of French 'adventurers,' and implored them to try an escalade. This succeeded, though the risk was great, and the fortress fell. When all was over, three landsknecht captains came to Bayard, and asked that their companies should have the double pay, for the place had been taken. 'But *you* didn't take it,' replied the angry knight, 'and I shall report you to the Duke of Suffolk, your commander, as impudent cowards.' Then followed such a murmur and tumult that Bayard had to call out a company of men-at-arms to protect himself. That evening a very drunk landsknecht burst in to a dinner-party of officers, and announced that he was come to kill Captain Bayard, the man who promised money and didn't give it. Fortunately he was so inebriate that he was put off by being given a long drink—at the end of it he forgot his purpose, and went off saying that no one should harm the captain, for he had very good wine indeed.[1]

Naturally these sort of people, who knew that they were theoretically traitors for being in arms against their liege lord Charles V, drifted everywhere—if not in the French service, you might find them in Poland, Sweden, or Denmark. Protector Somerset somehow collected a band of them to help in putting down Ket's Norfolk rebellion in 1549! If they were lucky, and had not got into trouble with Swiss, who never gave quarter to landsknechts, or with Turks, who beheaded all prisoners, they might drift back into the Imperial service again —from which they had probably deserted in the beginning.

The amount of these Germans serving, at one time and another, against the Kaiser's black eagle banner is surprising, when one looks at such statistics as come to hand. But in a French army they were also very numerous—we find 5000 in the Navarre campaign of 1521,[2] 8000 at Marignano,[3] 4000 at Pavia, five 'companies' under the Rhinegrave (perhaps 1500 men) in the fighting around Boulogne and Calais in 1544; Strozzi and Montluc had ten companies in the war of Siena, under George Reckrod of Hesse, a Protestant refugee[4] (probably 3000 men). In the last battle of the war, at Gravelines in 1558, we find in a French army, quite half of which was composed of foreigners, not only many companies of 'lanquesnets,' but

[1] Bayard, p. 214. [2] *Ibid.* p. 225.

[3] Du Bellay, i. 47. The Duke of Guelders was in command.

[4] Montluc, iii. p. 133.

(rather a novelty) four squadrons of German horse—'reiters,' as they were beginning to be called about this time.[1]

An amusing comment on the supposed relative values of the various national infantries towards the end of the war may be found in the Marshal de Brissac's debate with his captains before the Walls of Lanzo in 1522. 'Don Fernando has at Vercelli 3000 Germans, over and above his Italians and Spaniards, and I have with me neither Swiss nor lanquesnets to put in line against them.' To which Montluc says that he answered: 'Our Italians can pair off against their Italians. Give the Admiral d'Annebault and Captain Ysnard the French companies and put them opposite the Germans, the rest of our French infantry under M. de Bonnivet can face the Spaniards. You have a superiority of a third in cavalry over the enemy. If Ysnard would rather tackle Spaniards than Germans, Bonnivet and I will face the Germans.'[2] Ysnard and Bonnivet both said that they would leave it to the commander-in-chief. But no battle took place, and a page or two further on we find 'quelque temps après le Sieur Dom Ferrand dressa un camp, surpassant les forces de M. le Maréchal, car ledit seigneur n'avoit ni Suisses ni Allemands.' Common French military opinion, then, certainly made out the class-list: i. Swiss Infantry; ii. Lanquesnets; iii. Spaniards and French bracketed; iv. Italians.

The landsknecht companies were always accompanied by a certain proportion of men with firearms. The cross-bow seems to have been dropped in Germany long before it had disappeared among the French and Italians. The invention which disposed of the old hand-gun and turned it into an arquebus seems to have become known by the middle of the fifteenth century. This consisted in fixing a cock (*hahn*) and trigger on to the original simple tube of the hand-gun, the cock having a hole in it through which passed the string of a long coil of 'match,' which was kept continually smouldering. A small pan with 'touch-powder' was fitted on to the tube, and a hole in the pan communicated with the powder inside the tube. When the trigger was pulled, and the cock with the smouldering match clashed with the pan, the touch-powder was kindled and passed on its explosion into the main charge inside the barrel. This form of firearm, destined to a long career, had as its original name in Germany 'haken,' apparently from the

[1] Tavannes, chap. xiv. p. 6.

[2] Montluc, ii. suite, p. 3; *idem.* pp. 19, 20.

hook-shape of the projecting cock, when held back before letting it fall upon the pan. 'Hackbut,' the English form, is apparently merely *hackenbüchse*, 'hook-gun.' Arquebus, the French form, is one of the typical perversions of Teutonic words which are normal in all spheres of the Gallic vocabulary—the French *could not* get anywhere near a German gutteral sound.

The Emperor Maximilian was interested in firearms of all sorts, probably in artillery more than in smallarms, and possessed enormous quantities of hackbuts (or arquebuses, to use the form with which one is more familiar). In his armoury at Innsbrück in 1515 we find 469 of the best quality, and 1662 of smaller size, all of brass, and then in another list 1125 large and 665 small, of iron. In addition there were a few 'freaks,' double-barrelled guns, wall-pieces, and six primitive mitrailleuses (*streubüchsen*) consisting of several barrels fixed on to carts—the biggest had twelve barrels.[1] If the number of hackbuts—nearly 4000—looks large, it is only fair to say that there are no less than 31,527 landsknechts' pikes and 3381 halberds—which looks as if provision was made for only about one man in nine to be equipped with firearms. But of course it would not be fair to make any such deduction, as conceivably more hackbuts in proportion may have been out and in use than pikes, so that the stock may have been comparatively low at the moment. At the same time we may reflect that the hackbut was so much more expensive a weapon to manufacture that financial reasons may have restricted its output.

As might have been expected from their casual record of mutinies and desertions, the landsknechts were a terrible problem to Charles V at the end of each of his four wars with France. All the companies, save a minimum required on the Turkish frontier by Archduke Ferdinand, were naturally disbanded. At the same time the King of France would be throwing his German mercenaries into the street. If Charles V had to keep garrisons in Italy, he would use Spanish 'tercios,' and Francis I and Henry II preferred to keep Swiss in the way of permanent infantry corps. Hence masses of military unemployed, largely military unemployables, drifted back into Germany.

Some of them got service in Sweden, Denmark, or Poland; but bands great or small wandered about the country, practising

[1] For this list see Jähns, *Kriegswesen*, i. p. 423.

begging, gang-robbery, or blackmailing. Something of the kind had happened before in France, after the Treaty of Bretigny, when all the French and English mercenaries were turned loose by their employers, and went about frightening the Pope at Avignon, and the small princes of the frontiers. The difficulty in Germany was that state boundaries were so complicated, and state police so entirely unable to act outside their own limits, that a robber band might meander for months plundering villages and travelling merchants, and occasionally committing murders. This was only cured in parts of Germany by an agreement of January 1546 between the sovereigns of Electoral Saxony, Hesse, Lüneburg, Brunswick, and Hanover, the bishops of Münster, Magdeburg, and Hildesheim, the free towns of Cologne, Augsburg, and Goslar, and the counts of Mansfeld and Tecklenburg to take common police action against these 'sturdy and valiant beggars,' as Queen Elizabeth would have called them.[1] But wars inevitably broke out again, whenever the King of France saw his opportunity for another thrust at Charles V, and then recruiting officers, raising new landsknecht companies, raced round the realm, and swept together the greater part of the reprobates for a new venture in Italy or Flanders.

The German cavalry, of which during the earlier half of the Great Wars we find comparatively little mention, was trained in the fifteenth century to charge in deep many-ranked squadrons—entirely opposite tactics to those of the French gendarmerie, who always charged in line (*en haye*), as is recorded on many fields. There is an (unpublished) military manual for princes and commanders compiled by one Philip of Seldeneck, a Franconian knight, which dates itself, by allusion to the Swiss methods, as having been written not earlier than about 1480, and probably not much later. It gives not only good advice to princes about war and tactics in general, but very elaborate plans for the drawing up of cavalry in battle. Seldeneck is particularly enamoured of deep formations, even for the comparatively small body of men-at-arms of which a German petty prince was likely to have the command. The lowest depth which he is prepared to recommend for such a force is one with a minimum of nine ranks. With 200 riders to arrange, the author recommends a formation which he calls the *Spitz* or pointed order. There are to be only five men in

[1] See Delbrück, *Kriegskunst*, iv. 79.

the front rank, then four more such ranks, in the fourth of which is to be carried the princely banner, then the formation broadens out into nine ranks of 18 or 20 men apiece. For a corps of 1000 horsemen Seldeneck makes the leading rank only 7 men broad, to be followed by seven more such ranks, having respectively 9, 11, 13, 15, 17, 19, 21 men apiece. The banner would be in the seventh rank, and then the remaining 900 riders would come on in a broad column of 20 ranks. Care was to be taken to put the very best and most experienced men in the projecting front ranks—in the rear ones inferior material both in skill, armour, and horses might be stowed away.[1]

This looks a very clumsy and heavy formation, but such was undoubtedly the sixteenth-century German ideal, for after the men-at-arms began to decrease in numbers, the reiters who superseded them always advanced in very deep formation. It sins against the doctrine of shock tactics, for clearly when the front rank had closed with the enemy, and got into lance or sword fighting at a standstill, the rear ranks would be unable to push on, and would be forced to pull up their horses and come to a long halt The same would be the case in charging a firm front of pikes in square—the rear ranks obviously cannot assert any impetus. And if cannon could be brought to bear on the columns the results would be terrible. Fabrizio Colonna, whose cavalry had stood for some time in column under artillery fire, told his French captor that one cannon-ball had knocked over thirty-three of his men and horses, the fire being enfilading, and from across a river.[2]

In the last battle of the Middle Ages—or the first of the great Valois-Hapsburg struggle—Guinegate (1479)—the French heavy cavalry, more numerous than those of Maximilian, are recorded to have outflanked and completely routed their adversaries, pursuing them for many miles. This, like Rupert's charges at Edgehill and Naseby, led to the loss of the battle, for in the infantry fight Maximilian's Flemings beat the *francs archers* of the Marshal Des Cordes, and remained in possession of the field.

It is remarkable to find how little mention there is of German heavy cavalry during the Italian wars of the first period of the Hapsburg-Valois contest. Maximilian, when he co-operated in the League of Cambray and the Holy League, seems to have brought down into Lombardy, and to have laid his confidence

[1] See the analysis of Seldeneck in *Jähns*, pp 328–29.

[2] Bayard, L.S., p. 207.

in the landsknechts, not in German feudal horse, though he had a mass of it collected at the siege of Padua in 1509.[1] At Pavia the victorious army had with it only 500 transalpine horse, all 'Burgundian,' *i.e.* Netherland,[2] though there were 6000 landsknechts. In 1536 the Emperor had 2000 German horse in Italy, but they seem to be catalogued as light rather than heavy cavalry.[3] When he made his well-known invasion of Champagne in 1544 with an army almost exclusively German, his Spaniards and Italians being mostly occupied in Italy, he is said to have had with him 14,000 horse, light and heavy. But this disappointing campaign, from the historian's point of view, which might have provided some interesting tactical information, if only there had been a pitched battle, frittered itself away in sieges and manœuvres. King Francis had sent to Italy for d'Enghien and all the gendarmerie which he had there, but never offered battle—probably because he was decisively outnumbered. Indeed the only general action on the northern frontier of France in which cavalry was engaged on a large scale during the first forty years of the war, was the disgraceful Battle of the Spurs (August 16, 1513), of which more elsewhere. It does not appear again, and to any effect, till Renty (1554), St. Quentin (1557), and Gravelines (1558), and by that time considerable alterations had taken place in both armament and tactics—at least on the German side.

This came from the growing importance of the 'pistoleer,' or armoured horsemen carrying small firearms. Horse arquebusiers had been known for many years—they were the legitimate descendants of the horse-cross-bowmen of earlier decades. But neither cross-bow nor arquebus are easy to manage on horseback, since each of them requires the use of both hands, which is incompatible with proper riding, and the smouldering match used for the arquebus must have been particularly hard to manage for a trooper who wanted to keep his left hand for the bridle. I imagine that (in spite of some military drawings of the time) the horse-arquebusier must have halted in order to fire, and have been compelled to drop his reins on his horse's neck. He could only, therefore, have been used for exploration and skirmishing. For he had but one shot to fire, and, when that was spent, would have been quite helpless, as he would want to re-load, an operation taking

[1] Bayard, p. 100.

[2] Guicciardini, vi. p. 266.

[3] Army list of Charles V in 1536, printed in Clonard, iii., 326.

a long time and requiring the use of both hands. It would be impossible to reload an arquebus when in movement.

This disability was removed by the invention of the pistol, essentially a very small arquebus not placed to the shoulder but wielded by one hand, combined with the introduction of the wheel-lock, a device whereby fire could be struck without the need of a long smouldering match perpetually kept alight and pulled forward by degrees. The word 'pistol' has nothing to do with Pistoia, as was long believed, nor was it first manufactured in that small Tuscan town. It is apparently derived from the Bohemian word *pistala*, meaning a firearm, found as early as 1483 in the inventory of a Silesian armoury.[1] Why it was specialized about 1540 into meaning a small arquebus wielded by one hand is not clear; but the invention crops up first in east-German lands. The earliest mention of the weapon comes from 1543, and in a curious context. When the Austrian garrison had to surrender by capitulation the Hungarian fortress of Stühlweissenburg to Sultan Soliman, they were allowed to depart with arms and baggage. The Turk kept the terms honourably, with the exception that he confiscated certain wheel-lock pistols, which provoked his interest by their novelty and ingenious construction.[2] In the very next year, 1544, German cavalry are recorded to have provoked the surprise of the French in a skirmish in Champagne by firing on them with 'pistols,' 'which are little arquebuses, with barrels only a foot long, firing by a wheel-lock, wherewith were shot MM. Delaborde and Genly.'[3]

The wheel-lock, which did away with the need for the tiresome match, was a contrivance fixed to the barrel of a short arquebus, consisting of a toothed wheel which was wound up by a spring: the sharp teeth, when the trigger was pulled, and the wheel revolved, came violently against a lump of pyrites (*schweffelkiess*) fixed by the pan, and struck sparks from it, which, falling on the touch-powder in the pan, transmitted the flash to the charge in the barrel of the weapon. It would work for an arquebus only a foot long and easily carried in one hand. This the rider, having loaded, could put inside his holster till it was wanted: if wound up properly, it would keep for many hours in good order. Very soon it was discovered that he

[1] See Delbrück, *Kriegsgeschichte*, iv. p. 57, note.
[2] For this curious story, see Delbrück, *ibid.*
[3] See Du Bellay, book x. p. 138.

might carry three of these pistols—two in holsters, one in his right boot—which would give him three shots.

The pistol achieved immense popularity, and was adopted in every army of Christendom, and by the Turks as well. But it had many drawbacks. (1) The wheel was a delicate piece of machinery—if knocked against any obstacle and bent, it refused to revolve. A sword-cut, or even the rider's fall from his horse, might put it out of order. (2) If left wound-up too long, the spring got weak, and did not set the wheel going with effective strength. General Ludlow, the Parliamentarian leader, has a story of his being found in an unpleasant condition of disarmament in 1645, by leaving his pistols wound-up all night. A surprise at dawn being delivered, he found that the spring would not set the wheel whizzing,[1] and he had only his sword to depend upon. (3) The lump of pyrites, from which the sparks were struck, might be shaken out, or fail, since it wore away rapidly: flints, which were substituted later, were better, but even they got blunted or chipped off. (4) It is hard to take good aim when the pistoleer is going at full speed, pressed in between comrades on both sides, who are also riding hard. Cases were known where a man shot crooked, and hit his neighbour, or blew off the ears of his own horse. If men in the second rank started firing, the danger was of course much greater, and it was difficult to get nervous troopers to hold their fire. The only effective discharge was when the riders waited till the last moment, and did not blaze away till they could see the whites of their enemies' eyes.

The wheel-lock was tried for infantry weapons as well as for pistols, but was rejected after some experiments. Its real virtue was that it could be used with pistols wielded in one hand: foot-soldiers, who could use both hands, reverted to the match, which prevailed far into the seventeenth century, and was superseded by the flint-lock only in its second half, and not very long before the invention of the bayonet.

The most pernicious habit of the pistoleer was the 'caracole,' of which mention has already been made, the tactics by which each line fired and then swerved off to the rear, and formed up again to reload behind all the other ranks. This was a cause of disorder and confusion, unless the men were extraordinarily well trained, and all of good morale. The purpose of the 'caracole' was of course to deal with squares of pikemen,

[1] See Ludlow's *Memoirs*, i. 72.

impenetrable by a charge. The idea was to pour a succession of pistol-bullets into its front, an angle, or a part of a flank, till a gap was caused in the pike-front, or many gaps, into which the rear ranks of the cavalry could penetrate, using either pistol or sword.

The defensive tactics against such an attack were either to charge the column of pistoleers with other cavalry, or else to have arquebusiers so correlated to the pike-square that they could shoot down the front ranks of the pistoleers before they could get close enough in to do much damage. For the pistol was essentially a short-range weapon.

The pistoleers, entirely German in their origin, are soon found designated as *Schwartzreiters*, black horse, from a habit which they adopted of wearing black armour to impress the enemy (as we are told) by their grim appearance.[1] The term is found as early as 1548—the French called them, when first seen, 'diables noirs.'[2] Imitation is the best form of flattery, and Henry II—determining to utilize the idea—had a few pistoleers in his service as early as 1548, and could show as many as four squadrons of 'Reiters'—German deserters—in the campaign of Gravelines (1558). When first seen, some of them carried light lances: but this weapon was discarded early, as hampering the free use of the pistol and always getting in the way, since it had to be carried in some sort of a bucket while the firing was in progress and both hands were otherwise occupied—the one with the pistol and the other with the reins.

'Schwartzreiter' soon got shortened into 'reiter,' though this word might have meant any sort of horseman. But the man-at-arms with his barded horse, lance, and full armour was going out of date. By 1600 the lance and the complete cap-à-pied suit of plate had been discarded in all continental armies, save the Spanish, where they were considered somewhat of an archaism. The 'reiters' who appear so frequently in the chronicles of the French 'Wars of Religion' were all German mercenaries, enlisted freely by both sides, without much reference to creed or political tendencies. Indeed these horsemen had as bad a reputation as the landsknechts for desertion, mutiny, and general plunder. They went on strike, returned home, or pillaged friendly districts at large, whenever their

[1] See for all this Delbrück's chapter, pp. 141–49 of vol. iv. of his excellent *Kriegskünst*.

[2] Rabutin's *Commentaries*, vi. p. 620.

monthly pay got into arrears. Of their perverse and disorderly habits much will have to be said when dealing with the campaigns of the Huguenots, and also with the War of Independence in the Netherlands.

They are sometimes called cuirassiers (*kürissen*), for they wore cuirass and closed helm, though they had discarded leg-armour in favour of high leather boots. Thus, though originally derived from the cross-bowman, who was catalogued as a light-horseman, the 'reiter' had moved into the position of heavy cavalry by the end of the sixteenth century. The lighter troops, without much body armour, were stradiots, Croats, pandours, hussars,[1] or what you please in the general category of 'chevaux-legers'—they were predominantly foreign, and generally considered half-savage. They were reckoned even worse robbers and brigands than the 'Schwartzreiters.'

The first campaign in which the 'pistoleers' or 'reiters' were engaged freely on both sides was that of the Protestant League against Charles V, which ended at Mühlberg (1547), and a typical cavalry engagement on a large scale with this sort of cavalry was Sievershäusen (1553), where Maurice of Saxony got his death-wound leading his black riders against those of the marauding Albert of Hohenzollern-Kulmbach. And so he missed a career which bid fair to make him the dictator of Germany. He was mortally wounded by a pistol-bullet, dying two days after his victory.

[1] Hussars, borrowed from Hungary, are found in Austrian armies as early as 1548. Croats in the Venetian service are called savages and barely Christian in 1512. 'Stradiots,' originally Albanian refugees in the Venetian service (cf. Fornovo), were taken over early by French kings, and keep the name 'Stradiot' even to 1558.

CHAPTER VIII

THE ITALIANS—THE CONDOTTIERI AND THE THEORIES OF MACHIAVELLI

WHEN the ill-advised invitation of Ludovico of Milan confirmed Charles VIII of France in his plan for making Naples his first stopping-place on his wild journey to Constantinople and Jerusalem, the Italian states, small and great, were still—from the military point of view—in full fifteenth-century conditions. Though they had been enjoying what was, from the Italian point of view, a comparatively quiet time since the Peace of Lodi (1454), the old conditions of suppressed hostility and rival territorial ambitions still survived. And all the states, whether republican or monarchical in form, were depending, as they had been for the last two centuries, not on national armies but on hired mercenary bands, engaged and dismissed as financial or political occasion dictated. The time of the greater condottieri had gone by, still more so the age when the mercenary bands of Italy were formed of, and commanded by, foreigners. That was a fourteenth-century condition—in the fifteenth the bands and their organizers had become Italian. Some of the old adventurers, like the Sforzas at Milan, had made themselves into local princes, but this was exceptional. In the later fifteenth century the condottieri were persons of lesser importance, and their bands (as Machiavelli remarked) were numbered by hundreds rather than by thousands. But they retained the organization of the fourteenth century, and still were primarily companies of heavy cavalry, with no more than a moderate proportion of infantry of no great merit attached to them, for the purposes for which cavalry was obviously useless—mainly siegecraft and pioneering. The companies were still wholly without local or national ties, ready (as of old) to serve one employer in the spring and another, his enemy, in the autumn, if their contract was up, or their pay not forthcoming. Their comparatively unimportant captains could no longer aspire to become local tyrants.

But one of the most interesting points of Italian military organization during the Renaissance wars is that princes of families with a certain antiquity were now letting themselves out, with their personal levies, to serve as condottieri for larger powers. The Marquises of Saluzzo, the Gonzagas of Mantua, and the Estes of Ferrara all represented houses of some standing —the Estes had been at Ferrara since the thirteenth century, the Gonzagas at Mantua since 1329, the Saluzzos were a very old feudal house, holding from the Holy Roman Empire since time immemorial. But they are all found in the Great Wars acting like pure condottieri—hired to one side or another. It is therefore less surprising to find the great old baronial families of the Papal States, the Orsinis, Colonnas, etc., supplying generation after generation of mercenary captains in this age. And of course the 'tyrants' of the Romagna were natural partakers in such transactions.

The wars which the condottieri had been wont to wage on behalf of Milan, or Florence, or Venice, or the King of Naples, had been—as was pointed out in a previous volume—very scientific and bloodless affairs. They were games of chess in which checkmate was accepted with little acrimony, and still less bloodshed. Hence came the awful surprise to these old gamesters when the Swiss and the French came over the Alps, with their disgusting habit of storming towns by escalade and massacring the garrisons. It was generally felt in Italy that this was not 'playing the game.' And when the Italian states, appalled at the easiness of Charles VIII's triumphal march to Naples, felt that the establishment of a French empire in the peninsula must be prevented at all costs, the military caste sympathized, not—we may judge—without certain qualms of fear as to what fighting the French might mean to respectable soldiers of fortune. But skill might perhaps triumph over the brute force of the transalpine barbarian.

Their first experience was a dreadful one; at Fornovo all the mercenaries of Milan and Venice and some of the lesser tyrants were collected, with Giovanni Francesco Gonzaga of Mantua in command, accompanied by his personal following. The trap was set, with corps neatly arranged in supporting positions, while the French were to defile in line of march across the front of the waiting army. So they did—but, when the crash came, the gendarmerie of Charles VIII sent the Italians flying in every direction. And there was no quarter given—every horseman who fell was butchered. When Gonzaga sent a day after to negotiate about ransom for his uncle and three of his cousins, whom he supposed to be prisoners, he found that they were all dead—as were sixty more of his personal following.[1]

This must have been a nasty jar to a general skilled in the old customs of condottieri warfare. But the really astounding thing about the matter is that, five years after, Giovanni Francesco of Mantua made his peace with Louis XII, and actually entered the French service with his bands. Obviously he had made up his mind about the efficiency of 'barbarian' methods, and bore no lasting grudge for the unfortunate loss of his nearest relatives, for he was actually in command of a French army on the Neapolitan border in 1503, and nearly became involved in the disaster of the Garigliano. But he fell sick in November [very wisely], and left the charge of the doomed army to the Marquis of Saluzzo, another condottiere of the highest lineage, who was responsible for the December battle and the surrender of Gaeta.

The lesson of Fornovo to the Italians was fully appreciated, and they commenced at once to seek devices for bringing up their methods of war to the latest developments. The two obvious things were (1) that chessboard-manœuvre battles must cease, and that, if necessary, troops must fight seriously, and (2) that no army was complete without infantry—the best that could be improvised, if it were impossible (as it was) to get infantry that could compete with the Swiss. Shortly afterwards it was discovered that if it were impossible to get high-quality pikemen, it might be well to imitate the Spaniards, and to specialize on arquebusiers.

The first endeavour seems to have been to strengthen the cavalry arm with light-horse, as well as to train the existing men-at-arms to more serious fighting than they had hitherto expected to meet. But, in view of the predominance of the French gendarmerie, set battles with cavalry had better be avoided as much as possible. There ought to be no more Fornovos. Hence generals like the Count of Pitigliano (Niccolo Orsini), Bartolomeo de Alviano, Giovanni dei Medici, Gianpaolo Baglioni, Prosper Colonna, wanted to keep off if possible from fighting battles in the open, and preferred entrenched positions, raids, and manœuvres to cut the enemy's

[1] Comines, book vi. p. 221.

communications. The increased number of the light-horse is specially notable—horsed arquebusiers were numerous, developed by a specialist named Camillo Vitelli, and much used by the Venetians. But the 'stradiots' are much more frequently mentioned. Although they had wasted their chance at Fornovo by plundering the French baggage-train, they were in high esteem for many years after. These troops with the Byzantine name (*στρατιῶται*) had originally been the Albanian horsemen of the great Scanderbeg, the terror of the Turks.

When Albania was finally overrun by Mahomet the Second, the Venetians sheltered the refugees, who were for some years actually commanded by Ferrando Castriot, the heir of the old prince, who figures in Bayard's memoirs [1] as the 'Captain Scanderbrebec,' and died with the Venetian title of 'Conte de St. Angelo' at the battle of Pavia.[2] They looked, as we are told, much like the Turks with whom they had so long fought, save that they wore no turban round their caps, and they used the scimitar as their characteristic weapon.[3] They wore a mail shirt or a cuirass, but no other armour, and were mounted on small but very speedy and much-enduring horses, which could outmarch any other cavalry. Hence they were invaluable for desperate raids and surprises, and would often ride round the rear of the enemy, and cut up his convoys or his small detachments. It was rather useless to put them in line of battle, for they would not stand before a charge of gendarmerie ; but for all other purposes they were excellent. So much was this the case that the French soon raised squadrons of 'stradiots'—though how many of these can have been genuine Albanians is most doubtful. Certainly some of them were 'Greeks'—whatever that may mean—perhaps any sort of Christian inhabitant of the Balkan Peninsula—and others were Croats.

The other innovation which was taken up by the Italian captains after their first experience in the wars of 1495-1500 was the raising of respectable corps of infantry. Experiments with the pike-drill of the Swiss had commenced with the condottiere Vitellozo Vitelli (1497) of Citta di Castello, who had about 1000 men under arms—but he and his disappeared, when he was murdered by Cæsar Borgia. Perhaps some of his men,

[1] Bayard, xxxvi. p. 115.

[2] Where, of course, he was in command of the Imperialist 'stradiots.'

[3] Comines, viii. chap. ii., describes their scimitars as 'terrible weapons' and much to be dreaded.

however, may have been in the body of Romandiole infantry which the tyrant embodied after his general massacre of the small lords of the northern Marches. When Pitigliano and Alviano organized for the Venetians the mercenary army with which they were unwise enough to face the French at Agnadello (1509), Alviano had embodied several thousand well-drilled Romandiole pikemen under a commander called Brisighella,[1] who were all dressed in red slashed with white, 'gentils compagnons et nourriz aux armes,' says Bayard's chronicler.[2] They stood for a space against the French infantry in respectable fashion, but were taken in flank by a charge of gendarmerie and absolutely cut to pieces. 'Les rouges et blancs demoureurent sur le champ.' It was some time before any other solid body of Italian infantry could be formed by the Venetian government, and those that they did collect were mainly arquebusiers —pike-fighting was for a long time considered unlucky.

The celebrated 'Black Bands' with which Giovanni dei Medici—Giovan' delle Bande Nere as he was called—made some reputation, and were, it seems, mainly composed of infantry—of whom there were sometimes so many as 3000 embodied, with only 300 light-horse attached. The infantry appear to have been predominantly arquebusiers.

Of course it was not only the condottieri who were interested in the rapid transformation of warlike methods that began after Fornovo and Cerignola, and which went on developing in unsystematic and contradictory fashions for many a year after. The politicians and 'intellectuals' were doing their best to formulate general deductions, and to set them forth in logical form. The keenest intellect that was at work was, of course, that of Machiavelli, whose *Arte della Guerra*, '*Principe*,' and several minor works are not only technical but politically propagandic. He thought that he could apply his readings in classical military history, from Livy to Vegetius, to the needs of the day, and the establishment of a national army for his loved Florentine republic.

Many of his premises were quite sound, and his observation of recent facts was keen. But unfortunately for his reputation as a prophet in the military sphere, all his recommendations of a practical sort bear no relation whatever to the actual development of tactics and organization during the later years of the century. He 'backed the wrong horse' in almost every

[1] Hardly to be recognized in Bayard as 'Bresiguelz.' [2] Bayard, xxix. p. 92.

instance; he thought that artillery was going to continue negligible, that the day of cavalry in battle was quite over, that infantry was going to continue in very huge units, like the legion, and that the pike was destined to be put out of action by short weapons for close combat, like the sword of the ancient Roman or of the Spanish footmen of Gonsalvo de Cordova. In every case his forecast was hopelessly erroneous.

Machiavelli started with the thesis that the old condottiere armies of the fifteenth century, with their hired and corruptible generals, and their men-at-arms who fought battles without casualties, were contemptible, and must be replaced by national troops, interested in the defence of their own native states. Here, no doubt, he was quite obviously right: but his obsession for classical antiquity came in, and he envisaged the new army of Florence as organized on the model of the Roman Legion of the days of the old republic. It would be essentially a great body of infantry, with some minor trimmings in the way of cavalry and artillery, corresponding to the very modest allowance of *equites* in a Roman consular army, and its equally modest train of *catapults*, *balistae*, etc. There were to be 2000 pikemen, 3000 swordsmen with bucklers in the Spanish style, and 1000 'shot' in each unit. He was behind his times in the appreciation of the growing importance of firearms: thinking apparently of the slowly moving, slowly firing, heavy guns of the condottieri, which he had seen in his youth, he will allow no use for artillery save to get in a few salvos at the commencement of a battle,[1] and to scare 'peasants,' *i.e.* undisciplined militia. With the arquebus he is still more in error: he would give only one-sixth of his men firearms, and considers that arquebusiers are only useful to open a battle with skirmishing, or to fire from behind trenches or other cover. The real push of battle must be made with the mass of infantry, and after considering the exploits of the Swiss pike-column, which had ruled the field for a whole generation, he considers that the Spanish sword-and-buckler men are a better model for the future—largely, it is clear, because they are a sort of reproduction of the ancient Roman legionaries, who won battles with the sword, not (as some thought) with the pilum. His underrating of cavalry comes partly from his scornful memory of the condottieri, partly (one is forced to think) from reflecting how seldom the Romans won a battle by a cavalry stroke, and how

[1] In *Arte della Guerra*, iii. p. 660, the guns are allowed *two* discharges.

often they beat off the masses of horsemen put against them by Antiochus, Mithradates, or Tigranes. He has always got in mind the exploits of the legionary's sword at Heraclea, Pydna, or Cynocephalae against the Macedonian phalanx, which had such a close resemblance to the Swiss pike-column of his own day.

The theorizer dreams of battles fought in the fine old Roman way, and gloats over the success of his tactics. At the same time it must be borne in mind that when Machiavelli makes his mouthpiece, Fabrizio Colonna, discourse on war in the 'Arte,' we find survivals of the condottiere theories cropping up—as was indeed likely. Battles are risky things: nor must the general make it his first aim to find and fight at all costs the main army of the enemy. The good general will only fight when he pleases—that is to say, when he has found the opportune moment, and has manœuvred his enemy into an unfavourable position. Fabius Cunctator of old, who foiled Hannibal but never beat him in the open field, is cited with admiration. There is almost as much care spent on the study of entrenched camps, and of the best form for retreats and flank marches, as on the description of the ideal victorious battle. The list of advantages that a commander must have at his disposition, in order to be justified in fighting a general action is formidable. He ought to have favourable ground, time to marshal his troops well, and either the larger army, or one composed of more tried soldiers than the enemy's. He must not fight if his men's morale is low: on the other hand, if the troops are anxious and eager for battle, he must not be led into giving them their desire, unless all other conditions are satisfactory. The only occasions on which a general may be pardoned for fighting without having all advantages on his side, are when he is forced to do so by dire necessity—*e.g.* when all provisions are exhausted, when the army shows signs of breaking up for want of pay [this is a very characteristic symptom of the survival of condottiere morale], or when news comes that the enemy is to receive enormous reinforcements within a few days. If disaster is certain unless he takes the initiative, the general may be pardoned for gambling on the event—that is, if he is certain to be ruined if he does not strike, and has a fair chance of winning if he does. 'E meglio tentar la fortuna dove ella ti posse favorire, che non tentando vedere la tua certa ruina.'[1]

[1] *Arte della Guerra*, iv. p. 675.

But normally a commander should manœuvre, retreat, or entrench himself, if he does not like the general situation. This reminds one of a dictum of Pescara, 'better a hundred years of wars than one day of battle'—risks are too great. Napoleon's comments on such a pronouncement may easily be conceived!

As to cavalry, 'of course one must have some horsemen, but they are the secondary not the important section of the army. They are most useful for exploring purposes, for raiding and wasting hostile territory, to keep the enemy's army worried and always compelled to be under arms, and to prevent him from collecting provisions. But as to the clash of general actions, which are the critical things in war, and the end for which all armies are raised, cavalry are more useful for pursuing a routed enemy than for working in any other way, and they are sufficiently inferior to the infantry in merit.'[1] So cross-bowmen would certainly be more effective than men-at-arms! On the whole the cross-bow is preferable to the horsed arquebusier—who is only useful to make a noise.[2] And this is written with Marignano and Ivry, Tournhout and Rocroy, Naseby and Blenheim all to come! Cavalry as a battle force had still three centuries before it.

As to the poor arquebusier, he is barely mentioned; we do hear of *scopettieri*, but, along with cross-bowmen, they are assimilated to the *velites* attached to an ancient Roman legion; and in the ideal battle described in the third book of the *Arte della Guerra*, we find them thrown out for preliminary skirmishing, like the slingers and bowmen of the Romans, and expected to be out of the way when the decisive clash comes. Pavia was, of course, in the future when Machiavelli was writing, but, though he mentions it, Cerignola is quoted not as a victory of firearms but as an example of the benefit of entrenchments and hindrances.[3]

It is interesting to find that the author of the *Arte della Guerra* once had in his charge the drawing up of the organization of a national militia for the Florentines, when they had determined in 1506 to organize a citizen-force which should relieve them from the unhappy alternative of depending entirely on those hired soldiers whom he so much hated. In the *Provisione per Instituere Milizie Nationale*,[4] we find arrange-

[1] *Arte della Guerra*, ii. p. 636. [2] *Ibid.* ii. p. 653. [3] *Ibid.* iv. p. 671.
[4] Printed on pp. 721–51 of Machiavelli's work in the Milan edition.

ments for a conscription of persons of military age, which was estimated to produce nearly 20,000 men, divided into bands or companies of 800 apiece. The interesting thing is to find seven-tenths of the men of each band provided with the pike, one-tenth only with firearms, the remaining two-tenths being permitted to serve with halberds and other 'white weapons.' A supplementary 'Provisione' provided for the levying of 500 light horse, who were to be armed with arquebuses or cross-bows indifferently, as each man presented his weapon of choice. But as the ordinance was passed only a short time before the republic fell (March 23, 1511) it is probable that the corps was never raised.

The most notable thing about these 'Provisione' is that they show a curious dread of local influence, for no captain of a band was to be elected from the district where he was in residence. And on the 1st of November every year the captains were all to change bands, so that they should never be in charge of the same set of men for more than a year. Moreover, though a 'gonfalonier' was nominally in charge of the whole levy, he was under the control of several 'councils of war,' with numerous members, who were to be changed once a year. There was a special clause providing that captains should have no disciplinary power over their men, except when they were actually on service — in case they should become too authoritative.

It is difficult to conceive a worse system—the idiotic expression of democratic jealousy for any one who might show signs of gaining influence over the soldiers of the republic.[1]

The Florentine militia had five years of existence before a Spanish army came down on them, stormed Prato with circumstances of considerable atrocity, and restored the Medici to power. The new army made no great show, and the gonfalonier resigned (August 1512), giving place to Cardinal Giovanni dei Medici. But in 1527–30, when the republican party made its last bid for freedom, expelled Alessandro dei Medici, and installed the government of the 'Dieci de Liberta,' a much better fight was made. This was a forlorn struggle, for the Florentines had counted on the help of other Italian princes, and especially on that of Francis I of France, who deliberately abandoned them when he made the Treaty of Cambray

[1] For some excellent remarks on these 'provisions' and their folly, see Delbrück's *Kriegskunst*, iv. 118–20.

with the Emperor in 1528. Left to their own resources, they fought on for two years against an Imperial army commanded by Philibert, Prince of Orange. They raised sixteen companies of infantry from the city and thirty from the rural districts, making an army of 13,000 native troops—only about 1000 condottiere-infantry were enlisted and embodied with the rest. It is interesting to note that in the city contingent there were 1700 arquebusiers to 1000 pikemen and 300 halberdiers—the proportion shows clearly how the importance of firearms had risen in the few years since Machiavelli wrote the *Arte della Guerra.*

Florentine public opinion was much divided—the Medici had a strong body of partisans—and the republican enthusiasm was confined to the city and its immediate district, owing to the unwise policy which Florence had always pursued of denying full political rights to her subject-towns. Nevertheless the resistance to an overpowering enemy was much more obstinate than had been the case in 1512. There was no pitched battle, but the citizens rebuilt their walls in the modern style, and stood a siege of eight months, from October 1529 to August 1530. Even after the outlying fortresses had fallen, the Florentines made a most determined resistance, delivering many sallies—often successful—against the lines and camps of the besiegers. The commander-in-chief was an exiled prince or tyrant, Malatesta Baglioni, late of Perugia, who was accused of the usual condottiere habits of procrastination and irresolution; but the defence was kept up to a high standard by the 'Commissary General' Francesco Ferruccio, a Florentine, born of an ancient but impoverished family, who had served for some years in the 'Black Bands' of Giovanni dei Medici. Though a trained soldier, he was (a rare thing in sixteenth-century Italy) an enthusiastic patriot, and a born leader of enthusiasts.

He got better service out of the city bands than any one had ever expected, raided the enemy's communications, and relieved Volterra by a sudden march when it was about to fall. The inevitable only came to pass when he and his opponent, the Prince of Orange, met at Gavignana, in the hills above Pistoia, where both fell in a well-contested combat on August 2, 1530. Ferruccio was executing one of his usual raids round the rear of the enemy, when he was intercepted by several columns drawn out from the besieging lines, under the commander-in-chief himself. The Imperialist cavalry was routed, and Orange was killed by an arquebus ball while trying to rally it. But concentric forces, coming up from several sides, beset Ferruccio, who held the village of Gavignana for many hours against all attacks, but was in the end mortally wounded and his bands were cut to pieces. A Calabrian colonel, Fabrice Maramaldo, found him lying at the point of death, but ordered him to be stabbed, as a revenge for the death of Philibert of Orange. Ten days later the city surrendered, the commander-in-chief, Baglioni, having opened the gates to the enemy while negotiations were in progress for the capitulation, because he found out that a desperate party were still set on a hopeless resistance (August 12, 1530).

The defence of Florence by its citizens was one of the few proofs that there existed in some quarters of Italy a patriotism which could fight for the state, and thought no more of using hired arms. But this patriotism was local, and particularist—the Florentines to the last days of the republic were hard masters to their subject-cities, most especially to Pisa, whose very natural rebellion gave them much trouble. A similar outburst of desperate city-patriotism was shown by the Siennese twelve years after the fall of Florence. The important republic of Siena had hitherto escaped annexation: though, having taken the French side, it had been forced to make peace with the Emperor and accept a Spanish garrison into its walls. In 1552, insurgents, headed by the great Siennese house of the Piccolomini, expelled the Spanish troops and sent to ask for aid from France. Though from the strategical point of view a French lodgment in middle Italy, when both Naples and Florence were in the enemy's hands, was worthless, Henry II dispatched by sea the Florentine exile, Piero Strozzi, now a French marshal, with ten companies of landsknechts and in addition some French and Italian bands. The Siennese themselves raised a considerable force, and there were disturbances in other parts of Tuscany, which was now a 'grand duchy' under Cosimo dei Medici, whom the Emperor had imposed as sovereign on the conquered Florentines. Though attacked by the Viceroy of Naples from the south, and by the Marquis of Marignano with a Spanish-Imperialist army from the north, the Siennese made a resistance of unexpected obstinacy. There was only very intermittent help from France, since the sea was scoured by the Emperor's Genoese fleet, and it was impossible to send reinforcements by land. Nevertheless, the Siennese rising lasted for more than two years; and, after the outlying towns had been reduced, Siena itself stood a siege of eighteen months, and yielded only from starvation. The interesting point about this hopeless defence is that the Siennese themselves took a most gallant part in the fighting. Montluc, who was in command in the city during the greater part of the siege, gives high praise to their local companies, who—as he considers—were more trustworthy than his Germans and some of his French. But the game was lost from the beginning—an isolated force in central Italy was bound to succumb before an enemy who had command of the sea, and who ended by pouring in overwhelming reinforcements, when the Emperor had got to the end of the German troubles which were distracting him in 1552.

The Siennese rising was the last effort of what we may call the old republican Italy of the Middle Ages. After 1555 there were plenty of Italian troops in the field, but the rank and file were mercenaries, serving for their pay indifferently in the French, the Spanish, or the Imperialist armies. It is notable, as has been already remarked in another place, that a very large number of Italian professional soldiers served Francis I and Henry II. These were not only the descendants of the old condottieri; some were remnants of the old Angevin faction in Naples, others political exiles from the vanquished republican factions in many states. When Piero Strozzi came to France in 1543, he brought with him a band of no less than 300 persons, who had all been captains, lieutenants, or ensigns, mainly in the old Florentine army; they were all in full armour and mounted; a third of them had firearms.[1]

From a party such as this quite a number of Italian companies could have been officered and organized. While the Italian officers were many of them well esteemed, the rank and file seem to have inspired no great confidence, more especially the pikemen; the national bent seems to have been towards the arquebus. That these troops had no great reputation was the natural result of their being pure mercenaries, inspired neither by national feeling nor by loyalty to the employer whom they frequently changed. The officers had often old political connection with the French party, and many of them, like Strozzi, were exiles—mostly Tuscans or Neapolitans but some Milanese. But this, of course, was not the case with the men

[1] Du Bellay, x. p. 45.

in the ranks, who were hirelings and often deserters. They had no prejudice for or against the French or the Emperor to inspire them to any effort for one side or the other. Hence the way in which the French commander disliked putting them in line against landsknechts or Spaniards, as has been already noted.[1] They do not even, save in a few exceptional cases, seem to have had *esprit de corps* to inspire them ; the cases in which it is spoken of are those of the ' Black Bands,' which had been led for many years by Giovanni dei Medici, and Alviano's companies, who had grown old in the Venetian service, and wore mourning for their old leader long after his death. But usually these companies were hurriedly put together by adventurers, and had little more interest in their officers than they had in their native country. They were disbanded at every peace, and were casual and temporary formations for the most part. Some of them wandered far afield—Protector Somerset once hired an Italian band, under one Malaspina, for service against the West Country rebels,[2] and (more astonishing still) Aristotele Fioravanti organized a train of artillery for Ivan the Terrible, when he was not engaged in other less military business for the Czar.

[1] See above, p. 80.

[2] Montluc says that he saw a few Italian arquebusiers in the English ranks in the fighting around Boulogne in 1544. Book i. p. 342.

BOOK II

THE GREAT BATTLES OF THE ITALIAN WARS, AND THEIR TACTICAL MEANING

CHAPTER I

FORNOVO (JULY 6, 1495)

WE have already mentioned the tactical importance of several of the general actions of the Great Wars, and spoken of Fornovo and Marignano as cavalry victories, of Novara as the last triumph of the old Swiss pike-column, of Cerignola and Bicocca as fights in which entrenchments decided the day, and of Ravenna and Pavia as two early examples of the growing importance of gunpowder. But there are other fights to be taken into consideration, and of those already cited several have more aspects than one. It may be well to commence with a consideration of Fornovo—not only the first pitched battle in the long series of campaigns, but one which marked the end of an epoch—that of the old Italian condottiere tactics of elaborate cavalry manœuvres, where the side that was out-generalled used to give up the game, and acknowledge the checkmate. This one victory of Charles VIII, won against every rule of the old system, made an end of chessboard tactics.

Having accomplished his all-too-easy conquest of Naples, and left there his cousin of Montpensier with about half of his original army, King Charles had to start back northward, to open up his lost line of communication with his base, and to pick up Louis of Orleans with the French corps that had been left in Piedmont—his nearest possible reinforcements. The position was much like that of Macdonald in 1799, when the French army of Naples heard that the Austro-Russians had burst into Lombardy, and, after winning the battle of Magnano, had driven back the armies of the Republic into the Ligurian Alps. The problem in 1495 as in 1799 was to escape up the peninsula, and to get into touch with the friendly force without being intercepted on the way. To be intercepted would mean ruin and destruction. Maconald saved himself by shouldering off Suvaroff at the well-contested battle of the Trebbia (June 18–20, 1799), and getting through to join the army of Moreau at Genoa. King Charles fought his corresponding battle on the Taro, not fifty miles away from the spot where Macdonald and Suvaroff were to meet three centuries later, and reaching Asti after his victory got into touch with Orleans, who was at Novara. In both cases the Army of Naples saved itself, but the campaign ended with the loss of all northern Italy by the French, who had to abandon Lombardy, and to fall back into the mountains.

King Charles had no Russians and Austrians to fight, but a purely Italian enemy, the league of all the powers which had combined against him, after his conquest of Naples had filled the whole peninsula with terror. Milan and Venice, enemies for the last two centuries, had joined their forces, Pope Alexander VI was giving them the very doubtful advantage of his blessing, and the minor princes and tyrants of the north were in the league, while the King of Aragon was about to invade the Calabrias, and Emperor Maximilian in the far distance was raising troops for an Italian campaign.[1] Giovanni Francesco Gonzaga of Mantua, prince and condottiere, was in command, and had the planning of the campaign. At first the allies had thought of pushing down to Rome, and stopping the King there, but at the Pope's request they withdrew the advanced guard which they had sent thither, and resolved to set the trap in the Apennines, not on the Tiber.[2] The King of France, therefore, was able to march unmolested through Florence—an ally but a suspicious one—Pisa, and Lucca, and to enter the high pass of the Apennines where the railway now goes from Sarzana to Parma, crossing the watershed at Pontremoli, after ascending the valley of the Macra, and descending into Lombardy along the valley of the Taro.

The French captains had feared that the enemy would take position at the southern end of the difficult succession of defiles which they had to pass, perhaps at the Salto della Cerva beyond Pietrasanta 'where a few infantry and a couple of guns could have stopped us,' perhaps at the upland town of Pontremoli, 'town and castle so strong that if they had been well garrisoned they could never have been taken.'[3] But there were only 300 men in Pontremoli—who surrendered tamely, and as a reward were massacred by the King's Swiss, who broke

[1] His first contingent joined the Duke of Milan, but did not come down to Fornovo, but remained facing Orleans in the Milanese.

[2] Comines, book viii. chap. ii.

[3] *Ibid.* chap. v.

the articles of capitulation. This defile passed, the way down the broad valley of the Taro began to open up. This is not a pass but a barren shaly gap in the hills, with a narrow stream —generally nearly dry in July—running down the middle ; the country is all stony, the villages few and poor. The valley is often a mile broad, sometimes more, sometimes less ; it is not a defile of the forbidding sort, with overhanging cliffs, and there is generally a good deal of elbow-room.

The obvious reason for the choice of the valley of the Taro, near Fornovo, as the spot where the trap for King Charles was to be set, is that it shows ground on which cavalry can manœuvre. For an army of the old Italian sort, where the man-at-arms was the important person, the defiles of the Apennines were not suitable ground. Charles, as Comines remarks, might have been stopped before he had gone far into the defiles by infantry and guns blocking the road, and he would have had to turn back. But though the Marquis of Mantua had got some guns, he did not want an infantry fight, having no confidence in his men, and he was desirous not of blocking the French but of annihilating them. They were obliged to keep to the single track along the Taro, since their artillery and their considerable train of sumpter-beasts made it necessary for them to keep to main roads. They would have to march in a column of great length, and Gonzaga's plan was to fall upon them when they were moving slowly past the position on the other side of the shallow, broad, shaly bed of the Taro, where he had drawn up his army and laid out his artillery to play upon the long procession. Normally an army marching in column of route across the front of an enemy in position, ought to be knocked to pieces at the first shock, since it has to form front to flank in a hurry, and has no reserves—all the troops being strung out along the road.

The Marquis had pitched his camp on the eastern or right bank of the Taro, about two miles north of the village of Fornovo, some way back from the river, but had laid his guns so as to bear on the road, upon the other side of the water, along which the French must pass. He had with him 2400 men-at-arms, all old mercenaries of Venice, Milan, or the minor princes, 2000 light cavalry, mostly horsed cross-bowmen, but including 600 of the celebrated 'stradiots,' whom the Venetians had raised in the Balkan peninsula—mainly Albanians. Of infantry of one sort and another, there are said to have been more than 10,000, mostly cross-bowmen, but with some German pikes: but no one attached much importance to them, and many were no better than camp-followers. The tactical arrangement was that when the French, on the other side of the water, should have got level with the guns planted to annoy them, the whole of the cavalry, divided into seven bands, some in first line and some in support, should cross the shallow river, and charge in at various points in the French line of march. The stradiots, on the extreme right wing, were to get round the enemy's flank, fall upon his baggage-train and then cut in from behind into his main-battle.

The very elaborate array of the allies was marshalled in no less than nine corps or brigades. The extreme right was formed by a body of light horse, half Venetian stradiots under Pietro Duodo, half horsed crossbowmen under Alessio Beccacuto, who were ordered to cut across the head of the French marching column, and to stop it from further progress down the road. Next came the chief item of the allied right wing, entirely composed of Milanese troops—six hundred men-at-arms supported by 3000 infantry (some of the latter Germans) all under the Count of Caiazzo. This corps were ordered to attack the van of the French army, when it should have been checked by the stradiots and would be driving them away. Next in the front line was the division of the Marquis himself, 500 men-at-arms mainly of his own following, but including contingents of some of the smaller allies. They had with them 500 horsed cross-bowmen and 4000 infantry. The left 'battle' of the allies was composed of 500 Venetian mercenary horse under Fortebraccio de Montone—grandson of the famous condottiere general of the same name.[1] He was to attack the French rearguard. But in Italian tactics of the fifteenth century there was no mere front line, but an elaborate provision of supporting corps. On the right, Caiazzo's wing had behind it, at a discreet distance, Annibale Bentivoglio and Galeazzo Palavicino, with another body of Milanese horse. In the centre, Antonio, son of the Duke of Urbino, with troops from the Marches and the Papal States, was behind the commander-in-chief, with orders to support him when necessary. Lastly, a body of Venetian horse, under Gambara of Brescia and Benzone of Crema, were in support of Montone's corps on the

[1] Probably with infantry support, like the other two 'battles,' but it is not mentioned.

left. A detachment of horse and a very large body of the least useful infantry were left by the camp, and the place where the artillery had been posted.

The bulk of the enemy, formed in these two lines, was to cross the Taro and fall upon the French, when the head of the moving column should have reached the point, opposite the allied camp, where the artillery was stationed to play upon it. This crossing of the Taro, which looked child's play when the plan was drawn up, proved not so easy as had been calculated. For a desperate night of wind, storm, and rain not only put some water into what had been a dry bed, but made the rocky bottom of it as slippery as glass, and turned the sand on the opposite bank into soft mud.[1] It was not rendered impassable anywhere, but had become in the morning a hindrance, which it had not been overnight.

The King of France was perfectly well aware that he would have to cut his way past the allied army, if he was to reach the Lombard plain and his cousin of Orleans. His vanguard under the Marshal de Gié was already in touch with the enemy: a party of stradiots had driven in its fore-riders, and reached the skirts of its encampment on the 4th. They had killed several soldiers, and (this provoked much rage) cut off their heads in the Turkish fashion. They fled when fired on—being much afraid of falconets and arquebuses—taking away one prisoner, a Swiss captain who gave some most untrustworthy information to the Marquis of Mantua—as a good prisoner should.[2]

The King pushed on, slept at Fornovo on the 5th, and, whether with honest or with delusive intentions, sent Philip de Comines with a trumpet of truce to the allied outposts, with certain proposals to be laid before the Venetian commissaries.[3] But as he pushed on while the negotiations were in progress, and the discussion was interrupted by the first cannon-shot, it is probable that he had intended from the first to fight. The army was drawn up so that its line of march might be immediately converted into a line of battle by a simple 'front to flank.' First came the Marshal de Gié with the vanguard, which was very strong, 350 men-at-arms, 300 cross-bowmen, 200 of the

[1] Giovio, ii. p. 93.

[2] Comines, book viii. chap. viii.—the story of the Swiss captain is told by several other writers, *e.g.* Paolo Giovio, p. 85 of his *Istorie*.

[3] *Ibid.* chap. ix.

mounted cross-bowmen of the royal guard, who were told this day to fight on foot, and the 3000 Swiss pikemen who formed the core of the army: they were led this day by Engelbert of Cleves, brother to the Duke of Cleves, and the Bailli of Dijon, and they had the artillery with them. The main-battle, was commanded by the King, who had around him the various fractions of his guard and personal retinue, with Louis de la Tremouille as a sort of chief of the staff: the rear-ward battle was under the Count of Foix. Each had about 300 men-at-arms and half the French infantry—probably under 2500 cross-bowmen apiece. All the Swiss were in the front corps, which had to clear the road for the whole army to follow. It was therefore made somewhat stronger than either the 'main-battle' or the 'rearward.' The very considerable baggage-train was under charge of Foix's division, and got a little up the hillside on his left, so as to be covered by him, and not to fall to the rear. The whole force was estimated by an eye-witness, the condottiere Niccolo Orsini, Conte de Pitigliano, as 9000 men,[1] so was much inferior to the allies both in men-at-arms (900 against 2400) and in infantry (7000 against 10,000). But the enemy did not have 30,000 or even 20,000 men as some of the French chroniclers allege, but about 14,000.

When the French van reached the spot where Gonzaga's artillery could play upon it, the Italians opened fire, but Gié pushed on, though he dropped his guns to reply to the enemy. But before any appreciable harm had been done to either side, the Marquis gave the word of command for his whole army to cross the Taro and fall upon the enemy. They were seen to pass the river in three distinct bodies, but not simultaneously, for the slippery bed of the stream and the mud beyond it caused some lack of exact order. But this—as it turned out—did not much matter, as each corps carried out its intention reasonably well.[2]

The result was astounding—in a quarter of an hour the attack was made and the battle lost. To take the fighting in detail—at the north end of the engagement the stradiots did succeed in cutting across the road in front of the French van. But they then, instead of doing anything useful, rode along the hillside in rear of the enemy's line, where they fell upon the baggage, took some valuable plunder, including the

[1] See Comines, book x. chap. ii. He was present with the army as a hostage, not as a combatant.

[2] Comines, x. chap. ii.

King's private store of relics and jewels, and killed some hundred camp-followers.

But as the Count of Caiazzo's Milanese horse came within a hundred yards of de Gié's men-at-arms 'when they should have couched their lances, their hearts failed them, they halted and fell into disorder.' The Marshal charged them and drove them back over the river with loss. At the same time his Swiss charged the Milanese infantry, who had found part of Caiazzo's column, and made a great slaughter of them—especially of the

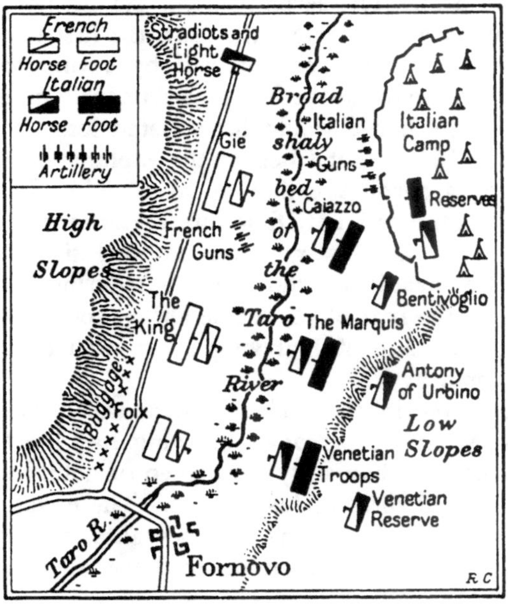

BATTLE OF FORNOVO, JULY 6, 1495.

Germans, who made some stand.[1] The Marshal then re-arranged his line, for he perceived another body of horse before him—this was Bentivoglio's brigade, which ought to have supported Caiazzo, but had no chance of doing so, since the corps in the first line had simply bolted. For the rest of the battle the Marshal kept still, observing Bentivoglio's horse and the garrison of the camp, which was visible behind it.

The real decisive action was in the centre and rear section of the line, where the Marquis of Mantua came into collision

[1] Guicciardini says that these Germans made for the French artillery, and were rolled over by the Swiss just as they were reaching it.

with King Charles, and the Venetian horse under Fortebraccio with the Count of Foix. 'They came on at a gentle gallop and to say the truth the charge was brisk,' but after the first crash and mêlée the ascendancy was obviously with the French—the Mantuans and Venetian mercenaries broke and fled, 'casting away their light lances, which lay thick on the ground—but they are good for little, being hollow and light and weighing no more than a javelin, though they are prettily painted.' The cavalry of the first line being routed, their attendant infantry gave ground also. The second line of Italian horse retreated without striking a blow, when they saw the first line dispersed. 'Antony of Urbino excused himself by saying that he had orders not to come in till the Marquis sent for him, and that he thought that the battle would have lasted long, according to the custom of Italy. But I believe the explanation was that he saw nothing to encourage him to advance.'[1] The whole of the Italian centre and left, first line and supports, fled back across the river, and many never stopped till they reached Parma. The stradiots, who were plundering the baggage meanwhile, in the rear of the whole affray, soon saw what had happened, and made off, carrying away such things as were not too hot or heavy.

The slaughter in the Italian centre was very great—the Marquis of Mantua's personal following had led the charge, and he lost no less than sixty gentlemen of note, including his uncle Ridolfo and three of his cousins. There was no quarter given, for the French grooms and servants, running in from the plundered baggage train, knocked every fallen man-at-arms on the head.

King Charles had charged very gallantly in the front rank, in front of his banner, and was in the thickest of the fray. When the clash was over, he halted, with hardly anyone left near him, for his guards had swept on in pursuit of the routed enemy, who had dispersed in all directions. He was in some little danger after the great charge, for a broken party of Italians trying to escape up the road, came down upon him when he had but one knight at his side. He was making good play with horse and sword when his friends rode in to his rescue and the enemy fled.

When the King and Foix had collected their scattered horsemen, some of whom had ridden a long way in reckless

[1] Comines, x. chap. xi.

pursuit of the enemy, the main body of the army moved on to join the Marshal de Gié, who stood with the vanguard opposite the enemy's camp. In this there could be seen collected not only the camp guard, but many of the horsemen of the reserve brigades of the Italian second line, who had never been properly engaged. It was debated whether it would not be well to fall upon this assembly, and the condottiere chiefs, Trivulzio and Vitelli, advised the King to charge in, saying that such troops would never stand. The French captains demurred, saying that enough had been done for one day, and that night was drawing on. Their opinion prevailed ; but no doubt Trivulzio was right—the enemy was astounded and demoralized.

So the army encamped, and marched on next day, leaving the Italians to rally, which they did. The King reached Borgo San Donnino and Fiorenzuola unmolested, and at Nizza got into the friendly territory of Montferrat. He was followed cautiously by the Conte de Caiazzo with some Milanese horse, but took no harm. The enemy, gradually getting together again, were soon in possession of all the land through which the King had marched.

This was a battle in which the losses on the two sides were wholly disproportionate. The French lost only 200 men,[1] of whom Julian de Bourganeuf, the Captain of the King's Gate, was the only person of importance. But, by an odd chance, the Bastard of Bourbon, pursuing recklessly, was taken prisoner by a band of horsemen whom he had chased alone and too far.[2] Nine of the King's Scottish archers of the guard are reckoned among the slain, and three or four captains of companies. The Italians had 350 men-at-arms killed, and some 3000 of their foot slain in the rout by the Swiss or the cavalry. Among the dead were Ridolfo Gonzaga, the uncle of the Marquis of Mantua, and Ranuccio and two others of his cousins, also four Venetian commanders, John Picinnino, heir of the famous condottiere of that name, Galeazzo de Coreggio, Alessandro Beraldo, and Roberto Strozzi.

This, as has been said before, was a terrible lesson to the Italians. Their veteran men-at-arms, trained in the old wars of the condottieri, had bolted before a desperate charge of the gendarmes, who meant business, and were not given to tactical manœuvres. The disgraceful part of the fight was

[1] Not certainly the 1000 given by Giovio, nor the twenty-two men-at-arms and seventy or eighty servants cited by Comines. I agree with Guicciardini.

[2] Like Maurice Berkeley at the battle of Poitiers.

that *all* the reserve brigades, set in due place to support the corps in the front line, had turned and fled, when they saw their comrades ridden down. What is the use of a good battle-array if the men do not intend to fight? And the stradiots, from whom much had been hoped, turned out to be savages who lost their heads when they saw plunder available, and forgot the purpose for which they had been told off. It was clear, after July 6, 1495, that the Italian states could not survive if defended by mercenary armies who fought on the old principles of much pay, no casualties, and the pleasant chance of rich ransoms. When the Marquis of Mantua found that all his relatives and personal retinue were dead on the field, instead of held to ransom, he saw that the end of an epoch had come. So did Machiavelli, studying the Art of War in distant Florence. This was a very different sort of war from that to which they had been accustomed.

CHAPTER II

BATTLE OF THE GARIGLIANO (December 29, 1503)

THERE were several encounters of interest between July 1495 and December 1503, but some of them were on a rather small scale, such as both the first and the second combats of Seminara, and the protracted fighting round Barletta. We have already given some account of the more important action of Cerignola (April 1503), the great triumph of Gonsalvo de Cordova's careful combination of entrenchment and the large-scale employment of firearms. This, as has already been pointed out, was the first of a long series of battles in the new method, of which he was the founder and the great exponent.[1]

But far more notable is his 'crowning mercy,' the rout of the Garigliano, whose story reads like that of a modern campaign in the best style, of which the Duke of Wellington or Robert Lee would have been proud.

The original French army of Naples had been beaten and dispersed at the fight of Cerignola in April, long before the reinforcements which Louis XII was preparing to dispatch to its aid had entered the Italian peninsula. When the army of succour at last appeared upon the scene, the main body of the wrecks of Cerignola had collected in the sea-girt fortress of Gaeta, a place on a projecting headland connected with the mainland by a low-lying spit. It is practically impregnable, if the Monte Orlando, a hill 500 feet high, alongside of which it is built, is held by the garrison; and this eminence had been surrounded with lines and taken into the periphery of the fortress. The officer in command was Yves d'Alégre, the rash captain whose taunts had stung Nemours into his unwise attack at Cerignola. He had with him a good proportion of the cavalry which had escaped from the recent battle, when the unlucky infantry had been so badly cut up. The only other places in the kingdom of Naples which were still in French

[1] See above, page 53.

hands were the distant fortress of Venosa in Apulia, where another fraction of the routed army had taken refuge, and Rossano in Calabria. The rest of the realm was in the hands of the victor, after the outlying French detachment in Calabria had been annihilated at the second combat of Seminara (April 21) and the forts around the city of Naples had been stormed or had surrendered in May and June.[1]

After setting things more or less to rights in Naples, and giving his army the rest which it needed, Gonsalvo directed his attention to the one task of importance which lay before him, the siege of Gaeta, which would obviously be a long and difficult operation. For not only was that fortress at the end of a long sand-spit, difficult to batter with the imperfect artillery of the sixteenth century, but it was in full communication by water with France, and the French were at the moment in possession of the sea, and had made Gaeta into a formidable naval base, into which food and munitions had been poured. Though there was a Spanish squadron in the Bay of Naples it was driven off by a superior French fleet, and a body of no less than 4000 infantry, under the Marquis of Saluzzo, a noted condottiere-prince, was shipped across from Genoa to strengthen the garrison. All the ingenuity of Pedro Navarro, the celebrated Spanish specialist in artillery and trench-work, could make no impression on such a strongly held place. An attempt to storm the fortified position of the French on the Monte Orlando, alongside of the walls, failed with loss. The batteries set up before the land-front were overpowered by the fire of the defences, and after some weeks of lying before Gaeta, Gonsalvo withdrew to Castellone, four or five miles inland, and turned the siege into a blockade. This was all the more necessary because news had come to him that the French army of succour was already in Lombardy, and moving southward. It would never do to have the enemy descending into his rear, and marching on Naples, while he was occupied in beleaguering an impregnable fortress on a strong headland.

Louis XII had been collecting troops even before the news of Cerignola arrived, and when that disaster was known, resolved to strengthen his expeditionary force. He subsidized no less than 8000 Swiss, raised new French infantry bands in Gascony and Normandy, collected some Italian mercenaries

[1] The Castel Nuovo was stormed on May 21; the sea-girt Castel d'Uovo capitulated in June.

in the Milanese, and called for volunteers for the cavalry arm to supplement the *compagnies d'ordonnance* of the gendarmerie who were available. In July the French part of the army was crossing the Alps, and in August it was joined at Parma by the Swiss and the Italian levies. The whole, according to Guicciardini, made up at least 20,000 men.

The charge of the army was given to a veteran soldier, the Marshal La Tremouille, who had served at St. Aubin de Cormier and Fornovo; the Marquis of Mantua, who had brought up the Italian auxiliaries, was second in command. The troops were starting from Parma when an unexpected event complicated the situation. Pope Alexander VI had died suddenly on August 18—poisoned by misadventure, as every one believed, with a draught that he had prepared for one of his cardinals.[1] No sooner had his blackened and unsightly body been huddled into a hasty grave, than the question of a Papal election cropped up. Now France had a candidate for the Tiara, the Cardinal George d'Amboise, who played at the court of Louis almost the same part that Wolsey was to play a few years later at that of Henry VIII. The moment that the news of the Pope's death reached Paris, he hurried down to Italy, with the King's high approval, and with permission to use the French army as a threat to the Papal conclave in support of his own candidature.

The army reached the vicinity of Rome early in September, but not its general—who had been left sick at Parma.[2] At the same time Gonsalvo pushed forward some light cavalry into the Papal States, as far as Nettuno and Rocca de Papa, so that the conclave seemed to be threatened from both sides—but the French threat was much the more dangerous. If d'Amboise had used his power ruthlessly he might have attained his end, but he was induced to move the army to a distance, by the assurance of false friends, who persuaded him that his election was certain, and that open display of force was both unseemly and unnecessary. When the French had drawn off, the conclave elected the moribund Cardinal of Siena, who took the title of Pius III (September 22, 1503), and died only twenty-six days later. But before his decease the tricked and

[1] So thought, at least, Guicciardini, Burchard, Giovio, Bembo, and everyone else who tells of the event, and of the curious disappointment of Cæsar Borgia, stricken down with illness (supposedly poisoned like his father) at the crucial moment.

[2] So Guicciardini—but Giovio says that he collapsed at Rome.

disgusted d'Amboise allowed the French army to start on its way to Naples, and during the short pontificate of Pius it crossed the frontier and got into touch with the Spaniards. The Roman intrigues had cost it a month of good weather, and in October the autumn rains had begun. La Tremouille, on falling sick, had handed over the command to the Marquis of Mantua, who did not enjoy the confidence of the French captains—some of them must have remembered his unlucky generalship at Fornovo eight years back. Still he was a trained soldier, and his army was much stronger than that of Gonsalvo.

There are two obvious lines of invasion for a force moving from Rome on Naples : it may take the inland route, the ancient Roman Via Latina, which goes by Tusculum, Anagni, Ceprano, San Germano, and Mignano on to Capua, following in all its middle course the valley of the Liris (Garigliano). Or it may take the shorter and more direct Via Appia which keeping close to the coast, cuts through the Pontine Marshes to Terracina, and then goes by Fondi, Itri, Formia, Minturnae, and Sessa to Capua. The former road has been much more commonly used by mediaeval and modern armies than the latter, probably because the Via Appia runs for so many miles along the Pontine Marshes, which were malaria-ridden and practically uninhabited, so that food was hard or impossible to obtain. On both roads there are obvious defiles, on which an army entrusted with the defence of Naples might array itself. On the inland route Ceprano and San Germano were old mustering-places of the troops of the Angevin and Aragonese kings of the fourteenth and fifteenth centuries, having in front the strong fortress of Rocca Secca, and above the towering cliffs of Monte Cassino—a monastery but also a stronghold. There is another defile somewhat nearer to Naples which is also very defensible—the pass over the watershed between the valleys of the Garigliano and the Volturno below the castle of Mignano.

On the coast road, the Via Appia, the obvious defensible defiles are the passages round the end of the Auruncan Mountains by Fondi and Itri, then, just behind, the narrow shore between hill and sea by Formia, next the marshy estuary of the lower Garigliano, and lastly the pass over the Monte Massico between Sessa and the sea—blocked in the sixteenth century by a castle which has now been destroyed.

While the army of the Marquis of Mantua committed

itself to the Via Latina, and began to descend the upper valley of the Garigliano, Gonsalvo saw that he must transfer himself from the position on the Appian Way from which he was blockading Gaeta. On October 6 he went inland, and took post at San Germano, under the shadow of the lofty rock of Monte Cassino, and with Rocca Secca in front of him. When he had departed from before Gaeta, the Marquis of Saluzzo came out of the fortress with 4000 men, occupied the surrounding defiles, and sent out light cavalry across the hills, which soon got into touch with the vanguard of the main army as it descended the valley of the Garigliano, and finally he went in person with a large detachment to report at headquarters. After the manner of mediaeval generals, who disliked leaving a fortress behind them, the Marquis of Mantua sat down for some days before Rocca Secca, whose governor, Villalba, not only treated with contempt a summons to surrender, but hanged the *parlementaire* when he threatened him that the garrison should have no quarter unless the gates were opened at once.[1] This act of ferocity made the Marquis resolve that the castle should be stormed at all costs, and Villalba punished, but at the same time inspired the garrison with desperate determination, since they were doomed to death if they flinched. The Marquis brought up his artillery, breached the walls, and made two desperate assaults, which failed. Meanwhile Gonsalvo sent out his vanguard from San Germano and threatened the camp of the besiegers, though he could not have dared to close, since he was inferior in numbers by about a half to the enemy.

Finding Rocca Secca too hard a nut to crack, and seeing the Spanish army on his flank, the Marquis of Mantua gave up the siege, and marched past the fortress to Aquino and Ponte Corvo. He is said to have been forced to move, since he had eaten up his provisions, and the countryside was bare. October was near its end when he settled down at Ponte Corvo, and harried the upper valley of the Garigliano in all directions to get food. Gonsalvo followed him with caution, and attacked his rearguard and baggage train, capturing his hospitals at Aquino. 'With singular piety'[2] he allowed no harm to be done to the sick, though on both sides there had recently been horrid mishandling and massacre of disabled men. But there was no question of attacking the French camp—the odds were too great.

[1] Giovio's *Life of Gonsalvo*, book ii. p. 136. [2] *Ibid.* 158.

After staying for seven days at Ponte Corvo the Marquis resolved that he would have to give up all idea of forcing his way by the inland route to the plain of Naples. The neighbourhood had been swept clear of food, the weather was abominable—such a November as had never been seen—rain never ceased—the roads were so deep in mud that the waggons and the artillery were always sticking fast, the cavalry could hardly move, and many of the infantry were dying of dysentery and cold. Gonzaga determined that he could not face the defiles about San Germano, and beyond it in the high hills of the watershed between the Garigliano and the Volturno. Accordingly he determined to make a short cut to the south by Pico and Itri, and to throw himself on to the coast road—the Appian Way—where he would have Gaeta at his back, and would be joined by the large force which its garrison could send out. Gaeta was full of provisions, and as a French fleet was lying in its harbour supplies would always be forthcoming.

Accordingly Gonsalvo de Cordova was surprised to find the French camp at Ponte Corvo evacuated, and discovered that the enemy had disappeared southward. He guessed what this must mean, and hurried off his cavalry to watch the passage of the lower Garigliano, following with his infantry and guns, moving by a difficult cross-road across the Auruncan Mountains parallel to the French. He then crossed to the east side of the Garigliano, and took ground there, so that by the 1st of November the two armies were again facing each other near the sea, with the swollen reaches of the Garigliano between them. The Via Appia had in old Roman days crossed the river by a long and solid bridge at the town of Minturnae. But Minturnae and the bridge had both disappeared in the Dark Ages. The bridge was gone and the inhabitants of the old town had moved uphill, to a new settlement, Trajetto,[1] so called from the ferry which had replaced the bridge. The ferry was not usable at the moment, for the river was in spate, and had overflowed the low-lying ground on both of its banks. Of this Gonzaga was probably aware before he started to get on to the Via Appia, but he intended to build a pontoon bridge on the spot where the approaches to the ferry were over reason-

[1] Like so many Italian towns, Trajetto has now adopted the name of the old Roman city which it represents, and is called Minturno, though it is three miles from the deserted site, where the remains of an amphitheatre stand in the middle of modern 'Scavi.'

ably good ground. As a preliminary operation he had to get rid of a little fort overlooking the point, the Torre de Garigliano, where Gonsalvo had left a small garrison. The officer in command capitulated rather than face artillery fire, and was allowed to take his men off. It is a strange sidelight on the ferocity of the times to find that, on the arrival of the detachment in Gonsalvo's camp, it was mobbed, assaulted, and mostly killed off by the soldiery—and that the general, though he disapproved of the riot, took no measures to punish its ringleaders, thinking apparently that the moral effect of the affair was rather profitable than otherwise. 'Other garrisons would take warning from it.'

The lower Garigliano, after emerging from the hills, reaches the sea by a very circuitous course with many loops and bends in a waterlogged plain. Its estuary is through a perfect bog—those marshes of Minturnae, in whose mud Marius once sought refuge from Sulla's assassins. Above the bog the higher ground is always on the western side, where the French were in occupation, and the place of the ferry above the Torre de Garigliano is the most obvious spot for a passage. This did not escape the eye of Gonsalvo, and his first care on reaching his new position was to order a long trench to be cut across the meadows facing the ferry, out of cannon-shot of the opposite side. The ground was so wet that the ditch at once filled itself with water, and the parapet behind it had to be built up with fascines, since the thrown-up mud would not hold itself upright. On this front he hoped to make a stand against any force attempting to cross the river.

The trial soon came. The Marquis collected beams and barrels and some small boats from Trajetto and the neighbouring villages, and, having placed heavy guns on the ground above the ferry to protect his workmen, started to build a bridge. The Spaniards annoyed the builders with arquebus fire, but were soon driven off by the artillery, to which they could not reply, as their own cannon could not be brought near the river. A broad bridge able to bear cavalry as well as infantry was completed on November 6, and the Marquis launched his assault over it. When about 1000 men had passed, and were going rapidly forward, Gonsalvo, who had collected his troops behind his trench, charged out and attacked them, while the bridge was jammed with reinforcements coming up in narrow file. There followed a very fierce fight of the hand-to-hand sort, in which the French were brought to a stand, as they could not deploy nor bring up more men, while their gunners across the river had to cease firing, lest they should hit their own friends. Finally the head of the column which had crossed the bridge was thrown back to the water's edge; those who could recoiled across its planks, but many were rolled into the river and drowned. When the assault had failed, the French guns opened again on the Spaniards massed at the bridge-head, and with such good effect that they had to draw back out of range.

The bridge had been ruined, partly by the back-rush of the retreating crowd, partly by misdirected cannon-balls aimed at the Spaniards during the heat of the struggle. But the Marquis was determined to force the passage, and sent back to the fleet at Gaeta for large ships' boats, which were brought upstream to the place of the ferry, and joined together with beams and ropes, while the heavy guns kept the Spaniards off. He then sent troops across, with orders not to press on recklessly, but to throw up a *tête-de-pont* entrenchment, capable of defence, just across the water. This was done, and a palisaded work was constructed, which was secure against any infantry assault (November 10 ?). But attempts to issue from it failed; the Spaniards checked three at their trench a few hundreds yards out.

The days in mid-November became worse than ever—perpetual rain and a few falls of snow. The Marquis resolved to put off further operations, and to wait for better weather. Even if he could deploy across the river from his bridge-head, the whole countryside in front was a morass, in which cavalry could not move, and artillery would sink up to its wheel-axles. He would wait till the river should fall, and other practicable spots for a crossing be discovered, supplementary to his bridge, and outflanking the Spanish trench which blocked the way.

There followed an extraordinary deadlock of six weeks, during which the situation of Gonsalvo's front line recalls descriptions of the trenches in Flanders during the winter of 1914–15. It was necessary to hold a line watching the whole lower course of the Garigliano. The pickets were up to their knees in mud, and it was only possible to get from place to place by 'duckboards' as we should call them—hurdles or plaited boughs laid in succession. The supporting companies made themselves hovels of wattled twigs—it was impossible to 'dig in,' for water answered the spade. Reliefs were difficult, and took an unconscionable time, the men were never dry, and food came up from the rear with intolerable delays. Many of his officers besought Gonsalvo to take back his whole army to the foot-hills four or five miles away, and to make his head-quarters at Sessa, the only town of any size that lay behind the position.[1] He refused to move, saying that a retreat would break the morale of the army, and that, if he surrendered the passage of the river, the French could not be stopped farther back—considering their overpowering numbers. To keep the men in heart he visited the front-line pickets every day, and lived in a hovel only a mile from the trenches, where he was always accessible. That he held a drenched and sullen army in hand for so many weeks is a marvellous testimony to his personal ascendancy over his veterans.

But there was something more than dogged obstinacy in his policy. He was beginning to have some hopes for the future. The most encouraging thing was that the French army, though better placed than his own as far as mere quarters went, was falling into demoralization from discontent and indiscipline. The sight of their ill-kept and undermanned outposts told him much, but more encouragement came from secret intelligence from their camps. The whole army was in a state of barely suppressed mutiny, the men were melting away, the officers forming cliques and cabals to remove the commander-in-chief. The Marquis was openly insulted, and privately accused not only of incapacity but of treachery. Comprehending the position, he developed an official attack of fever, turned over the command to the next senior officer, the Marquis of Saluzzo, and went home to his palace on the Mincio in high dudgeon.

The change of generals was of little profit—for the French officers still grumbled at being left in charge of an Italian, and though Ludovico of Saluzzo's record of service under Charles VIII and Louis XII was creditable, he did not pass as a military genius or an inspirer of enthusiasm. The army was out of hand, the captains neglected their rank and file, and sought warm quarters far from the front, or dropped into Gaeta for good company and good food: the men deserted wholesale. The state of the horses was growing so bad that Saluzzo sent back whole squadrons to get cover and forage in towns ten miles or even more from the front, at Itri, Formia, or even farther

[1] Giovio's *Life of Gonsalvo*, iii. p. 166.

afield. The whole outpost line was barely held, while the men sheltered in farms or villages far back from the river. Some companies made themselves rainproof shelters under the arches of the ruined amphitheatre of the old Roman town of Minturnae. The last thing that any one, from the commander-in-chief to the meanest cross-bowman, was expecting was a general attack at midwinter, by an enemy who was known to be in a great numerical inferiority and who had hitherto kept to a purely defensive policy. The main signs of life which the Spaniards gave in December were some attempts to break the French bridge—once by floating heavy logs down the river, which went aground or were shoved off by poles, on a second occasion by sending a fire-boat down the current. But it burnt itself out and exploded before reaching the bridge. Pedro de la Paz is also said to have executed a cavalry raid far upstream, but to no great effect.

Neverthless an assault on the French lines was what Gonsalvo was contemplating, when he discovered how carelessly they were being held. He had already received a considerable reinforcement. The whole Orsini faction had come to an agreement with the Spanish Government, and its chief, the great condottiere Bartolomeo de Alviano, turned up with 400 lances, 1000 light horse, and 4000 Italian foot. This brought up the Spanish army to a respectable figure, though its total was still much below that of the French. For some weeks Gonsalvo had been preparing materials for a bridge at the Castle of Mondragone, fifteen miles behind his front. They consisted of elaborate carpentry, short lengths of beam carefully measured, planks which would make into pontoons when fitted together, and long ropes. All had to be light and easily handled, since they were to be carried on the backs of mules or horses—the byroads were impassable for anything on wheels.

At Christmas there was something like a truce between the two armies—Gonsalvo appears to have let it be known that the 25th and 26th of December would be devoted to religious and convivial relaxation, and to have notified this to the French.[1] Once more we are reminded of the Flanders front in December 1914, and the 'fraternization' on Christmas Day. But on the 27th, while the enemy was still engaged in belated revelry, Gonsalvo began to move up from Mondragone his train of bridge material, and to shift the greater part of his troops up

[1] Havendo fatta tregua, quasi per una certa ragione, reposare dalle fattioni 'della guerra' (Giovio, iii. p. 160). This seems to imply an actual agreement.

to the northern end of his position, opposite the extreme left of the French lines, which rested on the village of Sujo some five miles above the bridge. He had chosen this point for the delivery of his blow, because the ground on each side of the river was less wet than lower down, and one bank did not command the other. Alviano, who had highly approved of the plan, and had suggested some of the details, was placed in command of the van and the train of bridge-timber. The 'main-battle,' under Gonsalvo himself with the Spanish infantry and the men-at-arms, was really to follow. Only the 'rear-battle,' under Fernando Andrada, was left in the trenches opposite the French bridge; he had a body of German infantry and part of the light horse, with orders to attack the bridge if the manœuvres upstream showed signs of success, or at any rate to check any attempt at a counter-attack which the enemy might make at this point.

The whole scheme worked with even greater speed and accuracy than might have been expected. The bridge opposite Sujo was laid at dawn on the 29th with no opposition—it was finished very rapidly, the engineering calculation being laudably correct. The French line here was held by some companies of Norman infantry, who were caught completely by surprise—they were not even under arms before the bridge was finished, were incomplete in numbers, and lacked many of their officers. Alviano's light horse fell upon and swept them away. They fled by the path along the river, spreading alarm, and were so closely followed by the enemy that no chance of making a front was given to the Swiss and other infantry, who were cantoned in Castelforte, San Cosmo, and other villages downstream. The cavalry were mostly far to the rear: the Marquis of Saluzzo could only collect a small body of horsemen who, under Yves d'Alégre, charged the Spanish van, but were ridden down and scattered. The enemy was coming on with such speed that the Marquis, vainly trying to collect a force at Trajetto, saw that he was in a hopeless position, ordered a general retreat to the narrow defile between the sea and the hills at Formia, and sent orders to destroy the bridge, and to bring off, if possible, the artillery which was lying there. This command was given too late, for the Spanish rear-battle under Andrada had begun to deliver its attack before the business could be completed. The French at the bridge cut its moorings and tried to embark the guns on the barges and ships' boats of which it was formed. All was in confusion, and the Spaniards were only checked for a short time; for they succeeded in capturing some of the boats and much material, and reconstructed an extemporary and imperfect bridge, over which they began to cross the river. At this point died obscurely Piero dei Medici, once lord of Florence and now only an officer of no importance in the French army. He had got a gun into one of the large boats, and was moving off downstream, when a party of routed soldiery jumped into the stern of the boat and caused it to capsize. Being in heavy armour, Piero went to the bottom, along with the gun and the fugitives. So much had he been forgotten that many chroniclers do not even mention the termination of his rather futile life. After crossing the repaired bridge, Andrada's column got possession of the bulk of the French artillery and of camps containing much valuable spoil.

Meanwhile a ten-mile chase, with no serious opposition met anywhere, brought the victorious Spaniards past Trajetto and Scauli to the narrow seaside defile of Formia, or Molo de Gaeta, as it was then usually named. Here a great block of French troops, horse and foot, had accumulated, and the Marquis of Saluzzo succeeded in stopping many of them, and in forming a front. This was the real battle of the Garigliano, for there was hard fighting for an hour on the narrow road, where a considerable body of the French gendarmerie had rallied, and under leaders such as d'Alégre, Bayard, the Bastard of Sandricourt, and the Genoese captain, Bernardo Adorno, held the bridge of Molo for some time.[1] They only gave way when Andrada's column came up, a little late, to lend its weight to the thrust. This cavalry rearguard having been at last driven in, the Spanish horse rode for miles among broken infantry and guns stuck in the mire, trampling down all opposition: they were not stopped till they reached the gates of Gaeta, and were fired on from the ramparts.

The French were surrendering on all sides—a body of horse which came up from distant cantonments, and was intercepted on the Itri road by Spanish light cavalry, which had taken a short cut across the hills,[2] yielded without striking

[1] Adorno was killed. Was this the occasion on which Bayard's 'Loyal Serviteur' speaks of him as having been captured and rescued by Sandricourt? (xxv. 81, 82). Probably not, as it is spoken of as an isolated skirmish at a bridge, not as part of a general action.

[2] Giovio's *Life of Gonsalvo*, iii. p. 172.

a blow, though it might easily have cut its way back along the Via Appia. The vanquished lost some 3000 or 4000 dead, all their guns and baggage, and several thousand prisoners. Yet it is probable that even at the end of the day they were

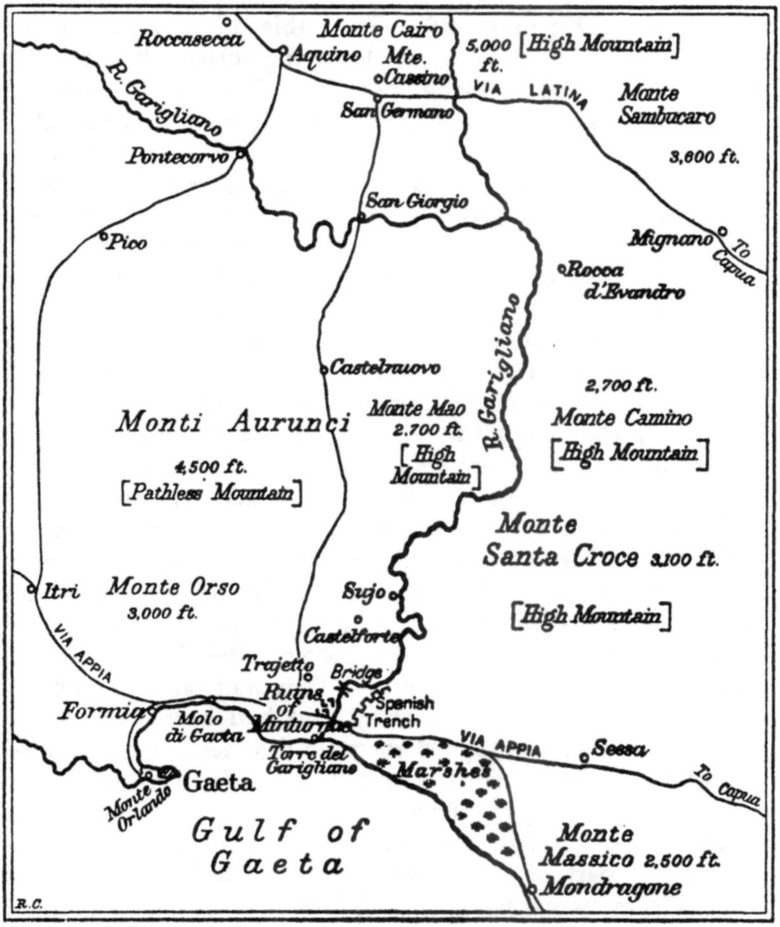

Campaign of the Garigliano, December 1503.

not outnumbered by the Spaniards, for their original force had been well over 20,000 men, and Gonsalvo had certainly not so many as 15,000.

Three days later the campaign ended, in a fashion which recalls to mind the Convention of Cintra. On the 30th,

Gonsalvo noted that the hill Monte Orlando, which overhangs Gaeta, and had been held as part of its *enceinte* in the early part of the year, seemed to be neglected by the demoralized enemy. He ordered it to be stormed, and found that its entrenchments were practically unmanned—they were carried without loss (December 30). From this position guns could have searched the whole of the low-lying fortress below. But there was no need to bring them up. For that evening the Marquis of Saluzzo behaved exactly as Junot behaved after the battle of Vimiero. He sent out a flag of truce, offering to surrender Gaeta, with all its artillery stores and munitions, in return for a free departure. As he could probably have got away the best part of his army on the numerous ships which lay in the harbour, and the siege might have been a long business if the enemy proved obstinate, Gonsalvo accepted the offer without hesitation. The French might depart by sea or land as they pleased, if they handed over the fortress intact with all its contents. A convention was drawn up and signed on January 1, 1504 the terms—like those which Junot got from Dalrymple in 1808, were very liberal, as Gonsalvo undertook to return his very numerous haul of prisoners without ransom.

He kept his word, save in one detail, for which he was much criticized by the French; among the captives were some Neapolitan barons of the old Angevin faction, who had been made prisoners at the second battle of Seminara, nine months back, and had been released on parole, after disowning their allegiance to the King of France. They had taken the opportunity to rejoin the French army, and had been captured in the field. Gonsalvo claimed them as deserters, not as honest prisoners, and sent them to the dungeons of the Castel Nuovo at Naples. Among them were Honorato and Alfonso of the great house of the Sanseverinos, and Matteo of Aquaviva.

A great part of the French army, including most of the chief officers, went away on shipboard: the rest and all the Italians and Swiss preferred to march homeward by way of Rome. It is said that only a fraction of either party returned in safety—the Marquis of Saluzzo and many more of those who went by sea died at Genoa, from marsh fever or exhaustion. The column that went by land suffered many afflictions on the way, being attacked by the peasantry whom they had plundered, and starved for want of any regular provision of food. Not only the hospitals but the sheds and stables of Rome were filled with moribund soldiers, who had barely staggered so far. The newly elected Pope, Julius II, was charitable enough to provide rations and clothing for many of them, but thousands are said to have been buried in Roman cemeteries.

Louis XII, enraged at the loss of such a fine army, sought for responsible persons to punish. Ludovico of Saluzzo was dead, but d'Alégre [1] and Sandricourt, who had been at the head of the cabal which got rid of the Marquis of Mantua, were banished. Sandricourt in a fit of despair committed suicide. The bailiff of Caen, who had been chief commissary, and was accused of wholesale frauds in the matter of supply, was condemned to the gallows.

On the whole this campaign is the most interesting of all those of the wars of Italy, and well justifies Gonsalvo's claim to the title of the 'Great Captain.' With an army of inferior numbers and heterogeneous composition, he had not only beaten a much superior force, but completely broken it, and driven it out of Italy by his skilful manœuvres, and his resolute optimism. His power to hold his men in hand in times of defensive warfare was equalled by his power to inspire them in the critical moment of taking the offensive, and the details of his passage of the Garigliano, followed by ruthless pursuit of the enemy whom he had taken by surprise, were models to be studied by all generals charged with the problem of attacking a superior army across an unfordable river. This was Gonsalvo's last campaign. After spending four more years as Viceroy of Naples, he was called back to Spain with all honour and ceremony. But his master, King Ferdinand, was decidedly jealous of his greatness, and never employed him again. He died in 1515, aged seventy-two.

[1] D'Alégre was afterwards pardoned, and survived to die a soldier's death at the battle of Ravenna.

CHAPTER III

RAVENNA (April 11, 1512)

It is obviously necessary to give a detailed account of this celebrated fight—mainly because it was the first general action during this series of wars in which the event of the day was settled by the force of artillery, but in a secondary way because its details offer several puzzling problems, to which no generally accepted solution has yet been found.

The preliminary campaign has already been alluded to in general terms.[1] The League of Cambray had come to an end, and most of the allies formerly combined against Venice had now reversed their policy, and had joined with Venice in an attack on France. The successes of the arms of Louis XII had frightened both the Pope, the King of Spain, the Swiss, and the Emperor. The first three were already at open war with France, and the fourth was on the eve of following their example, though the rupture had not yet taken place, and his armies were not yet in the field.

Gaston de Foix, Duke of Nemours, the very young but most capable commander to whom Louis XII had entrusted his army of Italy, had just driven the Spanish and Papal forces away from Bologna, which they had been besieging, and then turned to Lombardy, inflicted a severe defeat on the Venetians, and stormed Brescia. When he had gone off northward on this latter enterprise, the Spanish-Papal army descended from the Apennines, and reoccupied most of Romagna. Turning back upon them, Gaston came down the Via Aemilia, retook several small places, and then marched against Ravenna, the most important city which was in hostile hands. This was rather a dangerous move, as he left the enemy upon his flank, and exposed his communications with his base, Ferrara, whose Duke Alfonso of Este was the only trustworthy ally whom the French possessed in northern Italy. After a very short battering by a strong artillery—the Duke of Ferrara had brought his

[1] See above p. 25.

celebrated train of cannon to join the French—Ravenna was breached, and in some danger of being stormed, though its governor, Marcantonio Colonna beat off one serious assault.

Gaston, it is said, had, as one of his chief motives for besetting Ravenna, the notion that the enemy would not allow such an important place to fall into his hands without attempting to relieve it. And if they dared to come into his close neighbourhood, he hoped to force them to an open battle and beat them.[1] In this expectation he was justified, for Ramon de Cardona, the Viceroy of Naples, who was in command of the Spanish-Papal army, determined that he must do his best to succour Ravenna, though he knew that his total force was decidedly inferior to that of the French army. Accordingly, he came down from Imola, where he had been lying for some days, and advanced into the low country near Ravenna, between the rivers Ronco and Montone, intending so to place himself that he might disturb the siege operations without necessarily committing himself to battle. There was good reason for delay, because the Swiss and the Venetians were known to be stirring in Lombardy, and the latter had just begun to operate against Gaston's line of supply—an expedition in boats had come up the lower waterway of the Po, and captured a convoy which was moving from Ferrara to the camp of the French.[2] But, on the other hand, Ravenna was in serious danger, and so delay might lead to a disaster.

All the country about Ravenna is dead flat, water-logged, and cut into ditches and drains. The ground east and south of the city is recent alluvial silt—the sea has been retiring for a thousand years— leaving Ravenna, once a port in old Roman days, stranded nearly five miles inland. Two great churches Sta. Maria in Porto, two miles outside the present gates, and St. Apollinare in Classe, still farther out to the south, mark the situations of two successive short-lived harbours, both long deserted by the Adriatic. Between them and its present shore stretches the sandy waste of the Pineta. On each side of the city ran in 1512 two rivers called the Montone and the Ronco, of which the first flowed close under its northern wall, the second

[1] So Guicciardini, x. p. 269. 'Sperando que gl' inimici, per non diminir tanto de reputazione, non volessero lasciar perdere sugli occhi loro una tal città, e cosi avere occasione di combattere in luogo eguale.' Giovio says the same in his book (x. p. 209), of which only the epitome survives.

[2] Guicciardini, x. p. 271, and Bayard, iii. p. 194.

close under its southern front. They united some little way to the east of Ravenna, and went through the Pineta in a single stream. Unfortunately for topographers of to-day an energetic eighteenth-century cardinal named Corsini completely altered the hydrography of the region, by blocking the courses of both rivers, turning the Montone into the Ronco a mile south-west of the city, and then cutting a new joint course for both rivers somewhat south of their original estuary, and close to Sta Maria in Porto. This broad, artificial channel is called the Fiumi Uniti; into it drain many ditches from what was in 1512 mere marsh, but has now been turned into rice-fields and other patches of irrigated cultivation. The old lower courses of both Montone and Ronco are now mere ditches or canals.

Consequently Ravenna, instead of lying between two sluggish rivers, now stands without its old natural wet ditches, a mile north of the course of the artificial Fiumi Uniti.[1] But in 1512 Gaston de Foix, wishing to batter it, chose very reasonably its west front, that which then lay between the two parallel rivers, and was shielded by neither of them, from the Porta Adriana near the bank of the Montone to the Porta Aurea near the Ronco. But though he encamped between the rivers, he gave himself the chance of operating on either side, by casting a bridge over the Montone on the flank of his position.

Cardona, advancing cautiously from Imola, first halted on the south-east bank of the Ronco about four miles from Ravenna, but was then persuaded by his captains to move closer in, though Fabrizio Colonna, the leader of his cavalry, besought him not to risk a battle by pressing too near to the French line of investment, and to give himself room to manœuvre, and to retreat if necessary. But Pedro Navarro, the celebrated commander of engineers, induced the Viceroy to move close up to the enemy, by showing him a position which he considered almost impregnable, less than two miles from the city, between the Ronco on one flank and marshy ground on the other.[2] Granted that the time to throw up a trench across this short line was conceded by the enemy, Navarro guaranteed

[1] I spent some hours looking through sixteenth and seventeenth-century maps of the Ravenna district in the British Museum map-room. The early ones are all very careless, and do not in the least agree with each other as to the courses of the Montone and Ronco, and their way of reaching the sea. The most careful of them, that of a geographer named Magini, which shows the relations described above, seems to me the best.

[2] 'Da una parte l'aqua, d'altro il palude, locho forte.' Marino Sanuto, *Epistole* vol. xiv. p. 76.

that it could turn back any assault, and spoke with contempt of Colonna's suggestion to move into the Pineta, and get into touch with Ravenna on the side where it was not being attacked or battered. Arriving on this ground 'in the meadows of Sta Maria in Porto,' north of the mill called the Molinaccio, in the afternoon of April 10, the Spanish army dug a long trench from the Ronco on the left to the commencement of the marshy ground on their right, and encamped behind it.

The arrival was, of course, perceived by the French, and Gaston determined to force on a fight at all costs, even if it had to be made by a frontal attack, a thing which generals were beginning to dislike. But he was intending, as all his arrangements show, to try a new system, that of battering the enemy's weak points with overpowering artillery fire, before letting loose his infantry and cavalry for the assault. He had a train of exceptional strength—thirty of his own guns and twenty-four which the Duke of Ferrara had brought in. And during the night which the Spaniards had utilized for making their entrenchment, Gaston had set his pioneers to build a bridge across the Ronco, only a mile from the north-east angle of Cardona's camp, and to make ramps on each side of it, by which troops could come down easily to the river, which was quite fordable. The bridge indeed was mainly intended for the artillery, which could not be risked in the muddy bottom of the Ronco. No one doubted that there would be a general action next day. The commanders-in-chief had exchanged formal defiances by trumpet.[1]

The French prince had obvious reasons for pressing on a battle—his superiority in artillery, the approaching dearth of provisions, which must ensue from the Venetians having cut his line of communcation with Ferrara, and the knowledge that the Swiss were threatening to come down into Lombardy behind his back. But there was another and a secret reason. Jacob Empser, the veteran commander of the landsknechts, who formed the most valuable part of his infantry, had come to his tent to warn him that Kaiser Maximilian had sent letters to bid him withdraw his men at once from the French camp, as there was going to be war between the Empire and France within a few days. To depart on the eve of a battle would be heart-breaking to one who had served Louis XII

[1] About these defiances there is plenty both in the Spanish *Relacion* and in Florange.

long and faithfully, and he intended to put the letters of recall in his pocket for the moment. But more would come, undoubtedly, within the next few days, and all the Germans would get to know of the Emperor's orders. Wherefore it would be well to fight at once, before rumours got abroad.[1]

Gaston's army consisted of about 23,000 men. Of these he left 2000 Italian foot under Paris and Nicolas Scotti, two brothers, to guard his camp and trenches, and watch against sorties from Ravenna. The striking part of the army was divided theoretically into the old-fashioned vaward, main-battle, and rearward, but the denomination meant little in the arrangement of the fighting line, as will be seen. To the vaward, comprising about 900 lances under the Duke of Ferrara and the Sieur de La Palice, were told off the best of the infantry, 5000 landsknechts under Empser, and 3500 French foot, mostly Gascon cross-bowmen: it had charge of the artillery. The main-battle, which was supposed to be under Gaston himself—though he was moving about the field all day, not acting as a division-commander—was really led by the Senéchal of Normandy, by Odet de Lautrec, and those veterans of the Neapolitan campaign of 1503, Louis d'Ars and Bayard, with 780 lances. Its 3000 French foot, Picards and Gascons were under the Senéchal. The comparatively weak rearguard corps was led by d'Alégre—in favour again despite of his disgrace of 1503: it consisted of 4000 foot, mainly Italians under Federigo de Bozzolo, but including some French companies, and of only 300 horse. It left a strong detachment at the bridge, in case the garrison of Ravenna might sally out, and fall upon the camp-guard. The light cavalry was destined to take the extreme left of the line; it was headed by a Neapolitan noble, Gian Bernardo Caracciolo, one of the Sanseverino clan, and included the Duke of Ferrara's 300 horse-arquebusiers as well as the French mounted cross-bowmen and stradiots.[2] It was estimated at over 2000 men.

At dawn on April 11—it chanced to be the day before Easter—the columns came down to the bridge—the landsknechts, leading, formed up on the farther bank of the Ronco to cover the passage of the rest, more especially of the artillery.

[1] All this curious story from Bayard, lii. pp. 194–95.

[2] All this from the contemporary letters in *Marino Sanuto*, xiv. pp. 170–74, which I judge to be more trustworthy than later details and figures given by Guicciardini and others.

It had been intended that the French infantry should use the bridge, but the greater part of them forded the Ronco on their own initiative, led by a Gascon captain, who cried to his men that they must not let the Germans get ahead of them! The guns were then brought over the bridge, and the cavalry followed, first the 'vaward' and then the 'main-battle.' The horsemen of the 'rearward' remained for a time at the bridge-head, ready to retire to Ravenna if there should be trouble in that direction, or to join the fighting-line if there were not.

While the troops were crossing and deploying, Gaston de Foix rode out ahead to survey the Spanish lines, with an escort of only twenty men-at-arms—Bayard's chronicler says that he met and exchanged courtesies with a similar exploring party of Spaniards, under the well-known Pedro de la Paz, the chief of the 'genetours.' This story of chivalrous compliments may be as well founded as the tale of the salutes of the French and English guards at Fontenoy. It reads well in the pages of the 'Loyal Serviteur.' And it is borne out by the tale of polite defiances on the preceding day in the Spanish *Relacion* of the battle.

When the army was all on the south bank of the Ronco, it took up the formation which Gaston had prescribed. The cavalry of the vaward drew up next the river, then its infantry, then the infantry of the main-battle, lastly, on the left, the Italian infantry of the rearward, minus the detachment left at the bridge. Beyond them, and facing the extreme right of the Spanish position, were the 2000 light horse. The guns belonging to the French themselves were placed in the front of the right wing. The whole formation had a somewhat crescent-shaped appearance, for the infantry in the centre were drawn back, and the cavalry on the wings thrown forward. This may perhaps have been due to the Spanish trench having a convex front with flanks tending back. There is a definite statement that the cavalry of the 'main-battle' under the Senéchal of Normandy, were placed in second line behind the vaward and not far from the river.[1] This, though made by a good, if not by a contemporary, authority, Guicciardini, is difficult to reconcile with the details of some of the later fighting given

[1] Guicciardini, x. p. 275. 'Dictro a tutte questi squadrone . . . in sulla riva del fiume erano collocate le seicento lance della battaglia, guidate del La Palice.' But Bayard and the other French authorities give La Palice the command of the 'vaward,' not of the 'battle,' and give the Senéchal charge of the 'battle.'

by other writers. It would seem more likely that they were in the true centre of the array.[1]

The French army, having got into line, advanced to within cannon-shot of the Spanish entrenchments unmolested. This was not the fault of Fabrizio Colonna, who, while the enemy was in the process of crossing the river, besought leave from the Viceroy to charge out upon them with his horse, and attack the infantry before either the guns or the cavalry were across the water. It was undoubtedly a very plausible proposition, for the heads of columns engaged in a river-passage are most vulnerable, and cannot be easily supported by those in the rear. [The cases of Stirling Bridge (1297) and the Katzbach (1813) may be remembered.] But Cardona refused his leave for the sally, guided by the advice of Pedro Navarro, who insisted that this must be a purely defensive battle, and that the line must be kept intact.

When the French looked upon the Spanish position they were a little puzzled. The Spaniards, as exploring parties had reported, had all their infantry collected in one body, and their cavalry in two masses on the wings. But this morning the infantry were invisible; Pedro Navarro had withdrawn them behind the trench and made them lie down flat in the lowest ground.[2] The only things clearly visible were masses of cavalry on the right and left of the position, and guns placed behind the trench, mainly towards its left and apparently its centre. The Viceroy had trusted his engineer-general with the arrangements, and Navarro had prepared a system of traps for the enemy. The trench covered the whole front of the position, except for a gap of no more than 20 yards left between its northern end and the high bank of the Ronco, along which there was a raised footway, a thing seen in all Italian rivers that pass through land liable to floods. This gap was for cavalry to charge out in close column at the decisive moment. At the southern end we do not hear of a similar prepared exit, but (as the event proved) it was possible for light cavalry to get round 'tra fossi e pruni,' when needed for an advance.

The ground, however, was uninviting for an enemy desirous of turning the right flank of the position. Along the front of

[1] 'Ils ne faisoient que une troupe de tous leurs gens de pied, et deux de leurs gens à cheval, sur ça se fallirent ranger.' This was Bayard's report, after a sally into the Spanish front on the previous day. Bayard, liii. p. 202.

[2] But certainly not under the bank of the Ronco, or they could not have occupied the trench fast enough.

the entrenchment, mainly in its centre, however, Navarro had planted not only his thirty guns, but a number of devices which surprised his enemies. They are called 'carts' and are compared by several chroniclers to the scythed chariots of ancient Oriental history, but seem rather to have been inspired by a knowledge of the strange pictures which adorn some editions of Vegetius, and are found at the end of all copies of the *Notitia Dignitatum*. They were two-wheeled machines with a long projecting spear fixed in front, and two scythe-like blades placed at its sides, while above these were strapped several heavy arquebuses, with machinery for discharging them all at once—*mitrailleuse* fashion. They were fitted at the back with a long handle or pole, by which they could be either drawn about or shoved forward as occasion dictated. These devices seem to have had two purposes—they would break the regularity of the front of an attacking column, both by putting sundering obstacles between the files, and by delivering a volley at the critical moment. But they would also be useful protection for infantry attacked by cavalry—the tactician Urbicius, over a thousand years before, had advocated the use of some such machines.[1] To compare them to the *wagenburg* of the Hussites would be less apposite, as they could not have been formed into a square.[2] They would have been far too few—the numbers given for them vary from thirty in Guicciardini to a vague hundred in Bayard's 'Loyal Serviteur.'[3]

The array of Cardona's army, though elaborately described by more than one narrator, is not very easy to make out. There was certainly, next the river, and opposite the gap already mentioned, the cavalry of the 'vaward,' 670 Italian lances in the Papal service under Fabrizio Colonna. Then all down the front of the trench were four bodies of Spanish infantry in a long line, behind the guns and carts, with on the extreme right wing 1700 light horse—Spanish 'ginetes' and Italian horsed arquebusiers under the Marquis of Pescara, a young but a very distinguished officer. But the exact placing of the rest of the troops is difficult to make out. We are told that *behind* Fabrizio's brigade, and close to the river, was the 'battle' of horse, 565 lances under the Marquis of La Palude; on their

[1] See vol. i. p. 23. [2] See vol. ii. pp. 363–65.

[3] Who says that Navarro had 200 *arquebus à croc*—arquebuses mounted like swivels, with a 'cart' between each two of them. Bayard, liii. p. 206. He says that they were to be pushed forward against attacking infantry, to break their order.

front, and apparently forming a second line of infantry behind those at the trench, were 4000 Spanish infantry, with whom the Viceroy Cardona had placed himself. And finally there was apparently in third line or reserve, 490 Spanish horse under Carvajal, a veteran captain, with three Spanish foot regiments and the Italian foot-soldiers of the papal levy, only 2000 strong according to the best contemporary evidence, but stated by Guicciardini at 4000, which is probably an exaggeration. They were under a captain called Ramassot by Fabrizio Colonna and Bayard, but Cornelio Romaeo of Bologna by Köchlein—and Hernan Magote by the Spanish *Relacion*. He is otherwise unknown.

When a mediaeval army had deployed itself opposite its enemy, it was not usual for the attack to be long delayed. But the battle of Ravenna was the first example of a new sort of tactics. Having got his troops in line, Gaston halted them, and brought up his artillery, with which he commenced to pound the Spaniards for more than two hours. The bulk of the French guns was placed in front of his right wing and near the river, a position from which they played mainly upon the cavalry of Fabrizio Colonna, who were easily visible. The infantry in the neighbouring entrenchment were lying down, and could not be seen. But it was not on this flank only that Gaston was intending to apply artillery pressure. The Duke of Ferrara, a skilled artillerist, left his own men-at-arms in the vanguard division, but went off with his twenty-four select pieces, by a long circuit, to place himself on the extreme left wing of the French army, where the light cavalry were drawn up. Since we know that he had his heaviest gun, the 'Great Devil,' with him,[1] it is certain that he had to pick his route very carefully through ground full of ditches, where no 'point-to-point' movement was possible. But Cardinal Corsini's eighteenth-century meddling with the rivers has made all accurate detection of his route difficult or impossible. As his battery was working not very long after the artillery combat had begun at the other end of the line, it may be taken as certain

[1] Here comes in a ludicrous note. 'Coccinius' (Michael Köchlein), the pedantic German who wrote an account of this war, says, much to our surprise, that the 'Great Devil,' whom he takes for a French captain, was in command on the French left. He must have misunderstood his landsknecht friend, who described the battle to him, for the 'Great Devil' was undoubtedly Alfonso of Este's biggest and most beloved gun, which he showed to M. de Florange, who thought it 'la plus belle piéce que je vis jamais, et qui tiroit le mieux.' Florange, p. 77.

that the Duke with his guns acted as an independent factor in the fight, from the moment of the crossing of the Ronco, and never was involved in the cannonade at the right end of the battle. When he got into position, we are told that he brought an enfilading fire to bear on the Spanish entrenchments, which seems to imply that they were convex, and he certainly pounded the cavalry behind Cardona's right flank, *i.e.* Carvajal's 'rear-ward' squadron and Pescara's light horse and 'ginetes,' just as the French guns at the other end of the line were pounding Fabrizio Colonna's horsemen.

The extraordinary thing about this artillery preparation is to find that the Spanish guns in the entrenchment seem not to have been doing any harm to the French cavalry on the wings, but to have been playing most effectively on the German and Gascon infantry in Gaston's centre. By all accounts their work was excessively murderous; we have notes upon it both from a German source (Köchlein) and from two French sources (Bayard and Floranges, both eye-witnesses of the battle). The casualties are said to have numbered over 2000 before the final advance was ordered. The Gascons were so hard hit that they flinched, and fell back against the flanks of the landsknechts, who had to push them off with their pikes.[1] Philip of Friberg, second in command of the Germans, and the Gascon captain De Molard were cut in two by the same cannon-ball, as they were talking together between their lines.[2] We are told that the columns swerved a little to the left—perhaps to move off ground where the balls seemed to be beating with special fierceness. This swerving may have caused a gap to arise between the northern end of the Gascon infantry and the vanguard-cavalry on the banks of the Ronco.

As the French guns were playing upon the Spanish cavalry, while the Spanish guns were making havoc of the French infantry, this looked like a 'killing match.' Gaston was resolved not to charge in upon the entrenchments—Navarro was equally resolved not to come out of them. A tightening up of the artillery pressure seems to have been given by Yves d'Alégre, who pointed out to the commander-in-chief that still more destructive fire could be brought to bear upon the cavalry of the Spaniards near the river, if some guns were sent back across

[1] This comes from Coccinius, p. 563, who is rather propagandic in praising the steadiness of his compatriots, and depreciating that of the French.

[2] This from Coccinius, 564, and also from Bayard, liv. p. 206.

the bridge, and set to enfilade Fabrizio Colonna's division from the shortest possible distance, *i.e.* the other bank of the Ronco immediately facing his rear. Bayard's narrative says that it was this cross-fire of only two guns which made Colonna's position impossible. When a prisoner after the battle, Fabrizio said that he had seen one cannon-ball at close range knock over thirty-three men-at-arms! At any rate he made up his mind that he would have to charge at all costs: his troopers were raging, and asking for men instead of cannon-balls to fight. 'Cuerpo de Dios! somos matados del cielo: vamos combater hombres.'

But it would seem that the Duke of Ferrara's guns on the opposite wing were exercising an equally intolerable pressure on the cavalry of the Spanish right. For both Fabrizio and the long 'relacion' of the Spanish friar which is one of our most valuable sources, agree that it was Carvajal's corps, with Pescara's 'ginetes,' which made the first sally from the regular line, cutting round by the rough ground on the Spanish extreme right, and charging for the French light horse and the destructive guns which they were guarding. And apparently after an interval the Viceroy sent out also the Marquis of La Palude and the horse of the 'main-battle' to support Pescara and Carvajal. It is said that this corps had difficulty in joining the rest, charging 'tra fossi e pruni' with great disorder to its ranks.

Here the story begins to be difficult. With what foe did this considerable accumulation of Spanish cavalry fight? To our surprise we find that it was not only with the French flank-guard of light horse, but with the 'main-battle,' the gendarmerie under Gaston himself, Lautrec, and the Seneschal of Normandy. But this corps was, as we have seen, originally drawn up beside the French right centre infantry. I can only suppose that Gaston must have moved his central cavalry corps to the right, when the light horse were hard pressed. Otherwise it is difficult to see how it came into action.[1]

Fabrizio Colonna, according to his own narrative—most fortunately preserved—did not move till he saw the main body of the Spanish horse already in action at the other end of the

[1] Coccinius, for what he is worth, says that Colonna's division flinched under the artillery fire, and fell back for a moment on to the supports in its rear, but then charged. But what were their supports? The 'main-battle' had already moved away—perhaps Ramassot's infantry.

line. He then led his men, who had been eagerly demanding instant action, out from the gap, in between the Ronco and the trench, which has already been mentioned. The corps which he charged *must* have come from the French vaward cavalry under La Palice,[1] and not from the horse of the main-battle, who were engaged a long way off.

There was a stiff cavalry fight on the French left: the Spanish horse are given by all authorities a fine testimonial for their repeated charges. The *Relacion* tells how all their lances were broken before the joust was over, and they were sending for more to the Viceroy who could not supply them.[2] The French main-battle was also much exhausted, and we hear that Yves d'Alégre, who is found on this day not acting as the mere commander of the rearguard, but rather as a sort of chief of the staff to Gaston, had to hurry up certain units from La Palice's vaward division—the bands of Robert de la Marck (the Florange who has left us an account of the battle), and the archers of the royal guard under M. de Crussol. According to some authorities more reinforcements were also drawn from the bridge-guard in the rear.[3] These succours sufficed to turn the fortune of the day, and the Spaniards, after a very gallant struggle, broke up and turned to fly. 'Between the two ditches,'[4] says Bayard's scribe, 'there died 300 or 400 men-at-arms.' Colonna's brigade gave way, according to himself, a little later. Both the 'main-battle' and the 'vaward' cavalry were practically annihilated. Their casualty list chances to have been preserved in a Venetian document,[5] and shows, of 28 captains commanding bands that were present, 11 killed, 3 wounded and taken prisoners, 4 taken prisoners unwounded, and 10 simply wounded. Carvajal's and Pescara's units show less appalling statistics, but the former was wounded, and the latter both wounded and captured.

[1] It is Bayard's narrative which makes Fabrizio change the main-battle, not Fabrizio himself, who does not name his opponent. But, Bayard says that this charge, in which the 'main-battle' was assailed, was the *first* charge of the day (liv. p. 207), 'joyeux d'avoir le premier combat du jour,' and Colonna says that he started long after Carvajal was engaged, and did *not* lead the first charge. Bayard's 'serviteur' must have made a mistake as to which Spanish corps charged the main-battle. It *must* have been Carvajal and Pescara. But only Pandolfini among persons present, says definitely that Colonna fought La Palice, which seems to have been the fact.

[2] *Relacion*, p. 280.

[3] So Pandolfini.

[4] No doubt the ditch of the Spanish entrenchment, and the watercourse mentioned in Coccinius, which ran parallel to it, and which the Germans are noted to have been forced to scramble over.

[5] In Marino Sanuto's *Diarii*, xiv. p. 171–74.

The survivors fled in disorder down the Cesena road by the river south-westward. Fabrizio Colonna, with a mere handful of men whom he had kept together, fell back behind the left of the Spanish infantry in the entrenchment. The Viceroy, Ramon of Cardona—much to Colonna's disgust—left the field with the broken squadrons, while Pedro Navarro and the infantry were still making a gallant bid for victory.

The chase after the routed Spanish horse engaged a large part of the victorious French cavalry, and went on for a very long way. Bayard's chronicler notes that he rode on for six miles, till his charger was tired out. He says that before engaging in the hunt, Bayard besought his commander-in-chief to halt and rally scattered men, not to ride on recklessly, 'which he promised,' says the narrative, 'to do, but failed to keep his word, and evil befell him in consequence.'

At the moment when the Spanish cavalry were at last forced into leaving its positions and attacking, Gaston de Foix had determined that the time had come for setting in motion his infantry, who had so long been kept back, waiting for orders under a devastating fire.[1] We are told that the first command was given to 2000 Gascon cross-bowmen to press on towards the Spanish line, and play upon it with the purpose of making the enemy stand up and show himself, for all during the cannonade he had been lying flat! The Gascon captains asked for help of pikes, saying that they would be helpless if the Spaniards sallied out upon them. Accordingly 1000 Picard pikemen were sent up to strengthen them. But when they had come close to the trench and began to let fly over it on a high trajectory, the enemy arose and opened such a blistering fire of arquebuses and swivel guns [2] upon them that both Picards and Gascons melted away, after losing many of their officers, and fell back out of range. It seems that at this moment the Spanish second line of infantry moved forward and reinforced the first.

Matters were far more serious farther down the line, where the great column of German landsknechts under Jacob Empser delivered their assault. Braving the fire which was poured upon them, they rolled up to the ditch, after passing a water-cut (not part of the Spaniards' work) which was found to lie across

[1] For this triumph see Coccinius, p. 562. He is quite useful for the doings of the landsknechts, if not for other parts of the battle.

[2] *Arquebuses à croc*—heavy firearms fixed on pivots, perhaps those on the 'carts.

their path. Jacob Empser was shot through the body as he scrambled over the ditch, but one of his lieutenants, Fabian

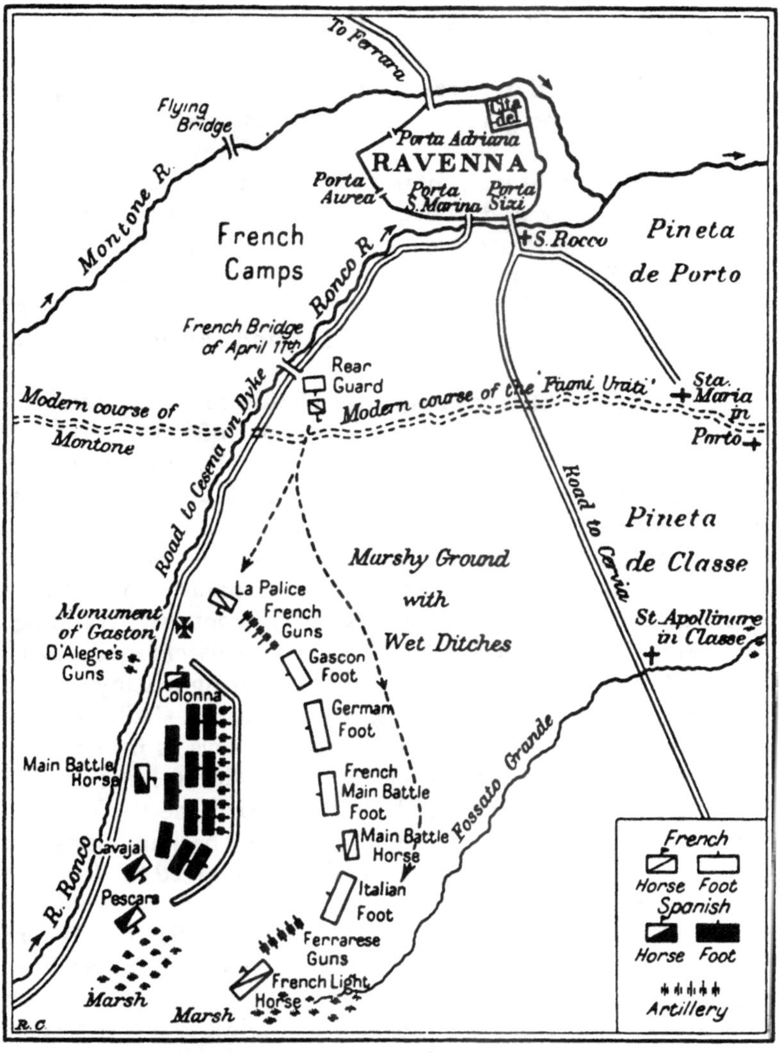

BATTLE OF RAVENNA, APRIL 11, 1512.

von Schlabrendorf, made a gap in the line of Spanish pikes by taking his own pike by the butt and using it like a flail, whereby he broke down to the ground a dozen hostile weapons,[1] and

[1] Coccinius, p. 564, and Bayard, liv. p. 210.

allowed his men to scramble in—though he himself (as was to be expected) was mortally wounded. At several points the Germans crossed the ditch, and got to hand-strokes with the Spaniards, but could make no great way forward. We are told that the sword-and-buckler men from the rear ranks slipped in among them, and did great damage with their short weapons —against which the pike was practically useless. After a long struggle the Germans were expelled from the works—it is said that they and the French infantry lost another 1200 casualties in this storm over and above the men killed in the previous cannonade.[1] The Spaniards raised a loud shout of victory, but as two French narrators remark, the battle was already lost to them, for their horse, at both ends of their line, were at this very moment breaking up and abandoning the field.

It was no doubt in consequence of this cavalry success on their flank that the French and German infantry were persuaded to try one more assault on the blood-stained trenches. This time both got over the ditch—M. de Maugiron, a Gascon captain, was slain actually on top of one of Pedro Navarro's carts [2]—nine of the twelve German landsknecht captains were killed or wounded. Thus once more the Spanish line held its own, but at the end of the clash, as the Gascons were recoiling, two companies (' enseignes ') of the Spaniards charged out of the position, and broke in among the retiring enemy. This column was to have a strange experience, as we shall see. At the same moment Fabrizio Colonna (as he says in his contemporary letter) came on the scene with a few dozen lances, saved from the rout of his vanguard, and fell upon the flank of the beaten troops. He alleges that if he had only been in possession of a couple of hundred horse he could have changed the fortunes of the day—but this no doubt is an optimistic view of the situation.[3] The beaten Gascons recoiled some distance, before they rallied upon other French infantry which had not yet been engaged—drawn no doubt from the main battle corps of which so little had been heard during the fight. The victorious Spanish companies found themselves isolated in the rear of the enemy's line, and cut off from their friends.

Things in this part of the field came again to a deadlock— the bulk of the French and Germans stood halted before the

[1] This from Florange, p. 96. [2] Florange, p. 95.
[3] See Fabrizio's letter in Sanuto, xiv. p. 180, and Coccinius, p. 563.

trench[1]—the Spaniards still lined it, save the two companies which had burst out in pursuit, and had broken a hole in the hostile front. At this moment the whole face of affairs was changed by the appearance in the flank and rear of masses of the victorious French cavalry, mainly from La Palice's 'vaward' division, which had come in by the gap in the trench near the water; others, we are told, had managed to scramble over it.[2] Fabrizio Colonna asserts that he had made some attempt to cover the Spanish infantry from the rear by bringing up Ramassot's Italian foot, but that it was a hopeless business, as they were overwhelmed not only by horse but by fresh French and Italian infantry—probably from the hitherto unengaged French left.[3]

The line of Spanish battalions along the entrenchment, charged by cavalry in flank and rear, and once more attacked in front by the rallied French and Gascon foot, had now come to the end of their powers. They broke, and were for the most part ridden over and trampled down. Only Pedro Navarro and a few captains were spared for their ransoms. At most a solid body, perhaps 2000 men, keeping closely packed together and pushing through the turmoil, succeeded in escaping down the Cesena road.

But the most extraordinary escape was made by the two isolated Spanish companies ('enseignes') of which we spoke some little time back. It will be remembered that they had broken right through the Gascon corps opposed to them, and had emerged at the back of the French line. Seeing the rest of the army ruined, it is said that they made the desperate resolve to get off in the direction of Ravenna, by passing round the Pineta, very far from the battlefield. But before they had gone a mile they found themselves intercepted by the French detachment of the rearward battle, which was guarding the bridge, under the Bastard du Fay, and turned back on to the narrow raised path along the river, passing by their old position in the entrenchment. The tide of war had swept along it some time before, and it was nearly empty of men.

[1] Florange is very positive that they did not attack again, but halted before the trench at the moment that the French cavalry charges began.

[2] Florange, p. 96.

[3] Guicciardini, ix. p. 285, makes much of this attempt, 'con somma laude' to hold back the Gascons near the river-road; but he is always ready to praise Italian valour, and Colonna does not think much of the effort. 'Ne volsero movere mai, si non a fugire' (M. Sanuto, xiv. p. 180).

As they were pushing along it in narrow column, with such speed as was possible, some stray French cross-bowmen pointed them out to the commander-in-chief, who was halted with his personal staff of some fifteen or twenty gentlemen, watching the last agonies of the Spanish infantry from a short distance. These broken men indicated the retreating column to Gaston, and cried to him that these were the Spaniards who had beaten them and cut through the line. Whereupon the young general forgot the promise that he had made to Bayard to keep out of mischief when the day was won, and, calling to the handful of gentlemen who were about him, forced his horse up on to the raised path, and galloped to intercept the narrow column of desperate men who were hurrying along it. They met his party with some arquebus shots, and then charged with levelled pikes upon the group of horsemen. Not one of the French riders escaped—Gaston's charger was hamstrung, and when he rose from the ground he was slashed down with deadly wounds in his head. His cousin Lautrec was left for dead, his aide-de-camp Vivarolz (the son of Yves d'Alégre) was rolled into the canal, and drowned there in his armour, as were several more. There was none of the staff left alive, we are told, to call for help or report the disaster, and for some time the captains were seeking in vain for the general, who had perished so obscurely and with so little necessity, in a backwater of a battle that was over.[1]

The two Spanish companies, pressing on with all speed, escaped down the road by the river. Six miles from the field they met Bayard and his party of men-at-arms, returning dead-beat from the pursuit of the Spanish horse. According to the tale of the 'Loyal Serviteur,' a Spanish captain cried to Bayard, 'Why meddle with us—you are not strong enough to break us—you know that you have won the battle and slaughtered our army—be content with your honour, and leave us alone, for by God's good will only have we escaped from this business.' The good knight thought that the Spaniard spoke sense, and the

[1] There is a weird tale of a presentiment in Florange, pp. 88–9. There was a certain bastard of Chimay who had second-sight. Gaston rallied him upon his powers of prophecy, while the original cannonade was going on, and asked if he was to win or lose the field. 'You are to win,' said the bastard, 'but take good care of yourself lest you stop on this field.' When all seemed to be over, Gaston, with his helmet off, asked the bastard whether he had taken enough care of himself, as all was now safely over. 'Do not say that all is over till all is quite over,' replied Chimay. Just then the messenger came to report the stray Spanish column. Gaston dashed on his helmet, and rode off to be killed.

two parties went by each other unfought. 'But if only he had known that these were the men who had killed Nemours, he would have died a thousand deaths rather than have failed to try to revenge the good duke.'[1]

So ended one of the most bloody battles on record, for the slaughter was not all on one side, as had so often been the case in the Middle Ages, but the victorious army suffered huge losses, if the vanquished army was almost exterminated. We are reminded of Eylau or Tagliacozzo, more than of Austerlitz or Agincourt. Of the Spanish infantry only a remnant escaped —of twelve colonels commanding units eleven were killed.[2] Of the cavalry of the van and the main-battle—as we have already mentioned, hardly an officer escaped. Indeed the Viceroy Cardona, Carvajal commanding the rearguard, and Pedro de la Paz, chief of the *genetes*, were the only persons of note who got off—and the second was wounded, while the last named was murdered by brigands in the Apennines during the retreat. Eight hundred men-at-arms out of 1800 in the field are said to have been slain, many scores more were prisoners, mostly wounded. The Viceroy, who never stopped his flight till he reached the Tronto, and the borders of the kingdom of Naples, could only collect 300 horse and 3000 foot out of some 16,000 who had stood behind his intrenchments on the dawn of April 11. Among the prisoners were Fabrizio Colonna, general of the horse, and Pedro Navarro, general of the foot, with a number of Spanish and Neapolitan nobles and captains: the Marquis of La Palude, who had commanded the cavalry of the 'main-battle'; Pescara, who had led the light horse; the Counts of Trajetto, Populo, Bitonto, and Concha, and Juan of Cardona, brother of the Viceroy—who died of his wounds a few days later. A more abnormal captive was the Pope's legate, Cardinal Giulio dei Medici,[3] who was found in the camp with more than 200 other clergy—one of whom was the author of the valuable Spanish *Relacion* of the battle. In addition to the casualties in action—probably 9000[4]—there

[1] Bayard, liv. p. 211.

[2] The author of the *Relacion* gives the names of all of them; the corps in the first line were those of Diaz, Velasquez, Chaves, and Luzan: then in the second line Pacheco, Samaniego, Paredes, and Salzado: in the third line were Arneja, Artiade, Arriaga, and Zamudio. Every one except Salzado was left dead that day, and Salzado died of his wounds!

[3] Afterwards to be Pope himself, as Clement VII.

[4] The author of the *Relacion* will only allow 3000 or 4000, but is 'tendencious.'

were many losses in the retreat, 'for seventy miles the roadside was strewn with the carcases of Spanish horses, ridden to death in the flight,' and the brigands of the Apennines—the 'mala gente,' as the Italians called them—waylaid, plundered, or killed many stragglers.[1]

The French losses, though severe enough, were of course less than those of the enemy—their own narratives allow that the unfortunate infantry had, first and last, some 3000, perhaps 4000, casualties, and that 80 of the gendarmes of the *compagnies d'ordonnance* were slain, besides seven gentlemen of the King's retinue, and nine archers of the royal guard. These figures do not allow for losses among the light horse—which must have been considerable in the early fighting. On the whole, it is probable that the French casualty list must have contained well over 4000 names. Besides the commander-in-chief who perished so needlessly, there died that unruly veteran, Yves d'Alégre, who was technically in command of the 'rearward,' but actually appears in all corners of the field, acting apparently as a sort of chief-of-the-staff, also his son, the sieur de Vivarolz, who was killed at Gaston's side, all the three senior officers of the German foot, Empser, Friberg, and Schlabrendorf, and nine of the twelve captains who served under them.[2] Among the French infantry, de Molard, the senior Gascon captain, was killed, also Montcarré commanding the Picard pikemen, and other officers in command of units, Maugiron, La Crotte, Grantmont, Bardassare, and several more.[3]

So ended the first Italian battle in which cannon settled the event of the day, and Gaston de Foix must be given every credit for seeing that the Spanish tactics of infantry entrenched, with good smallarms and a competent allowance of guns, could be beaten by superior artillery strength. His only mistake was in bringing his German and Gascon columns too close to the Spanish position, so that they suffered heavily before the cannonade had forced the Spanish horse into taking a hopeless offensive. Navarro's defensive tactics, founded no doubt on a

[1] *Relacion*, p. 293.

[2] Coccinius, p. 563. The command then fell to Jacob Empser's brother, Burckhard Empser.

[3] The casualty tables of different authorities are generally coloured by their party sympathies. The *Relacion* will only allow 3000 to 4000 Spanish losses! Bayard's 'serviteur' makes the estimate of 3100 French to over 9000 of the enemy. Coccinius gives much the same figure. Guicciardini gives the French casualties as half of the Spanish. Florange gets the French *infantry* losses alone up to 3200, which would make the total at least 3500.

memory of Cerignola, required that the army which stood entrenched should have a superiority of fire, which he had not —largely owing to the Duke of Ferrara's famous train of heavy guns. The Spanish cavalry was in such a cramped terrain, within a narrow entrenched space, that it could take cover nowhere. Every yard inside the position could be reached by cannon-balls. As a retreat along the narrow river-side road must have been equivalent to annihilation in detail, it may be said that the Spaniards were doomed, the moment that they discovered that they were 'out-gunned,' and had not strength enough to resist the French by 'counter-battery.' The victory would have been comparatively uncostly if Gaston had halted his infantry farther back, during the two-hours' cannonade. This error lost him at least 2000 casualties in his best foot-soldiery. As to Cardona he found that, by Navarro's advice, he had put his army into a sort of slaughter-pen, with a river behind, and no line of retreat save from one extreme end of the position—and that (like the rest of the ground) liable to be searched by artillery. If he could have silenced the French batteries he would have been secure—but he could not possibly do so.

A NOTE ON AUTHORITIES FOR RAVENNA

The most absolutely contemporary account, from a prominent actor in the battle, is Fabrizio Colonna's letter of April 28, written from a French prison. He is naturally full of reticences—presumably his letter was read by a 'censor' before being allowed to pass. He makes it clear that he did not (as Bayard supposed) lead the first charge, which was made by Carvajal and Pescara. And he says nothing clear about his own defeat by La Palice. His account of his service at the end of the day, along with the Spanish infantry, is corroborated from other sources. Marino Sanuto's very miscellaneous collection contains several other accounts of the battle, received within a few days; most of them (*e.g.* that on p. 111) are quite worthless rumours, *viz.* the Viceroy is drowned, and Lautrec and the Marquis of La Palude dead. The Spanish *Relacion* is very valuable, and is in pleasant literary style, but was written by a friar, not by a soldier, and he was praying in his tent while the battle was in progress. It is wholly 'tendencious,' putting all the blame on Italians—their cavalry is the first to fly—and mere camp-gossip comes in. When Navarro asks Colonna why he has brought so few horse to help the infantry, Colonna is made to reply that it is because his men are probably at Forli by now—so fast have they run. He will not concede that more than 3000 Spanish infantry have fallen—though he allows that 11 out of 12 colonels of the foot are killed. That any Spaniards got away at all was due to the fact that Alfonso of Ferrara shirked tackling a desperate Spaniard. Bayard's 'Loyal Serviteur' is very full, and generally accurate, but having written many years too late, has some slips of memory, *e.g.* the idea that Fabrizio Colonna led the first cavalry charge. But he is quite correct on the story of the isolated Spanish column which slew Gaston and his staff. Florange's story, shorter than Bayard's, and written in a Spanish prison about 1525, is also useful, and obviously genuine and honest stuff. Coccinius had clearly got his account from landsknecht officers present on the field, and can be trusted for the doings of the German column, but he is most 'tendencious,' against their French comrades, all of whose flinchings he is glad to detail, and is most ignorant on important points. Guicciardini and Giovio are both very second-hand, and full of errors of detail; the former is terribly 'tendencious,' in wanting to give much more credit to the Italian units than any eye-witness concedes. Fabrizio, though an Italian, speaks with contempt of the infantry whose efforts Guicciardini praises. Guicciardini, being accessible and easy reading, has had far too much influence with modern historians, who have not seen the *Relacion* or Colonna's letter. It is curious to find that Machiavelli, thinking over the fight, can only exalt the efficiency of the Spanish sword-and-buckler men, which suits his plea in behalf of Roman legionary tactics. He does not see that the whole battle is a testimony to the all-importance of the artillery which he so much disliked!

CHAPTER IV

NOVARA (JUNE 6, 1513)

THE victory of Ravenna, as we have seen, was the exhibition by a talented amateur, aged only twenty-three, of a new method of winning battles by the skilful use of an overpowering artillery force, and it served as a decided check to the Spanish theory of war, which had been started by Gonsalvo de Cordova—the idea that the receipt for victory was to get into a fortified position, well garnished with firearms great and small, and then to lure the enemy to attack you in it. This device reminds one, indeed, of the old English receipt for victory, the defensive combination of bow and lance, which had won great favour for more than a hundred years, but depended entirely for its efficiency on getting your enemy to be obliging enough to deliver a frontal attack.

We are surprised to find, only a year after Ravenna, a recurrence to another old system of tactics, that of the Swiss, which had had its triumphs, but was growing out of date, as Machiavelli clearly saw. But Novara was a repetition of the tactics of Morat and Granson, the last great example of that sort of victory—I mean the running down of an enemy by vast columns of pikemen, arranged in the échelon formation which the Swiss loved, and which, when successfully managed, had the effect of a tank or a steam-roller charging a barricade. In its most effective form the attack of the pike-column should be accompanied by surprise; its best successes had been against an enemy caught napping, with insufficient provision for scouting, resulting in over-late construction of a line of battle. And at Novara this important advantage was happily secured, the attack having been delivered at dawn, against a general who vainly thought that he had some hours to spare, and that a hostile army which had arrived on the ground deadbeat, after a forced march, would be unlikely to make a general advance and force on a battle, without taking a few hours of rest. In a way Novara was a psychological as well as a tactical triumph.

It may be asked why, when we left the French victorious at Ravenna, far to the south of the Po, we find the decisive battle of the next year fought many miles back, on the western side of the duchy of Milan. The reason was that Foix had won a battle, but that the fruits of his success were lost during the campaign that followed. The fighting round Ravenna had been pushed to a rapid close, because Gaston knew that there was another foe about to press in upon his back. The duchy of Milan had been left behind him lightly guarded by insufficient forces, under the old condottiere-captain Trivulzio. Descending from the Tyrol, an army of Swiss, having with them Maximilian Sforza, the heir of the captive Ludovico il Moro, fell in upon the western side of the Milanese, while a Venetian force, acting in conjunction with them, moved against the towns in the valley of the Po. La Palice, to whom the command of the French army had fallen on the death of Gaston de Foix, found himself much weakened by the departure of his landsknecht contingent, which the Emperor had recalled,[1] and was obliged to evacuate the recent conquests in the Romagna, in order to fall back to keep his line of communication open. He marched for Pavia, where there was heavy fighting with the Swiss, but learned that Trivulzio had been forced to evacuate Milan, except the citadel, and had retired westward. La Palice pushed on to join him at Alessandria—nothing remained to the French in the duchy save the citadels of Milan, Cremona, and Brescia—all the rest had been lost without serious fighting or a general action.[2] In the autumn La Palice thought it wise to draw the whole army back behind the Alps, as its numbers were low and its morale unsatisfactory.

It might have been expected that in the next spring Louis XII would have left Italy alone, after having seen one more campaign of disaster added to the list of French transalpine expeditions. For though he succeeded during the winter in concluding a truce for one year with Ferdinand of Spain, there was heavy trouble impending on his northern frontiers, where Henry VIII and the Emperor Maximilian were making unmistakable preparations for that campaign of invasion which was to leave the year 1513 notable for the Battle of the

[1] It will be remembered that the orders for recall had reached Jacob Empser a day before the battle of Ravenna, and that he kept them secret, and led his men to the fight. His brother Burckhard took back the survivors to Germany on getting further dispatches from Maximilian. (See above, p. 78.)

[2] See Bayard, lv. p. 216.

Spurs and the captures of Tournay and Thérouanne. But Louis was enraged at the idea of losing his duchy of Milan, where he had reigned for so many years, and he had succeeded in concluding an unnatural and unprofitable treaty of alliance with the Venetians, who had quarrelled once more with the Emperor Maximilian. Encouraged by this diplomatic success —but how could he expect the Venetians to prefer the King of France rather than a much weakened Sforza duke at Milan ?— he prepared to collect a new army in the Alps. La Palice served that year on the Flemish frontier—as did his friend Bayard—and the King selected for the Italian command the veteran La Tremouille, remembering rather his previous conquest of the Milanese from the sons of Ludovico il Moro in 1500, than his unlucky expedition of 1503. He was accompanied by Trivulzio, who held out great promises of being able to rouse a pro-French party in the duchy.[1]

The army destined for Italy was not on the largest scale— Venetian co-operation being (rather rashly) expected. It consisted of about 12,000 men, including a large landsknecht contingent collected by that old enemy of the Emperor, Robert de la Marck, Duke of Bouillon, which he had placed under his son Florange, the cheerful annalist, who has helped us at Ravenna. He himself brought 100 lances. We are surprised to find on this expedition the wandering Scottish prince John, Duke of Albany, as leader of a *compagnie d'ordonnance.*[2] There seem to have been in all about 6000 German pikes, 4000 other infantry—Gascons and Navarrese—about 1000 lances and 1000 light horse, including some stradiots, for the French had already many of these Albanians in their pay. As was usual with the armies of Louis XII, a good train of artillery had been provided.

Tremouille crossed the Alps unopposed, having marched from Grenoble earlier than the enemy expected, and came down upon Alessandria. The Swiss, who had sent out some detachments to watch for him, were taken by surprise, and collected at Novara, the first strong town on the Milanese border. They were in no great strength, as their main army had gone home, after having placed Maximilian Sforza on his father's throne in the preceding June. There seem to have been about 4000

[1] Florange, p. 116.

[2] Son of that Alexander, Duke of Albany, who had given so much trouble to his brother, James III.

only of them left in the duchy when the French struck. This being a force obviously insufficient to stop Tremouille, hasty messages were sent to the federal diet, asking for prompt help. These appeals required some time to work, and though the Cantons voted the raising of an army of succour without delay, it took several weeks to gather. The total appears to have been about seven or eight thousand pikes. Meanwhile the force already in Lombardy, unable to take the field for want of numbers, stationed itself at Novara, and was gradually outflanked on both sides by La Tremouille, cut off from its base at Milan, and finally blockaded in its positions. Duke Maximilian had joined it with a couple of hundred of his own horse—all that he could muster. Before the army of succour could appear the French captured Milan aided by treachery from within, for there was a strong anti-Sforza party among the citizens.

Many other towns sent in their prompt submission—Como was about the only exception. Having achieved this success, La Tremouille pressed in upon Novara, and opened trenches against it: he was aware that a siege would be a long business, for the garrison was large and the place strong. But he relied upon the power of his battering train, and thought that the Swiss (as in 1500) might possibly be ready to abandon the falling cause of the Sforzas.[1] Breaches had been opened, and preparations for a storm were being made, when news came that the Swiss army of relief was close at hand on the road from Ivrea, and on the morning of the 5th June its vanguard was in sight. La Tremouille resolved not to be caught in his siege-lines, but to fight on ground selected by himself, and drew off all his troops towards Trecate, a small town on a rising ground two miles east of Novara on the road to Milan. Here his army encamped that night, in no great order, for the artillery and train had got off late, and the landsknecht infantry escorting it reached Trecate after dark. Tremouille kept the troops under arms for an hour or two, thinking that the Swiss might have followed the retreating columns, but seeing no trace of a pursuit dismissed them to their bivouacs—'quils pouvoient dormir sensément, et faire bonne chére.'[2] He had come to the not unnatural conclusion that the enemy would take a night's rest after a forced march, and he had learnt that not the whole of

[1] For secret intrigues with certain Swiss captains, see Giovio, xi. p. 219.
[2] Florange, p. 131.

them had yet reached Novara, there being still columns which could only arrive next morning.

The French army encamped just where it stood: Tremouille himself took quarters in the town of Trecate with many of the horse. The artillery and the landsknechts lay outside, with some marshy ground between them and the town. Several chroniclers note that Robert de la Marck had brought with him, with great trouble, a provision of posts and palisades to form a defence against cavalry, also a number of 'arquebuses *à croc*'—swivel guns, so to speak—which were to be placed behind the woodwork. But all this stuff was not unpacked, and lying in the waggons when the blow came.[1] It was to be put up in front of a position, and the position was not yet chosen.

The conduct of the Swiss captains was wholly unexpected by their enemy. Instead of resting the troops and waiting for the missing columns to come up, they resolved to attack before dawn, by surprise, with such force as they had in hand, which (even including the old garrison) did not exceed eight or nine thousand pikes. There were only 200 horse, Maximilian Sforza's retinue—'few, but all gentlemen and the flower of the country'[2]—and of field artillery only eight falconets. According to the old Swiss practice the army formed up outside Novara in three blocks of pikemen, but of very unequal size. Two were small, and intended only to make demonstrations, which would keep the French cavalry in check; the third, made up to 6000 or 7000 men, was to strike at the exposed bivouacs of the landsknecht infantry, along with whom the French artillery was parked. After taking a meal and only three hours' sleep, the columns started off for Trecate.

The French were completely surprised, the first warning of what was impending being a rush of Swiss skirmishers (*enfants perdus*) against the pickets in front of the house where La Tremouille himself was lodged—he had just time to escape by the back door, and mount his horse only partly armed. The trumpets sounded and the men-at-arms began to call for their chargers and assemble, in great confusion. But the attack on the town was not, to their surprise, pressed forward.

[1] Both Du Bellay, p. 16, and Florange, p. 119, mention these contraptions, which were never used, and also Giovio, xi. p. 223.

[2] Florange, p. 132, no doubt exaggerates when he says that the Duke had 500 horse, but these were 'la fleur des gens de guerre du pays.' Giovio (p. 222) calls them 'molto pocchi ma tutti nobilissime.'

The Swiss captains, who must have been guided in the dark by persons accurately acquainted with the face of the country and all its byroads, had prepared a most elaborate scheme. On the extreme left was a small column consisting of only 1000 pikes and Maximilian Sforza's handful of 200 cavalry. It coasted right round Trecate, and fell upon the packed baggage of the French, which had been sent ahead, and was in the rear of the town, under charge of the Albanian stradiots and other light horse. Being behind La Tremouille's position they had taken no precautions for keeping watch, but just had time to assemble, and were engaged for some time with the turning column, but finally broke and abandoned the whole waggon-train to the enemy, who started plundering a rich booty.

The second column, a little stronger, some 2000 pikes, had advanced up the high road. In front of Trecate there was a wood through which the road passed: under cover of this they got quite close to the town. It was their 'forlorn hope.' composed of arquebusiers, which had alarmed La Tremouille and set the French horse mustering. But when they saw the enemy in confusion, and hastily forming some sort of a line of battle, the arquebusiers fell back on to the edge of the wood and continued shooting from its skirts. But the bulk of the column, under cover of the demonstration, swerved to the left behind the wood, and fell upon the French infantry—Navarrese and Gascons—who were encamped to the north of the town, and were still imperfectly drawn up. They were routed, and their commander, the Sieur de Beaumont, was killed: by all accounts they made no great stand. But the main blow of the Swiss, their left-hand column, at least 6000 strong, had been directed against the camp of the French landsknechts. We are told that they marched by bypaths through fields of standing corn, which so hid them that when the enemy's artillery was hastily put into action, most of the balls went over their heads.[1]

But Florange had got his landsknechts into some sort of a line before the clash came, and had the artillery drawn up in their front. The Swiss columns charged, somewhat flankwise, as we are told, and risked the cannon-balls—trusting to their speed in running in before more than two or three discharges could be made. This was expensive—the last salvo tore through the thickest of their ranks, and killed the com-

[1] If this is correct, and great detail is given (Giovio, book xi. p. 223), the dawn must have been breaking, to allow the Swiss column to be visible at all.

manders of the Bernese and Zurich contingents and many scores more. But the column closed up, overran and captured the guns, and fell upon the line of the landsknechts. There was a furious clash, but not a very long one, for the landsknechts were run down—all their captains save two were killed, and (as their commander Florange relates)[1] the whole front rank save eight men went down dead. Florange says that a body of halberdiers, distinct from the main column, fell upon his arquebusiers, whom he had placed on one side, and having routed them, turned against the flank of his landsknechts. 'To put it shortly, we lost the battle, and the landsknechts got no help: for of the French infantry not a man stood to fight when they saw the other Swiss column coming down upon them, so my landsknechts were broken and routed, and all the artillery was taken.' The only succour that came to hand was from Florange's father, Robert of Bouillon, who, seeing his son's corps beaten, came down with his own company of gendarmes from the main position—La Tremouille seems to have sent no help, and this was Bouillon's own inspiration. 'My father came down seeking after his sons, and found them in a very bad way. He first picked up my brother Jamets, who was on his horse trying to rally the flying landsknechts, and after he found me, lying among the dead, and quite unrecognizable, for I had forty-six cuts upon me, which took six weeks to heal. He put me upon a strong horse and sent me to the rear. When they sewed me up at Vercelli, the surgeons put in 72 (or was it 74?) stitches, and I could use neither arm, hand, leg, nor eye, and had to be carried on a litter. My father tried to rally the landsknechts twice or thrice, but the Swiss turned the French guns, which they had taken, upon them, and that broke their spirits, and all the best of them had been killed.'[2]

Meanwhile, what had become of the French cavalry, of whom we have no notice save of Robert of Bouillon's own company? It would appear that they were never properly engaged. Tremouille, hearing of the attack of the right-hand Swiss column upon his baggage in the rear, sent off some in that direction. But the main body, it would seem, were occupied in waiting for the false attack from the wood in

[1] Florange, p. 135.

[2] All this from Florange's interesting personal narrative, which must be followed in preference to the romantic narrative of some chroniclers, who say that Bouillon carried off his wounded son on his own horse, and broke through the Swiss in spite of the double load; *e.g.* Giovio, xi. p. 225.

front, where there were only a few Swiss skirmishers. French authorities, trying to find some excuse for the general, say that he was separated by marshy ground from the spot where the landsknechts were being cut up. But this is obviously a mere excuse, for Robert of Bouillon got across the swamp in his endeavour to save his sons. It would rather seem that La Tremouille was paralysed by the demonstrations, and never made any serious attempt to strike where a stroke was wanted. Italian authors accuse him of cowardice—which does not fit in at all with his military record, and add that the French men-at-arms fell not only into disorder but into actual panic. It is more probable that, having thoroughly misjudged the situation, and having taken a long time to get his squadrons together, La Tremouille waited for an attack that never came, while his unfortunate infantry was being cut up.

When it was certain that all his foot-soldiery, French and German, was routed and in flight, while behind him, on the Milan road, a Swiss force of uncertain strength was in possession of his baggage train and blocking the way, La Tremouille gave orders for a retreat in lateral directions, since he believed that he could not get to Milan. The cavalry split up into two columns on different roads, and made a rapid and disorderly retirement via Vercelli and Susa, which did not stop till the wrecks of the army had passed the Alps.[1] 'If the Swiss had possessed any cavalry there would have been a general slaughter.'[2] Fortunately for the French they had only to fear Maximilian Sforza's handful of men-at-arms, who were too few to do anything.

This was a disgraceful affair both for La Tremouille and for his gendarmerie—only forty French men-at-arms fell, though their infantry had been cut to pieces. The casualties included Louis de Beaumont, commander of the Navarrese foot; Monfalcone, commander of the stradiots; and practically all the officers of the landsknechts, except their wounded commander Florange, his brother Jamets, and two others. Of the 10,000 infantry more than half were left on the field. The Swiss loss was as much as 1300, all practically in the conflict with the landsknechts—the salvos of artillery poured into their attacking column are said to have killed 700 men in

[1] The Italian chroniclers insert bitter taunts made by Trivulzio and the Venetian deputy Gritti, who was with the army, to the flying French.

[2] Florange, p. 135.

three minutes. Among them were the captains of the cantonal units of Unterwalden, Zurich, and Bern.[1] On the news of the battle all the towns of the Milanese fell back into the hands of Maximilian Sforza, except a few which the Venetians had occupied—the French reconquest of the duchy had lasted for only three weeks.

This victory, the last great Swiss triumph, can only be ascribed to an almost impudent *tour de force*—the reliance on surprise for the discomfiture of a careless and improvident enemy. If La Tremouille had taken ordinary precautions, he should not have been beaten by an enemy practically destitute of cavalry and guns. But it was precisely the knowledge of his carelessness, and his reckless disposition of his army which made the Swiss try their bold game. If he had made up his mind to retreat a little earlier, had gone back a little farther, and taken a properly selected position, they must have tried some other method of attack. But their tactics show a real knowledge of the art of war—to mask and distract the most important part of the hostile army by trifling detachments, while striking with full force at its exposed wing, was a fine exhibition of military ability. Unfortunately for themselves the Swiss were only battle-folk—not capable of utilizing their own victory. They followed up the conquest of Milan by an invasion of France, and reached Dijon in force, but then allowed themselves to be bought off—taking a vast war indemnity from the French king in return for granting him peace. The Emperor Maximilian and Henry of England, who had been relying on Swiss co-operation from the east while executing their invasion of Picardy, called them sordid, petty-minded hucksters for making such a peace, and with some justification. For if Louis XII had difficulty in making head against his enemies in the north, what could he have done to meet the simultaneous invasion of the east by a large and victorious army? The explanation, presumably, is that the Swiss did not desire the destruction of the French monarchy, and looked to their own profit, with a serene disregard of treaty obligation to their allies.

[1] Giovio calls them Mottinus, Graff, and Antius. He has a story that Albert of Stein, commander of the column which arrived too late for the battle, hanged a captain, 'Vertio de Glarona (Werter of Glarus?), who had been intriguing with the French during the siege of Novara.

CHAPTER V

MARIGNANO (SEPTEMBER 13, 1515)

THE complete failure of Louis XII to re-establish his domination in northern Italy was marked not only by the rout of Novara (June 6, 1513) and the retreat of La Tremouille's army beyond the Alps, but by the defection of Genoa, where the party favourable to the French alliance was driven from power by internal conspiracies. These disasters were followed by the defeat of the French army of the north by Henry VIII of England at the Battle of the Spurs (August 16), and by the Swiss invasion of Burgundy in September. The Swiss, as we have seen, were bought off by an ignominious war indemnity, but the general situation of the French king was so dangerous that it is not surprising to find that, early in the following year, he sued for peace on such terms as he could get. He was fortunate in coming to a bargain with Spain and the Emperor at the Truce of Orleans (March 1514) and with Henry VIII, who chafed at being deserted by his allies and refused for some time to negotiate, at the Peace of London (July 10, 1514). Henry was bought off (like the Swiss, whose conduct he had so much condemned) by considerable monetary payments, and gave his young sister Mary in marriage to the French king—though Louis was elderly, invalidish, and only four months widowed of his excellent wife, Anne of Brittany.

Louis had bought peace not only by paying out vast sums of money, but by formally surrendering all his claims on Milan and Naples, and abandoning Navarre to the Spaniards—giving up his support of the French King and Queen of the little Transpyrenean kingdom, where his armies had fared so badly in 1513. These two pacifications of 1514 might well have been the end of the long wars of Italy, if France had only consented to see that the scheme for establishing an empire beyond the Alps was hopeless. For the misfortune of Christendom this reasonable expectation was quite vain, and peace was to prevail for little more than a year. Louis XII died on New Year's

Day, 1514, worn out (as the French chroniclers declared) by the interminable series of banquets, jousts, and processions which had attended his marriage with the lively Mary Tudor. His successor, Francis of Angoulême, from the first moment of his assumption of power, took up the old game which had led Charles VIII and Louis XII into so many ruinous enterprises, defying all the powers whose interests were incompatible with his own.

Unfortunately for Italy, and we may add for France and for Europe, Francis of Angoulême was as ambitious, and much more reckless, than Louis had ever been; he was possessed with the same chivalrous impulse to do deeds of personal daring and win glory that had inspired Charles VIII. It was hardly possible that at twenty-one he should be a good strategist, but he was determined to show that at least he was a good man-at-arms, and he was never so happy as when leading a cavalry charge, at the head of the gentlemen of his guard. The moment that he had been crowned at Rheims, and had made his state entry into Paris, he began to move his *compagnies d'ordonnance* and his royal train of artillery and his mercenary landsknechts towards Lyons and Grenoble, and to issue warrants to all captains of note for the raising of new units for a new war. Most important of all, he concluded a secret alliance with the Venetians, who were still at war with Ferdinand of Spain.

At the outset of his invasion of Italy, which started in June 1515, there was, for the first time during this series of wars, some chance that there might be a fight for the Alpine passes, for the Swiss masters of Milan had garrisoned Pinerolo and Saluzzo, and Prosper Colonna, the Pope's general, had a cavalry force at Carmagnola, ready to support them. But the probable did not occur. Francis passed the Alps, not where he was expected, but by the lofty Col d'Argentière, whose difficulties were supposed to be impracticable for anything more than a mule train. But the very real hindrances were levelled or swept away by thousands of pioneers, directed by the great engineer Pedro Navarro. This is the same man who had fought so obstinately for Spain at the battle of Ravenna, but, slighted by his lawful sovereign, he had become a renegade, and had entered the French service—a trick more suggestive of Italian condottieri than of Spanish generals. In a few days Francis crossed the Col d'Argentière with horse, foot, artillery, and baggage-train; he descended on to Turin, where he was 'honourably received' by his kinsman the Duke of Savoy.[1] His enemies, asking plaintively whether the French had flown over the Alps 'like birds,' had to evacuate the passes, lest they should be taken in the rear. This fate, indeed, did happen to Prosper Colonna, who, retreating too late, was surrounded and captured, with 700 of his horsemen, by La Palice and Bayard, who had got behind him and surprised him while he was at table in his lodgings at Villafranca.[2]

The enemy fell back on Milan—the main body was composed of Swiss, for Maximilian Sforza, as usual, had only a very moderate company of horsemen from his own retinue, and most of the Papal cavalry had been destroyed, when Prosper Colonna was captured at Villafranca. Help had been expected from the Spaniards and the Pope's infantry; but the Viceroy Cardona was distracted at the moment by his campaign against the Venetians, who had (as we have seen) rather unwisely formed an alliance with France, and had an army in the field. Cardona was operating against them on the Po when the French invasion washed up to the gates of Milan, and near him was the infantry of the Papal force under Lorenzo dei Medici at Piacenza. Both were solicited by Maximilian Sforza to join him at Milan, but neither appeared—more (it is said) from jealousy of each other than from fear of the Venetians.[3]

Francis had an army of formidable strength—2500 of his gendarmerie, 1500 light horse, 6000 Gascon foot under Pedro Navarro, 4000 other French infantry, and 8000 or 9000 German landsknechts who had been brought by Charles Duke of Gueldres—6000 of these were veteran troops known as the 'Bandes Noires'—'nourris et aguerris ensemble depuis vingt ans'[4]—also five or six companies of Italian mercenaries, and a train of 72 guns great and small. It is certain that he had nearly 30,000 men in all, and the Swiss having with them no allied troops save 500 of Maximilian Sforza's horse, must have been much weaker. French figures giving 40,000 men to them cannot possibly be accurate, when one reflects on what must

[1] Francis's mother, Louise of Savoy, was daughter of Philip, Duke of Savoy, and aunt to the reigning Duke Charles III.

[2] For amusing details, see Bayard, lix. pp. 240–42.

[3] Du Bellay, i. 55–56.

[4] Florange gives higher numbers: 10,000 in Navarro's corps, and 10,000 French infantry in addition.

have been the total fighting force of the Confederacy. And it was—as we shall see—not with their whole army that Francis I was about to fight.

The proceedings of the days which preceded the battle of Marignano form a very despicable page in Swiss history. Their council of captains took into serious consideration the idea of selling their vassal, Maximilian Sforza, to the French, as his father Ludovico had been sold fifteen years before, under similar conditions! The French king had opened secret negotiations with certain of his partisans in the Swiss camp, of whom the chief was Albert von Stein, the commander of the Bernese contingent. His offers were so tempting that they were made public, and at a general council of the captains the matter was discussed. Francis offered huge subsidies, including the payment of the balance of the indemnity promised at Dijon in 1513 by La Tremouille, 300,000 gold crowns more, and yearly pension to each of the Cantons, and promised that Maximilian Sforza (whom the Swiss were to betray) should be given honourable residence and a handsome yearly allowance in France. As a foretaste of his generosity he made a sweep of all the money and plate that could be found in his camp, and sent it to the Swiss, 'leaving to each gentleman only enough cash to keep him for eight days.'[1] In this way 50,000 francs were collected, and sent out to Galarate, where deputies from the Confederates were to confer with the Duke of Savoy and his brother, who acted as the agents of King Francis.

It is shameful to have to record that the five Swiss commissioners accepted the bargain. When it was reported to the council of captains there was lively discussion on the conflict between profit and honour. Finally the representatives of Bern, Fribourg, Soleure, and Valais declared that the Treaty of Galarate must be ratified, and marched away with their contingents, some 12,000 pikes. But Cardinal Mathias Schinner, bishop of Sion, the Pope's representative, made such vehement appeals to the captains of the eastern cantons, who had no love for Bern, the largest and most greedy of their confederates, that they refused to consider the bargain as tolerable by honest men. As they were still debating, the Cardinal contrived to bring on a trifling skirmish between some Milanese horse and some French exploring parties, and persuaded the Swiss captains that they were being attacked.

[1] Du Bellay, i. 54.

The council broke up, and the troops were hastily mustered in the square before Milan Cathedral, where the bishop harangued them from a pulpit—they were the soldiers of Holy Church, and the defenders of honour,[1] the French King had only part of his army with him, and victory would not be too difficult and very glorious.

Without making any delay, the army marched out of Milan, probably not half the strength of the French, when it is considered that such a large proportion of it had gone off homeward. It is hard to believe that it can have been more than 14,000 or 15,000 strong, even including the few hundred Milanese horse who accompanied it.[2] Cardinal Schinner rode some way with the vanguard, appealing to the officers and haranguing the men as he passed along the line of march. He then joined the Milanese cavalry in the van. The army was marshalled in the normal three masses of Swiss tradition, and when it came upon the French outposts deployed into the usual échelon of columns, the left leading, with the Italian cavalry on its outer flank.[3]

The French, encamped in and about the village of Marignano [4],had not been expecting to be attacked that day, since the King had come to a successful bargain with the Swiss deputies at Galarate, and was expecting to enter Milan in triumph next morning. The afternoon was well advanced, as the debate among the Swiss had wasted most of the daylight hours, and they had about ten miles to march from the gate of Milan to Marignano. It is curious to find that some of the French chroniclers speak of the ' bad faith ' of the Swiss in attacking—as if those who had been trying to corrupt the enemy's captains had any right to talk of morality.[5]

There was, however, no question of one of those surprises by which the Swiss, as at Morat or Novara, had often started a successful day of battle. For the French had light cavalry out in their front close in to Milan—the troops with whom the

[1] Florange, p. 190.

[2] Florange gives the Swiss 25,000 men (p. 192), and Du Bellay, i. p. 61, 35,000 ! —impossible figures.

[3] The right column contained the contingents of Zurich, Appenzell, Glarus, St. Gall. The main-battle consisted of the 'Old Cantons,' Uri, Schwytz, Unterwalden, Zug, and the Grisons. The left column was from Lucerne, Basle, Schaffhausen, etc.

[4] Now known as Melignano.

[5] 'Les Swisses furent par le Cardinal de Sion persuadez de rompir et fausser leur foy,' says Bellay (i. 56).

Cardinal had provoked a skirmish a couple of hours earlier. The army had been encamped in the usual administrative divisions of 'vaward,' 'main-battle,' and 'rearward.' The 'vaward' into which the leading Swiss column plunged at headlong speed, in their usual fashion, was in charge of the Constable Charles of Bourbon, who had with him a number of well-known captains, the old Italian condottiere Trivulzio, La Tremouille, the Admiral Bonnivet, the Marshal La Palice, with for infantry Pedro Navarro's 6000 Gascons and Basques, and 4000 other French infantry, while Florange and the Bailif of Dijon, who had been skirmishing up to the gates of Milan, were out in front, with 200 lances, retiring before the enemy as slowly as they might.

When the Swiss came close to the position of the vaward battle, Florange and his party turned and delivered a charge, not because they wanted to do so, but because they had been caught at disadvantage, not being able to pass over a ditch, behind which the French infantry were posted, before the Swiss were upon them. Many of them were unhorsed and fell, and so did some of the infantry, who tried to help them by coming round upon the flanks of the Swiss pikemen. But apparently the first impetus of the charge of the great column was broken by this diversion, and though the Swiss got over the ditch, and captured fifteen guns in a row behind it, they did not completely break through the French vaward division. It hung on to them, by means of cavalry charges, which forced them to halt and face outward, till the King and the main-battle came up in line with Bourbon's troops.

Francis happened to be in armour when the approach of the Swiss was announced—he was trying on a new suit which had been given him by Galeazzo de San Severino, and while his division was being arrayed, according to the lively narrative of Florange, he had a moment's leisure, which he utilized in a curious way—praying Bayard, as the best soldier in the army, to knight him there and then—which the 'chevalier sans peur et sans reproche' did with great dignity.[1] Apparently it was at about the same moment that the King and the 'main-battle,' and the second pike-column of the Swiss, the 'battle' of the Old Cantons, came up into line with their

[1] Florange, 193–94. Giovio says that this was *after* the battle, but Florange was on the spot. On the other hand, the *Loyal Serviteur*, lx. p. 346, might be, in his few modest words, taken to support Giovio rather than Florange.

respective vanguards. There was another great clash between the King's landsknechts of the 'Black Bands' of Gueldres, and the Swiss second battle—with results not dissimilar to that at the opening of the fight, for the Swiss burst in among the Germans, but did not quite break the line—the King and his bodyguard charging them again and again in the gathering dusk, and the fight became confused but indecisive. The King got his guns to bear in the intervals, and the Swiss, though they rushed in desperately upon them, could not reach them. The King was in the thick of the fray—he wrote to his mother next day that his bodyguard had made twenty-five charges, and that no one could for the future call the gendarmerie 'hares in armour.' He himself received several thrusts, and one which went through the buff coat which he wore beneath his armour. The mêlée went on long after dark, and the enemies were so intermingled that they could only distinguish each other by their national war-cries.

A curious incident related by Bayard to his 'loyal serviteur' may give some idea of the confusion. He had charged right through a mass of Swiss, and then hurtled in the dark into a vine-trellis, trained from tree to tree, which brought him to a stand, and forced him off his horse. He stripped off his helmet and greaves, and started to crawl on all-fours along a ditch, in the direction where he heard the cry of 'France' with mixed scuffling all around. By good luck he came on the Duke of Lorraine rallying a party of his men, borrowed a helmet and a horse (his third for the day) and charged again with the Lorrainers. It was ten at night before he struck his last stroke.[1]

After three hours' fighting in the dark the two armies fell back a little, in order to disentangle themselves, and the day ended indecisively. The King had kept his guns safe by constant charges of cavalry, which forced the enemy to halt whenever he got too close in. He caught some hurried moments of sleep on the limber of a gun, but they can have been but short, for after midnight he noted that the Swiss had lighted fires behind their line, whether to allow companies to see each other and form up, or to cook food, for the Duke of Milan sent them up carts of provisions of all sorts to the rear of the host.[2] These fires making the Swiss very visible, Francis ordered a full volley of cannon-shot to be discharged on them, which did much damage, and caused them to clear out of the neighbourhood of

[1] Bayard, lx. p. 245.

[2] Giovio, xv. p. 420.

the too-well-lighted spots.[1] They finally got into order, guided by the bellowing of their two great war-horns the 'Bull' of Uri and the 'Cow' of Unterwalden, whose deep moans contrasted with the sharp notes of the cavalry trumpets by which the French gendarmerie were guided in their reassembling.[2]

When day broke, the French were found to have reformed a close and orderly line of battle: Bourbon's 'vaward' and Alençon's 'rearward' divisions had been brought in close to the flanks of the King's 'main-battle.' The undaunted Swiss, however, reformed their array, and came on once more, 'if they had charged fiercely over-night, they charged still more fiercely in the morning.' This time, as we are told, they had put their main strength into two columns: the larger, with the banner of Zurich at its head, attacked the French centre; the lesser marched against Alençon and the rearward division. Bourbon and the 'vaward' had only a detachment left opposite them. The larger column suffered terribly from the King's artillery, but kept on and came to 'push of pike' with the landsknechts. They were thus charged in flank by Bourbon and Trivulzio from the 'vaward' and forced to form a double front, which brought them to a stop, with the guns still playing on them. One desperate band made a dash for the artillery, but only a single pikeman reached the goal, and was struck dead as he clapped his hand on the nearest gun. After this the battle in the centre stood still, with heavy loss to the Swiss increasing every moment. The small force left opposite the French right had not moved.[3]

The smaller or left-hand column of attack, which charged in upon Alençon and the rearward division, had at first considerable success—there seem to have been no guns on this part of the French front. Some of the troops gave way, spreading considerable dismay in the rear, and a disaster was only averted by a series of desperate and successive cavalry charges by the squadron of d'Aubigny—the Scottish condottiere—and Aymar de Prie. Affairs on this part of the field looked inconclusive. But the main attack on the centre had failed completely. By this time, however, doubts as to their victory began to come upon the Swiss. They had fought as well as their fathers, but 'ils visrent qu'ils n'estoient point le nombre

[1] Bayard lx. p. 246.

[2] Florange, p. 200.

[3] And was much vilified for its quiescence. Giovio, xv. p. 44.

de gens qu'il leur falloit'[1]—they were indeed hopelessly outnumbered, and their assault had only been inspired by an overweening confidence in their old tactics, confirmed by what had happened at Novara when they last met the French. In all earlier Swiss victories the success had been won in a very few hours, by the swiftness and vigour of their first charge. When the fighting on the afternoon of the 15th had failed to break the French line, there was little chance that the second day would bring better luck. The columns could not get forward if charged again and again by a gallant and self-sacrificing cavalry, which kept them halted and exposed them in the intervals of the charges to a devastating artillery fire. To this they could make little reply—they had brought up a few guns, and placed them on the high road to batter the French centre, but they were overpowered by the superior fire of the enemy.

Intermittent attacks on the French rearward still went on for some time, when they came to a sudden stop. A new hostile force was approaching on the left rear of the Swiss. This was the light horse at the head of the Venetian army, marching up from Lodi. On the preceding afternoon, just as the battle began, King Francis had asked Bartolomeo de Alviano, the Venetian general, for instant help, and it was given promptly. Riding to Lodi, Alviano started off his men on a night march, and by eight o'clock on the morning of the second day the head of the Venetian column began to show in the Swiss rear, and Alviano's light horse shouting 'San Marco!' were cutting up isolated parties and baggage guards. The road behind them showed long clouds of dust from the trampling of 12,000 horse and foot. The Venetian general had very properly ignored the unenterprising Spanish army, which lay some way from his immediate front, and had 'marched for the guns.' It is only 20½ miles from Lodi to Milan, which cavalry at least could easily cover in a single night. Alviano had chanced to be at the French Headquarters on the afternoon of the 13th, conferring with the King on future operations. When the advance of the Swiss was reported, he rode hard for Lodi, and must have reached it by seven o'clock. He roused his army, and started it on the Milan road at midnight. Travelling on a good highway, his light horse were in the rear of the Swiss by, I suppose, eight o'clock

[1] Florange, p. 200.

on the morning of the 14th. The heavy cavalry was not far behind, though the infantry must still have been some way off.[1]

The Swiss saw that the day was lost, and sullenly prepared for a retreat. Nothing did them more honour than the deliberate and soldierlike way in which it was executed. They picked up their wounded, pulled their few guns out of action, and formed a proper rearguard to cover their retirement down the Milan road. Two companies were detached to hold back the Venetian cavalry; they drove off its first squadron, killing its commander, Chiappino Orsini, son of the great condottiere, Orsini of Pitigliano, of whom so much has been told in earlier years. But being forced to continue their retreat, they were driven into a village, and battered by artillery till they perished —so died 400 pikemen of Zurich, doing their duty. Several other outlying parties of Swiss were cut up,[2] but the main body went on its way to Milan, observed rather than pursued by 300 lances commanded by the Admiral Bonnivet. The King, on the advice of his most trusted captains, refused to order close pursuit, for men and horses were exhausted, enough glory had been gained, and any lives lost in harrying such a desperate enemy would be wasted—like that of Gaston de Foix at Ravenna.

This very un-Napoleonic method of treating a beaten enemy was perhaps justified in this case. For the Swiss, instead of rallying and reorganizing, went straight home. These most astonishing and enigmatic people, after two days of most honourable fighting, suddenly relapsed into the mentality of petty hucksters or shameless blackmailers. The surviving captains presented themselves before the wretched Duke of Milan, and demanded three months' arrears of pay due to them. The unhappy Maximilian had, of course, nothing to give. Whereupon he was told that he had broken his financial contract, and the army marched off next morning for the passes, via Como and the Valtelline. The Duke was left in the 'Castello' of Milan, reckoned in those days impregnable, with a garrison of about 1700 much depressed adherents. Despairing of any help from the Emperor or the Spaniards,

[1] That only the Venetian cavalry were engaged, and that the infantry came too late for the battle, is shown by the letter in Marino Sanuto's *Diarii*, xxi. p. 14.

[2] Among others, a party which took refuge in a house which had been Bourbon's headquarters and were burned therein. Du Bellay, p. 61.

who had shown themselves such useless allies in his time of need, he surrendered early in the following month (October 4), when Pedro Navarro had begun to dig trenches, and threw up batteries in front of his stronghold. According to both French and Italian sources, he was thoroughly disgusted with his allies, and quite ready to take the terms which the Swiss had proposed for him at the conference at Galarate—a complete resignation of his claims to the duchy, in return for a handsome pension, and permission to reside anywhere in France that he might please. One chronicler puts into his mouth the words, 'I hold myself to be the luckiest man of my house, for while I was a titular duke I was a slave, the Swiss were my masters, and did with me whatever they pleased.' [1] Italian commentators say that he was a man of a poor spirit, and guided by councillors who wished to secure their own safety when the French should rule in Milan.[2] Maximilian lived for many years in France, obscurely but with a comfortable maintenance, and died at Paris on June 6, 1530. He had never married, but the Sforza claim on Milan did not expire with him, for his younger brother Francesco, safe with the Emperor in Germany, took the title of duke when he heard of Maximilian's tame abdication, and was destined some years later to recover the much-devastated duchy.

The losses on both sides at Marignano had been exceedingly heavy. But it is impossible to give much credence to the chroniclers—the French who say that the Swiss lost 14,000 or 15,000 men give a figure which would imply almost complete annihilation for the very moderate army of the Confederates. In all probability the casualties were not unequal on the two sides, and may have reached 5000 or 6000 apiece. On the French side the gendarmerie and the infantry had both been exhausted by long close fighting—often unfortunate—while the Swiss had been badly mauled by the artillery on both days of the battle, more especially on the second. The casualty list of nobles and captains among the French contains some notable names—among the dead were Francis of Bourbon, brother of the Constable; the Prince of Talmont, only son of La Tremouille; the Count of Sancerre; and the Sieur de Roye, younger son of the Duke of Bouillon; Jean de Mouy, who carried the royal standard that day; and d'Hymbercourt, Bussy d'Amboise, Beaumont, all noted captains. Claude of

[1] Florange, p. 210.

[2] Giovio, xv. p. 434.

Guise, who commanded the 'Black Bands' in the absence of the Duke of Gueldres, was trampled down and left for dead, but survived to fight on other fields—not so the German officers, who served under him, of whom a large proportion perished. The Swiss casualty-lists are full of names of leaders and banner-bearers of cantonal contingents, but owing to the strange anonymity of Swiss generalship, they mean little to the historian. The council of captains directed all operations, not any individual commander-in-chief.

Marignano, though the first great Swiss defeat, was not destined to be the last Swiss battle fought in the fine old ancestral style. It most certainly was not taken to be a refutation of the all-importance of the pike-column, or a condemnation of the offensive *à l'outrance* which had been the essential points in the normal Swiss tactics. Indeed the orderly retreat of the army after such a bloody day was considered as more honourable than some victories. The defeat could be put down, and with reason, to the desertion of the Bernese and their fellows on the eve of battle, to the hopeless disparity in numbers, and to the lack of auxiliary arms—there had been only eight guns in the field—some say only four. And the Italian horse had been not only few but nearly worthless—after some skirmishing on the first day, many of them had not appeared at all on the second, and had only turned up, when all was over, to escort the Cardinal of Sion from Milan on his way to the court of the Emperor.[1] Under proper conditions and with proper backing of horse and guns, the pike-column might yet assert its supremacy. The old game was tried once more, as we shall see, at Bicocca.

[1] Giovio, not at all given to disparaging Italian troops, says that after the first day many returned surreptitiously to Milan, 'la maggior parte, mossi da paura, o da disparatione, abandonati i capitani si ritornarano nella citta' (xv. p. 420). Indeed they were 'poco utile.'

CHAPTER VI

BICOCCA (APRIL 27, 1522)

MARIGNANO, as has already been observed, was not to be the last of the battles fought in the old Swiss style, in which the traditional columns of pikes were dashed recklessly at the enemy, whether he was caught by surprise, as at Novara, or allowed time to form his line of resistance, as at Marignano. But we have now reached the period in which the Swiss infantry was once more (as in the very earliest period of the wars) fighting on the side of the French king, and not (as at Novara and Marignano) against him. After Francis had got possession of Milan in 1515, the Confederates concluded with him the so-called 'Perpetual Peace' of Geneva, by which they engaged themselves to allow him to raise troops from among them in perpetuity, while they were at the same time not bound to declare war on the Emperor—still their suzerain in theory, though they were no longer sending representatives to the Imperial Diets. This was a most curious compromise—securing to the cantons a territorial neutrality in time of war, though their military resources might be purchased by the French whenever they were needed. That the Emperors never replied by an invasion of Switzerland—a most justifiable move under the circumstances—can only be explained by the fact that Austrian experiences in the Helvetic mountains had been uniformly unfortunate—and that if the Swiss would only stop their recently developed habit of incorporating Swabian towns in their confederacy, they had better be left alone. Not only in the time of Francis I, therefore, but for two centuries to come, the French always counted on being able to raise Swiss contingents, whenever there was money to pay them and a crisis on hand. On the few occasions on which Swiss corps are found acting on the Imperialist side,[1] they were mercenaries

[1] The Bishop of Pistoia in 1521 collected 2000, but was warned not to trust them in the Imperialist army, as they were mostly individual deserters from the French. (Guicciardini, xiv. pp. 60, and 28.) In 1522 Cardinal Schinner collected a similar force.

serving against the orders of their cantons, just as the landsknechts, so often found in the French ranks, were all rebels under the Emperor's ban, and technically traitors.

The general action in which the Swiss methods were to be tried for the last time on a grand scale was that of Bicocca, close to Milan, and not so very far from the old fields of Novara and Marignano.

When the Valois-Hapsburg contest recommenced on the largest possible scale, the initiative, as always, came from the side of France. Charles V was theoretically in possession not only of the Imperial crown but of all the immense heritage that had come to him from his grandfathers, Ferdinand and Maximilian. But he was in 1521 involved in two devastating wars—the first was the extraordinary rebellion of the 'Comuneros' in Spain, the unique outburst of discontented towns backed by many discontented nobles, who with the slogan of 'Santiago y Libertad' rose nominally in the cause of the mad Queen Juana, really with the intention of upsetting autocracy. This war dragged on for two years, and its last embers were not crushed out till the battle of Villalor (April 1523). At the same time the old trouble-maker, Robert de la Marck, Duke of Bouillon, after a temporary reconciliation with the Emperor, had reverted to his French alliance. With secret help from Francis (who yet pretended for some time to disown him) he had stirred up war on the Belgian frontier, and overran Luxemburg.

Without any open declaration of war against Charles, the French king sent a large army to invade Navarre and replace on his throne Henry d'Albret, the expelled French king of the little realm. Nominally the army was Navarrese, not 'French':[1] With the revolt of the 'Comuneros' in full fury the invaders found no one to oppose them, swept over the whole country and installed d'Albret once more in Pampeluna.[2] The conquest of Navarre had seemed so easy that L'Esparre, the French commander, took no care of the disposition of his troops, and

[1] It was impudent of Francis to pretend that he had no responsibility, when the army of invasion had in it 300 lances of his *compagnies d'ordonnance*, 6000 Gascon infantry, and was commanded by L'Esparre, brother of his favourite Lautrec, and of his mistress Madame de Chateaubriand.

[2] The celebrated Ignatius Loyola, the founder of the Jesuits, was a captain of infantry in the garrison of Pampeluna: he was wounded while defending the citadel, permanently crippled in the use of one leg, and turned to religion—with no small consequences to the world.

disbanded part of his infantry. He was turned out of the country almost as rapidly as he had come into it, by a hastily collected royalist Spanish army, which beat him at the combat of Esquiroz near Pampeluna (June 30, 1521) and swept him back over the frontier, with the loss of all his artillery.

This stab in the back would have been sufficient for a *casus belli* by itself; but King Francis succeeded in finding a counter-claim of violation of territory, when the Count of Nassau, having driven Robert of Bouillon out of his dominions, trespassed over the frontier, and laid siege to the neighbouring French towns of Mouzon and Meziéres. The former fell, the latter was saved by the valour and ingenuity of the great Bayard—the last successful exploit of his long military career. War was now lit up along all the borders of France, from Tournay to Bayonne. Its northern episodes—very tedious manœuvring for the most part—need little notice. But there was epoch-making business on hand in Italy, where the last great battle of the old style was to be fought.

Francis was in possession of the whole duchy of Milan, and was still allied (as in the days of Marignano) with the Venetians. In opposition were the German forces of Charles V on the north, in Tyrol, and his Neapolitan forces in the south. But the latter were weak at the moment—the war of the Comuneros was distracting all the resources of Spain to domestic dangers. And the duchy of Milan with the Venetian territory made a complete bar across Italy, to prevent any junction of the landsknechts of the north with Spanish horse and arquebusiers from the south.

The army with which the French Viceroy of Milan—the unlucky Lautrec[1]—had first to deal was a composite affair—it consisted of the Papal troops under the Marquis of Mantua, 'Captain General of the Church,' a Florentine contingent, and all the Spaniards that could be spared from Naples. The main problem was whether they could draw in the German contingent collected at Trent—it had to pass over the Monte Baldo through Venetian territory—and Venice was hostile. This difficult and dangerous movement, however, was effected. Crossing the narrow strip of Venetian land between Adige and Mincio, in two forced marches unmolested, the Germans passed the Mincio at Vallegio and got into friendly Mantuan territory. The French complained that they might have been intercepted,

[1] At the commencement of the war his brother Lescun was in actual command, Lautrec being on leave. He turned up in July.

if only common precautions had been taken.[1] Thus the allies were able to collect an army which could cope with Lautrec—without the Germans they would have been helpless. The general command was given to Prosper Colonna, the old condottiere of whom we have been hearing much in past years—to the great disgust of the Marquis of Mantua. In the two campaigns which followed, Prosper won the name of 'Cunctator,'[2] borrowed from the ancient Roman Fabius; but after much cautious manœuvring he put in a decisive battle—such as the real 'Cunctator' never won.

He was a typical specimen of the great condottieri, 'who fought with his brains more than with his sword, who grew famous by defending the state without ever exposing himself to disaster, and accepted battle only when it was needful, and when he had placed all advantages on his own side.' His first campaign passed off without a general engagement—he refused battle several times, raised the siege of Parma when it was almost in his hands, rather than risk anything, and wearied out Lautrec by retreats and flank movements. By the autumn the Frenchman was holding the line of the Adda with a diminished army, and had made a large detachment to garrison Cremona, through which he kept up his communication with the Venetians. 'All Saints Day had come, the nights grew long and cold, and rains fell continually—the bulk of our Swiss got sulky, and began to desert in bands without leave—it is true that they made excuse of their pay which was in arrears. By the end of October there were only 4000 of them left out of 20,000 with whom we started.'[3]

This was what Colonna had been waiting for—the circumstances had reproduced the situation of Gonsalvo's passage of the Garigliano in 1503.[4] Like Gonsalvo the old condottiere passed his army far upstream on the Adda, nearly to the foothills of the Alps, and pushed a vanguard across on boats at Vaprio, the very end of the French line. A small Italian detachment under Count Ugo of Pepoli, which was all that Lautrec had left in this direction, was crushed, and before the French could concentrate and march north, the whole of Colonna's army was over the Adda.

[1] Guicciardini, xiv. p. 34. [2] Guicciardini, xv. p. 212.

[3] Du Bellay, i. p. 181. Some of these deserters were picked up by Cardinal Schinner, and formed an unauthorized Swiss unit in the Imperialist army.

[4] See above, p. 122–3.

Lautrec, no longer anxious to fight, for he was now very weak in infantry, retired into Milan, thinking the campaign ended. But he was sadly mistaken, for on November 23, Colonna, having secret friends of the Sforza faction in Milan, and being well acquainted with the enemy's careless dispositions, risked a night-surprise, which had complete success. The Venetian infantry, which was holding a long segment of the defences by the Porta Romana, was seized with panic, and flinched before the escalade. The Imperialists swarmed into the town, and Lautrec, after some unskilful street-fighting, withdrew to Cremona with the wreck of his army, reduced to 500 men-at-arms and 5000 infantry of his own, with 400 horse and 6000 foot of the Venetians.

Both parties then went into winter quarters—but Alessandria, Pavia, and Como all passed into the hands of the Imperialists before the new year—the first two by popular risings, which drove out the French, the third after a formal siege. By the spring Lautrec had lost not only Milan but the greater part of the Milanese; but he was still hanging on in Lombardy, and hoped for great reinforcements, which King Francis, much irritated at the loss of the wealthy duchy, where he had been ruling for several years, proceeded to grant him. He requisitioned 16,000 'new' Swiss from the Cantons, sent over the Alps the Marshal de Foix and Bayard with some companies of gendarmerie, and Pedro Navarro with some French foot. At the same time he bought in Italy the services of the well-known condottiere Giovanni dei Medici, and his 'black band' of 3000 arquebusiers and light horse—Giovanni had been serving in the Papal army, but changed over on the light pretence that his contract had been made with Pope Leo X, who was just dead, and not with his successor. The Venetians were asked to come to the muster with their best troops—but proved (as often before) disappointing allies, for in the critical month they allowed a large German reinforcement for the enemy to slip past their fortress of Bergamo, by devious mountain-paths, and to join Colonna at Milan. However, several thousand of their mercenaries—horse and foot—reported at Lautrec's headquarters.

The newly collected French army decidedly outnumbered the forces of Prosper Colonna, even when he had been joined by the Germans, who brought with them Francesco Sforza to assume possession of his long-lost duchy of Milan. The

French, after retaking Novara, and picking up their Venetian contingent at Marignano, laid siege to Pavia, hoping to provoke the enemy by attacking such an important place. Almost immediately Lautrec discovered that his move had been so far successful that Prosper Colonna had come out into the field, and had fortified himself in the water-cuts and ditches of the famous Certosa monastery, ten miles outside Milan on the Pavia road. The French general knowing that he had now the larger army, was anxious to get the pitched battle which he had vainly sought for in the preceding year. But reconnaissances sent out towards the Certosa reported that the enemy had fortified himself in such an immensely strong position that it would be a very serious matter to turn him out of it.[1] On the other hand, it might be possible, by manœuvring, to threaten his communication with Milan, and to cut off his supplies, by raiding around both his flanks. In such a case he would have to move, however little he might like the idea of quitting an ideal defensive position, and he might perhaps be caught at a disadvantage while on the march. Accordingly the army went northward in a long sweep, and stationed itself at Monza, cutting the roads out of Milan toward the Alps.

Unfortunately for Lautrec, he found a most unexpected hindrance to this very rational strategy. The Swiss, who formed the largest contingent of his infantry, and without whom he could not hope to win his ends, went on strike, with their habitual perversity. They had come over the Alps, but they had not received any of their promised pay. Lautrec assured them in vain that it was coming, under convoy, to Monza. But they refused to listen to him, and a deputation of their captains, headed by the Bernese Albert von Stein, put before him their resolve that unless he attacked the enemy on the very next day, the whole of the cantonal contingents would march homeward. That they were not shirking danger would be proved by the fact that they were prepared to assault the enemy's position, if they were given the chance: but there must be no delays or manœuvring. Probably the captains did not put into words the boast that Guicciardini attributes to them—'We stormed the French positions at Novara, though we were outnumbered hopelessly, and we will storm this position, for we hold that the Spaniards may be more cunning, and more skilled in ambushes and deceit than the French, but we think

[1] Du Bellay, i. p. 217.

that they will fight no better, when they are tackled with desperate courage and soldierly skill.' [1]

Lautrec, yielding most unwillingly, committed himself to the venture: the Swiss, as was reasonable, undertook to bear the brunt of the battle which they demanded. Prosper Colonna had found outside Milan, about four miles to the north, another position as formidable as that of the Certosa. This was at the Park of Bicocca, a large country house surrounded by ornamental gardens and deep ditches. The neighbouring fields were full of irrigation-cuts, as is all the flat land round Milan. At the west end the position seems to have been too marshy to be practicable for military movements. At the east end was the high road from Milan, with a wet ditch along it, crossed some way to the rear by a narrow stone bridge. This looked the most accessible point in the neighbourhood. Prosper had been given sufficient time to strengthen a position already formidably strong. On the north front, where a sunk road leading to the manor-house ran along the front of the park enclosure, he had deepened the hollow, and made a rampart along it, with the cast-up earth, piled above the existing bank. In several places he had constructed high platforms in this rampart, and placed guns upon them, which would sweep not only the fields in front, but laterally the ditch immediately below.[2] The whole of his Spanish arquebusiers, in four-deep line, were placed behind the rampart, and in the rear of them solid continuous blocks of German and Spanish pikemen.

[1] Guicciardini. xiv, p. 124.

[2] The topography of the long-vanished manor and park of Bicocca is described by Du Bellay, Paolo Giovio, and Guicciardini in rather irreconcilable detail. Du Bellay calls it 'a gentleman's house surrounded by great ditches, with a circuit so large that it would hold 20,000 men. When the enemy reached it they deepened the ditches, and flanked them with large platforms, on which they placed guns. . . . The Marshal of Foix was told off to assault the most convenient spot, which had been ascertained by reconnaissance the day before. He searched about till he found a stone bridge, by which he could get inside the position.'

Giovio writes: 'It was a place very secure and convenient. On the right flank was the high road to Milan with a continuous wet ditch. The left and rear were covered by two curved water-courses (*emissariis amniculis*). In front was a road at right angles to the high road, broad enough for two carts to pass in it; it was sunk below the fields in front, about as deep as a man's middle. This road Pescara used as a ditch, and placed guns and infantry behind it. Prosper kept the cavalry to the rear. The high road was watched by Francesco Sforza and his men.' (*Vita Piscariae*, p. 339.)

Guiccardini's version is: 'Bicocca is a large mansion surrounded by spacious gardens, which are enclosed by deep ditches: the fields around it are full of springs and irrigation-cuts, such as the Lombards use to water their meadows. . . . Lescun, the Marshal, was told off to the left, with 300 lances and some foot, to proceed along

It will be remembered that large bodies of landsknechts, under George of Fründsberg, one of the most noted German captains, had recently got into Milan. The whole position was not more than 600 yards long, so that there were enough troops to pack it closely and to form reserves. The cavalry was kept far back (no doubt some of those present, and especially Pescara, remembered the lesson of Ravenna), and sent to the rear a detachment to watch the bridge on the Milan road. There was little danger of a repetition of Gaston de Foix's artillery preparation of 1512, since the ground in front was unsuitable for the moving of guns, being flat and cut up with irrigation ditches, besides being exposed to the artillery of the defence.

As if to give the enemy warning of his projected assault, Lautrec sent out in the evening of the 26th a strong reconnaissance—400 horse with infantry supports—to coast along the position as near as was possible. The officer commanding, the Sieur de Pontdormy, reported that the ground was very difficult, but this had no effect on the minds of the Swiss. His movement, however, was noted by Prosper Colonna, who at once sent into Milan, to beg Duke Francesco to come out with his personal retinue and such urban militia as he could collect, to watch the bridge on the high road upon the right rear. This was done—Sforza brought out 400 horse and 6000 indifferent foot to join the flank guard, and was in his place next day.

At dawn Lautrec delivered his assault, first sending out Giovanni dei Medici's light troops of the 'Black Bands' to press in Colonna's pickets—they got engaged with Spanish 'genitors' and finally cleared the ground in front of the position. We hear in the French accounts of the old formal division of the army into vaward, main-battle, and rearward, but the array showed no such simple features. Two dense Swiss

the Milan road, toward a bridge by which it was possible to penetrate into the rear of the enemy's position' (xiv. p. 125).

The French left is, of course, the Imperialist right. But we have to find what was Colonna's front line. Guicciardini and Du Bellay think of it as a ditch. Giovio calls it a sunk road, below the level of the fields. But a sunk road would not be a natural boundary for a gentleman's park, unless there was a wall or rampart above it, to keep out trespassers. I am bound to suppose that Pescara deepened the sunk road, and threw up the earth on its inner side to make the 'large platforms' on which the guns were placed. There must have been already some wall or bank above the sunk road, and this would be heightened by the upcast earth when (as Du Bellay says) Prosper proceeded 'relever les fossez.' This must have made a bastioned front, since the guns must have been mounted on 'cavaliers' dominating the rest of the rampart and the ditch.

columns of pikes were told off for the frontal assault, each 4000 strong, and some guns were attached to them. The general's brother Lescun (the Marshal of Foix) was to move out to the left, parallel with the Milan road, and to endeavour to force a way across the bridge in the enemy's right rear. In support of the Swiss columns of assault Lautrec himself had the 'battle' in hand, in which were the remainder of the Swiss, the French infantry, and the main body of the gendarmerie. Lastly, the Venetian army, under the Duke of Urbino, formed a third line, or reserve, somewhat apparently to the French right, and facing the marsh-protected part of the Imperialist position. The engineer Pedro Navarro with his sappers were attached to the Swiss, with orders to fill up ditches, and make the movement of artillery possible 'pour faire faire les esplanades.' But he got little chance of making himself useful—as we shall see. A small detachment of horse under Pontdormy was told off to watch Lescun's left. lest he should be taken in flank from the Milan direction.[1]

Technical command of the Swiss was given this day to the Sieur de Montmorency, a practised soldier, who marched with them on foot, as did several other French nobles, whose names need not be recapitulated. But Montmorency might as well have stayed behind, for when he had conducted the column up to near gunshot of the enemy, and bade the Swiss halt, to allow time to bring up the artillery, who were to batter the Imperialist entrenchments—as at Ravenna in 1512—he met with frank disobedience. The pikemen did not stop. Possibly the Swiss captains noted the way in which the Imperialist guns were entrenched and embrasured, and doubted the efficiency of artillery preparations. But it is more likely that (as French observers stated) they were inspired by blind pugnacity and self-confidence. At any rate they rushed straight at the Imperialist position: we are told that the two columns were in rivalry with each other—the one consisted of the contingents of the old rural cantons, under the command of Arnold Winkelried of Unterwalden, the other of troops of Bern and other urban cantons, under Albert von Stein—the captain who never forgave himself for having been a day late at Novara, and who had marched off before the battle at Marignano.[2]

The moment that the columns came in range of Colonna's guns they began to suffer heavily—the ground being level irrigated fields, and the distances having been marked out by

[1] All this from Du Bellay, i. p. 218.

[2] See p. 163.

the enemy's master-gunners. A thousand Confederates are said to have fallen before they reached the hollow road—and then they came to a dead stop—for the obstruction was so deep, and the rampart behind it so high, that the head of a pike could hardly be stretched from the bottom of the ditch to the crest of the earthwork.[1] And as its sides were cut perpendicularly, there was great difficulty in scrambling up the smooth

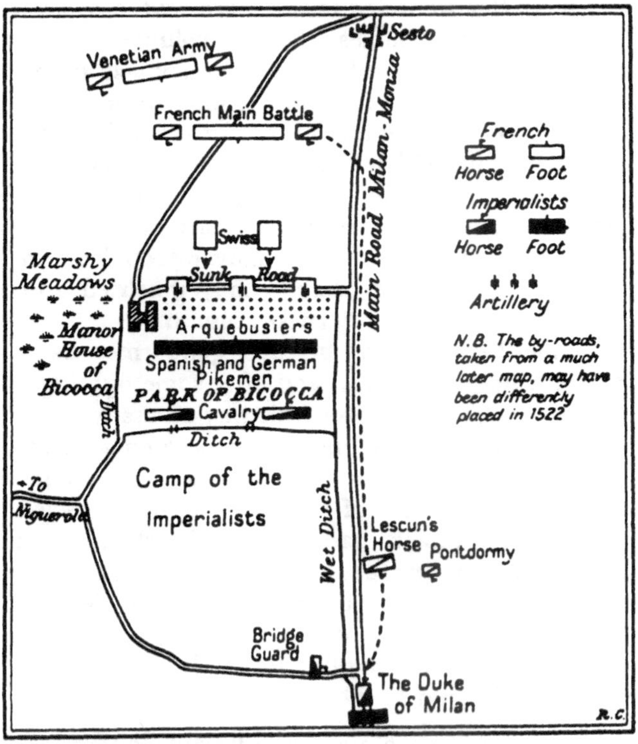

BATTLE OF BICOCCA, APRIL 27, 1522.

face. When the heads of the columns jumped down into the ditch, they were received by four successive volleys from Pescara's Spanish arquebusiers ; it is said that all the standards went down, and that the three or four first ranks perished wholesale.[2] But this did not quite check the onset : the Swiss were heard calling for the officers and double-pay men to come to the front, and the summons was not in vain. There was a desperate attempt to break into the position, and in several

[1] Du Bellay, i. 219. [2] This from Giovio, p. 339.

places groups, scrambling up over the dead and wounded, did reach the top of the rampart. But they were at once charged and tumbled back into the ditch by the German and Spanish pikemen, who came up through the arquebusiers. It is said that Arnold Winkelried, the captain of Unterwalden,[1] and George Fründsberg, the landsknecht leader, actually met hand to hand, and that the former was laid low by the German's pike, and the latter wounded in the thigh by the Swiss halberd. Almost all the French nobles who had volunteered to serve in the assault were killed : only Montmorency was dragged half dead out of the ditch by a faithful friend. After spending half an hour in desperate attempts to get forward, the survivors of the two columns turned back and retreated in a slow and sullen fashion—as at Marignano—to the ground from which they had started. They left before the rampart in the ditch, and on the field which they had first crossed, over 3000 dead, including 22 captains, among whom were both Winkelried and Albert von Stein, who had led the Bernese column.

Meanwhile Lescun with his 400 gendarmes had accomplished his turning movement, and after some fighting with the guard at the bridge,[2] had forced his way across it, and into the camp of the Imperialists behind the Bicocca enclosures. They thought for a moment that they had secured an important success, but—as a matter of fact—it was quite useless, since the main frontal attack had failed. Much confusion was caused in the camp, and panic-stricken messages came up to Prosper Colonna. But the Imperialist general detached some horse from his rear, under Antonio de Leyva, while the Duke of Milan, with his rather irregular force, came up to block the bridge behind Lescun.[3] The Milanese were held in check

[1] So Fründsberg, but Giovio, p. 339, says that Albert von Stein was the Swiss who fought him. It was perhaps Winkelried's desperate charge on this occasion which got his name into the legend of Sempach, as a man who sacrificed himself for his followers. See vol. ii. p. 249 of *Art of War in the Middle Ages*.

[2] Guicciardini (xiv. p. 125) has what looks like an improbable tale, that Lescun intended to capture the bridge by surprise, having told his men to mount red crosses instead of white, so as to be taken for Imperialists, but that Colonna, hearing of this device (one would like to know when and how), directed his troops to wear on their helmets that day a knot of corn-stalks or long grass, so that the French trick was detected at once.

[3] Du Bellay (i. p. 220) evidently exaggerates Lescun's success, but he is to a certain extent borne out by Paolo Giovio, who gives details as to the disorder caused in the camp by the raid (*Vita Marchionis Piscariae*, p. 339). He is also the only author who mentions Francesco Sforza's interference with Lescun's retreat, and the action of Pontdormy's squadron.

long enough by Pontdormy's detachment to enable Lescun to cut his way back across the narrow passage, and to retrace his steps to the main army. French narratives make much of this raid, and assert that it might have settled the fate of the day, if only the Swiss had held their own and attracted the main attention of the enemy, or if the Venetians had threatened the enemy's left flank.[1] But this is evidently unsound reasoning. The turning force was too weak to have accomplished much by itself, even though it had got well into the enemy's rear.

When the Swiss had been driven off, and Lescun had retreated, we are told that the Marquis of Pescara and other captains implored Colonna to deliver a counter-attack with his whole army, while the enemy was in disorder. And, apparently without Colonna's leave, some Spanish arquebusiers and light horse followed the routed Swiss for some little way, till they were driven back by Giovanni dei Medici's troops of the 'Black Band,' whom Lautrec threw out to cover the Swiss, and to see to the dragging off of the guns which had been allotted to them in the original plan of battle. But Colonna refused to move from his trenches, pointing out that the French second and third lines, including the bulk of their cavalry, were quite intact. He is said to have remarked that the enemy was so well beaten that next day the Swiss would break off and retire into the Alps.[2] George of Fründsberg agreed with him—*not* (we may probably believe) for the discreditable reason given by Paolo Giovio—namely, that his landsknechts called out that they must have a bonus (double pay) if they engaged in a second fight.[3]

Colonna's prophecy was quite well-grounded, for the Swiss refused to stay a moment longer with Lautrec, and went off home upon Tuesday, April 30. This departure left the French general so weak in infantry that he had to give up all hopes of further offensive operations, and crossing the Adda at Trezzo marched off eastward into Venetian territory. Why he did not rather retreat toward his own base, westward, is hard to make

[1] Du Bellay (i. p. 222) is hard on the Venetians, but it does not seem that they were in a position to approach with profit the water-logged west front of the entrenchments, or that Lautrec (whose reserve they formed) gave them orders to do so.

[2] Guicciardini, xiv. p. 127.

[3] Giovio (*Vita Marchionis Piscariæ*, p. 340) is always prone, like Guicciardini, to make scandalous allegations against non-Italian troops—German or French. But from other manifestations of the sordid mentality of the landsknecht (cf. p. 78), I should not like to be absolutely sure that there is nothing at the bottom of the story.

out. Probably he wished to keep in touch with his none-too-trustworthy allies, in order to prevent them from dispersing completely and abandoning the campaign. After conducting his army to Cremona, where he left his brother Lescun in command, he went off hastily to Paris, almost unescorted, to make his unhappy report to King Francis, and to receive in the end pardon for his mistakes—largely by the intercession of his sister, the King's favourite mistress.[1]

Only a small proportion of the native French troops had been engaged in serious fighting at this disastrous affair of Bicocca, their only serious loss being among the officers who had volunteered to accompany the Swiss: and the Imperialists had very few casualties indeed, having been under cover most of the time; the only person noted as having fallen in their ranks is Juan de Cardona, a kinsman of the ex-Viceroy of Naples, who was shot through the head by a chance ball.

But the importance of the day was that the Swiss were at last cured of their old pugnacity, which had been hardly affected at Marignano. They had lost not only their two chief captains, who had both been prominent in urging on the battle, but nearly all the officers who had charged in the two great columns. 'They went back to their mountains,' says Guicciardini, 'diminished in numbers, but much more diminished in audacity; for it is certain that the losses which they suffered at Bicocca so affected them that in the coming years they no longer displayed their wonted vigour.'[2] This is absolutely true: they never tried again the headlong assaults by which they had so often swept away enemies of superior strength. And when we next find them in action on a large scale, at Pavia in 1525, they surprised friends and enemies alike by their want of dash and initiative. They always remained good steady troops in battle, but were no longer inspired by the wild self-confident fury of earlier years.

This was the last incident in a long section of the Art of War—the Swiss methods had been tried for the last time, and

[1] There is a long story of court scandal concerning this pardon. The disaster said Lautrec, came from the money for the pay of the Swiss being denied him. The King demonstrated that the money had been provided, but the treasurer, Semblançay, pleaded that 400,000 *écus* had been requisitioned from him by the Duchess of Angoulême, the King's mother, to pay old debts of hers, just as he was about to send it to Italy. After many recriminations and legal proceedings, Semblançay was condemned to death years after his supposed offence. The whole matter is obscure, and Semblançay was apparently a scapegoat.

[2] Guicciardini, xiv. p. 128.

found wanting even by themselves. But it was also the first incident in a new period—that of the predominance of small firearms: for though the Imperialist artillery had worked well, it was the arquebus which had really won the battle. Also it was a notable vindication of the importance of field-fortification—Pedro Navarro, serving with the French that day, saw the justification of his own idea of defensive battle, which had failed at Ravenna only for want of a proper complement of artillery, and of sufficient cover for the cavalry arm. On this occasion the Imperialist horse were kept far to the rear, and quite out of cannon-shot. At Ravenna the position had not been sufficiently deep.

It had been known before that frontal attacks on an entrenched position were perilous—as witness Cerignola. After Bicocca no one tried them again during the whole course of the wars—the best storm-troops in the world had failed before the arquebus. For it was the bullet that stopped the assault: the pike only did the gleaning of the field—though no doubt it was necessary to have pikes at hand to finish the game.

CHAPTER VII

PAVIA (February 24, 1525)—THE CAMPAIGN AND THE BATTLE

THE disaster of Bicocca had not destroyed the French army of Italy, though it had been put out of action, and had withdrawn to the doubtful shelter of Venetian territory. Though Lautrec had allowed himself to be cut off from his direct line of communication with France, the French were still in possession of many towns of the Milanese, and there remained a formidable force in existence. King Francis, very unwisely, resolved that the game was not wholly lost, and prepared to send reinforcements across the Alps in the next spring. But everything went wrong during the months of delay. Since the French army was far away beyond the Adda, Pescara made a dash on Genoa, and with the aid of the anti-French party in the city surprised and destroyed the small garrison. Pedro Navarro was captured as he tried to organize resistance, and paid for his long service against his old master in the French ranks by a three-years' captivity in the Castle at Naples, which he had once captured for Spain by his explosive machines. On hearing of the fall of Genoa, Lescun, under a convention,[1] evacuated what he was still holding in the Milanese, and brought back the remains of his army over the Alps, to the great displeasure of the King, who had not given up his Italian ambitions.

It was not unnatural that the Venetians in the following spring (1523), seeing the French cause ruined, made peace with the Emperor, so that Francis was deprived of his only important ally beyond the Alps. The great condottiere, Giovanni dei Medici, of the 'Black Bands,' followed the example of the Venetians, and took service under the Emperor against his old employer. Nevertheless, with the most perverse

[1] By which the Castello of Milan, still in French hands after a siege of many months by the Duke Francesco and the townsfolk, surrendered, as not having been relieved in a term of months settled between Lescun and the Duke.

obstinacy, the French king adhered to his old project, and collected a considerable army at Lyons and in Dauphiny, hiring 10,000 Swiss (in spite of the memories of Bicocca) and 6000 German mercenaries, a force which, with the wrecks of the army of 1522, made up 1800 lances and 30,000 foot. He had been intending to take the command himself, but was distracted from his purpose by the surprising and desperate treason of the Constable Bourbon—the most important of the princes of the Blood Royal. Charles of Bourbon—with a frank disregard of loyalty and national feeling which recalls the worst examples of the fifteenth century—had planned a revolt in his own feudal lands, and entered into negotiations with the Emperor for the bringing in of a German force from the east to assist his rebellion. Bourbon's provocation had been considerable; he had been hardly treated by the King,[1] and was known to be discontented, but it was now forty years since a French prince had gone into rebellion against his sovereign, and his treachery was unexpected. His plans, however, were discovered before they were complete, and he was forced to fly for his life into Germany, while the captains of his castles of Moulins, Chantelles, and Carlat surrendered tamely at the first summons. Only some sixty or eighty gentlemen of his vassalage followed him across the frontier, so that he brought little practical aid to the Emperor.[2]

But at the moment when Bourbon's treason was discovered, it was not known how far his conspiracy had spread, and the King remained behind to deal with possible revolts, and handed over the conduct of the invasion of Italy to his favourite, Guillaume de Bonnivet, the Admiral of France, who had distinguished himself of late by his lucky conquest of Fontarabia on the Spanish frontier. The task was far beyond the Admiral's capacity, as it turned out, but he had a fortunate start. For Prosper Colonna, the old condottiere general who had discomfited Lautrec, was sickening with his last illness, and (what was as important) had made up his mind that, since

[1] Francis had refused him military commands which should naturally have fallen to the Constable. But the main grievance was an attempt made by the King's mother, the Countess of Angoulême, to claim from him great part of the Bourbon heritage, which had come to him from his wife and cousin Suzanne, on the ground that she was the next of kin to the deceased lady (being her first cousin), while Charles of Bourbon was only her second cousin and heir-male. The Chancellor Du Prat, backing the Countess, Bourbon made up his mind to rebel before the lawsuit was finally settled. See Du Bellay, i. p. 262.

[2] Curious details of the Constable's flight may be found in Du Bellay, i. pp. 265–70.

their loss of Genoa and their retreat behind the Alps, the French would make no move on the side of Italy for a long time. It was, indeed, a misguided venture, when war was loose on the Channel, and Anglo-Flemish armies were in Picardy. In consequence, the forces that might have been available against Bonnivet, if he had been expected, were much scattered. It was a very long time before the Viceroy of Naples, Lannoy, brought up reinforcements from the south, and the Venetians had always to collect a mercenary army before they could put it in the field. The Emperor's attention, too, was more attracted to the Spanish and Flemish frontiers than to Italy at this moment.

Hence Bonnivet had a good start—he turned Prosper Colonna off the line of the Ticino, which he tried to defend with a very inferior force, captured Novara without resistance, and—as several chroniclers assert—might have had Milan also, if he had tried to rush it, instead of occupying strategic positions around it, and endeavouring to starve the Duke Francesco and Colonna out of the great city, by planting troops at Monza, Chiaravalle, and San Cristofano, and cutting off the aqueducts. This scheme failed entirely, the months dragged on without a decision, and in November the Admiral, whose army had eaten up the countryside and was almost as famished as the Milanese, drew aside into winter quarters some miles back. The greater part of his Swiss mercenaries had melted away, and he discharged as useless many of his French infantry of the new levy.

Meanwhile Prosper Colonna died of his lingering disease in December, having by his obstinate persistence in Milan justified once more his nickname of 'Fabius Cunctator.' His place was taken by Charles de Lannoy, the Viceroy of Naples, who brought up tardily the much-needed reinforcements from the south. The Venetian army, raised at last, came in under the Duke of Urbino, and presently the Constable Bourbon descended from the Tyrol with 10,000 landsknechts, whom the Emperor had given over to his command. The balance of force had now gone over to the Imperialist side, and in March the united contingents marched out of Milan, and attacked Bonnivet in his cantonments. The Admiral was caught with his troops in dispersed positions, manœuvred in vain, then tried to retreat in order to concentrate, and was finally pursued and discomfited at the 'Rout of the Sesia'—where his rearguard of Swiss infantry was cut to pieces, most of his artillery captured, and he himself badly wounded (April 30, 1524). But a much more serious loss to France was that of the gallant Bayard, who was laid low by a ball from an arquebus in a hopeless charge against the pursuing Spaniards. It is *not* true that, as he lay dying under a tree, the Constable Bourbon came up and expressed his sorrow, to receive the cutting retort: 'I do not need your pity, who am dying like a gentleman—but *you* have need of *my* pity when I see you serving against your King, your country, and your oath of allegiance.' [1]

The Count of St. Pol, replacing the wounded Bonnivet, brought the remainder of the French army back across the Alps, where they were joined by the large garrisons of Lodi and Alessandria,[2] who obtained under a convention safe conduct to France, on condition that they left the fortresses intact—easy terms enough, but Lannoy and Duke Francesco wanted to clear the Milanese of the enemy without delay.

The reason for this hurry was that the Constable Bourbon was planning a counter-stroke, the invasion of France, and wished to have everything clear behind him. He had persuaded the Emperor that his own influence was so great in France, and that the allegiance of Provence—annexed only a generation back—was so unsteady, that he could count on a general revolt, if he showed himself at the head of a large force beyond the Alps. Accordingly he and the Marquis of Pescara, with the bulk of the Imperialist army, some 20,000 in all, crossed the mountains by the Col de Tende, and advanced as far as Marseilles, meeting little resistance, while a Genoese fleet coasted round and blockaded the city on the sea front.[3]

This was the first attack on France from the Italian side for many centuries—and like all that came after it was to prove disastrous. King Francis descended hastily from the north with all the forces that he could gather, and picked up the remains of his army of Italy, with several thousand newly enlisted Swiss. The Constable besieged Marseilles for a month, but failed to break in—a good garrison under the

[1] The 'Loyal Serviteur' gives no account of this, but speaks of the compliments made to the dying hero by the Marquis of Pescara, the Spanish commander; lxv. p. 261. But Du Bellay (ii. p. 342) gives the story.

[2] This convention, according to Guicciardini (xv. p. 227), restored 5000 infantry to the French king.

[3] Bourbon, it is said, wanted to march on Lyons, as nearer his own sphere of influence, but Pescara insisted on trying Provence, as a less risky direction.

Sieur de Brion and the Italian condottiere Lorenzo de Ceri held him at bay, with the enthusiastic support of the citizens. Presently it was reported that the King of France had reached Avignon with an army of overwhelming strength—nearly 40,000 men. Bourbon's prophecies of a general revolt had proved futile, and Pescara insisted that the Imperialists must give up the siege, and retreat, or King Francis would drop down behind them and cut them off from Italy. The siege-guns were packed off on the Genoese fleet, and the Imperialist army marched away by the road along the Riviera, which was then narrow and almost impracticable, in order to keep off as far as possible from the French.

It was not pursued : for Francis had conceived a fascinating strategical plan. The month of October had arrived, but the Alpine passes were still clear, and he resolved to strike at Milan, while Bourbon and Pescara were threading their way back to Genoa on bad roads.[1] Suddenly turning his army towards the Argentière pass, he descended on Saluzzo by forced marches. The Viceroy of Naples, who was in command in Lombardy, had far too small a force to resist him, and hastily retreated from Asti toward Milan, sending word to Bourbon and Pescara to join him at the earliest possible moment. But the defile along the coast of the Riviera by Fréjus, Monaco, Nice, and Finale had proved a very fatiguing and a very slow business; in many stretches of the road only mules could pass—nothing on wheels. The infantry, as we are told, were so worn out that they flung their pikes and corselets into the ravines, in order to be able to stagger along in a state of semi-starvation.[2] Pescara with the Spaniards had only reached Pavia, and the landsknechts under Bourbon were far behind, when the vanguard of the French army presented itself before the gates of Milan (October 24).

The Viceroy Lannoy then evacuated the city, for a strange and non-military reason. A violent outbreak of the Plague had declared itself in Milan in that autumn—the population were dying by hundreds, and if the Spanish and Italian troops whom the Viceroy had in hand were to have entered the narrow streets of the place, they would have caught the infection,

[1] It is said by Du Bellay that Bonnivet suggested the plan, and that it was opposed by La Tremouille, Chabannes, and Lescun, the Marshal of Foix (ii. p. 348).

[2] Du Bellay, ii. p. 354. Guicciardini, xv. p. 243.

and have perished like the citizens.[1] Lannoy therefore fell back on Pavia, where he awaited the coming up of the troops from the raid on France, leaving only the great Castello of Milan garrisoned.

The magistrates of Milan surrendered the plague-stricken city to the Marquis of Saluzzo, who commanded the French advanced-guard. But the King did not bring his army into the place—from the same dread of pestilence which had driven off Lannoy. He left a blocking detachment under La Tremouille before the Castello, and took the bulk of his force southward, in search of the Imperialist main army—a very right and proper course. But Lannoy refused to fight—his army was in a wretched state, and the Venetians, who ought to have joined him, had not put in an appearance: so he drew off eastward, after throwing a strong garrison under Antonio de Leyva into Pavia. The King of France, arriving in front of that important fortress, found that the Imperialist field-army had disappeared in the direction of Lodi and Crema.

The turning-point of this long and complicated campaign of 1524 came when Francis halted in front of Pavia on October 28. Some of his councillors advised him to pursue Lannoy and drive him right away into Venetian territory—there was a chance of catching up his dilapidated army on the way, and crushing it. But the King resolved to capture Pavia as a preliminary precaution, and though it failed to fall at once, took up a very rash plan—the maddest of all the strategical errors for which he and his two predecessors had been responsible, namely, a repetition of the old scheme for the recovery of the kingdom of Naples. Certain considerations impelled him toward this venture—the exiled Angevin nobles of the Regno, who were still numerous in the French service, were eager for it: the Viceroy and Pescara were now in Lombardy with their army, and were known to have left very small garrisons behind them. But perhaps the most tempting inducement was that the new Pope, Clement VII (elected in September 1523), had sent an agent to the camp before Pavia, with secret information that he had resolved to break with the Emperor, and to quit the 'Italian league' which had been opposed to France since 1522. He could promise the help of Florence and perhaps that of Venice also. This was surprising behaviour on the part of Clement, who, while he was only a cardinal, had been reckoned

[1] See Guicciardini, xv. p. 241, for the state of Milan.

a strong Imperialist. On December 12 a secret treaty was signed, by which the Pope undertook to welcome a French army at Rome, and to connive at the levying of mercenary troops for the King's service in his dominions.

Francis decided to act at once, and to send a great expedition against Naples, though Pavia was holding out stiffly, and midwinter was a mad time to choose for a march of 300 miles down the Peninsula. Clement had not expected to be taken at his word immediately, and, when he heard of the King's resolve, sent messages begging that the expedition should not move till the spring, when military operations would be more practicable. But Francis was set on the matter—the glamour of the reconquest of Naples clouded his view of all obvious dangers. He waited till he had recruited his army—6000 Swiss from the Grisons came in, 2000 from the Valais, and he had hired the 'Black Bands' of Giovanni dei Medici, who had just quarrelled with the Viceroy Lannoy, and was ready to change sides again at short notice. When these reinforcements had come to hand, he detached from his army nearly a third of its strength, giving over to John Stuart, Duke of Albany (just back from a feckless regency in Scotland),[1] 600 lances, 10,000 foot, 300 light horse, and a train of twelve guns.[2] With this force Albany was to march via Reggio, Florence, and Siena on Rome—he picked up at Lucca 3000 infantry landed from the French fleet at Pisa, and levies were being made for him in Rome. Altogether this was a considerable force—enough to reduce the King's army to a dangerous weakness, but not too much for an invasion of a kingdom so broad as Naples. But as a matter of fact, Albany never got farther than the Tiber, being delayed by bad roads and papal politics—he was induced to turn a pro-Spanish government out of Siena—and had not reached the borders of Naples when he got the appalling news that his master was beaten and a captive. He was lucky to get back to France on shipboard, with such of his troops as did not disband themselves when the news of Pavia reached them. The one important thing was that he lay near Rome with some 15,000 men when he was wanted in Lombardy.

[1] A queer choice, for Albany had twice showed himself unable to manage an army in Scotland.

[2] So Du Bellay; Guicciardini says only 200 lances and 4000 foot (xv. p. 251), which seems a small force for the work.

But to proceed with the King's unhappy campaign. When he arrived before Pavia on October 28 he made a first attempt to storm its western front, after a short battering with his siege guns. But the assault failed—the walls were injured, but Antonio de Leyva made a scientific use of barricades and cross-fire on the back of the breach, which completely checked the stormers. After commencing a more leisurely attack with zigzag approaches and 'parallels,' Francis was distracted from it by an ingenious plan of his engineers. The Ticino at Pavia flows in two branches, the main one under the city walls, the lesser one—called the Gravellone — 500 yards to the south; they rejoin before the Ticino enters the broader stream of the Po. It was demonstrated to the King that he might throw a dam across the larger branch, below the abbey of San Salvatore just above the town, and so force all the water into the Gravellone, which would leave the whole south side of Pavia accessible and practically defenceless, for there was no solid wall along the river bank. This idea seemed ingenious, and an immense amount of work was spent upon it in November. But when the dam was nearly finished, heavy winter rains in the Alps swelled the Ticino to twice its normal strength, and on one unlucky December night the dam was completely swept away, and the river resumed its normal course. Francis then fell back on simple blockade, and invested the city on all sides, with detachments guarding his bridges over the Ticino on the south side of the river.

Long winter sieges are destructive both to the morale and the personnel of an army, and some of his advisers begged Francis to give over the matter till spring, and to put his army into comfortable quarters till better weather arrived. But he refused, partly from obstinate dislike to abandon an enterprise, partly because he was convinced that the enemy could be starved out—he was informed (and the tale was true) that there was much discontent among the garrison—all mercenaries of the usual type—both Spaniards and landsknechts. He therefore persisted in the siege, which dragged on from October 28 to February 24 through all the worst months of the year. He and his officers were comfortably lodged in suburban abbeys and country houses, but the rank and file, in their tents or rough huts, murmured and deserted, or went sick, in appalling numbers. Nevertheless, he sent off Albany's great detachment and kept the rest of his army in a ring round the city.

The garrison suffered comparatively little, being all well housed in Pavia, though somewhat stinted in their diet. But Antonio de Leyva had no small difficulty in keeping them together; their pay was in arrears, and artillery munitions were growing scanty. Francis was quite correct in believing that discontent was rife—though the rumour current in his camp that the governor had secretly put to death the senior German captain [1] was apparently without foundation.

Meanwhile Lannoy and Pescara had collected all such troops as could be gathered together in north Italy, at or behind Lodi and Cremona. They could get no help from the Venetians, who were in communication with Pope Clement, and were abandoning the old 'League' for good. But the Constable Bourbon went off to Germany, and there scraped together some reinforcements by the aid of a small subsidy sent him by the Emperor—500 Netherland horse ('Burgundians,' as the chroniclers call them) and 6000 landsknechts, under the old George Fründsberg. They came to hand in January, when the Viceroy Lannoy, having now a respectable force under his command, proposed to send a detachment southward to Naples, to parry the thrust which Albany was threatening to deliver. Pescara succeeded in dissuading him, pointing out that it was all-important to be strong in Lombardy, and that all previous French raids on Naples had failed—Albany's army would get dissipated in garrisons and siege-work, if it ever reached its goal.

Accordingly the Imperialist generals resolved to attempt to raise the siege of Pavia; Antonio de Leyva had well deserved their assistance by holding out so long. On January 25 they concentrated at Lodi, and marched in the direction of the French, but set on manœuvring rather than a pitched battle, for they were still much inferior to Francis in numbers—more especially in cavalry.[2] It is doubtful whether they had so many as 1000 men-at-arms, and their infantry of all nations was not much over 17,000 strong. Their idea was that they might lure the King into abandoning the siege and seeking them in the open—when they would retreat: or else that, if he persisted in pressing Pavia, they could threaten his communications with Milan and France, and force him to dislodge in that fashion.

[1] Guicciardini mentions this (xv. p. 267), besides French authors.

[2] The Spanish cavalry, we are told, had been ruined by the retreat along the Riviera in October, and was still in poor order, and half unhorsed.

Their first move was to Marignano, on the Lodi-Milan road, giving an appearance that they might be threatening Milan itself. But the King was not moved by this diversion. Wherefore they swerved south, and beset Castel St. Angelo, half-way between Lodi and Pavia, where there was a garrison of 1000 mercenaries. It fell on the first day of battering, far more easily than they had hoped. Having destroyed this outpost, they advanced cautiously toward Pavia, still hoping that the King would abandon the siege and come out to assail them in the open—in which case they would have retreated and drawn him after them. But Francis took the opposite course—he merely altered the location of his army, and placed the bulk of it in a defensive position which would cover the siege against any attack from the east.

The line which he took was that of the Vernacula brook, a tributary of the Ticino, which runs into it about a mile below Pavia. At the same time he sent for the troops which were blockading the Castello of Milan, and drew them all in to his camp, save 2000 Italian foot and 300 horse under the old condottiere Trivulzio. The Vernacula position was very strong—the brook, though not broad, was deep-sunk and very muddy. When the Imperialists continued to draw near, the King threw up a complete line of contravallation, garnished with artillery, along the near side of the water ; it was over two miles long and practically impregnable. His headquarters up till this moment had been on the west side of Pavia, at the Abbey of San Lanfranco ; he now moved them to San Paolo to the north-east of the city, and brought up the bulk of his army to face eastward, leaving only the ' Black Bands ' of Giovanni dei Medici and 2000 Swiss infantry to blockade the western front. The south end of his new position rested on the Ticino, the north end on the wall of the Park of Mirabello, a ducal hunting-palace three miles north of Pavia ; but he took down the south side of the Park wall in order to allow free movement of troops in this direction.[1] His ' rearward ' division, under the Duke of Alençon, had as its headquarters the buildings of Mirabello ; his ' battle,' in front of San Paolo, was encamped behind the north end of the entrenchments along the Vernacula ;

[1] This important fact is mentioned only by Guicciardini, xv. p. 273. If we did not know of it, we should be wondering how Francis was able to advance into the Park without hindrance during the battle. The east and north sides remained intact, till the Imperialists broke in on the 24th–25th.

the 'vaward' was holding the rest of the entrenchments, and watching the eastern side of the besieged city, in a rather constricted position, between the besieged place on one side and the fortified line of contravallation on the other, reaching down to the Ticino.

Finding that Francis absolutely refused to move away from Pavia, the Imperialist generals took the bold step of advancing to the opposite side of the Vernacula, and entrenching themselves there, in a line parallel to the French contravallation (February 3). The brook was as impassable from the west as from the east bank, and there was no danger from a direct attack across it. Presently the Imperialists brought guns into their line, and for three weeks there was intermittent artillery-fire across the Vernacula, with no conclusive result. So close did the two fortified lines lie to each other that in more than one place there was only 40 yards of mud and water between the trenches—a situation that seems to recall episodes of the Great War of 1914–18. This deadlock lasted for three weeks—each side was waiting for the other to show signs of failing—and oddly enough mainly signs of economic failing. For Francis was aware that the Imperialist army was on the verge of mutiny owing to bankruptcy—the landsknechts were grumbling and were commencing to slink home, when no pay had come to hand for months. While the Imperialist generals were well informed as to the results in the way of sickness and desertion which had followed the tarrying of the French army for four months of rain and cold under canvas and in trenches. These mercenary hordes of the sixteenth century, whatever their nationality, were most unreliable material.

The final decision came when the Imperialists learned of two incalculable disasters which thinned the French army down to danger-point. On February 17, Giovanni dei Medici received a disabling wound[1] while repelling a sortie from the town; when he was sent to the rear his 'Black Bands' disbanded themselves—or at least two-thirds of them disappeared during the next few days. But a much greater blow came from an unexpected quarter. A roving band of Sforza's Milanese troops in the Valtelline captured the castle and town of Chiavenna, through which lay the main route into Italy from the Grisons. The Council of the 'Three Leagues' at once sent

[1] The bones of his foot were shattered by a bullet, and he retired to hospital at Piacenza.

orders to their soldiers in the French camp to return and clear their own borders. In spite of all the King's appeals to their mercenary honour—they were actually not in arrears of pay!—the whole 6000 Grisons troops marched off to the Valtelline on February 20. This was a terrible blow—several of the older French captains besought their master to abandon the siege, for the army had lost 8000 men in three days, and the companies

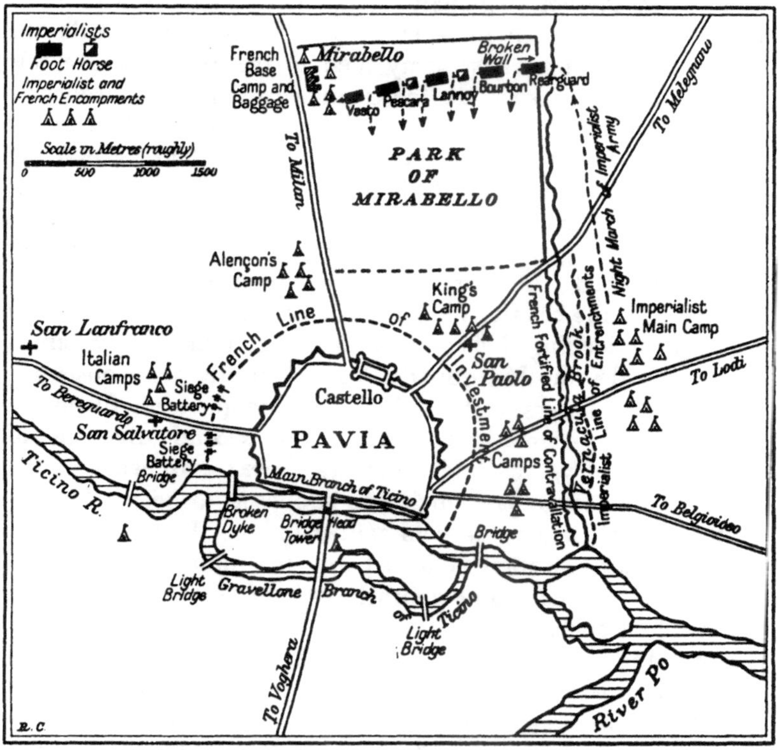

BATTLE OF PAVIA, FEBRUARY 24, 1525.

were so reduced in all the corps that he had with him now on more than 1300 men-at-arms, 4500 landsknechts, 5000 Swiss of various cantons, and perhaps 9000 miscellaneous French and Italian infantry. The whole had shrunk to about 20,000 of all arms—not much over the force which Lannoy and Bourbon had now in line against him; but the French cavalry was still some hundreds stronger than that of the Imperialists.

It was undoubtedly the change in the balance of numbers,

certified by deserters who were continually passing from camp to camp, which was the main reason that impelled Lannoy and Pescara to take a long-deferred offensive. But the chroniclers assure us that a strong contributory cause was the semi-mutinous attitude of their own troops, who (like the Swiss before Bicocca) were demanding pay long due, or battle as an alternative. On the night of the 23rd–24th of February the artillery in the entrenchments along the Vernacula was set to bombard the French lines by way of distraction, while the whole of the Imperialist army filed to the right, save a few companies left to keep up the demonstration in the trenches. The project was to turn the flank of the French lines, by passing the Vernacula two miles upstream, when it had become an insignificant obstacle, and to break in on the north-east corner of the Park of Mirabello, despite of its formidable wall, which the French King regarded as a secure protection, and had left hardly watched. The army marched in five divisions of infantry, with the cavalry on each side of the third division. In front of the leading division (composed of Italian and Spanish infantry under the Marquis del Vasto—mostly arquebusiers—with 200 light horse attached) were 2000 pioneers furnished with battering-rams and picks. The second division, of Spanish infantry, was under Pescara; next marched half the cavalry, then the mass which was supposed to be the 'main-battle' (here Lannoy as commander-in-chief had his post)—German landsknechts—followed by the other half of the cavalry, after which came another corps of Germans under the Constable Bourbon, with a rear column composed partly of Spanish and partly of Italian foot. The movement was carried out in darkness, on a night of rain and wind, and to keep themselves in touch and distinguish themselves from the enemy the troops were told to wear shirts over their armour—this was to be a 'camisade' in sixteenth-century parlance.

The leading column arrived absolutely unseen in front of the north-east corner of the Park walls, where the sappers attacked with their rams and picks three lengths of the solid brickwork—artillery would have worked quicker, but was not used for fear of rousing the attention of the enemy. Broad sections of the walls were cast down, after delays which filled the Imperialist commanders with great anxiety, for every moment before dawn was valuable. But at last the breaches were made, and the columns pressed through them. The surprise was complete, no hostile force was on guard, save a few Italian foot under a Genoese captain named Giustiniani, who had failed to keep any watch, and were swept away at once. The leading division—Del Vasto's light horse and Spanish and Italian infantry—made for the buildings of Mirabello, where they found hundreds of huts and pavilions full of servants, camp-followers, and sutlers, but no formed troops. They took an enormous booty, and captured the Pope's legate and many other civilians, but the Marquis kept them from dispersing for plunder (a hard task), and pushed southward, to come into line with the rest of the army, which was forming up in succession in the northern half of the Park. He left only his handful of light horse at Mirabello. The ground was partly open rides and lawns, such as ducal pleasaunces would naturally show, but was intersected with avenues and copses, shelters for the game. On the whole it was not unsuited for manœuvres on a limited scale, and there were many stretches where cavalry could act.

King Francis had been taken absolutely by surprise—before dawn his flank had been turned, and the enemy was deploying in a fairly formed line behind his headquarters. Naturally he sent orders in all directions for the various divisions of his scattered army to come up and get into array as quickly as possible. He himself with the main body of his gendarmerie, who were encamped around him, were ready long before the rest—there were expected on his right hand the 'Black Bands' of German landsknechts under the Duke of Suffolk and Francis of Lorraine, on his left the Swiss under Diesbach. The greater part of the native French infantry, then holding the southern section of the contravallation, towards the mouth of the Vernacula, were left there for a time, in case there should be an attack on this part of the line—as the artillery cannonade seemed to threaten. Bussy d'Amboise, who was in charge there, had orders to come up only if no trouble seemed to be brewing opposite him. The Italian infantry, both the wrecks of Giovanni dei Medici's people and the other mercenary companies, had been watching the west side of the walls of Pavia. They were distracted by a sally made by the governor, who had received orders (through an emissary who penetrated the French lines on the previous day) [1] to co-operate according

[1] Giovio, *Vita Piscariae*, p. 399, says that he disguised himself with French badges, and slipped through the line, saying that he was one of Medici's Italians, carrying a message to the front trenches.

to the best of his ability. Another fraction of the French army hardly got into action at all—this was the force which technically was called the 'rearward' of the general organization of corps, *i.e.* 300 lances and some companies of French and Italian foot which had lain south of the buildings of Mirabello, opposite the north side of Pavia—their leader, the Duke of Alençon, got them together very late, and never struck in against Del Vasto's division as he should have done—only, as it seems, watching the small body of light horse which had been attached to it. Another isolated force which was quite out of the game was the detachment under the Count of Clermont, which was guarding the works at the south end of the main bridge of Pavia and the minor bridges by which the French kept touch with the farther bank of the Ticino. Altogether, as the chronicler who relates the fight in the fullest details observes, 'nunquam dissipatius et confusius ullae copiae dimicaverunt.' [1]

The King, seeing the Imperialist army getting into line in front of him, while his own scattered troops were trailing up from their cantonments, resolved to strike at once with what he had in hand, while the rest were arriving—this was indeed necessary, or his whole army would have been overrun before it had got together. Accordingly he advanced with his cavalry and a small artillery reserve which had been encamped near his headquarters,[2] and found himself facing the 'rearward' and left centre of the Imperialists, close to the east wall of the Park. His guns opened upon the last column of the enemy, which had only just passed through the broken walls, and did it much harm—it is said that this Spanish-Italian division gave way entirely when charged by two squadrons of gendarmerie and fled back as far as the Vernacula, leaving behind it several pieces of artillery which it was guarding.[3] The King himself then charged—across the front of his own guns, and masking their fire—at the hindmost of the two Imperialist cavalry

[1] Giovio, *ibid.* p. 400.

[2] The guns under Galliot, the 'Grand Master of the Artillery,' cannot have been taken out of the 'lines of contravallation'—there would have been no time to get them out of their emplacements—so must have been, I imagine, an artillery reserve. We are told by Du Bellay that Galliot got them up and placed them 'advantageously' (ii. p. 390).

[3] I am sure that Giovio is right in saying that these victorious squadrons were those of Federigo de Bozzolo and the Sieur De Biron, and that Du Bellay errs in giving the credit to the Duke of Alençon, who was in another part of the field.

brigades in the centre of the hostile line. It was composed of 'Burgundian' and Austrian[1] horse, under Ferrante Castriot, Marquis of St. Angelo, the last descendant of the great Scanderbeg. He was killed—not by the King's own hand as some French chroniclers assert—and his corps scattered. This made a gap in the left centre of the Imperialist army, and Francis thought that the day had started well. He turned sideways to attack the neighbouring infantry columns, could not break in, but put them in some disorder; his own gendarmerie also began to fall into disarray after repeated charges.

Meanwhile the remainder of the Imperialist army stood fast, and was in good condition to receive the belated attack of the French infantry, who only came up when the King's stroke had lost its first effect. There were two isolated attacks delivered, the first by the Swiss contingent, which got engaged with Del Vasto and the Imperialist right wing, and afterwards drew upon itself the attention of Pescara's division. By all accounts the reputation of the Swiss suffered much from their conduct on this day. Harrassed by the volleys of the arquebusiers, who formed the greater part of Del Vasto's corps, they came on slowly, hardly got to 'push of pike,' and finally went off to their left and retreated *en masse* on the Milan road, with their commander Diesbach at their head. None of the chroniclers either on the French or the Imperialist side have a good word to say for their conduct; indeed the French writers attribute the loss of the day entirely to their fault—which is probably unjust, for other causes were operating.[2]

The 'Black Bands' of the French landsknechts made a very different advance—they plunged headlong, when they got on to the field, into the column of their countrymen which Lannoy commanded, and waged a desperate conflict with it, till they were taken in flank by the other corps of Imperialist landsknechts under Fründsberg, and were exterminated, fighting hard and never recoiling. Both their commanders,

[1] Giovio, with his usual perverse classical phraseology, calls them 'Noric' horse, lent by the Archduke Ferdinand—Noricum being the ancient name of archducal Austria in Roman days.

[2] On which side of the King's cavalry did the Swiss operate? If it was on the *right*, as Du Bellay says (ii. p. 391), how did they manage to escape on the Milan road? For to reach it they would have had to go right through the Imperialist line. I am constrained to believe that they must have been on the left, a position from which such a retreat would have been possible.

the exiled Duke of Suffolk, the 'White Rose of York,' and Francis of Lorraine were killed, and nearly all of their captains. This was the end of the old German companies, originally raised by the dukes of Gueldres and Bouillon, of which we have heard in so many Italian battles. Florange, who had commanded them in many earlier campaigns, but was serving this day with the cavalry, was taken prisoner. 'If only our Swiss had behaved like our landsknechts, what a different battle it would have been,' exclaims Du Bellay.[1]

The King meanwhile was charging right and left with his gendarmes in the centre of the Imperialist line—losing men every moment by the fire of the enemy's arquebusiers, who, sheltered in the copses and hedges of the Park, could not easily be got at with the lance. The final disaster came when Bussy d'Amboise, and the French infantry, who had been holding the lines of contravallation, came up, much too late, and were overwhelmed by the same columns of landsknechts which had exterminated the 'Black Bands.' The King and the wrecks of his chivalry were left alone on the field and were surrounded and crushed. Francis himself fought fiercely, even after his horse had been killed, was thrown down slightly wounded, and would have been killed by a Spanish foot-soldier if M. de Pomperant, one of the Constable Bourbon's exiled partisans, had not arrived in time to save his life. He was taken before the Viceroy Lannoy, who received his sword, treated him with all courtesy, and sent him to be imprisoned in the castle of Pizzighettone.

This was the end of the battle. Alençon, whose conduct throughout the day is inexplicable, had never joined in the main action, and after some feeble skirmishing with Del Vasto's light horse, had remained out on the left, without any attempt to close in. He may have been distracted by the sally which De Leyva and the garrison made against the western end of the line of investment, but must have known, after a time, that this was a comparatively unimportant part of the day's operations. He finally went off to his left, picking up on the way the remains of the Italian troops who had been holding the section of the lines about San Salvatore and San Lanfranco. Another corps which escaped was that with which the Count of Clermont was holding the bridge-heads across the Ticino. When the day was obviously lost, he collected his detachments,

[1] Du Bellay, ii. p. 392.

broke the bridges, and got away to Mortara. There he was joined by Trivulzio, who with 2000 men had been holding the trenches before the blockaded Castello of Milan: when the fugitive Swiss divulged the fate of the day, he packed up without delay, and went off westward in haste.

But only fractions of Francis's once formidable army saved themselves. The actual loss in action is said to have been 8000 men—some of them drowned in the Ticino while trying to swim it after the bridges had been broken. These, I suppose, must have been fugitives from Bussy d'Amboise's infantry, and some of the wrecks of the gendarmerie. The Imperialists did not refuse quarter—or the slaughter would have been still greater. On the day after the battle the Constable Bourbon paroled all the prisoners who were obviously incapable of paying a ransom—they had (of course) been well stripped by the soldiery before they came before the Constable. Montluc, then a simple man-at-arms, was one who profited by this complaisance on the part of the victor 'n'ayant pas grande finance.'[1] One asks oneself whether this was a chivalrous impulse, or whether the Constable (like Bernadotte dealing with French prisoners in 1813) was hopeful of making himself friends in view of future political ambitions.

The slaughter among the French noblesse at Pavia was only paralleled by that which took place at Agincourt 110 years before. A very large number of the commanders whose names are familiar to us from earlier campaigns, perished around their king. The oldest of them was La Tremouille, who had reached his sixty-sixth year, and had fought his first fight against the Bretons at St. Aubin de Cormier in 1488. Of the unlucky leaders of troops in the pastthere died the Admiral Guillaume de Bonnivet, Jacques de Chabannes (the Marshal de La Palice), Lescun, brother of Lautrec (the Marshal de Foix), the two great Angevin nobles from Naples, Galeazzo de San Severino, Master of the Horse to King Francis, and his brother, the lord chamberlain, Richard de la Pole, the exiled Duke of Suffolk, Francis of Lorraine, Réné Bastard of Savoy, Master of the Cross-bows, the Count of Tonnerre, Bussy d'Amboise who had commanded the French infantry on the right, Chaumont, the son of the celebrated Charles d'Amboise, Louis d'Ars, and the Milanese Galeazzo Visconti. The lists of lesser captains would fill a page. Among the numerous prisoners were Henry

[1] Montluc, i. p. 53.

d'Albret, the dispossessed King of Navarre, the Count of St. Pol, Florange, the son of the Duke of Bouillon, the Marshal Montmorency, the Prince of Talmont, Louis Count of Nevers, Francis of Bourbon, nephew to the renegade Constable, and the Italian condottiere Federigo de Bozzolo. Hardly a man of mark who had been present at the battle escaped death or capture—the only prominent names are the Duke of Alençon, and the Scottish veteran Robert Stuart of Aubigny, a scion of the house of Lennox, who spent his whole life in the French service.[1] Alençon died a few months later, overwhelmed—it is said—by the universal contempt and insult with which he was received in France on his return.

The Imperialists reported their loss at only 700 men, largely in the cavalry corps which King Francis had ridden down at the commencement of the action. Ferrante of St. Angelo, its commander, was the only person of importance killed; but Pescara had received two wounds, and Antonio de Leyva, while leading the sortie of the garrison, got an arquebus ball in his leg. Both these distinguished officers were on foot again within a few days.

From the point of view of tactics Pavia was a most interesting battle. Both armies being on the move, the clash was irregular, as each division got to the front, and found itself in action with hostile troops coming up in successive corps. The Imperialists were not quite in line when Francis charged in with his cavalry: the French never got into a well-ordered array, and fought as each corps straggled up. Indeed on both sides there were detachments which were not seriously engaged—not only did Alençon's 'rearward' never close in, but we are told of troops on the Imperialist side which did not move up, owing to misconceived direction, and the reluctance of a divisional commander to take responsibility and to strike in without orders. This is said to have been the case with some of the Spanish horse.[2]

The main tactical deduction made by contemporary observers was that the importance of firearms was still on the increase. Not that the artillery counted for much. Francis masked his own guns by his original cavalry charge, and the

[1] Guicciardini and others wrongly place him in the casualty list as 'd'Obigni.' He survived to go through several more campaigns, and died in 1536. In the left-hand lower corner of the great picture of Pavia in the Ashmolean, he and Alençon are shown riding off together in safety. He is labelled 'M. d'Aubigni, Escossois.'

[2] See Giovio's life of Pescara, p. 399.

few pieces which the Imperialist rearguard was convoying were overrun and taken by the French. But the arquebus fire is said to have been absolutely decisive, not only in delaying the advance of the Swiss and riddling the flanks of the French landsknecht column, but most of all in thinning and disabling the squadrons of the gendarmerie, when they split up after their first charge. Sheltered in copses or behind hedges, the arquebusiers kept up a rolling fire, which brought down horse and man in detail, yet they could not be got at by the lances of their opponents, and no corresponding force of French light-infantry seems to have been sent against them. Indeed the French had few arquebusiers,[1] and cross-bowmen were ineffective.

This was the last regular pitched battle that was fought in Italy for many a year, and we may add that it was the last general action since Marignano in which both sides were ready to take the offensive. The Imperialists turned the left flank of King Francis's prepared position with the object of forcing him to come out of his lines, under penalty of seeing himself cut off from his all-important base at Milan. This move compelled the King to attack them, as they defiled past his front: he gladly accepted the arbitrament of battle, having been long desirous of getting at an enemy who had hitherto covered himself with trenches, behind that formidable obstacle the deep-sunk bed of the Vernacula. Accordingly there supervened a 'battle of manœuvre,' and the side which fought in the better-formed line, and co-ordinated the use of the various arms, naturally had the advantage. The strength of the French cavalry was spent before their infantry came to the front. In a way Pavia may also be called a 'victory by surprise,' since the original advantage gained by the Imperialists through their having formed their line in the dark, across the flank of the French, was never lost. If Francis had not been taken by surprise, he could have utilized all his forces, and have attacked in a regular formation. As it was, his corps fought piecemeal, and one important detachment—Alençon's wing—was practically out of the battle altogether, being cut off from the main body, and lacking orders as well as initiative.

Pavia might have ended the war if the Emperor Charles had been contented to impose less onerous terms of peace upon his captured adversary. It was the project for dismembering

[1] See p. 48.

France, imposed by the treaty of 1526, which goaded Francis into breaking his oath and renewing the war in desperation.

NOTE ON THE MAP OF THE BATTLE OF PAVIA

There are two contemporary pictures of the battle, one in a German pamphlet published at Nuremberg shortly after the event, which gives a good representation of Pavia, the island opposite it and the four French bridges. It is useless for the troops, except that it shows the point where the Imperialists broke through the Park wall to have been towards its north-eastern corner. They are represented in two droves—the infantry passing through the gap, the cavalry waiting behind to follow. The French army is not shown, save for an ill-seen clump of spears at the south end of the Park. The other representation is a very large panoramic picture in the University Galleries at Oxford, which purports to give the field at the moment of the general rout of the French and the capture of the King. Little labels are placed over the head of each corps, and each important officer, giving their names in French, 'Les Suisses du Roi,' 'Les Italiens,' 'Les Lanquesnets,' 'Le Viceroy de Naples,' etc. The French line is broken, a heap of dead knights lie in front of the advancing Imperialists, 'M. L'Admiral,' 'Le Duc de Sufforck dit Blanc Rose,' 'Le Vicomte Galeas,' 'Le Bastard de Savoie,' etc., as the labels on them show. The Count of Clermont's men look on helplessly from the other side of the Ticino: Alençon and 'M. d'Aubigny, Escossois,' are riding off in the lower left-hand corner. The perspective is almost non-existent, and the figures are of a mediaeval rather than of a Renaissance execution. One might guess that it was drawn, from a description, in the French-speaking Netherlands (to account for the labels all in French) rather than in Italy. The author has no good conception of the true topography, and has got the hunting-lodge of Mirabello wrongly oriented.

These two contemporary productions are of some use, but the battle has mainly to be constructed from the very numerous written sources, French, Italian, Spanish, and German, which are often contradictory in their narrative. It is impossible to guess the exact extent of the Park of Mirabello in 1525. The little German picture makes it narrow, and covering little

THE BATTLE OF PAVIA, FEB. 21, 1525

From the Painting in the Ashmolean Museum, Oxford

more than the front of the Castello: an Italian map of 1650, on the other hand, makes it stretch all along the north front of the city, from the Milan road to the Melegnano road. I have adopted this version in my map. Neither the German picture nor that at Oxford show the course of the Vernacula brook, though the former gives the French and Imperialist trenches facing each other very clearly, without indicating that there was water between them.

BOOK III

THE LATER YEARS OF THE GREAT WARS

(1527–59)

CHAPTER I

CHANGES IN TACTICS, FORTIFICATION, AND ARMAMENT

IT might have seemed reasonable, as we have already observed, that the battle of Pavia should have brought the Great Wars of Italy to an end; the attempt of the French kings to establish a domination in the Peninsula having finally failed, and the third monarch who had continued the long endeavour being left at the mercy of the Emperor. His last army was destroyed, his person consigned to a Spanish prison: the terms of peace were obviously to be dictated by the enemy. Charles V made them too hard: it was quite reasonable that Francis should resign all his claims on Naples, Milan, and Genoa. Nor was there anything contrary to equity in the demand that the old French suzerainty over Flanders and Artois—which was a relic of the antiquated division of the empire of Charlemagne among the three sons of Louis the Pious, seven hundred years back—should be consigned to oblivion. Flanders had for long centuries been part of the block of states which formed the Netherland group, though by this odd survival its count was reckoned a vassal of France, while all his neighbours in Brabant, Namur, Hainault, Luxemburg, Holland, etc., were vassals of the Holy Roman Empire. To make legal fiction square with political fact was entirely justifiable. Nor was there anything unreasonable in the compulsion put upon Francis to abandon his clients—the King of Navarre and the dukes of Guelders and Bouillon—all of whose dominions were at this moment actually in the hands of Imperialist armies. To restore the Constable Bourbon to all his territorial possessions was, no doubt, a bitter pill; and he would be for the future a declared enemy set down in the middle of the French realm, a triumphant traitor. But there was one clause in the Treaty of Madrid which no French King, and indeed no Frenchman of any rank, could be expected to accept for a moment longer than he was compelled to do so by actual force.

This was the mutilation of the realm by the cession to Charles of the ancient heritage of Burgundy, not only the duchy, but all its dependencies, the Nivernois, the county of Charolais, Château Chinon, Auxonne, Auxerre, and other smaller patches. This was an enormous block of territory, cutting into the middle of the Seine valley, and reaching to within eighty miles of Paris at one point. Francis signed and swore to the treaty, and as a pledge of his obligation undertook to marry the Emperor's widowed sister Eleanor. But oaths made under duress are flimsy bonds, even to men of a conscientious turn of mind, and Francis, with all his external parade of chivalry, was anything rather than conscientious. It would have been too much to expect him to keep the terms of the Treaty of Madrid for a moment longer than was necessary, when once he had got himself free, even though he left his hostage-sons in the Emperor's hands. As soon as Charles was involved in new troubles—the disastrous conquest of Hungary by Sultan Soliman, which brought the Turks up against the borders of the Austrian hereditary states (1526) and the 'Holy League' of the Pope and the minor Italian states against the Spanish domination—the King of France repudiated his oaths (he got a dispensation from Clement VII to salve his conscience) and once more declared war on the old enemy.

The war of 1527–29 was the first of four separate attempts of Francis and his son Henry to realize the old ambition of mastering Italy, a futile and disastrous infatuation. As has been remarked already, each of these wars was a deliberate attempt to fall upon the Emperor when he should be distracted by external dangers. If the resumption of hostilities in 1527 had been intended merely to undo the monstrous annexation of Burgundy at the Treaty of Madrid, there would have been considerable justification for it. But Francis plunged at once into the wild old scheme of conquering the kingdom of Naples, and sent out one more army to perish in the south.

Since we are about to deal with the four wars of 1527 to 1559 under one heading, it may be best to give a synopsis of them, before descending into detail.

I. *The War of 1527–29*.

Francis, repudiating the Treaty of Madrid, renewed the attempt to reconquer Naples. The expedition under Lautrec ended in complete disaster (1527), and in the following year the

French army of Lombardy, under St. Pol, was destroyed at the rout of Landriano (1528). Francis then opened negotiations, and Charles granted peace, because the Turks were besieging Vienna, and a Protestant rebellion was threatening in Germany. Hence the Treaty of Cambray (1529) by which the Emperor gave up his claims on Burgundy; but Francis once more surrendered his old ambitions in Naples, Milan, and Genoa, ratified again the cession of the French suzerainty over Flanders, and paid a considerable war-indemnity. He was lucky to get off so cheaply from the struggle. The peace was a reasonable one.

II. *The War of 1536–38.*

The Emperor being very busy with the Turks on the Danube (1533–34), and afterwards with his expedition to Tunis (1535), Francis declared war very suddenly,[1] overran Savoy and Piedmont before the enemy could mobilize against him, but failed to reach Milan (1536). The Emperor, when he had collected his armies, retaliated by an invasion of Provence and an attack on Picardy, both of which failed, Harassed by his still-continuing Turkish War the Emperor began negotiations which ended in the Truce of Nice (1538), not a peace but a suspension of arms on the *uti possidetis* principle, which left Francis in possession of Savoy and part of Piedmont, in occupation but not in ownership. Thus the terms were somewhat better for Francis than those of 1529.

III. *The War of 1543–44.*

Charles having utilized the time of truce for a formidable but disastrous invasion of Algeria (1541) was left in such a depressed condition, that Francis, forming a close alliance with Sultan Soliman, thought the opportunity favourable for a resumption of hostilities. The Emperor's attempt to drive the French out of Piedmont failed at the battle of Ceresole (April 14, 1544), but his invasion of northern France, combined with an attack of Henry VIII of England on Boulogne,

[1] The *casus belli* was wholly inadequate—execution of a French secret agent (Maraviglia) for murder, by the Duke of Milan's judges. Francis declares that he was an ambassador and immune from justice, though a Milanese born. The Emperor supported the Duke's action. The invasion of Savoy was much like the German invasion of Belgium in 1914—a demand for armed passage, and when this was refused sudden attack with no legitimate justification.

forced Francis to sue for peace. The Emperor, who had trouble behind him, an exhausted army in Champagne, and no real help from the English, granted it: hence the Peace of Crépy (September 18, 1544), made very much on the same lines as the Truce of Nice. Francis kept his hold on Savoy and part of Piedmont, but by a separate and later agreement ceded Boulogne to the English King, who had besieged and captured it, while giving no help at all to his Imperial ally.

IV. *The War of 1551–59.*

Francis I had died (like Henry VIII) in 1447, but his son, Henry II, made himself heir to his father's policy. The Emperor being engaged in dangerous wars with the German Protestants, and with the Sultan of Turkey, the opportunity for attacking him once more seemed very favourable. In his first rush Henry II captured Metz, Toul, and Verdun (April 1552). The Emperor, having patched up a peace with the Protestants at Passau (August 1552), was able to turn his attention to the French frontier, but after a long siege (October–December 1552) failed to recapture Metz. Continuous fighting followed on the side of Flanders, but came to an end without any very important engagement or definite results. On the side of Italy the French held on to Piedmont, but failed in their two last attempts to attack southern Italy—the invasion of Tuscany (1553–55) ended with the fall of Siena, long defended by Montluc: the still more hopeless invasion of Naples ('cymitière des Français,' says Tavannes), the last of these vain attempts to get possession of the old Angevin inheritance, failed in 1557, the Duke of Guise bringing back only the wreck of an army. King Henry's disasters continued into the loss of the battles of St. Quentin (August 10, 1557) and Gravelines (July 13, 1558). The only counter-advantage won by his arms was Guise's surprise of Calais (January 6, 1558), left unguarded by the maladministration of Mary Tudor. It was not surprising, then, that King Henry signed in April 1559 the Peace of Cateau Cambrésis, by which he gave up all that he was holding in Italy save the Alpine Marquisate of Saluzzo, and all the scattered places still in his power on the borders of the Empire, save the old conquests of 1552—Metz, Toul, and Verdun. Calais was also retained, as Elizabeth wisely 'cut her sister's losses'—though she continued to haggle about this old English *tête-de-pont* in France in many later negotiations. There had been a promise of monetary compensation at the Peace of Cateau Cambrésis, which could always be raised when convenient—and was (naturally) always evaded by the French.

Henry II died by an accident at a tournament, only three months after the treaty (July 10, 1559), leaving behind him a distressed and shifty Queen Dowager, with four young boys under her charge, much hampered by over-great subjects—the Duke of Guise and his kin. The wars of Italy had at last come to an end—to be followed almost immediately by the 'Wars of Religion' in France itself. The battle of Dreux (December 19, 1562)—the first of the series of general actions between the Catholic and the Huguenot factions—came only three years after the death of Henry II. Of them and their tactical interest we shall have much to say hereafter.

It will be noted that between 1521 and 1559 there were eighteen years of peace interpolated among the years of fighting, and that at each rupture it was France that took the initiative, invariably because the Emperor was known to be hampered at the moment by dangers which had nothing to do with the Italian wars, but were caused by troubles with the Turk or the Germans. The first war (1521–25) synchronized with the outbreak of a revolt against Charles in Navarre, and a rising of the Dukes of Guelders and Bouillon against their suzerain. The fact that lay below the outbreak of the second war (1527) was the sudden conquest of Hungary by the Turks (1526) which compelled the Emperor to turn all his attention to the Danube, and the defence of the old Hapsburg hereditary dominions. At the third declaration of hostilities (1536) Francis took advantage of Charles having turned off to a serious African war, and having transferred to Tunis the flower of his army of Italy. The fourth war, similarly, was a repercussion of the dreadful disaster suffered by the Emperor before Algiers, and was accompanied by a close alliance with Sultan Soliman. The last and longest struggle (1551–59) was deliberately begun when the ageing Charles was in the midst of his German troubles, which engaged his main attention all through the last years of his reign: and Henry II was in secret communication with the treacherous Maurice of Saxony and with the Turk before he moved.

Even when we have allowed for the distractions caused to Charles by the ever-pressing Turkish danger in the rear, and for the more intermittent trouble caused by the German Protestants, it still remains difficult to discover the causes which prevented him from making an end of the French, during the periods when he had a comparatively free hand, and was able to turn his main attention to the indomitable enemy in the west, whose stabs in the back had been delivered with such accurate attention to his difficulties in other quarters. Not once but on three or four separate occasions it looked as if the time for a decisive blow had arrived. The first was in 1536, when he had crossed the Alps and overrun Provence. The French King, devastating the land behind him, had retreated to Avignon, where he stood at bay and waited to be attacked. Charles declined to accept battle, though he had advanced as far as Aix, only forty miles from Avignon, and had scoured the countryside up to his enemy's outposts. He hung back and finally returned to Italy, with an army somewhat depleted by privations, much as the Constable Bourbon had done some twelve years back. The second opportunity for a final settlement by a pitched battle was in October 1543, during the operations on the frontiers of Picardy, when the Emperor was besieging Landrécies. Two great armies faced each other for several weeks—with the tame ending that both marched back and went into winter quarters, though on each side the more enterprising officers held that 'il estoit plus honorable d'aller chercher l'ennemy que de tourner autour du pot.' [1] The third occasion when decisive action was evaded was in the autumn of the next year, 1544, when the Emperor had forced his way deep into Champagne, and lay at Epernay and Château Thierry, with his scouting parties pushing so near to Paris as Soissons and Meaux. There was absolute panic in the French capital, and many citizens were moving away to Normandy or Orleans. But the Emperor refused to make the final push, and instead of a battle we come to the indecisive Peace of Crépy. Incidentally, it may be noted, Charles thereby betrayed his ally, Henry of England, who was left involved in a dangerous war without the aid that had been promised him. The English King was lucky to get out of the struggle with Boulogne to his credit—which the French had vainly endeavoured to recover just after the Peace of Crépy.

The fourth disappointing deadlock, followed by no decisive

[1] Du Bellay, x. p. 71.

battle, befell in 1554, when the Emperor, having patched up matters with the German Protestants at the Diet of Passau, endeavoured to repair his check before Metz in the preceding winter, by one more large-scale attack on the northern frontier of France. After taking Thérouanne and Hesdin, he was brought up in front of the main army of Henry II, and after the cavalry fight before Renty—where the French had somewhat the advantage—refused to commit himself to a general action, and allowed the enemy to make off hardly molested.

This series of inconclusive ends to campaigns which seemed to require a definite decision would have shocked Napoleon, or any other master of battle-science. There was obviously a lack of initiative and offensive spirit at the root of the matter. But various characteristics of the warfare of the time may suggest partial explanation.

In the later years of the war, when one combatant had taken up a very strong position—perhaps partly entrenched—it had become the habit of the other to avoid frontal attacks, and to try to manœuvre the enemy out of his chosen ground. He might distract his adversary by false attacks, while aiming a real one at some unguarded point. Or he might try to cut his communication by cavalry raids, or he might simply wait for the effect of starvation, on an army which had eaten up the local supplies. Or he might plan to fall upon the enemy when he was shifting his ground, for movement when two armies are face to face is very dangerous to the party which first goes off from its fixed position, turning a battle-front into columns of route. The commencement of a retreat gives opportunities to a watchful foe. Very often none of those chances came off, and cautious opponents simply blocked each other.

It is interesting, therefore, to find on occasion a problem in the way of retreats that very much reminds us of certain episodes of the Great War of our own day. When two armies have entrenched opposite each other, with a very moderate 'No Man's Land' between them, how is the one which finds itself compelled (for one reason or another) to move off, to secure a retreat, when the opponent is so near that he can rush in, the moment that he knows that the trenches opposite him are growing undermanned? A rearguard left on the old position may be annihilated at once, and the main body which has evacuated the position may be caught in marching order not in line of battle, and overwhelmed in detail—like Lee's army of 1865 retiring from the Petersburg Lines. One remembers two very successful retreats of the kind in the late War, both the result of most elaborate and thoughtful planning—the withdrawal of the British army from Gallipoli in the winter of 1915, and the large-scale retirement of the Germans from the projecting bulge of their front in Picardy in the spring of 1917.

We are therefore somewhat interested when we find a precisely similar problem badly solved in 1553, far down in the centre of Italy. An Imperialist army under the Marquis of Marignano had been forced to raise the siege of Siena, but, retiring a few miles, had taken up a good position in front of the village of Marciano, with three 'strong points' and artillery properly placed. Piero Strozzi with the Franco-Sienese army, pressing hotly, pushed in so closely to the Imperialist position that he was only brought up on coming to a long hollow road a hundred and fifty yards in front of the enemy's trenches. He saw that Marignano was waiting for him, halted, and hastily 'dug himself in' on the hollow road. Their numbers were about equal, perhaps 12,000 a side, and both weak in cavalry. Montluc's narrative explains the situation :

'Now M. de Strozzi had pressed forward so far hoping to attack the Marquis, if he could only get him out of his entrenchments. But he found this impossible, and there they stayed seven or eight days, each waiting for the other to dislodge. The Marquis knew that if he moved first Strozzi would attack, and he had received orders [from the Emperor's G.H.Q.] to risk nothing. Now between the two armies there was only a single field 150 yards broad, across which there was "sniping" by the infantry all day, and our men were having the worse casualties, because the enemy used to fire on us from guns in his strong points. Indeed he had found out our watering-places, which we could only use at night, and bombarded them in the dark with some effect. M. de Strozzi at last determined that he must get out of his position, but knew that, the moment he had begun to do so, the enemy would charge in, and he hesitated about giving the Marquis the advantage. I sent him urgent prayers not to retreat by day, but to go off with all precautions in the night. And he did send off his heavy guns at 1 a.m., but thought that his columns of horse and foot would get mixed and disordered, if he went off across the fields in the dark. Accordingly he retreated at noon—with the most dreadful results—the enemy sallied out, when he saw the troops moving off, and caught them on the march. At five o'clock M. de Combas, colonel of the French infantry, rode into Siena, to tell me that the battle was lost, and Strozzi mortally wounded. At dawn the wrecks of the infantry began to straggle in—the cavalry had gone off into the open country, carrying the wounded general with them. We had lost many officers and 4000 or 5000 men. The cavalry had got mixed up with the baggage train, the infantry had been cut up on the roads. Strozzi was wounded early, and there was no attempt to do the only possible thing—to sacrifice a rearguard—300 or 400 arquebusiers, perhaps, and the cavalry—in order to let the main body get off (August 2, 1553).'

'Wherefore let me reiterate my conclusion. Retreats in daylight, under the eyes of the enemy, are so dangerous that they must always be avoided if possible. If it is not possible, it would be better to hazard everything, and commit oneself to a general action.' [1]

Machiavelli, writing long years before the combat of Marciano, had recommended that a general should try to retain his liberty of movement, by keeping far enough off from the enemy to allow himself time to move at leisure. It was the want of such elbow-room which had ruined the Spaniards at Ravenna. But this, of course, was by no means always possible. We do detect, however, many cases, like those quoted above, where the adversaries were so resolved not to give the enemy a chance, or to be forced into a battle at a disadvantage, that they behaved with what looks like excessive caution. Generals who were incautious sometimes came to sad grief—like St. Pol at Landriano, Montmorency at St. Quentin, and Les Thermes at Gravelines—all forced to fight when they did not intend to do so. Of this more shall be said later. Charles V was never caught in this way.

But in accounting for battles that were never fought, there are other causes which may serve as partial explanations. The first was the thoroughly undisciplined *morale* of the mercenary troops who served in the sixteenth-century armies. The landsknechts were as bad as the Swiss on the other side.. They melted away or mutinied when pay was not forthcoming, and the monetary resources of Charles were much more limited than might have been expected from the vast extent of his

[1] Montluc, iii. pp. 235–45. I have, of course, condensed the verbiage somewhat.

dominions. The treasures of Mexico and Peru, which were so invaluable to his son in the next generation, were only beginning to drift across the Atlantic. Spain and Germany, even the Low Countries, gave subsidies grudgingly: Naples and Milan were almost drained dry by the end of his long reign. It seems surprising, but is actually true, that Charles was often brought to a halt by mere lack of money. Without it large mercenary armies, uninspired either by loyalty or by patriotism, could not be kept together.

The second consideration which has to be borne in mind was the difficulty of supplies in an offensive campaign. An army could only live for a short time on the plunder of an invaded district, more especially when, as in the Provence campaign of 1536, the enemy had harried his own country before the advancing invader, had driven off the live stock, and burned the standing crops and even the mills. This old system, as we have seen in an earlier volume, had served the Scots well, in the time of their wars with the three first Edwards. Even in a richly-cultivated region an invader could not live for more than a few days on mere plunder—all the more so because mercenary bands were most wasteful pillagers, and spoiled in one week resources that should have lasted for three. When an army was in friendly territory magazines could be formed not too far to the rear, and convoys brought up, even over the bad roads of the sixteenth century. But to get food to the front when an invasion had been pushed deep into hostile regions, and the countryside had been wasted, was a very hard matter. Convoys from a distant base would take an unreasonably long time to arrive, and might be intercepted by angry peasantry, or roving parties of hostile light cavalry. At the best, they would have to be guarded by very large escorts.

The primary difficulty for an invader who had pushed far forward was the question of keeping the roads behind him safe. If he left fortified places behind him, he must block them by detachments, whose loss would enfeeble his striking force. If he preferred to destroy them before going on, the siege of even a petty town or castle might cause delays which hindered the advance for an unconscionable time. Some fortresses were almost inaccessible from natural strength, others had been brought up to the new level of military architecture, and defied inferior artillery. But many were so weak or so antiquated as to tempt a general to turn against them, in the hope that they would fall within a few days: this hope was often deceived—and the course of a campaign might be unreasonably protracted. We are sometimes reminded of Wellington's unexpected and unintended delay before Burgos in the autumn of 1812. Good examples of this sort of hindrance may be seen in Charles's sieges of Fossano (June 1536), which preceded his unlucky invasion of Provence, and of St. Dizier (which place held out for forty days), at the commencement of his advance into Champagne in 1544. If a general, set on the solid conquest of a region, allowed himself to be distracted into besieging every fortified place within it, the matter might be interminable.

Montluc, in one of his divagations into general principles, which lie scattered about among his tales of personal adventure, has a remark to this effect, which is worth pondering upon. He is speaking of Charles's position in Champagne in 1544, when Paris seemed in imminent danger. 'France if united cannot be conquered even by a dozen battles, considering the patriotism of its noblesse and the number of its fortresses. I hold those to be in error who said that if Paris fell, France was lost. No doubt Paris is the treasury of the whole realm, and its sack would be a matter of inestimable loss. But there are so many other cities and fortresses in this realm, that it would take thirty armies to capture and garrison them all. To hold down what he had won, the conqueror would have to unpeople his whole kingdom, which is obviously impossible. While the invader was taking one place, he would be losing another, for want of garrisons which he could not provide.' [1] Montluc was no doubt thinking of the history of his country in the times of Joan of Arc, which rather bears out his thesis, for undoubtedly the English attempt to subdue France failed, in the long run, from the impossibility of garrisoning the whole country, even when there was a French faction allied to the invader. Though Henry V and Bedford won battles and captured cities, there was never any chance of ultimate success, because they could not provide both a field army and a garrison for every city. Any place left ungarrisoned, or inadequately garrisoned, might revolt at the first chance, national patriotism being one of the *imponderabilia* which cannot be dealt with in strategical theses.

No doubt Charles V possessed military resources far greater than those of which Henry V of England and his brother the regent Bedford could dispose, but even with these it would be a

[1] Montluc, ii. p. 302.

hard business to garrison every conquered city and region, and the field army would have dwindled sadly as each detachment was left behind. It is interesting to compare the problem with those which faced Napoleon two hundred and seventy years later. It may not be generally realized that, when the French Emperor overran a hostile country, he left behind very numerous *étape* troops to keep open his communications, and to besiege any fortresses which might be holding out behind him. The corps so employed were generally foreign contingents, or recently levied and inexperienced units. In 1806 he so utilized Dutch and Spanish auxiliaries ; in 1812 Polish and German divisions were dropped along his route to Moscow. Similarly when the Allies in 1813 drove Napoleon out of Germany, and found that he had left countless garrisons behind him, the blocking of these places was turned over to Prussian and Russian third-line levies, or insurrectionary corps of Hanoverian and other German patriots. But troops of some sort had to be found. During the Peninsular War half the enormous French force which lay beyond the Pyrenees—often 250,000 strong—was tied up in *étape* and garrison duty, and the field armies were consequently enfeebled. The same problem existed in the sixteenth century, when hostile territory was won : the garrisons swallowed up the striking force, till it dwindled to a comparatively moderate strength. During the later years of the great Italian wars, both the French and the Imperialist armies in Piedmont and Lombardy had to hold so many places that it was hard for them to concentrate more than 12,000 or 15,000 men for field operations—hence Brissac, Del Vasto, and the Duke of Alva spent whole summers without any large enterprise being taken in hand, and employed themselves in minor operations, about sieges and reliefs, which could be conducted with comparatively small numbers.

As has been already mentioned, there were only three general actions of importance during the later wars (1543–1559) between France and the Hapsburgs. And of these only one, Ceresole (1544), was a pitched battle, to which each side came with intent to arrive at a decision. At St Quentin (1557) the Constable Montmorency was manœuvring, with the object of succouring a besieged town, and was forced into an undesired fight by the sudden offensive of Philibert of Savoy. At Gravelines (1558) the Marshal des Thermes, raiding the Flemish coast, was surprised and brought to action by Egmont before he could take measures for a prudent retreat. It is well worth while to describe the tactics of each of these actions.

In the thirty-two years which elapsed between 1527 and 1559 there are many other points to be noted, which show us that we are far away from the days of Fornovo and Novara. One is the immense development of scientific fortification—in the field this had already started with Gonsalvo de Cordova—as witness Cerignola : we have seen its development at Ravenna and Bicocca. But for city walls it was also on the move. Old political centres like Milan and Verona saw their time-honoured mediaeval ramparts superseded by elaborate enceintes, with bastions and *fausse-brayes* and glacis, covered ways, and horn works. New foundations like Hesdin, Vitry, and Havre were of course built in the modern style. Indeed the scientific engineer was once more getting the better of the artillerist, and sweeping conquests of whole rows of antiquated fortresses by gunfire—like those which the brothers Bureau had carried out in Normandy in the preceding century [1]—were no longer possible. Instead of rearing high walls visible to the besiegers' gunners, the military architect 'dug himself in' and lay low. Sieges tended to grow longer as the wars went on—though the culmination of the tendency was reserved for another half-century at the interminable siege of Ostend, which ran on into three summers. As will be shown in a later chapter, the Dutch-Spanish wars of the Netherlands ultimately became one long series of siege operations, with no battles in the field for long periods of years. The spade got the mastery of the cannon-ball.

If, however, the larger firearms were foiled by low-lying fortifications all through the later sixteenth century, in the field the smaller firearms were victorious, alike over the old methods of the Swiss pike-column, and the charge *en haye* of the French gendarmerie. The arquebus, and its legitimate issue the musket, continued to extrude all the earlier shooting-weapons. Even in England, where the long-bow survived all former continental rivals, it had finally to be abandoned by the ordinance of 1595—as will be told in the chapter dealing with the English military organization. It will be remembered—from Montluc's interesting notice already quoted [2]—that down to 1525 the French native infantry had hardly any arquebuses or calivers, and was cumbered with thousands of cross-bows, while after the triumph

[1] See my *Art of War*, ii. p. 226. [2] See above, pp. 47, 48.

of small arms at Pavia every captain raising a band strove, with more or less success, to collect as many arquebusiers as possible [1] —which led in later years to an over-production of firearms and the neglect of the pike—its necessary complement, so long as the arquebusier could not defend himself against cavalry.

La Noue in his ' paradoxes ' gives examples of the way in which the combined force could defy very superior numbers of horsemen.[2] That the unprotected arquebusier was helpless was sufficiently shown in many cases, for example, in the rout of the Huguenot infantry of Guienne in 1569, of which mention will presently be made. The pikeman, on the other hand, though he could beat off cavalry easily enough, was very vulnerable to firearms, small or great, when stopped by a physical obstacle, *e.g.* trenches manned by arquebusiers (as at Bicocca), or when artillery could be brought to play upon him while he was perforce halted and detained by real (or even by simulated) attacks by horsemen—as at Marignano, and in much later years at Moncontour.[3] It was the deficiency of reliable ' corselets,' as the French officers called their pikemen, which led to the persistence of Swiss regiments in the' Wars of Religion,' long after the old idea that the national French infantry was practically worthless had been abandoned. For, as we shall see at Ceresole, it had become a fixed notion that though the French arquebusier had established a firm reputation, the French pikeman could only be induced to face Spanish or German infantry by very capable leaders, who could inspire their men with an unwonted confidence.[4]

The firearms of infantry were steadily improving during the second half of the century. The arquebus or caliver was gradually superseded by the musket,[5] a weapon with a much longer range, though with the countervailing fault of being slower in discharge, and heavier to carry. But in fire-fight a force with a proportion of long-range muskets could begin to plague an enemy armed only with arquebuses or calivers, before the latter came within effective range. We shall note what seems to be an early example of the useful employment

[1] And, if possible, to arm them with good Italian arquebuses, not with the inferior article produced by French smiths in the middle of the century. See above, pp. 44.

[2] See above, p. 44. [3] See p. 166 above. [4] See p. 80 above.

[5] The actual word ' musket ' seems to be first found as early as 1527, but I cannot find any certain mention of its use by bodies of troops for many years after that date. Mühlberg, if correct, is an early case.

of muskets at Mühlberg (1547)[1] when the Saxons, who thought themselves beyond the risk of molestation, found that musket-balls could carry across a river which had been judged to be sufficiently broad to form a complete protection. The musket had to be worked with a fork or rest to support its excessive weight, and the tiresome management of this auxiliary tool made necessary very elaborate drill-arrangements, and consequently great slowness of fire. It was observed that muskets would have been more practical if only Nature had endowed mankind with three hands instead of two.

The matter of loading was very much complicated by the existence of the fork—wanting both hands for the use of musket and ramrod, the soldier had to get rid of the fork somehow—either by hanging it from his elbow by a cord, or by sticking it into the ground, both highly inconvenient devices. Musketeers were still rare in the second half of the sixteenth century—the Duke of Alva's celebrated Spanish companies had only 15 musketeers to 100 arquebusiers or more. The weapon was exceptional, and required well-drilled specialists to manage it, for many years. But by the end of the century it had become normal; the Marshal Strozzi is said to have been its first great patron in the French army.[2] In England, always late to make changes, the first mention of musketeers in a company is as late as 1587. In the last years of the Great Wars of the first half of the century the musket was still a new invention. It may possibly have been, when first employed, suggested by the arquebus *à croc*, a heavy wall piece fixed on a peg or pivot, and used only from walls, where its support was immovable. But obviously if such a weapon was made much lighter, and supported by a strong man and a movable rest, instead of by an immovable pivot clamped into a wall, it could be converted into a musket for service in the field. Ronsard has a satiric note in one of his sonnets against those who think it too heavy to carry.

In the cavalry arm the most notable development was the long controversy between the exponents of the benefits of the deep order for squadrons, and those who held to the old practice of the charge in line, *en haye* as the French called it. Much could be said on each side—the line might easily fall into disorder by

[1] See below, p. 249.

[2] He was a great musket-shooter himself, and could bring down an individual enemy at 300 yards' range, according to Brantôme.

breaches in it caused by early casualties, or even by ill-trained cavaliers failing to keep touch with their neighbours on each flank. On the other hand, in the deep order the rear ranks might be practically useless; if the first rank failed to break through, and came to a standstill, the ranks behind could only check and wait. If the whole mass was moving rapidly, this halt would infallibly result in jamming and confusion at the sudden cessation of the power to push forward. If two resolute hostile squadrons in the deep order met with level fronts, and neither flinched, it is obvious that the leading ranks would clash, and while many horses and men would go down as casualties, the survivors in front would get locked in sword-combat at a standstill: the rear of each mass would be immobilized. As a matter of fact in cavalry charges one side or the other generally flinches before contact, and is routed. But if both parties mean business, and neither side is ridden down by heavier weight at once, the result might be a mere fencing match. The rear ranks are blocked. (Cf. the charge of Scarlett's heavy cavalry brigade at Balaclava against a deep column of Russian horse.)

La Noue, who devoted much thought to this problem in his comfortless prison in the Netherlands, made decision in favour of the deep order, if cavalry met cavalry and both sides charged home. But by his time there had arisen the unhappy system of the 'caracole,' already spoken of,[1] by which cavalry in deep formation delivered fire from successive ranks, instead of charging home, each rank going to the rear and re-forming after it had discharged its weapon. Used against an enemy who disregarded initial losses, and charged in with sword or lance, this device was certain to lead to disaster, since the sight of rank after rank filing off to the rear was most demoralizing to the rest of the column, and often led to a general *débandade*. If two cavalry forces, both using the 'caracole,' should meet, there would be prolonged rolling fire, but no decisive result till the morale of one side or the other gave way. But La Noue regarded the 'caracole' only as an abuse of the deep order, not as an essential feature of it.

By the middle of the century the Germans had all become 'reiters,' mailed cavalry using firearms, and moving in deep formation.[2] The French long preserved their preference for the linear formation and the lance, but took to employing as an

[1] See above, pp. 41–42.

[2] See above, pp. 84–85.

auxiliary arm bodies of reiters (*reitres* as they called them) or ' pistoleers.' They were already to be found in the armies of Henry II, and were invariably present in those of the ' Wars of Religion '—especially on the side of the Huguenots, who imported German horse—Protestant allies in theory—on a very large scale.

But these reiters were heavy cavalry. Quite separate from them were light cavalry in all national armies, whose tactical ancestors were the ' stradiots ' used by the Venetians and then by the French,[1] and the ' genitors ' of the old Spanish hosts. These were originally armed with sword or javelin alone, and were normally not employed in battle-line, but only for raids and surprises, like the English ' prickers ' and Border Horse. When firearms became common, the mounted arquebusier begin to appear, who differed from stradiots or genitors in that he carried a weapon with which he could do harm at a distance. Normally he would have to dismount in order to use it effectively, and his original efficiency consisted in his power of rapid movement for the occupation of defiles, bridges, etc., where an enemy could be checked for a time till he brought up a serious infantry force. The mounted arquebusier was the lineal ancestor of the ' dragoon ' of the seventeenth century, who is so prevalent in narratives of the Thirty Years' War and the civil strife of Charles I of England and his Parliamentary enemies. In the French army the steel-clad *compagnies d'ordonnance* came to form a less numerous proportion of the mounted men as the years rolled on, both because complete armour became more and more costly, and because the stock of big horses capable of carrying a fully equipped man-at-arms was gradually dwindling. The *compagnies* continued to exist, and are found still in line at the commencement of the ' Wars of Religion ' in 1562. But lighter cavalry, not unfitted to take part in a general action, began to appear—they may be compared to the ' demi-lances ' of an English army of the time of Henry VIII. The best squadrons of these *chevaux légers* had corselet and helm, but wore high leather boots instead of the old leg-armour of the Middle Ages. They carried sword and lance, and were ready to charge home, unlike the ' pistoleer ' or ' reiter ' who depended on his firearms in close combat.

But there were other light cavalry in most continental armies, not merely the arquebusiers on horseback, who were

[1] See above, p. 92.

really mounted infantry, but troops carrying firearms of the smaller sort, which they could use without dismounting. Such were the Spanish *Herrueleros*, who had short carbines or pistols but little or no armour, and were intended (like the stradiots or genitors of earlier generations) to skirmish and to raid, not to fight in battle-line. In the French armies we occasionally hear of *Argoulets* as light horse, apparently of rather inferior quality—for the word passed into common parlance to mean a paltry fellow. Whether they carried firearms in their later days is uncertain—they are occasionally reckoned along with the stradiots, who certainly did not, being equipped with sabre and light lance only. In Imperialist armies we sometimes find Hussars—Hungarians lent to the Emperor by his brother, King Ferdinand—*e.g.* in the Mühlberg campaign. These were essentially raiding troops, without firearms, trained originally to fight the plundering Turkish horse of the East, not to meddle with heavy cavalry of the Western type.

But Herrueleros, Argoulets, or Hussars were but auxiliary troops, often useful in a campaign, but not decisive for a battle. The real military problem between horse and foot consisted in the trial of the composite regiment of pikes and arquebuses against heavy cavalry, whether of the old type of the French gendarmerie or the new type of the German reiter.

CHAPTER II

CERESOLE (CERISSOLES) (APRIL 11, 1544)

THE tactics of the later battles of the Great Wars show so little resemblance to those of straightforward affairs like Ravenna, Novara, or Bicocca that they well deserve study.

The only one, as we have already observed, in which both sides met with a fixed intention to fight, and with forces drawn up in a careful array, was Ceresole. In the winter of 1443–44 the French, under the Sieur des Boutières, were in possession of central Piedmont, both the plain, the foot-hills of the Alps, and the passes over them. They held Turin as their central base, and Pignerolo, Carmagnola, Savigliano, Susa, Moncalieri, Villanova, Chivasso, and many smaller places. The Marquis Del Vasto, the Emperor's lieutenant-general in Lombardy, was in possession of a semicircle of fortresses outside the solid block of the French holding, Mondovi (which he had captured only lately), Asti, Casale, Vercelli, and Ivrea. These two commanders had been of late executing partial enterprises against each other's outlying places of strength. Boutières had battered and stormed San Germano (near Vercelli), but had failed before Ivrea. Del Vasto, striking at a more vital point in his enemy's line, had seized Carignano, only fifteen miles south of Turin, right in the middle of the French line, had thrown into it a heavy garrison of his best Spanish troops, and had strengthened its fortifications. These, by the work of several weeks of his whole army, were turned into a modern-type stronghold, with five bastions, good curtains between them, and a deep ditch.[1] Boutières could not collect for the moment a force sufficient to disturb these operations, his army being much dispersed, and largely tied up in garrisons. Del Vasto then went back to winter quarters, leaving Carignano well victualled, 'a thorn in the foot of the French general.'[2]

King Francis, dissatisfied with the operations of his representative in Italy, displaced him, and sent in his stead a

[1] Du Bellay, x. pp. 90–91. [2] *Ibid.* x. p. 10.

prince of the royal blood, Francis of Bourbon, lord of Enghien,[1] a young man of known energy, though as yet untried in high command. And recognizing that his force in Piedmont was none too great to cope with the Imperialists, he sent thither 200 lances of the old gendarmerie with 3000 or 4000 Swiss—or quasi-Swiss—of a new levy, and ordered ten or twelve companies of French infantry to be embodied in Dauphiny and Languedoc and sent to this front. These 'Swiss' were raised by the Count of Gruyères, a subject-ally of Canton Fribourg, from his own uplands; they were trained in the Swiss pike tactics, but much disappointed their employers.[2] The new French companies were not utilized in the approaching battle, and were probably left on blockading duty when it came.

Enghien laid siege to Carignano before the winter was well over, and rightly was of opinion that Del Vasto would come up to relieve the place, into which he had thrown some of his best troops, under a resolute commander, Pietro Colonna (known as 'Pyrrhus of Epirus' for his daring). This would undoubtedly lead to a battle, unless the French army should abandon the siege in a tame fashion. But general actions were by now recognized as very serious and dangerous tests, and Enghien took the curious step of sending an officer to Paris, to ask leave from King Francis to commit himself to the decision by arms. The emissary sent was Montluc, the self-assertive Gascon whose autobiography so often helps us to details of interest.[3] His narrative cannot always be trusted, since he sees himself in the limelight at every crisis. But if he may believe him, the King's advisers, headed by the Count of St. Pol, were against giving Enghien permission to fight, on the ground that if the army of Italy were badly beaten, Del Vasto would invade France, and find nothing left in his front, since the King would have taken every available man to the North to resist the impending invasion of Picardy by the

[1] Not *Duke* of Enghien, as many modern historians, thinking of the nineteenth-century Enghien of unhappy memory, have called him.

[2] Paolo Giovio defines these people as having been 'raised from all regions by the Upper Rhone and the Lake of Geneva.'

[3] Du Bellay carefully avoids saying that Montluc was the messenger, having apparently a personal dislike for him. And he differs from the Gascon by saying that Francis only gave leave for a battle if the captains in Italy should approve. It seems hardly credible that Montluc could have invented his whole graphic tale of the dispute at the council board, and his own impassioned plea for action. Cf. Bellay, ix. pp. 102–3, where 'un gentilhomme,' anonymous, is sent to Paris, with the ten vigorous pages of debate in Montluc, i. pp. 245–55.

Emperor and the King of England combined. Be this as it may, Francis heard Montluc's arguments, if we may believe his tale, and sent him back with the message that a battle might be risked, if Enghien's captains agreed that it was prudent. Here follows a typical anecdote of the days of chivalry—nearly a hundred young men of enterprise about the court, hearing that there was to be heavy fighting in Piedmont, posted off at full speed and rode for Turin, in order to be on the spot where glory was to be won—among them was Gaspard de Coligny—the staid Huguenot hero of twenty years after, but now still in his hot youth. He was among many young knights of fame, Dampierre, St. André, Vendôme, Rochefort, Jarnac. Du Bellay has a page full of their names.

Del Vasto, as Enghien had calculated, was determined to relieve Carignano, and moved forward from Asti so soon as he had received some reinforcements from Germany, which the Emperor had promised, 7000 landsknechts picked from old regiments, and all furnished with corselets—full armour was rare among them, as it was also among the Swiss. The Marquis was rather short of cavalry, but, as we are told, considered that he had a more than compensating advantage in the steadiness of his German infantry, and the very large proportion of arquebusiers among his Spanish and Italian foot companies.[1] His total force consisted of 7000 landsknechts, 6000 Italian foot, and a body of 5000 veterans, Spaniards and Germans, who had served in the Emperor's last African expedition; there were under 1000 cavalry, of whom 300 were Florentine light horse under Ridolfo Baglioni—lent by Duke Cosimo—300 Neapolitans under Philip de Lannoy, prince of Sulmona, and only 200 Spanish and Italian heavy armed gendarmerie.[2]

Enghien, leaving a small blocking force in front of Carignano, had assembled his army at Carmagnola, a junction of roads two miles west of the Po, which commanded all possible

[1] We are told that he made this observation after the battle to his prisoner M. des Thermes. He said that after Pavia the Spanish officers had come to think little of the French gendarmerie, and believed that arquebusiers would always get the better of them, if properly covered. He also thought that his landsknechts, all picked men, were as good as, and more numerous than, the Swiss, on whom Enghien relied as his mainstay. Des Thermes was in bed, wounded in the leg, when the Spanish general came and sat beside him and made these interesting confessions. Cf. Montluc, ii. pp. 267 and 282.

[2] These figures are lower than those given by Du Bellay and Montluc—but the victor always exaggerates the number of the vanquished.

routes by which the enemy might try to approach the besieged town, and only left Del Vasto the choice between a frontal attack and a circuit by which he would expose his flank to the French, while marching past them. The cavalry of Enghien kept close touch with him, and found that he was not making a long sweep by Sommariva and Raconigi, as he might have done, but courting a decisive fight, by aiming straight at the French position at Carmagnola. On the night of April 10 he encamped at Ceresole d'Alba, a village on the high road five miles or less south-east from Enghien's headquarters. The French prince had been watching him all day, at the head of a strong force of light cavalry, and was besought by some of his officers to call up his whole army, and attack headlong, while the enemy was in line of march along a countryside destitute of hedges or other obstructions. But he refused, and probably with wisdom, for his own troops would have been coming up in successive detachments, and the day was far spent. He had determined to fight in good order, and on his own chosen ground. Accordingly the units were roused before dawn, and marched out to a position three miles in front of Carmagnola, where there was a long hillside of no great elevation and with gentle slopes, with in front of it a dip, beyond which there rose a similar slope. The ground was bare, save for two farms with their outbuildings between the two hillsides, and a wood on the right. The village of Ceresole was out of sight, two miles behind the enemy's slope.

The battle moment came luckily for Enghien, for his troops had been in a state of incipient mutiny for want of pay. The Swiss were (as before Bicocca) threatening to march home, and the Gascons were sending angry deputations to headquarters. Forty thousand *écus*, which was all that the King had been able to send, was a wholly insufficient sum to discharge even a month's arrears. But on the news of imminent battle the mutineers fell into their ranks again.

Enghien's forces consisted, so far as can be estimated from a comparison of sources,[1] of 4000 old Swiss infantry—the core of the army—4000 French infantry of the 'old bands'—mostly Gascons, 3000 of the 'Gruyères' contingent, with 2000 Italian

[1] I give figures somewhat lower than Montluc's (ii. p. 272) and somewhat higher than Du Bellay's (x. p. 122). Tavannes (i. p. 65) takes a good deal from Du Bellay, but allows for 5000 French infantry and 800 men-at-arms, which are heavier figures than Du Bellay's.

foot, 600 light horse, French and Italian, and 800 of the *compagnies d'ordonnance*, including their 'archers,' with as a supplement the 100 volunteers from the court who had just

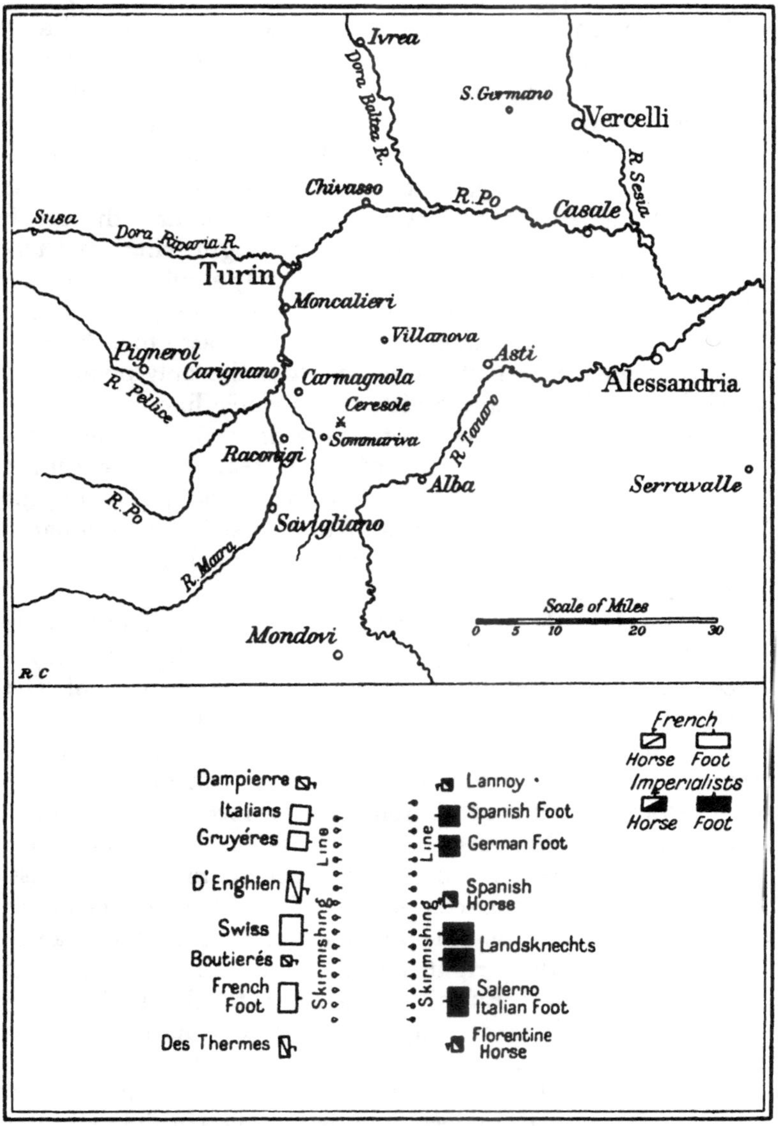

Above: LOCALITIES OF THE BATTLE OF CERESOLE.
Below: DISPOSITION OF TROOPS.

reached the camp. His artillery consisted of twenty guns, a number about equal to that which Del Vasto had brought with him. It would seem that the new French foot-companies, which had been raised in the winter, were the force left behind to blockade Carignano. Altogether the total was probably several thousands weaker than the Imperialists in the way of infantry, but as many hundreds stronger in the mounted arm, which was notoriously deficient in Del Vasto's army.

The position which Enghien had selected was on the crest at the head of the long, gentle slope. Its central section was somewhat higher than its two flanks, which had the result that the left wing was invisible from the right wing, and *vice versa*. There seems to have been an exactly similar disposition of ground on the opposite side of the intervening dip, as we are told that a central knoll, on which Del Vasto took his post at the commencement of the action, was the only point from which he could command a view of his whole line.

The French army was theoretically divided, as of old, into the three corps, 'battle,' 'vaward,' and 'rearward,' *i.e.* centre, right, and left wings. Enghien himself took post with the 'battle,' de Boutières (whom he had superseded in command) had the 'vaward,' and Dampierre the 'rearward.' There was no second line or reserve.

On the right the 4000 French foot of the 'old bands' were flanked in their outer wing by the bulk of the light cavalry—three companies, but incomplete in numbers—those of Des Thermes, Bernadino, and Mauré. If the ranks had been full, there should have been 650 troopers, but apparently the actual strength was 450 or 500. On the inner flank of the French foot, de Boutières placed himself, with an incomplete squadron of gendarmerie—80 men-at-arms, who should have been 100.

The centre was formed by the pike-column of the Swiss, all old troops, thirteen companies, of which seven belonged to William Frülich of Soleure, and six to a captain named St. Julian. On their right was de Boutiéres' detachment of horse, on their left Enghien himself, with the bulk of his heavy cavalry, three compagnies of gendarmerie, not quite complete in numbers,[1] and the 100 volunteers from Paris, besides 150 light horse. The whole must have been at least 450 strong.

The left under Dampierre had as its infantry the two pike-columns of the Gruyères people and of the Italian bands. The

[1] Companies of Crusol, d'Accier, Montravel, and d'Ossun's company of light horse.

former were about 3000 strong; their commander, the Count of Gruyère, and some of their companies were not yet arrived from Dauphiny. They were commanded on this day by a Sieur Descroz, who was also in charge of the six Italian companies which were ranged alongside with them. Outside this infantry and on the extreme left of the line were 400 mounted archers, detached from the companies of gendarmerie, and acting this day as light cavalry. Dampierre, who was entrusted with the whole 'rearward' corps, took his post with them.

When the Imperialists came in sight, and began to deploy along the slope opposite the French position, the two armies were neither of them fully visible to each other, for there was some attempt to keep under cover till the decisive moment. The Swiss, we are told, lay down flat behind the crest, and of the enemy only the left wing was at first fully in sight. Del Vasto opened the fight by sending out a cloud of arquebusiers, with whom he was well provided, to feel for the French, and if possible to find and turn their flanks. Enghien replied by sending out 800 arquebusiers from his French and Italian companies under Montluc to hold them back. This skirmishing lasted nearly four hours. 'A pretty sight,' says Du Bellay, who was present, 'for anyone who was in a safe place and unemployed, for they played off on each other all the ruses and stratagems of petty war.' Sometimes one side and sometimes the other gave back, and the skirmish flickered up and down the slopes between the two lines. Presently each side brought up its artillery, when it had discovered the lie of the hostile army. Del Vasto produced his twenty guns and placed them beside the two farms which lay in front of his line, one half in front of his right wing, the other in front of the centre. Enghien replied by displaying two batteries of similar force, one in front of his left wing, where the Gruyères contingent lay, the other in front of the Swiss of his centre. The interchange of cannon-balls seems to have had little effect—the distances being great and the troops largely under cover.

This ineffective shooting went on for hours, till Del Vasto showed signs of sending out cavalry to take the French arquebusiers in flank, which caused Montluc to draw back his skirmishers, and ask for help from des Thermes' light horse on the right wing. When this came up, and more serious fighting seemed likely to begin, the Marquis took the general offensive, and showed his whole fighting line, crossing the dip between the two armies. It appeared that his 'rearward' or left wing was composed of his Italian foot, under the Prince of Salerno, flanked by Ridolfo Baglioni's 300 Florentine horse. The centre, or 'main-battle,' was composed of the 7000 armoured landsknechts lately arrived from Germany, under Aldobrand Madrazzo, the brother of the Bishop of Trent. Del Vasto was on their right, near his small body of gendarmerie—some 200 lances, under Carlo Gonzaga—more or less opposite to Enghien's own cavalry. The 'vaward' or right wing consisted of the 5000 Spanish and German veterans from the African wars under Ramon de Cardona, flanked by 300 Italian and Neapolitan horse under Philip de Lannoy, the Prince of Sulmona.

The clash began apparently at the south end of the two lines, where des Thermes' light horse broke at once the Florentine troopers of Baglioni, with such force that they drove them back into the pikes of the Prince of Salerno's infantry. Hoping to take advantage of the confusion, des Thermes charged right into the Italian foot, but though he put them into some disorder they stood their ground. He himself, cutting in deep into the column,[1] was wounded, unhorsed, and taken prisoner by an Italian soldier named Napoliello, 'thinking that he would have been better followed.' This charge, though unsuccessful, was of immense profit to the French, for the Prince of Salerno took some time to get his men into order again, and before he was ready to resume the offensive, matters had gone very badly for his comrades in the centre.

The French foot-bands in the right centre had started downhill to attack Salerno's corps, which had been opposite to them, when it became obvious that the Italians were stationary, in some disorder, and still far away, while the columns of the Imperialist centre had come a long way up the hill, and were exposing their flank unguarded. It was, according to his own narrative, Montluc who pointed out to de Tais, the commander of these Gascon bands, that he would be wasting his opportunity if he went off to his right to attack the Italians, and that he could easily strike at the side of the vast column of the landsknechts, who were much closer to him. The suggestion was taken, and the Gascon companies swerved somewhat to their left, and went down the hill diagonally against the enemy's

[1] Du Bellay, x. p. 129; Montluc, ii. p. 282.

main column, who were not yet engaged with the Swiss, for the latter were still ranged on the crest, waiting for them.

It says much for the tactical skill of the commander of the landsknechts that when he saw a second force making its way toward him, he was able to break up his great mass of 7000 men into two sections, of which one swerved to face the French foot, while the other went on against the Swiss—and there must have been more manœuvring power among his companies than one would have expected. But there is clear evidence that the division was made. 'Seeing that the French had changed their plan,' says Du Bellay, 'the Imperialists made a parallel change, and of their great battalion made two, one to fight the Swiss, the other the French, yet so close to each other that seen sideways they still looked one great mass.[1]

On both sides there was a tremendous crash when the columns hurtled together. The French foot were practising what they considered a new device—behind the first rank of the pikemen were a rank of arquebusiers, who were ordered not to fire till the moment of impact had arrived. 'In this way,' says Montluc (who claimed to have been the inventor of the idea) 'we should kill all their captains in the front rank.[2] But we found that they were as ingenious as ourselves, for behind their first line of pikes they had put pistoleers. Neither side fired till we were touching—and then there was a wholesale slaughter: every shot told: the whole front rank on each side went down.' The second and third ranks met over the corpses of their comrades in front, the rear ranks pushing them forward. 'And as we pushed harder, the enemy tumbled over. I was never so active and adroit in my life,' says Montluc, 'and I had good need to be so, for thrice I was brought to my knees on the ground,' and to rise again for a man in armour was no small task.

The Swiss, who faced the other half of the landsknechts, waited till they saw the French within twelve pike-lengths of their immediate adversaries, and only then rose to their feet and rushed in. The clash here was a little later than further to the right, for the hillside was steeper, and the enemy climbed more slowly. Along the whole centre of both armies there was 'push of pike' for some minutes, each side being full of *esprit de corps* and obstinacy. The matter was apparently settled by the charge of de Boutières' small corps of gendarmerie,

[1] Du Bellay, x. p. 128. [2] Montluc, ii. p. 286.

who struck in at the landsknechts and took them in flank, giving the final thrust which sent them reeling down the hill.

The French accounts say nothing of what happened to the small body of Imperialist men-at-arms, who were on the right of the landsknechts, and ought to have helped them, as de Boutières did the Swiss. The most detailed narrator from the other side says that they disgraced themselves ; that the Marquis ordered them to charge the Swiss in flank, but they that recoiled so promptly from the pikes that many supposed that they were merely practising a manœuvre, and were about to come on again. But they never came about, and rode to the rear, carrying the Marquis with them, who had got an arquebus-bullet in his thigh.[1] Their commander, Carlo Gonzaga, was taken prisoner, while vainly trying to bring them back to the front.

The victorious Swiss and Gascons made a horrid slaughter of the landsknechts, who were so jammed together in their dense order that only the rear ranks had the chance of throwing down their pikes, and making a run for individual safety. For over a mile, as far as the entrance to the village of Ceresole, the road was strewn thick with corpses—the Swiss gave no quarter, having to revenge a recent incident of ' bad war,' as it was called, when the Germans had mishandled the Swiss garrison of Mondovi. The figures given for their slain are no doubt exaggerated—it was said that they were practically exterminated, and a loss of 5000 out of 7000 is indicated. Practically all their captains were killed, as Giovio's casualty list shows. Some of des Thermes' light horse, who had been driven off by the Italian foot, joined in the pursuit and slaughter.

The Prince of Salerno, seeing this disaster in the centre, and judging it irretrievable, marched off the field at once, and reached Asti with his column intact, picking up on the way the survivors of Ridolfo Baglioni's Florentine light horse, who had got together, since they had not been pursued. Meanwhile, on the other end of the battlefield, matters had gone in a very different fashion. The Spanish-German column on the right wing of Del Vasto's line had gone forward against the Gruyère and Italian infantry on Enghien's left with a fine impetus—the pikes, as we are told, flanked by a strong swarm of arque-

[1] Paolo Giovio, xliii. p. 329. It looked, he says, at first as if they were practising the Moorish trick of wheeling after a feigned attack, in order to lure the enemy to pursue—but it was not tactics, but panic.

busiers. To the horror of Enghien, who was watching this part of the fight from his point of view on the highest ground of the position, the pseudo-Swiss broke without making any serious resistance—leaving their captains in the front rank to be trampled down and slain; and the Italian bands next them in the line followed their example—the commanders Descroz and du Dros were both killed. The Imperialists pursued, slaughtering the hindmost freely, and took no notice of the fact that Dampierre's cavalry on their flank had beaten and routed Lannoy's Neapolitan light horse, and had galloped after them in pursuit.

Seeing his left wing swept completely away, Enghien lost all count of how the battle was progressing elsewhere—probably the landsknecht columns were not yet defeated—and forgetting the duties of a commander-in-chief called upon all the cavalry of his main-battle to follow him in a charge to stop the rout on the left.[1] The whole of his fine gendarmerie and volunteers trotted after him, and he drew them up for an attack on Cardona's victorious infantry, who were now well behind the original French position. As it chanced, the lie of the ground was such that from this part of the rear-slope no view of the rest of the battlefield was available, and the engagement went on without any reference to what was occurring elsewhere.[2]

The story of this part of the fight, purely a matter of infantry *versus* cavalry, recalls the record of Marignano. Enghien's first charge broke in a corner of the Imperialist column, which had halted to receive him—he pushed right through to their rear, losing many of the volunteers on the way—fourteen or fifteen gentlemen of mark were killed. But the enemy closed up, and he had to deliver a second charge, under a tiresome arquebus-fire from the flank; again he rode in, but failed to break up the Imperialists, and according to Du Bellay the second charge was even more costly than the first.[3] Enghien drew up his shattered ranks and was apparently joined at this moment by Dampierre and some of his light horse, returning from the pursuit of Lannoy's Neapolitans. A third charge was made—according to one French narrator the Prince was encouraged by the Sieur de Tavannes, who cried to him that

[1] That Enghien saw the rout and came down *at once* to save matters is proved by Paolo Giovio's narrative, I think.

[2] See Du Bellay, x. p. 132, for this particularity of the ground.

[3] Montluc corroborates Du Bellay as to the slaughter in the second charge.

the cup must be drained to the dregs: 'Monsieur, il faût boire cette calice' [1]—but this charge, too, failed to be decisive, and barely a hundred men-at-arms formed up when it was over. Montluc says that Enghien at this moment thought that all was lost, knowing nothing of the rest of the battle, that he cursed the day when he was born, and tore open his gorget, intending to stab himself in his fury, a 'thing which ancient Romans might do, but not good Christians.' [2] Be this as it may—Montluc loves a tragic scene—at this very moment St. Julian, the Swiss captain, rode up to report to him that the Imperialist centre was routed, and the battle won.

At the same time the Imperialist column, which had halted to receive the three charges, was observed to draw itself together and to be moving off hastily towards a wood in their rear. The news of the defeat of their main body had just reached them, as it had reached Enghien. The Prince followed them with the wreck of his cavalry marching parallel with them, and was at last able to do them some harm, for he had just been reinforced by a company of Italian horsed arquebusiers, who had been stationed that morning eight miles away, to hold the fords of the Maira River at Raconigi, and had very properly 'marched for the guns,' when they heard the artillery fire beginning at ten o'clock. Dismounting, firing, and remounting, they kept alongside of the retreating column detaining it. But the fatal blow to the Imperialists was given by the victorious troops of the French centre—'when we heard at Ceresole,' says Montluc, 'that M. d'Enghien wanted us, both the Swiss and we Gascons turned toward him—I never saw two battalions form up so quick—we got into rank again actually as we ran along, side by side. The enemy was going off at quick march, firing salvos of arquebuses, and keeping off our horse, when we saw them. And when they descried us only 400 paces away, and our cavalry making ready to charge, they threw down their pikes and surrendered to the horsemen. You might see fifteen or twenty of them round a man-at-arms, pressing about him and asking for quarter, for fear of us of the infantry, who were wanting to cut all their throats.' A great many—perhaps half—got killed, the rest were accepted as prisoners.' [3] Some got away, including the Baron of Seisneck, who commanded the German companies in the column.[4]

[1] Tavannes, i. p. 68.

[2] Montluc, ii. p. 284. Brantôme also gives the anecdote.

[3] Montluc, ii. p. 289.

[4] This we get from Paolo Giovio, xliv. p. 329.

So ended the day—Enghien found that he had 2530 Germans and 630 Spaniards alive in his hands. The dead, no quarter having been given in the centre, were much more numerous, but 5000 or 6000 (the least numbers given) are rather more probable figures than the 10,000 or 12,000 mentioned in French accounts. The Imperialist army was, after all, only 20,000 strong, and the whole left wing had got off intact, while the cavalry had been more frightened than hurt. Among the prisoners were Ramon de Cardona, commanding the left wing, and Carlo Gonzaga, who led the heavy cavalry; Aldobrand Madrazzo, who had chief charge of the landsknechts, was taken up for dead, but survived his wounds. Of his captains nearly all perished—Giovio gives a long list of them—the heir of Fürstenberg, the Baron of Gunstein, two brothers Scaliger—Christopher and Brenno—Michael Preussinger, Jacob Figer, etc. etc. There were several Spanish captains among the prisoners, but apparently no Germans of note save Madrazzo.

The French loss is obviously understated at 500 men (or less) by their chroniclers; it is clear that the Gascon foot must have suffered severely in their contest with the landsknechts—five captains of bands fell—la Molle, Passin, Barberan, Moncault, and St. Geneviève, of whom the first is said to have been killed in personal combat with Madrazzo, whom he wounded [1]—the casualties among their rank and file must have been very heavy. Moreover, *all* the captains of the unfortunate Gruyères folk are said to have fallen, along with Descroz who led them, and Charles du Dros, late governor of Mondovi, who was with the Italian foot; hundreds of their followers must have been caught and slain, for the men of a broken column cannot get away. Enghien's following of gendarmerie was, as we have seen, cut to pieces—among the dead were his two body-squires, and of the volunteers d'Accier, D'Oyn, Montsallais, de Glaive, governor of Cahors, Rochechouart, Courville, and some dozen more of the noblesse. The only prisoner lost was des Thermes, carried off by the Italians, among whom he had cut in too deep. But it is hard to believe that the total French loss can have been under 1500 or 2000, considering the toughness of the fighting, and the complete rout of their left wing.

The tactics of this battle have their interest. We note that cavalry, when opposed alone to steady infantry, failed to break

[1] This, a doubtful tale, comes from Paolo Giovio.

them, but could pin them to the spot by repeated charges. On the other hand, infantry already engaged with other infantry could be cut up by horsemen. Del Vasto was wrong in holding that a hopeless inferiority in cavalry—both in numbers and quality—could be compensated by a superiority in arquebusiers. For when all his cavalry had been driven off the field, the combination of hostile foot and horse was too much for infantry unsupported by the other arm. The loss of the battle was due in the main to the fact that both of the Imperialist wings were paralysed by cavalry charges, which held them occupied, while the landsknechts' columns in the centre were attacked not only by the Swiss but by the Gascon foot, who ought to have been 'contained' by the Italians of Del Vasto's left wing. If Baglioni's light horse could have kept des Thermes occupied, the Prince of Salerno would have got to business with the Gascons and given them occupation. In that case the landsknechts would have had nearly double numbers against the Swiss in the centre, and things might have gone very differently. The ineffective results of the long artillery engagement, before the clash came, would have been gratifying to Machiavelli. It cannot be said that either of the generals in command got much credit from the day. Del Vasto miscalculated the hopeless drawback caused by the want of efficient cavalry. Enghien lost his head completely, and forgot the main engagement, while he was leading cavalry charges on one wing. It is said that he was only saved from absolutely suicidal self-sacrifice by officers who reminded him of Gaston de Foix's useless death at Ravenna thirty years before.

The result of the battle of Ceresole ought to have been the conquest of the Milanese by the French. But nothing of the kind happened. The King sent orders that Carignano must be taken before the army moved on. The governor, Pietro Colonna, held out with desperate resolution for several weeks after the battle, and just when he had been starved out, and had capitulated, orders came from Paris that Enghien must at once detach the best part of his army, to join the force in Picardy which was holding back the dangerous inroad of the Emperor on the northern frontier. Twenty-three companies of the old Gascon and Italian foot and half the gendarmes marched over the Alps, leaving Enghien nothing but his Swiss, his light horse, and some newly raised south French units. Before the summer was far advanced, Del Vasto got some small revenge for

Ceresole, by cutting up at Serravalle, near Alba, a large body of newly raised Italian mercenaries, whom Pietro Strozzi and the Count of Pitigliano were bringing up to reinforce Enghien (June 2). Oddly enough this success was won by the Imperialist cavalry, who had so little distinguished themselves in the recent pitched battle. Strozzi had very few horsemen with him, and some squadrons which Enghien was to have sent him never turned up. He and the greater part of his men were taken prisoners. This was another proof that infantry could not yet stand alone against even an inferior force of both arms.

NOTE.—There are excellent and lengthy narratives by Montluc and Du Bellay, both eye-witnesses, and something may be gleaned from the memoirs of Tavannes. Paolo Giovio has a very full account of the day, from the Imperialist side, but has got oddly wrong, placing the Swiss on the French right, and the Gascons in the centre. He puts the Swiss opposite Salerno's wing, and makes them (not the Gascons) deliver the flank-attack on the landsknecht-column. Nevertheless he gives valuable notes on points neglected by all the French narrators—especially on the disgraceful conduct of Carlo Gonzaga's cavalry, who are not even mentioned by Montluc and Du Bellay.

CHAPTER III

THE GERMAN CIVIL WARS (1546–53)—MÜHLBERG AND SIEVERSHÄUSEN

IT is certainly necessary to make a note on the internal troubles in Germany, sometimes swelling up into actual campaigns, which distracted the unfortunate Charles V at several crises, when his attention should have been concentrated on the French or the Turkish enemy. In only one case, that of 1546–47, can it be said that he was responsible for the complications, for on that occasion the initiative was his own; and at a moment when he thought himself free from his usual wars to east and west, he made an endeavour—apparently successful for the moment—to put an end to a situation in Germany which he regarded as intolerable.

The record of these wars is as uninteresting from the military as it is depressing from the moral point of view. For, save in one incident, the tactical lessons appear to be negligible, while the strategical lessons turn not on generalship but on motives of apathy or suspicion or selfish particularism. There was no battle of a notable kind, for the Emperor's victory at Mühlberg (April 24, 1547) was hardly a battle at all, but only the destruction of a small army by a larger one, in a surprise and a running fight. And all the many manœuvres of 1546 or 1552 depended in the main on political psychology of a discreditable sort, *i.e.* intrigue and selfishness, apathy or treachery, not on the ostensible position or numbers of the armies which the combatants could put into the field.

The Schmalkaldic League, founded in 1530 and crushed in 1547, was an intolerable thing from the point of view of the suzerain of the Empire in which it sprang up, though eminently comprehensible from the point of view of princes or free towns determined to fend off an armed Catholic reaction. It is comparable, of course, to the associations formed by the Huguenots in France, or by the Netherlanders in the 'Burgundian' dominion of Philip II a generation later. In all these cases the associated malcontents persisted for a long time in declaring their loyalty to their sovereign, while in arms against him. The 'Diffidation' by which the Schmalkaldic princes disowned their allegiance to Charles in 1546 was an exceptional phenomenon—the Huguenots and the Dutch kept up the farce of loyalty for a much longer space of time.

While Charles was oppressed with French and Turkish wars he had to tolerate the Schmalkaldic Confederacy, even when it took such extraordinary steps as the expulsion of the Austrian governors from the sequestrated duchy of Würtemberg, and the restoration of the exiled Duke Ulrich (1534) by force of arms, a provocation to open war. In compensation he had got the appearance of contingents from the Protestant states for the great army which he gathered at Vienna, to ward off Soliman's second great invasion of Hungary, which ended so tamely at Güns and Gratz (1532). From 1534 and the Truce or Treaty of Cadan there was a sort of compromise between Charles and the league of the professedly loyal Protestants of the north, which neither of them regarded as a permanent settlement. The Emperor had vain hopes of procuring, by means of the Council of Trent, to which Protestants were invited to come, a general reunion of Europe under a reformed Catholic Church. The Schmalkaldic leaguers, determined never to submit to a Papacy, however much reformed, had hopes of drawing all the German states, one after another, into an anti-Catholic union—the annexation of Church lands by their lay neighbours would be a useful attraction. The ideas were incompatible, but the actual solution which was to come, a sufficiently illogical one by an unsatisfactory compromise, could not be arrived at without fighting.

In 1544 Charles had concluded with France the Treaty of Crépy, and in 1546, after protracted negotiations, both he and his brother Ferdinand had obtained a long truce with Sultan Soliman. His hands being free, for the first time for many years, he pressed the German problem to a definite issue. He was perfectly sincere in his notion that the ills of Christendom could be cured by a reform of Church organization and the abolition of scandals. That he was up against something much more fundamental than a demand for the redress of acknowledged abuses he did not understand—any more than did the contemporary Popes. And many of his difficulties came from the Popes themselves, who feared Emperors and

General Councils, and failed to understand that the Protestant movement was an intellectual and moral revolt against the whole system of Catholicism, not an artificial agitation organized by some heretical thinkers and a number of very greedy princes, desirous of absorbing Church lands.

When Charles, whose ideals were genuine though his methods were tortuous and diplomatic, pressed matters to an issue, the question arose among the Protestants whether he could be trusted. Many thought that he was well-intentioned, and remembered that he was after all their suzerain, and that to take arms against him was high treason. He talked about General Councils, the reform of Christendom, and the reconciliation of dissidents. But what did this mean? He had started the Council of Trent, and that body was already showing itself intransigent and apparently incorrigible. The more resolute among the Schmalkaldic Confederates feared the worst, and resolved to fight. A much greater number of them hung back, opened private negotiations with the Emperor, and sent no contingents, or very small ones, to the Schmalkaldic army. One, and he the most enterprising and ablest of the Protestant princes, sold himself for good consideration to Charles, and presently attacked his confederates from the rear. German commentators have striven to find palliatives or excuses for the conduct of Maurice of Saxony, Margrave of Misnia and lord of Dresden, but cool selfish ambition is not only the most obvious but the most logical explanation of his conduct, when his whole career is surveyed.

In July 1546, Charles, unaware of the lengths to which the more desperate of his Protestant subjects were prepared to go, waited at Ratisbon, where he had summoned the Diet of the Empire to meet. He was professing the most pacific intentions, but at the same time preparing to use force, if his invitations to the Council of Trent were refused. He was unprepared for an open rupture, and had hardly any troops with him, when the Council of the Schmalkaldic League published an open defiance of his authority, and a large force headed by John Frederic, Elector of Saxony, and Philip of Hesse marched upon him from the north, while another, mainly composed of the levies of the free cities of Suabia and Franconia, came up from the south-west, and showed signs of cutting him off from the Tyrol and Austria. His brother Ferdinand seemed unable to give prompt help, because his own Bohemian subjects were showing open sympathy with the Schmalkaldic party. The Catholic Duke of Bavaria, from whom the nearest assistance could have been expected, seemed wavering toward neutrality.

The Emperor would have been in a parlous strait if the Protestant confederates had used their opportunity, but they negotiated instead of striking hard—many of them were uncomfortable at the idea of using violence against their suzerain. Hence while his enemies dallied, Charles succeeded in collecting an army from all the quarters of his wide-scattered dominions —large contingents from the Imperial garrisons of Milan and Naples, all Spanish and Italian veterans, and a very large force from the Netherlands, under his trusted general, Maximilian Egmont van Buren, which included 4000 heavy men-at-arms. The Catholic princes of Germany rallied around him somewhat tardily, and the Pope, anxious about the future of the Council of Trent, contributed an appreciable body of mercenaries, though he was no true friend to an Emperor whose political aims he distrusted. With these and such troops as the more loyal sections of the old Hapsburg lands (Tyrol, etc.) could send in, Charles gathered by September an army able to face the Confederates, and engaged in an indecisive campaign with them in southern Bavaria, round Ingoldstadt, where the adversaries cannonaded each other ineffectively across the Danube.

The decision in the war was then given by a blow at the rear of the Schmalkaldic armies. Maurice of Misnia, in accordance with his secret treaty with the Emperor, suddenly declared war on his cousin the Elector of Saxony, and overran, almost unopposed, nearly all the dominions of John Frederic; indeed he got possession of almost everything save Wittenberg, the electoral capital, Gotha, and Eisenach. This treacherous stab in the back, by a prince who had always emphasized his Protestant views, ruined the cause of the Schmalkaldic confederates. John Frederic marched north with the bulk of their army to evict Maurice, leaving only an inadequate 'containing force' in front of the Emperor under Philip of Hesse. Charles, turning against this nearer opponent, drove him away northward, after some insignificant skirmishes, and then marching through Swabia and Franconia received the submission of the Duke of Würtemberg and most of the free cities, who made no serious resistance. They were fined, and compelled to repeat abject formulæ of homage, but not badly handled—Charles aimed at conciliation rather than at chastisement (November–December 1546).

Meanwhile the Elector John Frederic, arriving in his own dominions with the largest part of the Schmalkaldic army, had expelled his cousin Maurice from Thuringia, and driven him back into his own Misnian dominions. But this success did not compensate for the complete ruin of the Protestant cause in the south, though it had the incidental result of encouraging King Ferdinand's Bohemian subjects to take sides openly with the Schmalkaldic party against their master. Their rising, however, had no decisive effect, and gave no practical help to John Frederic. Nor was assistance from France, promised by several secret messengers, ever forthcoming; though (as usual) there was encouragement sent to rebels when the Emperor was known to be in difficulties. But Francis I was near his end—he died in April 1547—and his failing health appears to have kept him from utilizing opportunities as of yore—the Elector got out of him 100,000 gold *écus*, but no military help. The last piece of good luck that fell to John Frederic was that early in March he surprised and destroyed at Rochlitz, near Leipzig, a reinforcement of 7000 men, whom the Emperor had sent before him to join the Misnian margrave—capturing its leader the turbulent Hohenzollern prince, Albert of Culmbach, who was destined a few years later to destroy Maurice, whom he now came to help.

But Charles himself was now on the move with his main body, though short of the Pope's Italian contingent, which had been recalled. Coming up by Nuremberg and Eger, he there picked up his brother, King Ferdinand, who had brought up many Hungarian hussars, since he could not count on his Bohemian subjects. When Maurice also came in to join him—from a raid which he made toward Prague to keep the Bohemians quiet—their united force amounted to some 30,000 men. The Elector was caught unawares—he had sent troops to join the Bohemian rebels, and had detached companies to garrison his capital, Wittenberg, Gotha, and Meissen, which he had just captured. Unable for the moment to concentrate more than 3000 horse and perhaps 12,000 foot, of whom only 6000 were old soldiers, he went behind the Elbe, whose course he intended to defend, having Mühlberg, whose bridge he had broken, on his left hand, and a broad stretch of river, which was believed to be unfordable, in his front. He had detachments watching its course, lest the Emperor might build a bridge somewhere, and he encamped three miles back from the river with his main body. The only princes with him in this campaign appear to have been his son and namesake, John Frederic 'the Magnanimous,' Ernest of Lüneburg, Wolfrad of Mansfeldt, and Ernest, son of Duke Philip of Brunswick-Grübenhagen. There seems to have been no proper concentration of the Protestant forces to oppose the Emperor—one large body, consisting of the contingents of Hamburg, Lubeck, and other Hanse towns, was at Bremen; Philip of Hesse was watching the borders of his own dominions, and carrying on futile negotiations, when he ought to have been co-operating with the Elector.

The Emperor was seeking a rapid decision while the enemy was weak, for he had received news (false as it happened [1]) that the large detachment of the Elector's army which had been sent toward the Bohemians, was not far off, and was hastening up. To cross a river 200 yards wide and reported unfordable was a hazardous business, even in face of an enemy of inferior strength—as Macdonald found at the Katzbach in 1813, and Burnside at Fredericksburg in 1862. Charles had collected a number of skiffs and punts on his own side of the Elbe, with which he hoped to make a bridge of boats, if fortune favoured. But his main reliance for what looked like a venturesome stroke, was that a discontented peasant—whose farm had been sacked by the Elector's men—had informed him of a ford no less than 300 feet broad, which was passable by horse, at the level of the river as it then stood, and even by infantry wading up to their waists or shoulders. The Duke of Alva urged that the risk was too great, but his master resolved to make the experiment.

On a very dark and foggy morning he moved his whole army down to the river, and pushed forward into a thin line of willows and rushes, at the very water's edge, a thousand Spanish infantry, some armed with the long-range musket which was just coming into use.[2] These skirmishers opened fire on the Saxon pickets on the other side, and threw them into disorder by repeated volleys—the Elector's men had wrongly supposed that

[1] 'Ut in bello multa inania ferunter,' says Lambert Hortensius, vi. p. 1649.

[2] Some speak of arquebusiers, some of musketeers. The musket was only just coming into favour.

they were out of range.[1] After making an ineffectual return fire the pickets melted away, abandoning some boats which were moored alongside of their own bank.[2] The Spanish infantry then plunged into the water, some wading shoulder high, others actually swimming, and got a footing on the opposite bank. Seeing them established there and not repulsed, Charles put the whole of his cavalry at the ford, the light horse and Hungarian hussars in the van, the heavy cavalry following. He himself rode at the head of the German reiter squadrons, in full armour, as he is represented in Titian's famous equestrian portrait in the Prado Gallery. Meanwhile the punts and boats which had been already collected on the hither bank were made into a bridge—some help being also got from the skiffs which the Saxons had abandoned, so that infantry was beginning to form up on the other bank before the last of the cavalry had passed. There was, of course, no chance of getting heavy guns across the improvised bridge of boats : only six very small pieces were brought over, and attached to the light cavalry in the van.

The Elector had been taken entirely by surprise when the Emperor threw his advance guard across what was believed to be an unfordable river, in the darkness of an exceptionally foggy dawn. John Frederic's troops were at breakfast when the first spurt of firing at the ford broke out, and he himself is said to have been at morning prayers—'sacrae concioni in castris vacabat.' His only possible chance would have been to have hurried on each unit as quickly as it could be formed up, to oppose the passage of the Imperialists : no doubt there would have been some disorder, and some regiments would have got to the point of danger later than others. But it was also obvious that the Imperialists would be crossing very slowly, especially the infantry, who had to make use of a single extemporized bridge. As the Saxon camp was only three miles from the ford, it is clear that cavalry at least might have been at the front, and holding back the enemy by a detaining fight, if proper diligence had been used when the first sound of firing had been heard.

But John Frederic made no attempt to counter-attack, or even to throw out a fighting rearguard. He packed off the

[1] The water being extremely shallow for many yards out on the Imperialist side of the river, the arquebusiers are said to have waded up to their knees, or even higher, and to have continued their fire actually standing in the river-bed, shortening the range by many scores of yards.

[2] Said in the rather untrustworthy anonymous *Pugna apud Mühlberg* to have been part of the Elector's bridge destroyed on the previous day.

whole army on the high road to Wittenberg, perhaps a little delayed by the fact that he sent ahead his guns and his baggage before starting his infantry on the march. Finding that there was no opposition to the passage of his troops across the ford and bridge, the Emperor sent on the Duke of Alva and the Margrave Maurice, with the light horse and the Hungarian hussars, to seek for and detain the enemy, even though the infantry was only half across the river. The light horsemen, we are told, each took up an arquebusier behind him, in order that, when the collision came, fire-support might be forthcoming.

The Protestant army had only gone some three miles when its rear began to be seriously worried by the Imperialist vanguard, and the heavier cavalry were visible coming up rapidly. The Elector evidently realized the danger to which an army retreating in haste along a single road is exposed, when its rear is assailed and detained. Possibly he may have remembered cases in recent wars when destruction had started with a running fight, such as Bonnivet's unhappy retreat to the Sesia (1523), St. Pol's affair at Landriano (1529), and (most disastrous of all) the annihilation of Katzianer's Austrian army by the Turks at Valpo (1534). At any rate he halted, and drew up his whole army in front of the wood of Lochau, the infantry and fifteen guns in the centre, on ground which had marshy fields in front and the trees close behind, with the cavalry on the wings. He hoped, it is said, to hold the enemy back for the afternoon, and to retreat to Wittenberg under cover of the night.

The Emperor had time to bring up his whole army before he needed to attack. As was natural, he held back for a time from a frontal assault upon the well-posted Saxon infantry, even when his own had come up, and drawn itself out on a parallel line. But he employed cavalry onsets against the weaker squadrons on the Elector's flanks. The Saxon horse of the left wing was completely beaten by charges headed by the Duke of Alva's men-at-arms. The right wing, pestered for some time by swarms of mounted arquebusiers sent forward by Maurice of Misnia, gave way also when charged in earnest. The Elector 'inclinante jam victoria' sent his guns to the rear, but held out firmly with his infantry. Attacked on both flanks by horse, and after a creditable resistance which lasted for some hours, the Saxon foot finally gave back in disorder into the woods.

Meanwhile the victorious Imperialist light horse had circled round the forest, and intercepted the broken enemy as they came out of cover into the Wittenberg road. There was considerable slaughter both in the wood and at its debouches. Sixteen standards of foot regiments were captured, and 59 'cornets' of cavalry units, also Duke Ernest of Brunswick and many officers. The Elector himself, riding off almost alone, was intercepted by eight Spanish and Italian troopers: he showed fight, was wounded in the face by a sword-cut, and brought as a prisoner to the presence of the Emperor after dark. Only 400 Saxon horse are said to have escaped to Wittenberg by the high road: a larger body with the Elector's son, John Frederick 'the Magnanimous,' who was badly hurt, cut their way out westward and reached Gotha. But the Protestant army was completely dispersed and practically annihilated.

The Emperor treated the captive Elector with asperity. The prince was led to him bleeding, and so exhausted that he sat huddled upon his saddle-bow, and could not dismount. He had only strength to mutter that he hoped for clemency from the Emperor. His captor made the sarcastic reply that he was hailed as Emperor now, but remembered the 'diffidation' in which John Frederick had styled him only 'Charles of Ghent.' He ordered the Elector to be arrested and placed in prison. He was actually tried for high-treason by a court-martial and condemned to lose his head—an unconstitutional proceeding, for he should have been brought before a proper Imperial court. But it is pretty certain that the Emperor never intended to proceed to this extremity, and was only intending to make John Frederick order his wife to surrender Wittenberg and his other Saxon fortresses, and to secure the complete dispersion of all his surviving troops. The Elector was sent as a prisoner to Oudenarde, in Flanders, with no promise of release, and was forced to give up all his dominions. Charles made over his electoral dignity to Maurice of Misnia, who had served him so well and so treacherously, and gave along with it Wittenberg and the old Saxon duchy, with John Frederic's Thuringian and other lands, all save Gotha, Weimar, and Eisenach, which might go to his heirs.[1] Thus the 'Elector of Saxony' means for the future Maurice of Meissen and his successors, and the old 'Ernestine' electoral house became petty dukes of scattered scraps in Thuringia.

[1] To which Altenburg was afterwards added. The Ernestine house did not even keep its scraps together under one hand: after John Frederic 'the Magnanimous' they broke up into three or four lines, condemned to political impotence.

On the news of the battle of Mühlberg the various princes and free cities that were still in arms for the Schmalkaldic cause hastened to make what terms they could with the victor. Philip of Hesse, the greatest offender save John Frederick in the Emperor's eyes, found himself, greatly to his disappointment, committed to an imprisonment that seemed—like the ex-Elector's—to be limitless. Of how Charles dealt with Germany, with the 'Interim' and its implications, there is no occasion to tell—the story is political and psychological, not military. Nor is there any occasion to tell in detail the strange tergiversation of Elector Maurice's second set of campaigns, when he betrayed the Emperor just as he had betrayed the Schmalkaldic Leaguers in 1546, and led against his suzerain an army levied in that suzerain's name for the siege of Magdeburg. He leagued himself with the French, and forced on the reluctant Charles the inconclusive treaty or pacification of Passau (August 2, 1552). How far Maurice would have prospered in his design of making himself the dominant personage in Germany can never be known. For he was killed most ingloriously and inopportunely, while hunting down the ever-turbulent raider, Albert of Hohenzollern-Kulmbach, who was plundering the lands of the Franconian bishops. A chance bullet in the cavalry skirmish at Sievershausen (July 9, 1553) gave him a mortal wound, and ended his ambitions, which were as limitless as they were unscrupulous. His influence only survived in the long continuance of the unsatisfactory agreement of Passau—which merely put off the decision of religious strife in Germany for another two generations. But he had broken the nerve of the old Emperor, who never recovered from the ignominy of the compromise of Passau, and the disastrous failure of the siege of Metz.

[*P.S.*—In this account of Mühlberg I have followed Lambert Hortensius's *De Bello Germanico*, vii., rather than the very divergent account of the anonymous '*Descriptio Pugnae non procul ab oppido Mühlberg*'—both printed in the Augsburg 1574 collection of chronicles of 'Quae in Imperio Caroli V. Caesaris inciderunt.' Many details differ—those in the *De Bello Germanico* look more reliable. Nothing intelligible is to be made out of Thomas Winzer's long metrical narrative of the combat of Sievershausen, which is pure literary trash. I cannot make out the details of this cavalry skirmish on a large scale.]

CHAPTER IV

THE BATTLE OF ST. QUENTIN (AUGUST 10, 1557)

AFTER the rupture of the short-lived truce of Vaucelles (February–November 1556), the final episode of the great war between Hapsburg and Valois began—destined to last for two years. Charles V had abdicated, a broken man since the siege of Metz and the last inconclusive campaigns in Picardy (October 1555), and the director of policy on the Hapsburg side was now his son Philip II, King of Spain and lord of the Netherlands, but not Emperor as his father had been. The German domains of the house, and the Imperial crown, had passed to his uncle Ferdinand, and with them the burden of the Turkish War on the Danube, and the problem of the disruptive tendencies of the Protestant Reformation. Philip could count on the co-operation of the Emperor, but not on his enthusiastic support; Ferdinand was wholly unable to prevent the appearance in French armies of thousands of mercenary landsknechts, not Protestant zealots but simple mercenaries, destitute of all national feeling. Such hirelings had been numerous in the days of Francis I; we have already had to mention them time after time. In the reign of Henry II they formed a still more important part of the French infantry—even outnumbering the Swiss. Both at St. Quentin and at Gravelines there were more German than native companies in the beaten army. On the other hand, Philip could count on a much more effective help from England than his father had enjoyed, for after he had forced his unfortunate wife Queen Mary to declare war on France (June 1557) the considerable English contingent on the Continent acted simply as an auxiliary division in his army, and was never used for cross-purposes and purely English objects, as had always been the case when Henry VIII had been in league with Charles V.

In the campaign of 1557 Philip disposed of an army as considerable as those which his father had brought into France in the days of St. Dizier, Renty, and Landrécies, apparently about 50,000 men of all arms, including a solid Spanish contingent, a very large levy from the Netherlands under the Duke of Aerschot and the Counts of Egmont [1] and Horn—the scaffold-companions of 1568—many thousand Germans lent by his uncle, under the Dukes Henry and Ernest of Brunswick and the Counts of Mansfeldt and Schwartzburg, with the English contingent of 7000 men under the Earl of Pembroke. The King was theoretically in command, but hardly showed himself before the army during the whole campaign. The charge of operations was really in the hands of a rather young but already experienced commander, Philibert, Duke of Savoy, a bitter enemy of the French, who were still occupying Turin and the greater part of his Italian states. Hence his nickname of the 'Landless,' changed afterwards, when he had won back everything by hard fighting, into Philibert 'Testa de Ferro.' The duke's plan of campaign was to demonstrate against the frontier of Champagne, and when the enemy were attracted towards Mezières and Rocroy, to pass westward, and throw himself into Picardy. This he did with success, and, after threatening Guise with a false attack, suddenly sent forward his cavalry to encircle St. Quentin, which was at the point where he intended to make his breach into northern France.

Henry II was in a very bad position at the moment—for he had sent the best of his troops, under the Duke of Guise, to carry out the last of those mad raids into southern Italy, which had always been so fatal to his predecessors, Francis I and Louis XII. Lured into this enterprise by the anti-Spanish Pope, Paul IV—whose military resources turned out to be negligible—Guise, with the larger half of the gendarmerie, the Swiss auxiliaries, and the 'old bands of Piedmont,' was bickering with the Duke of Alva on the borders of Naples, when he was wanted in Picardy. The Constable Montmorency, to whom the defence of the northern frontier was entrusted, had concentrated at Attigny (behind Mezières) an army which was intended to co-operate in the protection of Champagne. It did not amount to much more than half the number of which the Duke of Savoy could dispose, consisting of 20,000 foot—the landsknechts were somewhat more numerous than the native French, legionaries

[1] Lamoral of Egmont (1522–68) must not be confused with his distant relative, Maximilian of Egmont-Buren, who had co-operated with Henry VIII in the Boulogne campaign of 1544 and died in 1548. The latter left no male heir, and his daughter married William the Silent.

and others—and some 6000 horse of all kinds—gendarmerie, *chevaux légers*, arquebusiers on horseback, and German pistoleers (' Reitres '). Other new French companies were being raised, but were not ready for service in July.

The sudden swerve westward of the enemy compelled Montmorency to make a similar movement, which brought him to La Fère, on the Oise, which formed his base of operations during the succeeding campaign. His first duty was to provide for the defence of St. Quentin, which had been caught undermanned and low in munitions. It was only saved from instant capture by the energy of the Admiral Gaspard de Coligny—the Huguenot hero of future years—who was in charge of the government of Picardy, and cut his way into the place on August 2, before the investment was complete, at the head of three companies of gendarmes and two of infantry.[1] The resource and obstinacy which he put into the defence detained the enemy in front of it for many days.

St. Quentin slopes up steeply from the north bank of the Somme, which there (as in so many parts of its upper course) does not run in a single, well-defined stream, but in many small channels concealed in broad marshes. The low-lying south side of the town lay above a wide expanse of bog and mud, with patches of shallow open water showing among it. This makes it practically impregnable on that front, but also inaccessible to an army coming from the south. The only exit from it on its south and south-east side was by a causeway crossing the marsh, which carried the road to La Fère and Chauny: this passed by means of bridges over two low islands. Of these the larger one was occupied by a suburb, the Faubourg de l'Isle, which was so far from the main line of defence that the besiegers got hold of it without much trouble, and placed in it a garrison of Spanish infantry, which stockaded itself so strongly that it completely blocked the great road southward, the natural line of approach for a relieving army.

The Duke of Savoy distributed his divisions around the three

[1] He entered by the Ham road, on the south-west corner of the town, which was only watched by a mere cordon of German horse, because this part of the line of investment had been destined for the English contingent, which (marching up from Calais to join the main army) had not yet arrived. The whole of Coligny's force did not succeed in entering—a quarter of the gendarmes and more than half the infantry lost their way in the dark, and returned to Ham, as did some light horsemen who had been attached to the force.

remaining fronts of the town, the Spaniards on the eastern side, communicating with their detachment in the Faubourg de l'Isle by a ford [1] upstream, the Netherlanders on the north-east side, the Germans on the north-west, and the late-arriving English on the west across the Ham road, the only other possible line by which succour could arrive. He commenced to dig trenches and erect 'mounts' for batteries, against the northern and higher front of the town, but it was some days before his heavy guns began to arrive, as they moved slowly and the army had outmarched them.

The Constable Montmorency was most anxious to do what he could to succour St. Quentin, all the more because Coligny was his sister's son, and attached to his political faction—that which was opposed to the growing influence of the Guise family. But it was obvious that he could not dare to cross the Somme and offer battle to an army almost twice the strength of his own. Having established himself at La Fère, he first attempted to reinforce the garrison, which was still inadequate. He sent a considerable detachment to Ham—south-west from St. Quentin, under the Marshal St. André, with orders to make an attempt to throw a body of infantry into the town by night, while other troops should make demonstrations against different points of the besieger's lines. Coligny had succeeded in sending out emissaries from the town, by tracks across the marsh known only to fishermen and fowlers, who were to indicate a point north of the Ham chaussée where the hostile lines were thin.

This plan was tried, but failed disastrously. The enemy, apprised, it is said, of the design,[2] shifted his posts, and dug a trench, well garnished with arquebusiers, across the front which was to be attacked. When Dandelot, the brother of Coligny, charged at midnight with 2000 foot at the appointed spot, he met with such a salvo of shot that almost the whole of the head

[1] Or at least so I presume, for there was no way to the Faubourg across the marsh and open water near it, and the somewhat distant Rouvray ford is the only possible way of approach.

[2] Bussy Rabutin, book ix. p. 28, says that the failure was due to the fact that some Englishmen—political exiles from Mary's tyranny or mere adventurers—who were serving among the French light horse, were taken prisoners that afternoon in a skirmish, and, to save themselves from being hung as traitors, gave away the plan, in time for precautions to be taken against it. That there were English exiles in the Constable's ranks is incidentally mentioned by Melville (p. 24), who in his lively account of the battle of August 10, mention that when wounded he was looked after by his friend Mr. Henry Killegrew—a very well-known person in Elizabeth's reign.

of the column was laid low. Of the rest many were taken prisoners, and the rest dispersed in the darkness (August 3).

The Constable still persisted in his design of bringing help to St. Quentin. His next plan was a most ill-advised one. Knowing, by means of Coligny's emissaries, that there were tracks—if difficult ones—through the marsh on the south side of the town, and that there was a stretch of open water to the west of the Faubourg de l'Isle, he determined to bring his army up to the edge of the marsh, and to deploy it there, to protect his operation, while boats should be launched on the open water and carry relays of infantrymen to its north bank, under the walls of the town, where Coligny would have guides ready for them, to show them causeways and posterns by which they could cross the ditch and enter the place. The enemy, it was calculated, could not disturb the proceedings, as all his forces were on the north bank, save the Spanish garrison in the Faubourg de l'Isle, which could easily be blocked up. It was believed that there was no point for many miles where he could pass the river-marsh in force. There was, according to local advices, a ford at Rouvray, two miles upstream, but it was said to be difficult, so narrow that only four men could cross abreast, and to be unguarded. It was considered impossible that the enemy could use it for a turning movement in force, and its exit could be blocked by a trifling detachment.

On August 8 the Constable went out in person with a large body of cavalry to make a 'reconnaissance in force' of the ground on which he intended to operate. He found no outlying pickets in his way, till his observation-officers reached the neighbourhood of the marsh, and discovered a couple of Spanish companies holding a mill, above the causeway leading to the Faubourg de l'Isle. The camps of the enemy were visible for many furlongs, stretching uphill on the other side of the water. But there was no cavalry-screen (as might have been expected) on the south side of the Somme.

On the following night Montmorency marched out from La Fère in two columns, the cavalry through the town on the right, the infantry on the left by bridges thrown across the Oise downstream, by which they got on to the Chauny-St. Quentin road, which runs for some way parallel to the La Fère-St. Quentin chaussée. The infantry—22 German bands and 16 French bands—brought with them 15 guns and six or seven large boats 'of the sort that are commonly used in camps to pass an army over rivers and waters.'[1] Marching all night, the French appeared at eight in the morning, somewhat tired but absolutely unopposed, opposite the south-east front of the marsh. The Spanish arquebusiers in the mill above the causeway were easily driven in, and fell back on the Fauburg de l'Isle, where the garrison stood on the defensive, and could not well be attacked across a narrow bridge.

The Constable drew up his army on both sides of the chaussée, and ordered the boats to the front, to be launched on the shallow water on his left—his follower Melville describes it as a 'little loch.'[2] Meanwhile he sent an observation-officer, M. d'Eschenets, to report on the ford of Rouvray, who brought back word that there was no guard there,[3] but that it would be well to block it with a hundred arquebusiers—obviously he judged the point of no importance. The Constable, urging that foot-soldiers move slowly, sent off instead a squadron of German 'reiters' or pistoleers.

There followed a couple of hours of considerable worry for the French. It was of no use that they amused themselves by placing some guns on the bank by the mill, which made long shots at the Spanish camp across the water, and caused some confusion there.[4] This was futile—the main thing of importance was that the boat-scheme turned out impracticable. It was difficult to get them down to the water, and so many soldiers rushed into them and crowded them, that it was hard to shove them off, because of the mud and slush. Some stuck fast, others could not be steered to the appointed landing-places. And on getting out, the men could not find the dry paths in the mud, so that many lost their way and wandered into patches of marsh, from which they could not extricate themselves, and were smothered or drowned.[5] At the end of a couple of hours only Dandelot, Coligny's brother, and about 450 men had got into the town, and the boats were sticking here and there in the mud.

Meanwhile the Duke of Savoy had been granted time to

[1] Sir James Melville, p. 23. [2] *Ibid.* p. 23.

[3] 'Qu'il n'y avoit vue encore personne passer, n'y prendre le chemin pour y venir. Bussy Rabutin, ix. p. 38.

[4] Some one said that the Duke of Savoy's own tent was hit, and that he was seen to run without his armour, and to make for Egmont's camp up the hill. The distance makes it impossible that individuals could be identified. See Bussy Rabutin, ix. p. 38.

[5] This interesting description is from Bussy Rabutin, ix. p. 39.

think out a plan of operations, and had come to a very bold conclusion. It would be impossible to get his whole army across the Somme and its marshes in any reasonable time; but his cavalry could move rapidly, and he was aware that the ford of Rouvray was no mere defile, but a broad passage, over which horsemen could pass thirty abreast. He knew that he had immense superiority in the mounted arm, and could drive in any cavalry-screen that the French might throw out against him. There would be no need to attack the Constable's infantry, if it stood to fight; but it was certain that Montmorency would not wait in position for the arrival of the overpowering forces of foot-soldiery of which his enemy could dispose, but would retreat. And when he was on the march, probably in disorder, all manner of opportunities might occur for cutting him up. The initial movement was obviously possible, because it could be seen that there was only a handful of 'reiters' guarding the all-important ford of Rouvray.

The French officers, watching the hasty musterings, and falling-in to order, of the troops on the opposite hillsides, presently detected that there was a general move towards the east with cavalry at its head. The Constable therefore detached the Duke of Nevers with three companies of gendarmerie, to reinforce the picket of Germans at the ford of Rouvray. But on approaching it Nevers met the 'reiters' flying in disorder, and saw that Netherland horse were pouring across it on a broad front and in great numbers; there were already 1500 of them over the water, and the hill opposite was crowded with more approaching squadrons. Some of Nevers' captains advised him to charge the head of the column at all costs, and to endeavour to thrust it back and block the passage. But the Duke held that, considering the odds, this would be mere madness, and fell back towards the mill near the chaussée, where the Prince of Condé and the bulk of the French light horse were drawn up.[1]

On getting news from Nevers that the enemy was across the ford in great strength, the Constable determined on immediate retreat, sent off his infantry and guns on the Chauny road, and ordered the boats to be abandoned. He himself with the remainder of his cavalry stopped behind, to support Nevers and Condé, and to allow the slow-moving infantry to

[1] Bussy Rabutin, ix. pp. 41–2.

get a fair start. 'These horsemen come but to stay us, till their foot be advanced,' he remarked to his staff.[1]

The enemy, however, had other designs. When the whole of the cavalry were across the ford they began to deploy on a long front, completely outflanking Nevers and Condé, who had to retreat in haste in order to escape envelopment. Though

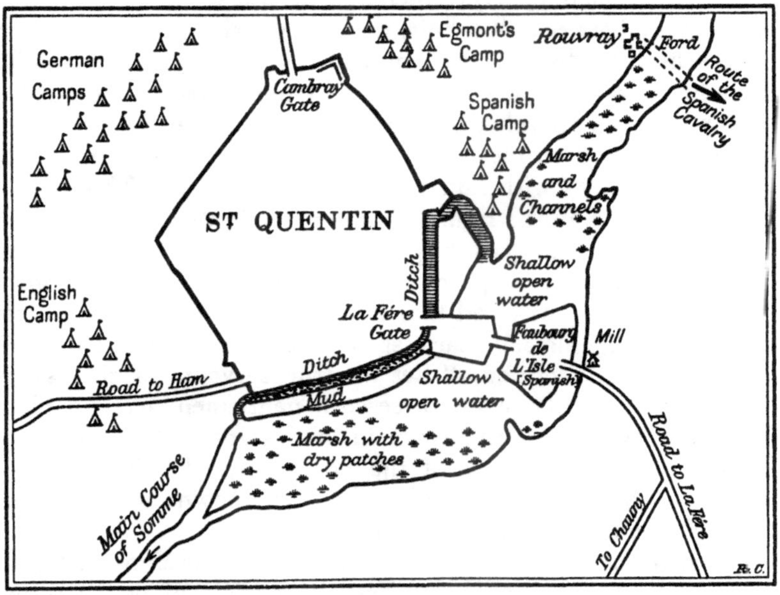

ENVIRONS OF ST. QUENTIN.
(From Tassin's Map of 1631.)

N.B.—It is unfortunately impossible to give the precise condition of the fortifications of St. Quentin in 1557, as the earliest accessible plan, that of Tassin, dates from 1631 only, and shows many bastions and several outworks added since the year of the battle. I have, therefore, only been able to give the general outline of the place.

one brigade of reiters under the Count of Schwarzburg was left to guard the ford, the rest came on, at least 4000 or 5000 strong, in eight well-ordered brigades.

As their pace was not very great, the Constable had time to draw up for a detaining action on a hillside some mile south of the Somme marsh, between a village and a wood. It was intended that Condé and Nevers should fall in upon his left flank. But it was impossible to improvise a line of battle,

[1] Melville, p. 24.

and complete disaster followed. Both Melville, our best eye-witness, and Bussy Rabutin state that panic started with a bolt to the rear of a body of non-combatants, paymasters ('trésoriers'), grooms,[1] and others, who went off in a mass. It is very difficult to stop cavalry who have been retreating for some time, as had been seen at the Battle of the Spurs, forty years back.

'The Constable,' says Melville, who was at his side, 'being in a valley between two hills, moving toward the strait place where he intended to stay, spurred forward up the little hill, that he might see how to resist and put order in his battle, which gave an apprehension to some that he was flying. And when he turned on the top of the hill, to behold the onset, no man would tarry with him for all his commands. Though he ever cried, "Return, return!" their heads were homewards, and their hearts also, as it appeared. Then his master of the horse brought him a speedy Turkish horse, to run away with the rest, but he answered that it was against his profession to fly, addressing himself fearlessly against the greatest troop of the enemy, and crying, "Let all true servants of the King follow me." Only some threescore gentlemen accompanied him, who were all overthrown in an instant. He desired to be killed, but his master of the horse cried, "It is the Constable; kill him not." Before he was known, he was shot through the thigh, and then taken prisoner.' [2]

The Netherland and German squadrons, therefore, charged in upon a disordered line of French cavalry, of which some troops fought, but more fled. Nevers and Condé, on the left, got off with a fair number of their men, by scattering across the hillsides. But the greater part of the French horse were either ridden down or taken prisoners. Nor was this the end of the day; pushing on in pursuit of the fugitives, Egmont and the Brunswick dukes came upon the Constable's unfortunate infantry, hurrying in one long column southward along the Chauny road. Taken in flank, the separate bands formed up

[1] 'There are as many servants as masters among the French men-at-arms,' remarks Melville, and adds that it was the Marshal St. André who told them all to clear off as quick as possible, as they were a mere impediment to the fighting men.

[2] Melville continues with his own adventure—he was hurt by a stroke on the head which knocked off his helmet, and stopped by two enemies in a narrow way. 'My horse ran between them against my will; they cut at me, but I leaped a dyke, and ran through the village, where so many of our foot were slain, after which there was room enough to escape. So I came safe to La Fère, where my old friend Mr. Killegrew held my horse, while my head was being dressed.' Melville, p. 24.

in small squares [1] but could not keep the cavalry out, each being beset on all sides, for the enemy burst in between band and band. They broke up, and the rush of fugitives got blocked in the street of the village of Essigny le Grand, where there was 'a great butchery.' But the victors finally gave quarter, and gathered up some thousands of prisoners. All the guns were taken except two, which were at the head of the marching column, and were abandoned by the wayside near the village of La Justice, not far outside La Fère.

If the pursuit had been pushed to the extreme—and there were German squadrons which had been little engaged, and were comparatively fresh—the whole French army might have perished. But Philibert of Savoy, content with such a victory as had not been seen since Pavia, halted his victorious cavalry at dusk without reaching La Fère. His infantry had not fired a shot all day, having been left far behind, and it would seem that Pembroke's English contingent, which had the farthest way to march of all the army, never reached the battle front at all.

The Duke of Nevers, taking command, as senior surviving officer, of the wreck of the French army, retired to Laon, where on a muster being held it was found that there remained some 1200 French horse, light and heavy, a couple of hundred German pistoleers, the equivalent of four out of the sixteen French bands of infantry, and some 3000 or 4000 men, of the twenty-two strong German bands, which had mustered 12,000 before the battle.[2] Altogether the loss must have come to about two-thirds of the army, but many thousands were prisoners not dead,[3] and many more had simply disbanded themselves. Among the killed were a Bourbon Prince, John, Lord of Enghien (heir of the victor of Ceresole), the Vicomte de Turenne, and many other noblemen. The still more numerous list of prisoners included, beside the wounded Constable, the Duke of Montpensier—another Bourbon—the Duke of Longueville, the Marshal St. André, the Count of Rochefoucault, Aubigny (heir of the old Scottish Marshal), Montbéron, son of the

[1] 'Se serraient en battaillons quarrez et bien joints,' says Bussy, ix. p. 49.

[2] Bussy Rabutin, ix. p. 68.

[3] The rank and file were dismissed on easy terms, after having been paraded before King Philip. The Germans merely had to go home after swearing not to serve the King of France for six months! Any one who could pay a ransom was, of course, kept in hand—there was much bargaining and sales of wealthy captives between the captors.

Constable, Ludovico Gonzaga of Mantua, and the Rheingrave Philip, who had commanded the German infantry. The victors were said to have lost only fifty men—an exaggeration no doubt, but they can hardly have lost 500, and no officer of mark was killed or wounded.

King Philip came up to St. Quentin a day or two after the battle, and had presented to him in triumph some fifty standards of horse and foot, all the French guns, some hundreds of noble prisoners, and some thousands of the dilapidated rank and file, who were dismissed on parole. There was a general expectation that he would order a march on Paris, which was now covered by no army, and was in a state of panic. But that cautious monarch refused to allow Philibert of Savoy to strike at the French capital, and directed him to complete the siege of St. Quentin, and after that to deal with the smaller neighbouring fortresses. Coligny held out desperately for seventeen days after the battle, and the place had to be stormed; the English contingent suffered heavily at the breaches—the only fighting that it got during the campaign [1] (August 27). The army then dispersed itself to the sieges of Ham and le Catelet, and after their fall took Noyon and Chauny. Then to the surprise of all Europe, it turned back in September, and broke up at Cambray; the King returned to Brussels, the German contingents were all disbanded, the Netherlanders went to their homes, and the Spaniards were put into garrisons.

So ended a campaign which bid fair to be the ruin of France—its termination was even more surprising than that of Charles V's invasion of Champagne in 1544, for on that occasion the French army was still intact, while in 1557 it had been practically destroyed at St. Quentin. The old Emperor, in his monastic retirement at Juste, is said to have received his son's letter concerning the victory, and to have remarked that the next dispatch would bring news of the occupation of Paris. So thought every one else who was following the course of the campaign. The explanation of Philip's refusal of such an opportunity may be referred to many causes. One was his own proverbial dilatory habit of mind; he liked to finish one job before he started on another, and stayed for a fortnight before St. Quentin, which was obviously doomed to fall, when he might have left 10,000 men to blockade it and have marched with 40,000 on Paris. But after the fall

[1] Holingshed, vol. iv. p. 89.

of St. Quentin (August 27) there was still a good month of campaigning season left him. He used it merely to sweep up trifling garrisons at Catelet, Ham, and Noyon, while the French were rallying, and calling in reserves from all quarters.

One reason for dallying that was alleged is the fact that money was running short—the army was ruinously large from the paymaster's point of view, and the German contingents were already beginning to get troublesome about pay in arrear, and the approaching end of their contracts. The English had been lent for a fixed term of months, and were due to return in the end of September to Calais—as they did. There had been some unpleasant riots and affrays between the Spanish and the German rank and file, and the officers were quarrelling about the ransom-money of rich prisoners—whom they had been selling about among themselves. But all these were trifling hindrances for a great king—and, if Philip had been a prompt and vigorous general, he would have tried the grand enterprise, and could certainly have been before Paris in a week after the battle, while the French were still hopelessly scattered.

Meanwhile King Henry, surprised at his own good fortune, called back Guise and the wrecks of his army from Italy, drew in many bands from his eastern fortresses, bought the services of 14,000 'new' Swiss, and issued commissions for the levying of many French companies. By October he had again an army at his disposition—but he ought never to have been given the time to collect it.[1] And at midwinter Guise, now in chief command, took the offensive, and performed the unexpected and surprising feat of capturing Calais—a success which wiped out the memory of St. Quentin, and prolonged the war for another year.

As to the events of August 10, 1557, the main fact that strikes the military observer is the rashness of Montmorency in bringing his army into a position where it could be beset by an enemy of overpowering strength. The blame must be shared by his staff, for he had several times sent out exploring officers to make out the lie of the country. In the heart of Picardy this should not have been a difficult task—but it is evident that they reported to their general that the water in

[1] Guise came back, in haste, by sea from Ostia. But the bulk of his army, under Tavannes, had to make the long march across the Apennines and the Alps, before it reached France.

front of St. Quentin was fit for the passage of boats, and—what was worse—that the ford of Rouvray was narrow and difficult, and only required to be watched by a small force. If its real character had been known, 5000 men and some guns could have been sent to block it, and would have sufficed.

The next most important deduction that must be made from the details of the battle is that unsupported infantry, on the march, when its cavalry had been driven away, was still helpless against a flank attack by a very large mass of hostile squadrons. There are other examples of the fate of infantry deserted by its horse—Gemblours (1578) and Tournhout (1598) were similar cases in the next great war—that of the Insurrection of the Netherlands. Even in more modern days examples could be quoted—the fate of Del Parque's army at Alba de Tormes (November 28, 1809), and of Henry O'Donnell's at the combat of Margalef a year later (April 3, 1810). On each of these occasions the French horsemen achieved a brilliant success without any infantry help, when the trifling Spanish cavalry had been brushed away. I am not sure that Allenby's immense cavalry raid into the rear of the Turkish army of Palestine in 1918 might not be quoted as offering some analogies for complete triumph of the mounted arm when properly used.

CHAPTER V

THE FALL OF CALAIS (JANUARY 7, 1558)

THE tame ending of the campaign of St. Quentin had surprised all Europe, and though in August it seemed as if France was ruined, and Paris in desperate danger, by the end of the autumn Henry II had once more a formidable army at his disposition, while the heterogeneous forces of Philip II had dispersed into winter quarters or disbanded themselves.

It was the complete dislocation of the army victorious at St. Quentin, which emboldened the Duke of Guise to undertake what had always been a rare thing during these wars—a mid-winter campaign; and its conduct adds considerable credit to his military reputation. Its aim was no less ambitious than the capture of Calais, which he knew to be undermanned, and neglected by the English Government. French armies had raided the Terre de l'Oie, the low-lying district in front of Calais, several times during the earlier wars—most notably in 1544 and 1545—but there had been no regular attack on this abnormal English stronghold since that of 1436, when Humphrey of Gloucester and Richard of Warwick scared away the beleaguering army. According to French chroniclers, the first idea of a dash at Calais had been conceived by Senarpont, the governor of Boulogne, who had communicated it, long before the battle of St. Quentin, both to Coligny and to the Constable Montmorency, who had approved it, but put it off to some future season, when external conditions should be more favourable.[1]

No better moment could have been chosen than New Year's Day, 1558, for after the expeditionary force which the Earl of Pembroke had taken to St. Quentin had gone home, Calais was disgracefully undermanned. To guard the town and its complicated system of outlying defences, Lord Wentworth had in hand only 500 men, even when to the depleted regular garrison there were added some belated scraps of Pembroke's

[1] Bussy Rabutin, x. pp. 144–5. Tavannes, xiii. p. 172.

force. He could count on 200 or 300 armed citizens to supplement this meagre total.[1] There was a separate garrison at Guisnes under Lord Grey, which amounted to some 150 men more. For the two towns and their subsidiary castle at Hammes there were about 1000 men in all disposable. Moreover, the fleet had disbanded for the winter, and could not be reassembled save after long notice. The English Government had not the least notion that mischief was at hand, when Lord Grey on December 22 and Lord Wentworth on December 26 sent intelligence to London that there was a concentration of French warships at Boulogne, and a mass of at least 12,000 men at Abbeville. By the time that the Council received and began to ponder over this news, it was too late to do anything, for Guise had appeared in great force before Calais on January 1, and a French squadron from Boulogne had simultaneously slipped round to the Straits, and was blockading the place from the sea side.

Guise had taken preliminary measures to distract the Spanish forces in the Netherlands—a fortnight before marching on Calais he sent a large detachment under the Duke of Nevers through northern Champagne, to demonstrate against the fortresses on the Meuse and in the Ardennes. When it had displayed itself, and attracted attention in this direction, it made forced marches back to Picardy, and joined the main army.[2]

Meanwhile he moved rapidly forward himself with 25,000 men by Abbeville and Ardres, and presented himself on January 1, 1558, before the fort of Newnham Bridge on the Boulogne Road, which covered the south-western approach to Calais above the marshes. In front of it he threw up a battery. Though this was an important outwork, Lord Wentworth ordered it to be evacuated at the first cannon-shot,[3] on account of his penury of troops. He could not afford to lose even 100 men, and drew them back into Calais itself. On the same day (January 3) that Newnham Bridge was abandoned, another French column cut in between Calais and Guisnes, isolating the latter place completely, while a third marched close to the shore along the sandy Dunes on the north side of the narrow

[1] Holingshed, iv. p. 92. Compare the letter of the Venetian Ambassador, written only a month before the disaster.

[2] Bussy Rabutin, x. p. 146.

[3] Which shot, as it chanced, took off the head of Henry Horseley, the master gunner of Calais. Holingshed, iv. p. 92.

harbour of Calais. On January 4, the commanders of this column, the Duke of Aumâle and Marshal Strozzi, carried out a very bold enterprise, the capture of the fort of the Rysbank, which commanded the northern side of the harbour. It was on the extreme point of the long sandy extension of the Dunes which lies opposite the town, guarded by a ditch not well filled when the tide was out. The French battered it from the side of the Dunes, and then stormed it at low tide, having easily made a breach in its not over-strong wall. The garrison, very weak in numbers,[1] surrendered at discretion, being unable to get away, as their comrades at Newnham Bridge had done. A number of ships, laid up for the winter, under the Rysbank shore, fell into the hands of the enemy. For the future all entry of reinforcements from Dover into the harbour would be made hazardous by the guns on the captured fort. But no reinforcements came!

The French now established themselves close under the walls of Calais, on ground of which much might have been inundated if Lord Wentworth had been given time to cut the dykes. Guise concentrated his attack on two points, putting up batteries, the one on the Dunes opposite the north side of the town, to play on the water-gate and the harbour front facing the Rysbank, the other opposite the castle, on the north-west angle of the place, which was built close over the foreshore, looking down into tidal mud. The first-named attack was intended mainly to distract the garrison. But the real attack was on the castle, at the west end of the town, which had been reported to Guise as its weakest, not its strongest, point.

It was an unimproved fourteenth-century building, without any outworks or even ramparts.[2] After two days' bombardment (January 4–5) its seaward angle crumbled into a long breach, which was stormed on the night of the 5th, the assailants wading through the slush at the head of the harbour and then through the castle-ditch. Wentworth had ordered some powder-bags and fougasses to be placed under the more accessible points of the breach: but they failed to work, and the French got in with little loss. It was only when the enemy were actually inside the castle that the governor resolved that it

[1] Though mounting over 40 guns, the fort had a garrison of under 150 men.

[2] Bussy Rabutin calls it 'creux et sec, et sans ancuns ramparts,' x. p. 152. By 'creux' I suppose that he means 'four square' with an open court in the middle, and no cross-walls or traverses.

must be retaken at all costs—two assaults were made upon its inner front from the streets—the second with the aid of cannon, man-handled forward to blow open the gates. But the garrison was too weak and was easily beaten back, with the loss of Sir Anthony Agar, the Marshal of Calais, and 80 more killed—according to English accounts—the French say 200 or 300, which would mean nearly half the available strength of the troops in the place.[1] On the evening of the 6th, Wentworth

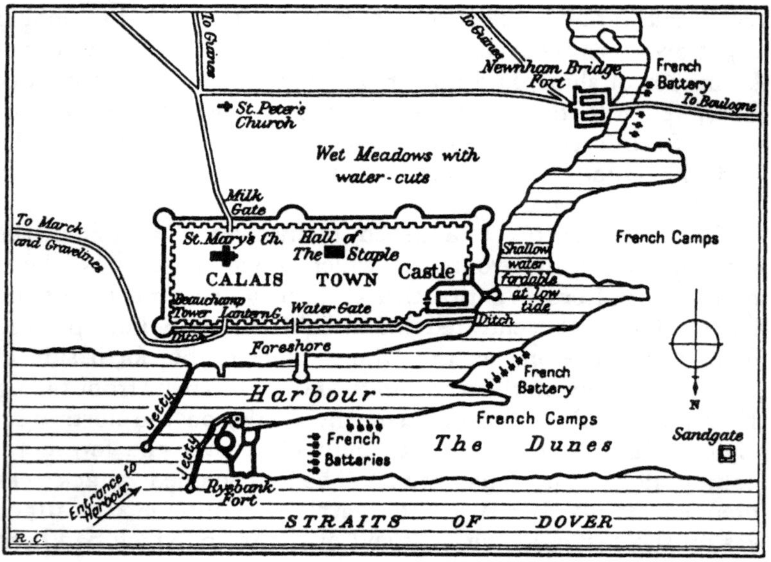

CALAIS.
(Mainly from the map of 'Stephen the Almayne,' 1549.)

offered to capitulate. Guise, anxious to have the matter finished before any reinforcements could arrive from Flanders or Dover, granted free exit to the garrison, save Wentworth and his chief officers, who were held to ransom. The civilians were to come out without bag or baggage, 'leaving behind everything that was worth more than a groat.' Four thousand two hundred of them were escorted to Gravelines, the nearest town in Flanders, by three companies of Scottish light horse, who (as we are told) treated them with much consideration and kindness—a rare thing in those wars.[2]

[1] So Holingshed and Bussy Rabutin respectively. Probably the one understates and the other overstates the casualties.

[2] Holingshed, iv. p. 92.

On the morning of the 8th, several ships appeared off the harbour from Dover, carrying hastily assembled reinforcements from Kent—but the French flag was seen flying everywhere, wherefore they turned back. It is hard to see that the Queen's Council could have made better haste—the first definite alarm had reached them only on December 28, and it had seemed incredible that a fortress reported impregnable, which had defied all attacks for two hundred years, could fall in seven days. The fault lay in the 'wilful negligence and lack of credit of earlier warnings'[1] which had left this large and important place hopelessly undergarrisoned. With 4000 men inside, instead of 500, it might have held out for a long time, even against modern artillery, for the outlying defences—such as Newnham Bridge—could have been held for purposes of delay, instead of evacuated, and the dykes might have been cut everywhere.

Guise then turned against Guisnes, which was so far inland that there was no possibility of succour from the sea. It had been observed during the siege of Calais by a large detachment, lest help might come from Flanders. But the governors of the Netherlands were taken as much by surprise as the Council in London, and their attention had been drawn off eastward by Nevers' demonstration on the side of the Ardennes. Lord Grey got the help of one company of Spanish hackbuteers from the garrison of Gravelines, under an officer named Montdragon, who distinguished himself greatly in after years in the Dutch wars. Without this aid he would have had as short shrift as Wentworth in Calais. But, evacuating the town of Guisnes, he retired into the castle, and there held out for eight days of bombardment, and withstood several assaults. On January 20, the walls being in ruins, he capitulated—the garrison marching out on honourable terms, though Grey himself with his son and his captains were, of course, held to ransom.[2] This was a creditable defence, contrasting happily with the miserable show which (by no great fault of his own) Wentworth had made at Calais. Guise blew up the castle of Guisnes, considering that, with Ardres so near, it served no good purpose. Useful to the English when they were holding Calais, it would be of no importance when Calais was lost.

[1] Holingshed, iv. p. 93.

[2] There is a good account of the siege of Guisnes in Lord Grey's *Life*, by his son, published as a separate tractate by the Camden Society.

The last holding of the English 'pale' in France was the little fort of Hammes, in the marshes between Guisnes and Calais—a place of no strength, but hard to approach. On learning of the fall of Guisnes, the captain in command there, seeing himself completely encircled, and deprived of all hope of succour, took the sensible but inglorious step of absconding by night by a circuitous path through the marshes, leaving many spiked guns and some munitions to the enemy. It would have been impossible to burn or blow up the castle, as noise or fire would have attracted the attention of the enemy, who would have caught the garrison on its retreat. As it was, they reached the Spanish fortress of Gravelines unmolested.

Guise's device of battering the Castle of Calais from across the shallow tidal water at the head of the harbour, and then sending attacking columns to mount the breach by wading through the slush, bears an interesting resemblance to Wellington's dealings with St. Sebastian in 1813. There, too, the fortress was shattered by fire from the other side of tidal water, and the hour of assault was fixed at the time of the lowest ebb. If Guise got in with little trouble and signally few casualties, the cause must be sought in the fact that Wentworth had only some 700 or 800 men to defend a large flat town, while Rey at St. Sebastian had over 3000, and an inexpugnable hill-fort behind him, into which the garrison could retire when the breaches were forced. Hence the difference in the length of the two sieges, and in the losses of the assaulting columns.

Thus England lost her *tête de pont* on the Continent, through which so many armies had marched into France between 1347 and 1557. The blow to national pride was bitterly resented: every one knows the tale how Queen Mary, though she survived for many months after the disaster, felt that her reputation and her policy had received a deadly wound, and died muttering that 'Calais would be found written on her heart.' But the loss was from the strategical and political point of view rather a benefit than a disaster to England. Though Elizabeth haggled again and again with the French, for terms by which she might recover the old stronghold, she never made any serious attempt to regain it. English expeditionary forces were seen at Havre and Rouen during the Huguenot wars, but never before Calais. It was, in effect, much too near to the Flemish frontier of Philip II, and if it had been in English hands at any time between 1560 and

1588, it is probable that the war with Spain, which Elizabeth so craftily avoided for many a year, would have been precipitated. But not only in her time, but in the whole period down to the nineteenth century, an English Calais would have been the inevitable cause of continental wars, which were happily avoided. It could not have been maintained against the strength of the French monarchy, with which Tudor and Stuart England would have been wholly unable to compete.

CHAPTER VI

THE BATTLE OF GRAVELINES (JULY 13, 1558) AND THE PEACE OF CATEAU CAMBRÉSIS (APRIL 2, 1559)

GUISE, after having dismantled Guisnes and repaired the breaches of Calais, went back in March to the Ardennes frontier, leaving about a third of his field army, under the Marshal des Thermes, to protect the newly conquered region from a probable counter-offensive on the part of King Philip's force in Flanders. It is surprising that no such attempt was made—and the English were furious with their ally, holding (rather unfairly) that the loss of Calais was due to him rather than to their own neglect. They refused to send an army to Flanders, and instead dispatched the fleet—mobilized all too late—to execute a futile descent on the coast of Brittany, which failed completely. Philip had disbanded a large part of his army for the winter ; by the time that he was beginning to collect a competent force, the stress of the campaign had passed away to the side of the Ardennes, when Guise laid siege to Thionville (May 1558). The place fell by assault, after a stout resistance, before the King's army was ready to relieve it. But Guise had then to move westward, for at last a formidable Spanish-Netherland-German army had been collected, and was advancing into Picardy once more, as in the previous year, leaving Calais unthreatened. By midsummer one of those surprising deadlocks which had been seen in previous years occurred. The two armies faced each other behind the Somme, dug themselves in, and each refused to take the offensive (July–August 1558).[1]

Meanwhile, however, the last, in a small way one of the most decisive, battle of this long series of wars was fought, on a side-issue, and by a detached force on either side. The Marshal Des Thermes made a sudden incursion into west Flanders, with the

[1] There is an excellent account of this deadlock in Boivin's *Journal*, under date September 2.

not unimportant detachment which had been left under his command at Calais. We are assured that this was not done with the reasonable object of distracting the enemy's main army by threatening his northern flank, by way of a diversion, but purely because Des Thermes was aware that the garrisons of Flanders had been depleted, and that there was a fine opportunity for raiding a wealthy region, full of ill-fortified or unfortified towns. The special opportunity offered was that Dunkirk was full of valuable merchandise, everything that used to pass from England to Flanders having been shifted in this direction since Calais fell into French hands.[1] To get to Dunkirk he would have to march past Gravelines, the Spanish frontier stronghold, which was properly fortified and not undermanned like other places. But he resolved to take this risk, knowing that the garrison would be too weak to meddle with an army: and, if he prospered in this raid, he may have intended to make a dash at Gravelines on his way back to Calais.[2] Des Thermes has left a very unconvincing report of his campaign in a dispatch addressed to the Duke of Guise.[3] He alleges that it had been his first intention to lay formal siege to Gravelines, though there were ten companies of foot in the garrison—nearly 4000 men—but that he dropped the idea, because he had received orders from Picardy not to tie himself up in lengthy operations, but to hold his army in hand, so that it could easily be recalled. 'Wherefore,' he writes, though he could more honestly have written 'in spite of which orders,' he proceeded to do precisely what he had been told not to do.

He had marched from Calais on June 30 with 500 gendarmes, three companies of light horse, and eighteen German and fourteen French foot companies—1500 horse and 9000 foot. He went past Gravelines, leaving it untouched, and spent twelve days in his raid, taking Dunkirk 'which,' as he owns, 'was no fortress,' after two days' bombardment of its feeble defences. The place was sacked and proved a most rich prize. Des Thermes then turned against the wealthy open town of Bergues,

[1] This is expressly stated by Bussy Rabutin: 'Etant adverti que la pluspart des villes selon la Coste de la Mer estoient mal pourveuës et garnies de gens de guerre, il fit l'enterprise de s'aller amperer d'une fort belle ville appellée Duinkerke, ou il y a fort beau et riche port' (xi. p. 237).

[2] 'Déliberant encore d'entreprendre et essayer Gravelines, s'il en offroit quelque occasion' (Bussy, *ibid.*).

[3] This may be found *in extenso* in the end of the edition of Bussy Rabutin's *Memoirs*, pp. 339–52.

from which the inhabitants fled in the night, leaving their goods behind them. Here also much plunder was taken. 'The soldiers in our camp before Dunkirk,' says the Marshal, 'were so clogged up with the spoil that they had got that we could not keep them to their standards, so to prevent general disbandment, we had to allow the booty to be taken back to Calais, which was done under the escort of M. de Senarpont and a body of cavalry. Such good haste was made that they were back to our camp at Dunkirk in two days.' There was some idea of going forward again to Nieuport, but this would only have led 'to getting so much more plunder for the soldiers that they would have again become unserviceable,' so on the 10th of July he turned back to take a look at Gravelines.[1]

Meanwhile he was, so he says, smitten with a violent access of gout, which kept him on his couch for two days, during which his officers reconnoitred Gravelines. They reported that it might be breached, but that the garrison was so strong that to get into the place would be a difficult and lengthy business. Wherefore Des Thermes resolved to return to Calais, crossing the river Aa, which flows from St. Omer past Gravelines, by the fords between that fortress and the sea, which were good and easy for many hours of the day. On the morning of July 13 the army began to cross the Aa, the cavalry leading, the infantry following, while Des Thermes brought up the rear with 500 arquebusiers and two companies of gendarmes, escorting the baggage: he could, as he explains, mount his horse that day. He took this post for fear that the garrison might make a sally, when they saw the waggons defiling before their walls.

But the garrison of Gravelines was not the thing to be feared. Though Des Thermes seems to have taken no thought of the possibility, his twelve-day raid into Flanders had given time for the enemy to get together a field force, with which he would have to deal. And the enterprising commander of this force was not intending to drive the French out of Flanders by

[1] 'Il y avait ceux qui estoient d'advis d'aller jusqu'àu Nieuport. Toutesfois il y avait une difficulté, qui estoist que d'aller la ce n'estoient qu'enricher nos soldats, pour n'en tirer après aucun service, et nous laisser en un inconvenient.' Des Thermes' dispatch, p. 346. Can any better evidence be wanted of the condition of the army, and of the fact that the whole matter was a mere raid for plunder?

In this account of Des Thermes' campaign, I have used his own dispatch as the main authority for dates and designs. It is a disingenuous document, but can only excuse facts, not misdate them. From it I contradict frequently-made statements that the French took Bergues before Dunkirk and advanced as far as Nieuport.

frontal attack on the seacoast route, but to cut in between them and Calais, and to force them to fight with the water at their backs and no possible way of retreat.

Lamoral of Egmont, who had so much distinguished himself in the cavalry fighting at St. Quentin, was now commanding in Flanders. Without weakening the main army in Picardy, he gathered together a miscellaneous force from all the garrisons of Flanders and Artois, of which a great part was composed of burgher-militia from the big Flemish towns. But there was a considerable cavalry contingent, including several German 'pistoleer' or 'reiter' squadrons, and some of the old Burgundian (*i.e.* Netherland) gendarme companies, which had owed their origin to Charles the Bold, and which Charles V had kept regularly in service. Probably the units were fragmentary, since they had been gathered together out of garrisons, but the mounted arm was formidable, though the infantry was heterogeneous and of unequal value. The whole force, however, was numerous, and must have outnumbered Des Thermes' raiders by several thousands—though not by two to one, as he alleged. The whole may have mustered 2000 horse and 10,000 or 11,000 foot.

The French commander writes, in his dispatch to Guise, that his first alarm about an enemy from the inland being in his neighbourhood, was when his camp before Gravelines was raided and his outposts driven in on the night of the 12th. On asking next morning why he had been left uninformed of this, 'the captains answered that it was because there was no order in the army, and no one had given them any commands.'[1] Moreover, enemy forces could be seen crossing the Aa by fords above the town, which were dry from the recoil of the tide, a good hour before the passages at the mouth of the estuary were practicable. It was evident that the cavalry at the head of these marching columns was aiming at getting ahead of the French on the road along the shore to Calais. Des Thermes, therefore, hurried across the ford in person, leaving the baggage and the rearguard to cross where they could, and took command of the gendarmerie, who were at the head of the French column of route.

Soon after he arrived the enemy's horse came into action, two bodies, one of German reiters, the other of heavy cavalry, cutting into the road in front of the French. They were

[1] Des Thermes' dispatch, p. 349.

charged and repulsed, so that for a moment Des Thermes thought that he was having the best of the day.[1] But the enemy's main body, his reserve of horse and his infantry, were now so close up that there was no hope of renewing the march on Calais, and getting across the front of the approaching troops. Des Thermes was constrained to halt, and form up in line of battle—the German infantry next the sea, the French infantry more inland, seven guns in the centre, the horse on the left flank. The rearguard from across the water seems to have got on the ground very late, possibly after the crisis of the battle was over. To be forced to fight with his back to the Aa and Gravelines, the sea on one flank, and the enemy on the other, meant ruin, unless a complete victory were obtained, and the enemy thrust quite out of the way.

Possibly Egmont's first cavalry attack had been made only with the object of pinning the French to the ground, and forcing them to draw up in battle array. For his main body, though approaching, was not available at the moment of this initial clash. The French had taken up a position with no natural advantages—as the map shows—the low sandy dunes are only fifteen or twenty feet above the sea, and their very gently rolling surface is not broken by any marked dips or gullies behind which a line could be formed with any utility.

Egmont, having rallied the squadrons which had taken part in the early unlucky charges, put himself at the head of his reserve of gendarmes, with the rest of the cavalry behind him, and waited till his infantry had come up and formed themselves ready to join in the general advance. He then attacked *à l'outrance*, broke the much inferior French horse, of which some units are accused of having failed to support Des Thermes at the critical moment,[2] and then fell with all arms upon the French infantry. There was some hot fighting, in which Egmont himself was unhorsed, for the native French companies on the left, who had covered their flank hastily with wagons full of hackbutteers, held out resolutely for some time in solid blocks. But the landsknecht companies, who formed more than half of the army, made little resistance; they threw up their pikes instead of levelling them, in sign of

[1] Des Thermes' dispatch, p. 351; cf. Bussy, x. p. 240.

[2] Bussy Rabutin says that Des Thermes was 'mal secouru du surplus de notre gendarmerie, et on accuse certains des chefs et quelques compagnies d'avoir fait mauvais devoir.' Book xi. pp. 240–1.

surrender, and broke up—'which was thought strange and disgraceful,' [1] for the reputation of this German mercenary infantry in the French service had hitherto been high. The collapse of this wing of the French army is explained by a fact mentioned by Holingshed [2] and (oddly enough) by Guicciardini, though it is not cited either by Des Thermes in the meagre end of his dispatch, nor by Bussy, the best French narrator of the fight. There was a small squadron of the English warships, which (though the fleet had gone to Brittany) was hanging about the Flemish coast, and endeavouring to communicate with the garrison of Gravelines. This it had not been able to do, for that fortress is a mile from the sea, and the

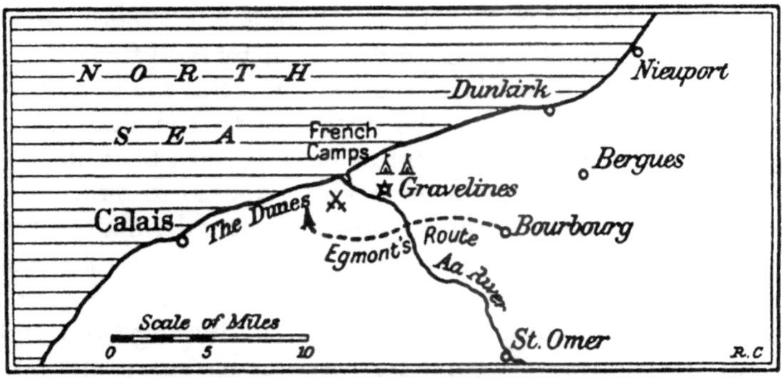

LOCALITIES OF THE CAMPAIGN OF GRAVELINES, JUNE 31–JULY 13, 1558.

French were in force on the estuary of the Aa. But seeing and hearing a battle in progress on the low dunes, the commander of this squadron (his name is unfortunately lost) pushed as far toward the land as the shallowness of the water permitted, and began trying long shots at the wing of the French army which rested on the sea. This was done, no doubt, from a memory of what happened at Pinkie, where the English ships had so bombarded the Scottish left wing that it flinched inland and got out of range. Cannon-balls, from the unexpected side of the sea, plunged into the massive blocks of German pikemen at the water's edge, and so demoralized them, at the moment of Egmont's general attack on the French line, that they threw

[1] 'Les allemands ne firent aucun resistence, se rompant d'eux mesmes, et hausserent leur picques, ce qu'ón a trouvè étrange et mauvais.' *Ibid.* p. 241.

[2] Who was writing in 1577, only nineteen years after the battle, and must have been familiar with the facts. See iv. p. 119.

up their pikes and asked for quarter, when attacked by the hostile infantry.

The French army was absolutely annihilated. Only a few individual horsemen, who had somehow got round the flank of the mêlée, escaped to bring the bad news to Calais. Des Thermes himself, wounded in the body, Senarpont, the governor of Boulogne, and all the other French commanders, d'Annebault, Villebon, Morvilliers, and de Chaubrun, were taken prisoners—a valuable asset in ransom money—and of the rank and file half were dead and the other half captives. Holingshed remarks that if Egmont and the commander of the English ships had agreed to move on Calais next morning, they might have captured it. For only a skeleton garrison had been left behind when Des Thermes went out: there were no French ships in the Channel, and succour by land from Picardy would have taken some days to come up.[1] But the opportunity was missed—probably, in the end, not to the detriment of England. Egmont was content at having, with his improvised army, completely destroyed the enemy opposed to him.

It is impossible to speak in sufficiently condemnatory terms of Des Thermes' little campaign. He had been told (as he confessed himself) not to risk his army. Instead, he went for a twelve days' raid into a country which promised good plunder, leaving Gravelines—a strong fortress—in his rear, and taking no precautions against interruptions from the east. Nothing was easier than to cut off his way of return, and then, unless he should win a victory over superior numbers, he and all his army were doomed to destruction. Egmont won a deserved reputation for the soundness of his strategy, and the resolution with which he had pushed matters to a quick issue. But for his preliminary cavalry attack the enemy might possibly have pushed past his front, and have reached Calais with some portion at least of his army intact.

This disaster had no small part in bringing the interminable war of Hapsburg and Valois to an end. King Philip and King Henry, entrenched opposite each other on the Somme, and each refusing to attack, were both at the end of their financial resources, and disinclined to hazard any general action. Philip was no soldier, but a fanatic intent on the suppression of Protestantism throughout Europe. Henry had, if we trust his admirer, De la Noue, tardily come to the con-

[1] Holingshed, iv. p. 120.

clusion that military glory was not a safe career, and was already somewhat failing in health and energy.[1] Gravelines coming on top of St. Quentin indisposed him for further risks, and long negotiations were already in progress between the two royal camps before the autumn was out. The intentions of the two parties were sufficiently shown by the fact that both the French and the Spanish kings disbanded part of their armies in November, and sent the rest into garrisons far to the rear. The peace negotiations, however, dragged on for many months, for there was unlimited opportunity for haggling over scraps of land. The French were in possession of the three bishoprics in Lorraine, of Thionville, of Calais, and of a large holding in Piedmont and Savoy. On the other hand, Philip, besides the complete possession of the long-disputed Naples and Milan, had a number of the strong towns of Picardy in his hands. In the end, Henry gave up all his claims on the Italian lands, and evacuated Savoy and Piedmont, retaining in the south only the narrow derelict Alpine marquisate of Saluzzo, when the reigning house had died out in 1548. On the northern side of France Henry kept Metz, Toul, Verdun, and Calais, while Philip retired from the occupied Picard fortresses. He was no longer interested in the fate of Calais, since his wife, the English Queen, had died in November 1558, and the claims of her enigmatic successor, Elizabeth, whom he had every reason to distrust, were not likely to influence him overmuch. The English ambassadors at the Congress of Cateau Cambrésis were put off with some vague promises of a future restoration of Calais, or a war indemnity, neither of which did they really expect to turn into actualities. Elizabeth, in later years, found them useful points on which to haggle with the French government at one time and another, without any very serious anticipation of success.

Thus the dream of a French Empire in Italy vanished, and a real Spanish domination was established there, which was to last for a century and a half, for the Pope and the minor Italian princes were helpless before the master of Milan and Naples. On the other hand, the three bishoprics and Calais, though they look small patches on the map, were of the highest importance to France. Not only was all future danger from English invasion in the north removed, but a strangle-hold was secured

[1] Boivin in his amusing account of the deadlock on the Somme, tells of the king being riddled with chills and gout in his hands, and compelled to lie up in his tent.

on the duchy of Lorraine, whose territory was completely dominated by Metz, Toul, and Verdun, in French hands. The only marvel is that it did not fall into the power of France till many generations had passed away. Strategically it was doomed after 1559.

Henry II was undoubtedly wise to acquiesce in the terms settled at Cateau Cambrésis—yet his professional soldiers, who loved the long wars, grumbled fiercely, with Guise as their spokesman, and murmured that France was giving up thrice as much to the enemy as Philip was giving up to France.[1] They even proffered the fatuous explanation that the peace was the work of the Constable Montmorency, who wanted to escape from his captivity, and to see his rival Guise deprived of further opportunities of winning military glory! The real governing cause was that Henry was failing in health and spirit, and financially ruined—the very large army that he had collected for his final effort was far too expensive for his resources. Whether after a few years of rest he would have started again on his earlier policy must remain an unsolved question—for, only a few months after he had signed the treaty, he was accidentally killed in a tournament (September 1559), and France, deprived of a governing hand, lapsed into faction-fights and religious wars. There were those who, like Coligny, thought that domestic strife might be brought to an end by luring the whole nation into one more attack on Philip II, when he had tied himself up in his Netherland troubles. But this was not to be the solution—the Catholics and the Huguenots had to settle their contest, before France as a whole was fit once more to plunge into foreign wars, and make her weight felt in European politics.

Meanwhile Philip of Spain, with his hands free for the repression of Protestantism in all quarters, was able to work out his own ruin, without having the whole force of France turned against him—as it had been against his father the Emperor—whose troubles had been both more complicated and less self-sought. Charles failed because France was always distracting him by sudden and usually improvised attacks—Philip because he engineered his own disasters without external assistance.

[1] See, for example, Montluc's wholly unconvincing remarks on a 'disgraceful peace.'

BOOK IV

MILITARY HISTORY OF ENGLAND UNDER THE TUDORS

CHAPTER I

THE EARLY CONTINENTAL WARS OF HENRY VIII (1512–13)—THE BATTLE OF THE SPURS (August 16, 1513)

HENRY VIII had to face the complete transformation of strategy, tactics, and fortification, which had spread all over western Europe since the commencement of the Great Wars of Italy in his father's time. When he came to the throne in 1509, he found the old English organization and armament still based on the ideas of the fifteenth century. Henry VII had succeeded in keeping out of any serious intervention in the continental struggle, though he had once led an army for a few weeks to Picardy in 1492, and had lent small expeditionary forces to his allies on several occasions. An English army was still composed of the traditional 'bows and bills,' with hardly any provision of cavalry, and was a force raised for a short campaign, with no permanent embodied units; it expected to be dismissed when winter cold set in. There was no survival of the veteran bands which had garrisoned France in the time of Henry V and Henry VI—a few hundred bowmen, billmen, and gunners at Calais were the only standing force which the King possessed. If an army was wanted, it would have to be created by the combination of contracts with nobles and military adventurers, and of unwilling shire-levies collected by the Commissioners of Array. The old triumphs of the fourteenth and fifteenth centuries had been won, and the bloody battles of the Wars of the Roses fought out, by armies marshalled in the old style—masses of billmen and spears, flanked by large wings of archery. In no case had they been decided by cavalry; in very few of them had artillery played any part.

The smaller firearms had been seen,[1] but merely to be

[1] As at second St. Albans, where the Burgundian hand-gun men played a very poor part. See vol. ii. p. 402. Henry of Richmond may have had a few French mercenaries of the same sort at Bosworth in his foreign contingent, but they are not mentioned in the fight, though their captain, Philibert de Chaundé, was rewarded.

despised: the bow was still the king of battle, and was held to have justified its old reputation against the landsknechts of Martin Schwartz at Stoke,[1] and the Flemings and French at Dixmuide. But its predominance was about to decline, and though Henry VIII was a believer in the bow in his early days, and himself a great archer, by the middle of his reign he was, probably with reluctance, yielding to the tendency of the age, and subsidizing the purchase of hackbuts and calivers. The national 'bill' was also destined to give way to the pike: billmen were more numerous than pikemen even at the siege of Boulogne in 1544, but Henry's recognition of the fact that the pike was better than the bill for holding off cavalry charges is shown by the facts that already in 1513 he was hiring 'Almaynes,' *i.e.* landsknechts, for his first invasion of France, and in 1544 was ready to enlist as many of them as he could get, because his English pikemen were all too few, even at the end of his reign.

It will be remembered that a similar change had come over the armament of the Swiss in the previous century. Originally the halberd—the continental equivalent of the bill—had been their national weapon: but by the time of their Burgundian wars they were wedded to the pike, and in the Italian campaigns, which we have been discussing in previous chapters, only a very small proportion of each Swiss column retained the old weapon—for the chances of close combat. The explanation of the triumph of the longer weapon over the shorter was, of course, the increased importance of cavalry—an eighteen-foot weapon could keep off horsemen in a way that was impossible for a six-foot weapon.

The old 'bill and bow' combination was by this time somewhat despised by the professional soldiers of the Continent, and perhaps not without reason. Florange, the unexpectedly literary son of the rebel Duke of Bouillon, had come to the conclusion that the English were good steady troops to hold a position, but for the rest no better than other men.[2] The amusing Gascon braggart Montluc waxes quite insolent when discussing them. He had been brought up, he said, to believe that they were almost invincible, never retreated, and died on their ground. He had heard all about Poitiers and Agincourt. But after experience in the Boulogne campaign of 1544 he had come to the conclusion that if the English were so formidable

[1] So Molinet, *Chroniques*, iii. p. 158.

[2] Florange, p. 144.

in previous centuries it was because they were then 'demi-Gascons'—they held Aquitaine, had many Gascons in their armies, and had married for centuries into Gascon families. When this connection was cut off, they were much like other troops, would retreat if their flank was turned, or if they fell into an ambush, and had been sometimes seen in panic. He was probably piqued by the memory of one unfortunate night, when at the celebrated 'Camisade of Boulogne' he had to run for a mile uphill, with three arrows in his roundel and another in his sleeve, and the foe in hot chase.[1] But he was justified in his conclusion that the legendary superiority of 'bow and bill' over all other weapons had disappeared. And an English army, as it stood in 1544, was not the instrument by which France could be overrun in the old style of Edward III or Henry V.

Henry VIII invaded France in force thrice during his long reign. The first was when he led his own army to the capture of Thérouanne and Tournay in 1513, and had his never-forgotten and glorious experience of chasing the chivalry of France in the Battle of the Spurs. The second was in 1523, when his bluff brother-in-law, Charles Brandon, led a raiding army as far as Montdidier and the Somme, and turned back, when autumn came, with a famished force riddled with mutiny. The third invasion was that of the 'Enterprise of Paris,' which dwindled into the 'Enterprise of Boulogne' in 1544, led by the King himself, who though unwieldy and diseased—a very different figure from the active adventurer of 1513—thought that his own presence would achieve marvels such as no duke could compass. He turned back, betrayed by his allies, and bankrupt, with the capture of one town to his credit.

The military record of the reign is, indeed, not an inspiring one. Henry's unhappy delusion that he could be the arbiter of Europe, by diplomacy or by arms, was an instance of obvious megalomania. He was clever, and quite as unscrupulous as the French, German, and Spanish monarchs with whom he had from time to time to deal. But though he had a loyal people at his back, and more money at his disposal than was usually available to Maximilian or Ferdinand, Francis I or Charles V, he was not really strong enough to compete either with the new French monarchy, or with the immense territorial accumulation of power that ultimately came into the hands of

[1] For his description of his adventures in this disaster see below, pp. 347-8.

the Hapsburgs. Courted when his assistance could be useful, he was invariably cheated when it was no longer necessary. He left his realm in a dire state of exhaustion—the direct result of his wilfully personal ambition, and his inability to appreciate the fact that he was not able to direct the general course of European politics.

Undoubtedly the want of a permanent national army had a great deal to do with his failures. When all other contemporary sovereigns were building up standing armies on a greater or a smaller scale, Henry never rose to such an idea. This was fortunate for his subjects—with *compagnies d'ordonnance* or *tercios* ready to his hand, his rule would have been even more arbitrary than was actually the case. More than once he had to restrain himself, when he discovered that the general feeling of his subjects was against him. As the Pilgrimage of Grace showed, great bodies of malcontents might flare up in arms, and he had no sufficient military force to oppose to them. His 'gentlemen pensioners' and his yeomen of the guard were but a handful, and bills or bows were in every farm and cottage. It is curious to note that among the many results of the growing importance of firearms was the fact that popular risings became progressively more impotent against trained soldiery, from the mere question of armament. The last and most complete demonstration of the fact was reserved for the next century, and the field of Sedgemoor, but there were examples of the same sort to be seen in Tudor times—especially in the suppression of both the eastern and the western insurgents in the third year of Edward VI, when the hired bands of Protector Somerset crushed all resistance. But King Henry never let matters come to the last extreme, or turned mercenaries loose upon seditious assemblies. He hired Almaynes and Burgundians, even Italians, freely enough for his continental campaigns, but was never obliged—thanks to his tact—to bring them across the Channel in any numbers.

From the first of his military enterprises to the last we find Henry taking into his pay foreign troops—the only engagements during his reign at which they were not present were the two affairs with the Scots, Flodden and Solway Moss, which were fought by extemporized armies of northern levies. Even in the comparatively small expedition to Biscay under Dorset in 1512 there were 500 'Almaynes' under a Flemish captain, one Guyot.[1] In later enterprises the number was progressively greater, down to the Boulogne campaign of 1544, when the King was ready to hire them by the five thousand at a time. These foreigners were, at the commencement of Henry's reign, all pikes and heavy cavalry; in his latest years he was also enlisting arquebusiers, or 'hackbuteers' as the English called them, Spanish and Italian companies, while the pikes and horsemen were always Germans and Burgundians.

The armament of these strangers sufficiently indicates the deficiencies in Henry's own national army—'bills and bows' were proved not to be the only contingents necessary in an army that was to face the French, who were now organized in a very different fashion from their fifteenth-century ancestors, and could not be trusted to make headlong frontal assaults on English armies comfortably established in defensive positions. For open fighting, with cavalry employed on a large scale, and manœuvring necessary, it was impossible to rely solely on blocks of billmen flanked with wings of archery.

Heavy cavalry was particularly necessary to meet the French gendarmerie, and England in early Tudor times could supply very little of it. The number of 'men-at-arms' fully equipped was very low, although there was a quite competent supply of what we may call light cavalry—the people who appear in the muster-rolls of Henry VIII as 'javelins,' or 'northern horse.' These were useful troops for raids, skirmishing, and exploration, equivalent to the 'stradiots' or 'Albanians' whom the French were employing. But of fully equipped men-at-arms there never seem to have been more than a few hundreds available—the English upper military classes had ceased for the most part to fight on horseback in the fifteenth century, and had not recovered the habit. Many gentlemen enrolled themselves as 'spears afoot' in 1513. At most the King could find 'demi-lances,' half-armoured men on unbarded horses, in considerable numbers, but only nobles and knights appeared in the full equipment. Henry VIII had a nucleus of real heavy cavalry in his 'gentlemen pensioners,' but outside their ranks full men-at-arms were scarce, and the units composed of them were few and of very modest numbers. This fact it was which

[1] Hall, *sub anno* iii. fol. xvi. *S.P.D.* vol. i. gives only 400 Almaynes (No. 3231). Guyot and his 'Almaynes' took part in the defence of St. Sebastian, when the rest of Dorset's army had gone home. *S.P.D.* i. p. 451.

made Henry so anxious to enlist as much foreign heavy cavalry as he could procure. 'Burgundians' is the usual name for them, *i.e.* troops raised in the Netherland states which the Emperor Maximilian had christened the 'Circle of Burgundy'—*Burgunder Kreis*—perpetuating the name of his father-in-law's inheritance, which had come into the Hapsburg hands by his marriage with the short-lived heiress Mary of Burgundy. Their spiritual ancestors were the gendarmerie of the *compagnies d'ordonnance* which Charles the Bold had maintained in rivalry with the regular cavalry of the King of France. Considerable numbers of them were hired and welcomed.

With the expedition that went under the unlucky Marquis of Dorset to Biscay in 1512, it would seem that no cavalry of any sort was sent—probably horse-transports were hard to procure. Ferdinand of Aragon had promised to provide not only draught-beasts but a competent force of cavalry for the projected invasion of Guienne; neither were forthcoming, and when the French of Bayonne came down to the Bidassoa, the English had to form up 'all on foot for lack of horsemen.'[1] As the King's secret agent, John Stile, wrote to his master: 'As for horsemen, verily my sovereign lord, if your army but had the horsemen, Bayonne and the country thereabout had byn gotten, or put in great distress by this day.'[2] But no Spanish horse ever appeared, and Dorset's army wasted away into sickness and mutiny, and finally returned to England without having advanced a step farther into France than St. Jean de Luz, some ten miles from the frontier! Ferdinand had never designed that it should—being in no wise desirous to see the English in possession of Guienne, and well content that the threat of their invasion should immobilize the French force at Bayonne, while he himself was overrunning the much-coveted Navarre.

When King Henry himself took over to Calais, in the following year, the army which captured Thérouanne and Tournay, and won the Battle of the Spurs, it is very difficult to calculate its exact strength and composition, for of the several statistical documents which deal with it, no two agree. Putting aside ridiculous estimates of 80,000 men and of 52,000,[3] and another of 34,000, which is still far too high, there seems some reason to accept the figures of 8000 for the 'vaward,'[4] 6000 for

[1] Hall, *Henry VIII*, *anno* iv, p. 17. [2] *S.P.D.* i. pp. 398–99.
[3] Du Bellay, book i. p. 7. [4] See *S.P.D.* 575, for the composition of the 'Vaward.'

the 'rearward,' and 9000 for the main-battle and the King's retinue, which would give 23,000 in all, plus about 1000 more of gunners, artificers, and pioneers.[1] Of mounted men there were at most 3000, including not only northern light horse and 'javelins,' but also mounted archers, who were apparently always intended to act along with the cavalry. There may have been 1000 'demi-lances,' though only the contingents ascribed to very important persons, such as the Earls of Northumberland and Derby, and the great Welsh captain Sir Rees ap Thomas, show any. Knights and barons of the smaller sort contribute only foot-soldiers in their companies of 50, 100, or 200 men. Of these foot-soldiers the vast majority were 'bills and bows,' but there were a certain small proportion of English pikemen—the King was anxious to increase their numbers: they might some day make it unnecessary to hire so many landsknechts. But here again, as with the demi-lances, it is only some few contingents that show 'morris-pikes' among the foot. Why the long pike was called Moorish is hard to make out, possibly because the weapon was, by tradition, known to have been used by the Moors, before the Swiss and the landsknechts made it more usual and famous. In the army of 1513 pikes appear in the companies of the Duke of Buckingham, and the Lords Lisle and Abergavenny,[2] but are exceptional.

The real strength of pikemen came from the hired 'Almaynes,' of whom there were several thousand in the host—the King's 'main-battle' alone had 2000,[3] and there were 2500 others with the vaward.[4] Some authorities quote the total at 6000, which may not be much of an exaggeration. These were to have borne the stress of cavalry charges, in the event of a general engagement—but no such great clash was to come during the campaign of 1513. What did happen was a purely horseman's battle—if battle it may be called, for in the combat of Guinegate (August 16) no infantry was engaged on either side. As we shall see, the Battle of the Spurs was to have a curious resemblance to the battle of St. Quentin, which forty years later (1557) showed, at the very end of this period of interminable wars, exactly the same characteristic faults in French generalship and French cavalry discipline. In each case a

[1] Which is rather a forecast of what *ought* to be got together, drawn up before the expedition started; *S.P.D.* 576.

[2] *S.P.D.* i. p. 631, No. 4306.

[3] *Ibid.* 4306.

[4] *Ibid.* No. 4070, p. 575.

commander, who had intended to make no more than a demonstration, got involved in a general action by venturing too close to his enemy, and when he tried—too late—to draw off, saw what was intended for a retreat turn into a panic—a reckless canter to the rear.

Henry, with perhaps 24,000 English and 7000 foreign troops, had formed the siege of Thérouanne. The main body of the regular French forces was at this time in Italy—the disastrous fight of Novara had been lost a few weeks back. To face King Henry's invasion of Picardy the French could only oppose an army insufficient in numbers to court a pitched battle, but strong enough to molest the besiegers by raiding on their line of communications—one big convoy from Calais had been cut up with severe loss on June 27. On July 28 a minor affront had been suffered. One of King Henry's biggest siege-guns—a bombard called 'St. John,' one of a set of twelve named after the Twelve Apostles—while on its way to the front had become embogged in a marsh near St. Omer. It was captured by a raiding party, and only rescued on the next day by a detachment of 700 horse, sent out hastily from the camp under the Earl of Essex and Sir Rees ap Thomas.[1] Another raid by the Boulogne garrison came to a bad end, as the English clamped their waggons together, and fought successfully by archery from behind them, driving off the French with heavy loss.[2]

The French observing army was estimated by those who could judge it best at 4000 horse, heavy and light, including several companies of 'stradiots' and 11,000 foot, partly French, partly hired landsknechts. The Marshal La Palice—a familiar name from the wars of Italy—was in command in June, but was superceded in August by the Duke of Vendôme. With them were several other officers whom we have met elsewhere—the Duke of Longueville, the Counts of St. Pol and Clermont, Florange the son of Robert of Bouillon (newly escaped from death at Novara)[3] the Chevalier Bayard, with his friend Richard de la Pole of the White Rose, 'the great traitor of England'—last hope of the House of York. Thérouanne being in desperate straits not only from battering and mines, but from exhaustion of food and especially of munitions, Vendôme thought out an elaborate demonstration, by which he would so distract the besieging army in its trenches, that a rapidly moving detach-

[1] Du Bellay, i. 8 tells of the capture of 'St. John,' but not of its recovery.

[2] *Ibid.* p. 9.

[3] See p. 157.

ment might pierce the line at a weak point, and throw relief into the place.

For this desperate venture a company of the stradiots, under the Sieur de Fontrailles, was told off, each with a side of bacon at his saddlebow and a powder-sack behind him—a nasty charge when firearms were about. Meanwhile two separate demonstrations were to be made against opposite points of the siege-lines, the lesser one by the Duke of Alençon with the Count of St. Pol against the west front, where the Earl of Shrewsbury was in charge, the stronger under La Palice, the Duke of Longueville, and the Count of Clermont against the south side of the lines. The French infantry were left in camp at Blangy many miles away: they had just been joined by 7500 landsknechts under Florange and Richard de la Pole,[1] since a rapid cavalry operation only was in view. The two bodies of horse moved out before dawn and drew near the lines, but not unobserved, as the English 'border prickers' were all about the countryside, and had reported 'a great plump of spears' behind the church tower of Guinegate.[2] Apprised of the approach of at any rate the larger division of the French horse, whose numbers were ludicrously exaggerated,[3] the King drew out a field force from his lines, sending out 1100 horse, English and foreign, as a vanguard, while a mass of infantry 10,000 or 12,000 strong came a mile behind. The Emperor Maximilian, who had ridden over from his fortress of Aire with a small escort of 30 men-at-arms, to visit his ally, went out along with him, and as a matter of courtesy mounted the Red Cross of St. George that day—a compliment much appreciated.

The English scouts came in sight of La Palice's corps on the hillside above Guinegate by a village called Bomy, four or five miles outside Thérouanne, on a plateau 'between a low wood and a fallow field.'[4] The French had been hoping to find the enemy unprepared, and checked themselves on the edge of the hillside, when they saw that Henry's army was alert, and out in force. The Albanians, however, started out on their rather forlorn venture, riding in a circuit far from the main body. La

[1] Florange, p. 146. Du Bellay, i. p. 20. Blangy, it may be noted, is only four miles from the famous battlefield of Agincourt.

[2] Hall, *Henry VIII*, v. folio 32.

[3] The Northern scouts said that they had seen 6000 horse! *Ibid.* p. 93, Du Bellay says that there were 1400, plus the Albanians; Bayard's 'loyal serviteur' only allows 1200 (p. 342). I should guess that 2000 would be about the probable figure.

[4] Hall, folio 33.

Palice was inclined to give up the enterprise, and to retire, but waited longer than he should have done, partly to give the stradiots time for their stroke, partly because some of his men-at-arms had gone forward before the main body, and were skirmishing already with the English vanguard of horse.

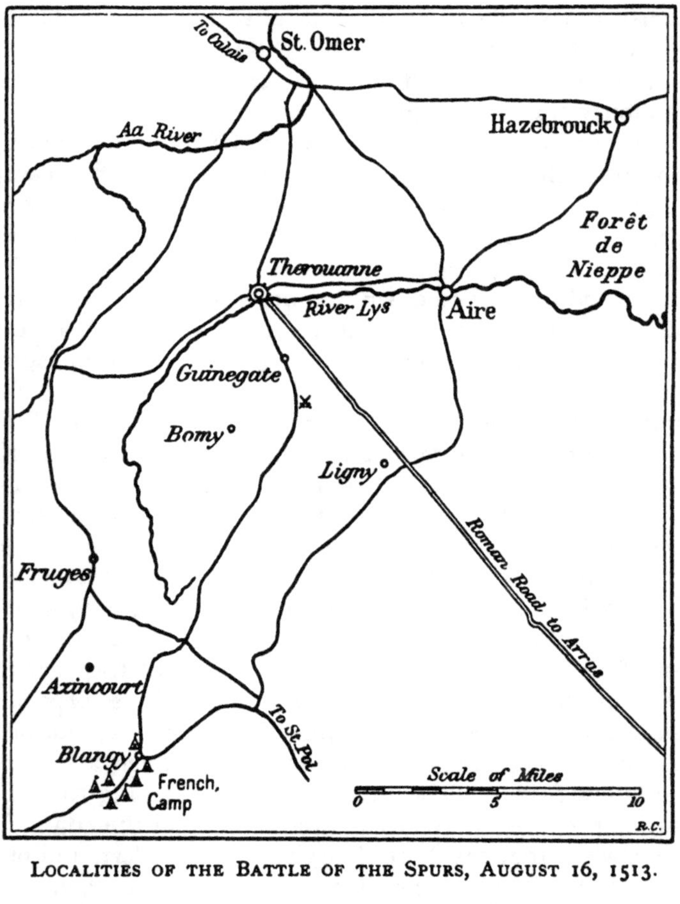

LOCALITIES OF THE BATTLE OF THE SPURS, AUGUST 16, 1513.

There were some lances broken, and the main body of the English horse was waiting to be attacked, with sure knowledge that their infantry in full force was close behind. The mounted archers got off their horses, and lined some hedgerows on the flank, from which there was good shooting to be done into the mass of cavalry opposite them, 'thirty-six deep around more than thirty guidons and banners.' Then La Palice at last took note that the English infantry was coming up, in good order under the King's great banner, outflanking his position on both sides. He wisely, but too late, gave orders for a retreat. Hall says that at this moment Clarencieux Herald, who was at the front, rode up to the Earl of Essex and Sir Thomas Peche, and cried, ' In God's name set forward, the victory is yours, for I see by them that they will not abide, and I will go with you in my tabard.' The English horse charged, and the gendarmes, caught at the moment when they were just about to move off, were thrown into disorder. This was turned into absolute panic by an unlucky chance—the body of stradiots which had been trying to get into the town of Thérouanne had been driven off by a couple of cannon-shot. Retreating hastily across the hillside, they ran in confusion against the flank of the heavy cavalry, and upset the whole array. The Burgundian horse pressed in sideways against them, while the English were pushing straight ahead against the main body. La Palice and Bayard made an attempt to keep the rearguard steady, to cover the rout of the rest, but in vain. The whole body galloped away for their camp at Blangy, throwing down their standards and their heavy lances : some men-at-arms cut off the ' bardings ' of their horses, which were strewn all along the road. King Henry was not fortunate enough to have been in the first charge, but rode up from the head of the infantry, with his personal bodyguard, to join in the exhilarating chase, which went on for many miles. The disordered fugitives finally drew up before the French infantry ranged in order in front of their camp at Blangy.[1] Florange opined that if Henry had waited for his infantry to come up, and put his cavalry in order again, he could have destroyed the whole French army.[2]

The casualties among the defeated force were few—not more than forty dead, as we are told. But nine standards were taken, and more than 120 prisoners of note, including the Duke of Longueville, the Count of Clermont, the Chevalier Bayard, de Piennes, governor of Picardy, the Lord of Hymbercourt, and several of the gentlemen of the French King's chamber and

[1] The best account, undoubtedly, is that of Hall. One can use to supplement it : the letter of Giles to the Earl of Devon (*S.P.D.* i. p. 664), John Taylor's Diary (*ibid.* p. 625), and useful narratives of Florange and Du Bellay. Bayard's ' Loyal Serviteur ' has some curious errors, *e.g.* he describes the capture of the English great gun ' St. John ' on this day and has a very doubtful story of how the Chevalier got off paying ransom.

[2] Florange, p. 146.

members of his guard. La Palice was captured by a Burgundian knight, but somehow got away in the confusion.[1]

While this combat and rout was going on, the smaller body of French cavalry under St. Pol, which had demonstrated against the Earl of Shrewsbury's part of the siege lines, was driven away without much trouble—Sir Rees ap Thomas, pressing them, captured four standards. The garrison of Thérouanne surrendered on the 22nd, being deprived of all hope of succour, and were allowed to march out with the honours of war—there were still over 2500 of them surviving.

The Battle of the Spurs has received more notice than it merits by contemporary historians—undoubtedly because it was the only action at which Henry VIII was present in person, and he loved to hear his own praises. Moreover, it was the only fight in the open field on the Continent during the whole of four English wars with France. In Suffolk's invasion of Picardy in 1523, and in the King's own 'enterprise of Boulogne' in 1544, there was not a single serious clash, barely a skirmish or two. There were some cavalry affairs outside Boulogne, after the King had gone home in the autumn of 1544, but the numbers engaged were not great, and the tactical lessons were few. Indeed the only large-scale and interesting fight with the French during all these wars was the 'Camisade of Boulogne'—of which more will be told in its proper place. This was no battle, though a good many thousand combatants were engaged, but the midnight escalade of a fortified town, followed by desperate street fighting, which ended by the eviction of the storming parties, who had in their first onset gained a solid footing in Basse Boulogne. Meanwhile it is necessary, and also instructive, to look at the last general action ever fought by an English army consisting only of the traditional 'bows and bills'—the bloody fight on Branxton Hill, better known as 'Flodden Field,' which took place only three weeks after the Battle of the Spurs—September 9, 1573.

[1] Not without some suspicion of collusion on the part of his captor. Hall, folio 33.

CHAPTER II

FLODDEN FIELD (September 9, 1513)

WHILE King Henry was conducting his leisurely operations around Thérouanne and Tournay, and winning a not very glorious victory at the Battle of the Spurs, a campaign of a much more interesting sort was being fought on the Northumbrian border. That the French King would try to stir up the Scots to attack Henry in the rear, was an almost certain expectation. This had always been the policy of France when vexed by an English invasion, and there was (as usual) a sufficient number of grievances on the part of the Scots to make a plausible *casus belli*, if James IV could be induced to intervene. The political situation exactly reproduced that of 1346, when Philip of Valois persuaded David II to cross the Tweed—'weening that there was nothing left in England save millers and mass-priests.' And the result was to be precisely the same—just as Crécy (August 26) was followed by the Scottish disaster of Neville's Cross (October 17) in 1346, so in 1513 the Battle of the Spurs (August 16) was followed by the much more woeful Scottish slaughter at Flodden (September 9). The diversion which the French King had planned to distract the English invader ended in the destruction of his too confiding northern ally.

For some time King James had been arming: he had collected a train of heavy artillery such as had never before been seen in Scotland. English agents in the Low Countries reported that enormous quantities of the long landsknechts' pikes were being shipped across from Zealand to Leith—which, so long as the Scots were neutral, could not be stopped by the Emperor's Custom-house officers. In the winter of 1512–13 it was reported to King Henry that the arsenals in Edinburgh were hard at work on the making of gunpowder, lance-staves, and military carts. Small sendings of munitions from France might have been complimentary gifts, meaning little, but the arrival of a complete military mission of 40 trained officers under a captain named d'Aussi at Dumbarton in May 1513 [1] could have but one interpretation: if James had not been meaning war, he would not have allowed them to land. They were utilized for the training of the Scots in the modern pike-tactics—eight of them, we are told, were assigned to each of the columns in the new style which manœuvred at Flodden. And the formation in a number of moderate sized corps, instead of in the vast blocks of men which were marshalled in earlier battles, was an obvious lesson from the Continent. The formal offensive and defensive alliance with France was apparently ratified on May 22, binding each party not to make peace with King Henry unless both states were comprehended.[2] A confirmation of it signed at Edinburgh on July 10 had the provision that if the King of either country should die, his ally was to make no attempt to interfere in the matter of the succession.[3] This was curiously prophetic—James had only three months to live, though he was in the flower of his strength and energy. Meanwhile nothing certain concerning the alliance was known at London. On June 4, a person so well informed as Bishop Ruthal, a member of the Privy Council, was writing to Wolsey that the intention of King James was still doubtful. Probably he may make no open war, 'but rob and spoil the English both by land and sea—more especially on the water.'[4] The actual declaration of war was apparently drafted on July 16, but only delivered to King Henry in his camp before Thérouanne by Lyon King-of-Arms on August 11,[5] but the general summons to the national levy of Scotland to mobilize at Edinburgh had been issued long before it was possible for the herald to reach Flanders.

This document was a command that within twenty days all recipients should be at the Borough Muir outside Edinburgh, the normal mustering-place for royal armies. Apparently the day on which the levies were to reach their appointed destina-

[1] Venetian State Papers, vol. i. No. 316. The statement in this same letter that the French brought many hundreds of arquebuses is odd, as these were not common in French armies till after Pavia (see above, pp. 43–4). The name Dunbar is an obvious error for Dumbarton, as it is described as a port on the *west* coast of Scotland.

[2] On April 22, James writes to his uncle, the King of Denmark, that he has made proffers of alliance to France, and suggests that John of Denmark may do the same. Letters and Papers, No. 3138. Cf. Letters and Papers, No. 3218.

[3] *Ibid.* 3303.

[4] *Ibid.* 5757.

[5] The text so dated is in Letters and Papers, No. 4351.

tion was August 13.[1] Each man was to bring forty days' provisions with him, counting from the day of his appearance in the camp. Twenty days' notice seems not a moment too long, when it is realized that the messengers went to the remotest Highlands—the chiefs of the Mackenzies and Macleans and the Earl of Caithness were all killed at Flodden. The time required for the summons to be passed to the extreme north, for a chief to muster the contingent of his clan, equip it for long service, and bring it to Edinburgh from, let us say, Ross-shire, would be very considerable. We happen to know that the King's large train of artillery marched for the Border in three divisions on the 18th, 19th, and 20th of August, and that the whole army crossed the Tweed on the 22nd of that month. Twenty days, counting backward from August 18, would take us to July 30, but it is clear that the artillery would not have started till mobilization was fairly complete. It seems certain that the royal summons to the remoter contingents cannot have been issued later than the 20th July—*i.e.* James had definitely shown his hand by that date, and the peace party in Scotland (which, of course, included the Queen) must have passed the news on to Berwick without delay. Possibly their last effort to turn the king from his purpose were the weird warnings in Linlithgow church, of which Pitscottie gives such a striking description. Very probably the date of the drafting of the declaration of war (July 16) was the same as that of the summons to the more distant contingents.[2] It will be noted that it was on July 22 that the Earl of Surrey collected his retinue, and began to move forward from London to the north—as if he had just received intelligence that trouble was certain on the Border, though no formal declaration of hostilities was to come to hand for many days.

It is clear that the army collected outside Edinburgh on August 13 was the largest that Scotland ever put in the field. For once in a way the kingdom was in good order, and not vexed with faction or civil strife. The King was very popular among all classes of his subjects—again a rare thing in Scotland. The mere list of earls, lords, and chiefs of clans slain at Flodden shows that the remoter Highlands and Isles were well represented in the host. The whole nobility of the realm,

[1] This, at any rate, is the day given by Leslie on his page 346.

[2] The Scottish Treasurer's accounts show that July 24 was the day on which Fife, Angus, and Bute were called up.

save individuals too young or too old (as, for example, the famous Angus, 'Bell-the-Cat') to serve, were present. It is, of course, impossible to calculate with any accuracy the strength of the army. Scottish and English chroniclers of the next generation speak freely of 100,000 men—Pitscottie is as liberal in his figures as Hall. We may surmise that 40,000 is about the highest total that can be conceded. But English as well as Scottish authorities agree that a much smaller number was present at Flodden. There had been a considerable leakage during the short three weeks of the campaign; a contemporary English document—Bishop Ruthal's letter to Wolsey—acknowledges this,[1] no less than the Scots, who are anxious to reduce their national army to a lower figure than that of the enemy. Pitscottie gets the fighting force of September 9 down to 28,000 men—a not impossible figure—out of his original 100,000.[2] There was no obvious reason for a desertion of three-fourths of the host—though no doubt Borderers or Highlanders must have amassed considerable plunder of cattle during an unopposed stay of three weeks in Northumberland, and would be anxious to place it in safety. Pitscottie says that some of the remoter contingents were already near the end of their food; this may be so, but the Scots' camp was found full of victuals on the night after the battle. It may be that an original force of 40,000 might have been reduced to 28,000 or 30,000. But the army, like the King, must have been eager for a clash with the ancient enemy, and Surrey was known to be coming up. Desertion cannot have been wholesale.

James had no intention of making a spectacular march on Newcastle or York; he was operating as the ally of Louis XII, with the sole purpose of distracting King Henry's offensive in northern France, of forcing him to turn back, or at least to detach a large part of his field army to defend his northern borders. On consideration the only thing that surprises us is that he did not start the campaign by laying siege to Berwick, an important place, Scottish of old and intermittently in Scottish hands within the memory of aged men. Instead, he began the leaguer of the great castle of Norham, the bishop of Durham's property, which was second only to Berwick as a protection to the English border. Bishop Ruthal had been sending munitions of all sorts thither, the moment that the declaration of

[1] Ruthal to Wolsey, Letters and Papers, No. 4462, of September 20.

[2] See Pitscottie, pp. 268 and 276.

war by King James came to hand,[1] and was convinced that the place was in a good state of defence. Herein he erred, for Norham, though a first-class castle, was not fit to cope with modern artillery, of which King James had good store. By only five days of battering great parts of its outer defences were ruined, though the donjon and the inner ward were not injured beyond possibility of repair. The governor, John Ainslie, got fair terms of surrender, and was sent as a prisoner to Falkland ; the well-furnished residential parts of the castle yielded some attractive loot in the way of furniture, arras,[2] and wine (August 29).

After sacking Norham, the King made a leisurely advance southward, and took by force of artillery the castles of Etal, Ford, and Chillingham, after which he established himself in an entrenched camp on Flodden Edge, the long hill above Ford, where he waited for the inside of a week, intending to accept battle in this almost impregnable position, when the English should come up against him. He must have known by now that the enemy was collecting opposite him—Surrey (as will be seen) had reached Newcastle on September 1. Nothing could have been stronger than the ground which James had chosen, a steep hill 300 feet high, the river Till covering its eastern flank, and a marsh its left. 'There was but one narrow field for any man to ascend up to the said hill, and at the foot of that lay the King's ordnance. It was too strong to be approached on any side, unless that the Englishmen should have temerariously run in upon the ordnance.'[3] Indeed the only defect of the position was that it was too strong for any reasonably-minded enemy to attack. Here James lay, with his pavilions stretching for a mile along the crest, waiting for Surrey to appear. Scottish historians of a later day, not understanding his quiescence, attributed it to amorous dallying with Lady Heron of Ford.[4] But the poor lady was only attempting to drive a bargain with the King, by which her husband should be released, and her castle not blown up, in return for the handing over of Lord Johnstone and Sir Alexander Hume, then prisoners in England. Meanwhile the Earl of Surrey and his army were drawing nearer each day. The King welcomed their approach.

[1] Letters and Papers, No. 4388, August 4.
[2] Cf. Exchequer Rolls of Scotland, xiv. 38, g.
[3] Hall, *Henry VIII*, v. folio 41.
[4] Especially Pitscottie, p. 268.

It may be worth while to illustrate the extraordinary lack of heavy cavalry in English armies of the period, by pointing out that Surrey's personal retinue, round which the rest of his forces mobilized, when the Scots declared open war and crossed the Tweed, included in a total of 500 men, all in uniforms of white and green, precisely *one* man-at-arms fully equipped—his name was Avery of Berwick—and 22 demi-lances. All the rest were infantry, 462 bows and bills under ten 'petty captains'; there were two surgeons and a trumpeter attached.[1]

The levies from the north which poured in to join Surrey on his march, included numerous 'prickers,' Border horse useful for raids and exploration, but we are told very definitely that they dismounted to fight.[2] And of the infantry many were 'horsed and harnessed' both billmen and bowmen; but, of course, their mounts were country hacks, which were sent to the rear in time of battle. Records at Bolton Abbey show that the villages of the Craven region sent about nine men apiece—all horsed and harnessed—but no doubt there must have been contingents not so well furnished with mounts.[3]

The army, therefore, at Flodden fought all on foot. Sir Walter Scott's picture in *Marmion*, when 'charging knights like whirlwinds go,' is a romantic misconception. The engagement was entirely an infantry battle, like Homildon, the last occasion on which English and Scots had met for a general action. And in this respect it was entirely unlike Pinkie—the one great fight that was yet to come—in which cavalry charges played as great a part as at Bannockburn—though with very different results.

When King Henry sailed for France on June 29, he had warned the Earl of Surrey that if trouble should come from the Scottish side, he was to have charge of the defence of the north.[4] But that James IV would declare war was by no means certain at that date; it was not indeed made formally clear till on August 11 Lyon King of Arms delivered his master's ultimatum to King Henry in his camp before Thérouanne. This document was drafted at Edinburgh on July 26. Its general tenure must have been known some days before, by

[1] Pay-roll. for August 20, in Letters and Papers, i. No. 4376. Avery, as a full man-at-arms, got 1s. 6d. a day as salary, the twenty-two demi-lances only 9d. apiece.

[2] Ruthal to Wolsey, *ibid.* 4462. 'The borderers are not to be trusted. They never lighted from their horses till the battles joined, and they plundered both sides.'

[3] See *Flodden Field*, ed. Federer, pp. 155–6.

[4] See Hall, *Henry VIII*, *anno* v. folio xxxvii.

information from the peace-party who were numerous in the Scottish court. At any rate, on July 22 Surrey had moved forward from London with the 500 men who formed his personal retinue, and had reached Doncaster on the first day of August, from which place he sent out letters of warning to the sheriffs and nobility of the northern counties, bidding them to be in readiness, as trouble was probably at hand. Surrey remained some days at Doncaster and Pontefract, till he got news that probability had become certainty. For on August 13 Lord Home, the Scottish Warden of the East March, made a destructive raid into Northumberland, before the delivery of the ultimatum, or the English king's answer to it, could be known.[1] He burned seven villages, but returning with much spoil of cattle was set upon by Sir William Bulmer, who had collected some local levies, at Broomhouse, near Wooler, routed and stripped of his plunder. After this there could be no doubt that war had come, and Surrey issued orders for general mobilization. Meanwhile he commanded Sir Nicholas Appleyard, his Master of the Ordnance, to push forward to Newcastle with his guns, and put them safe within its walls, till the rest of the contingents should come up to that town, which was designated as the general mustering-place for September 1. The captains of the frontier fortresses of Berwick and Norham were warned to expect assaults at short notice.

On August 25 Surrey was at York; he got notice that King James had crossed the Tweed on August 22 with the full force of Scotland, and had laid siege not to Berwick—as had been thought likely—but to the Bishop of Durham's great castle of Norham. On the 26th the Earl, having gathered in the levies of southern and central Yorkshire, marched for Durham, where he picked up on August 30 the contingents of the bishopric, and took down from above the high altar of the Minster the banner of St. Cuthbert, to serve as his chief ensign. He reached Newcastle on September 1, having heard meanwhile that Norham had fallen after a siege of only five days. He found there many outlying detachments already arrived—Lord Dacre with the men of the Western March, and Sir Edward Stanley who had brought up across the Pennine Chain, via Skipton-in-Craven, the levy of Cheshire and Lancashire, under the banner of the Earl of Derby.

On September 2, Surrey moved forward from Newcastle to

[1] The date is given only by Leslie, p. 347.

Alnwick, where the contingents of the Earl of Northumberland were already awaiting him. This he did, we are told, because Newcastle was overcrowded already, and other detachments were still due there, which would cumber the town and eat out its provisions.[1] He was at Alnwick on the 3rd and 4th September, much troubled by foul weather, which made the roads very difficult for the remoter contingents that were coming up from the rear. He was, however, joined on the 4th by all the troops that he expected, and some of whom he had not been sure. He had sent word to his son Thomas, the Admiral of England, who was on the high seas, to bring him to Newcastle any landing force that could be spared from the fleet. The Admiral, though much buffeted by the weather, reached Newcastle on the 3rd, and brought up a thousand men—marines, as they would have been called in later centuries. It would seem that the Derbyshire contingent, of which we hear mention, must also have been late, and perhaps the large and solid bands sent by the Bishop of Ely, James Stanley,[2] all bearing on their breasts the three crowns of the see coat-of-arms combined with the eagle's legs of his family shield. The Ely men brought with them the banner of St. Audrey, which had not seen the field for many a year.

Having now in hand all the troops that he could expect to gather, Surrey marched out from Alnwick to Bolton-in-Glendale, where he divided up the host into its fighting formations. Hitherto it was an undigested mass of contingents. Instead of distributing the army into the ancient three divisions —vaward, main-battle, and rearward—the Earl adopted a new order of battle, two main divisions each furnished with 'wings.' It is possible that he originally intended to fight in two lines—but, in the actual battle that was to come, the two main corps with their four subsidiary divisions all drew up in one continuous line—whether this was designed or no.

We are fortunately able to reconstruct from one source and another most of the details of the marshalling of the host. The right-hand main division, often called the 'vaward' in the old style, was under the Admiral Thomas Howard, who displayed the banner of St. Cuthbert, and had under him his

[1] Hall, *Henry VIII*, v. folio xxxviii.

[2] This Bishop Stanley was a typical pre-reformation political prelate, active and sumptuous. He was often the mark for Protestant cynicisms, as he had at least three natural children, of whom one, John, served in the Flodden contingent, and was knighted for good service in the battle.

own marines, the Durham contingents under William Bulmer, and the retinues of a number of Yorkshire and Northumberland peers—Conyers, Latimer, Scrope of Upsall, Ogle, Lumley, and the Westmoreland contingent of the Cliffords. Its right-hand 'wing' under the Admiral's brother, Edmund Howard, consisted of a thousand Cheshire and 500 Lancashire levies, under Sir John Booth and Sir Richard Cholmondely, with the citizens of Hull, the tenants of the royal lands of Hatfield Chase, and the small contingents of a number of gentry—Fitz-williams, Warcops, Savages, Stapletons, Harbottles, Tunstalls. The left 'wing' under Sir Marmaduke Constable, a very aged Yorkshire knight, included his own kinsmen and their tenants, and a Northumberland corps under Sir William Percy, cousin of the Earl of that county, with a thousand men of Lancashire.

The Earl of Surrey took his own stand in the main central left-hand corps, called indifferently the 'main-battle' or the 'rearward.' He had with him his own retinue, with the Ely men under the banner of St. Awdrey, the tenants of the Abbot of Whitby, the citizens of York, the contingents of the Lords Scrope of Bolton and Darcy, and of many Yorkshire knights—Barkeley, Pickering, Tempest, Dawnay, Clapham, Gascoigne, Willoughby, Tilney, Radcliffe, etc. Its right 'wing' under Lord Dacre included the Cumberland marchmen, and the local levies of Tynedale [1] and Bamboroughshire from the Northumberland borders. In his account of his doings Dacre says that his own contingent being judged too small for a battle-unit, these Northumbrians were added to him—to his very small profit, for they behaved badly. The left wing of the main-battle under Sir Edward Stanley was made up of the bulk of the Cheshire men—minus the detachment lent to Edmund Howard—and of those of the Lancashire levies who had not been told off to Howard's and Constable's units.

The whole army is said to have been 26,000 strong—perhaps somewhat of an overstatement, but the whole of the northern counties had marched very willingly against the ancient enemy, and the number does not appear impossible for what amounted to a *levée en masse* to defend the home-lands.

On the 5th, Surrey had sent the King, by the herald *Rouge Croix*, a sort of defiance. James, it was declared, had entered England and slain English subjects without justice or reason—he was therefore challenged to try the decision of right by

[1] Hall, *Henry VIII*, v. folio 40.

force of arms before Friday next, the 9th of September. Incidentally Surrey mentioned that he approved of Lady Heron's proposed exchange of Lord Johnstone and Sir Alexander Hume for her captive husband Sir William and his castle, and would undertake to hand over the two Scottish prisoners. Undoubtedly *Rouge Croix* was intended to get a sight of the Scottish army and position, rather than to carry out a successful negotiation. And Surrey suspected that the enemy would be trying the same game. 'For it was thought by the Earl that the said King would be fain to send a herald of his own, only to view and oversee the manner and order of the royal army, ordnance, and artillery.' He was quite correct in this hypothesis, for James detained *Rouge Croix*, and sent out Islay herald on September 6, with the message that he wished to be assured that the defiance addressed to him was rightly delivered; if so, he accepted the challenge for Friday next. As to the Heron affair, he would consider the matter. Whereupon just as James had detained *Rouge Croix*, so Surrey detained Islay. On the next day (September 7) the two heralds were exchanged—each (no doubt) having made out what he could of the enemy's dispositions. The result of *Rouge Croix's* observations was manifested in the most absurd fashion.

He was sent back to the Scottish camp with the preposterous proposal 'that the King of his noble courage would descend the hill whereon he lay, for the place is no indifferent ground for armies to fight upon.' James naturally refused to see the herald, and sent him back with the remark, delivered by a servant, that 'it beseemed not an earl to handle a king after that manner.' It was really ludicrous that a war-hardened soldier of seventy should use pseudo-chivalrous verbiage to lure his adversary out of a good position, and James thought that this exchange of compliments should come to an end. Let Surrey turn him out of the entrenched camp on Flodden Edge if he could! Probably the French captains in the Scottish host had made him comprehend the full advantage of field fortifications strengthened with artillery.

Surrey now found himself in a most tiresome situation. The English army was already short of provisions, and in a few days would have exhausted all its scanty stores. It is recorded as a special sign of trouble that beer had run completely out on the 6th, and that for three days following no one had anything but water to drink ! [1] It was necessary to fight at once, for there was no adequate supply of food to be got nearer than Newcastle or Berwick. The Earl had counted on a battle in the open field with a willing adversary, and was disappointed when he found himself invited to storm a strong entrenched position garnished with artillery. He was driven by sheer pride and privation to adopt the daring and risky—almost desperate—expedient of manœuvring, which was to bring him an astounding victory. If James would not move out of his chosen ground of his own accord, he might be made to do so, by taking the English army by a semicircular flank march on the other side of the Till river, and placing it between Flodden Edge and the Scottish border. If his communications with his base were thus cut, the King would be compelled to come down, and to drive off the intercepting force. But to make a flank march round a watchful enemy is terribly dangerous—especially when there is a river to be crossed, which must for a time divide the army which is on the move into two parts : the one which has crossed and the one which has yet to do so. And if the battle went wrong, there would be the Tweed and hostile Scotland behind a beaten army.

On the 8th, Surrey moved his camp from Wooler Haugh to Barmoor Wood, facing the east end of the Scottish position, at a distance of about two miles ; between them was the Till and a low range of hills, which made their march only partially visible from the Flodden summit. On the same evening the Earl issued his final orders, which were certain to bring on a fight. His 'vaward' under the Admiral, with its wings, was to move northward and cross the Till by Twizel Bridge, close to the junction of that little stream with the Tweed. The Admiral had to take the artillery with him, as his father intended to ford the Till at the Millford (or Sandyford), where guns could not pass. The troops had eaten their last ration, and rode from five in the morning—when they left Barmoor—to four in the afternoon, when the battle began, without further food. This day the Scots remained quiet in their camp, save that parties of Border horse went out southward, and burned some villages.

As long as the English kept to the right bank of the Till, their movement might be construed as a march toward Scotland, with the object of crossing the Tweed and ravaging the

[1] Ruthal to Wolsey, Letters and Papers, No. 4462.

lands of the Merse, and so of drawing the King after them. Or it might mean a move on Berwick, to get a new base and provisions. It was not till the Admiral's advanced guard began to cross Twizel Bridge that any other intention was suspected. King James was viewing those partly-visible operations from the high ground at the east end of Flodden Edge; no doubt cavalry scouts must have been out, watching the course of the Till. When it became clear that a part at least of the English were crossing the bridge—at some time late in the morning—a new review of the situation had to be made. What was Surrey's object? It might be to cut the Scottish line of communications, or it might be to invite battle on the 9th—as his challenge had promised, by attacking the King's position from the rear, since its front was impregnable. It is said by both English and Scottish authorities that Giles Musgrave,[1] an English outlaw, on whose local knowledge James much relied, pointed out to him that Surrey's object must be to seize Branxton Hill, the eminence behind Flodden Edge, separated from it by a dip only, and only a few feet less in height. Established on this position, Surrey would be asking to be attacked, and would have some advantage of ground. It would be well therefore to face front to rear, occupy Branxton Hill, and force the English (if they intended to fight) to attack with the slope against them. There was time to execute the change of position, since the Admiral's vanguard was still a long way off.

The King had several hours before him to carry out this suggestion, which struck him as most reasonable. He resolved to seize Branxton Hill, transfer his artillery to it, and draw up his whole army facing north, dominating the low and marshy ground over which the English would have to advance. This move must have taken several hours, as the heavy guns had to be dragged up from their original position on the east side of Flodden Edge, and the fighting-order of the army would have to be inverted; the corps originally intended for the right wing now becoming the left, and *vice versa*. According to one account James even ordered his pavilions to be struck on Flodden

[1] Leslie (p. 348) has got the matter clearly, saying that Musgrave pointed out that Surrey was not only going to cut communications, but to seize Branxton and force on a fight. Hall (*Henry VIII*, v. folio 42) does not put things so well—Musgrave 'did it for a policie, to cause him to come down from the hill, and the king caused his tentes to be removed to another hill in great haste, lest the Englishmen should have taken that other hill.'

Edge, and brought across to be erected on Branxton, which would have been a very complicated business. When one

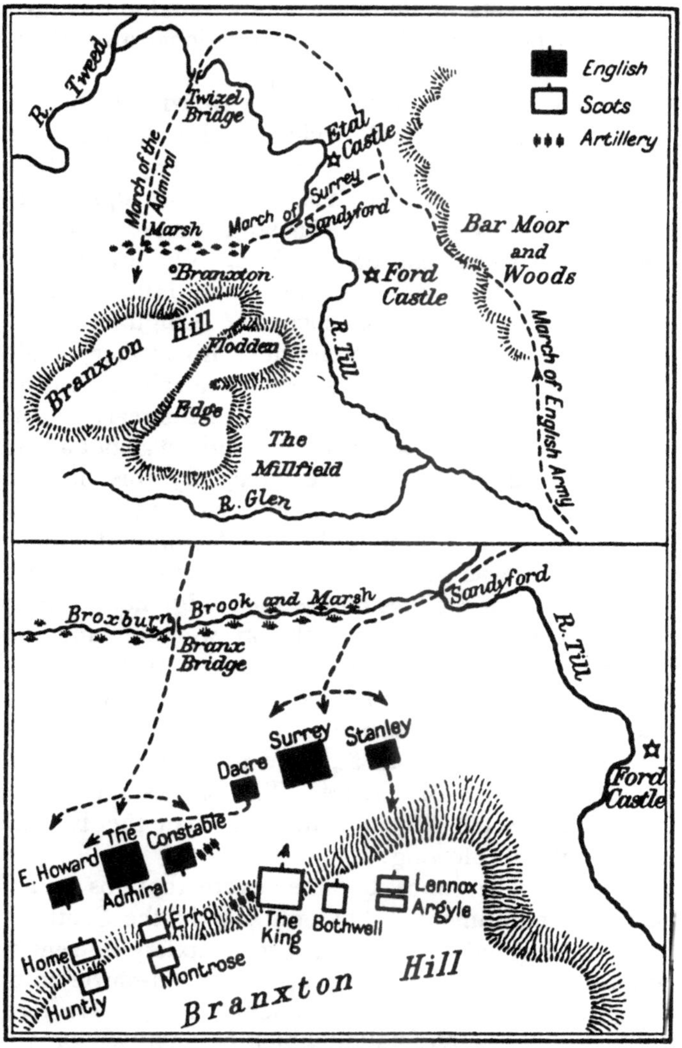

FLODDEN FIELD, SEPTEMBER 9, 1513.
Above : Topography. *Below :* Order of Battle.

reflects that there must have been at least ten thousand nags of sorts, and a mass of horse-boys, gillies, and suttlers in the camp,[1]

[1] In the curious picture of the Scottish camp to be found in the first (and illustrated) edition of Holingshed, suttlers are conspicuous.

it seems incredible that, with a battle impending within the next few hours, James should have shifted his impedimenta and non-combatants on to what was to be fighting ground. But no doubt he gave orders for preparations to break up the camp.

Meanwhile the Admiral with his 'vaward' and wings and the guns had crossed Twizel Bridge, deployed, and advanced in line southward. Somewhat later his father, with the other half of the army, forded the Till some miles upstream, and began to come forward in échelon with the Admiral, but (as was inevitable) somewhat behind him. On arriving at the marshy dip a mile in front of Branxton Hill, down which trickles the insignificant Broxburn, the Admiral found that there was only one point in it on which his artillery could get forward, perhaps the causeway known in later times as 'Branx Bridge.' [1] The troops sent their horses to the rear, and waded through the swamp—the water in the middle was but a 'tailor's yard broad'—but the guns were late.[2] By this time it was nearly four o'clock, and the Admiral began to make out the Scots already on Branxton Hill, and forming up for fight. But clear vision was for the moment difficult, for, as we are told, the servants and gillies on Flodden Edge had, in preparation for a move, raked together the straw, litter, and rubbish of the camp, and set it on fire 'according to their custom.' A south-east wind took a broad cloud of smoke across the hillside and hid the armies from each other for half an hour. This was not an early example of an advance under cover of a smoke screen, but a matter of chance. The smoke must have incommoded the Scots only a trifle less than the English, in whose faces it was blowing.

The Admiral at once became aware that his own force was not only outnumbered but outflanked; the Scottish host extended far beyond his left. In great anxiety he sent riders to implore his father to bring up the main-battle into line with the 'vaward' without a moment's delay. He tore, we are told, the *Agnus Dei* badge which hung at his breast and sent it to the Earl as a token of desperate need.

Surrey had just time, and not a moment more, to hurry up and deploy on his son's left. Indeed the battle seems to have

[1] See Bates's *Flodden Field*, p. 12, for local topography.

[2] As we find them in action at the beginning of the battle, they cannot have been 'left behind,' as some accounts say.

begun on the extreme right before his men were quite level with the vaward. For King James, seeing the two halves of the English army separated, charged downhill with his army in an échelon of columns, the left leading, apparently in the hope that he might smash the Admiral's corps before the Earl's was fully in line.

The array of the Scots on this day was obviously governed by the advice given by the officers of the French military mission, for (as has been already mentioned) its order was unlike that of old Scottish practice, and was evidently inspired by continental tactics. The troops were drawn up in five columns,[1] each apparently composed of two brigades, with intervals of a bow shot (200 yards ?) between each column. All were armed with the long pike of the landsknecht type, of which many had been imported, and many manufactured in the Edinburgh arsenal. The brigading, as one Scottish chronicler points out, was regional—the men of each Scottish district being massed under the most prominent noblemen of the neighbourhood. The left-hand column was composed of the Borderers under the Chamberlain, Lord Home, with in support the Highlanders of the north-east under Alexander, Earl of Huntly. The second column, in two brigades, under respectively the Earl of Errol, Constable of Scotland, and the Earls of Crawford and Montrose, consisted of the levies of Fife, Angus, Perthshire, and the small counties about the head of the Firth of Forth. The King had the third or central column, which was formed of his own personal following and the levies of the western Lowlands, under the Earls of Cassilis and Glencairn and the Lords Herries and Maxwell. The fourth consisted of the full force of the Lothians under the Earl of Bothwell, the fifth comprised all the western and northern Highlanders, under the Earls of Lennox and Argyle. The King's column is said to have been larger than any of the others—if the whole army counted 28,000 or 30,000 spears, it may have

[1] I follow Surrey's official report in speaking of *five* columns, some authorities give only four, as does Hall, who is generally accurate, and Geo. Buchanan, who is often the reverse. He gives a front line of three columns, and Bothwell with the men of Lothian in reserve. Hall mysteriously speaks of four Scottish columns which *fought*, and 'two battles which never came to handstrokes' (folio 43). I can only suppose them to mean Home and Huntly's brigades, which after their first successful clash with Edmund Howard, hung off, and did not join in the main decisive central engagement. Bothwell's corps may have been lost to sight when the Scots were first seen and counted, as we are told that it got into a 'cleugh' and came late into action.

had 8000 or 9000 men, the rest 4000 or 5000 each. English observers have left on record that some of the columns were 'quadrant,' *i.e.* regular squares, others 'pike-shaped,' *i.e.* with a front narrower than the depth. All came down the slope 'in the Almayne fashion,' very orderly and with no shouting.

Some time before any of the other columns closed, Home's Borderers clashed in upon Edmund Howard's 'wing' of the Admiral's corps. It was completely routed, largely because the Cheshire contingent flinched—their countryman, the author of *Scottish Ffielde*, gives the very inadequate excuse that they resented being put under a Howard, when they were always wont to follow the Stanleys.[1] The unfortunate gentlemen whose retinues had been brigaded with them suffered terrible slaughter, Brian Tunstall of Westmoreland, the first man who struck a blow, was run over and speared. Sir John Booth, commanding the Lancaster band, Sir William Warcop, Sir William Fitzwilliam, Sir John Lawrence, Christopher Savage, and Sir Richard Harbottle[2] were all killed. Sir Henry Grey and Sir Humphrey Lisle were taken prisoners. Edmund Howard himself was thrice felled, and only got away to the Admiral's main body by the help of John, the Bastard of Heron, according to Hall, though he slew Sir David Home, the Chamberlain's brother, who tried to intercept him.

The complete *débâcle* of Edmund Howard's wing might have been fatal to the English cause, but for the utter indiscipline of the victorious Borderers and their neighbours, the Aberdeenshire caterans (the 'Cathericks,' as the Cheshire bard calls them), who, instead of turning in upon the Admiral's flank, devoted themselves to the plunder of the dead and the pursuit of the fugitives. They got the sort of loot that suited them best by capturing the parked horses of the enemy, left on the other side of the marsh. Home, if we may believe Pitscottie's anecdote, made no effort to meddle with the rest of the battle. Huntly, it is said, suggested to him that they ought to collect their men, and strike again; but the Borderer replied, 'He does well that does for himself; we have fought their vanguards and won the same; let the rest do their part as well as we.'[3] This tale may be an invention, but it seems

[1] *Scottish Ffielde*, line 265. Letters and Papers, No. 4520.

[2] Richard Harbottle was certainly killed here, and not in the centre, while fighting personally with King James, as the *Rotta dei Scozzesi*—indulging in Homeric narrative—relates.

[3] Pitscottie, p. 278.

that for the rest of the evening Home allowed himself to be 'contained' by a detachment from the English centre. Edmund Howard's disaster happened just as Surrey's main body was coming into line; seeing the rout, the Earl detached his leading 'wing' under Lord Dacre to cover the Admiral's flank. And Dacre, 'like a true and hardy knight,' saved the situation.[1]

We chance to have a contemporary letter from Dacre narrating his experiences. Just as he was ordered off to the right, the Scottish artillery on the slope above opened on his detachment. Whereupon the Bamboroughshire and Tynedale Borderers bolted! 'They fled at the first shot of the Scottish guns, as my Lord Admiral can report.'[2] But he brought his own Cumbrian contingent round to the exposed flank, where he and Home seem to have paired off for the rest of the evening. He says nothing about an incident reported by a late Scottish chronicler, that he got engaged with some of Huntly's men and drove them off. A cruel slander by the Cheshire bard to the effect that 'Cheshire fled away and Dacre followed after'[3] is a complete misrepresentation. Cheshire fled first, and Dacre came on the spot after the initial disaster. His absconding Border contingent stole a number of the horses of the main body, as they fled from the field; but he himself held his ground.

Farther eastward along the line everything went in a very different fashion. The column under Errol and Crawford came into full frontal collision, 500 paces from the foot of Branxton Hill, with the Admiral's main body, and its surviving 'wing' that of Sir Marmaduke Constable. The fighting here seems to have been heavy and prolonged; but we have little information concerning it. Bishop Ruthal, in a contemporary letter, gives great credit to his own Durham contingent under Sir William Bulmer, who fought splendidly under the banner of St. Cuthbert.[4] But this may be local pride. So may be the claim of old Sir Marmaduke Constable to have 'corageously advanced himself above others, though being of the age of three score and ten.'[5] What is certain is that this Scottish

[1] The 'Treue Encounter' gives him a handsome testimonial for saving Edmund Howard.

[2] See Dacre's letter in Letters and Papers, No. 2386.

[3] *Floddon Fielde*, line 267.

[4] Ruthal to Wolsey, Letters and Papers, No. 4461.

[5] See Bates's *Flodden Field*, p. 7.

column was decisively beaten, and all its leaders, the Constable Errol, Crawford, Montrose, and Rothes slain. The Admiral is said to have given orders that no quarter should be granted.[1] This fight was over long before the battle in the centre was finished, and Thomas Howard was able to help his father at the critical point.

Going eastward up the battle-line, the clash came perceptibly later than that between Home and Edmund Howard, for we are told that it was the sight of Home's success which caused the King to launch his own column against the enemy's line,[2] not yet, perhaps, quite firmly compacted. But there was time for a little artillery practice. The Scottish guns, as we have seen, scared Dacre's borderers, though we are assured that they made few hits, because downhill fire is difficult. The English guns, placed, presumably, in front of Constable's wing of the Admiral's corps, are said to have replied effectively to the Scottish—though it is *not* true (as Hall relates) that they killed Robert Borthwick, the master-gunner of Scotland. He was alive years after. But they also sent some balls well into the front of the King's infantry column, which caused him to order the pace downhill to be quickened, in order to save casualties. But there can have been only a few rounds fired on either side—as was so often the case in battles of this age—Machiavelli (it will be remembered) lays stress on the brief share of artillery in general actions.

James led his column straight against that of the Earl of Surrey, and delivered a fair frontal shock. We have good evidence that the English archery, so often fatal to Scottish onsets in earlier actions, had on this occasion no decisive effect. The front ranks were 'most assuredly harnessed,' *i.e.* in complete armour, 'and abode the most dangerous shot of arrows, which sore them annoyed, but yet except it hit them in some bare place, did them no hurt.'[3] Bishop Ruthal wrote, only ten days after the battle, that 'they were so well cased in armour that the arrows did them no harm, and were such large and stout men that one would not fall when four or five bills struck them.' And the contemporary 'Treue Encounter,' declares that 'few of them were slain with arrows: the bills did hew and beat them down.'

[1] 'Neminem, quantumvis nobilem, Scotum, etiam si rex esset, captivum facere: sed occidere' is in the Latin letter to Cardinal Bainbrigge at Rome, printed in the *Rotta dei Scozzesi*.

[2] Leslie, p. 348.

[3] Hall, folio xliii.

The effect of the charge of a column of pikes was intended to be like that of a steam-roller, or a modern tank, to crush down opposition by sheer impetus. But if the first crash failed to disrupt the enemy's line, and only brought the assailants into a stationary combat, after bearing down one or two ranks of the hostile front, the effect of the impetus was lost, and the pike was a clumsy weapon at very close quarters. The long and tough fight between the King's 'battle' and Surrey's resolved itself into a contest between lance and halberd—the thing that had been seen at Sempach in 1386.[1] For the landsknecht pike was but a very long lance, and the bill was but the English form of the halberd. Apparently the halberd had the advantage, when once the pikes were jammed. 'The bills disappointed the Scots of their long spears, on which they had relied,' says Bishop Ruthal, in his summary of the fight.[2]

Since both sides were very much in earnest, and did not give way, there was a long standing mêlée. The Scots, we are told, found their pikes hewn down, and were forced to cast away the shafts and fight with their sword. Casualties were heavy on both sides, without any decisive result. When this strife had been going on for some time Bothwell brought up the Lothian column, which (as we are told) had been at first out of sight in a cleft of the hillside, to the King's assistance. At the same time, or perhaps a little later, the Admiral—having completely finished off Errol's and Crawford's division—came in to the assistance of his father, striking (as is reported) the King's division in flank and rear. The mêlée went on for some time still—the final decision being given, as it would seem, partly by the superiority of bill over spear in close fight, partly by the demoralizing effect of the death of nearly all the Scottish leaders. They had fought in the front ranks—like landsknecht captains—while the English officers had taken positions from which orders could be best given and manœuvres directed.[3] It would have been absurd for Surrey, a man of seventy, who generally travelled in a coach because of his rheumatism, to fight in the forefront of the battle. But King James had chosen to lead his column—and his corpse was found at the farthest point to which the Scottish charge had penetrated, within a few yards of

[1] Ruthal to Wolsey, Letters and Papers, No. 4461.

[2] See *Art of War*, vol. ii. pp. 250–2.

[3] The Scottish Regency wrote an account of the battle to the King of Denmark, in which they accused the English commanders of 'lurking in the rear, unlike Scots lords.'

the banner of St. Audrey, which waved over the centre of the English main-battle. All his personal retinue had fallen with him, save Sir William Scott, his secretary, and Sir John Forman, his sergeant-porter, the only prisoners of importance taken that day. The Earls of Glencairn, Cassillis, Morton, and Bothwell; the Lords Maxwell, Hay, Sinclair, Sempill, and Herries, with Lauder the Provost of Edinburgh, all perished in this part of the field—also the King's bastard son Alexander Stuart, who by a scandalous appointment was holding the archbishopric of St. Andrew's at the age of twenty-one. Along with this young prelate fell other clergy, who might have been better employed that day—the Dean of Glasgow and the Abbot of Kilwinning. The wrecks of the Scottish centre fled southward, to escape home by devious paths over the 'dry marches.'

It was apparently after the decision of the fight between Surrey and the King that the entirely separate engagement between the extreme eastern wings of the two armies came to an end. On each side these corps must apparently have been in échelon to the rest of their line, and came late into action. The fifth Scottish column, that composed of the Highlanders of Lennox and Argyle, was still far up on the ridge of Branxton when the King made his downhill charge. For to attack it, Edward Stanley's Cheshire and Lancashire 'wing' of Surrey's corps had to mount the slope, which was so steep that we are told in *Floddon Field* that they

> 'could hardly fast their feet,
> But forced on hands and feet to creep,
> At last the mountain top they wan.' [1]

When near the crest of the hill they were charged in disorderly fashion by the Highlanders. Why these last were so late to join in the general advance, we cannot say. One Scottish account says that they started in a loose swarm, another that a French officer tried in vain to array them properly,[2] and that the extreme end of Stanley's force overlapped them, and took them in flank. It is probable—but not stated by any contemporary authority—that the archery played on them with effect, since only a few of their chiefs were wearing the full armour which had made Surrey's bowmen so useless in the main clash. It is at any rate

[1] *Floddon Field*, p. 390.

[2] George Buchanan, says Lamotte, but it was more probably d'Aussi. Lamotte was not killed as several English accounts say: he was alive some years after.

clear that they made a feeble fight, and that their chiefs sacrificed themselves in vain, in trying to keep them together. Lennox and his son Lord Darnley, Argyle, the Earl of Caithness, the Lord of Lorn, the Master of Lovat, the chiefs of the Macleans and Mackenzies, M'Ian of Ardnamurchan, head of the Macdonalds, Campbell of Glenorchy, and two clerics, the bishops of Caithness and of the Isles, are recorded as having fallen. Stanley's Lancashire and Cheshire men followed them along the slopes, and—as Hall tells us—found themselves presently passing over the ground where Surrey had already cut the King's column to pieces. 'They found themselves among the Scots who were by the Earl's battail slain before, and suddenly left the chase and fell a spoiling, and stripped the King of Scots, and many that were slain with him: by reason of which some said that he had been slain by that wing. But this could not be true, for the prisoners of Scotland testified that the King's battle fought only with the Earl's battle.' [1]

The pursuit of the English carried them no farther than the Scottish camp on Flodden Edge, which was found full of al manner of provisions, a boon to a starving host which had tasted no food since yesterday. There was abundance of meat of looted cattle, and 'no man would have believed that Scottish beer was so good, had it not been tasted and viewed by our folks to their great refreshing, who had got nothing but water to drink for three days.' [2] In the camp was found much valuable gear, including some silver plate belonging to the King and the nobles, and (as we are assured) several thousand feather beds. Next morning Home and his Borderers were seen hovering in the distance 'the Scottish battles that never fought' as Hall styles them—they were dispersed by a few cannon-shot. When the English sought their own camp by Barmoor Wood, they were disgusted to find that during the battle their own fugitive borderers had stolen many horses beside other stuff.

> 'Many a wye wanted his horse, and wandered home on foote—
> All was long of the Marchmen, a mischief them happen!' [3]

So complains Legh of Baggaley, the Cheshire bard. Bishop Ruthal corroborates: 'Borderers are not to be trusted: they

[1] Hall, folio xliv.
[2] Bishop Ruthal, Letters and Papers, No. 4461.
[3] *Scottish Ffielde*, pp. 414–15.

have done much harm, and plundered both sides.'[1] Lord Dacre raged against these thieves as Warden of the Marches, and extracted 287 horses and mares which were restored to their owners on their book-oath[2]—but what had been the total stolen?

The casualties on the Scottish side almost pass belief. The most moderate Scottish chronicle puts them at 5000—English authorities double that figure. All such figures are proverbially untrustworthy. But it is worth noticing that of twenty-one Earls on the Scottish peerage roll of 1513, ten were killed at Flodden—Argyle, Bothwell, Caithness, Cassillis, Crawford, Errol, Glencairn, Lennox, Morton, and Rothes. Of the remaining eleven Huntly escaped from the field; Arran was at sea with King James's navy; Sutherland was insane; Angus, Eglinton, and the Earl Marshal were elderly, Buchan under age. This pretty well accounts for the whole class! Similarly, of twenty-nine barons in the list, thirteen fell at Flodden—Avondale, Crichton, Elphinston, Erskine, Fleming, Hay of Yester, Lorne (Innermeath), Maxwell, Ross, Sempill, Seton, and Sinclair, with Darnley, son of Lennox, who was not a peer of parliament. Only Home and Lindsay escaped from the field of those who had been present. Of the remaining barons one was insane, three very elderly, two very young. Of peers who do not appear in the casualty list several are represented in it by their elder sons, as was old Angus, 'Bell-the-Cat,—others by younger sons. The gap made by Flodden in the Scottish peerage is unparalleled, yet (oddly enough) none of the titles actually died out—families were numerous in those days. No sufficient tables can be compiled to calculate the corresponding losses of the landed gentry, or the chiefs of Highland clans, but I am under the impression that their casualty list was proportionately heavy among them, and that the statement that ninety heads and 400 other members of gentle houses were slain is not impossible. The published lists in the English authorities are hopelessly insufficient, but anyone looking through Scottish genealogical tables soon finds that a whole generation was cut off simultaneously—occasionally three or four brothers and cousins figure together.[3] It must be re-

[1] Ruthal to Wolsey, Letters and Papers, No. 4462.

[2] See Bates's *Flodden Fielde*, p. 22.

[3] Going through my own wife's family tree, I found three Flodden casualties in as many ascending lines—Boswell, Stewart of Morphie, and Fergusson. There may be more.

membered that all fought on foot and in heavy armour, so that flight, especially for a man weary with much weapon-play, would be slow and difficult. And very little quarter was given, as is shown by the fact that only the two knights, Scott and Forman, and 400 other prisoners were taken alive.[1] If 28,000 or 30,000 Scots were present, I imagine that the whole casualty list may have been a good deal over the 5000 allowed by Buchanan and others.[2]

Hall says that the 'books of wages by which the soldiours were paid' showed a loss of 1500 on the English side, which, considering the rout of Edmund Howard's division and the protracted hand-to-hand fighting in the centre, appears a very probable estimate. Bishop Ruthal of Durham, in his contemporary letter, guesses at 1000. All other authorities give lower estimates, but with no certainty. There was no peer slain—only about ten were present—Surrey, the two Scropes, Latimer, Conyers, Clifford, Dacre, Ogle, Lumley, Darcy. On the other hand, about ten knights of some note fell, nearly all in the rout of Edmund Howard's corps—Sir John Booth commanding the Lancashire contingent, Fitzwilliam, Warcop, Harbottle, Savage, Lawson, and one or two more.

The fate of the unfortunate James of Scotland was discovered with some difficulty, for his body had been stripped like those of all his company. He had an arrow wound, which did not much matter, but several deadly blows from bills—his head was so hacked that he was recognized with some difficulty by Scott and Forman, the two surviving prisoners from his retinue. Surrey caused the body to be embalmed, and sent it in a leaden coffin to the Queen-Regent, who stowed it away at the Nunnery of Sheen, without formal burial, because James had died excommunicate. There it lay for many years, much neglected and forgotten, till the time of the Dissolution of the Monasteries, and finally was lost. Stow has a sad story of the squalid end of the corpse of this romantic if erratic sovereign: it was gone before his descendant James VI came to the English

[1] The 400 prisoners come from Bishop Ruthal's already quoted letter. The 'Treue Encounter' adds one more prisoner—unidentifiable—'Sir John Coolchome.' What name is this? Possibly Colquhoun.

[2] Most of the printed lists in contemporary documents are hopelessly inaccurate, both in omission and mention of names; *e.g.* Hall wrongly gives the Earl of Athol and the Lords Lovat and Forbes as casualties, and adds as peers several persons who were only heads of knightly houses. He omits, on the other hand, Crichton, Hay, and Fleming—all peers—and murders the names of Elphinston and Erskine into Elveston and Arskyll.

throne. A sword, ring, and dagger purporting to have been taken from his body and sent to the Queen-Regent by Surrey, are preserved in the Heralds College.[1]

Both from the tactical and the strategical point of view this was a most interesting battle. It is the last on record of the fights purely between men on foot which had been so common in the fifteenth century. But it differs from its predecessors in that archery played no important part in the decision: the contest was hand to hand between pike and bill—a most exceptional phenomenon. The bill won, but the result had no effect on English military organization. Henry VIII continued to purchase more pikes, and to collect troops carrying firearms; moreover, he endeavoured to get together more cavalry, even if they had to be hired abroad. In the next great clash between English and Scots, at Pinkie, the army of Henry's successor in military authority, his brother-in-law Somerset, contained an immense proportion of horsemen, and a good many arquebusiers. In a way, then, Flodden marks the end of an epoch. But its tactics differed from those of earlier Anglo-Scottish battles, in that both armies fought with smaller and more numerous units—probably under the influence of recent continental experience—and not in the great blocks of 'vaward,' 'main-battle,' and 'rearward,' of earlier days. This precluded 'jamming' such as had sometimes been seen before, and was to be seen once again at Pinkie on the Scottish side, but perhaps it facilitated dispersion, and want of proper combined action. King James's columns gave each other little help. This may, perhaps, have been due in part to simple indiscipline: Home deliberately fought his own battle on one wing, while on the other Argyle and Lennox were obviously late in starting, and were still high up on Branxton Hill while Home and Errol had got to its very bottom.

Indiscipline and feeble fighting are indeed obvious on both sides: Edmund Howard's Cheshire contingent, and Dacre's Northumbrian marchmen disgraced themselves, even more than Home's plundering Borderers. Yet the main bodies on each side fought with most obstinate courage, though both were raw troops who had never seen a pitched battle.

[1] It is hardly necessary to mention two old Scottish legends which appear in Pitscottie, Leslie, and Buchanan—the one that James escaped the battle, but was kidnapped and murdered by Lord Home; the other that he went as a pilgrim to Palestine!

Considering the untried quality of his army, the Earl of Surrey indulged in a most hazardous experiment when he made his semicircular march, and placed himself between King James and Scotland. One asks what would have happened if more of his units had copied the behaviour of those which bolted, or again if Home and his Borderers had turned against the Admiral's flank, instead of scattering for selfish plunder. King James can, perhaps, hardly be blamed for the fact that his five columns fought dispersedly—but is liable to grave criticism for not marching at once to attack the English, when they were only half across the Till. Probably the explanation was that he wished, as he had all along, to fight with his army well arrayed on favourable ground, not to bring it up piecemeal against an enemy who was on the move. If success can justify a desperate manœuvre, Surrey must be given high credit. But he fought as he did mainly because his army was starving, and victory was the one possible alternative to a disastrous retreat, and the melting away of an undisciplined multitude.

CHAPTER III

THE SECOND ENGLISH INVASION OF FRANCE (1522–23)

CONSIDERED from the military no less than from the political point of view, the second war of Henry VIII with France, which started in 1522, is uninteresting. Its only noteworthy feature is that it led to the furthest irruption of an English army into France that was to be seen till the days of Wellington. As a judicious modern historian has observed, 'It was one of the most purposeless and unnecessary contests in which this country has ever been engaged—a war waged to the accompaniment of solemn religious pretexts, while the Turks were overflowing the plains of Hungary and beating down the heroic defence of the Knights of St. John at Rhodes. It was a war of fruitless raids and ravages, framed upon a scheme as disturbing to the balance of power in the West, as it was fatal to the interests of Christendom in the East.' [1]

At bottom this most ill-conceived and ill-managed war had its origin, like the earlier contest of 1513 and the later contest of 1544, in King Henry's misplaced desire to play the part of the Arbiter of Europe—a part beyond possibility considering his limited military resources. And the King's infatuation was helped at this moment by Cardinal Wolsey's equally misguided designs upon the Papal tiara, which no continental prince or prelate intended that he should obtain. Some of his suggestions concerning Papal elections, made to foreign negotiators, were as misplaced and unscrupulous as any of his master's proposals.

Right down to the end of his reign the King was still hankering after the renewal of the old claims of Edward III and Henry V to the Crown of France. This was obvious in 1544 no less than in 1522. His agreement with Charles V was the same at both periods—France was to be crushed, her provinces split up between the victors. But if this consummation failed to

[1] Fisher, *History of England*, 1485–1547, p. 240.

materialize, Henry was ready to take any lesser gains that could be got—if not Guienne and Gascony, then even such a modest prize as the County of Boulogne. But success—small or great—could only be obtained with the aid of the Emperor, and Henry had yet to learn that Charles V was as little anxious to gratify his ally's ambitions as his grandfather Maximilian had been in 1513. In this particular war of 1522, Charles merely made use of the English army, which was sent across the Channel into Picardy, as a covering force, to protect the frontier of his Netherland dominions, while he was turning his main strength against the French in Italy. The succours which he sent to join the expeditionary force were inadequate, and their co-operation half-hearted.

Henry's attention was in 1522 distracted, just as it had been in 1513, by the probability—which soon turned into a certainty—of an intervention of the Scots in his rear. The Duke of Albany, who had overturned the old regency of Scotland, and established himself as 'Protector,' had been born and bred in France, and was entirely 'Francophil' in his sentiments. He was ready to take up the old policy which had led to the slaughter of Flodden. Fortunately for King Henry, Albany was a fickle and inconsequent person, who succeeded in disgusting the Scottish nobility, and his military career was to be contemptible. But this Henry could not know in June 1522, when he started on his wholly unnecessary French war, and he had to make provision for an attack in his rear, which might prove more fortunate than that of James IV. Hence, probably, came the fact that he did not go himself on the expedition to France, and only sent first the Earl of Surrey and later his own brother-in-law, Charles Brandon, Duke of Suffolk, to take across the Channel a moderate force. When Suffolk in 1523 had brought over all the reinforcements, the whole came to less than half the strength of the army which Henry himself had led to Thérouanne and Tournay in 1513. A much larger force was ear-marked for service against the Scots, including all the levies of the six northern counties and Cheshire, Derbyshire, and Notts. But it never had to be employed, for Albany barely crossed the Border, and took back a mutinous army to Edinburgh.

Surrey in the autumn of 1522 confined himself to raids in Picardy and Artois, ending in an abortive blockade of Hesdin, which was abandoned for lack of battering artillery. In the next year, when Suffolk brought over larger contingents, a serious invasion of France began. But it could only be effective if very strong auxiliary forces should be lent by the Emperor, for an army of 10,000 or 11,000 English was obviously too small to face such strength as the enemy might conceivably gather. But the Netherlanders, whom Henry was to hire, came up late and in wholly insufficient numbers. This caused the King to halt in his scheme, and he ordered Suffolk to turn aside and lay siege to Boulogne, a useful but unambitious undertaking. From this plan he was dissuaded by the Emperor and by Wolsey, who was anxious to propitiate Charles, because a vacancy in the much-coveted Papal chair was impending. An invasion of Picardy actually took place during the autumn of 1523—which penetrated far into the land—even beyond the Somme. But since the promised Imperialist succours never turned up in any sufficient strength, it was doomed to become a mere mischievous march of destruction, like those which Edward III and his sons had several times carried out in the fourteenth century.

We have unfortunately much less information as to the composition of Suffolk's army than is forthcoming for the French expeditions of 1513 and 1543. The total of the duke's English force was 10,688 infantry and cavalry, and 1648 men belonging to the ordnance, transport, and other auxiliary services. There were hardly any fully equipped men-at-arms, but about 600 'demilances' and some hundreds of horsed archers. The deficiency in heavy cavalry was (as in 1513) to be made up by the Emperor's 'Burgundian'—*i.e.* Netherland—gendarmes. The infantry was all 'bows and bills,' the latter somewhat more numerous than the archers. We hear of no men equipped with firearms, though Henry was buying hand-culverins (*i.e.* arquebuses) in that same autumn,[1] nor of any pikemen, though there had been some few in the army of 1513. But the Emperor had promised to send a large contingent of landsknechts, who would supply this deficiency. The artillery, on the other hand, was numerous and heavy—the want of it had been the cause of the futility of Surrey's operations in the preceding autumn, and had now been repaired. When some drafts from the permanent garrisons of Calais and Guisnes had been added, the whole force came to about 13,000 men.

[1] Letters and Papers for 1522–3, No. 3494. Armour and lance-staves were also being purchased.

There were no contingents from the North—the Scottish war naturally absorbed them—but a larger levy than usual of Welsh, who appear in the chronicle of the campaign as a particularly tiresome and disorderly company. The army was organized with a 'vaward' and 'rearward' in the old style, but in addition (as at Flodden) there were two 'wings' to the 'main-battle,' which of course was led by Suffolk himself. The arrangement of the minor units seems to have been in companies of 100 commanded by 'captains,' with their 'petty captains,' into which the 'retinues' brought by individuals were drafted; a large retinue making one or more companies, many small ones being combined with others.[1] The peers with 'retinues,' smaller or greater, were Abergavenny, Daubeny, Dacre of the South, Berkeley, Ferrers, Fitzwarren, Marney, Mountjoy, Monteagle, Sandes, and Willoughby.[2] Of this organization more will be told in the chapter on 1543, when many more details are forthcoming.[3]

Suffolk was delayed at Calais till September, waiting for the promised Imperialist contingents. By the end of the month only 500 heavy cavalry and 3000 landsknechts, under Floris of Ysselstein, Count of Egmont-Buren, had turned up. The Emperor's excuse was that he had made over the main body of the landsknechts who had been raised to the renegade Constable Bourbon, who would advance with a separate army, and co-operate by an invasion of north-eastern France, in which he would be supported by all the strength of the Netherlands. The English army was invited to meet Bourbon, and Compiègne was suggested as the joining-place of the two forces. Very unwisely, King Henry consented to this hazardous strategical combination, which was to fail, even as a more ambitious one was to fail twenty years later in 1543. Suffolk was ordered to give up the idea of a mere attack on Boulogne, which had been contemplated when the exiguity of the imperial auxiliary contingent was discovered, and to strike for the heart of northern France, where Bourbon would meet him at Compiègne—perhaps for a blow at Paris. The season was now late and October at hand.

Accordingly Suffolk, leaving Boulogne, Abbeville, and Hesdin and Montreuil untouched upon his right or western

[1] Suffolk's own return had fourteen captains for just under 1400 men. Letters and Papers, No. 3288.

[2] *Ibid.* Hall's list is incomplete. [3] See p. 335.

flank, marched for the middle Somme; his foragers, especially the so-called adventurers,[1] wasted the country for miles around his route. He took Bray by force of artillery, with which he was well provided, and Roye by capitulation, and reached his limit before Montdidier, where he battered the walls so effectively that the garrison of 2000 men surrendered on terms (October 28). He was now within a few miles of the frontier of the Isle de France, and had seen no field-force opposed to him. But he sought in vain for news of Bourbon and his army—and not without reason. For the Constable's host of landsknechts had fallen into mutiny in Champagne and dispersed, leaving their general with nothing but a few hundred lances, so that he had been fain to turn back out of France.

After the capture of Montdidier the duke's captains began to murmur that the season was late, and the object of the campaign lost. Suffolk sent Lord Sandes post-haste to England, to ask the King's leave to retreat. Meanwhile he stayed, ravaging the counties of Santerre and Vermandois, and captured Nesle (November 2) and Bouchain (November 15). But in November the weather became abominable: 'After great rains had fallen, came a fervent frost, so sore that many a soldier died for cold. Some lost fingers and some toes, and many the nails of their hands, which was a great grief.'[2] To move the great guns and baggage waggons the duke had to contrive hurdles or sleds, for wheels would not work. The Welsh contingent went on strike, and cried, 'Home, home!' The spirit of the whole army was so demoralized that Suffolk was constrained to order a retreat into the Emperor's dominions, and came to Valenciennes in Hainault early in December. Unfortunately King Henry had refused to listen to the report brought by Lord Sandes, and had sent back orders that the army was to keep the field, and to wait for reinforcements. When the royal message arrived, 'the duke did what he might to restrain the soldiers. But it might not be, for many Englishmen shipped themselves at Antwerp, and many at Sluys and Nieuport, and other havens, and the duke, seeing that he could not bring his army together, or scarcely its fourth part, licensed the remainder to depart.'[3] He himself came to Calais (December 12) where he received a thundering reproof from the King for evacuating Picardy and breaking up his army. The report of the privations suffered by the troops had

[1] For whom see below, p. 328. [2] Hall, *Henry VIII*, xvi. folio 122. [3] *Ibid.*

evidently no great effect on the mind of Henry, who was comfortably domiciled in winter quarters at Windsor. But he finally forgave his brother-in-law, who was never very long in disgrace. Brandon seems to have been the only person for whom Henry had a genuine personal liking, and could count on comparative immunity when others would have suffered dreadful penalties.

So ended a very futile campaign, which illustrates a few characteristic tendencies. Henry was still unconvinced that small fire-arms and pikes were the weapons of the future, and contented himself with hiring landsknechts and heavy horse from the Emperor. It was only in the end of 1523 [1] that he started on heavy purchases of lance-staves and armour for pikemen in Germany. At the same time, he ordered calivers or arquebuses. But he was still of the same mind as Francis I; it required the disaster of Bicocca, and still more that of Pavia, to convince the French king that small firearms were indispensable. It will be remembered that Montluc lays stress on the way in which the arquebus was practically unknown to the French down to this period of the Great Wars.[2]

Obviously Suffolk started too late from Calais—which was none of his fault—and was set a plan of operations which rested on a wholly uncertain junction with a force moving from a distant base. He put garrisons in Roye and other places which he had taken, but had to withdraw them when he gave up the campaign and retreated into the Emperor's dominions. Otherwise they would have been surrounded and starved out during the winter. If he had dispersed his army into garrisons, there would have been no striking force left, when he had taken six or eight considerable places. At the end of the campaign only Bouchain was left of all his gains, and this he handed over to the Emperor's governor in Hainault, because it was close to the frontier.

The army was obviously ill-disciplined—if a trifle less so than the force which the Marquis of Dorset had taken to Biscay in 1512. It did not desert wholesale, as did that strange company, but gradually melted away: and there was still a nucleus left when the Duke gave the word for disbandment. The organization was, of course, deplorable—there was no intermediate unit between the company of 100 men and the

[1] Letters and Papers, No. 3494. The date was October 1923, when Suffolk's army was quite spent.

[2] See above, p. 43.

great mass of the vaward or the main-battle. Captains wrangled with each other for want of regimental or brigade control. The rank and file of a company acknowledged no obedience to any but their own officers. And in the case where a company was made up from many small retinues, it must be doubtful whether the numerous sections of which it was composed had much care for a captain who was a stranger to most of them. But we may defer criticism on this hopeless form of organization till we come to the army of 1543, concerning whose internal order much more information is forthcoming than can be gleaned concerning Suffolk's unlucky force.

A special source of trouble were the 'adventurers,' *alias* creckers or crackers,[1] a band of ne'er-do-weels who were not receiving the King's pay, or embodied in any corps. They had come to Calais in the hopes of being 'taken on the strength,' but Suffolk had refused to have anything to do with them. Under leaders chosen by themselves they preceded or flanked the march of the army, plundering right and left, and irritating the peasantry into desperate antagonism to the invaders. It was in vain that Suffolk tried to keep discipline, and to buy food as he went. The 'adventurers' sometimes fought desperately when they came into contact with French regular troops, but they just as often absconded. Much incendiarism and some petty massacres are ascribed to them by the French chroniclers—no doubt with justice. Suffolk would have done well to arrest them all, and hang some of their chiefs. He thought apparently that they were useful enough in the way of bringing in information, and purveying looted cattle for the camp, to justify him in tolerating their reckless and abominable dealings among the country folk. They taught the men of the regular companies all the worst habits of the soldier of fortune, engaged in habitual brawls (especially with the Welsh, as we are told), and in general were a focus of indiscipline and desertion. But, no doubt, landsknechts and French or Italian mercenaries, when their pay had not come in, were as great nuisances to their commanders and to the peasantry as the English 'crackers'—the origin of whose nickname I have been unable to trace.

Suffolk's expedition to Montdidier was the last regular campaign of an English army on the Continent for just twenty years. Though King Henry meddled in every league or counter-

[1] Hall, *Henry VIII*, 16th folio.

league that was devised, his matrimonial infelicities and his religious controversies gave him plenty of employment at home. These troublous times have no record of battles, but there was hardly a year when strife might not have flared up—generally abroad, occasionally at home, as in 1536, when there seemed every probability that the Pilgrimage of Grace might have ended in a lively civil war. Meanwhile, in those dreary days Protestants were being burned for heresy and Catholics hung or beheaded for treason, and Englishmen were vainly trying to make out exactly what 'the King's Religion' might be.

CHAPTER IV

THE LAST FRENCH WAR OF HENRY VIII (1543–45) —THE 'ENTERPRISE OF PARIS,' AND THE 'ENTERPRISE OF BOULOGNE'

FAR more interesting than the campaigns of invasion in Henry's earlier years, which had been led by Dorset in 1512, by the King himself in 1513, and by his brother-in-law of Suffolk in 1523, was the last military venture of the reign, the great scheme which started as a very ambitious project, the 'Enterprise of Paris,' a plan for crushing France in conjunction with the armies of the Emperor Charles, led by Charles himself, and which dwindled down into the siege of two isolated fortresses only a few miles within the French frontier.

This was the last instance of the appearance of a king of England in person at the head of an army preponderantly English in composition, using as his ostensible purpose the revindication of the time-worn claims of Edward III and Henry V to the French crown. That claim had been made so often before, and then bartered, for provinces ceded or moneys paid, by Edward III at Bretigny (1360), by Edward IV at Picquigny (1475), by Henry VII at Etaples (1492), by Henry VIII himself at the peace of 1514, and again at the peace of 1525—that one wonders at its making one more appearance as a very unconvincing *casus belli*. But it seems clear that the now elderly King was at least aiming at the complete disruption of France, and imagined that the aid of the Emperor Charles—obviously necessary for the task—could be obtained with more security than in 1523, when his ally had left him in the lurch and stranded an English army in Picardy, unaided by the promised Netherland and German succours. It turned out that he was to be deceived in 1543 just as he had been in 1523, though the Emperor himself took the field, and seemed to be set on crushing completely the French king who had foiled so many of his earlier designs.

Unwarned by former experience Henry faced the ruin of

his finances, debased his currency, recklessly sold away the spoils of the confiscated abbeys, and committed himself to a scheme of extravagant expense that only an impossible success could have justified. He emerged from two years of ruinous war, with the poor reward of one frontier fortress. The magnificent 'Enterprise of Paris' was to decay with the modest 'Enterprise of Boulogne,' and to end with the half-hearted peace of June 1546.

Inspired by belated ambition, though his health was failing, and he could barely drag his corpulent body on to the saddle of his war-horse, Henry determined to direct a great invasion himself, more effectively than his first adventure of 1513, and crossed the narrow seas at the head of such a completely equipped army as had never before landed at Calais. The army of 1544, of whose organization such elaborate details are preserved in the *Calendar of State Papers Domestic* for that year,[1] was undoubtedly the most formidable force which ever crossed the Channel during the existence of the old English monarchy. It was thrice or four times as strong as the army of Crécy or Agincourt, and double that of Suffolk's expedition of 1522. It was provided with all those subsidiary corps which modern armies know so well, and mediaeval armies habitually lacked. Not only were there pioneers, but a transport section, a commissariat section, an armourers section, a munitions section, companies of bakers with portable ovens, and even of brewers, a chaplain's department, and a provision of mounted scouts for exploration and intelligence.

The only thing which would strike a modern soldier as hopelessly neglected was the army medical corps ; sickness was always prevalent, owing to the hopeless ignorance of the time in the matter of contagious diseases, sanitation, and foul water. We know that by the end of a few weeks of campaigning there was always a mass of sick to be dealt with, besides any wounded that might come to hand from the normal chances of war. Yet we have no mention save of a few individual surgeons, with no staff of any sort attached to them [2]—except that the well-known Dr. Butts, the King's physician, seems to have had a personal following. There is obviously no regular provision for the distribution of medical attendance among the units of the army. Under modern conditions this would

[1] See *State Papers Domestic*, 1544, pp. 141–66.

[2] Blounte's and Salablanca's bands had each a surgeon.

account for the altogether abnormal number of sick reported in the later months of the campaign. But when we call to mind the dreadful and drastic methods of sixteenth-century surgery, we must perhaps reflect that the wounded were not altogether to be pitied for the absence of a medical staff.

The projected total for the army of invasion, as laid down in documents of March–April 1544, was about 42,000 men of all arms, including not only 32,389 English—4000 horse and 28,000 foot [1]—but a contingent of 'Almaynes and Burgundians' to be supplied by the Emperor. The King was under the impression—quite rightly—that his subjects had not been keeping touch with the important changes in the Art of War that had developed during the long struggle in Italy between Hapsburg and Valois. His army ought no longer to be formed solely in masses of 'bows and bills' as in the old days. What he lacked was men-at-arms, arquebusiers, and pikemen. All these he hoped to get from his allies. Maximilian, Count of Egmont van Buren [2]—heir of the Floris of Egmont Buren of 1523—was to bring, it was hoped, 2000 heavy cavalry and 4000 landsknechts. A German condottiere named Landenberg was to collect more levies on the Rhine. Altogether some 10,000 of the Emperor's subjects were to be taken into the King of England's pay. These were to join the army of invasion so soon as it should have crossed the Channel.

Meanwhile the main body was collected on the South Coast. It was to be divided into the old traditionary units, 'vaward, main-battle, and rearward,' but these are described in rather surprising fashion as 'three regiments' [3]—the first time that this familiar military word is found in an English document. It obviously has no technical meaning, and only describes administrative units, as these three 'regiments' are respectively 13,000, 16,000, and 13,000 strong. There was great lack of heavy cavalry—nearly all the horse being either 'demi-lances,' *i.e.* equipped only with helm, backplate, breastplate, and tassets, or 'javelins and light staves,' *i.e.* light horse with lances, or 'northern horse,' *i.e.* Border moss-troopers, or

[1] A variant *S.P.D.*, p. 156, has 3600 horse and 31,000 foot.

[2] The 'M. de Buren' or 'Count de Bures' of various chronicles and documents. Norfolk called him Bewers!

[3] *S.P.D.*, 1544, p. 157.

arquebusiers and archers on horseback.[1] It is rather surprising to find these last reckoned as cavalry at all, as they were really only mounted infantry. Similarly there was still a sad deficiency of infantry with firearms. In one assembly of foot, told off to the main-battle, there were 807 archers, 1073 bills, 380 pikes, to 181 hackbuteers (arquebusiers). Lord Wriothsley's contingent comprised 20 demi-lances, 20 light horse ('javelins'), 50 archers, 50 bills, 40 pikes, and 20 hackbuteers. The Duke of Suffolk certified for 100 horse, 100 archers, 300 bills, but no 'shot' at all. Sir Antony Knyvet had 100 hackbuteers, but this was an altogether exceptional contribution. It would seem doubtful whether in the whole 28,000 foot of the army of invasion there can have been anything like 2000 men with firearms.[2] This explains the eagerness with which the King enlisted mercenary companies of foreign arquebusiers, when he got to France. By the end of the campaign he had four of them in his pay, Italians under Giovanni de Salerno, and Spaniards under Alonzo Salablanca, Antonio Pompeo, and Alejandro Moreno.

The 'Burgundian' horse of Egmont van Buren were somewhat of a disappointment to the King, as he found that many of them were not men-at-arms but 'pistoleers' of the sort that was growing popular in Germany, armed not with a lance but with a 'boarspear' and a 'little hand arquebus,' *i.e.* a pistol. The landsknechts, however, seem to have given satisfaction, and their 4000 pikes were a useful contribution to an army wherein bills outnumbered pikes by three or four to one.

The mobilization of the army must have been a difficult matter, as there was need to divide these men into units much smaller than the vast 'vaward,' 'main-battle,' and 'rearward.' The rank and file came in as personal contingents contributed by individuals of the most diverse importance, ranging down from the 500 men of the Duke of Suffolk to the 'two hackbuteers afoot' of Roger Lygon, the 'one tall billman' of Lewis Ap Richard, and the one demi-lance sent by the Warden and

[1] In a list (*S.P.D.*, p. 169) of cavalry 'appointed to the master of the Horse,' 514 in all, we find 147 demi-lances, 139 'javelins,' 72 'Northern Staves,' 46 hackbuteers on horseback, and 110 archers on horseback, with only 6 'men-at-arms.' Of course, in the King's personal following there was a larger proportion of fully armed men, such as the 'gentlemen pensioners.'

[2] Norfolk, when blockading Montreuil, speaks of having 400 or 500 hackbuteers with his 'vaward' corps (*S.P.D.*, p. 543), and afterwards complained that when his Spanish company had been taken away from him he would have only some 150.

fellows of All Souls College, Oxford. The organization adopted appears to have been that of dividing all the mass of individuals into bodies of about 100, otherwise 'companies,' 'wards' or 'bands,' under captains. Apparently each body was to consist of a mixture of men armed with missile weapons, whether bows or hackbuts, and of billmen or pikemen. For the Council drawing up the 'order of battle' are 'to appoint how many of the traict (missile-bearing men) and how many of each other weapon shall be in every hundred, and whether those of the traict should be harnessed (*i.e.* wear armour) or no.'[1] That the band would be rather under than over 100 men seems to be indicated by provision being made for 52 captains of horse, and 337 of foot—which would give about 75 men to the horse unit and 83 to the foot unit, if the whole 32,000 men of the original scheme were raised. But it must be remembered that many of these captains would be detailed to what we should consider special or staff work,[2] outside the 'bands' of the line of battle troops. So the use of a 'hundred' for a unit may have been justified after all, and we shall see that there were many bodies of just about that strength, a little under or a little over.

Each unit seems to have had only two officers, a captain (at 6s. a day for horse and 4s. a day for foot) and a 'petty captain,' who got half that pay. There is no mention of non-commissioned officers unless we count the 'whiffler' as one: corporal is as yet an unknown word, and 'sergeant' was still used only in its mediaeval sense. 'Lieutenant' has not yet been turned to use for the captain's deputy: it is only employed for very important officers—such as the Lord Lieutenant of Ireland, or the Lieutenant of the Tower, or the King's Lieutenant in France,' who were really locum-tenentes of some great one. Applied to the second in command of a foot company it does not appear till Elizabethan days.

It seems surprising that we find no intermediate unit between the 'vaward' and 'main-battle' and 'rearward' and the small band of 100 men. It was forty years since the Spaniards had been combining companies with 'colonelcies,' and ten or so since the 'tercio' had come into existence. The

[1] *S.P.D.*, 1544, p. 142.

[2] *E.g.* the officers commanding the trifling units of 25 hackbuteers and 35 archers of the King's Yeomen of the Guard were captains, and Suffolk had four captains (obviously staff) attached to his personal retinue of only some 60 people, when acting as the King's lieutenant.

French 'enseignes' or companies of foot had varied from 400 upward, and the 'legion' recently invented by King Francis was an affair of 6000 men divided into six bands of 1000 each. We cannot trace any such large units in the English army of 1544. Yet it seems impossible to conceive of a 'vaward' of 13,000 men with no division save into 130 companies of a hundred each. And as a matter of fact we occasionally find mention of units of a much greater strength. For example, in the muster-roll of the garrison of Boulogne in December 1544 we have, along with 22 units of from 76 to 113 men—the normal sized bands—several of much greater strength: Lord Lisle, 270 foot; Sir Thomas Poynings, 500; Sir William Blounte, 374; Thomas Wyatt, 370; Edmund Rouse, 195; Sir John Luttrell, 208. What were these? Double, treble, and quadruple hundreds placed under a single commander apparently. It would obviously be natural for the very large contingents, furnished by such persons as the Duke of Norfolk or the Duke of Suffolk, to be kept together, though they might have to be divided into four or five units of 100 apiece. And if they acted together, we can only suppose that the senior captain and company-commander would be some knight of importance, named by the provider of the contingent, and exercising some authority over the other associated companies. We have at least one obvious example of this when, on October 30, Suffolk appointed Sir William Blounte to be 'Captain of 400 men sent to Boulogne taken out of our band (*i.e.* apparently out of Suffolk's personal contingent)[1] with 4 petty-captains, 394 footmen, and a chaplain, surgeon, standard-bearer, drum, fife, and 'whiffler.'[2] These are obviously identical with William Blounte's unit of 374 men found in the Boulogne garrison two months later.[3] Other notices look like less permanent organizations, *e.g.* on July 21, Suffolk, before Boulogne, 'sent Sir Edward Baynton to have chief rule of other captains, namely, Lord John Grey, Mr. Broughton, Mr. Cavendish, Francis Askew, and Edmund Hall, with 1500 men. This Monday the shooting was very hot on both sides.'[4]

That the cavalry unit continued to be the original 100 men, under a 'guydon' and a squadron standard, seems to be

[1] *S.P.D.*, 1544, p. 524.

[2] The 'whiffler' was not a musician, as might have been supposed, but a sort of non-commissioned officer charged with the conduct of marching and quartering.

[3] *S.P.D.*, 1544, p. 468.

[4] *Ibid.* p. 485.

proved by the pay rolls, which account for just such numbers—of 'demi-lances' under Sir Thomas Speke, Sir William Herbert, Sir Henry Knyvet, etc., of light horsemen under Throgmorton.[1] Hobby and Paston, etc., of the Northern horse under Sir Richard Manners, etc. Maurice Berkeley, for some reason, had the exceptionally full unit of 115 light horse.[2] The only indentures for a body of heavy cavalry are payments to Edward Rogers, 'captain of the men-at-arms,' for himself, his standard-bearer and '121 men-at-arms, each with footmen attendant upon them.' Probably these, together with the King's 'Gentlemen Pensioners' under Sir George Carew, may have been almost the only units of fully armed cavalry in the army. The Pensioners count up to 75 lances if the muster-roll may be trusted. There were, however, a few individual men-at-arms, some of them peers and knights not attributed to any unit, who turn up on occasion—*e.g.* there were four, attached apparently as staff to Lord Lisle, in the garrison of Boulogne in December 1544.[3]

It would seem probable that there were not much more than a couple of hundred of completely equipped heavy cavalry in the whole army, though of demi-lances and light horse of all varieties—lances, 'javelins,' and northern Borderers—there may have been over 2000—the rest of the 4000 mounted men of the muster-rolls being the archers on horseback and arquebusiers on horseback, who were not really cavalry at all. This accounts for the King's extreme anxiety to have as many Burgundian men-at-arms sent with Egmont van Buren as possible, and his disappointment at finding that many of the Imperialist contingent were 'pistoleers' and 'reiters' and such sort of people.

While there can be no doubt that in the ordinary band of 100 foot there was a mixture of bows and bills, it would seem that many of the arquebusiers may have been mustered into separate companies. We certainly hear of one John Hubberden, 'captain of a hundred hackbutteers' who was slain at the siege of Boulogne.[4] And the hired foreigners were apparently on the same organization—Alonzo Salablanca's unit were all 'hagbutiers' save his drum, whiffler, surgeon, and priest.

[1] An interesting note of Throgmorton's (*ibid.* p. 307) shows that his 100 men were made up of individuals drawn from six contingents—those of the Earl of Essex (40), Knyvet (20), William Willoughby (20), Sir Richard Long (13), Robert Barwick (3), and John Baker (1).

[2] *S.P.D.*, 1544, pp. 308–9. [3] *Ibid.* ., p. 161. [4] *Ibid.* Appendix, p. 485.

John Ap Richard, 'captain of the hacquebutiers on horseback' had a rather weak unit of only 74 men and one 'petty captain.'

It was certainly a strangely organized army, the devices of enregimenting and brigading being far less advanced as yet than they were in the French and the Imperialist armies. It is significant that no word has yet been invented for the officer who commands a permanent union of several or many bodies of a hundred men, though the Spaniards had been talking of 'Colonels' and 'Colonelcies' for some forty years, and the French were accustomed to deal with bodies of 1000 men as sections of the legion, and had come to call the officer who was in permanent charge of several 'enseignes' or 'compagnies' a colonel. But in the English army of 1544 it looks as if the 100-man unit remained the base of organization, and the commander of 'vaward' or 'rearward' worked them together as he pleased, sometimes giving several over for a longer or shorter time to a 'grand captain'—the term is known—who would take command over all the junior captains. But he has no permanent rank, and no special title, or—apparently—extra pay.

The campaign of 1544, like that of 1523, started as being a joint scheme of Henry VIII and the Emperor for the complete destruction of the French power. Their armies were to operate simultaneously and for a common end, the plan is more than once called in English documents 'the Enterprise of Paris.' And the union of intent is shown by the fact that Charles undertook to lend Henry the large body of Netherland auxiliaries under Maximilian of Egmont van Buren, one of his most trusted officers, and gave him permission to draw on Flanders and Brabant for carts and waggoners for his provision train. The disputes about these vehicles, and the payment for draught animals and their teamsters, between the King's commissioners and the Queen-Regent of the Netherlands, take up many weary entries in the *Calendar of State Papers Domestic*.

The two armies were both very large for the time: Henry's when the Netherland contingent is counted in, should have reached 40,000 men. That of the Emperor was still stronger in numbers. The original plan was that the King should force his way over the Somme to Compiègne, while the Emperor should sweep across Champagne to Soissons and Meaux. It was reckoned that Francis I could not put in the field forces capable of coping with both at once—all the more so because a large detachment of his gendarmerie and his invaluable Swiss was busy in Piedmont against Del Vasto. Such a plan involved either the rapid capture, or the blockading by detachments of the French frontier fortresses. Henry had to look after Boulogne and Montreuil: Charles after Luxemburg and St. Dizier. But it was calculated that they ought to meet on the Marne in August with a joint force of 80,000 foot and 18,000 horse—such a muster as had not been seen in the whole of modern history—and these are not the vague figures of mediaeval chroniclers, but are drawn from elaborately compiled muster-rolls which still survive. They allowed for marginal deductions, where there is some doubt whether contingents and corps may not be a little overstated, owing to the commanders not bringing up quite their full contracted numbers.[1] If all the units had been honestly produced by every responsible officer, the figures would be still higher.

But the 'Enterprise of Paris' dwindled gradually into two separate and leisurely advances into France, and the English and Imperialist armies never met. The Emperor, having mobilized his very considerable forces in May at Speier on the Rhine, where he had much talk with the English commissioner as to the conduct of the campaign, sent forward 12,000 men to besiege Luxemburg, which surrendered on June 6, not so much from battering as from starvation. This frontier fortress being secured, the Emperor started out, and was at Metz on June 19. After capturing Ligny and some other small places, he sat down in front of St. Dizier on July 6. The long and obstinate defence of this place by the Count of Sancerre, who held out till August 18,[2] was the essential cause of the failure of the 'Enterprise of Paris,' for after it had fallen the idea of a rapid dash at the French capital was obviously out of date. The French King had been given two whole months (June 19–August 18) between the muster of the Imperialist army at Metz and the capitulation of St. Dizier, in which he could reorganize his army, call up his much-cherished Swiss con-

[1] In the Emperor's service the landsknecht colonels were notorious for bringing up short numbers, and claiming 'mort pay' for deficient individuals. Landenberg, arguing with Henry's commissioners, urged that the Emperor had always to submit to this on a greater or lesser scale. The English commissioners denied the existence of the custom so far as England was concerned.

[2] The capitulation was really on August 8, but contained a clause giving Sancerre an option to wait a week for possible relief from outside. As no relief came, he opened the gates on August 18.

tingents from Italy and their own homes, and strengthen all his fortifications, including those of Paris.

Charles was obviously flagging from his original purpose, and had some good excuses to make, by pointing out that the English part of the joint invasion was hanging back, and making no serious attempt to strike into the heart of France. The Duke of Norfolk with the 'vaward'—perhaps 13,000 men—had landed at Calais on June 6 and the immediately succeeding days. On the 20th he had only got as far as Marquise, half-way between Calais and Boulogne, and was waiting for the appearance of the 'rearward' under Lord Russell. This corps arrived piecemeal at Calais on the 18th–20th. But the King with the 'main-battle' was much behind time, and did not cross the Channel till July 15, though part of his contingent under his brother-in-law, the Duke of Suffolk, had begun to drop in to Calais early in July. Some of the troops had been shipped from some harbours inconveniently remote—such as Ipswich and Harwich—and came in unpunctually.

Norfolk and Russell had over 20,000 men collected by June 20, but found plausible reasons for hanging back near Calais—the transport train from the Netherlands was inadequate, food scarce, and—most important of all—they were not certain whether their master wished them to march straight for the Somme, or to make sure first of the French first-line fortresses—Ardres and the much more important Boulogne and Montreuil. They had a long discussion with the Count of Roeulx, a sort of liaison officer sent over by the Emperor, to discuss the main trend of the campaign. He said that the English *might* march straight for Bray on the Somme, leaving Boulogne and Montreuil behind them—but did not insist on this course; Boulogne, perhaps, might be masked, but if the Marshal de Biez, French commander in Picardy, should throw himself into Montreuil with a considerable field-force, he did not like the idea of marching by him, and leaving him to cut communications. If Montreuil received a garrison of 7000 or 8000 men it could not be disregarded, and would have to be besieged, and this would be a matter of many weeks before success could be expected. But if de Biez left Montreuil to itself, with its normal garrison only, it might be had with small trouble. Boulogne, on its lofty hill, would certainly take a long time to capture, but might be masked and disregarded—Montreuil was much more important. De Roeulx did not like the idea of crossing the Somme and plunging into France till this place had been disposed of.[1]

This was not very helpful advice—Norfolk resolved that he must find out what was the state of affairs at Montreuil, and risked leaving Boulogne behind him—the place was strong but the garrison known not to be very large. Accordingly, he marched out from Beaulieu (not far from Calais) on June 22, and, moving with extreme slowness and caution, appeared before Montreuil on July 4.[2] Meanwhile Russell, with the 'rearward,' who had landed at Calais on the 20th, pushed forward and joined Norfolk on July 4, while Egmont van Buren and a large part of the promised Netherland contingent, including many cavalry, came in from the east. There was by the first week in July an accumulation of some 30,000 men before the fortress. Only the King and the 'main-battle' were wanting. But without the King's presence Norfolk and Russell felt loth to take responsibilities; they felt (very rightly) that whatever steps they took would be criticized, and that misadventures would be reckoned as crimes.

It was found that the Marshal de Biez had thrown his field-force into Montreuil—which must have had a garrison of at least 6000 men—and he had been busy for three weeks in strengthening the place with earthworks, 'cavaliers,' and external defences. It could not be had by a rush, as might possibly have been the case in early June. Meanwhile there was not only de Biez to be reckoned with. Despite of his danger from the Emperor's side, the King of France had sent another important field force into Artois—the Count of Vendôme was said to have 500 lances and 10,000 foot, somewhere about Hesdin. This force might try to unblock Boulogne or Ardres, or to start operations against the rear of the army besieging Montreuil. On the whole, Norfolk reported to his master that he should deprecate the 'Enterprise of Paris,' and thought that it would be much safer to confine the operation of the army to the capture of Ardres, Montreuil, and Boulogne—which might well require a couple of months for completion[3]—'a great conquest and profitable.' If they were won in good time, it might be possible to 'put Abbeville in jeopardy' or even to pass the Somme. But if they could not be 'won in

[1] See Norfolk's interesting letters of June 11 and June 14, on his conferences with de Roeulx in *S.P.D.* for 1544, i. pp. 420 and 435.

[2] Norfolk to the Council, *S.P.D.* ii. p. 524.

[3] *Ibid.*

good time,' *i.e.* if the sieges lasted well into autumn, Norfolk is silent as to what should be done. Obviously he did not believe in the 'Enterprise of Paris.'

No more did King Henry, for when he had landed in mid-July with the 'main-battle,' he applied himself to the siege of Boulogne, though he had been warned that the place was strong and might take two months to master. His corps sat down before it on July 19, leaving Norfolk and Russell to deal with Montreuil. Meanwhile they were acting as a sort of covering-corps to his own operations, and observing Vendôme's force, whose strength was obviously exaggerated—it cannot have amounted to half the united numbers of the English 'vaward' and 'rearward,' plus van Buren's Netherland contingent.

The siege, if so it may be called, of Montreuil appears to have been a most tame and ill-managed business. The town was strong and well garrisoned, on a high-lying plateau above the river Canche. But this does not explain the fact why a large army lay before it for more than two months without making any serious attempt to deal with it drastically, though a 'mount' was built and trenches opened. As Russell remarked in a confidential report to the King, it was the queerest siege that man ever heard of, for though such a large force was assembled, it never attempted to enclose the whole fortress, but left the front above the river and the gates in it unwatched.[1] Norfolk's excuse was that he had tried to get van Buren and the Netherlanders to pass the Canche, and encamp across the unwatched road, but had been unable to induce them to take up a position in which they would be separated by the river from the rest of the army, and exposed to attacks by Vendôme's field force.[2] But if the Duke thought that the place ought to be completely encircled, he could have detached some of his own considerable contingents, instead of applying to van Buren to complete the investment. Russell wrote to the King that he had offered to

[1] 'The king should send some one to report to him the way the place is being besieged. We lie so far apart that on a sudden attack we could not succour each other. As gates are left open the place cannot be won, for men and victuals go in at pleasure. On last Saturday 100 horse came in over the causeway, and so may they come in at the Boulogne gate.' *S.P.D.*, 1544, i. p. 601.

[2] 'Two gates are left open and a third might be used by night: we are enforced to give this liberty because that the Burgundians will in no wise lie in any quarter but near unto us.' Norfolk to Suffolk, *S.P.D.*, 1544, i. p. 560. One convoy with a Burgundian escort was cut up by Vendôme, on its way from Arras, with some loss.

take up the job himself, but that Norfolk said that 'the place is too dangerous.'[1] Altogether the chronicle of the siege is depressing, and one is left with a strong impression that Norfolk was not anxious that it should succeed, for if the place fell he would be committed to the 'Enterprise of Paris,' which (as he rightly supposed) was beginning to fade away from his master's mind.

This was undoubtedly the fact. Both the Emperor and the King of England were at this moment being visited by French envoys, proffering terms of peace. And each, though he did not wholly conceal the intrigues from his ally, did not refuse to listen to the proposals. They corresponded with each other, seeking to discover the minimum which could be accepted. Henry signified that, though he had published the old monstrous claims that went back to Edward III, privately he would not be averse to dropping all wider demands—the old pretensions to Aquitaine and Normandy—in return for an enormous payment in cash and the acquisition of Boulogne, which he was determined to get. Charles recapitulated the old terms inflicted on France after Pavia, but (in confidence) thought that he might abandon his demand for Burgundy, if the French retired altogether from Italy, gave up all ideas of Milan and Naples, and evacuated their holding in Savoy.

Meanwhile, Henry's siege of Boulogne proved, as had been prophesied, as long an affair as the Emperor's siege of St. Dizier. Though the place had been invested on July 19, and though the King himself encamped in front of it in great state on July 26, it did not surrender till September 14. The lower town, along the harbour, was defended by only a slight wall, and was captured with ease on July 21, the French evacuating it on the first serious assault. But the high-lying quadrangular upper town was a very strong fortress, with steep slopes below it on three sides, and only one practicable line of approach on its north-eastern front, where the ground is more or less on a level with the walls. Boulogne had good bastioned ramparts, surrounded by a *fausse-braye*, or low outer circuit of wall, commanded by the main line of defence. There was a dominating citadel with three keeps at its northern and most exposed angle. The only outlying defence was a post established in the block of the old Roman watch-tower or beacon, the *Tour de l'Odre*, on the height above the sea, a quarter of a mile

[1] Russell to the King, *S.P.D.*, 1544, i. p. 565.

outside the walls. This was surrendered by its trifling garrison on July 22, the moment that a gun had been dragged up to batter it. The French troops in Boulogne were a trifle under 2000 strong, composed of one company of horse, a thousand men of the 'Legion of Picardy,' a strong company of Italian hackbuteers, and some local levies. The governor was a rather young officer, Jacques de Coucy, Sieur de Vervins, a cousin of the Constable Montmorency and a son-in-law of the Marshal de Biez. He may be reckoned among the list of unlucky and ill-used commanders, for though he made a creditable defence of seven weeks, and does not seem to have failed in his duty, he was tried for cowardice and executed in the following winter—a predecessor of the unfortunate Lally de Tollendal.

The English very properly confined their approaches to the north-eastern front, from the castle westward, on the higher ground, leaving the steep sides of the place alone—so much so that the French grazed sheep upon them till July 27, when they were swept off by a party of the besiegers. The soil was shallow above the solid under-surface—not more than nine feet deep anywhere, and it turned out to be a slow business to hew trenches in the hard ground, and still more difficult to scrape together masses of earth to form a 'mount' opposite the chosen point of attack, which was in the curtain beside the castle. The mount, raised to a level with the walls and well gabioned in front, was furnished with artillery, which opened fire on August 3. But though much harm was done both to the walls and the town—the spire of the great church collapsed on August 19—there was no definite breach made, and the enemy still held the *fausse-braye*, outside the main ramparts. Much disappointed at the delay, the King threw up two additional batteries on each flank of the mount, and continued to pound the walls. The trenches being pushed farther forward—not without considerable loss of life—and the defences continually battered, an assault was made on the *fausse-braye* on September 1, and a long section of it was gained, and linked up with the trench system. This brought the besiegers to the actual foot of the rather shaken main wall, and to the flank of the castle, which was also much damaged. A general assault both on the curtain and the castle above was made on September 2, and failed completely. 'Our men in the Braye broke open certain doors, but found much resistence of men.

hailshot, and ramparts within of stones and earth, so that they could not enter, and at the breach of the castle the enemy cast down great abundance of fire and stones, so that our men were fain to recoil, many of them burnt and hurt at both places.' [1] The casualties were numerous, and the enemy's main line of defence still intact.

The King, in no wise turned from his purpose, now ordered mines to be dug under several points, starting from parts of the *fausse-braye* which were in his hands, both against the castle, the curtain beside it, and a bastion called the Flemings' Tower more to the south-east. The last-named mine, fired on September 4, ' rove the tower very sore,' but did not actually bring it down. On September 11 the great mine under the castle was exploded, and sent many stones flying, some of which killed Englishmen in the trenches. After it had worked, three assaults were made at various points, and some lodgment was made in the works, but no entry secured. The Italian engineer Jeronimo, who had been in technical charge of the operations, was killed, ' which would save the King five thousand pounds,' as a cynical correspondent remarks [2] in a letter home. On the following afternoon the Governor sent out two officers to treat for a surrender. ' The cause,' writes the Emperor's agent in the English camp, ' seems to be their inability to endure any longer the battery of the artillery, which has fired more than 100,000 shots, and moreover the English have footing in the wall in divers places. They had not been required to sustain the siege for more than six weeks by the King of France, and have sustained it for eight. As for artillery and victuals, they have plenty of wine, salt meat, and flour, but little powder.' [3]

On the 13th terms were easily arrived at, for the King had good reason for wanting the town at once, and granted the governor leave to march out with his men in honourable fashion, carrying all their private property, and for the civilian population to go forth with their carts and horses, saving any who would consent to swear allegiance to the crown of England. The interesting muster-roll of the garrison as it marched out gives 67 horse, 1563 foot, French and Italian, of whom 800 were

[1] All this from the Diary of the Siege, Cotton MS., Caligala E. iv., in the British Museum.

[2] Mason to Homings, *S.P.D.*, 1544, ii. p. 115. Because, no doubt, the King would regard the wages and bonus due to the engineer as cancelled by his death. Who would search for the heirs of a foreign adventurer ?

[3] Chapuys to the Queen-Regent at Brussels. *S.P.D.*, 1544, ii. p. 124.

hackbuteers, and 87 transportable wounded.'[1] There were about 2000 civilians following them, with horses and more than 100 carts, 'poor women leading their beasts, very lean and meagre.[2] The garrison must have lost several hundred men during the long siege, but the only casualty of importance recorded is that of Philip the Corsican, captain of the Italian hackbut company. The English loss must have been much greater, but is nowhere accurately recorded.

Henry's anxiety to get Boulogne as early as possible is explained by the fact that he had news that the Emperor was negotiating with the King of France—as indeed he was doing himself, for the Cardinal du Bellay and other envoys had come into his lines with elaborate and in some ways tempting proposals from King Francis. It was most important to be in possession of Boulogne when the terms were being settled, for if the Emperor came to an agreement (as was feared) before Boulogne fell, there would be little chance of obtaining the place by negotiation.

In this conclusion Henry was perfectly correct, for Charles signed the Peace of Crépy on September 18, leaving his ally in the lurch, five days after the surrender of Boulogne. His excuse was that Henry had completely failed to carry out the 'Enterprise of Paris,' and so far from having penetrated into central France to join him, had sat down before two fortresses only a few miles within the frontier of Picardy. As has been noted elsewhere, the terms at Crépy were much less favourable to the Emperor than might have been expected, and included a project (which Charles can hardly have intended to carry out) for a marriage between the second son of France and one of two Imperial princesses, who should bring as a dowry a rich duchy either in Italy or in the Low Countries. The Emperor pledged himself, indeed, to procure reasonable terms for his English ally as part of a general pacification. But naturally the ideas of the two sovereigns as to what constituted reasonable terms could not easily be made to agree.

Henry was determined to keep Boulogne, his only gain from an expedition on which so much money had been spent. The French were as determined not to surrender it, though they offered very large sums by way of a war-indemnity. The moment that the Emperor had come to terms, and begun to

[1] Holingshed, ii. p. 964.

[2] Diary of the siege quoted above from Cotton MS.

move his army homeward, Francis saw his opportunity, and detached the greater part of the large force which had been facing the Germans. The Dauphin marched into Picardy with a mass of cavalry, all the Gascon and Italian bands that had been withdrawn from Italy, the landsknechts of the Rhinegrave Philip, and 6000 Swiss. He had orders to give battle if the English would accept it, but if (as was more likely) they should refuse to fight, then to recapture Boulogne at all costs, which should not be a difficult task, since the walls were in a battered state, and there had been little time to repair them.

The English King did not hesitate for a moment when he got news of the Peace of Crépy. He would not be strong enough to resist the whole force of France, more especially as Egmont's considerable auxiliary corps from the Low Countries was to be withdrawn from him.[1] Accordingly Norfolk and Russell were ordered to raise the siege of Montreuil, and to retreat on Boulogne (September 26) to join the King and the 'main-battle.' On September 30 the King left for Calais and returned to England by sea, and on October 3 Norfolk and the whole army followed him, leaving only a garrison of 4000 men in Boulogne, under Dudley Lord Lisle—the Protector Northumberland of later days.[2]

The Dauphin appeared in the vicinity of Boulogne with some 30,000 men only four days later, while the English army was melting away at Calais, slipping home by detachments—their numbers were much thinned, as we are told, by a plague of dysentery, caused by lying out and marching in wet weather. The King wrote angry letters to Norfolk blaming him for sending so many men home, but the force at Calais continued to dribble away—some got off as convalescents, others with passports, many without them.

Meanwhile the Dauphin, after making demonstrations in the direction of Guisnes and Hammes, to distract Norfolk's attention, carried out the great venture of the year, the disastrous 'Camisade of Boulogne,' whose event almost consoled the English for the failure of the previous campaign. The situation was a tempting one: much of the English heavy artillery

[1] Egmont expressed genuine grief at the Emperor's conduct, and instead of marching off separately to Artois, accompanied Norfolk's corps till it was in safety. He said that he never wished to wear harness again in the service of such a master. Norfolk to the King, *S.P.D.*, 1544, ii. p. 163.

[2] Henry sent a most ungracious letter to Norfolk to reproach him for retreating so promptly. *S.P.D.*, 1544, ii. p. 219.

and munitions were still in the trenches of September and the camps behind them. The bulk of the stores were in Basse Boulogne, protected only by the flimsy and breached walls of the lower town, which had never been repaired. The French ascertained by a careful reconnaissance at night the condition of affairs—Montluc, who formed one of the party, says that he got far into the English lines, without being challenged by the three or four sentinels between whom he slunk forward among the trenches.

On the report of the reconnoitring party the Dauphin determined to deliver a 'camisade' at midnight on October 9, telling off for it the whole of the old French and Italian bands from Italy—23 companies commanded by de Tais, who had led the right wing at Ceresole. In support of them were the 6000 Swiss, who were to halt on the heights beyond the *Tour de l'Odre*, and to give help where needed. The leading companies wore white shirts over their armour, to distinguish them from the enemy in the darkness.

The 'camisade' commenced with a complete success. The English outposts seem to have been disgracefully negligent, considering that the foe was known to be not far off. Advancing in three columns, with the French companies under de Fouquessoles leading, and the Italian companies in support,[1] the assailants swept over the slopes where the artillery lay parked, hurling away a few pickets, and then charged into the lower town, by no less than four breaches which they found open. The English were taken completely by surprise; many were cut down as they ran to arms, others shut themselves up in houses, and defended themselves so long as they could. But these were mostly killed off—Montluc says that he slew some 200, some in white and red cassocks, some in black and yellow. But the bulk of the garrison of Basse Boulogne ran up the steep road to the south gate of the upper town, where they reformed, and were joined by a company of 200 men under Sir Thomas Poinings. Thus reinforced, they charged down into the lower town, where all was now in confusion in the dark.

For the enemy, thinking the affair over, had turned to miscellaneous pillage; the town was full of desirable things, and the ships moored at the quay also. The ensigns planted in the market-place, and before the great church, had not twenty

[1] Under an officer whom Montluc calls the Count of Pedemarie. I cannot identify him.

men apiece around them, according to Montluc, and the captains had completely lost control over their men. Charging into the midst of a crowd of plunderers, the English from the upper town scattered every one. Fouquessoles, who had led the storming party, was killed;[1] de Tais, in chief command, was wounded by an arrow; Teligny—famous afterwards in the Huguenot wars—was taken prisoner. The disordered bands streamed out of the breaches by which they had entered two hours before, some passing through the English camps, where they were saluted by gunfire from the ramparts of the upper town, more by the beach along the mouth of the port.[2] Their total loss was about 800 killed and prisoners. The former were the more numerous, for the English had been irritated by the fact that the stormers had given no quarter, when first they entered the town. The Swiss reserves, left on the heights beyond the *Tour de l'Odre*, had made no attempt to reinforce the van-guard—it is said that in the darkness of a rainy night they never discovered what was going on in the town until a rush of fugitives ran in upon them. The Dauphin's advisers were much criticized for having held this large force idle; but it is obvious that in all night operations on rough ground it is almost impossible to ensure accurate co-operation between separate corps.[3]

The general moral of the 'Camisade of Boulogne' might perhaps be compared with that of Graham's unhappy attempt to storm Bergen-op-Zoom in March 1814. In that case, too, the assailants entered the town at several points, seemed for a moment to be completely successful, and then allowed themselves to be driven out by a sudden offensive of the rallied garrison, who were not superior to themselves in numbers. Want of order and of strong leadership was fatal.

Much dashed in spirits by this disaster, the Dauphin led his army, which was suffering from short rations and bad weather, back to Montreuil, where he disbanded his Swiss and landsknechts, and left the Marshal de Biez to watch Boulogne, with the French and Italian infantry and a com-

[1] An English narrator gets his name as Foxhills!

[2] Montluc says that, making a sort of rearguard action, he escaped uphill close under the upper town. Holingshed says that the main flight was over the sands, and then uphill to where the Dauphin stood.

[3] Montluc's account of his personal adventures during the 'camisade' is most interesting, but he loses the general trend of events while discoursing of his clashes with small bodies of English in by-streets. *Memoirs*, ii. pp. 310–20. He is absurdly depreciative when speaking of the English fighting—yet has to own that his friends suffered a disgraceful rout.

petent proportion of cavalry. The recovery of the fortress was to be postponed to the next campaigning season, but King Francis was determined to have it back at all costs, and no peace with England could be made so long as this aspiration endured.

Meanwhile de Biez took advantage of the suspension of open fighting by commencing to build an elaborate fort on the south side of the Liane, where its tidal estuary faces Basse Boulogne. It was to be furnished with heavy guns, which were to prevent ships from entering the port—but the Italian engineer, Antonio Mellone, who was charged with the drawing up of the plans, seeking solid earth, chose ground so far from the sea and the entry of the harbour that the farthest range of the French cannon could not command the lower portion of the quays, and English vessels passed in and out freely with victuals for the garrison. The batteries of this fort of ' Outreau ' (over-the-water), when finished in 1545, could molest Basse Boulogne very seriously, but failed utterly to make any impression on the fortification of the upper town. The adversaries cannonaded each other across the estuary of the Liane and its tidal waters for many months, without any effective result whatever.[1]

And so matters remained throughout the year 1545, during which the interest of the war shifted away from land operations in Picardy to the French naval offensive in the Channel, described in the next chapter. When this came to an end after d'Annebault's ineffective sea-fighting in front of Portsmouth (August 1545), the French king resigned himself to the loss of Boulogne, and consented to the peace of 1546, whereby he not only surrendered that much-disputed fortress but granted large ' pensions '—really a war indemnity—in order to be quit of the strife. Henry was equally ready to make peace—he was in a bankrupt condition, could not find money to raise a new army for continental strife, and had been frightened most thoroughly by the appearance of a dominating French fleet in the Channel. It was this new phenomenon which set him to the fortification of all the ports and likely landing-places of the south coast—a vast and expensive system whose details have now to be studied.

[1] There is an admirable contemporary engraving in the British Museum, taken by an artist standing by the foot of Outreau, which shows all the English works on the other side of the water, each with its St. George's Cross flag. An important addition has been made to the defences by erecting a large fort by the *Tour de l'Odre*.

CHAPTER V

NAVAL WAR IN THE CHANNEL, AND THE FORTIFICATION OF THE SOUTH COAST (1545–46)

WE have seen that, all through his reign, Henry was with no great success, endeavouring to take hints from the progress of the Art of War abroad. Hence his intermittent efforts to introduce the employment of small firearms—which had so little effect that even in the campaign of 1544 he was gladly hiring Spanish and Italian 'hackbuteers,' no less than Burgundian men-at-arms. Yet he offered advantages to native 'shot,' as was seen in his Charter to the 'Royal Artillery Company,' which was granted in 1537. And in 1539 in the general review of the London Train-bands—said to have been 15,000 strong in all—there were several companies of arquebusiers or caliver men, who were given precedence of the bowmen in the marshalling of each ward, and discharged *feux-de-joie* as they passed the King's platform erected in front of the Palace of Westminster. What proportion they bore to the bowmen is not stated—but from records of later years it was probably very small. And undoubtedly firearms were more prominent in London and the south-eastern counties than in any other part of England. We have already noted Norfolk's complaint that he had practically no arquebusiers in the large levy raised against the 'Pilgrimage of Grace,' and have seen that in 1544 there were probably not 2000—even including foreigners—in the army that went out on the 'Enterprise of Boulogne.'

With the larger firearms Henry made much better progress; Edward IV had already possessed a movable train of artillery, and there were always trained artillerists available, no such prejudice existing against the gunner as there was against the hackbuteer—from memories of the early failures of the hand-gunmen from abroad, who showed so poorly in the Wars of the Roses.[1] Gun-founderies had existed for some time—

[1] See vol. ii. p. 229.

worked perhaps very intermittently; but between 1520 and 1530 they seem to have become permanent and important establishments, and the buying of cannon from abroad became exceptional. One obvious cause of the necessity for a large development of gun-founderies was King Henry's interest in his fleet. For the new ships of large dimensions which he built, pieces of the heaviest sort were necessary—the *Great Harry* had four 'great cannons,' 60-pounders, 8 feet 6 inches long, besides a considerable number of 'demi-cannons'—32-pounders, and other smaller guns. Before Henry's reign was over, the tendency to furnish English ships with artillery of larger proportion to their tonnage than was common abroad can be detected. This idea of making the warship a machine destined to operate by force of gunnery, rather than a fort with a garrison of soldiers, who were to board the enemy in close combat, was a cardinal change in naval psychology. When the days of the Great Armada fights of 1588 arrived, the English invariably tried to lie off and batter the enemy—the Spaniards wished to close and fight a land-battle on the water.

But guns for use on shore were still the more important products of the royal founderies. Every army that took the field had its train, and the Master of the Ordnance was a personage of high rank—a knight, as a rule, under Henry, though later in the century he was sometimes a peer. Though at Flodden and Guinegate the artillery was of no decisive use, we shall see that at Pinkie—only a year after Henry's death—it settled the whole course of the battle. The first important gun-founder in the reign seems to have been a certain John Owen, but in the later years the King introduced two German specialists into his great foundry and arsenal at Greenwich. These were Peter Bawde and Peter of Cologne, who are said to have invented in 1543 the first explosive shells for use from mortars. Stow describes them as 'hollow shot of cast-iron stuffed with fireworks, fitted with screws of iron to receive a match to carry fire kindled, that the fireworks might be set on fire, to break in pieces the same hollow shot, whereof the smallest piece hitting any man did kill or spoil him.' This invention was before its time, like the occasional breach-loading cannon of the fifteenth century, of which we had to speak in a preceding volume.[1] Probably the uncertainty of explosion

[1] See vol. ii. p. 231.

from the screwed match kept the invention from achieving the success that shells were to have in later generations.

The greatest change in the employment of artillery which we find in the reign of Henry VIII is concerned with a new development in fortification. Guns of position for defensive purposes at strategical points of the coast are a new phenomenon. In earlier centuries the English fleet had generally been dominant in Home Waters—the command of the sea which had been won by Edward III at Sluys was seldom lost. But there had been times of adversity, when French (and occasionally Spanish) [1] fleets raided the south coast and burnt small seaports, both in the reign of Edward III and Richard II, and later (after an interval) in that of Henry VI. Such coast-defence as was devised to cope with such dangers consisted in the occasional building of castles such as Bodiham and Cowling to cover likely points of attack, or in closing harbour mouths by chains with strong towers at their ends, as at Dartmouth and Fowey. But such measures were quite exceptional.

We come to a perfectly different state of affairs under the early Tudors, owing to the development of artillery. Hostile ships could now not only land raiding parties, but batter open towns along the shore from a distance, and level by their fire any antiquated shore defences built in the old style of lofty towers. In his wars with France, Henry VIII saw the command of the narrow seas challenged, not by raiding fleets of the old sort, but by ships carrying many guns, and also by galleys brought round from the Mediterranean, and formidable by their facility for manœuvring in a dead calm, which was fatal to the moving power of sailing vessels.

It may be argued that the best way of facing this new danger was to build a navy of overpowering strength, which would be able to sweep away any French fleet from the Channel. This would be an obvious remedy under modern conditions. But in the sixteenth century wind and weather had an importance very different from that which they possess in the days of steam. It was possible for an English fleet to be wind-bound in Portsmouth, Plymouth, or the Dover Straits, while a French fleet from Brest or from the group of ports at the Seine-Mouth could get to sea and do much damage—all depended on the blind chance of the weather. This condition of insecurity

[1] See the curious narrative of Don Pero Niho's raiding along the Hants and Dorset coasts in 1405 in 'El Victorial.'

went on right into the eighteenth century. The most obvious example of the danger befell as late as 1797, when the French Brest fleet, putting out in rough weather, reached the coast of Ireland unmolested, and might have thrown 20,000 men ashore in Bantry Bay, to start an Irish insurrection, but for a series of entirely fortuitous mischances.[1] The British fleet under Lord Bridport was never seen by the French; in consequence of the bad weather it had never stirred from the Channel, the admiral had taken it for granted that the French would make no experiments in a stormy December. He did not get out of Spithead—so perverse were the winds—till January 3; the French fleet was in Bantry Bay by December 21!

King Henry, therefore, was not so misguided as might be at first supposed, when he took in hand the protection by well-gunned forts of vulnerable points on the seaboard facing France. Open towns along the shore, the narrow entrances of harbours, and stretches of beach offering opportunities for easy landing, were the points which he selected for fortification, always by low-lying heavily gunned works facing seaward, but with a sufficient rear-defence to protect them against assaults on their backs by landing parties. They were closed works, not mere shore-batteries.

Military architecture, as we have already had occasion to remark,[2] had now been completely transformed, its end in the future was to present fronts hard to hit, not lofty and visible, but low and sunk, if possible, behind some sort of a *glacis*, and they must be furnished with very heavy guns, sufficient to silence shipfire from vessels brought close inshore. As every one now knows, shore-batteries have the advantage over fleets, since they may be made as thick as the builder chooses, with many feet of stone or earth in their front, and no vulnerable points except the gun-embrasures, while the guns on a ship have but a comparatively thin protection of wood (or iron in later days) to shield them, and a hit on any part of the ship counted and might disable or even sink it. In a duel between

[1] The main piece of luck for the British in 1797 was that the ship carrying both Hoche, the French commander-in-chief, and Morand de Galles, the admiral of the fleet, had been blown right out into the Atlantic, and ultimately returned to Brest nearly disabled. Grouchy, the senior military officer at Bantry Bay, refused to use his initiative—heads came off easily from revolutionary generals—and took back the 15,000 men who had reached Ireland without attempting a landing. This was the same Grouchy as in 1815!

[2] See above, p. 223.

guns on shore and guns afloat, the only advantage of the ship is that it is a moving target, while the shore battery is a fixed target, and therefore easier to aim at. This counted for a good deal in later days, but did not mean much when engagements took place at very close quarters, not at a distance of 6000 or 10,000 yards. In the sixteenth century a ship wishing to use its artillery effectively had to go very close inshore, for its guns would not carry far. And when it came within 400 or 600 yards of a shore-battery it was a large mark, not a mere distant speck, like a modern ship engaging at a distance to be measured by miles. The ship then was very vulnerable—the shore battery much less so, because it was low-built and had a thick protection of stone or earth. A well-built fort hardly showed over the skyline, while a ship on the surface of the water could not hide itself, and was eminently visible at sixteenth-century range distances. Modern naval history shows that in the matter of duels between ships and forts the former have no chance, unless they have great superiority both in the weight and in the number of their guns. Cases like Lord Exmouth's successful bombardment of Algiers in 1816, or the crushing of the Alexandria sea-forts in 1884, are quite exceptional. The normal phenomenon is what happened at Sebastopol in 1854—complete failure to get any effective result, with appreciable damage suffered by the fleet.

Henry's scheme for the fortification of vulnerable points along the south coast, facing France, was enormous, and its expense was quite unprecedented. To many of his works the name of 'castle' has been applied, but they are better described as closed forts, with accommodation for small permanent garrisons of gunners and watchmen, which might be brought up to any strength from the local levies of the neighbouring shires, when danger threatened from a French fleet loose in the Channel.

The line extended from the Straits of Dover to the end of Cornwall, and there were even outlying fortifications farther out—at the mouth of the Medway and the outlet of Milford Haven. At the narrows of the Channel, facing Calais and Boulogne, there were no less than three castles, each guarding a likely landing-place, at Sandgate, Deal, and Walmer. Dover had its old high-perched castle, but this would not protect the shore-town from bombardment, so Henry added a 'bulwark' or low-lying shore-battery down by the harbour. Farther along the coast was the lonely Camber, a forlorn circular work covering the marshy flat between Rye and Winchelsea, where French landings had repeatedly been made in old days. Four new forts on the Hampshire coast took up the line, of which each, it is melancholy to remember, drew its stones from a great and recently disestablished abbey. Netley, with Calshot, which faces it across the mouth of Southampton Water, were built with the material drawn from Netley Abbey. Farther down the Solent, Cowes Castle on the Isle of Wight, and Hurst Castle opposite it on a projecting spit of the Hampshire coast, drew on the ruins of Beaulieu for their stones. There were already some works, apparently, at the mouth of Portsmouth Harbour, but Henry supplemented them by extra batteries, and in addition threw up Southsea Castle on the low beach at the east end of Spithead.

These defences of the Solent were the only works of Henry which proved of actual use during his last long war with France, when the French had obtained a momentary predominance in the Channel, and the English fleet had retired to the narrow waters round its great base-arsenal at Portsmouth. The French admiral d'Annebault was inclined to press for a general action, following the enemy into the narrows—but his pilots and captains put in an absolute veto, saying that the shore-batteries would blow them to pieces at short range, if they pressed in, while the English fleet was making a front with both flanks covered by the land guns. Hence d'Annebault lay outside, skirmishing a little with his galleys in the more open water, but never venturing to come in, and finally retired, after making some fruitless land-raids on the Isle of Wight and the Sussex coast, which were easily driven off (August 18–24, 1545).

Henry's castles farther west never had the test of war, like those which covered Portsmouth and the Solent. They must, however, be noted—Portland and Sandsfoot covered the two ends of the excellent roadstead made by the projection of Portland Bill. There was a 'work' or 'bulwark' low down on the mouth of the Exe, for the protection of its estuary near Powderham. Dartmouth, and the long fiord which runs up with navigable water as far as Totnes, were already guarded (but *not* protected from artillery) by mediaeval towers at Kingswear and Dartmouth itself. Henry enlarged both, and gave them casemated batteries at the water's edge. The next protective fort was at Salcombe, at the mouth of the Kingsbridge estuary, built in the usual style and very strong—it stood a long siege in the civil wars of Charles I. There was some building at Plymouth, whose scope is hard to discern, as the front was thoroughly reconstructed in the seventeenth century, and the older fortifications have disappeared.

Passing into Cornwall, Fowey—a very important place in the sixteenth century—had down to Henry's time only a chain laid between towers at the mouth of its long creek. The King, regarding them as wholly insufficient protection, built a fort called St. Catherine's at the seaward entrance of the harbour, on high ground completely commanding the neighbouring waters. The great estuary of the Fal—no town of Falmouth yet existed, but the harbourage was well known—was the scene of some most effective, as well as picturesque, building. At the mouth of the haven the King placed the two most handsome of all his forts—St. Mawes and Pendennis. The former lies on comparatively low ground on the eastern side of the narrows, the latter on a much higher and more commanding site, on the very point of the rocky peninsula which juts out into the Channel.

St. Mawes is the best preserved, as well as the most artistic, of all these 'castles'—the outside is lavishly decorated with large heraldic shields and inscriptions in fantastic labels glorifying the King's name—'Semper honos, Henrice, tuus laudesque manebunt'—and such like screeds. The fort is a circular structure of no great height, with a three-lobed battery for eleven heavy guns in its lower story, and ample provision for smaller ordinance and swivels above. The embrasures are immensely deep, very narrow at the mouths, but splayed backward, so as to give ample space for the working of the guns. Of the three lobes of the battery one, with three embrasures, looks out to sea; the second, with five, commands the narrowest point of the haven entrance; the third for three guns looks up into the harbour.

Pendennis, terribly pulled about in recent centuries, survives in much less perfect form, for it was used as an artillery station down to 1922, and Henry's original circular fort was recast again and again to fit modern guns. But the Tudor dragon and lion may still be detected over its main entrance.

It is odd to find that the King's long series of splendid shore-batteries have hardly any warlike record. Save for the operations around Portsmouth in July 1545, and one hostile raid by a Dutch squadron against St. Catherine's at Fowey in 1666, they never saw an enemy from the side of the sea. On the other hand, nearly all of them were attacked from the land, during the wars of Charles I and his Parliament. They had not been built to face serious attacks from the rear, but nevertheless their solid masonry gave them much resisting power, even when battered from behind. Sandsfoot, Portland, Salcombe, and Pendennis all made long and honourable resistance to the guns of the Roundheads—though they all had the structural fault of not having a high *glacis* on their landward side, to protect them from siege guns working at short range. But attack from the rear was not what Henry VIII and his architects were thinking about in 1540, but only assault from the sea front—and this never came!

English military architecture had practically come to an end in the fifteenth century, when castle building ceased, and the great landowners built themselves spacious manor-houses with every convenience for comfortable living, instead of places of strength in the old style. At the most, the fifteenth or sixteenth century residence of a great peer showed some slight trace of fortification in a gate-house or a moat. It was Henry VIII who revived military architecture in England, and his structures—as we have seen—were quite in the modern style: low lying, heavily gunned, not in the least suited for occupation for a great civil household, but only for a moderate garrison. They were often handsome and picturesque—as every one who has seen St. Mawes will concede—but they were not residential, but purely military in their purpose. Domestic architecture developed in the splendid Tudor manors of which so many still survive—military architecture followed the tendencies of the sixteenth century, though on a modest scale. England never became a land of fortified cities in the style of Italy or the Netherlands. When the civil wars of Charles I and his refractory Parliamentarian adversaries broke out, neither side had any proper provision of modern strongholds—the combatants found themselves, almost invariably, besieging old-fashioned castles, or manor-houses roughly patched up, or city walls dating from the Middle Ages. The military engineer had to make sudden improvisations, like the lines which encircled Oxford or Bristol, or the never-tested entrenchments of Turnham Green, which were to cover the western approach to London. The sea-coast forts of Henry VIII were the only places in the modern style which were attacked.

CHAPTER VI

THE BATTLE OF PINKIE (SEPTEMBER 10, 1547)

ON the death of Henry VIII (January 28, 1547) the conduct of the foreign policy, which he had managed with so much perverse energy and so little success, passed into the hands of the councillors whom he had left behind him as the guardians of his nine-year-old son Edward VI. They had been for the most part unscrupulous parasites of an awe-inspiring master, selfish opportunists who had found their profit in carrying out his orders, however much they may have doubted their wisdom. On the King's death they came to an agreement, putting aside for the time many personal enmities, and, after decorating themselves with new and more gorgeous titles of nobility, proceeded to take stock of the position that lay before them. Edward Seymour, the little King's uncle, now newly made Duke of Somerset, was decidedly the least objectionable of this group of upstarts. He was not without laudable ambitions, quite sincere in his religious views, moral in his personal life, a good master, and full of schemes for the relief of the poor and the abatement of social injustice. That his ambition was greater than his ability justified, and that he loved to accumulate lands and wealth is certain. He came to an unhappy end because he had scruples, and was surrounded by men who had none. Meanwhile, as Lord Protector he proceeded, so long as he was able, to reverse his late master's religious policy, but to carry out his secular policy. The main thought that had been obsessing King Henry's mind during his last months was his plan for the union of England and Scotland, by the marriage of the nine-year-old Edward VI to his cousin the five-year-old Mary Queen of Scots.

The idea was good, and there were many in Scotland who were not averse to it—notably among those who inclined to the principles of the Reformation. Somerset repudiated all idea of conquest, and offered complete equality, the preservation of all Scottish laws and institutions, freedom of trade, and common citizenship. There should be a personal union of the Crowns, but no claim to suzerainty of the one over the other —the old pretensions of Edward I should be abandoned.

Unfortunately the offer was urged without tact, and offended patriotic sentiment — moreover, the Catholic party was in power in Scotland, and Somerset had already declared himself a zealous Reformer. The Queen-Dowager, who led the ruling regency, was closely allied with France, and a French squadron had just captured the Castle of St. Andrews, where a forlorn band of Scottish Protestants had been holding out (July 31). Among the prisoners taken away in French galleys was the famous John Knox.

Somerset's plan for pushing on the marriage project was the application of military force—a very unwise one, for it was in the power of the Scots to send away the little Queen to France—as they did somewhat later. On August 22 the Lord Protector started from London, and by September 1 had mobilized his army at Newcastle and crossed the Tweed. His force was not of overwhelming numbers, about 16,000 in all, but it differed from all other English invading armies of earlier generations by the great strength of the mounted arm. Here the experience of continental wars was working — Somerset had noted the deficiency of cavalry in all his master's French campaigns. He had hired a body of cosmopolitan foreign men-at-arms under a captain named Malatesta, had summoned across from France the 'Bulleners,' or horse of the garrison of Boulogne, 500 strong, and had brought up the 'Gentlemen Pensioners' of the Royal bodyguard. Adding the full muster of the 'Northern Horse,' the 'prickers' of the Border Marches, and a troop of the new sort, 'harquebusiers mounted,' *i.e.* cavalry with firearms, under a Spanish condottiere named Pedro de Gamboa, he had no less than 4000 horsemen. The main body of the infantry were still the 'bows and bills' of the old familiar sort—there were only 600 native English with firearms, and perhaps a company or two of foreign arquebusiers. A considerable fleet under Lord Clinton coasted along the shore, keeping close touch with the army, and reporting daily to receive orders.

This tactless invasion forced the greater part of the Scottish nobility into the field, lest they should appear to be open enemies of their country; but a certain number failed to appear at the muster—among them the Earls of Lennox, Bothwell, and

Glencairn, who had all been in private negotiation with Somerset. Only forty horsemen under the Laird of Mangerton unobtrusively joined the invaders.[1] The army, which mobilized on the Borough Muir, outside Edinburgh, was about 23,000 strong—but only 1500 were cavalry, and these mostly Border moss-troopers, unfit to contend in line with heavy men-at-arms.

Somerset understood that the Scottish regency intended to fight, as he had received no reply to a request that they should send commissioners to confer with him at Berwick. He expected to be faced at the defile of Cockburn's Path, where the ravine of the Pease Burn offers a splendid position for resisting an army approaching by the coast-road to Edinburgh ; but it was found unoccupied. It was on the night before the expected battle that the Lord Protector had the curious dream, which Patten, his confidential scribe, has recorded. He dreamed that he was being received by the King at Windsor in high triumph —but suddenly found himself wondering why he should be so honoured, for he could not remember that he had done anything at all ! Patten sagely remarks that dreams often should be read in reverse—' one may dream one thing and the exact opposite come to pass.' But the sleeping Somerset was not so far wrong—his whole campaign was to be absolutely without result, despite of a very spectacular victory.

The English light horse presently brought news that the enemy was prepared to defend a position as strong, if not so precipitous, as that of Cockburn's Path, the line of the river Esk, which falls into the Firth of Forth between Musselburgh and Inveresk. They had taken up a long front from the seashore to a morass on the middle course of the Esk, and had fortified its northern end, next the beach, by a long earthwork containing guns, intended to give cover from any fire coming from the English fleet, of whose presence the Scots were well aware. The Esk was fordable in most places, but the ridge above it, Edmonston Edge, was a steep slope with a good *glacis*, which any troops fording the river would have to mount in face of opposition. The position could not, of course, be turned on the side of the sea, and the other flank was so well covered by the morass, that an immense circular march into a rugged country would be required to get round it. Somerset

[1] Probably John Ramsay, the Scottish gentleman who drew the six plans of Pinkie which I discovered in the Bodleian, was one of them. He was certainly present on the English side, as his curious little pictures show.

resolved that he was committed to a frontal attack, and after lying a night before the Scottish camp, and executing careful reconnaissances, made out his plans. His cavalry, scouting inland in force, fell in with the main body of the Scottish horse, and routed it with loss—numbers and weight being both in favour of the invaders. The Scots had heavy casualties, and on the battle-day their troopers were both thinned in numbers and demoralized in spirit.

Somerset's plan was apparently inspired by memories of the record of Ravenna—an enemy on the defensive, partly

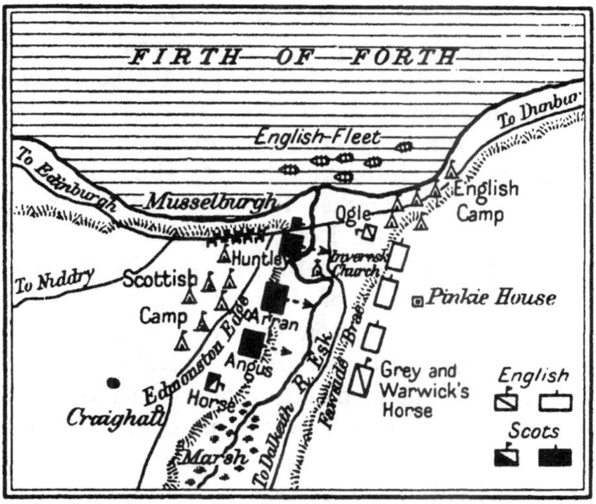

BATTLE OF PINKIE, SEPTEMBER 10, 1547.

covered by entrenchments, hopelessly weak in cavalry, and inferior in guns, should be tackled by intensive artillery fire, if possible flank fire as well as frontal, on a selected part of his array; this should be followed by a joint attack of cavalry and infantry on the battered and shaken section of his lines. In front of the Scottish left, on the opposite bank of the Esk, rose a knoll on which was the church of Inveresk. This knoll deflects the course of the river, which has to find its way round three sides of it in order to reach the sea. It almost projects into the Scottish lines, and is of a greater height than the ground immediately opposite.

Not dreaming that the enemy would fight anything but a defensive battle, Somerset resolved to occupy this knoll, place all his heavy guns upon it, and so to batter the Scottish left, while at the same time his fleet was to run inshore and open a flanking fire upon the rear of this section of the hostile lines. The superiority of guns would be so great that the few Scottish pieces in the earthwork above the shore might be disregarded. The fleet's cannonade would overpower them. When the Scots were shaken, the infantry and cavalry should ford the Esk above and below the church and its knoll.

As it turned out, this was precisely the battle that was *not* going to be fought. For the Scots, contrary to all expectation, took the offensive. The Regent Arran was watching the English lines, when he saw the whole army, drawn up on the slopes of Fawside Brae, begin to move northward towards the sea and the high road to England. By some astounding perversity of interpretation, he took this to mean a general retreat, judging that Somerset, having viewed the Scottish position and found it impregnable, was about to turn back to Berwick. The whole English force was on the move, first a body of light horse under Lord Ogle, then in succession the three infantry divisions—vaward, main-battle, and rearward—and last the main body of cavalry in two heavy brigades. The guns could also be seen on the move in the intervals between the infantry corps. Arran suddenly came to the conclusion that he had a splendid opportunity of attacking an enemy in flank, as he moved across his own front—a surprise blow against an unsuspecting adversary. Accordingly he sent word to his lieutenants to leave their ground on Edmonston Edge, ford the Esk, and strike the English marching columns in flank. It is said that the Earl of Angus, who led the vaward division, remonstrated against the proposal, but was told to obey orders on pain of treason.

The Scottish array was in the normal three columns of pikes—the left, or 'rearward,' under the Earl of Huntly, next the sea, was flanked by a body of Highland bowmen [1]—the only light troops which the Scots put in line. The centre, under Arran himself, was accompanied by some artillery—not very numerous, and drawn (it is said) by teams of men, not horses. The 'vaward,' or right wing, under Angus, also had

[1] Patten calls them 'Irish,' because they wore kilts and talked in a Celtic tongue. But what he means is sufficiently shown by his mention of 'Mr. Campbell, an Irish gentleman, close kin to the Earl of Argyle.' Pitscottie, who was not so ignorant, calls them 'Macleods and Macgregors and all the Islesmen of Scotland.' But he gives a most confused account of the array.

with it a few guns, and on its flank the remains of the cavalry which had been cut up in the skirmish of September 8. The royal standard of Scotland floated over the centre: among many other banners, one was noted bearing a picture of the Church kneeling before Christ, with the motto, 'Ne Obliviscaris Domine Sponsae Afflictae'—a palladium of the Old Faith.

Somerset was taken much by surprise at this sudden advance of the Scots, and had to form 'front to flank' in great haste, for the enemy came down the slope, and across the river, at a quick pace, more like horse than foot as his chronicler observes. It seems that the English had some confusion on getting into line: but in each infantry division the harquebusiers and bowmen formed up on either flank of the billmen, the guns were brought forward as fast as could be managed, and the cavalry placed itself on the wings. If the Scottish attack had gone on with the same rapidity with which it started, there might have been a serious congestion in the centre, where the guns were moving among the infantry.

But fate was from the first against the Scots. Their 'rearward,' which crossed the river between the Church-Knoll of Inveresk and the sea, was suddenly fired upon with effect by the warships in the bay. The first salvo entirely dispersed the Highland archers on Huntly's flank, who broke and fled back—they never came into action again. The same volleys also struck into the mass of pikemen, and killed the Master of Grey and other gentlemen, whereupon Huntly shifted his column inland, to get out of range of the ships' guns, and moving along the front of the Inveresk knoll, pushed up to fall into line with the 'main-battle,' which had crossed the Esk farther south, to the right (Scottish) of the knoll. Huntly's column was in some disorder, and seems to have actually coalesced with the Regent's division into one great mass. They continued, however, to advance, crossing the flat land beyond the Esk, and moving up the opposite slope, where the English were now getting into battle order. Angus's column, the right or 'vaward,' seems to have been a little late in getting forward, but ultimately came into touch with the main-battle on the other flank.[1]

[1] Here I am following the very clear representation in Ramsay's map, which shows the centre decidedly in advance of the two wings, Angus's corps being markedly to the rear, about to pass the Esk when the Regent and Huntly are well across. I am bound to say that one would have the opposite impression after reading Pitscottie's narrative, ii. p. 496, who makes the Regent cross the river covered by Angus corps already on the far bank. But I follow Patten and Ramsay's picture.

All were possibly delayed by the difficulty of getting their man-hauled guns across rough ground. If our best contemporary picture can be trusted, Arran's four pieces took their place on Huntly's left, when the line was reformed. The Scottish horse hung back, and remained écheloned far to the right rear, obviously fearing contact with the much larger English cavalry brigades opposite it.

The whole Scottish army had thus become one great mass of pikes, with its flanks wholly uncovered, since the Highlanders had disappeared, and the horse were far to the rear. Seeing this, Somerset resolved to stop their further advance by what I may call the tactics of Marignano—*i.e.* to cause them to halt by successive cavalry charges, while his artillery and his infantry armed with missile weapons, should beat upon the flank of the stationary body, and on such parts of its front as were not covered by the charging squadrons.

This was expensive work for the English cavalry—who, like the gendarmerie of Francis I at Marignano, were destined to suffer heavily in their self-sacrificing task. But the tactics were successful.

The Scottish host was half-way up the slope of Fawside Brae when the charges began, in the fields of Pinkie Cleugh—newly reaped stubble, with low earth-banks around it, but not sufficiently cut up to prevent easy motion both of horse and foot. The first charge was made on the southern flank of the Scottish columns, *i.e.* apparently on Huntly's 'rearward,' by Lord Grey with 1800 horse, including the 'Bulleners and the 'bands' (as they are called) of Lord Grey himself, Warwick, and Somerset. The first clash is described as having been a bloody business. 'The Scots' says Patten, 'stood at defence, shoulders nigh together, the fore-rank stooping low before, their fellows behind holding their pikes in both hands, the one end of the pike against the right foot, the other against the enemy's breast, so nigh as place and space might suffer. So thick were they that a bare finger should as easily pierce through the bristles of a hedgehog as any man encounter the front of pikes.'

From this formidable line of steel, Grey's men-at-arms, though they charged with good courage, were thrown back with ease, like the knights of Edward II at Bannockburn. The leading ranks of the riders went down *en masse* six feet in front of the Scottish line, and the obstruction of dead and wounded men and horses must have been inconvenient both to the Scots, who would have to advance over them, and to any further English charge against this part of the line. Shelley, the commander of the 'Bulleners,' was killed.

But the Scots had been forced to halt, and Somerset was determined to keep them halted, while his guns were being brought up, for despite of the failure of the first assault he delivered a second with the 'Gentlemen Pensioners'—the royal bodyguard—Sir Ralph Vane's 'band,' and the demi-lances of Lord Fitzwalter—1600 in all, as we are told.[1] The charge was directed against the Scottish 'main-battle,' the Regent's own division. This attack was beaten off with loss, like the first. Darcy, the commander of the Pensioners, was struck down by a cannon-ball from a gun upon the Scottish right,[2] and their standard was in danger, for Sir Andrew Flammock, who bore it, was beset by Scots who laid hold of its staff—which broke in the wrestling, and left only the spike and the banner in the knight's hands as he pushed back out of the mêlée. The total loss of the English horse in the two charges is given at about 200 men—but no leader of note was killed save Shelley, though Lord Grey and many others were wounded.

Meanwhile the Earl of Warwick had got the guns of the English right wing, and apparently those of the centre also, into position to bear upon the Scottish left (Huntly's division and parts of the right (Angus). The hitherto victorious pike-men found themselves suddenly assailed by salvos of round-shot fired into the mass from a distance of only 200 or 300 yards. At the same time the English hackbuteers and bowmen swarmed down in the intervals, and began shooting into the stationary columns. The mercenary horsed-arquebusiers of Pedro de Gamboa are said to have made an exhibition of continental tactics, by riding up close to the pikes and firing in succession as they galloped by.

The only chance for the Scots would have been to continue their uphill original advance before the cannon had completely shattered their array. But this was practically impossible—the ranks had shrunk into a serried mass under the impact of the cavalry charge, and there was a bank of dead and wounded

[1] Pitscottie has a silly story that Somerset, at this moment, resolved to fly, and was only prevented from doing so by Warwick.

[2] But not killed—only sore bruised through his armour and with the fingers of his right hand smashed.

English men and horses in front of them. There was no directive order given, for the Regent Arran left the field, called for his horse, and rode away to Edinburgh, crying that 'he was betrayed.' [1] The divisional commanders were left to their own guidance; Angus on the right concluded that it was necessary at all costs to get out of cannon-range, by falling back down the hillside. But the retreat which he ordered soon changed into a rout, because the harassed crowd of pikemen, when once started on a backward movement, would not stop, or try to reform at the foot of the slope, but broke their order and became a mere crowd of fugitives. The retrograde movement spread all along the line—Huntly's flank had been even worse used than the 'vaward' by the English artillery, and a general thrust to the rear was evident, which became a wild stampede when the English infantry closed, and Grey's rallied horsemen charged again. The Scots cast down their pikes and made off—the ground at the foot of the hill looked like a wood-yard, says Patten, from the immense number of long staves lying about. The Scottish horse got off with no damage, having never closed, but the infantry was sadly maltreated by the pursuing English horse, who had their early repulses to avenge.

The slaughter of the vanquished was great, according to all accounts English and Scottish, the pursuit having been over bare hillsides which were easy going for cavalry. The 10,000 casualties—even 14,000—spoken of by chroniclers are no doubt stated too high, and Somerset gave orders for quarter to be granted, not ineffectively, for 1500 prisoners were brought in. The Scots record only one peer slain—the Lord Fleming—but a number of young heirs to titles—the 'Masters' of Grey, Buchan, Erskine, Graham, Livingstone, and Ogilvie. Among the prisoners were the Earl of Huntly (slightly wounded), the Lords Wemyss and Yester, the Master of Sempil, Sir Halbert Hamilton, brother to the Earl of Cassilis, and Mr. Campbell (an 'Irish' gentleman), close kin to the Earl of Argyle. The English lost only about 250 men, all in the cavalry, for the infantry were hardly engaged. Shelley, commanding the 'Bulleners,' was the only important person slain; but many were wounded, including Lord Grey, Sir Thomas Darcy, chief

[1] Pitscottie says that his cry of 'Fy, fy, treason!' came early, when he saw the Highlanders quit the field, attributing their departure to treachery, not to panic, and he adds that the Border horse on the right wing followed the example of the Highlanders, and 'left early.' But he cannot be trusted, being set on finding excuses for a rout.

THE BATTLE OF PINKIE, SEPT. 10, 1547. THE LAST PHASE

From the contemporary engraving founded on John Ramsay's drawings in the Bodleian Library

captain of the 'Pensioners,' Calverley, standard-bearer of the men-at-arms, Clement Paston, and Pedro de Gamboa, the Spanish condottiere.

The battle, though such a complete victory, had no political results whatever, as the Scots still refused to listen to Somerset's peace-proposals, and ended by sending their young Queen to France, where she was wedded to the Dauphin Francis—to the entire wreck of all the long-conceived projects of Henry VIII. Tactically it forms the last link in the long chain of battles in which the Scottish pike-column proved itself ineffective against a combination of weapons on the English side. But Pinkie differs from all the other disasters, save Falkirk, in that it was won by the use of cavalry and artillery together—like Marignano—not by the union of bowmen and dismounted men-at-arms as were Dupplin, Halidon Hill, Neville's Cross, and Homildon. Flodden, as we have shown elsewhere, was a variant from the old series, as we are told that it was not won by the bow, but rather by the bill against the pike. The recrudescence of cavalry as the mainstay of battle, already well developed on the Continent, was demonstrated by the great strength of Somerset's mounted force, which amounted to a good third of his infantry, a thing never before seen in English armies—at least since the thirteenth century.

It is one of the curious chances of history that no battle of a similar sort was to be fought on this side of the Channel for nearly a century. The Scottish wars came to an end, England did not interfere as the chief combatant in the struggles of the Huguenots, of Dutch Independence, and of the Thirty Years' War, and the series of English battles in which cavalry played the prominent part only recommences at Edgehill in 1642.

N.B.—My chief guides in this chapter are Patten and the wonderful set of tactical sketches by John Ramsay, which I discovered in the Bodleian forty years ago, and published in the *Archaeological Journal* in 1933. The illustration appended to this chapter is from the contemporary engraving compiled from Ramsay's drawings. It shows Huntly's column blasted by artillery fire, while the last English cavalry charge is detaining Arran's 'main-battle.' The Scots are beginning to drift to the rear. The 'church' is that of Inveresk.

CHAPTER VII

EDWARD VI, MARY, AND ELIZABETH (1547–1603)—ORGANIZATION AND ARMAMENT—THE LAST DAYS OF THE LONG-BOW

WE have no pitched battle of primary importance to record after Pinkie in the annals of the Tudors. Wars and rumours of wars there were in plenty, but we have little to study in the way of strategy and tactics, though much of interest may be found in the matters of armament and organization. Somerset sold Boulogne back to the French, rather than take up a war with Henry II. Queen Mary threw in her lot with Spain, and sent a contingent to St. Quentin (1557) which had no part in the battle. Her only notable share in the last French war of the century was to lose Calais by miserable neglect—the tale of this disgrace has been told elsewhere. Elizabeth on many occasions lent auxiliaries both to the Huguenots in France and to the Protestant rebels in the Netherlands, but to little effect. On no occasion did they form the main part of an army, or give the decisive blow in a general action. When she employed a purely English army outside the bounds of her realm, it so chanced that no great fighting occurred—as at the siege of Leith (1560) and the 'Journey of Portugal' (1589). Obscure strife in Ireland against an irregular enemy leaves behind it certain lessons as to guerilla warfare, but no record of military importance. The whole period is singularly dull from the point of view of the historian of the Art of War. Its interest is political, diplomatic, commercial, religious, literary, exploratory, but not military.

In 1549 we get curious political sidelights—the Lord Protector Somerset is found dealing with public discontents in a fashion very different from that of his brother-in-law Henry VIII. He employed foreign mercenaries on an unprecedented scale against the Devonshire and Norfolk insurgents—obviously because no shire-levies from their immediate neighbourhood could be trusted in either case. But it is a curious instance of the irony of history that against the Devonshire Catholics, who were fighting for the restoration of the old religion, Somerset used bands of Italian arquebusiers under a captain named Spinola, while the rebellious reformers of Norfolk were being put down by the Earl of Warwick with forces that included not only some Italians but 1400 Protestant German landsknechts. As long as the mercenary was duly paid, it mattered nothing to him whether the rebels against whom he was employed were of his own religious opinion or not.

How prudent Somerset and Warwick were in employing masses of foreign auxiliaries in 1549 was sufficiently shown by events of a few years later. Local levies, newly raised, and embodied under official captains whom they knew not, were too liable to be swayed by public opinion, if it was against the government in whose name they had been levied. They might think for themselves, instead of obeying orders blindly. This was well shown when, on the death of Edward VI, Northumberland found his men melting away from the royal standard, when they were led out in the name of Queen Jane to put down the adherents of the Lady Mary, whom almost every man held to be the rightful heir to the Crown. Whatever the will of Edward VI might have directed, or whatever Northumberland and his Council might have proclaimed, no one would fight against the late King's sister, to defend the usurpation of his cousin. It is curious to reflect on what might have happened if Northumberland had been well furnished with bands of foreign mercenaries in 1553, as Somerset had been in 1549. Undoubtedly there would have been fighting, for the soldiers of fortune would have made some effort for the hand that paid them. We feel some surprise that, during the long weeks of the boy King's fatal sickness, the dictator had not hastened to make some provision of foreign pikes and arquebuses to guard against possible trouble. Apparently it was a defect in his mentality—he was so unscrupulous himself that he had forgotten that others might have scruples.

The same danger that comes to a government that depends on hasty local levies, and has no standing army, was experienced by Queen Mary in turn less than a year later. When Sir Thomas Wyatt and his friends raised the Kentish rebellion, with the ostensible purpose of preventing the Queen from making her highly unpopular match with Philip of Spain, he was playing for the support of public opinion. Whatever his ultimate intentions may have been, he professed unbounded loyalty to Mary, and purported to have taken arms for the deliverance of his mistress from evil councillors, who were trying to lure her into a marriage against which the House of Commons had protested. It was this appeal to national sentiment which gave Wyatt his initial success. When the old Duke of Norfolk led out against him the levies of London, a great part of them went over to the rebels, crying, 'We are all Englishmen'; and even the Yeomen of the Guard threw down their arms and fled, as did the rest of Norfolk's array.

The ultimate fate of the rebellion was not settled by the petty and spasmodic skirmishing that took place, but by the fact that the Queen made a vehement personal appeal to the fundamental loyalty to the Crown, which was still the strongest motive in the mind of nearly every Englishman. It was a matter of popular psychology whether she would succeed or not. When Wyatt, desperate at the discovery that if few were opposing him, still fewer were joining him, made his final dash at London with a dwindling host, there was a chance that he might win. Many of the Queen's levies showed a poor spirit; a panic seized the men stationed at Charing Cross, who fled down Whitehall shouting, 'Treason!' and were only rallied by the Queen herself. Would London rise and open its gates? Wyatt pushed up the long defile of the Strand almost unopposed. But when he reached Ludgate, discovered the gates locked, and was fired upon from the walls, he found himself entrapped. The long narrow column of his followers was enclosed in a *cul de sac*, and large forces had blocked his way back to Charing Cross, and came up from the side-streets to the north. There was very little fighting—neither side wanted to kill, and Wyatt, disheartened by his failure to find supporters, finally surrendered, on a vague appeal to stop unnecessary bloodshed. Never was a rebellion so entirely settled by the public opinion of the masses—and this opinion was wavering almost down to the last moment (February 1554).

Incidentally we may mention a fact that has some military interest in the way of organization. The London militia was by this time properly organized into companies with a fixed uniform, the white coat and red cross. Many men had deserted to Wyatt during the preliminary fighting in Kent; when the final scuffle about Temple Bar occurred, much confusion was caused by the fact that both the deserters and the companies who had adhered to the Queen were clothed alike. We are told that the only way in which they could be distinguished was that Wyatt's men had been marching all night from Kingston over rain-sodden roads, and were splashed from head to foot with mud, while the royalists came to the fight from comfortable housing under cover. Hence they raised the cry of ' Down with the draggle-tails,' on the principle that every muddy man was a rebel—which answered fairly enough as a test.

Mary succeeded in getting rid of the popularity which had attended her first months, in resentment at Northumberland's impudent attempt to change the succession to the Crown. It was waning at the time of Wyatt's rebellion in 1554 but still effective. By the following summer she had by means of her Spanish marriage lost her hold on the nation, and when the systematic burning of Protestants started in February 1555, she rapidly earned its disapproval—for the number of those who approved her policy, even among Catholics, was not considerable. She became an object of hate. Hence her well-remembered sobriquet of ' Bloody Mary.'

The military events of the unhappy Queen's interference in her husband's French war have been dealt with elsewhere—the appearance of an English contingent which won no particular distinction at St. Quentin (1557), the more effective intervention of an English squadron on the shore-battle at Gravelines (July 1558), and above all the details of the wretched affair of the loss of Calais. That old stronghold was doomed to fall some day, considering the strength of the new French monarchy. But it need not have fallen after a siege of less than a week, owing to an inadequate garrison, and the tardy dispatch of succour that never arrived.

The rolls of the contingent of 5000 men which went to the campaign of St. Quentin have been preserved. The whole is called a ' regiment '—a word used not in the modern sense, but meaning a force of all arms under a single commander, as it had been employed in King Henry's expedition of 1544.[1] The ' regiment ' consisted of 4000 foot and 1000 horse, with some pioneers; in the cavalry there were some ' lances '—apparently men-at-arms of the old sort,[2] but more ' demi-

[1] See above, p. 332.

[2] Rather more than we should have expected, considering the immense difficulty with which Henry VIII collected a few hundred fully armed heavy horsemen.

lances' and light horse. The foot were in forty companies of 100 men, each under a captain; there is still (as in the time of Henry VIII) no division into large subsidiary units of the battalion size. The contingent would have been better employed in strengthening the wholly inadequate garrison that had been left in Calais, but it was disbanded in the autumn.

Elizabeth had ascended the throne a few months before the last of the great wars of Hapsburg and Valois came to an end at the Peace of Cateau Cambresis. Her diplomats were present at the negotiation, and were put off with vague promises of an indemnity, if Calais were not restored within eight years. This was all that Philip II would do for the new English Queen—who (as he was well aware) detested him—when he was making his final settlement with the French. Wise men knew that eight years was a long time, and the half-million *écus* spoken of never likely to materialize. Peace was necessary, and the humiliation of a peace that did not include the restoration of Calais could be pushed off on to the detested memory of Queen Mary.

So Elizabeth after 1559 was free of the perils of the European wars in which her father—of his own perverse will, and her sister, to please her husband—had entangled themselves to so little profit. She could turn to her own policy, and start on that endless series of intrigues—just avoiding open war—which occupied the first twenty years of her reign. Like Mr. Gladstone in 1882, she was always able to plead that though English troops might be 'engaged in military operations' in France, Scotland, or the Low Countries, she was not technically at war with France or with Spain, still less with Scotland. And since not at war, she never needed to create a standing army—an expensive instrument, against whose cost her parsimonious soul revolted. Even the standing navy, which she knew that she must keep up, if England was to survive as an independent state, suffered sadly from consistent neglect, and had always to be supplemented by the hired vessels of adventurers in time of crisis. Its storeyards and arsenals were never full, and its rations were generally found to be rotten—cheap stuff which would not keep.

From time to time Elizabeth put quite considerable forces in the field, both for service abroad, and for home defence in time of rebellion, or of threatened invasion from abroad. The war in Scotland in 1559–60 which centred round the siege of

Leith employed at least 8000 men, the expeditions to Havre and Rouen in 1563, for the aid of the Huguenots, must have used up a somewhat larger number, when the reinforcing drafts are added to the original force. Leicester took to Holland in 1585 a contingent raised for longer service than usual—8000 foot and 1000 horse, beside some companies destined for the garrisons of the three 'cautionary towns,' Flushing, Brill, and Rammekins. The expeditionary force which went out on the fleet for the 'Journey of Portugal' in 1589 was at least 9000 strong; Lord Willoughby took 4000 to Normandy in the same year to join the Huguenot army of Henry of Navarre; and Sir John Norris and the Earl of Essex must have had between them a somewhat larger strength at St. Malo and Rouen in 1591. The landing force of the Cadiz expedition of 1596 was also considerable.

But of course much greater strength was concentrated for home service, during the periods of alarm concerning rebellion or foreign invasion: Elizabeth had 20,000 men under arms to suppress the Catholic 'Rising in the North' in 1569, though few of them were actively employed. And at least thrice that number were concentrated, in one place or another, to resist the expected Spanish landing in the Armada year 1588. The celebrated camp at Tilbury was only the greatest of many camps arranged strategically for the covering of the south coast.

It must be confessed that all these forces were very unsatisfactory gatherings of haphazard material, and that organization was terribly deficient and military training usually contemptible. A force sent abroad had to be knocked together into ordinary discipline, and instructed in rudimentary drill after it had landed in France or Holland. A home-defence army was a casual collection of country-folk irregularly armed—bow and bill were still normal, and pike and arquebus exceptional right down to the middle of Elizabeth's reign. Theoretically the supreme military organization of the realm had been recast, just before Elizabeth came to the throne, by an Act of the last Parliament of Philip and Mary, which did away with the time-honoured Statute of Winchester of Edward I, concerning the general military obligation of the Queen's lieges. It abolished with the ancient duties of the sheriff, and the more modern practice of levying the shire-force by 'Commissioners of Array,' and created a new supreme authority, the Lord-Lieutenant, in each county. He was charged with the appointment of the officers of the militia, the assessing (with the aid of local commissioners) of the number of men to be supplied by each hundred and parish, and the supervision of their armament. In the event of domestic rebellion or the alarm of foreign invasion he would lead the levy into the field. But the full levy was very seldom called out, only, indeed, on the occasion of such abnormal events as the 'Rising in the North' of 1569, and the approach of the Armada in 1588. It was as a rule only a moderate force, a few companies from each shire, that the Queen required for service either at home or abroad, and she nominated the chief commander for such employ, though the Lord-Lieutenant designated the subordinate officers from captain downward.

The designated contingent of a shire was not procured by ordering certain local units to take the field, but by authorizing the Lord-Lieutenant to draw the required number from his whole shire, with the aid of his commissioners, who would go into the details. A certain amount of volunteers were always to be had—restless, ambitious, or impecunious persons, of whom some were keen for adventure, others had good reasons for wishing to leave home—such as poaching, debt, or scandal. But except in a few cases when popular interest was stirred—*e.g.* in that of the first companies that went to help the Dutch Protestants in 1572—the volunteers had to be supplemented by pressed men, designated by the commissioners in each hundred.

The system did not work well, and produced contingents whose training was *nil*, their discipline poor, and their morale doubtful—months were required to lick the company into something like military shape—or at least so much of it as had not deserted, or died from privation or ill-living. Sometimes the men starved, and at others they drank themselves to death—as in the 'Journey of Portugal' in 1589. The officers were either young nominees of the Lord-Lieutenant, who sought adventure—restless cadets of old families on the look out for thrills—or old professionals who had served in foreign armies, and knew how to make soldiering pay by corrupt devices. When Shakespeare drew the levying of Sir John Falstaff's Cotswold company under the superintendence of those egregious commissioners, Justices Shallow and Silence,[1] he was (of course) giving his memories of a session for recruits in 1588, not in 1405. Sir John, in raising his ragged company, gets a very

[1] *Henry IV, Part II*, scene 2.

unsatisfactory selection together, by allowing the able-bodied rustics to buy themselves off, and collecting a residuum of tramps and weaklings.

In this most humorous scene Falstaff first gathers in Mouldy, a young farmer, and Bullcalf, a large-limbed shirker, who pretends that he has a cough which makes him a 'diseased man,' though he roars like his name-beast. While the captain goes off to dine with the two foolish justices, the pressed men offer his minion Bardolph two pounds each to get them off. Bardolph passes on three pounds to Falstaff, keeping the balance himself, whereupon the old reprobate releases Bullcalf and Mouldy, and takes instead Shadow, 'a half-faced fellow that presents no mark to the enemy,' and Wart, a professional tramp with his tattered clothes all pinned together. They and the rest are given coats and calivers (the calivers show that we are in Elizabeth's time, not in that of Henry IV), and are marched off, much to the surprise of Shallow and Silence, who would fain have served Falstaff with 'the likeliest men.'

In another scene[1] this typical Elizabethan professional soldier explains his whole system. He will not march his company through Coventry, because he is ashamed to show them. 'I have misused the King's press damnably: I have got, in exchange for a hundred and fifty soldiers, three hundred and odd pounds. I press me none but good householders and yeomen's sons, who would as soon hear the devil as a drum. I press me none but toasts and butters, with hearts in their bellies, and they have bought out their services. And my whole charge consists now of slaves as ragged as Lazarus, discarded unjust servingmen, revolted tapsters, and ostlers trade-fallen; the cankers of a calm world and a long peace. You would think that I had a hundred and fifty tattered prodigals, lately come from swine-feeding and eating husks. A mad fellow met me by the way, and asked me if I had unloaded the gibbets and pressed the corpses. No man hath seen such scarecrows: there is but a shirt and a half in the company.' 'Pitiful rascals,' says Prince Hal, as he passes by—but 'food for powder,' replies Falstaff—'they'll fill a grave as well as better than they: mortal men, Hal; mortal men.' So thought their villainous captain, when he led them at Shrewsbury Field where they were 'well peppered,'—the survivors 'mostly fit to beg for life at the town's end.'

[1] *Henry IV, Part I*, act iv. scene 2.

It is sadly true that the picture is a recapitulation of solid grievances to be verified from state documents. While some of the captains of companies were well-intentioned amateurs, who had got a commission by favour of the Lord-Lieutenant, others were systematic swindlers, who held back their men's pay, failed to report deaths and desertions, in order to pocket the money due to non-existent persons, and issued damaged rations, bought at low prices as spoiled goods, to the rank and file. The poor wretches thereupon contracted dysentery or Walcheren fever, and went to hospital, where they died. On several occasions we read of the county gaols being cleared out of all but the most serious offenders; poachers, thieves, and drunkards were drafted into some company of recruits. Many of them made good fighting men in the end—but they were naturally addicted to marauding and desertion.

One of the most ruinous contributory causes to the inefficiency of Elizabethan expeditions was that the companies raised for any temporary crisis were disbanded when it was over. Officers and men alike were thrown on the street, till there was another political difficulty. The only exceptions were in cases where a campaign extended itself over several years, and the units were kept on foot for a long time. This happened frequently in times of Irish rebellion, but the longest term of service was that of the expeditionary force sent under Leicester to the Netherlands in 1585, of which the surviving remnant—kept going by intermittent batches of recruits—was transferred to the service of the Dutch States in 1598—save some companies reserved for the garrisons of the 'Cautionary Towns.' By this time the force was a much more reputable body than Leicester's original levy. The feeble had died, the scoundrels had deserted—the Spaniards had a whole regiment of English renegades under William Stanley, the man who betrayed Deventer in 1586. The officers were mostly English Catholic refugees [1]—the men the scum of the earth.

But the survivors of the companies that had gone over with Leicester, and had been recruited at intervals ever since, were a tough body, after they had been for some years under the hands of Norris and the Veres, and had shed their weaker and more disreputable elements. All the more so, because they had been trained in the new Dutch system of pike and musket drill,

[1] Apparently Guy Fawkes served as an ensign in the regiment—and there were other similar fanatics in it.

invented by Maurice of Nassau, the professional-soldier prince who had succeeded his father, William the Silent, as the head of the Protestant cause in the Netherlands. William had been a politician and organizer rather than a soldier—his son was a first-rate tactician and disciplinarian—the forerunner of Gustavus Adolphus. His English companies came to be the best of his army—more esteemed than his German foot and still more than his native Dutch foot, for among the Hollanders all the best men went to the navy, and the army was the secondary service. Its numbers could not have been kept up, but for its English, Scottish, Huguenot, and German auxiliaries.

The nomenclature of the military hierarchy underwent a complete change during the reign of Elizabeth. We have seen, when dealing with the army of Henry VIII in 1522 and 1544, that of all the modern officer-names that of 'captain' alone for the commander of a company was in existence. His senior subaltern was known as the 'petty captain' right into Elizabeth's time—certainly as late as 1563; after which the name 'lieutenant,' hitherto used for all sorts of *locum tenentes* (like the Lord-Lieutenant of Ireland, or the Lieutenant of the Tower), begins to appear as the proper designation of the captain's deputy, who would take his place in his absence. The standard-bearer of the company, known as ensign (often written illiterately as 'ancient,' just as the standard itself *insigne* or *ensign* appears as an 'ancient') serves regularly as the junior company officer of commissioned rank, not as a mere bearer of the company flag, by the middle of Elizabeth's reign. The name of Sergeant (*serviens*), like lieutenant, had been known from time immemorial for petty officers of all sorts, not merely military, but like the King's sergeants-at-law, also civil. The first sign of a technical use of it in army organization comes from a royal ordinance of 1528, directing that each company or band should have a sergeant, who was to be an expert in drill—which was gradually growing more important all through the century. He was also supposed to have charge of the inspection of weapons. Corporal (an Italian title, *caporale*) does not appear till 1585, when we find that each company of a shire-levy shall be given three under-officers called corporals, who should be responsible for the instruction as well as the discipline of the three squads or platoons into which a company of 100 or 150 men would be divided. Apparently they were under the sergeant, to whom their duties had previously been allotted. The 'whiffler' of the time of Henry VIII has disappeared—his function had gone to the sergeant.

It is not till 1587 that the name 'Colonel' becomes fixed, though the Spaniards had been speaking of colonels (*coronels*) and their 'colonelcies' as early as the time of the battle of Ravenna, sixty years back.[1] And English writers, dealing with foreign affairs, had been using the word as a known term, for commanders of units containing many companies, for some time.[2] It was employed in such an unscientific way—much as 'regiment' had been,[3] that we find, for example, Lord Clinton called 'colonel of all the foot' in the contingent which went to St. Quentin in 1557; [4] though this was a very large body of 4000 men in forty companies. In 1587 only was it recognized that a definite title was needed for officers in charge of several companies permanently joined together as a unit, notably in the army which went to Holland with Leicester, and stayed there for many years. 'Regiment' for a body of ten or twelve companies kept together, and 'colonel' (more usually *coronel*, in Spanish fashion) for its commander, became stabilized terms in the later days of Elizabeth's long reign.

The history of the 'major' is even more curious. He is originally always 'sergeant-major'—a term which was destined to sink to much less importance. It will be remembered that sergeants had come to be company non-commissioned officers in charge of drill, armament, and to a certain degree discipline. When the permanent union of many companies into a single unit came into being, there was an officer appointed who was to be above all these sergeants and to supervise their activities. He was called the sergeant-major, was a regimental, not a company, officer, and enjoyed high precedence, being above all captains, so much so that if the colonel was sick or absent he assumed command of the whole regiment. In 1589 we find Sir Francis Vere with the odd title of sergeant-major-general of all the army in Holland. In the early seventeenth century these officers are still called sergeant-majors of a corps, but they ended by dropping the sergeant half of the title, and were called majors only. No doubt this was from dislike of the word 'sergeant,' which had come to mean a non-commissioned officer,

[1] See above, p. 57.

[2] Cf. *Oxford Dictionary* under the word 'colonel'—1548 seems the earliest year for the word—and this of Germans only.

[3] See above, p. 332.

[4] Holingshed, p. 1133.

while this important person was eminently a commissioned officer of high rank. Presently the term appeared again for the senior non-commissioned officer, who was above other sergeants, but, of course, was under the 'major' in his new nomenclature, as was every one else save the colonel.

There is another mysterious officer to be taken into consideration, whose queer title of the 'captain-lieutenant' appears by the end of Elizabeth's reign, and endures far into the seventeenth century. The colonel was supposed to be the owner of the senior company of the regiment, which carried his personal heraldic ensign: he could not, of course, command it tactically himself, but employed for its management a senior lieutenant, who discharged the duty of captain, and drew a pay above that of other lieutenants. He ranked with captains, though not technically holding the commission of one—hence his curious title.

As every company had a standard, borne by its ensign, and bearing its captain's arms, or badge (some captains not being of gentle birth had no armorial bearings), the aspect of an Elizabethan regiment was dominated by its ten or a dozen flags of diverse colours, a gay and parti-coloured show. The purpose of using such a profusion of standards was to make rallying easier, at a time when each company was composed of men armed differently, and serving in different ways—the arquebusiers were always getting apart from the pikemen, when they went out to skirmish, and right down to the time of Leicester's expedition to Holland there were still archers and halberdiers in the company. Readers of ballad-poetry will remember 'Brave Lord Willoughby's' adjuration to his men:

> 'Stand to it, noble pikemen, and look you round about,
> And shoot you right, you bowmen, and we will keep them out,
> You musket[1] and caliver-men, do you prove true to me,
> And I'll be foremost in the fight, says brave Lord Willoughby.'

A company of 250 men was at this time composed of some such proportions as eighty men with firearms, forty bows, forty halberds, and eighty pikes. This, at least, is the scale laid out for a company raised in Lancashire in 1584. When such a mixed force was put in line, the halberds took post in the centre to guard the standard, the pikes on each side of them, then

[1] Note that the musketeer appears among the 'shot,' though his weapon was much later than the arquebus, and is first mentioned in an English company in 1584. See 'musket' in the *Oxford Dictionary*.

the bowmen in two halves on each side of the pikes, and finally the musketeers and caliver-men at the ends of the line. But in the beginning of an action all the men with missile weapons, whether bows or firearms, might be thrown out to skirmish, and retire to take shelter with the pikemen when battle was joined, and (as the phrase of the day was) 'it came to push of pike.' The retreat of the skirmishers was often disorderly of necessity, and they had difficulty in forming up again, sometimes on the flank of the pikes, sometimes (it would seem) even by throwing themselves down in front of the long weapons) at least battle pictures of the time occasionally represent them as doing so—or taking refuge inside the square which the pikes formed in the case of a cavalry charge. For, of course, the arquebusiers were quite incapable of stopping by their fire a cavalry attack pressed home—though (as had been shown at Pavia long ago) they could do much harm to horsemen operating on ground broken by walls, thickets, or hedges, in which there was cover to be got.

The bow was going out of use, though very slowly, all through Elizabeth's reign, not without much controversy between the admirers of the old weapon and the modernists who condemned it. We have seen that as late as 1584 a company was expected to have a certain number of archers. In 1569 at the time of the 'Rising in the North,' a Yorkshire levy, called out by the Earl of Sussex and Lord Hunsdon, could produce only 60 arquebusiers out of over 2000 men who had come together, though a force from the Midlands, collected at the same time at Newark, had 300 arquebusiers and 300 pikemen. But Sussex reported 'great default of armour and weapons, no spears for the horse, no arquebuses or powder for shot; all that the people had for armour was plate-coats, jacks, and sallets, with black bills, bows, and arrows, and there was nought else to be got, until my Lord-Lieutenant got spears, pikes, and powder from Newcastle.' He only made up his 'shot' by borrowing 300 arquebusiers from the garrison of Berwick, who were trained men.[1] To get them to Durham he set them on horses, requisitioned in haste in the County Palatine. In the following year the Queen made a strongly worded appeal to the deputy-lieutenants of Kent, Essex, and Hampshire to provide arquebusiers, retained permanently to be in readiness for any crisis—their weapons to be furnished at the

[1] See *State Papers Domestic, Addenda* under 1569, pp. 121–3, and 136.

common expense of the county, and committed to the charge of skilful captains, who should train the men selected for their service. 'We cannot prescribe the number, yet the greater the better, above 600 for that part of the country near the sea and the Thames.' [1]

This seems a very modest figure for three shires, where the use of firearms was much better known than in the North or the Midlands. A few years later the Queen's ideas as to the numbers that could be provided were rising to much heavier proportions.[2] In 1573 out of the train-bands of London no less than 3000 men were selected to be trained in the use of the arquebus or caliver, apparently about half the total levy of the city on this occasion.

By 1588 the bow was distinctly obsolescent. There were still many archers in the great mass of shire-levies that was called out on the threat of Spanish invasion, the approach of the Invincible Armada. But by this time there were some regions where of all the men embodied there were no bowmen—only pikes and arquebusiers or caliver-men. This was the case in Cambridgeshire, Huntingdonshire, and Wiltshire. In London not one man of the regular train-bands had a bow, but among the 4000 supplementary untrained men 800 were scheduled as archers. In the Midlands and North the proportion of bows to firearms ranged down from one-third to one-fifth, but was always in the minority. Only in the two forest counties of Buckinghamshire and Oxfordshire is there a slight preponderance of archers over arquebusiers. The reason for this is not clear—possibly they were poaching counties, where the silent bow was dearer to the wood-haunter than the noisy self-revealing firearm.

It was seven years after the disbandment of the great musters at Tilbury and elsewhere, collected to face the Spanish army that never came, that the Privy Council issued a most important order, which marked the end of archery as the national speciality of English war. This ordinance decreed

[1] See *State Papers Domestic, Addenda* under 1570, pp. 305–6.

[2] In 1569 there had been a general order that the Commissions of Musters under the Lords-Lieutenant were to encourage the use of firearms, by providing that every government servant or official with a salary of 50 marks or more should be compelled to find an arquebusier, also every parson with pluralities worth £20 per annum, and every justice of the peace, and if such a justice had lands in more than one parish he must provide two. The weapons were to be stored in 'artillery houses,' not taken home by the arquebusiers, who were to meet at these houses, for practice with ball, twice a month. *State Papers Domestic, Addenda* under 1569, pp. 78–80.

that for the future the bow should not be passed as a competent weapon for any man enrolled in the regular train-bands of any shire, and that all the 'shot' must have arquebuses, calivers, or the more modern and heavier musket. This was the death-knell of the old weapon that had served England so well in every battle for three hundred years, from Falkirk and Halidon down to Flodden.

This decision was not reached without much controversy. There was a lively discussion in progress all through the later years of Elizabeth, when military books (rather scarce in earlier generations) began to appear in considerable numbers from the English press.[1] Some of them were infected with Renaissance classicalism, and go back to Aelian and Frontinus for general principles of war; others borrow wholesale from Machiavelli, the excellent La Noue, Du Bellay, and the Spaniard Mendoza. But there are purely English controversies, turning on the old dispute between the bowman and the arquebusier.

The most important advocate of the ancient weapon was Sir John Smyth, as good an old soldier as his opponents Sir Roger Williams and Humphrey Barwyck. Smyth grounds his preference for the bow on six practical points:

(1) The trained archer takes a better aim than the arquebusier—who can only shoot accurately at very close range—point-blank—while a good bowman can hit accurately at 150 or 200 yards—distances at which the arquebusier can only fire wildly at anything over 100 yards.

(2) The bow is a simple weapon—firearms very complicated things, which can get out of order in many ways; the piece clogs and fouls easily, is liable to breakage, and can only be repaired by a skilled gunsmith. Wet weather spoils the powder, windy weather blows out the match, or sends its sparks flying among the powder of horns or bandoliers.

(3) In the excitement of battle all but the oldest and steadiest troops are liable to mishandle their weapons. One man, in his haste, forgets the wadding between powder and ball, another the

[1] A complete bibliography of early English military books can be found in Cokayne's excellent work.

wadding to keep the ball down. Smyth has seen bullets trickle out of the mouth of a caliver, for want of wadding, when the soldier was holding it with the muzzle depressed. 'This is why when musketeers of a raw sort shoot point-blank at whole battalions, sometimes only few are seen to fall.'

(4) Arquebusiers can only stand two deep, archers are effective even six deep, when the rear ranks shoot with a high trajectory.

(5) The arquebus is a very heavy weapon compared with the bow, and tires out soldiers on the march; their arms grow unsteady after half an hour's rapid firing.

(6) Most of all important is the old and effective argument on rapidity of discharge. A bowman can let off six aimed shots in a minute, an arquebusier only one in two or three minutes, when he has gone through all his manual exercise carefully.

Smyth will only allow advantage for firearms in accurate shooting from 'bulwarks, mounts, and the ramparts of a fortress,' when the soldier is under cover, and supporting his piece on a wall or a 'rest' of the modern sort.

Humphrey Barwyck replies in detail to all of Smyth's six points, as follows:

(1) Archers are no longer accurate shooters at long ranges—their art is much decayed.

(2) If bad weather is pernicious to firearms, it is no less so to bows. Rain makes bowstrings slack, and after a march in the wet arrow-feathers flake off.

(3) Archers can be as nervous in battle as arquebusiers. Barwyck has seen them, when excited, fail to draw the arrow to the head, and shoot wildly without aim, in order to let off as many shafts per minute as possible.

(4) When archers stand more than two deep, the rear ranks are taking no real aim, but only shooting at hazard into the air.

(5) The bowman is dependent for real efficiency on his

bodily strength much more than the arquebusier. ' If he have not his three meals a day, as is his custom at home, nor lies warm at nights, he presently waxes benumbed and feeble, and cannot draw so as to shoot long shots.'

(6) The improvement of firearms, and the practice of constant drill, is making old soldiers capable of discharging many more shots in a fixed time than was possible a few years back. They can now shoot off forty times in an hour, and the rate will continue to improve.

The celebrated Sir Roger Williams, one of the hardiest veterans of the wars in the Netherlands, also took up the controversy with Smyth on the side of firearms *versus* bow. He would rather, he says, have with him in the field 500 good musketeers than 1500 bowmen. Archers, nowadays, are of such mixed quality that out of 5000 of them only some 1500 will ' shoot strong shoots.' And supporting Barwyck's argument about bad weather, he opines that after three months' campaigning in winter, or even in exceptionally cold spring or autumn seasons, not one archer in ten can keep up his proper bodily strength. ' Few or none of these will then do any great hurt at twelve or fourteen score ' (*i.e.* at 240 or 280 yards). The arquebus, on the other hand, will shoot as strong as ever if the soldier has strength enough to touch off his piece. Sir Roger, we must confess, omits to mention the fact that though the piece may go off all right, the aim of a tired, half-starved, or rheumatical arquebusier is not likely to be very good.

The modernists had their way—and the verdict was finally given against the bow, when the Privy Council by their Ordinance of 1595 decided that archers should no longer be enrolled in the train-bands as efficient soldiers, but only arquebusiers, caliver-men, or musketeers. The musket, which was becoming more prevalent as the century closed, was a heavier weapon, longer, and always fired from a ' rest,' a forked staff on which the man supported the barrel of his weapon. Its introduction complicated drill—adding several new movements to the manual exercise, but made accurate aiming much more possible. The arquebus used to ' wobble ' when its bearer grew tired. The first demand for musketeers as opposed to arquebusiers or hackbuteers in England goes back only to 1587.[1]

In the matter of clothing, as of armament, there was much change during the reign of Elizabeth. The white coats usual in the time of Henry VIII and Mary were no longer universal. While it was usual for the whole company, or regiment, after regiments became normal, to have the same uniform, different levies from different shires seem to have been clothed in colours chosen by the Lord-Lieutenant and his commissioners, or by the colonel. Red was not uncommon, but a Lancashire levy in the middle of the reign is ordered to have blue coats, and an interesting document of 1584, for raising troops for Ireland, has the curious phrase that the men are to be dressed 'in motley, or other sad green colour, or in russet.' That 'motley' should be reckoned a 'sad green colour' seems odd—presumably it meant stuff in which there was a pattern, with lines of different hues in it, something like a modern tweed. The choice of this particular clothing was an early instance of 'adaptation to environment'—the troops were destined for wood-fighting, for Ireland was then mainly a forest country, and white or red coats would have been too visible, while motley or russet or sad green would not be easy for the lurking enemy to make out among the trees. In short, this was an experiment like our modern adaptation of khaki, which began with the days of the Indian Mutiny.

The most distinguishing feature in the military equipment of the Elizabethan age was the 'morion,' a high pointed head-dress, with a brim sharply cocked before and behind, and usually a ridge-crest. This form was superseding the round-topped steel cap of infantry by the end of the reign of Henry VIII; a few figures wearing it are to be noted in the great military pictures of the King's army at Portsmouth and Boulogne, which are now preserved only in copies, the originals having perished at Cowdray. In the second half of the century it was almost universal among all nations of Europe both for pikemen and 'shot'; but towards the last years of Elizabeth some of the arquebusiers are found dropping it in favour of ordinary felt hats—the heavy metal morion proving inconvenient for skirmishing, though invaluable to the pikeman, who might have to face a cavalry charge, and downward slashes

[1] But in the ballad of 'Lord Willoughby,' quoted on p. 379, it will be noted that musketeers are mustered along with caliver-men.

from the sword, while the musketeer had only to 'run for it' if horsemen came down on him. The pikeman also wore both breastplate and backplate, with 'tassets,' broad curved plates of steel, hanging from the breastplate to protect his thighs. These the man with firearms never possessed, though he seems for some time to have had a breastplate—which was finally given up in favour of a buff jerkin. By the reign of James I military opinion had come to the conclusion that all protection of head armour or body armour for skirmishers was a mistake, and that lightness of movement was the one desideratum. Wherefore felt hats and leather coats became the only wear of the musketeer, though his comrade with the pike continued to sheath himself in steel.

As to the horseman, it may be said that the 'demi-lance' completely superseded the fully armoured man by the end of the century. Only generals and superior officers continued for some time to wear leg-armour, which in all other ranks was dropped in favour of high leather boots. It may be doubted whether the greaves and steel shoes, seen in some portraits of late sixteenth-century magnates, were not really 'armour of parade,' or for tilting, and not used on active service. But the cuishes covering the thigh continued to be worn by all heavy horse. It will be remembered that Sir Philip Sydney is said to have owed his death at the skirmish by Zutphen to his having lent his cuishes to Sir William Pelham in the hurry of arming. A ball, which would almost certainly have been deflected by the steel, penetrated into his unprotected thigh, and smashed the bone. In the same skirmish it is noted that Lord North was wearing *boots*, *i.e.* no armour below the knee. The lance was in general use down till the last years of the century, both by heavy and light horse, but the continental preference for pistol and sword rather than lance was making itself felt on this side of the Channel, though it did not become dominant till close on 1600. As early as 1569 the Earl of Sussex, speaking of the horse of the northern rebels, says that 'some of them are pistoliers,' which shows that they must have been making long preparation and buying weapons from abroad.[1] The pistol superseded the arquebus for cavalry using firearms, and we no longer hear of 'arquebusiers horsed' such as were seen at Pinkie. Its adoption led to the pernicious system of the 'caracole' mentioned already in an earlier chapter, which

[1] *State Papers Domestic, Addenda*, under year 1569, p. 121.

involved discharges by successive ranks, instead of the much more effective charge home without any halt for firing. The English were the latest of all the military nations save the Scots and Spaniards to drop the lance.[1] Among the French gendarmerie it was obsolescent at the end of the Civil Wars, while the Germans long before this had specialized on to the reiter with his pistols.

On all occasions when large forces were called out in the reign of Elizabeth we find complaints from the Lords-Lieutenant as to the inefficiency of the squadrons of horse which they had to raise. Many gentlemen and others, scheduled for the obligation to find one or more demi-lances, either came with imperfect provision of armour, or with weak horses, or did not come at all, but offered to serve on foot as officers for the shire-levy. Those who sent substitutes mounted, are often accused of supplying men neither skilled in horsemanship nor in the management of weapons.[2] The same complaint, a suggestive fact, is found both in France and in the Spanish viceroyalty of Milan, where a governor complained that the local levy of horse was mostly made up of the varlets of gentlemen or officials, who in time of peace only came to the standard on the day of a review, and in time of war slunk home on the fourth day of a campaign.[3] The few who were not of this class were bad characters, whose only object was to be free to carry weapons, which made them the terror of the neighbourhood.

Holland is a country particularly unsuited to cavalry operations, and the 1000 demi-lances and light horse who came over with Leicester in 1585 had small opportunity of acting in large units, or serving in line of battle. They found themselves involved in sieges, or in narrow operations on dykes, or on roads between canals and morasses, where charges were difficult. Indeed there had not been a pitched battle in the Low Countries since Gemblours (1578), and was not to be another till Tournhout (1597). The only occasion on which we find the English horse seriously engaged was in the abortive attack on the Spanish convoy in front of Zutphen in 1586, when Sir Philip Sydney met his death. And here they made a desperate on-

[1] Spanish cavalry with the lance are still found in Parma's campaigns, though he was given to adding men with firearms to his heavy horse. It seems that only Philip III ordered the disuse of the lance officially.

[2] The gentlemen due to serve as demi-lances had always some good excuse for deficiency in horse or arms; cf. *S.P.D. Addenda* under 1573, p. 454.

[3] Cf. the Conde de Clonard, *Historia Organica de caballeria y Infanteria*, iv. p. 273

slaught, led by a group of gentlemen volunteers, but did not win a victory. The Dutch wars, indeed, show a marked contrast to the contemporary struggle between the Huguenots and the Catholics in France, where, in a realm full of stretches of good 'cavalry country,' there were an immense number of fierce contests of horse, from Dreux down to Ivry. In the Huguenot armies at least (less so in the armies of their opponents) the mounted men were the really important part of the host. But of this we must speak elsewhere.

On the whole, the reign of Elizabeth makes a very depressing chapter in the history of the English Art of War. Things might have gone otherwise if the Queen had risen to the idea of raising and maintaining a permanent standing army. But her resolute policy of stringent economy made this impossible to her, and a hand-to-mouth system of perpetual levies, followed by perpetual disbandments, produced very unsatisfactory armies. Brilliant in every other respect, the Elizabethan epoch makes but a sorry show when military operations on land are in question. Occasionally one feels acute humiliation when reading of the conduct of English troops—and officers.

The worst case of all was the celebrated 'Journey of Portugal' in 1589, when the considerable force which, after a buccaneering raid on the ports of Galicia, landed at Peniche to support the claims of Don Antonio, went home depleted and disorganized without having accomplished anything, though its leaders had a splendid opportunity before them. Lisbon was practically ungarrisoned—the old 'tercio of Portugal' had gone off in the Armada, and had perished almost to a man. In the city there were only a depot and recruits—the nearest Spanish troops were miles away at Badajoz. Though the Archduke-Governor had arrested many suspected persons, the populace was ready to rise if only a lead had been given. Indeed, very many Lisbon citizens were hanged for conspiracy after the English had left the country. But the English leaders—Drake seems to have been the most responsible—quarrelled with each other, never attacked the place seriously, and hung about waiting for a general insurrection, which did not take place. Indeed their conduct hardly invited it, for their soldiers plundered the countryside and got hopelessly drunk. Drake, who was in command of the naval side of the expedition, refused to bring his ships up to the level of the city, and waited at Cascaes, by the Tagus mouth, exchanging recriminations with Norris.

It seems probable that a joint sea and land attack on Lisbon must have succeeded, and that a general rising of the Portuguese, who had a sincere hatred for all Spaniards, would have followed. But instead the general and admiral waited till half their men were sick, and Spanish forces began to show on the side of the inland—when they went off, having fired hardly a shot, and were met at home by the Queen's displeasure and the well-merited reproaches of their countrymen. It is a sorry tale.

But there were other incidents in Ireland and also in the Netherlands of which the memory is distressing. The Elizabethan age produced some fine soldiers, like the Veres and Morgan and Roger Williams—but not a respectable army, and rarely, if ever, a man capable of handling one. The best service done by English troops was as auxiliaries to Henry of Navarre or Maurice of Nassau, and the one officer who seems to have had a real capacity for high command was Lord Mountjoy, whose talents were wasted in the distressful wars of Ireland.

BOOK V

THE WARS OF RELIGION IN FRANCE (1562–98)

CHAPTER I

GENERAL CONSIDERATIONS: POLITICAL AND MILITARY—THE CATHOLIC AND THE HUGUENOT ARMIES

THERE is, as has often been remarked, an immense difference between the almost continuous strife between the King of Spain and his rebellious subjects in the Netherlands (1568–1607), with which we have presently to deal, and the contemporary struggle in France (1562–98) between the Catholic and Protestant parties, which can be broken up into nine separate civil wars, interrupted by longer or shorter periods of insincere pacification. The fundamental cause of the divergency was that in the Low Countries the Crown was represented by an autolatrous fanatic, equally set on asserting his own will to govern as he pleased, and on extirpating heresy, while in France the Crown meant a family of opportunists, Catherine de Medici and her sons, quite unscrupulous, indeed perfectly 'amoral,' whose main object was to survive, and to keep some hold on power during the clash of two contending parties among their subjects. The Queen-Dowager, the dominating spirit at Court during the great part of this period, was no doubt a Catholic so far as personal inclinations went, but she disliked the Guise family, the heads of the extreme Catholic party, quite as much as she disliked the turbulent leaders of the Huguenot faction.

She desired peace, even peace secured by giving toleration to the Protestants, if by peace alone could the royal power and the survival of her own family be secured. Her name goes down in the history with the red stain of the Massacre of St. Bartholomew smirching it, but it is only fair to her memory to recall the fact that this was a strange and desperate aberration from her general policy, and that half a dozen times she made energetic efforts for peace and toleration, not (of course) because she liked toleration, but because she desired peace for the salvation of her family and the monarchy. Her mental position has been not unjustly compared to that of Elizabeth of England who, like her, sought peace and pursued it, with the most lamentable disregard of pledges and promises, and with a perfect readiness to sacrifice individuals. The English Queen had her difficulties, but they were not complicated by the existence of a family of perverse and degenerate sons, nor of a fierce and reckless feudal noblesse demoralized by fifty years of the old Italian wars of plunder and aggression. Elizabeth's father and grandfather had worked for her in cutting down such of the ancient English houses as had survived the Wars of the Roses. While she could always count on the support of the bulk of the nation, Catherine de Medici and her sons were only sure of the moderate central faction, the 'Politiques' as they were called during the latter part of the period of civil war, who cared more for the safety of the realm and the preservation of the monarchy than for either Catholic Orthodoxy or Protestant Reform. But such men as the Chancellor l'Hôpital were not too common, and appeals to loyalty or national sentiment had only an intermittent effect on zealous Catholics or Protestants—though some on both sides were not unsusceptible to them on occasion. Even Coligny and La Noue could yield to them for a moment, till some provocative incident occurred.[1] It must be confessed that neither Catherine herself nor her wretched sons were personages from whose mouths appeals to loyalty or national sentiment had a very convincing effect. It was not till Henry of Navarre came to the front that any member of the French royal family could command enthusiasm, even from a section of the nation—and he was but a genial adventurer, not an unselfish hero: he recalled Francis I rather than St. Louis. All his display of chivalry—genuine enough in a way—was compatible with a most cynical outlook on mankind—and womankind—and he was entirely self-centred, though capable of fits of ostentatious generosity, when the circumstances demanded it. But to find a King who was capable, alert, a jovial companion, recklessly brave, an excellent cavalry officer if not a strategical genius, was a splendid discovery for a France that had lived for thirty years under the degenerate sons of Henry II.

[1] As an interesting study of Coligny's reluctance, see d'Aubigné, book III. chap. iii. pp. 131–33. For La Noue, his conduct when mediating between the Rochellois and the King is a sufficient proof of his honest hesitation between the calls of loyalty to his monarch, and loyalty to his co-religionists and his conscience.

The Crown being without lustre or appeal to national sentiment, internal trouble was inevitable. The particular shape which it took was settled by the outstanding problem of the sixteenth century, the clash between Orthodoxy and Protestant Reform, which had already shaken so many of the other countries of Europe, but was late in its appearance in France. Nothing can be more surprising than the fact that the sect which under Francis I and Henry II had seemed insignificant, attracting intermittent persecution and supplying but a moderate record of martyrs, suddenly flares up within two years of the death of Henry II as a great military force, challenging the control of the realm, and putting large armies in the field.

It looked in 1559 as if the accession of the young and weakly Francis II was to be followed by little more than the usual contest for power between ambitious coteries of princes and nobles, which normally accompanied a minority. The Guises, who had got control of the boy, and had supplied him with a wife—Mary Queen of Scots—from their own clan, were naturally hateful to other people who, like the Constable Montmorency, Antony of Navarre, or Louis of Condé, thought themselves equally entitled to a share in the governance of the realm. The Conspiracy of Amboise, put down with such efficiency and so much bloodshed, had little to do with religion, it was simply a *coup d'état* which failed, and left the Guises more firmly established in power than before. They lost it suddenly by the decease of their sickly protégé on December 5, 1560, which placed the crown on the head of his ten-year-old brother Charles IX, and the regency in the hands of the Queen-Dowager Catherine de Medici, who tried to confirm her difficult position, by compromises and edicts of toleration for the Protestants whom the Guises had been persecuting. Her policy provoked the wrath of the Orthodox, and open trouble came when the military retinue of Francis of Guise fell upon and slaughtered a Calvinist congregation at the Massacre of Vassy, March 1, 1562. This led to a general rising of the Huguenot party all over France, and to the dismay of the Orthodox these sectaries displayed a surprising strength, headed as they now were by Louis of Condé, a prince of the royal blood.

The traditions of old civil war, the strife of the Armagnacs and Burgundians, the Praguerie, and the 'League for the Public Weal' of 1465, were not forgotten in France. But three generations had passed away without any serious recrudescence of such phenomena. The Constable Bourbon's attempt to stir up rebellion on the old feudal lines in 1524 had been a complete failure—no one followed him save a few personal retainers. How came it then that from the time of the Conspiracy of Amboise (1560) civil war was once more endemic for more than thirty years ?

It is not sufficient to attribute the miserable record of the reigns of the later Valois to the groundswell of the Reformation, affecting France later than it affected Germany or Great Britain. The 'Wars of Religion' were not altogether religious in their essence. That admirable puritan and philosopher, La Noue, turned his mind on to the problem in his dreary years of imprisonment in a Spanish fortress, and found the solution in the moral decadence of all the governing classes in his native land, caused by the long wars of aggression and plunder in Italy—wars absolutely unjustifiable. 'The King of France rules a mighty state—two hundred leagues long from Metz to Bayonne, two hundred and fifty broad from Morlaix to Antibes, fertile in every product needed for human use, well peopled, with ample revenues, and a most gallant noblesse. Why then, the foreigner may ask, try to enlarge such a state by alien annexations rather than to perfect it by internal reform ? . . . We French must remember that the time of great annexations is over, and that it is no mean feat to keep what we have in good order. The dream of plunder and glory should come to an end. The young have been reading too many romances of reckless adventure, full of *amours déshonnêtes* and objectless fighting. The old have been reading and rereading Machiavelli, who saps all fundamental ideas of honour and justice. It is hard for the nation to settle down after so many nerve-wracking years.' [1]

More especially was this so because France was full of old soldiers and old officers, thrown upon the street by the almost complete disbandment of the army of Henry II after the Peace of Cateau Cambrésis. The officers were numbered by thousands, the men by tens of thousands. They had been engaged in wars which were the worst possible school for civic morality. And something worse was to come, a war of religion, which

[1] All this is reduced from the long chap. xx. of his *Discourse*, pp. 502–19. I have already had occasion to quote these most interesting paragraphs in the first preliminary chapter of this volume. I know no other passage which so well expresses the spirit of the times, as envisaged by a staid and religious-minded observer.

brings out all the worst forms of savagery—fanaticism that leads to massacre. La Noue has much to say on the way in which the Crown might have dealt with these masses of reckless veterans. He leaves the actual criminals to the gallows, but is much more interested in the majority, whom he regards as reclaimable citizens. One great class consisted of young men of good family, who had been drawn to Italy by the spirit of adventure, and when Italy was relinquished, and quiet times should have come, instinctively took sides with one party or another, from disgust at the tenets or practices of the rival faction, or not unfrequently from old family quarrels with their nearest neighbours at home. Most of them were no great dogmatists on one side or another, like that Duke of Bouillon who frankly declared that he did not know very much about the Bible, but that he could not endure to see decent bourgeois sent to the stake about some matter of the Real Presence in the sacramental bread. On the other side there were slmple Catholics who were driven mad by the sight of iconoclastic outrages by wild sectaries, who hacked down crucifixes, and blew up high altars. If only such people would be persuaded that toleration might be made the order of the day, and that their *bêtes noires* on the other side would be muzzled and kept from provocative atrocities, they might settle down to peace.

The case of the large mass of *roturier* or non-noble officers who had held captaincies in the wars, and disliked going back to civil life, as a step down in social status, was quite as difficult as that of the disbanded cadet of some old family. La Noue begged such folks not to be blinded by the example of Spain, where it was considered infamous for a man who had served in the regular army to take up a trade, but to think of Switzerland and Flanders, where an old officer easily turned to merchandise or farming. It is far less humiliating to open a shop than to hang on as a casual retainer of some lord, or a 'gentleman's gentleman,' expected to do the dirty work of his patron in return for the crumbs that fell from his table—the life of an Italian *bravo*.

La Noue, alas! preached toleration, and a return to long-forgotten peaceful avocations, to a generation that was deaf to his pleading. The poor gentleman did not relish the return to his few paternal acres, nor the professional soldier the idea of living what he considered the life of a huckster. They formed the ready material for the armed bands of civil war, and their old rank and file were strewn about the country—largely as vagabonds—craving for the sound of the drum. Hence the readiness with which armies sprang from the ground after 1562—it was only three years since Cateau Cambrésis had turned the veterans loose. The inflammable material was ready to blaze up when the match was applied.

That the conflagration took the shape of a war of religion was due to the honest revulsion among large sections of every class of society against the corruption of the times—in matters religious no less than matters of morals and manners. The courts of Francis I and Henry II had been sufficiently scandalous—less provocative, however, to a right-minded Christian than the sight of Jean of Guise, the first of the four Cardinals of Lorraine, a notorious evil liver,[1] who was holding in plurality the Archbishoprics of Rheims, Lyons, and Narbonne, the Bishoprics of Metz, Toul, Luçon, Thérouanne, and Valence, with some ancient and wealthy abbacies such as Cluny, Fécamp, and Marmoutiers,[2] and sharing in the royal orgies—he beat Wolsey all round both in see-grabbing and in ostentation. The state of the Gallican Church was as hateful to Calvin as that of the German Church had been to Luther a generation earlier. The French revulsion against the old régime took a much more Puritan shape than that of the German revulsion. The logically-minded Gaul formulated the church organization of Geneva, and pressed to its extreme the old doctrines of Original Sin and Predestination. The duty of the Elect to contend despite of all odds against the Reprobate—God's enemy—is the fundamental inspiration of all the desperate rallies of the Huguenots—but with what strange allies were the Elect sometimes to be found co-operating! Broken men, men with a grudge against some old enemy, ambitious adventurers, impecunious cadets who wanted excitement at all costs!

The first head of the armed rising of the Huguenots was Louis of Bourbon, Prince of Condé, who had been at the back of the Conspiracy of Amboise, and had nearly lost his head

[1] See Brantôme, *Dames Galantes*, vii. p. 321.

[2] It is easy to forget that there were no less than four Cardinals of Lorraine in this period: (1) Jean (died 1550), spoken of above, uncle of the great Francis of the Civil Wars; (2) Charles, brother of Francis and nephew of Jean, whom he succeeded in the Archbishopric of Rheims—this is the Cardinal of the conspiracy of Amboise period—he died in 1574; (3) Louis, younger brother of Charles, Archbishop of Sens, bishop of Metz, Alby, etc., died 1576; (4) Louis II, also Archbishop of Rheims, nephew of (2) and (3), and brother of the 'Balafré,' with whom he perished at Blois in 1588, at the hands of the minions of King Henry III.

on account of it, though his responsibility had not been legally proved. Condé had proclaimed his adhesion to the Reforming party; how far he mastered its creed and principles it boots not to inquire—he may be credited with a genuine dislike for the existing church régime in France and its corruption. His own life was not that of a Puritan, but that of a dashing cavalry officer; he had won many laurels in the old war against Charles V. But behind him were not only a miscellaneous collection of political malcontents, but a grim band of Calvinist zealots smarting under many years of persecution, and inspired with the same enthusiasm that was to be found a century later among the soldiers of Cromwell. When once they had taken arms, and realized their own strength, it became a matter of 'the sword of the Lord and of Gideon'—accompanied with the same outburst of iconoclasm and violence that was seen on this side of the Channel in 1642–46. The churches of France show in ruined shrines and mutilated statuary the same record as those of England. The blowing up of the high altar and lantern of the historic cathedral of Orleans, by means of many barrels of gunpowder, was only one among numerous specimens of wanton destruction. Many of the Huguenot leaders were religious enthusiasts but not vandals, and did their best to restrain their subordinates—such were the two Chatillon brothers, the great Admiral Coligny and Dandelot, and the admirable and philosophic La Noue, whose comments on the wars fill the reader with sympathy for a good man thrown in evil times. But mixed with such personages were mere blood-thirsty adventurers like that Baron des Adrets, whose atrocities go some way to excuse the corresponding excesses of Catholic leaders. But, as has been remarked by a contemporary, who started the game? The Huguenots had been counting their martyrs for years—and the first massacres, from Vassy onward, had been perpetuated on their unarmed co-religionists.

One of La Noue's most interesting chapters is dedicated to a description of the way in which the first Huguenot army of 1562 began as a band of Puritans, but ended by becoming a band of pillagers.

'When this war started there were some chiefs and captains who spoke of military discipline, but much more effective were the sermons in which we were admonished not to oppress the poor people, and the religious zeal which had brought most of us into the field. Without any constraint we tried to bridle ourselves, as no fear of punishment could have bridled us. The noblesse showed itself in these early days quite worthy of its name, marching in the open country, where the temptation to live by plundering is much more obvious than in towns, it lifted nothing, never maltreated a peasant, and was content with poor fare. The leaders and those who had any money in their pockets paid honestly for everything they got. There were no complaints, the villagers did not abscond from their dwellings. If any soldier was guilty of violence he was banished or put in custody, and his own companions made no attempt to excuse him. Among such a large gathering we heard no one blaspheme the name of God; persistent swearers tried to drop the habit, and were reproved with anger if they failed. You could not have found a box of dice or a pack of cards in the camp. Women were not allowed to enter the quarters—such women as haunt the camp are always the source of debauchery. No one was allowed to quit the standards to forage for himself; the men had to be contented with their rations and the small amount of pay that was forthcoming. Morning and evening, when the guards were set, there were public prayers, and psalms were sung. We noted piety in many from whom it could not have been expected—old soldiers of former wars. One day my brother-in-law Teligny and I were praising the conduct of the army to the Admiral Coligny. "Very fine," he said, "provided that it lasts. But I have my fears that our people will shed their virtue in the course of two months, and have nothing left but their bad qualities. As an old infantry colonel I cannot but remember the proverb, "Young hermits may become old devils." We smiled at his saying, but it was only too true.

'The first disorders were at Beaugency, which was stormed by some Provençal companies. They gave themselves up to pillage, and mishandled poor Protestant townsmen, who had not been able to get away, worse than the Catholic garrison. This example provoked the Gascon companies, who soon after showed that they would not be left behind in violence. But the North-French regiment of M. d'Yvoy would have got the first prize, if one were awarded for misbehaviour. Our infantry lost its original virginity, and allied itself to Miss Pillage (Mademoiselle La Picorée), who, as the war went on, would become Princess Pillage. Indiscipline spread to the noblesse, some of whom, after a first taste of the dish of stolen goods, refused to sup on any other meat. And so the evil commenced on a small scale but became general. This was no fault of the Admiral, who tried strong measures to stop the disease. In Normandy he hung a captain of irregular horse, who had sacked a village, with four or five of his men around him, all booted and spurred, with a pile of their plunder at their feet, and a written notice above, warning others of the same sort. The effect lasted about a month! I am bound to say that the Catholics during the first few weeks of the war also behaved with comparative decency—the gentlemen especially—but they went the same way as our people in full sail. It sometimes made me laugh in bitterness to think how "soldier" meant the same thing as "brigand" in our days of the Troubles.' [1]

Comparing the French wars of religion with the English war between King and Parliament that came eighty years later, we find a notable if surprising parallel. The Huguenot armies were much like those of Charles I, their strength lay in a numerous and chivalrous cavalry, which represented the better part of the gentry of the realm—their infantry, like the royalist infantry in England, was always relatively weak and untrustworthy. There were too many foreign landsknechts, and old professional soldiers of doubtful enthusiasm for the cause. True, there was a strong leaven of real Puritans among the Huguenot foot-soldiery, but not enough to leaven the whole mass. Where the Puritan element shows, it was in the defence of towns by their burgher levies, as at Sancerre in 1573, when the citizens were fighting for hearth and family as well as for the religion. The same thing—as has been shown elsewhere—was the case in the Revolt of the Netherlands. The Huguenot cavalry was formidable in its charge, capable of rallying again after desperate defeats; but it was most undisciplined, liable to melt away at critical moments, because some distant local danger suddenly distracted the gentlemen of a province. Subordinate leaders could never be trusted to carry out with strictness the orders of their commander-in-chief; and commanders-in-chief often had to swerve from their strategical intent, because their more important lieutenants refused to concur,[2] or even left the field in petulant disobedience.

Again and again we get complaints that these feudal gentry

[1] All this paragraph is contracted from the much longer passage in La Noue's commentaries, *Premiers Troubles*, pp. 818–24.

[2] This was the case with Coligny before Moncontour.

were desperately careless—they despised the dull work of reconnaissance and guard-setting. They could never be trusted to keep the exact day, when a concentration of scattered forces was needed. They quarrelled with each other on old family grudges and jealousies. But the most persistent fault was 'localism'; they disliked to be drawn far away from their own province for more than a few days, because their homes, during their absence, might be in danger from raids by their Catholic neighbours. Charles I in the English Civil War was ruined by precisely the same drawback; the Northern Horse or the levies of Cornwall and Devon hated to be involved in operations in the Midlands, and could only be lured with difficulty and on rare occasions to co-operate with the King's main army. So it was with the Huguenot commanders—Condé, Coligny, or Henry of Navarre could never be quite certain whether they would have 3000 or only 1500 horse in line on the day of a battle, if ominous rumours had come in from the South or the West.[1]

Hence a persistent lack of continuity and strategical unity in all these tedious wars. Matters were not finally settled by decisive actions between concentrated armies, for petty campaigns were going on in corners of the realm, whose progress and results were independent of the central struggle. This is precisely similar to the condition of affairs in England in 1642–46, just as the contest between the Fairfaxes and Newcastle can be studied almost without reference to the operations of Essex and King Charles, so the campaigns of that surprising local commander-in-chief Lesdiguières, in Dauphiny and Provence, went on as an independent side-show. And as Naseby did not settle the English Civil War, which dragged on for more than a year after the royal cause had become hopeless, so neither Ivry nor the surrender of Paris in 1594 put a stop to the French war of religion; the last Leaguers did not surrender till 1598.

From the very first the Huguenots were in the habit of calling in German auxiliary troops, especially the 'reiters,' the armoured pistoleers so much in favour at the time. La Noue thought them practically superior in tactics, if not in morale, to the French noblesse, who were still fighting in line—*en haye*—with the lance, as in the old wars of Italy.[2] But valuable

[1] The habitual carelessness of Huguenot armies in the way of keeping a good look-out on the countryside is not only stressed by Catholic authors like Montluc and Castelnau, but acknowledged by La Noue, Béza, and Agrippa d'Aubigné.

[2] See his 'First Paradox': 'Qu'un escadron de Reitres doit battre un escadron de lances.' *Discours*, pp. 438–48.

though they might be, these mercenaries were liable to the same fatal fault which they showed in the wars of the Netherlands. They were no enthusiastic Protestants, but professional hirelings, and they went on strike, or even returned *en masse* to Germany when their pay was not forthcoming. The Huguenot war-chest was always precariously sustained, and often empty—it had no grip on the regular revenues of the realm. Bankruptcy was periodical, and this entailed the departure of the reiter. Even German princes of note could not keep their squadrons together in France; William of Orange himself in 1568 saw his army melt down to 1200 men. And in the same year Wolfgang of Zweibrücken, of the Palatine House, led a very considerable force into the heart of France—as far as La Charité on the Loire—but when he died his successor Mansfeldt could not keep it together, and joined the Huguenots for the battle of Moncontour with a much reduced contingent.

If the armies of the French Protestants may be compared in many respects to those of the English royalists, those of the French Catholics had a certain resemblance to those of the English Parliamentarians. They had at their back the old civil machinery of the realm, the *Parlements* and the royal governors of provinces, also, what was most important, the power of taxation on the ancient lines, all the constitutional advantages. Moreover, they could use the royal name without displaying the obvious hypocrisy shown by the English Houses. The all-important Paris gave them a consistent and invaluable support, such as London gave to the Puritans. The strength of their armies lay in their infantry—all the old regiments save that of Languedoc (which broke up) adhered to them. Their command of money made it possible for them to keep on foot, and to pay (which was the important thing) the large Swiss contingents which were furnished to them under the 'Perpetual Peace' of 1520. From the start the King of Spain lent them some of his old *tercios* from the Netherlands; in the first battle of the war, Dreux, there were already 3000 Spanish foot in the Catholic army. And they drew on the Catholic German states for reiters and landsknechts on many occasions, headed by princes like the Rheingrave and the Margrave of Baden. Their infantry often counted twice or thrice as many pikes and arquebuses as that of the Huguenots. On the other hand, their cavalry was not at all superior to that of their enemies—Catherine de Medici once complained that the better half of the noblesse was in the field with the rebels, though the old *compagnies d'ordonnance*, the regular squadrons, adhered to the Crown. But both Protestant and Catholic contemporaries state that the *compagnies* were not what they had been in the old days—the proportion of gentlemen was smaller, that of professional hirelings larger—the morale no higher than that of the Huguenot cavalry, though the discipline might be better. Still the Catholic armies were not such hastily improvised and amateur forces as those of their enemies—the similarity to the English armies of the Parliamentary party is obvious. The 'League,' however, did not supply any troops of such obviously superior quality as the 'New Model' of 1644. And in the chaotic provincial warfare which went on in outlying parts of the realm, the Catholic bands were of exactly the same composition and quality as the Huguenot bands, with which they fought, and exchanged atrocities.

The general desire of the Guises and their supporters was the extinction of heresy; this ambition was always being foiled by the policy of the Crown—*i.e.* of Catherine de Medici—which craved for peace, and so much disliked a predominance of the Guise family that, again and again, a compromise was patched up, when the Catholics would have preferred a war of extermination. It is this, and not merely military considerations, which explains many of the surprises of the Time of Troubles. But one must also take note of the tendency of the generals of the day to waste the fruits of victory in the field on interminable sieges—a tendency which we shall also have to note in the Netherlands. This mistaken policy was not altogether unperceived by the keener minds of the time—as is witnessed by a curious dialogue recorded by La Noue.

When he was a prisoner after Jarnac and Moncontour, he met at Tours the Cardinal of Lorraine—the second of the name, not the victim of Blois. Rather to the prisoner's surprise the Cardinal sent for him, and started a conversation on military matters, 'for he was a prince who was by no means ignorant of them.' 'The Admiral Coligny and your friends,' said the Cardinal, 'ruined themselves by besieging Poitiers, when you had a much bigger chance before you. The royal army at the moment was weak and not yet concentrated; if pushed it would have had to fall right back to Paris. But you gave us time to get together and strengthen ourselves, instead of striking when we were half defeated.' 'Well,' said La Noue, 'this fault of ours ought to warn you never to do the same thing.' 'We will take care not to do so,' said the Cardinal. 'Nevertheless, events showed that to know our fault profited them little. They stumbled over the same stone, wasting their victory on the siege of St. Jean d'Angély, which went on for two months. That siege in a way cancelled the effect of Moncontour.' [1]

The greater number of the general actions in the French Wars of Religion were 'cavalry battles'; when the cavalry of one side had beaten that of the other out of the field, the infantry belonging to the routed cavalry was generally surrounded and cut to pieces. This was no doubt largely the result of demoralization—the foot regiments felt themselves deserted, and broke or yielded. This we shall presently see in the Netherlands at Gemblours, Tournhout, and Nieuport. There were exceptional cases, as at Dreux, where the victorious cavalry wasted itself in pursuit, and the infantry of the other side held its ground, and remained in possession of the field, though it was incapable of turning success to a decisive victory. Something of the same sort was seen in the English civil war at Edgehill, where the reckless dispersion of the Royalist cavalry enabled the Parliamentarian infantry to hold its own, and to push the King's infantry up the hill. But, like Dreux, Edgehill was not a real victory: and Coligny after Dreux, like King Charles after Edgehill, retained his liberty to manœuvre, though his army had been thinned.

The proportion of horse to foot in these wars was always heavy—much more so on the Huguenot side than on the Catholic side. In battle order the game usually commenced with cavalry charges—the Huguenots in particular hoping to make short work of the hostile infantry, if they had beaten the gendarmerie off the ground. The Catholic order of battle was usually in masses of infantry, with cavalry interspersed. If the ground permitted, the front was covered by swarms of arquebusiers placed behind hedges or ditches, *enfants perdus*, who had to run for it if the enemy attacked in force, but who could do some damage if he did not strike hard, or if there were points, *e.g.* spinneys, or houses where they could get cover, and avoid being ridden over. In some battles the infantry on both sides hardly got into the thick of the fighting, *e.g.* at St. Denis, which was almost entirely a cavalry battle—the Huguenots having only engaged a few hundred arquebusiers, while

[1] *Discours*, pp. 993–5, shortened to save space.

the Catholic infantry hung back, and for the most part retired without coming into action. This was an exceptionally ill-fought battle on both sides, as will be shown in the proper place. It is hard to find in history more perfect examples of military incapacity continued over a considerable period of time than those of Montmorency and Condé, both of whom were regularly beaten in action. At Dreux, by a comic and unique instance of righteous nemesis, each was taken prisoner by the other's army, because both chose to act as cavalry brigadiers rather than as responsible commanders-in-chief.

It was notable throughout the war that the Huguenots were generally short of pikemen to make their infantry array solid, while the Catholics had always a strong Swiss contingent, also the remains of the old royal foot-regiments, of which pikemen were a necessary portion. The Huguenots occasionally produced the pikes of landsknecht mercenary regiments—which had all the faults mentioned above; but when the Germans were not forthcoming their provision of pikes was always insufficient. Sometimes indeed they had none at all—as in the case of the side-show near Riberac in 1568, where the Huguenot arquebusiers had to make what d'Aubigné calls a 'bataillon de parade,' *i.e.* the mere show of a fighting formation, because they had no pikes to keep off cavalry, and were ridden down wholesale.

It is most difficult to calculate the strength of the armies of this period, for though sometimes the chronicler gives round and comprehensible figures—which are nevertheless not to be trusted—in most cases we have merely lists of units—so many 'cornettes' of horse or 'enseignes' of foot, and these formations varied hopelessly in strength. The only definite fixed standard was that of the old *compagnies d'ordonnance*, 50 lances or a very occasional 'double company' of 100. At the battle of St. Denis, Condé had 1500 horse divided into 18 cornets, but some were 100 strong, others down to 45.[1] It is impossible to

[1] See the curious evidence of a prisoner in d'Aubigné, iii. p. 215. 'There are 18 cornets supposed to average 80 troopers apiece. In my own there were 75: some are much larger; but the one supporting mine had only 40 or 45. The ranks behind the flag of this unit are filled up with men who have only buff coats and pistols; of gentlemen wearing cuirasses and closed helms, and with horses worth 50 gold crowns, there are, except in the very largest cornets, not more than ten or a dozen. These are called the "gens de combat," and these decide the day.' La Noue makes similar assertion about the rear ranks of a German reiter squadron, which were poor stuff compared with the gentlemen or old soldiers who rode in front. Hence the inferior effect of the 'caracoling' when the turn of the rear ranks came.

make accurate deductions from narratives stating that an army had 20 ' good ' cornets, or 20 ' poor ' ones. Occasionally the unit raised and led by a popular officer ran up to 200 men—more frequently five weak cornets might make no greater show.

The same was the case with the infantry—the ' enseigne ' was somewhat larger than the cavalry ' cornet,' but like the unit in the other arm might vary indefinitely. It was often about 200 strong, and this might be normal in a well-equipped army, but very frequently it seems to have had little more than 100 pikes and arquebusiers. The latter were always far more numerous than their comrades with the long weapon, except in German or Swiss corps : even in the French ' old ' regiments surviving from the time of Henry II, the arquebus predominated. In the new and transitory units, raised locally, the pikes were often a negligible quantity. French authors never tire of insisting on the fact that firearms and incessant activity suited the spirit of their countrymen, and that pike service was more congenial to stolid Swiss and German.

As in the old days of the Italian wars, it was still quite uncertain what a ' regiment ' might mean, save in the case of the old regular infantry. When a local commander had raised a good number of ' enseignes,' he might form them into a regiment—*e.g.* we hear in Dauphiny of a ' fine ' regiment of eighteen ' enseignes ' or 2000 men. But a very much smaller number of ' enseignes ' might be made into a regiment—*e.g.* if two local commanders joined forces, each with only seven or eight of these companies under his flag, they would certainly not be amalgamated into one corps, from reasons of personal and local *amour propre*, but would work as two small regiments. On the other hand, an official commander-in-chief, Catholic or Huguenot, who had gathered together a large number of unattached units, might enregiment them for convenience, with little regard to the wishes of mere captains—though tact would have to be employed, or desertion and perhaps mutiny might ensue. Strangers and old neighbours were equally ' touchy ' about being placed under the command of an officer whom they might consider no better than themselves.

Altogether the armies of the Huguenots were assemblies of amateurs, all through the early period of the war—though there were old soldiers like Coligny and Dandelot among the commanders, and a sprinkling of veterans in the ranks. But the mentality of the noblesse predominated for many a year, courage and indiscipline were fatally connected. They were always coming to grief, says one contemporary, because they had neither *maréchaux de camp* nor *sergeants* to keep them in order. By the former term was meant technically trained superior officers—the term in later years was sometimes equivalent only to a colonel of many companies, but came to mean something like a brigadier, and in the seventeenth century is found as a practical equivalent for major-general. By 'sergeants' were meant not petty non-commissioned officers, but the sergeant-majors of those days, commissioned officers charged with the discipline and internal organization of a corps, who afterwards came to be called majors only, and had rank immediately after the colonel.[1] The remark made above really meant that a Huguenot army lacked professional officers of high rank conversant with the moving of troops, and also regimental officers to look after the general discipline of a hastily compacted unit, and see to such matters as billeting and the setting of piquets.

Sieges, though frequent enough in the history of the Huguenot wars, are not so all-important as in the wars of the Netherlands. All French towns of any importance were, of course, walled; but few of them had the advantage of strong water-protections, which was so decisive in the Netherland wars. Indeed, La Rochelle was almost the only place which could be compared to a typical Dutch stronghold. The cities of the inland were still in 1560 mainly dependent on their mediaeval *enceinte* walls, but the frontier-fortresses and the ports had been brought more or less up to date during the long wars of Francis I and Henry II. Marseilles, for example, had its fortifications rebuilt, and Havre was a new-style fortress, built by Henry II to supersede Harfleur and Honfleur, the old guards of the Seine estuary. Naturally attention had been directed to the frontier-fortresses of the northern border—Verdun, Metz, St. Dizier, St. Quentin, Peronne, Mouzon, Toul, Hesdin, Landrécies, several of which were entirely rebuilt after celebrated sieges which had revealed their original insufficiency. Navarrens on the Pyrenean frontier was also a

[1] See the note on 'majors' and 'sergeant-majors' in the chapter on the English army under the Tudors, pp. 377–8. Brantôme says: 'le sergeant de bataille est à cheval pour aller par les rangs, par le devant, par le derrière, ou par les côtes, afin de mettre promptement ordre à ce qui est necessaire.' (*Discours sur les Colonels*.)

modern place. Francis I moved the whole town of Vitry and reconstructed it on better ground, after a defence which had been gallant but unfortunate in 1544. Calais, too, had been largely reconstructed after its capture by Guise in 1558. Boulogne, when taken over from the English by treaty in 1550, was found to have been much improved by them, particularly in the way of outlying redoubts.

But the large majority of the inland towns of France had still no more than mediaeval walls. When the Wars of Religion broke out, they were forced to improvise new defences. In the intervals of pacification governors rebuilt whole enceintes, when they had time and money, as Tavannes did at Dijon, with outlying intrenchments and palisades, redoubts placed on rising ground too near the old enceinte, glacis contrived around the old ditch, and the building up of earthworks, with platforms for cannon, against the back of the original ramparts. The more frequent hasty patching up of antiquated places of strength was sometimes called 'fortification à l'Huguenotte,' because in most cases it was the rebels who made their towns of refuge into formidable strongholds, which could offer long resistance to the artillery of the day. These extemporized improvements were often very effective, and led to sieges of great length, when the burgher-guard (like the Dutch) were fighting for hearth and home. Such was the case at the already mentioned siege of Sancerre in Berry, where a town-levy, commanded by an advocate, held out after the Massacre of St. Bartholomew, in the extremity of starvation, till it obtained the benefit of the Pacification of La Rochelle (August 1573).[1] Many less notable instances of obstinate resistance might be granted in each of the successive 'Times of Trouble' which fill up the intervals of semi-peace between 1562 and 1596. The case of La Rochelle itself after the St. Bartholomew Massacre is the most notable. In a way it was this long-maintained obstinate local resistance by the sectaries which brought the wars to an end, for the enemy had finally to admit that the Huguenots were unconquerable, though the Huguenots had to surrender their dream of 'smiting the Amalekites' and placing a Protestant king on the throne of France.

[1] The gallant lawyer-governor Joanneau was privately murdered by enemies after the hostilities had ended, but the survivors of the citizens got their 'liberty of conscience,' though their well-defended walls were ordered to be breached.

CHAPTER II

THE BATTLE OF DREUX (December 19, 1562)

THE Catholic 'Triumvirate'—the old Constable Montmorency, the Duke of Guise, and the Marshal St. André—had on March 29 got possession of the royal family—much to the disgust of the Queen-Dowager, who in 1562, as always, wished to pursue a policy of her own of an opportunist sort, and to avoid war. But the Triumvirs, having her and all her children in their hands, could, and did, use the royal name effectively. On April 2, Condé and Coligny seized Orleans at the head of a large party of Huguenot gentlemen, and sent out their appeal to their co-religionists throughout France. It was followed by sporadic risings all over the country, and many large towns such as Lyons, Rouen, and Bourges fell into the hands of the insurgents. Their triumph was accompanied in many places by gross vandalism in the way of destruction of shrines, statues, and historic relics—even the monument to Joan of Arc at Orleans was thrown into the Loire. The Catholic party was naturally infuriated, but insincere negotiation went on for some time before the actual outbreak of war, when on July 4 the Huguenots stormed Beaugency, the nearest Catholic stronghold to their headquarters; and on the same day their enemies seized Blois.

For three autumn months there was irregular fighting, accompanied by countless atrocities on both sides, all over southern and western France. On the whole the balance of success lay on the side of the Catholics, whose greatest exploit was the storm of Rouen on October 26.[1] The Huguenots, however, did not regard their cause as lost, as they had success-

[1] Here was mortally wounded on October 13 the worthless Antony of Navarre, the father of Henry IV, and the head of the house of Bourbon. He was a thriftless waverer, who was at this moment serving with the Catholic army. His death was of importance, as throwing the control of the Bourbon power into the hands of his very Protestant widow, Jeanne d'Albret, who brought up her son Henry to be the head of the Huguenot party. It was only much later that it became evident that he would also be heir to the French crown.

fully won over to their side important foreign allies. They had hired, with the help of the Elector Palatine, a considerable force of reiters and landsknechts, who were coming up from the side of the Rhine, commanded by the Marshal of Hesse, and guided by Dandelot, the brother of Coligny. Also, what was equally profitable, they had bought the alliance of Elizabeth of England by the Treaty of Hampton Court (September 20), by which they handed over Havre to her, to be held as a pledge for the restoration of the much-regretted Calais. The Earl of Warwick took over 3000 hastily-levied English troops to Havre, and £100,000 pounds in gold. Both further soldiers and further subsidies were to be forthcoming ; but it was too late to save Rouen, which fell not long after Warwick's landing.

The Huguenot chiefs at Orleans did not feel themselves strong enough to move, even to save Rouen, till they had been joined in the first days of November by the large auxiliary force from Germany which Dandelot brought to them, after having eluded an attempt of the Marshal St. André to intercept him in Champagne. They had also picked up a few thousand south-French Huguenots, the wrecks of several beaten contingents, which had been brought up to them by the Count of Rochefoucault. Having now some 4000 horse and eight or nine thousand foot, they took the field, marched on Paris, and challenged the Catholics to a battle. The enemy would not come out—they were expecting the arrival of 3000 Spaniards from Flanders, and of forces from the south—the troops that had been of late maltreating the Huguenots of Guyenne.

On December 9, Condé and Coligny abandoned their obviously futile demonstration against Paris, and moved off towards Normandy, intending to pick up the English men and money that Warwick had brought to Havre. The roads and the weather were bad, and they wasted some days in trying to reduce Chartres, and in devastating the countryside on the way. On the morning of December 19 they were marching on the high road Chartres–Dreux–Rouen, with some hope of seizing Dreux, when they found the Catholic army deployed across their route, between the villages of Epinay and Blainville, three miles south of Dreux. We are told that they had neglected all distant reconnaissances, and had no notion that the enemy was so near. The Catholics, on the other hand, were guided by the old Constable, who knew something about the usefulness of scouting, and was perfectly aware of the position of the enemy. On the previous night he had heard their drums beating on the Chartres road, and had discovered their camp. Secure of being undisturbed, he took his army across the Eure, which ran between the two armies, by two easy fords south of Dreux, and was comfortably in position to await Condé's approach on the following morning. His line was formed across the high road, with the right wing resting on the village of Epinay, and the left on the village of Blainville. Outside each of these places there were thick woods. The Bois de la Marmousse extends west of Epinay, with the little river Blaise beyond it. The Bois de la Place and Bois de Chanteloup touched Blainville, with the more important stream of the Eure on their eastern side. Thus the position could not be turned, having thick timber on each flank: and the only ground open to attack was the space of 2500 yards between the two villages, on each side of the high road. Montmorency had placed his modest provision of artillery partly in front of Epinay, partly across the high road just west of Blainville.

We have various details about the strength of the Catholic army, which are not quite easy to reconcile. On the one hand, we are told that they had little more than 2000 cavalry, but at least fifteen or sixteen thousand infantry: on the other hand that they had 42 companies of heavy and 8 of light horse, with of infantry 44 'enseignes' of French, 28 of Swiss, 10 of German, and 13 of Spaniards. Now giving 50 lances apiece to the gendarme companies—of which we know that some at least were 'double companies,' *e.g.* those of Guise himself and of Montmorency—there must have been much more than 2000 heavy cavalry, besides the rather negligible 500 or 400 light horse. The whole must have been over 3000 strong. As to the infantry, 95 'enseignes' at 200 apiece would make up more than the 16,000 men spoken of by Catholic narrators—the Huguenots would count them as more like 20,000.[1]

Montmorency was in command—Guise, though the greater captain, had refused to take up any charge, for political and personal reasons: he was both anxious to conciliate the old Constable, and reluctant to assume any responsibility, when co-operating with a general of such well-tried incapacity. Mont-

[1] D'Aubigné gives the figures quoted above, Castelnau, a well-informed contemporary Catholic, gives 35 companies of gendarmerie, 8 of light horse, 36 'enseignes' of French foot, 22 of Swiss, 14 of Spaniards, and 10 of German mercenary lands-knechts—82 'enseignes' in all. De Thou copies Castelnau exactly. Davila gives no detailed figures.

morency had divided the host into two, not into the usual three, corps: he himself had the 'main-battle,' the Marshal St. André the 'vaward,' in which Guise appeared in a very minor capacity, as captain of his own *compagnie d'ordonnance* and leader of his personal retainers. The 'vaward' was composed of 17 companies of gendarmerie, twenty 'enseignes' of French foot, the 14 Spanish companies, and ten 'enseignes' of mercenary landsknechts. This French infantry was largely composed of the 'old bands' of Piedmont, brought up from the south within the last few days. The Constable's 'main-

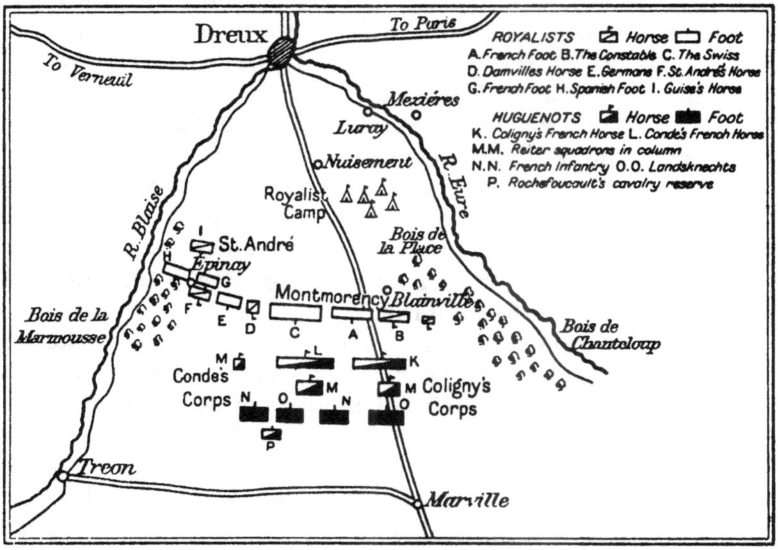

Battle of Dreux, December 19, 1562.

battle' had 18 *compagnies d'ordonnance*, the eight companies of light horse, and for infantry the largest unit in the army, the 28 (or 22) companies of Swiss, with 20 (or 16) companies of French infantry, Picards and Bretons for the most part. There were 22 cannon with the army, of which the Constable had 8, the Marshal 14.

The array in which the Catholics fought was a single line, with very trifling reserves. The Constable occupied the ground from Blainville to within a short distance of Epinay. On his extreme left was Sansac with the companies of light horse, then the Constable himself with his gendarmerie, then the village of Blainville, which was barricaded and had light guns in front of it: then came the mass of French infantry, and ending the line of the 'main-battle' the solid phalanx of 5000 or 6000 Swiss pikes.

In St. André's wing some companies of gendarmerie under Damville, the Constable's son, were next the Swiss: then came in succession the 10 German foot companies and the twenty French foot companies: between them was St. André himself with the main body of his gendarmerie, and on the extreme right the Spanish infantry. The greater part of St. André's troops were not in the open, but screened by the houses of Epinay, which lies on high ground and was barricaded, and by the wood adjacent.[1] Guise with 200 horse, including some noble 'volunteers,' was in second line or reserve, behind the village. We are told that by his advice the Marshal kept his men well under cover, the cavalry dismounted, the infantry with pikes lowered, so that among the houses and trees they were almost invisible. Indeed one authority states, probably in error, that the Huguenots never discovered them till the battle was well begun, and made no allowance for their existence. The guns of St. André's wing were placed east of Epinay, bearing on the high road.

On the morning of the 19th the Huguenot army broke up from its camps at Ormoy and Neron, and marching towards Dreux suddenly, after passing the village of Marville, ran into the unexpected enemy in position. The Constable's artillery fired a few long shots, which dispersed the screen of light horse which headed Condé's column of route, and compelled the leading squadrons of his main body to seek shelter in a dip of the ground. The whole army halted, and the Prince and the Admiral rode out to the front, to reconnoitre the hostile lines, which were not too clearly visible. We are told that Coligny made up his mind that the Constable would act on the defensive only, and would not quit his position. He advised that the Catholics should not be attacked, but that the army should turn off by the Treon road, and march westward, avoiding any general action. He so far prevailed that the baggage of the army, and all the guns save five light pieces, were set off in this direction. Meanwhile the Huguenots deployed from column of march into a fighting front, in case the enemy should take

[1] We are told that Montmorency's corps was so strong that it filled up almost the whole space between Blainville and Epinay, so that St. André's corps showed little of its front.

the opportunity of making a sudden attack on an army moving across his front and exposing its flank. Condé argued that the two hosts were so close that an action was inevitable, and that if the Huguenots swerved off towards Treon, the Constable would descend and fall upon them. In this view he was right—the fatal mistake had been made when the army was allowed to march blindfold into close contact with the enemy.

Meanwhile the Huguenots deployed, facing the Constable's army. Like the enemy they were in two corps, but they had a second line, and reserves. The Admiral with the 'vaward' faced Montmorency, Condé with the 'main-battle' should have faced the corps of St. André, but as a matter of fact only placed himself opposite its eastern third. Detailed narratives from the Huguenot side say that Coligny had some 400 French heavy cavalry and four companies of German reiters—say 1000 pistoleers—with eleven 'enseignes' of French and six of German infantry—some 3500 foot it would appear. On the right Condé had slightly larger numbers—500 gendarmes, six companies of reiters, then 400 argoulets and horse-arquebusiers, twelve 'enseignes' of French and six of German infantry. The whole army made something near 4000 horse and 7000 or 9000 foot.[1] La Noue, an eye-witness, says that the Catholics were about as strong as the Huguenots in the mounted arm, but had thrice as many foot. This would seem to be a not unnatural miscalculation of numbers by the beaten side. Probably the Catholics were decidedly inferior in cavalry, and rather more than double their enemies in infantry. The German reiter-squadrons were much stronger than French 'cornettes' or 'companies' of horse, and appear to have averaged over 200 apiece, as ten of them seem to have amounted to 2400 troopers. The Marshal of Hesse had brought 3000 to Orleans in November—normal wear and tear would have reduced them to 2400 by the 19th December. Similarly the German foot companies were much larger than the French, as twelve of them seem to have made up 4000 men, while 23 French 'enseignes' came to only 3000—little over a hundred apiece.[2] It will be seen that more than half the Huguenot army consisted of these foreign auxiliaries, who

[1] Castlenau, a Catholic witness, rates the Huguenot infantry at this figure, which is the same as that given by Huguenot narrators. Tavannes says that the Admiral and the Prince had each 1500 French infantry.

[2] Throckmorton, the English diplomatist, who was with the army, says 7000 foot in all, the 3000 French very ill equipped and armed.

seem to have numbered about 7500 in all, as against 4500 native French.

The two armies being in battle order, with a mile of 'no man's land' between them, there was a pause of nearly two hours about noon, a fact commented on both by La Noue and Mergey from the Huguenot side. The former attributes it, with his usual psychological insight, to the fact that this was the first battle between French armies for a century, that every one had friends or brothers in the opposite ranks, and that there was a noble reluctance to shed the blood of kinsmen, and to open the first act of an inevitable tragedy. More materialistic observers opined that the Constable did not want to come out of his strong position, and that the Huguenots were thinking of the perils of the first blast of artillery fire that they would have to endure, if they took the offensive, before they could get to handstrokes with the enemy.

After two hours, about noon, spent in this indecisive fashion, Condé came to the conclusion that the Constable would not 'come out of his fort' on any condition, and resolved to follow Coligny's original advice, and to move westward to get on to the Treon road, following the artillery and baggage.[1] Orders were sent along the line to turn to the left, and march, still in battle-order, in that direction. But the Huguenots had not gone 200 yards when they saw the Constable's line begin to come downhill, to fall upon their flank as they moved. Whereupon they faced to the front once more, and commenced the general offensive advance which had been Condé's original intention. There was a long clash from Blainville westward as the Protestants charged, but not in and about Epinay, where the bulk of St. André's corps was lodged, for Condé did not attack the entrenched village or the artillery beside it, but left in observation only the infantry of his column and two squadrons of reiters as a detaining force. But he attacked, and this was a novelty, in two lines—French cavalry in front, reiters in support, infantry with a small cavalry reserve [2] in rear.

In all the eastern half of the battlefield the Protestant charge was completely successful: the Admiral's French horse, supported by his reiters, rode down the Constable's cavalry, which must have been somewhat inferior to them in numbers.

[1] Mergey, p. 63.

[2] One hundred French lances under Rochefoucault, who stood by the landsknecht infantry.

Montmorency, fighting in the front line in spite of his seventy years, was unhorsed, slightly wounded, and taken prisoner. The gendarmerie, under the Prince of Porcien, turned on the mass of French infantry and on Sansac's light horse about Blainville, and completely routed them, capturing the village and the guns in front of it. The Admiral's horse then seem to have got out of hand, many pursued the flying Catholics as far as their camp on the banks of the Eure, where they took much plunder, including the baggage and silver plate of the Duke of Guise. It was some time before Coligny could collect a nucleus of rallied men, with which to intervene in the struggle on the other wing of the battle. Of what became of his 3000 infantry we have, oddly enough, no information: possibly they co-operated in the defeat of the Constable's companies of French foot: but they are found, when next we hear of them, not far from their original position in front of Blainville.

The casualties among Montmorency's following had been very heavy: we have a list of many notable persons killed and wounded. The young Duke of Nevers received a pistol-bullet in his head and died of it.[1] The Count of Rochefort, Annebault, the son of the general of the old Italian wars, and Givry were killed outright. Some of the fugitives, largely from Sansac's light horse companies,[2] fled as far as Paris, where they spread the news that the Constable was taken and his whole army routed. Whereupon the Queen-Dowager, who was ready to take things as they came, is said to have observed, 'In that case we shall have to learn to say our prayers in French.' One of her letters mentions that she was for twenty-four hours in ignorance of how the day had really ended.

Meanwhile on the western side of the battlefield things had gone differently. Condé had facing him the big phalanx of Swiss infantry, and to right of it St. André's whole corps, largely screened by the fortified village of Epinay. Apparently he made up his mind that the proper tactics were to break down the Swiss, the best soldiers in the Catholic army, and then, having pierced the hostile centre, to fall on St. André from flank and rear. All Protestant critics blame him for wasting his

[1] According to d'Aubigné and other authorities, he was shot by the clumsiness of his mentor the Sieur Desbordes, who got himself killed on seeing what he had done. Nevers is said to have been a concealed Huguenot, and to have bitterly regretted taking the field against his conscience. Some say that Desbordes shot himself—others that he ran headlong into the mêlée to be killed at all costs.

[2] See Vielville's *Memoirs*, viii. p. 53.

whole strength in this way, and leaving the hostile left wing in and about the village unmolested. All that he did was to place opposite it some of his foot and a couple of reiter squadrons.

The honours of the day were to remain with the Swiss. Condé first charged them in front with the gendarme companies of De Muy and Avaret, who cleft a way right through the column and came out on the other side, without causing it to break up. The Prince then brought up the main body of his horse, both French and reiters, and charged the Swiss in flank. They were forced to recoil, and left the artillery, which had been in their charge, to be captured. But they held together, only falling back perforce towards St. André's line. The Marshal sent out to relieve them part of his gendarmerie, three companies under Damville, son of the Constable, and Aumâle the son of Guise. But this succour failed—the gendarmes were routed by two squadrons of reiters, whom Condé detached against them. Aumâle was unhorsed and wounded, Montbéron, the Constable's youngest son, was killed. The remains of the companies fell back and rallied upon Guise's body of reserve.

Condé then seems to have lost his head—somewhat like Ney during the great French cavalry charges at Waterloo. He resolved to break the Swiss at any cost, and failing to do so with the troops already in hand, sent in the six landsknecht foot-companies which had hitherto been observing St Andre's front. When the Swiss saw them, their old national hatred for landsknechts blazed up, and so far from recoiling they advanced to meet them, and to the disgust of Condé, drove them off with ease. In wild rage the Prince then ordered his last cavalry reserve, the 100 lances of Rochefoucault, to charge in—they had no better success than the others—' il est malaisé d'enfoncer tels herissons,' writes Mergey, one of Rochefoucault's followers.[1]

All the Huguenot units had drawn back from this much battered but invincible phalanx of pikes, when Guise advised St. André to charge at last, as the time had now come. The Marshal had suggested to him, more than once, that it had arrived already, but the Duke put him off till he had seen the whole of the Prince's cavalry used up. He then came out of the woods with 200 gendarmes in his centre, 500 arquebusiers, *enfants perdus* in front of them, the Spanish infantry on his

[1] *Memoirs of Mergey*, p. 63.

left and the French infantry of the 'old bands' on his right. St. André charged at the same time with what he had left of cavalry, and the ten companies of Catholic landsknechts. Guise's move out of the wooded position came as a sort of flank attack on the Huguenot line, where there was nothing left save a reiter squadron or two and the two squares of Rohan-Fontenay's south-French infantry. Attacked by horse and foot, these raw troops broke at once—most of them were cut to pieces, their colonel was one of the few who escaped. Guise and St. André then fell upon the German foot companies who had just been beaten off by the Swiss, and routed them with unexpected ease. 'These were the most cowardly lot of landsknechts who came into France during the forty years of war,' says an angry Huguenot historian, 'though they looked fine men enough.' The main block of them threw up their pikes and surrendered, on promise of their lives. The scattered and exhausted squadrons of Condé's gendarmes and reiters had little fight left in them when they saw all their infantry collapse. The Prince in vain tried to rally them, failed to gather more than a handful, and charged with this into the front of Damville's gendarmes; his horse was killed and he himself taken prisoner. It was the act of a hero of romance, not of the general of an army.

For the battle was not nearly over — though Guise and St. André thought for half an hour that they had no more to do. But between four and five o'clock, as December dusk was falling, they saw a large body of cavalry coming in upon them from the south. At first, it is said, they thought that it was reiters coming in to surrender, as the landsknechts had done. But when the horsemen got close the white scarves became visible, and they found that they had to fight the Admiral.

Coligny was collecting his scattered horsemen when he received news that matters had gone badly on the other wing. He appointed a general rallying-point behind a long hedge somewhere south of his original position, and there got together some hundreds of gendarmes and reiters; he was joined by many of those who had made off when Condé's line collapsed. A very interesting narrative by one of Rochefoucault's men says that about thirty or forty of his *cornette* turned up behind the hedge, and found many others dropping in. Rochefoucault, the Prince of Porcien (who had commanded the gendarmes in Coligny's wing), and the Admiral himself had each got together some hundreds of men, part reiters, part French.

The gendarmes, we are told, had for the most part broken their lances, and had only their swords. The numbers of the rallied horse are given at different figures by each Huguenot witness—apparently there may have been 1000 reiters and 300 gendarmes.[1]

Having arrayed this force in three troops, Coligny brought them out in good order against Guise and St. André, who were taken somewhat by surprise, and had to form a new front in and about Blainville. The fighting lasted into the dark hours, and was most confused. The Huguenots rode down both Guise's and St. André's horse. Des Brosses, the Duke's lieutenant, was killed, St. André taken prisoner—to end villainously a few minutes later. But Coligny's horse could not dispose of the hostile infantry, who were largely under cover among houses and trees, and held their own. After many vain attempts to dislodge them, the Admiral had to give the signal for retreat, which was made in good order, with a proper rearguard. Guise made little or no attempt to pursue, and Coligny got off, and encamped at Neuville, only three miles off the battle-ground—bringing with him two of the five cannon which the Huguenots had brought to the field and a number of captured standards. He was joined there by some relics of his infantry. In a dispatch, which he wrote a few days later, he claimed that he had only lost 140 horse, 2200 foot, with 1500 landsknechts who had basely surrendered. This undoubtedly was an understatement of his casualties. But the Catholic allegation that the enemy left 6000 men on the field was ano verstatement equally incorrect. Retaining complete control of his movements, Coligny sent his surviving infantry, under his brother Dandelot, to strengthen the garrison of Orleans, taking with them the captured Constable, and rode with his cavalry into Normandy, where he picked up some local levies, and captured Caen and several other places. The battle had been by no means decisive; and the losses on the two sides were probably not unequal.

Guise, who by the capture of Montmorency and the death of St. André,[2] found himself the only surviving 'Triumvir,'

[1] So De Thou. Mergey, who was present, says only 700. D'Aubigné gives 800 reiters and 250 gendarmes.

[2] St. André's death is told shortly by several narrators—at great length by Vielville on one side and d'Aubigné on the other. All agree that he was killed after surrender by a man whom he had wronged in the past. According to the long version the father of Baubigny of Mezières, a very wealthy man, had placed his son in St. André's household, and gone surety for a number of bonds for which the Marshal was respon-

assumed command of the army, and sent arrogant letters to Paris, in which he intimated to the Queen that he had distributed honours and offices to those who had served well, and had taken over responsibility for the war. Catherine thought that she had fallen into the hands of a dictator, but was freed from her fears in a few weeks. Guise had laid siege to Orleans, instead of following Coligny into Normandy. While he was riding behind the siege-works on February 18, with a single companion, he was shot by a gentleman named Poltrot de Méry, a fanatic who had deserted from the Huguenot ranks a few days before, pretending to have abandoned the rebels. Apparently he was a deliberate assassin, who came over with the one object of killing Guise, just as Balthazar Gerard, who shot William the Silent, presented himself at Delft as a refugee with the sole purpose of murdering that prince. Poltrot escaped for a moment by the swiftness of his horse, but—as Catholics opined by the just vengeance of heaven—rode in a circle all night on cross-roads, and was captured next morning almost on the spot where he had shot the Duke. Tortured, after the fashion of the times, he first said that he had been suborned by Coligny and the preacher Béza and then, when racked harder, declared that the Queen Mother was at the bottom of the matter. He withdrew the charge, then repeated it, and then withdrew it again, contradicting himself twice. When on the scaffold he exclaimed that he would do the deed again gladly if it were possible. There is no reason to suppose that either Coligny or Catherine, whom some have suspected as an instigator, had any real connection with the affair; Poltrot was a homicidal political maniac like Balthazar Gerard. But on the principle of *cui bono* the Queen was the person who got most profit.

She induced the two great prisoners, Montmorency and Condé, to get into negotiations for a pacification, for which each was naturally eager, and by their aid patched up a reconciliation, which gave the Huguenots most of the items of toleration which they claimed. Hence the 'Peace of Amboise'

sible. Some years later St. André, always a spendthrift, failed to repay the money when Baubigny kept pressing for it: hence friction, in which Baubigny killed, in a duel or a scuffle, one of the Marshal's squires. St. André, who was then all-powerful with the King, got Baubigny tried for murder, executed in effigy, and stripped of his estates. When he was taken prisoner by his enemy, and had handed over his sword and helm, the Prince of Porcien rode by: the Marshal tried to surrender to him, in order to get into less unkindly hands. But the Prince said, 'Every man's prisoner is his own,' and passed on. Whereupon Baubigny cried, 'Always trying to defraud me,' and shot St. André through the head.

(March 1563)—which was to hold for more than four years. The person who came badly out of the business was Elizabeth of England: for when Catholics and Huguenots were reconciled they attacked Havre in common, and forced the Earl of Warwick to surrender (July 28, 1563). His garrison had caught typhus fever, and a fleet sent for his succour had been wind-bound in the Channel. Elizabeth never forgot the way in which she had been tricked by the Huguenots, by whose aid she had intended to get back the old stronghold of Calais in exchange for Havre. In late years she often sent help to the Protestant party in France, but always with a conviction that they might play her false—in which she was not far wrong. Henry of Navarre was a far more unscrupulous person than Coligny or Condé, whose exculpations for their conduct ring most false! They had only French patriotism to plead for disowning their bond—a plea not likely to appeal to Elizabeth.

De la Noue, philosophizing in his usual acute style, says that there were seven things to note about the battle of Dreux:

(1) It opened without any preliminary skirmishing, which was due to the reluctance of the combatants to start a civil war.
(2) The resistance of the Swiss infantry passed all heroism seen in other wars, and settled the day.
(3) The extraordinary self-restraint of Guise, in holding back so long, and not striking till Condé's horse was absolutely exhausted, was very notable.
(4) The battle lasted five hours—well into the dark. Usually battles were settled in a short clash in these wars.
(5) That the two commanders-in-chief were both taken prisoners is an incident unique in history.
(6) The army that lost the battle retreated in good order, unpursued.
(7) The exceptional courtesy with which Guise treated his prisoner Condé, whom he invited to supper, and harboured in his own bedroom at Dreux, instead of putting him in chains, was a testimony to the Duke's essential magnanimity, which even enemies were forced to recognize. The only parallel was the treatment of John of France by the Black Prince after Poitiers in 1356.

One might add a few of one's own glimpses of the obvious, namely :

(1) Commanders-in-chief should not act like cavalry brigadiers.
(2) It is criminal for an army with superior cavalry not to know that an enemy is within an easy day's march.
(3) Steady infantry can win a victory, but cannot utilize it when the enemy has a superior cavalry force still in the field, and can block pursuit.
(4) Reserves of both arms are necessary; simple linear tactics dangerous.

CHAPTER III

BATTLE OF ST. DENIS (NOVEMBER 10, 1567)

BETWEEN the Peace of Amboise (March 19, 1563) and the outbreak of the 'Second Troubles' in September 1567, there was a period of more than four years of ostensible pacification, during which Catherine de Medici succeeded in maintaining her position, playing off against each other the two Catholic factions headed by the old Constable Montmorency and the Cardinal of Lorraine, who had succeeded his murdered brother as the head of the house of Guise. She persuaded both that her policy was definitely anti-Protestant, as most undoubtedly it was, and so held them back from open attacks upon her predominance—which was now screened by the arrival of her son Charles IX at the year of his nominal majority. But the young King was really still at her disposition—it was only some time later that he began to show signs of an ill-regulated desire to assert his own personality. During the four years of nominal peace the concessions made to the Huguenots were very imperfectly observed, and provocative incidents were continually vexing them. But their main apprehension came from the ostentatious friendship which Catherine was displaying for Philip of Spain, the husband of her eldest daughter. And if they had known all that was discussed between the Queen-Dowager and the Duke of Alva at the conferences at Bayonne, they would have found all their suspicions confirmed. The passage of Alva's army along the frontier of France, through Savoy, Bresse, and the Franche Comté in 1567, seems to have been the final cause of the second Protestant rising, which had as its object nothing less than to kidnap the young King, then at Meaux with his court, and to get possession of his person and his authority. One is reminded of the Conspiracy of Amboise of seven years before.

On Michaelmas Day 1567, Condé and Coligny, having secretly collected 500 gentlemen of their party, tried to surprise the court, but were foiled by the arrival at Meaux of a body of 5000 Swiss, newly hired by the Queen, who escorted Charles IX, his mother, and his brothers into Paris. But the kidnapping plot was only part of their plans—on the same day Huguenot bands took possession of Orleans, Auxerre, Vienne, Valence, Nîmes, Montpellier, and Montaubon. Civil war had broken out again, and from many directions small bands of cavalry flocked in to join Condé before the walls of Paris, to whose gates he had pursued the King and his Swiss escort.

On October 2, Condé established his headquarters at St. Denis, waiting for reinforcements, while the old Constable, who had assumed command of the Royalists, kept quiet within the walls of Paris with the same object. A whole month was spent in this curious deadlock; meanwhile sporadic fighting of a minor sort went on all over France, and absorbed a great part of the succours which each side was expecting. All through October vain negotiations went on—the Huguenots demanding sincere compliance with the terms of the old treaty of 1563, and many additional concessions, such as equal partition of office between members of the two religions, and the expulsion of the Queen's Italian favourites and ministers. All this was purely dilatory—they were really waiting for the arrival of a large corps of German reiters, which was being raised for them by their zealous co-religionist the Elector Palatine, and was due to arrive about Christmas. The Constable was equally inclined to await the approach of a Spanish contingent from Flanders, which Alva had promised him.

Presuming too much on the procrastination of the old Constable, Condé began to spread his none too numerous force all round Paris, detaching Coligny's brother Dandelot with 800 horse and 2000 foot to the other bank of the Seine, to a distance from which they could not easily be recalled, and another force under Montgomery [1] to seize Pontoise, the key of the Isle-de-France.

Hearing of this, and urged to do something by the enthusiastic mob of Paris—always fanatically Catholic—the Constable suddenly marched out, on November 10, from the northern gates of the city against the enemy's headquarters at St. Denis, intending to take Condé by surprise—as in a way he did. For though the Huguenot prince had time to get his troops into battle-array, he had no chance of recalling Dandelot,

[1] The involuntary slayer of Henry II at the famous tournament of 1559.

much less Montgomery. But Condé nevertheless resolved to fight—he was always rash in the extreme, full of over-confidence, and thinking too much of his own prestige. He had an intellectual contempt for the old Constable, and a great belief in his own dexterity, and in the invincible courage of the noblesse which followed him. As La Noue remarks, we may sometimes admire the pluck of a general, while seeing that he was attempting the impossible.

Although he had under his hand at the moment no more than some 1500 of his cavalry, and little over 2000 foot,[1] Condé refused to move when the Constable came out against him. His outposts were in front of St. Ouen and Aubervilliers, his main body lodged in St. Denis. On the approach of the enemy he took up a prepared position. The cavalry were in three equal bodies—one under Coligny by St. Ouen, with a trench filled with arquebusiers covering its right or outer flank ; the Prince himself, with the centre, was deployed in front of St. Denis ; the left wing, under Genlis, was drawn up beside Aubervilliers, flanked by a trench of arquebusiers on its left, just as Coligny's squadron was on the outer flank. These two bodies of ' shot ' were of no more than 400 men apiece. The poor remainder of Condé's infantry, 1000 pikes and a few arquebusiers, was drawn up far to the rear, in front of St. Denis.[2] The Prince's resolve was to allow the enemy to come up on to his chosen ground, and to charge in, when they should find themselves surprised by the unexpected fire of his hidden arquebusiers. His force was so small that he arrayed his cavalry *en haye* in the old style—not that he did not know of the advantages of a deeper order, says La Noue, but simply because it was necessary to make a long front, in order to cover the broad space between St. Ouen and Aubervilliers.

The royalist army took an unconscionable time in emerging from the Porte St. Denis and Porte St. Martin, and forming up outside the walls of Paris, for the long march across the waste fields and market-gardens, which are now covered by the sordid streets of the ' Banlieu.' Montmartre and its church were on the left of the army, which was drawn up for a linear attack on St. Ouen, St. Denis, and Aubervilliers, where the enemy were

[1] La Noue gives even smaller figures—only about 1000 of each arm. *Memoirs*, p. 888.

[2] It is curious that the Tavannes Memoirs mention this, when so many better authorities do not. But undoubtedly the fact is correct.

known to lie. The Constable had arrayed his army in a long front, with four masses of infantry in column; in the gaps in front of them and on their outer flanks were placed his cavalry in five divisions. The right-hand infantry column was composed of French foot under the Duke of Montpensier, with companies of horse under Biron on its outer, and under the Marshal de Cossé, on its inner flank. Next to the left were the 5000 Swiss, who were in charge of the artillery, which was drawn out in a single line, and opened a distant fire on Genlis's troops near the village of Aubervilliers. Beyond the Swiss, in the right centre, were companies of gendarmerie under the Marshal

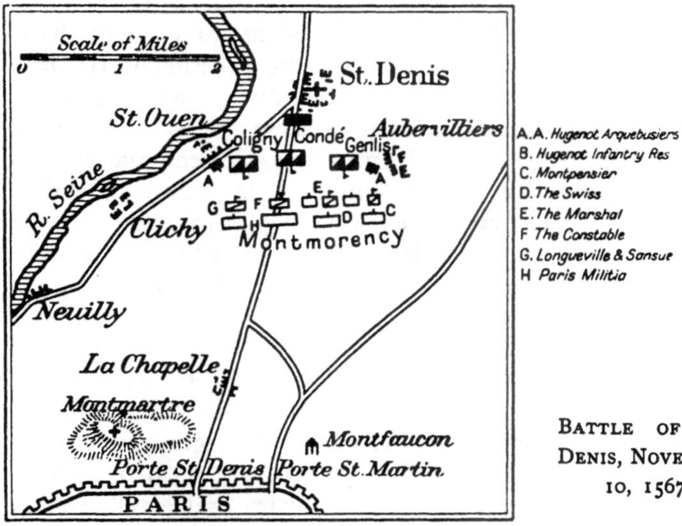

BATTLE OF ST. DENIS, NOVEMBER 10, 1567.

Montmorency, the old Constable's eldest son. Next to them in the left centre, were two regiments of French infantry under Strozzi and Brissac, beyond whom rode the Constable himself with his own *compagnies d'ordonnance*. Behind him was the great body of Paris militia infantry. Finally, forming the extreme left wing, were companies of heavy and light horse under Sansac and the Duke of Longueville, facing toward St. Ouen. There were small cavalry reserves—one under D'Aumâle, the younger Guise, behind the right wing, and the other under Damville, the Constable's second son, behind the left.

The advance of such a long array over ground not entirely level, and with one or two groups of houses in its centre, was necessarily very slow, and it was afternoon before the armies came into active collision, when the royalist horse on the extreme left and right wings attempted to circle round and outflank the stationary Huguenot squadrons of Coligny and Genlis. The horsemen of the latter suffered some loss from distant artillery fire. But in each case the Catholics fell into the trap, and were received at fifty yards distance by smashing arquebus volleys, from the hidden marksmen in the trenches, which threw them into disorder. Coligny and Genlis charged at once, and drove their assailants in disorder back against their infantry—Montpensier's column on the royalist right, the Paris militia on the left. The former closed up, held firm, and were soon relieved by de Cossé's horse, so that Genlis was completely checked. But the Parisians mixed with the flying cavalry, gave back and retreated in the direction of Montmartre.

Seeing his flank corps successfully engaged, Condé then charged with his central body of horse against the Constable's gendarmes, and broke them. The old commander-in-chief was fighting—as at Dreux—in his own front line, though he was now seventy-four. He was ridden down and summoned to surrender by a Scottish adventurer, Robert Stuart,[1] to whom he replied by a buffet with his sword-hilt which knocked out three of the Scot's front teeth. The answer was a pistol-bullet in the side, which inflicted a mortal wound. Condé's men were making havoc with the Constable's followers, when they were charged in flank by the companies of the Marshal de Montmorency, who came diagonally from the right, too late to save his father, but in time to pick up his bleeding body, which was borne to the rear. A furious cavalry mêlée followed in the centre, in which neither party had much advantage; but Condé was finally compelled to give the signal for retreat, because his left wing under Genlis had failed in its attack on Montpensier, and had then been overpowered by Aumâle's companies of gendarmerie from the reserve and Biron's rallied squadrons. Coligny on the right, who had got far into the rear of the central fighting, and become engaged with Damville's companies from the reserve, had to cut his way backward through the rear of the royalist cavalry, but got off without much loss, save among the arquebusiers who had served him so well at the beginning of the fight. Damville is said to have shown little spirit in pursuing him.

[1] The same man who is said to have shot the President Minard for burning Anne du Bourg, the Huguenot judge, eight years back.

The Battle of St. Denis, Nov. 10, 1567

From the Tapestry in the Musée de Cluny

All the three Huguenot corps rallied on their small clump of pikes in front of St. Denis, and were not pursued. The whole affair had taken only three-quarters of an hour, but dusk was beginning to fall, and the Catholics, deprived of their commander-in-chief, did not reorganize themselves for a new advance, and drew up on the battlefield. They were contented to have won a technical victory, and to have the corpses of the slain, and a certain amount of captured standards. The losses of their cavalry were certainly no less than those of the Huguenots ; their infantry had hardly been engaged.[1]

This was a mad exploit on the part of Condé and his men—quite unjustifiable but attended with no ruinous results. It is said that the Turkish ambassador, who had watched the fight from the height of Montmartre, exclaimed to those about him that 'if his master the Sultan had only a thousand of those "white coats" to put at the head of each of his armies, he could become master of the whole universe.'[2] Certainly the moral result of the fight was to inspire the Royalists with a wholesome dread of the Huguenot horse. Next day Condé, having been rejoined by Dandelot and his corps, offered battle outside the gates of Paris, and burned the village of La Chapelle and a number of windmills.

No enemy came out against him—the excuse given being that all the royalist generals were gathered around the death-bed of the old Constable, who expired on November 12. In reality they were disputing among themselves for the succession to his command-in-chief—which was claimed not only by his son the Marshal Montmorency, but by several others. The Queen-Dowager adroitly turned their rivalry to her own profit, by appointing the King's younger brother Henry, Duke of Anjou, as 'lieutenant-general of the kingdom,' though he was only seventeen years of age. He was her favourite child, the only one for whom she showed any real affection, her other sons and daughters being little more than pawns in her game. King Charles was showing some signs of becoming restive, but was still reduced to subjection, whenever she exercised her well-tried ascendency over him.

In spite of his reckless display of courage at St. Denis,

[1] The Huguenot casualties included the Counts of Sault and Suse, and Picquiny, the Vidame of Amiens. The Catholics, beside the Constable, lost the Count of Chavannes and d'Anton, the Constable's nephew.

[2] D'Aubigné, i. p. 307.

Condé was beginning to grow anxious as to his position. The reinforcements which he was expecting from the south had not turned up, and it was obvious that he could not blockade Paris with 6000 men, more especially as Spanish troops sent by Alva were on their way. The only hope was to move east, and join the large German force which the Elector Palatine was raising for him on the Rhine. On November 4, only four days after the battle, Condé decamped, and after marching first to Montereau, where he picked up the levies from Poitou and Orleans, which had failed to join him before Paris, moved slowly through Champagne and Lorraine, living on the hostile countryside. The royalist army followed him with caution, and could not prevent his junction with the Germans at Pont-à-Mousson soon after the New Year of 1568.[1] Being once more at the head of an army strong enough to seek battle, Condé turned east and laid siege to Chartres, before whose walls he was joined by a great body of Gascon and other south-country Huguenots, who had hitherto been fighting their own campaign on the Garonne (February 24).

Then followed one of the unexpected and insincere pacifications which make such frequent appearance in the annals of the Wars of Religion. The Queen offered the Huguenots all that they could ask—the restoration of the Treaty of Amboise, and the removal of all restrictions which had been made since its ratification. She feared a general action, though the Catholic army was now in great force, and fell back on intrigue —hoping to resume her old policy when once the rebels should have dispersed, and have sent home their formidable German auxiliaries. The terms were too tempting to be refused, and were duly accepted, though Coligny warned his friends that they would not be carried out honestly. Hence the Peace of Longjumeau, signed on March 23, 1568, which ended the 'Second Time of Troubles.'

So closed a campaign mainly to be remembered by the astonishing battle of St. Denis, where 3500 Huguenots fought an army of five times their own strength, without suffering the destruction that Condé's reckless courage had well earned. It is, of course, one more testimony to the power of the cavalry arm in these days—and (we may add) to the extraordinary

[1] The Germans are said, probably with exaggeration, to have numbered no less than 8000 horse. This seems an incredible figure, but no doubt the force was very large, as it terrified the Royalists.

incapacity of the leaders on both sides, which we have already marked at Dreux. At the fight in 1562 Condé had at least some chance of victory, which he threw away; at St. Denis there could be none whatever, considering the odds. The younger Tavannes, compiling the memoirs of his father long after the wars were over, hazards the opinion that Condé was only intending to deliver a sharp blow, and then to draw off, without pushing things to extremes. Their dispositions, he says, 'montrent qui'ils pensaient bien â la retraite.' But it is doubtful whether the headstrong Condé had any such subtle intentions. He exposed himself to dangers which should have proved fatal, if the enemy had any tactical skill whatever. His psychology must be studied with reference both to Dreux and to Jarnac.

CHAPTER IV

BATTLE OF JARNAC (MARCH 13, 1569)

THE pacification made at Longjumeau in March 1568 was never effective, since local hostilities continued in many corners of the realm: but (such as it was) it was officially in operation for just six months. Queen Catherine had made her surprising surrender to the Protestants merely in order to secure the disbandment of their formidable army, and—most of all—the return of the great mass of German auxiliaries to their own country. The King actually advanced 300,000 gold *écus* to pay them off.

The royal troops, on the other hand, were not dismissed, and the Queen made contracts for several thousand Italian mercenaries, to be levied by the Pope and the Duke of Tuscany, —who were not destined to arrive till next year. In August the mask was dropped, and an attempt was made to kidnap Condé and Coligny, who were staying in the country at the Prince's château of Noyers. Warned in time, they fled with their families to La Rochelle, and sent a protest to Paris against the general non-execution of the Treaty of Longjumeau. The Queen replied by dismissing the Chancellor l'Hôpital, the one genuine advocate of peace, and by causing the Parliament of Paris to publish an edict declaring toleration at an end, depriving all Protestant officials of their posts, and ordering all ministers of the Reformed religion to quit France in fifteen days (September 28).

War had come again, and all over the south and west of the land the Huguenots once more rushed to arms; this time the struggle was complicated with the first serious outbreak of insurrection in the Spanish Netherlands—Louis of Nassau had entered Friesland, to be victorious at Heiligerlee and sadly beaten at Jemmingen (July 21). His elder brother, William the Silent, who had invaded Brabant with a large German mercenary force, as will be told elsewhere,[1] was conducting his

[1] See below, p. 558.

unlucky campaign with the aid of many Huguenots of the north, who joined him under Genlis, the man who had led the left wing at the battle of St. Denis in the preceding year. Though not beaten in battle by Alva, he was foiled and outmanœuvred, and finally forced across the French frontier in November; his reiters mutinied and deserted him, and he reached Strasburg with a mere remnant of his once formidable army, mainly consisting of Genlis's Huguenots and his own personal retainers. All that had been accomplished by the Netherland insurrection was to distract Alva from interfering in French affairs during the autumn and winter of 1568. It is true that William and his brother Louis, untamed by their disasters, set themselves to gather new levies in Germany, with the aid of their ever-constant ally the Elector Palatine. But these could not be ready till the spring of 1569, and meanwhile Condé and Coligny had to do what they could with their own resources.

These were greater than might have been expected—not only did three-quarters of Poitou, Angoumois, and Saintonge fall into the hands of their local supporters, but a very large body of friends from the south-east forced their way through many dangers, and joined Condé in upper Poitou in the end of October. This, unlike most of the Huguenot armies, was a popular levy from regions where Protestantism prevailed in all classes of society—it is said to have numbered 16,000 foot to only 500 horse, and of the foot the large majority were arquebusiers, and the pikemen very few.[1] It resulted—a thing uncommon in these wars—that the Huguenots, for once, had an army equal in numbers to the royal troops for the moment, though the bulk of it was composed of raw, untrained infantry. hastily embodied and thrown into new regiments.

The Queen-Dowager had been taken somewhat by surprise at the consequences of her projected *coup-d'état* against the Protestant leaders. Not only had they escaped, but a third of France had flared up into insurrection. Her troops were scattered, and her treasury empty; there was no help to be got from Alva in Flanders, who had his own hands full. It was only by the middle of November that a royal army had been collected on the Loire, under the nominal command of

[1] It was to a strong column of this force that the disaster happened, which is mentioned on p. 406, when 3000 arquebusiers, caught in the open without pikes, were exterminated by the royalist cavalry, when they tried to form a *bataillon de parade*.

the eighteen-year-old Henry of Anjou, but really directed by Gaspard de Tavannes, an experienced and unscrupulous soldier, of military talent much superior to old Montmorency, who had led the earlier royalist armies. It amounted to 6000 Swiss and 10,000 French infantry, with some 4000 horse. The Italian mercenaries had not yet come up, and no Spaniards could be got from Flanders. But the troops under the Duke of Montpensier, which had failed to intercept the Huguenot reinforcements from the south, had moved up to join the main body.

Then followed a very indecisive campaign between Poitiers and Saumur, in which neither party would commit itself to a general action, when it discovered the enemy in a good position, though there were several lively affairs of outposts. Winter weather of the most bitter sort—the hardest December of the century, it is said—having set in, the two armies stood blocked opposite each other 'with a fierce and equal countenance,' till 'the universal murmurs of the soldiers on both sides no longer permitted the generals to hold the open field among the ice and the blasts, and they separated on the fourth day. The Duke of Anjou drew back to Chinon, and dispersed his army in the Limousin, while the Prince and the Admiral retired to Niort.'[1] The retreat was no easy matter—the sleet was continuous, the foot-soldier fell down every three paces, the cavalry had to dismount, because their horses slipped on every puddle. Many men died of sheer cold, more of dysentery caused by chills. The royalist army suffered, if possible, more than the Huguenots, because they were worse fed.[2] There was an immense amount of desertion on both sides—young soldiers quitting the ranks and making for the nearest place where they could find shelter and food, after which they went home. It was calculated that each army lost 6000 or 7000 men before the retreat and the campaign were over.

Neither side could pull itself together for an effort till the hard frost came to an end in February. But there were minor events in the depth of winter—the successful defence of Sancerre by its citizens ended in January; a new Huguenot army was raised by the 'Seven Vicomtes of Quercy' in Guyenne, and paralysed the royalist forces in the south. Elizabeth of England, warned by her mistake as to Havre in 1562, sent the

[1] Castelnau's Memoirs, v. p. 117.

[2] La Noue, pp. 956–7.

insurgents not auxiliary troops but £100,000 in gold angels, delivered at La Rochelle. The royalist army was reinforced by 2000 Catholic reiters, raised in the ecclesiastical electorates of Germany by the Rhinegrave Philibert and a Margrave of Baden.

Early in March, Anjou, or rather his mentor Tavannes, took the offensive, intending to cut off Condé and Coligny from the levies of their co-religionists in Guyenne, and crossed to the south bank of the Charente to the east of Angoulême. The Huguenot generals occupied the opposite bank, and had all the bridges in their hands. Tavannes, intent on bringing them to action, succeeded in seizing the little town of Chateauneuf, which lies on the south side of the river, at the end of a long loop of its course, secretly repaired its broken bridge, and laid another of boats close to it, thus having good means of crossing. The Huguenots were dispersed along the other bank, guarding many passages. Coligny's corps was nearer Chateauneuf, Condé's a good many miles away at Cognac. Tavannes next made an ostentatious forced march along the south bank of the Charente towards Cognac, in order to attract the enemy's attention, and then brought back his army after dark, and passed it all successfully over the two bridges at Chateauneuf. Coligny had left 200 horse and two regiments of infantry to watch the passage, but the officers in command,[1] finding the river banks cold and inhospitable, very improperly took up quarters in villages some miles back from the water, and were surprised at dawn on the 13th by finding that the Royalists were across the river in overpowering force, without having had to fire a shot.[2] It was only when their vanguard under Biron was a mile forward on the opposite bank that fifty Huguenot horse discovered it, and turned back in haste to warn Coligny. This was a typical instance of carelessness among officers of the noblesse, of which contemporary critics so often make mention. The troops, horse and foot, which ought to have watched Chateauneuf, made an attempt to fight a 'detaining action' at Vibrac, but were soon driven off with loss.

Coligny had to collect his troops in haste from scattered billets, in order to make a front, and to send the disastrous

[1] Apparently La Louë was the person mainly responsible: he was a hard fighter, if incautious.

[2] La Noue, p. 557.

news to Condé at Cognac, begging for instant support. Meanwhile he took up a position behind a rivulet, the Guirlande, at right angles to the Charente, with his headquarters at Bassac; he had intended to go back still farther and join the Prince, but, before he had got all his troops together, found himself so closely pressed that he had to turn and fight, rather than be caught in full retreat.[1]

The rivulet which falls into the Charente in front of Bassac had a limited number of points of easy passage owing to pools and marshy spots, and offered a fair ground for a detaining action; but the Catholics pressed on with all possible speed and in overpowering numbers, being aware that they had less than half the Huguenot army before them, and might dispose of it before Condé could come upon the field. The infantry of their vanguard crossed at one point and drove the enemy's arquebusiers out of Triac, and entrenched itself there. The cavalry of the vanguard under Martigues, Cossé-Brissac, and the young Duke of Guise—who was then seeing his first field at the age of eighteen—crossed the rivulet at another point, and rode down four cornets of Huguenot horse under the famous La Noue, who was taken prisoner.[2] They then got engaged with the Admiral's main body. Meanwhile the rest of Anjou's army had come upon the field and deployed—in three great bodies of infantry—French regiments on the flanks, the Swiss in the centre with the artillery. Between the Swiss and the right-hand column of infantry was the 'main-battle' of Anjou's horse; the 2000 German reiters under the Rhinegrave formed the extreme southern end of the line; some companies of light horse under La Valette covered the other flank, rather in front of the infantry of the northern column.

The Admiral's troops, both horse and foot, were giving way, though still maintaining a desperate resistance, when Condé came upon the field, not with the whole of his corps, but with the bulk of his cavalry. The infantry were still miles behind on the road, returning from their unlucky march toward Cognac. The Prince was at a height of spiritual exaltation—a Scot would have said that he was 'fey.' When

[1] One of his columns, 3000 South-French infantry, guarding his cannon and baggage, had been started off early for retreat, and got so far forward on the road to Angoulême that it did not get back in time for the battle.

[2] He was 'exchanged' soon after, against Lessac, the lieutenant of the Duke of Guise.

he was putting on his helm, the horse of his brother-in-law, the Count of Rochefoucault, reared against him, and broke one of the bones of his leg by an unlucky blow of its hoof. Instead of dismounting, he raised himself in his saddle and cried, 'Gentlemen of France, the long-expected hour has come, see in what a state Louis of Bourbon can go into battle, fighting for Christ and fatherland,' and with that he spurred into the fray, with his personal retinue—twenty-five gentlemen of the house of La Vergne—immediately behind him, and some score of companies of gendarmerie—under Rochefoucault, the Count of Choisy, Rosny (the father of the famous Sully), and others at his heels. Coligny at the same time rallied his horse, and advanced on the Prince's left—the Catholic chronicler says that his charge was *assez molle*, with tired troops.

But the thrust which Condé gave to the battle was for the moment decisive; the royalist vanguard was turned to complete rout, and driven back against the cavalry of the main body. At the same time the Huguenot light horse under Fontrailles drove in the royalist light horse of La Valette to the north end of the battle, and thrust it against the infantry column of Anjou's right wing. But the numerical odds were too great—on the one flank Fontrailles' squadrons failed to break the royalist infantry—he himself was unhorsed and made prisoner among the pikes. His followers broke and left the field. Coligny's charge 'petered out' on the other flank.

In the centre Condé, fighting furiously and not unsuccessfully with the royalist gendarmerie, was suddenly charged in flank by the 2000 reiters of the Rhinegrave, from Anjou's left wing. This settled the day—the Huguenots were ridden down, though they fought most desperately and refused to fly.[1] Fifteen of Condé's bodyguard of twenty-five were killed, and many scores of the best of the noblesse with them. The Prince himself was unhorsed, and being unable to stand because of his broken leg, was dragged up by a common soldier. He saw a gentleman named d'Argens riding by, and surrendered to him, handing over his sword and gauntlet. D'Argens had him carried to the rear, and while he was conversing with this officer and another named St. Jean, Montesquiou, a Gascon adventurer, captain of the guard to the Duke of Anjou, came up and deliberately shot him dead from behind, with a pistol ball through his head. Whether this was done

[1] Castelnau, vii. p. 135.

by Anjou's orders, or merely to curry favour with him by this disgusting exploit, is uncertain. Anjou is said to have given sinister hints to some of his confidants,[1] and his whole life speaks for itself; on the other hand, Montesquiou was a ruffian quite capable of acting on his own inspiration, if he thought a murder likely to win him promotion.[2]

Thus ended a most genial and inspiring leader, but a very bad general. The records of his three fights, Dreux, St. Denis, and Jarnac, show that he had no idea of winning save by headlong cavalry charges, led by himself in person. In this last campaign he had been completely outgeneralled by Tavannes, who lured him to fight with a half-assembled army against very superior numbers. But after the event of St. Denis, Condé may have thought that numbers did not much matter—he had on that occasion brought off his troops with little loss after engaging against fivefold odds. His personal qualities—his wild courage, and gay and chivalrous bearing—had made him the most popular figure in France; even the Catholic chroniclers speak of him with respect and regret, and only marvel that he should have taken up the Huguenot cause. They can only explain it by personal ambition, not making any allowance for a genuine dislike to the state of the Roman Church as then existing, or for a wish to foil the Queen-Dowager's Machiavellian policy as fatal to France.

Condé's cavalry having been cut to pieces, the Admiral drew off his own troops, foot and horse, as best he could, and retreated to St. Jean d'Angely, where he picked up such of his own corps as had not arrived in time for the battle. He was not pursued with any vigour. Condé's infantry and the shattered remains of his gendarmerie rallied at Cognac. The loss to the Huguenots had been 'more in quality than in quantity,' as de Thou remarks—of 400 dead half were gentlemen of high quality—the *élite* of the party—and the prisoners (mostly wounded) were numerous also. Castelnau gives long lists of both; among the prisoners were La Noue, Rosny, with the ex-bishop of Cominges, who had thrown off his mitre and joined the Protestants, and Fontrailles who had commanded the light horse companies. Robert Stuart, the Scot who was credited with the deaths of the President Minard and the Constable Montmorency, was shot in cold blood after the

[1] Brantôme alleges this, but is not the best of authorities.

[2] He was killed not long after at the siege of St. Jean d'Angely.

battle, as Anjou confessed in a letter, 'for his crimes.'[1] The loss of the Catholics was small—200 to 300 only—almost all in the cavalry, and included only a few persons of note.

The course of the campaign reflects much credit on Tavannes—we need not give any of the glory to the Duke of Anjou, though he was, of course, hailed officially as the victor, and could pose as a Catholic hero. His whole character and career suffice to show that neither courage nor decision were his strong points; though he could plot a murder, or conduct an

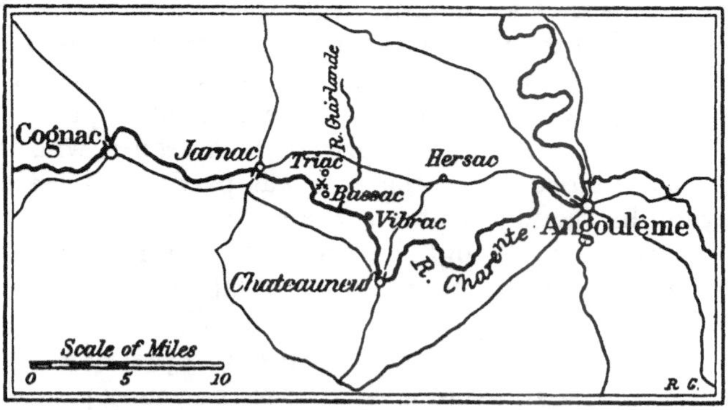

TOPOGRAPHY OF THE BATTLE OF JARNAC, MARCH 13, 1569.

intrigue, or make a sudden display of piety or cruelty, he was no general.

N.B.—The best authorities for the battle are La Noue on the Protestant and Castelnau on the Catholic side, but incidental help can be got from many minor sources. I found the plan-panorama in the Cluny Museum, and Perissot and Tourterelle's engraving most useful; they name each unit in both armies.

[1] It is not *quite* certain that Stuart killed either of them—though it was generally so believed. Some of the contemporary chroniclers say that, at St. Denis, when the Constable had delivered his blow at Stuart, *some one* shot him straightway. And the assassination of Minard was never legally brought home to this wandering Scot, though he was held to be the culprit. Anjou, in his letter which mentions Stuart's execution, calls him the man guilty of the murder of Minard, 'and of other crimes.'

CHAPTER V

BATTLE OF MONCONTOUR (OCTOBER 3, 1569)

THE military effect of the battle of Jarnac, strange as it may appear, was almost negligible—save that it threw the command of the defeated Huguenot army into the hands of the Admiral Coligny, a much more cautious man of war than the headlong Condé. It might have seemed important that the Protestants were now deprived of a leader who was a prince of the blood. But this loss was partly repaired by the display of his young nephew, Henry of Navarre, as titular head of the party, though he was only fifteen years of age. Henry's indomitable mother, Jeanne d'Albret, brought her son to the Huguenot camp, along with his slightly older cousin Henry, the heir of the murdered Condé, and presented them to the captains, who all swore allegiance to the young prince.

Meanwhile Anjou, or rather Tavannes, had done little to exploit the victory of Jarnac. Two days after the battle the royalist army presented itself before Cognac, whither the bulk of Condé's infantry—who had not been in the battle—had retired. Under the idea that the Huguenots were demoralized, the young Duke of Guise made a dash at the place, but was repulsed by a vigorous sortie of 1200 men, who sent him flying. 'On ne prend pas tels chats sans mitaines,' remarks La Noue, alluding to the unbroken spirit of his friends. The Royalists then made an equally futile demonstration against Angoulême, and had to content themselves with capturing Mucidan, Aubeterre, and Bergerac—'attacking little places because they could not tackle large ones, for want of the train of artillery expected from Paris.[1] They then placed themselves across the roads leading from the Charente district towards Guyenne, in order to prevent the Admiral from getting any succour from the south, where his friends were in force both in Quercy and in Béarn.[2]

Coligny, therefore, was able to reorganize the Huguenot

[1] La Noue, p. 962

[2] Tavannes, xxii. p. 156.

army beaten at Jarnac at his leisure: it was little diminished in numbers, though so many of the best of the noblesse had fallen with Condé. His hope now rested on the arrival of the great levy of Germans which was being made by the Elector Palatine on the Rhine, with which were to be found both William and Louis of Nassau, with the wrecks of the French force under Genlis which had fought in Brabant in the previous year.[1] But the command was in the hands of Wolfgang, Duke of Zweibrücken, not in that of either of the Nassau princes. It was a powerful corps of six thousand reiters and nearly as many landsknechts, and there were some 2000 Huguenots in their company, under Morvilliers, the successor of Genlis. But it is 250 miles from Alsace to La Rochelle, and the Admiral had many doubts as to whether this great succour would ever reach him.[2] It was destined to do so, however: Duke Wolfgang seems to have shown considerable military ability, for he shouldered off, on three separate occasions, large royalist forces which were sent to intercept him. The Queen-Dowager would not entrust this task to the Guises alone, but gave two separate corps to Aumâle—the representative of the house of Lorraine—and to the Duke of Nemours. They appear to have been jealous of each other, and to have failed to co-operate, though their joint forces would have outnumbered the Germans. At any rate, the Duke of Zweibrücken fended each of them off on various occasions, and forced his way through Burgundy and the Nivernais, though both these regions were unfriendly country, with no Huguenot faction to bring help.

On May 20 he stormed La Charité on the Upper Loire, which Nemours had left undergarrisoned, and began to get into touch with Coligny, though the main royalist army might still have slipped between them. On the next morning Duke Wolfgang died—'of a surfeit,' it is said: after a hard day's fighting he indulged in over-deep potations and a hearty meal, and ended with an apoplectic seizure.[3] Command was taken over by Wolfrad, Count of Mansfeldt, his lieutenant, though both the

[1] Genlis had died of fever at Strasburg during the winter, and this refugee Huguenot force was now under Morvilliers.

[2] So says La Noue, p. 968.

[3] Some Protestant chroniclers say that he was poisoned by a physician. The same allegation is made about the death of Coligny's brother Dandelot, who died about the same time at Saintes. But all unexpected deaths in these times were put down to poison. The Tavannes Memoirs—though a Catholic source—make Zweibrücken poisoned by 'a doctor from Avalon.' who gave him drugged wine.

Nassau princes were still with the army. The new general succeeded in evading Anjou, though the Prince came up with his rearguard, but the royalist reiters, it is said, refused to charge their Protestant countrymen, and allowed them to slip off. They made the excuse that they were famishing, and that their horses were absolutely exhausted—but it was in reality a case of 'dog does not eat dog.'

On June 10, Mansfeldt crossed the Vienne, and joined Coligny's army at Chaluz,[1] near Limoges—after finishing one of the longest and most dangerous forced marches recorded in these wars. He should never have been allowed to get away from the various royalist forces which beset him: but did so. Meanwhile Aumâle and Nemours joined Anjou and Tavannes—their joint strength had a numerical superiority over the enemy, even after the Germans had got into touch with the Admiral's army. But as two long dispatches from Tavannes to the Queen-Dowager show, 'directed against his calumniators,' there was a strong feeling of discontent and 'défaitisme' in the royalist army. It was felt that the victory of Jarnac had not been properly exploited, and that the march of the Germans ought to have been intercepted. For the first-named grievance Tavannes was held responsible, for the second the Council at Paris, *i.e.* the Queen-Dowager. But an even more serious matter was the war-weariness of the Catholic noblesse—they had been called to arms in September 1568, and had been subjected to the rigours of an exceptionally hard winter. After an unmistakable victory in March they were still held to the standard in June, and nothing had been settled. Tavannes wrote to the Queen: 'Those gentlemen have become so downhearted that they ignore the "rassemblez vous." They want a month or six weeks' leave, to see whether their manors have not been burned or sacked. Many slink home. If these fine gentlemen play us the trick which they have already tried once or twice this year, *videlicet*, to go off, without saying a word, just as we are getting near the enemy, the whole business may end in a general upset.[2] Things are at this moment so depressing that I dare not write down what I feel. I hear, by letters sent to the Duke of Anjou, that the Spanish ambassador has been proposing that Alva should be sent for, and placed in supreme command, as pedagogue to His Royal Highness. If he comes,

[1] Near Limoges—the castle fatal to Richard Cœur-de-Lion in 1199.

[2] 'Tout ne peut aller à fin que sens dessus dessous.'

it will be with a very large army, and the crown of France will be like a wafer nipped between two irons. It might almost be better to listen to the Huguenots, when they ask for negotiations in favour of a general peace—a truce would be better than nothing.' [1] Aumâle's corps on joining the main army was in a disgraceful state: of his own company of gendarmes, his lieutenant and all his lances except ten had deserted!

The Cardinal of Lorraine, as has been mentioned in an earlier page, told La Noue, then a prisoner at Tours, that if the Admiral and the Germans had only made a push in July, the royalist army was '*sans vigeur et à demi dissipée*,' and would have given way without making a serious stand.[2]

This psychological condition of the royalist army in June 1569 is the real explanation of the fact that the campaign went on for three months, after the Germans had joined Coligny, without any definite result, the general action at Moncontour only coming in October, when it might have been expected in July. Tavannes only plucked up courage to take the offensive after many weeks, when he had been joined by the Italian mercenaries under the Count of Santa Fiore sent by the Pope, by the long expected train of heavy artillery from Paris, and by some Spanish horse detached to his help by Alva, under Peter Ernest of Mansfeldt—a person to be carefully distinguished from his kinsman Wolfrad of Mansfeldt, who was leading the Protestant reiters in Coligny's army.[3] The morale of the Royalists gradually went up, when it was seen that the Huguenots were getting no profit out of the strategical situation, and many absentees returned to the front. By September, Tavannes could write '*la chance se tourne: qui fuyoit la bataille la cherche maintenant.*'

That the Huguenots failed to take a resolute offensive after they had been joined by their German auxiliaries is at first

[1] All this is boiled down from Tavannes' two very long letters to Catherine, of July 20 and August 11. The Cardinal of Lorraine and the Guise party had been pressing for the summons to Alva, which, of course, would have put Tavannes 'out of his job.' Coligny had sent a gentleman named l'Estrange to Anjou, with a request for a safe-conduct for him to go to Paris, and negotiate for a general pacification. The Prince refused the safe-conduct, but Tavannes has come round to the conclusion that it might be worth while to grant it—a truce would be useful at the moment. But, of course, neither Anjou, Tavannes, nor the Queen-Dowager wanted a real peace.

[2] See above, p. 441.

[3] This is the Mansfeldt who was in command of the Spanish army in the Netherlands twenty years after, and was outmanœuvred by Maurice of Nassau. See pp. 572-3.

sight somewhat surprising, as several incidents encouraged them at this time. The first was a successful affair of outposts at Roche l'Abeille on June 25, in which the young Duke of Guise and the Italian Colonel Strozzi, who with a mixed force had advanced without orders too close to the Admiral's outposts, were suddenly set upon by 4000 arquebusiers and a few cavalry, and routed with loss. Strozzi was taken prisoner, forty captains and 500 of his regiment were slain, Guise abandoned the infantry and fled. He incurred the most bitter reproof from the commander-in-chief: '*il vous fust esté plus louable de mourir que faire ce que vous avez faict.*' Tavannes thought for a moment of sending him to the rear—'Limogé' or 'Stellenbosched' in modern phraseology, but, considering his political importance, detached him at the head of a body of light horse, to keep him away from headquarters.

Tavannes thought that if the Huguenots had made a general push, after routing Guise and Strozzi, they might have secured a complete victory.[1] La Noue explains that the hour was late, that heavy rain was falling, and that the royalist position was strong; the risk was too great.[2] Both armies drew away, to get better quarters in lands that were not thoroughly devastated.

Coligny was at this juncture of opinion that the proper policy was to march on the Loire and seize Saumur as a good crossing-place; such a move would tempt Tavannes to come north also, in order to cover Paris. He was forced, however, contrary to his desire, to remain in Poitou. A very large proportion of his cavalry was composed of the Huguenot noblesse of the west, and they insisted that no advance toward Paris ought to be made till Poitiers, the one great fortified city in the hands of the Royalists in their native region, had been reduced. Unwillingly the Admiral fell in with their desire, and, after capturing several small outlying places, laid siege to Poitiers on July 24. This move, as all the contemporary critics allow, was a fatal one for his cause. The place was well garrisoned already, and the young Duke of Guise threw himself into it, with the flying corps of cavalry which Tavannes had given him. Apparently he did so without orders, and much vexed his chief, but the move turned out a fortunate one, as it inspired the garrison with a confidence which they had not felt before.

[1] Tavannes, Memoirs, xv. p. 165. [2] La Noue, p. 971.

The siege of Poitiers lasted for six weeks (July 24–September 7) and made no progress. The Huguenots had little siege artillery, the place was covered for the greater part of its circumference by the rivers Clain and Boivre, across which no approaches were possible. On the S.E., the unprotected front, the ground was highly unfavourable, owing to flanking fire from high points of the defences. On the one occasion when a breach was made, and the outer wall forced, the stormers found in front of them not only a second new entrenchment, but the garrison massed in such force that progress seemed impossible, while a heavy flank fire was being directed upon them from dominating points. They abandoned the attack on this front, and shifted their siege batteries, with little profit. It was here that the younger Guise made his military reputation—his earlier efforts at Jarnac, Cognac, and Roche l'Abeille had been by no means to his credit ; but at Poitiers a very enterprising and effective defence was made.

During these fatal six weeks the army of Tavannes and the Duke of Anjou was shaking itself into better order, had received its heavy artillery, and been joined by reinforcements—the *arrière ban* was called out—and by many repentant deserters. By the beginning of September it was reorganized and ready to move. It was no profit to the Huguenots that during the course of these weeks they achieved a great success in the far south. The main part of the royalist army in Gascony had developed an invasion of Béarn and Navarre under Terride, the lieutenant of Montluc. After some successes—the most notable of which was the capture of Pau, the capital of Jeanne D'Albret's possessions—it was destroyed at Orthez by Montgomery, one of Coligny's most daring lieutenants.[1] Detached with a handful of cavalry by the Admiral, he had crossed the Garonne,[2] rallied the Protestants of Béarn, and fallen on Terride, who was forced to surrender with most of his troops (August 24). By the orders of Queen Jeanne those of the Catholic nobles of her State who were taken prisoners were tried for high treason and beheaded. This was an unwise act,

[1] The man who had killed Henry II by accident at the famous tilting-match in 1559.

[2] Montluc has an interesting explanatory chapter as to how he let Montgomery slip through. He was first told that the Huguenot had a large force, then that he had a mere handful of horse—which was true—and so made up his mind that he was a negligible quantity, and that Terride would deal with him. He omitted to reckon on the rally of the Béarnois Huguenots.

for all the Huguenots of France were at the moment guilty of high treason, and the provocation to retaliate was obvious. About the same time William of Orange, who had been following Coligny's camp without any official command since June, made his perilous ride right across France with only twenty gentlemen as his guard, and reached Germany in safety, from whence he returned to the Netherlands. He left his brother Louis with the Huguenots; Coligny seems to have employed him as a corps-commander, though Wolfrad of Nassau still had charge of the great mass of reiter horse.

Early in September Tavannes thought himself strong enough to take the offensive, and moved against Chatelherault, the enemy's supply-base and general hospital, judging that this would compel Coligny to abandon the siege of Poitiers, and come out into the field to rescue his depot and sick. It is said that the manœuvre was pressed upon him by the Cardinal of Lorraine, who feared that Poitiers might fall, and his nephew Guise be taken prisoner. At any rate the plan succeeded—Coligny abandoned his siege lines, and moved out toward the enemy; it is said that he was only too glad to quit his position before Poitiers, having a good excuse for doing so—'one may raise a siege without dishonour in order to fight a battle.' He had lost several thousand men before the place, and the whole army was growing sick of the fruitless and expensive leaguer, though of course it was the infantry which had suffered the casualties.

On the very day on which Coligny raised the siege of Poitiers, the Royalists had made an attempt to storm Chatelherault, which was repulsed with great loss among the Italian infantry who headed the column of assault. The Huguenots of the garrison, according to La Noue, took a special pleasure in trouncing the hirelings of the Pope. On receiving news of Coligny's approach, Anjou and Tavannes raised the siege of Chatelherault and drew off towards Chinon. There followed a campaign of manœuvres—each army trying to catch the enemy at disadvantage, but refusing to attack strong positions. This led for some time to nothing but partial rearguard actions; at Port des Pilles, on the Creuze, the Admiral almost caught the Royalists passing a defile, but just missed his chance by halting before a rearguard which slipped away in the night. Next day he refused to assail the Royalists drawn up in order and in a strong position,[1] and put his army into cantonments

[1] Castelnau, vii., vi. p. 40.

at Faye la Vineuse, near Richelieu (September 3). A fortnight later the Catholics took the offensive from the side of Loudun, and another *rencontre* took place at St. Cler, both armies being on the move without any proper knowledge of each other's position. The Catholic vanguard under Biron came in unexpectedly upon the Huguenot rearguard under De Muy, rode it down, and nearly threw into confusion the whole of the marching column in front of it. The Admiral had to come back and draw up a line of battle, behind a rivulet with narrow fords, along which there were several lively cavalry charges.[1] Next morning the two armies were opposite each other—so close, says La Noue, that it seemed impossible for them to avoid a battle.[2] But though the royalist artillery did much damage to the Huguenot squadrons, Tavannes would not risk a general assault, and Coligny decamped under cover of the night (September 30).

The Admiral's tactical object was now to take up his ground beside Moncontour, a small fortified town already in his hands, when there was good forage to be had, and strong ground between woods and the Dive River, which he thought afforded a position where he could accept a defensive battle with advantage. Tavannes sent out a long-distance reconnaissance, which reported that Moncontour was surrounded by old walls, and covered on one side by the river, wherefore he resolved not to press the Huguenots back on to their chosen ground, but to make a flank movement, going round almost to the sources of the Dive, at a place called La Grimaudière, and crossing it there, so as to force Coligny to fight, if he chose to risk a general action, on a different front. The move had also the advantage of placing the Catholic army in a position threatening the enemy's natural line of retreat towards his bases in Poitou.

The Admiral, inadvisedly, did not keep good touch with the enemy on the day after the skirmish of St. Cler (October 1), but waited for further developments—it would seem to have been only on the second day that he detected the meaning of Tavannes' neglect to press in upon him, and resolved to fall back from the Moncontour position, to one behind Ervault,[3] where he would be covered by the Thouë River. He sent out a cavalry detachment to seize Ervault, only just in time, for a

[1] On this day Charles of Mansfeldt, brother of Wolfrad, the commander of the Germans, was killed by a cannon ball. The Protestant loss was quite considerable.

[2] La Noue, p. 984.

[3] Airvault on modern maps.

royalist exploring party came up and reconnoitred the place just after it had been occupied. He had intended to follow, by making a night march, but many of his captains protested against it, urging the very real dangers of moving large bodies of men in the dark, when columns frequently get astray, block each other, or straggle. Coligny gave way, and postponed the retreat till dawn. He then came upon one of those maddening delays which often exasperated generals in command of mercenary troops. When orders to march were given, Mansfeldt's landsknechts, and five cornets of his reiters, went on strike, asking for pay long overdue, and refusing to move off. Such things, as will be remembered, were common in the wars of the Netherlands no less than in France. It was nearly two hours before the Admiral succeeded in pacifying them with promises—there was, as usual, no money forthcoming.

The month was October and morning light came late, hence the Huguenot army lost much precious time, and failed to get forward to their position by Ervault, which the Admiral desired to reach. Its head had only got half-way towards the Thouë when the outlying pickets announced that the royalist army was coming down in battle-formation on the flank of the columns of march. There was just time, but not more than enough time, to deploy into line to face the enemy, on a rolling sandy plain with little advantage of position, except that there was some dead-ground before certain parts of the front, which would give protection from distant cannon-fire.[1] The Huguenots faced south-east in front of the Moncontour–Ervault road. Their left wing was more or less covered by the Bois de Maire, a considerable wood.

The forces of the two armies at Moncontour are stated with somewhat varying figures by all the narrators of the battle. The only point on which they agree is that the Royalists were in somewhat superior numbers; but natural party prepossessions always lead to diminution of one's own strength and exaggeration of that of the enemy. Castelnau, by far the best authority on the Catholic side, gives Anjou and Tavannes 7000 horse and 18,000 foot with 15 field-pieces; the Tavannes' Memoirs credit the Admiral with 7000 horse, 16,000 foot, and 11 guns. Pro-

[1] All this from La Noue, 987. It is most unfortunate that his narrative of the actual battle is very succinct, and lacking in explanatory detail. Possibly he was wishing to spare the military reputation of the Admiral, to whom he was personally devoted. He himself was in the thick of the matter, and taken prisoner—for the second time in the year!

THE BATTLE OF MONCONTOUR

From Perissot and Torterelle's contemporary plan

ROYALISTS

A. Infantry of the Royalist 'Avant-Garde,' a regiment of Swiss flanked by two Italian and five French units.
B. Montpensier's Horse.
C. Martigue's Horse.
D. The Rhinegrave's Reiters.
E. Two bodies of Italian Horse.
F. La Valette's Horse.
G. Guise's Horse.
H. Biron's 'Ost de Reserve.'
K. Anjou's Gendarmerie.
L. Reiters of Philibert of Baden.
M. Thoré's Horse.
N. Main body of Royalist Swiss.
O. Ernest of Mansfeldt's Horse.
P. Cossé's 'Ost de Reserve.'
Q. Carnavalet's troop.
S. 'Enfants perdus.'
T. Isolated farmstead.

HUGUENOTS

1. Huguenot Landsknechts, with five regiments of French foot.
2. Reiters in front line.
3. 'Enfants perdus.'
4. De Muy's Horse.
5. Renel's Horse.
6. Briquemont's Horse.
7. Wolfrad of Mansfeldt's Reiters.
8. The Admiral's reserve of Horse.
9. Reiters under Louis of Nassau and French foot in reserve.
10. Five regiments of French foot of the 'bataille.'
11. Choisy's Horse.
12. Laverdun's Horse.
13. Tracy's Horse.
14, 15, 17. Corps of Reiters.
16. Escort of the Princes, about to leave the field.

[*N.B.*—Half of the Landsknechts were really brigaded with the French foot of No. 10 in the 'bataille.']

The Battle of Moncontour, Oct. 3, 1569

testant writers acknowledge only 18,000 men on their own side, 6000 horse and 12,000 foot, and put the Catholic cavalry up to 8000 in all. Both armies had a surprising proportion of foreign troops—the Royalists had 6000 Swiss infantry, 3000 or 4000 Italian infantry, and of mounted men the 3000 Catholic reiters under the Rhinegrave and the Margrave of Baden, 800 Italians under the Pope's general, Santa Fiore, and a few hundred Walloons under Ernest Mansfeldt, lent by Alva. The Huguenots were still relying on 4500 reiters—the remains of the 6000 originally brought by Zweibrücken, and some 4000 or a little more of his landsknechts. It appears that there was much more native French cavalry on this occasion in the Catholic than in the Protestant ranks—an exception in these wars. The élite of the Huguenot noblesse had been sadly thinned at Jarnac: the Tavannes Memoirs say that the companies had been filled up with 'bourgeois'—who, of course, were less formidable because less well horsed and armoured.[1]

There was a considerable delay between the moment at which the two armies got into touch with each other, and the commencement of the action—Castelnau says as much as four hours, for the Huguenots halted and formed up at about eleven o'clock in the morning, and all the fighting was in the afternoon. Coligny was only prepared to receive an attack, Tavannes was not quite sure whether he could dare to take the offensive. He rode on a Spanish horse all along the front of the Admiral's line, out of cannon-shot, looking for useful indications. His memoirs (embodying his recollections but not written by his pen) say that he noted that some of the Huguenot cavalry was not well closed up, and that the pikes of the landsknechts wavered in a disorderly fashion. If he refused to attack, Coligny would draw off at night, and the campaign might continue for an indefinite time. Winter was drawing nigh, which always meant widespread desertion on the part of French troops.[2] Accordingly he took the decision

[1] Tavannes, p. 175.

[2] There is an interesting story in La Noue, p. 985, that at dusk on October 2 two gentlemen from the royalist army came secretly to the Huguenot outposts, and shouted across a ditch that if the Admiral would move off and refuse battle, hundreds of the royalists' horse would be melting away from the standards within the next few days, and Tavannes would be brought to a standstill. Coligny was inclined to accept the advice, but the majority of his captains took it for a ruse, and thought the enemy wished to induce the Huguenots to make off, because they feared a general action. Some openly demanded battle. The tale is repeated by De Thou and other writers, obviously copying La Noue.

of advising Anjou to advance—the chance of a battle might not come again.

Both armies were divided, as at Dreux, into two corps, not as of old into three. Though these were called 'Avant-Garde' and 'Bataille' there was no 'Arrière-Garde.' And in each case there was no great difference in strength between the two halves of the army. Both fought not in a single line, but in several; and in addition there was a reserve on the royalist side kept quite to the rear, an 'ost de reserve,' squadrons of horse behind the second line, under the Marshals Cossé and Biron. In Coligny's host the 'bataille' formed the right wing, and the 'avant-garde' the left, in that of Tavannes the order was reversed, so that the two 'avant-gardes' faced each other. This may have been, in the case of the Huguenots, because Coligny thought that the 'avant-garde,' of which he took charge, would have to cover the 'bataille,' and prevent it from being thrust off the road to Ervault, along which it was intended to get away when possible. He handed over the 'bataille' not (as might have been expected) to Wolfrad of Mansfeldt, whom he kept at his own side along with the best of the reiter squadrons, but to Louis of Nassau—an improvised corps-commander, as it is said, for Coligny's old friend Rochefoucault was to have had the charge, but had fallen sick.

On the royalist side the 'avant-garde' or right wing was led by the Duke of Montpensier, the 'bataille' by the Duke of Anjou himself. They were not dissimilar in strength. Montpensier had a front line of cavalry, the 800 Italian horse of Santa Fiore on the extreme right, Martigue's French horse in the centre, and La Vallette's light horse on the left. In second line were infantry, a strong Swiss regiment in the centre, with two Italian regiments on its right, and five French foot-regiments upon the other flank. In reserve, behind the infantry, were Montpensier's own companies of gendarmes, behind the Swiss foot, twelve 'cornets' of Catholic reiters (under the Rhinegrave Philip, the Landgrave of Hesse and the Count of Schaumberg) behind the Italians, with the Duke of Guise's squadrons behind La Vallette. The whole is reckoned at 3500 horse and 8000 foot. Far to the rear was a 'ost de reserve' of horse, under the Marshal Biron.

The main body of the royalist Swiss under their 'Colonel' Pfeiffer formed the core and central block of the 'bataille.' Attached to them were four French foot regiments, and behind them the gendarme squadrons of Thoré, the son of the old Constable Montmorency. To the right of the Swiss was the largest corps of French gendarmerie, which was led by Anjou in person: in front of him was a flying vanguard of 50 gentlemen on barded horses, led by the Breton Carnavalet, told off for his special protection. On the extreme left of the royalist front line were Peter-Ernest of Mansfeldt's five squadrons of Walloon (or 'Burgundian,' in the nomenclature of the day) horsemen. This made up over 4000 horse and 9000 foot. Behind the whole of the 'bataille' was a small 'ost de reserve' of French gendarme-companies under the Marshal Cossé.

In the opposing line of the Huguenots the Admiral had the left wing or 'avant-garde' which lay opposite Montpensier. He had thrown out some arquebusiers into a large farm (grange) or small village some way to his front—'enfants perdus,' intended merely to hold off the first demonstration of the enemy, and in a very perilous position. His front line was composed of French squadrons *en haye* under de Muy, with reiters, in column, on their right. There were arquebusiers in front of them. In support of de Muy were French horse under the Marquis of Renel and Briquemont on the left; then many squadrons of reiters, under Wolfrad of Mansfeldt; and then infantry, half the landsknechts under the Baron of Geroldseck, and five regiments of French foot. The third line or reserve consisted of Coligny's best companies of gendarmes, commanded by his son-in-law Teligny, d'Acier, and Puy Greffier—the last two east-country Protestants—also of two squadrons of reiters. Here, like a good general, Coligny himself took part—not in any front line.

There were five guns placed far out on rising ground to the left, intended to enfilade the enemy when he should advance—perhaps on the knoll by the village of Douson.[1] The 'bataille' had of infantry five regiments of French foot—almost destitute of pikes as we are told, which is no doubt the reason why the other half of the landsknechts, under the Alsacian Count of Pfirt were placed among them for support.[2] The striking force of the 'bataille' consisted of French horse, three lines all *en haye*, under the Count of Choisy, Laverdun, and Tracy.

[1] Tortorel's big contemporary (1570) plan is apparently wrong in putting them in the centre, before Geroldseck's landsknechts.

[2] There can be no mistaking the statement of the royal official report that the landsknechts were in two bodies, though Tortorel's great battle-plan puts them all in one. Castelnau and d'Aubigné, the former an eye-witness, agree.

In the rear of the infantry were two heavy bodies of German horse, and out on the right, guarding the flank, two more reiter-corps which must have faced Peter-Ernest of Mansfeldt's Netherland auxiliaries: they had with them six guns and some 'enfants perdus' for skirmishing.

During the nerve-racking pause of four hours before the royalist attack began, Coligny gave two orders which have been much criticized. The first one was to send to the rear, for safety's sake, the two young princes, Henry of Navarre and Henry of Condé, with an escort of 150 lances. 'This was too early or too late'—their departure shook the confidence of some who saw it: Catholic narrators say that there were timid souls who joined themselves to the cortége without leave. The second order, given just before the fight opened, was to direct Louis of Nassau, who commanded the 'bataille,' to send over some 'cornets' of his best reiters to strengthen the avant-garde, because the force opposite looked very heavy. Louis committed an astonishing fault, by coming across in person with three picked squadrons. The battle commenced just after he arrived, and he became involved in the central fighting, leaving his own corps without a proper general. Thus the three senior officers of the army—Coligny, Count Wolfrad, and Louis, were all with the avant-garde, while the 'bataille' had no one in charge of it, 'ne sachant comment se gouverner.'[1] This may have been one of the main causes of the loss of the day.

It was on the side of the two avant-gardes that the fight began. Tavannes ordered Montpensier to advance—he had to repeat the order four times, for the Duke, remembering the story of Montmorency and Guise at the battle of Dreux, did not want to get engaged before the other half of the army was on the move. He was at last started off, and when he got under fire of the Huguenot guns on his flank, was obliged to charge in, '*boire la calice*,' in order to get out of the way of the cannon balls.[2] The first clash came when the horsemen of Martigues, in his centre, cleared off the Huguenot 'enfants perdus' who were skirmishing in front of the lines. Whereupon De Muy's squadron, and his cornets of reiters out of the Admiral's front line, came out to cover the fugitives, and were beaten off—the reiters are said to have behaved badly. Thereupon Coligny launched the rest of his second line cavalry, French and German,

[1] So La Noue, p. 988.

[2] Tavannes, Memoirs, p. 175.

upon Martigues, and Montpensier replied by sending up his Italian, German, and French squadrons, against whom from the other side Coligny charged in person at the head of his third line reserve of French horse. He came into collision with the reiters of the Rhinegrave and Schaumberg, and was getting dangerously cut off from the rest of his party, when Wolfrad of Mansfeldt came in upon the mêlée, and broke his Catholic compatriots. The Rhinegrave was killed: Coligny received a pistol-ball in his cheek: his wound bled so profusely that he had to be taken to the rear, and responsibility for the command passed to Louis of Nassau and Count Wolfrad,[1] not (as the Huguenot narrators complain) to the benefit of their side.

Meanwhile the charges had spread all along the front. Anjou's cavalry of the royalist 'bataille,' Thoré's horse, and Peter-Ernest of Mansfeldt's Spanish contingent getting engaged with the Huguenot gendarmerie *en haye* and the reiters in column opposite them. La Noue observes that if only there had been some one in command on the right, things might have gone well: 'the said corps without any leader and without order, so shook the troops of Monsieur, that one can but guess what would have happened if there had been some one to guide them to a great effort.'[2] Anjou's gendarmerie were driven back, he himself was unhorsed, and only saved by members of his personal guard, the *cornette blanche*. The royalist second line cavalry under the Margrave of Baden, Thoré, and the Duke of Aumâle was brought up, but did not turn the tide. The Margrave was killed, Aumâle 'had sufficient trouble to disentangle himself.'

Matters were quite undecided, the field covered with broken squadrons rallying in disorder, and making inconclusive partial charges, when Tavannes brought up his last reserves—the mass of Swiss infantry and the 'ost de reserve' of Cossé and Biron. It was unusual for foot to charge horse, but Pfeiffer's Swiss are said to have come up 'at a trot' and to have thrust themselves into the mêlée, giving the royalist cavalry a solid nucleus on which to get into line. Louis of Nassau had still some reiter squadrons in hand, fresh troops according to Catholic narrators, rallied units according to his own friends. With them he

[1] I can hardly believe the theatrical tale in d'Aubigné that the Admiral and the Rhinegrave met in front of their squadrons, and exchanged shots—Coligny scoring a 'bull's-eye' and the Rhinegrave an inner.

[2] La Noue, p. 988.

charged the Swiss, but failed to break them, being held off by heavy fire from the arquebusiers who flanked the pikes. According to the Tavannes Memoirs the Germans executed the regular manœuvre of the reiters, the 'caracole'—'faisant leur limaçon accoustomé' round the flank of the Swiss column, when they were charged by Cossé and Biron with the royalist last reserve of horse—the only intact body of that arm left on either side. The reiters swerved off and retired, with heavy loss.

This last effort settled the day, for the disordered squadrons all along the Huguenot front, when they saw their last militant unit give way in flight, threw up the game, and left the field, abandoning their unfortunate infantry, which had been allowed to take so little part in this essentially cavalry battle. Louis of Nassau gathered the reiters from both wings into a formidable mass, and went off in some order, the wrecks of many French squadrons with them. The few parties of the royalist horse which had the pluck to pursue them were repeatedly checked when they tried to close, for the reiters—and some of the French too—turned back and formed up a steady rearguard. They reached Ervault little molested. Other fractions of the French horse scattered more widely, and turned up in unexpected quarters.

The large majority, however, of the Catholic horse had stopped upon the battlefield to help to make an end of the deserted foot-regiments of the enemy. Mansfeldt's landsknechts from both wings had coalesced into one solid square, if we may trust Perissot's famous plan-panorama of Moncontour, and made face in all directions. They would not have been averse to a surrender, but this was made impossible by the fury of their old professional rivals the Swiss, who ran in upon them with the shout of 'Remember Roche-Abeille!' and 'Remember St. Colombe!'[1] while the royalist horse charged them in the rear. The killing was done in cold blood. La Popelinière says that many of the Germans fell on their knees crying, 'bon Papiste, bon Papiste, moy!'[2] but no quarter was given. With them perished many of the French infantry, though these scattered more, and did not form up to be

[1] At La Roche-Abeille the Royalists held that there had been much killing of prisoners who asked quarter. (See above, p. 444.) St. Colombe was the chief of the Navarre gentlemen who had been put to death after surrender, when Montgomery took Orthez. (See above, p. 445.)

[2] See Popelinière, xix. p. 140.

slaughtered in a square. Anjou is said to have interfered somewhat late, and to have saved the lives of 500 men, as well as those of d'Acier and La Noue, whom Montpensier wished to kill—they had been captured in the early cavalry fighting. This is one of the very few creditable episodes in the prince's contemptible life: La Noue vouches for it.

About half the Huguenot infantry must have been massacred—the remainder got off, some into the woods behind the battle-line. Of the royalist infantry very few fell, but as to the cavalry, the curious fact emerges that the victors lost more than the vanquished. The total is given, even by first-hand Catholic sources like Castelnau, as 500 against 400 Huguenots. Of the latter there fell the Count of Pfirt,[1] in command of landsknechts; Biron, brother of the Catholic Marshal, but his bitter enemy; and Puy-Greffier, a very old captain of high repute.[2] La Noue and Crusol d'Acier were taken prisoners—the Admiral (as we have seen) was wounded. Among the Catholics both the reiter-commanders Philip the Rhinegrave and Philibert of Baden were slain, and Clermont-Tallard the chief of the Catholic noblesse of Dauphiné; also the Italian Count of Sasatello. Guise, Peter-Ernest of Mansfeldt, Schaumberg, Bassompierre, and two Italian counts were wounded. It would seem that the main body of the Catholic horse of the 'bataille' was very badly cut up: the Swiss infantry saved the day.

This awful disaster of the Protestant army ought to have ended the war, but did not! This was mainly owing to political intrigue. King Charles IX was bitterly jealous of his brother Anjou, and came up to the front to supersede him in supreme command. Tavannes, in high disgust at losing his post as mentor, pleaded ill-health, and retired to his home. The Catholic army sat down before St. Jean d'Angely, in which large numbers of the fugitive infantry had taken refuge, and lay before it for six weeks. The defence by the Huguenot captain, Armand de Piles, was skilful and desperate—he was allowed to surrender on very easy terms on December 2, because winter had come, and the royal army was melting away. All that Moncontour had given to the victors was possession of

[1] Count of La Ferette in all French narratives—but his holding was the Alsatian county of Pfirt, in the Sundgau.

[2] He is said to have sacrificed himself and his squadron in protecting the retreat of 1000 broken Huguenot infantry into the Bois de Maire, which lay behind the battlefield.

most of Poitou—but La Rochelle was not seriously threatened, nor even Angoulême and Cognac.

Meanwhile the Admiral, when he had recovered from his wound, carried out an astounding diversion. Still followed by the bulk of Count Wolfrad's reiters, he turned south and gained Guyenne, where Montgomery's victorious army was in force, and where the Huguenots in November had taken Nimes, the second most important city in Languedoc. He spent the winter in the south, practically unmolested, and in the spring undertook the most extraordinary raid on record during all these wars. With 3000 horse, mainly reiters, and 3000 arquebusiers mounted on country nags, so as to be able to keep up with the cavalry, he made between April and June an astounding march from Nimes to the Upper Seine, through Languedoc, Forez, Burgundy, and Champagne, and repulsed at Arnay-le-Duc (near Autun, in Burgundy) an army under the Marshal de Cossé, sent to intercept him (June 26). Having picked up at Sancerre and La Charité on the Loire—places which had held out all through the winter—guns and reinforcements, and having been joined by scattered Huguenot parties from all quarters—one considerable body came from Geneva—he was actually threatening Paris at midsummer. His force was inadequate for any real offensive, but his position had immense psychological advantages.

For the Queen-Dowager was in one of her depressed and nervous phases of mind—her son, the King, was getting out of hand and threatening her favourite boy, the worthless Henry of Anjou. The Guises on the one side and the numerous Montmorency clan (the Marshal and his brothers, Damville, Thoré, and Meru) on the other were cavilling at every proposal laid before the Council. Catherine was also, as Tavannes had admonished her, cherishing suspicions of the King of Spain. Moreover, there had been a series of petty disasters in the west, where sallies from La Rochelle had cut up a considerable royalist force, and retaken several towns in Poitou.

The Admiral had been for several months making overtures for a pacification, on such terms as he could get for his co-religionists: and, to the disgust of the more extreme Catholics, the Queen ended by accepting his proposals, and signed, on August 8, 1570, the 'Peace of St. Germain,' which granted the Huguenots liberty of conscience everywhere, free exercise of their cult in two cities of every province, and in all other places where it was actually established at the moment, including the castles and manors of the noblesse. Protestants might be admitted to judicial and administrative posts, and finally—a most surprising concession—they might keep garrisons in four of their strongholds—La Rochelle, Cognac, Montauban, and La Charité on the Upper Loire. The tenure of these places was theoretically to be for two years only; when matters should have settled down, they were to be restored to the King.

This peace was as insincere as any other that Catherine de Medici ever signed; it helped her over a crisis whose danger she possibly exaggerated, and she had no more intention of abiding by it honestly than she had when the Peace of Amboise or the Peace of Longjumeau were proclaimed. The game was to be resumed, again and again, when she could see her way to repudiating her pledge.

NOTE ON THE BATTLE OF MONCONTOUR

The details of this fight are most difficult to reconstruct, not for want of evidence but from perplexing contradictions in the narratives which have to be studied. The contemporary documents are the royalist official report and the panorama-plans of Perissot and Tortorel, published in 1570. These last are obviously derived from the official report, but are defective in scale and topography and have some obvious blunders. De La Noue, who was present, gives comments rather than a narrative. Castelnau, also present, but writing many years after the battle, seems to have been affected by Perissot's plans, which he must obviously have seen. De Mergey, also an eye-witness, gives only his personal adventures. The Tavannes Memoirs are valuable, as giving the reminiscences of Anjou's 'chief of the staff,' but they are not from his own pen, though doubtless they contain his impressions. In one or two details they are obviously inaccurate. D'Aubigné is, as always, clear and comprehensible, but belongs to the next generation; he has, however, much good material from the Protestant point of view. Davila is also much too late, and has little original evidence, having obviously studied his predecessors. De Thou and Popelinière are also worth consulting.

Most modern accounts of Moncontour seem to me to have got their topography wrong—they make Anjou (or rather

Tavannes) absolutely cut the Huguenot line of retreat to Ervault. This is clearly an error, as Mergey—a combatant—says that the mass of the retreating Protestants got to that place, and Castelnau allows that the few Catholics who pursued only followed as far as Ervault (vii. p. 52). If the Royalists had got across the line of retreat, this would have been impossible. That the battlefield was not immediately south of Moncontour, but some way down the high road, is clear from the fact that the large wood—Bois de Maire—covers the terrain in this quarter. It is only mentioned by d'Aubigné, who says that some flying Huguenot infantry got shelter there after the rout. Obviously it was some way to the rear of the Admiral's left wing. I utilize as a plan Perissot's diagram, herewith reproduced: it is invaluable for the general disposition of the armies, though there are the errors noted above as to the Huguenot guns, and the disposition of the landsknechts.

CHAPTER VI

CHARACTERISTICS OF THE LATER YEARS OF THE FRENCH WARS OF RELIGION — HENRY OF NAVARRE—1572–96

FROM the Peace of St. Germain (August 8, 1570) to the Peace of Fleix (November 29, 1580) the weary chronicler has to record four more 'Times of Trouble' and four more insincere pacifications. After a few years of nominal truce the eighth civil war broke out, with most complicated permutations and combinations of political parties, and dragged on for more than ten years before the last malcontents owned Henry of Navarre to be also Henry IV, King of France. The whole dismal period seems a nightmare of blood and perjury—the main mystery being to discover how anyone could ever attach any reliance to promises made by Catherine de Medici, Charles IX, Henry III, or (for the matter of that) by the younger brother of the kings, the wretched Alençon, who so deceived the expectations of the Netherlanders, and was so egregiously fooled by Elizabeth of England.

The most astonishing phenomenon of the period is that it was possible to patch up even temporary and unsatisfactory agreements between the Crown and the Huguenots, after the Massacre of St. Bartholomew (August 24, 1572). Even when we grant that this atrocity was not long-premeditated, but an explosion of sudden panic on the part of the Queen and her advisers, who feared that the unstable Charles IX might slip out of their control, it remains astounding that the Huguenot party could ever again have signed a treaty with such people, even when 'sanctions' of all kinds were granted to them.

The explanation would seem to be that the party had grown conscious of its own desperate position, and was glad to see that the enemy was not aware of it. Down to 1572 there had been the dream of conquering France for 'the Religion,' of launching a regenerated nation against Spanish and Papal tyranny. After 1572 the Huguenots were fighting for survival, not for domination. It was a relief to find that the Crown had proved unable to exterminate them, even when the Queen-Dowager and Anjou and their gang had murdered Coligny and so many of his old captains, and had engineered mass-slaughter in dozens of French cities. The interminable siege of La Rochelle had proved the military incapacity of the government, even more clearly than its failure to trample out the sporadic risings of the South. Though leaderless, these indomitable sectaries had put up such a resistance that the enemy stood foiled. When the Crown offered peace on not impossible terms, it was possible to negotiate, even with the memory of St. Bartholomew in mind. The day had passed when complete military and spiritual victory could have been hoped for; there remained the chance of accepting the terms offered by the demoralized enemy. One might trust in Providence, and wait for the vengeance of Heaven to fall ere long on the Jezebel and Ahab of the day. Some bold spirits were driven into anti-monarchical theories, tending toward republicanism, since nothing could be hoped from the treacherous and bloody house of Valois. This was always an underlying idea, till the strange chance came which smote the royal family with death and the failure of male heirs, and made Henry of Bourbon and Navarre the destined successor to his distant and degenerate kinsmen. But this possibility did not arise till twelve years had passed since St. Bartholomew's Day.

But meanwhile the Huguenots, instead of being crusaders for the triumph of 'the Religion,' became merely one of several parties contending for control of the Crown, and were forced into touch with other factions of doubtful political morality in that struggle. The governing facts of the period were two—the first was the growing estrangement from the Crown of the more irreconcilable Catholics, to whom all projects of toleration were hateful, and who found their leaders in the ambitious house of Guise. The second was the development of the 'Politique' faction, at the other end of the Catholic party—the existence of a powerful group which so disliked the extremists that it was willing to tolerate Huguenots, even to co-operate with them actively—if thereby it could gain its ends. The Crown was represented by Henry III, a contemptible Heliogabalus, with an occasional dash of Nero, who was making royalty ridiculous as well as odious. Behind him, but not dominating him so much as she had hoped, was his mother, whose methods were sufficiently well known to all parties. The game of the Crown was to keep matters going by turning one party against the other with no relation to principles. The royal prestige was sinking so low that it was hopeless to dream of controlling the realm by its ancient authority. Hence from time to time Henry III is found cajoling 'Politiques,' Protestants, and Leaguers with equal insincerity. It was morally unfortunate for the Huguenots that they were utilized as allies by such contemptible creatures as Alençon, and such blatant egoists as Damville, when they fell into line with the 'Politiques.'

And when Henry of Navarre escaped from the four years of degrading abjuration and servitude which followed St. Bartholomew, and resumed the creed of his boyhood and the headship of the Reformed party, his followers had gradually to discover that his personal interests had more weight with him than the 'Pure Gospel.' It was rather absurd to find, as the successor of the austere and puritanical Coligny, a cynically humorous prince of deplorable manners, whose geniality and quick courage could hardly be considered to atone for his lack of appreciation of the binding force of the moral law. When once Henry of Navarre became the accepted head of the Huguenot party, there came an end to idealistic conceptions of its duty and its aims. It had become a party among parties, guided by a very astute and self-centred politician, who was set on getting a fair deal for his followers, but had no intention of sacrificing his own chances for their tenets and theories. When the assassin's hand made him in 1589 the legitimate king of France, his abjuration of the Reformed religion was inevitable—the long delay in its consummation came not from doctrinal doubts in the mind of the catechumen, but from careful consideration of the precise balance of interests at the moment. When would the unavoidable disgust at his relapse, on the part of his old Huguenot followers, be much more than counterbalanced by the reluctant acceptance of him by Catholics shocked at the prospect of falling completely under the yoke of Spain?

The most perplexing problem in the chronicle of these years is to realize exactly how the balance of parties stands at the particular moment which is being studied. Is the Crown, with the assistance of all other parties, at war with the Huguenots? Or are the Huguenots and 'Politiques,' more or less united, attacking the Crown? And if so, are the Catholic extremists backing the Crown or playing a game of their own? 'Three-cornered duels' are by no means unknown. The only incredible and impossible combination was that of the Huguenots and of the 'League' under the Guises; but all others were possible. The 'Politiques' might be in arms for or against the King—so might the Leaguers. A clear definition never came till, in 1588, Henry III murdered the Duke of Guise and his brother the Cardinal at Blois, under circumstances of the vilest treachery, which made any subsequent alliance between the Crown and the extreme Catholics impossible, and threw the assassin into forced connection with his heretic cousin the King of Navarre. What would have happened if the worthless life of Henry III had been prolonged for some time it is difficult to judge: but his death at the hands of a Catholic fanatic, only six months after the crime of Blois, clarified the situation, though it left it apparently insoluble. The 'Politiques' had to recognize the rights of the Huguenot heir—or to face the much worse alternative of throwing over the sacred 'Salic Law,' and accepting the late King's niece, the daughter of the hated Philip II, as next heir—so that France would be absorbed in the Spanish Empire. The Leaguers, for their part, disliked the Spanish succession almost as much as did the 'Politiques,' both for personal and for patriotic reasons, but were being inevitably pressed toward it for four years, as they found no prospect of worsting the Navarrese by their own power, and grew more dependent on Spanish aid for survival. On the other side, Henry of Navarre, in spite of his legal title, and in spite of the support of the greater part of the 'Politiques,' had failed to subdue Paris, and was not master of half the realm of France; his campaigns had brought no definite victory.

Hence the ultimate compromise—Henry, with no great mental pangs, abjured Protestantism and conformed—many of his own Huguenot supporters condoned the act for reasons of patriotism, knowing well that their master was not, and never had been, a conscientious Calvinist, and that spiritual sympathy would be wasted on him. The Leaguers felt that a Catholic King of any sort was better than a Spanish Queen, and allowed themselves to be bought over, one by one, by titles, governorships, and vast gifts of money. It was not a very high-principled affair on either side. But it saved France.

The study of intrigue and policy in all these unhappy years

is much more engrossing than that of the military art. No epoch-making changes or inventions are to be chronicled, and in the three or four general actions which occurred there is nothing fundamentally new to be found. But certain marked developments may be noted as the years slip by. The regular use of the lance, which had still distinguished the old gendarmerie at St. Denis and Moncontour, and which had not entirely died out by the day of Coutras, had very nearly disappeared from French armies by the date of Ivry (1590) by which time the Spaniards alone seem to have been constant to the long weapon. Henry of Navarre's cavalry were all pistoleers in full armour, as the reiters had been, and like the reiters they were charging in deep formation. The charge *en haye* had been found too shallow—a few casualties in the long front of a squadron so arrayed produced such gaps at the moment of contact that unity of command was lost, and an enemy working in a deeper order would get through somewhere. At least so thought La Noue, when he expounded in his *First Paradox* that with equal numbers the reiter would beat the gendarme. He does not, oddly enough, allow for the fundamental defect of reiter-tactics working by the 'caracole'—the successive pistol-discharge of different ranks. For he regards this tactical device as an error on the part of the reiters, not as their proper system.[1] A reiter-squadron caught using the 'caracole' by an enemy riding also in deep formation, and not *en haye*, ought to be broken by the onset of such men in the hostile front ranks as had not been shot, falling upon a mass of individuals making to the rear in order to recharge their weapons. Apparently the French heavy cavalry, when they had dropped the lance, and had taken to the pistol and to charging in a fairly deep formation, were expected to use the sword after the first moment of clash with the enemy's front. La Noue was of opinion that the best formation was six or seven deep, so that a company of about a hundred riders would have a front of about fourteen or fifteen.[2] He is careful to provide

[1] 'La première faulte est, qu'estant a vingt pas des ennemis, ils leur tournent le flanc, et déchargent sur eux leur salve de pistolets, pour ce que (disait-ils) plus de gens peuvent tirer que s'ils heurteraient par teste. Et sans doubte si lesdits ennemis s'estonnent et tournent le dos ils les accousteront mal. Mais s'ils tiennent ferme et les reitres vont refaire un grand circuit pour recharger leurs pistoles, et si les autres suivent chaudement, ils vont prendre la carrière tout du long' (*i.e.* the reiters gallop away right ahead if charged and pursued). *Paradoxe*, i. pp. 445–6.

[2] *Discours Militaire*, pp. 415–16.

that the best men should ride in the two or three front ranks; recruits and men with insufficient armour or weak horses must be relegated to the rear of the squadron.

It is worth noting that Henry of Navarre won his first victory in the open field at Coutras (October 20, 1587) precisely with this cavalry formation. The royalist commander, the King's favourite, Joyeuse, advanced with his main body of horse in a long line, two deep only. Henry, with his cousins Condé and Soissons, who faced the attack, had the Huguenot horse, decidedly inferior in number, drawn up in three solid squadrons six deep. There was some garnishing of arquebusiers between the three bodies, but these were for defence, not for attack. The important thing was that each of the three Huguenot squadrons completely broke through the line opposed to it, and then turned sideways, and rolled up the broken fragments on each edge of the gaps by flank charges. There was a great slaughter among the Catholics: Joyeuse had been hanging his prisoners of late, and the victors replied by cutting down or pistoling him and some 400 other royalist gentlemen, who would have meant good ransom money. What happened in other parts of the field had not mattered, when Joyeuse and his chivalry were exterminated.

The cavalry on all the more important battlefields of the later Wars of Religion was very numerous in proportion to the infantry. At Coutras the Catholics had 2500 horse to 5000 foot; the Protestants 1500 horse to 4000 foot. At Arques the army of Mayenne was nearly thrice the strength of that of Henry IV—perhaps 22,000 to 9000, but the proportion of cavalry in each was much the same, about one in three. At Ivry, Henry had 3000 cavalry out of a total force of 10,000; Mayenne 4000 out of a total of 12,000. The idea of creating a lightly moving force, every man horsed, for a long raid, which had been seen in 1570, when Coligny mounted 3000 arquebusiers on country nags to accompany his cavalry in the sweep from Nimes to the Upper Seine, was repeated by Henry of Navarre in 1592. Alexander of Parma having marched from the Netherlands to raise the siege of Rouen, then hard pressed, the King left all his infantry in the lines of circumvallation, and rode into Picardy with 3000 German and 2000 French horse accompanied by 2000 arquebusiers provided with mounts. His design was to hang about Parma's line of march, blocking defiles and cutting off detachments, without committing himself to an action. He failed completely—the Duke so conducting his campaign that no chance was found of taking him at advantage. Henry always had the worse when he came up against this great commander—who left as his verdict that he had supposed that he had to deal with a general, but found that it was only with a captain of mounted infantry.'[1] Rouen was duly relieved by the Spaniards.

Infantry was still, as in the earlier wars, organized on the system by which each unit was composed of the clumsy combination of pikes and firearms. Sometimes the unit in battle array was the regiment—now of comparatively small size—ten or twelve companies and about 1000 of all ranks being the supposed average. But occasionally several regiments were massed together into great blocks of 3000 or 5000 men—as had been seen at Dreux and Moncontour. If an army was large, there would probably be a front with cavalry units between each mass of infantry: but on the other hand it was not unusual for cavalry to be flanked with detachments of arquebusiers, who were intended to help them against hostile cavalry not so provided, as at St. Denis and Coutras. The fact that these 'pelotons' were commonly called 'enfants perdus' is a sufficient proof that the danger of their position was realized. If their comrades on horseback were routed, they were in dire peril of being cut up. If the ground permitted it, these daring skirmishers were expected to utilize hedges or thickets for self-protection; but obviously this was not always possible. Coligny is said to have been specially prone to this formation, and Henry of Navarre certainly employed it. But this may have been in part due to the habitual lack of an adequate proportion of pikemen in the Huguenot regiments: there was a desire to utilize the over-plus of arquebusiers in some fashion. To leave them in line of battle of regiments, with each regiment sadly short of the necessary stiffening of pikes, was not a promising formation.

By the end of the century there was a growing proportion of musketeers, as opposed to arquebusiers, in the 'shot' of each regiment. The musket, of which the Marshal Strozzi was the first patron in France, was so heavy to handle that it had to be worked with a fork, to support its muzzle at the

[1] Parma called Henry a 'carabin,' this being the name used at that time for mounted arquebusiers—the force that afterwards developed into dragoons. Dragoon is a seventeenth-century designation.

moment of firing; it was decidedly a more clumsy weapon than the arquebus. Drill-books show the extraordinary complication that ensued in the handling of musket and fork while loading was going on. And rapidity of fire must certainly have been hampered. Nevertheless, the musket, owing to its longer range and the greater penetrative power of its balls, continued to grow in popularity. Even in the Swiss infantry, always most conservative, and noted for the small proportion of men with firearms in its regiments, there were by the end of the century 30 musketeers to 30 arquebusiers in a company, of which the majority was composed of some 200 pikemen. Of course in a Spanish, and still more in a French, regiment the proportion of the 'shot' was much greater. When Henry of Navarre had become Henry of France after 1589, he was not so habitually short of pikemen as in his earlier days, when his landsknecht units alone were properly equipped with the long weapon. This came from the fact that, when he inherited the forces which had belonged to his kinsman, Henry III, he took over the old Royal regiments and Swiss, which were properly organized. He was, therefore, no longer hampered in fighting power by having such a large proportion of south-French Huguenot corps, in which the pike was always so greatly lacking.

Henry was a genial leader and a master of men, but he undoubtedly belonged to the same school of thought as his uncle, the elder Condé, and believed in winning battles by smashing cavalry charges, in which he himself should lead the decisive attack. He was not of the more modern way of thinking, which held that the proper place for the commander-in-chief was to be between two solid blocks of infantry in the reserve line, the '*experts envieux qui disoient que le général ne devoit estre à la teste*.' But that he must have had an uneasy feeling that there was something in the idea is sufficiently shown by the remark of his friend, the elder Biron, after Ivry. Henry had led the main front line of horse, and left the Marshal in command of the reserve. '*Sire*,' said the latter when the fight was over, '*vous avez faict aujourd'hui le devoir du Maréchal de Biron, et le Maréchal de Biron a faict ce que devoit faire le roi*.' And the remark was not resented.

His most trusted officers several times took him to task for what they did hesitate to call his vanity, in risking his life without reason in unnecessary exploits. Sully records two occasions on which he dared to speak out, one after the repulse of a sortie during the siege of Rouen, the other at the combat of Aumâle (1592) when the King took charge of the last squadron covering a retreat, and saw half his men cut up, while he was doing the work of a mere captain.[1] Henry is said to have endeavoured to justify himself. Half his army, all the Catholics, he said, were following him not because they liked his faith, or thought much of his constitutional rights to the throne, but because he imposed upon their imagination and their sense of honour. His reputation and his personal influence was everything: 'ainsi hazardez-je tous les jours ma vie, et endurez-je mille choses qui me faschent bien fort, pour maintenir ma reputation, puis qu'il m'est beaucoup meilleur de mourir les armes à la main que de voir disperser mon royaume, remettant en Dieu moy et mes affaires.' But what would have become of France if he had stopped a fatal pistol-bullet in one of these exploits, like Maurice of Saxony at Sievershausen, or Gustavus Adolphus at Lützen? His death would have meant the ruin of France, as he was well aware. The fact is that he enjoyed these thrills!

No doubt Henry did get the best effort out of his Catholic gentlemen by leading them in person—but obviously he was thinking of his own triumph, not of the future of France, or of French Royalty (still less of French Protestantism), when he led one of these exhilarating charges. If he should be killed (which was quite possible) these problems would not concern him. Providence must take the responsibility! It is difficult to understand the idealized view of him which has been held by French historians—essentially he was a genial, self-centred adventurer, liable to be distracted from his obvious duties, down to the last, by purely personal indulgence. One has only to think of his scandalous pursuit of the Princess of Condé, which he made into a matter of high politics, when he was fifty-five, grey-headed, and gouty. But the 'vert galant' was such a relief to France after thirty years of the degenerate Valois brothers, that one must make allowances for popular enthusiasm in his own day, and literary enthusiasm in succeeding generations.

As to his military talent, he obviously belonged to the same class of leaders as Pappenheim or Murat—he was a cavalry specialist, not reaching up to the standard of Cromwell or

[1] Sully, *Oeconomies Royales*, ii. pp. 36 and 49.

Charles XII, who could use all arms in battle—much less to that of the great masters, Turenne and Gustavus Adolphus of Sweden. When we find him employing his pelotons of arquebusiers, or the fire of his very modest train of artillery, we can see nothing that differentiates him from Coligny and other predecessors. Invariably he won by his cavalry charge, not by the collaboration of the three arms.

One thing we must concede to him—that he got the best work (as he claimed himself) out of a very difficult team of cavaliers. Enthusiastic attachment to his person had replaced Protestant enthusiasm, which was waning after thirty years of lost effort, or traditional loyalty to the Crown, which the odious Valois had done their best to kill, or the military pride of the professional soldier, who after all expected to be paid, and seldom was. No one else could have kept his armies together. He himself could barely succeed in holding his horsemen to their standards in hard times, when the squire wanted to go home to his (possibly ruined) manor, and the reiter-captain was a full year in arrears, and had been given no chance of late to sack a fat town. The strength of his armies varied from month to month, and more than once his purely French contingent dwindled to a nucleus, round which were gathered landsknechts, Swiss, or English, contingents of doubtful enthusiasm. But with a cheerful countenance, and a ready fund of raillery and humour, he persuaded his adherents to follow him in the darkest days.

Two of his battles show that he had at any rate the power to utilize a position, though when he had taken it up he intended to win purely by one of his smashing cavalry charges. The first was at Coutras—the victory where he made the remark that 'After this it will be impossible to say any longer that Protestants can *never* win a battle.' The second was at Arques, two years later, when he blocked an infinitely superior army by holding a defile where numbers were of little avail. But a study of his campaigns against Parma show that he had no such skill in open warfare—grand scale manœuvres—as his adversary. And the general study of his operations would seem to prove that he was sadly wanting in the Napoleonic faculty of utilizing a complete victory. After Coutras and Ivry he remained for long days without making any decisive move —in the former case he 'took a week off,' and went to Béarn to lay 22 captured standards at the feet of his mistress, the

Countess of Grammont,[1] a fact which would seem incredible if we had not for it the evidence of the best of the contemporary historians. All through these wars there appears astonishing inconsequence in the movements of most of the commanders —not to be explained by lack of money and munitions, nor by the habitual desertion common to both Royalist and Huguenot armies, nor by bad weather. Henry of Navarre was no exception to this rule. On the other hand, he was almost as frequently a quick mover; it is often difficult to make out why one mood was on him rather than the other. His energy was freakish and intermittent—the one certain thing was that when the active inspiration was on, he would be capable of very daring and risky improvisations, and of any amount of feats of personal valour. As his more cool-blooded adversary, Parma, is said to have remarked, 'He can make a splendid retreat; but why does he get into situations where such a retreat is necessary? I never do.'[2]

[1] D'Aubigné's criticisms, i. xv. p. 58, seem perfectly justified.

[2] This story only comes from Péréfixe, writing Henry's life in 1661, but he is full of good matter, if somewhat anecdotic, and knew his Henry well. He was born in the reign, but was too young to remember anything about it himself.

CHAPTER VII

THE BATTLE OF COUTRAS (October 20, 1587)

THE eighth and longest of the 'Wars of Religion' had started in 1586, when the miserable Henry III, after a feeble attempt to defend himself against Guise and the Leaguers, had submitted himself entirely to them at the Treaty of Nemours (July 7) and declared open war on the Protestants, by publishing an Edict which withdrew all toleration, and gave fifteen days' grace for all who would not accept the Catholic faith to quit the kingdom. Naturally the Huguenots flew to arms in all directions, and since they had great tracts of the West and South in their interest, it was clear that the Edict of Nemours was a declaration of war rather than an effective document. As in 1569 they called the Protestant princes of Germany to their aid, and the Elector Palatine, John Casimir, promised to raise once more hordes of reiters and landsknechts for an invasion of France in the next year. But the best security for the Huguenots was that the King was more interested in shaking off the yoke of the Guises than in extirpating heresy, and his forces and those of the League never co-operated. While sporadic warfare spread over most districts of southern France, the King let other local campaigns fare as they might, but took as his own main share of the war an invasion of the South by a royalist army under his profligate and expensive favourite Joyeuse, whom he had married to his own sister-in-law. The forthcoming German irruption would have to be looked after by the Guises and their friends. Joyeuse's first operations were futile, but in September 1587 he was marching from Saumur across Poitou and Angoumois, with the definite intention of joining another royalist force which held Bordeaux, and of shutting off the Protestants of the west from their co-religionists of Languedoc and Gascony. In this manœuvre he was marching across the front of the King of Navarre, who was collecting forces in the La Rochelle-Saintes district, with the object of carrying out precisely the manœuvre which

Joyeuse was wishing to prevent, namely, that of dropping down through Perigord and Languedoc, with the ultimate intention of joining hands with the succours expected from Germany. This was indeed a repetition of the plan which Coligny and the Duke of Zweibrücken had carried out nearly twenty years before.[1]

The King had lavished money and reinforcements on Joyeuse—according to some because he really wished him to make an end of Henry of Navarre. But according to others he wanted to get Joyeuse away from Paris and the Court, because he was becoming besotted on his other ' mignon ' d'Epernon, and suspected his elder favourite of intriguing with the Leaguers.[2] Joyeuse had with him about 8000 foot [3] and 2000 horse—his cavalry included a great number of courtiers of his entourage, little experienced in war but magnificently equipped—they were said to be the most showy body of cavaliers that had been seen for many a year. In all, the Duke had 24 companies of gendarmerie, four of light horse, and four of ' stradiots ' or ' Argoulets '—still called ' Albanians ' (the last time that this name is found used) under an Italian named Mercurio Bua. Of foot he had four strong regiments, one of them the old ' Picardie,' the senior infantry corps of the French army, besides some independent companies. But of cannon he only brought two to the field—perhaps because he was set on a forced march. He had sent orders to the Marshal Matignon, who lay at Bordeaux with 4000 men of all arms, to come out via Libourne and join him on the Dordogne.

Henry of Navarre and his cousin Condé had collected a rather haphazard assembly of troops in Saintonge, leaving much infantry in garrison, and bringing with them regiments that were small, because they had dropped behind all their ineffectives : many of the arquebusiers had got themselves nags, in order to be able to keep up with the cavalry. Eight foot-regiments only made up 5000 men—they did not average half the strength of the units in Joyeuse's army, but they were formidable veterans. The cavalry was about 1300 strong—more Poitevins and other western men than Gascons. A small reinforcement of 200 horse had got through from beyond the

[1] See above, pp. 441.

[2] See d'Aubigné, i. p. 47. If Joyeuse won, it was said, the King would score a point : if he was beaten, his master would be ' deffait d'un ingrat trop eslevé.'

[3] Sully's figures, not Davila's, but d'Aubigné says that the number of the foot is too high, and that it was only 5000.

Loire—scattered Huguenots from Anjou, Touraine, and Maine, under Charles of Bourbon, Count of Soissons, the younger brother of Condé. Though reared a Catholic[1] he stuck to his family connection on this occasion, from pure jealousy of the Leaguers. He was a good soldier, but a most untrustworthy friend to his kinsmen: his intrigues were a plague to the King of Navarre for many a year.

The Princes had marched from the Charente southwards, and had been joined by scattered Huguenot reinforcements at Archiac and Montlieu, from which they marched toward the Dronne River, intending to cross it at Coutras, and then to make for Bergerac, where they expected to pick up friends from Guyenne. This route, as Henry of Navarre was aware, would take them between Joyeuse's army marching south from Poitiers by Chateauneuf and Barbezieux, and Matignon's force moving up from Bordeaux. But he thought that he could pass between them, and was ready to fight either, if he met them before they united. As a matter of fact he succeeded in cutting in before Joyeuse could get south of him, and was at Montguyon on October 18, while the Duke's army was at Chalais, a long day's march north-east of him. From thence he marched for Coutras on the Dronne, with a strong cavalry vanguard under La Tremouille, one of his most trusted officers, to clear the way. Joyeuse had also an exploring party many miles in front of his army—his 'Albanians' under Mercurio: the two cavalry detachments met in the long street of Coutras, in the dusk of October 19, and the Albanians made off, knowing that their own supports were a long way to the rear, and believing that La Tremouille had the Huguenot main army close behind him. Joyeuse had halted and encamped at La Roche Chalais, ten miles from Coutras, and about the same distance from his last stage at Chalais town, on the same afternoon.

Henry was thus able to cross the fords of Coutras, and to take up a position in front of it, between the Dronne on his left hand, and the Isle, which joins the Dronne just south of Coutras, on his right. It was beautiful defensive ground, an open space only 700 yards broad alongside of the high road from Chalais,

[1] At the St. Bartholomew, when Condé was imprisoned, his younger brothers, Conti and Soissons, were mere children. They were kidnapped—Conti's guardian was killed with the boy in his arms—and handed over to Catholic relatives. Reared in the old faith, they never reverted to Protestantism, as did Condé and Henry of Navarre, but they became very untrustworthy 'Politiques,' sometimes fighting the League—sometimes the King.

with woods on each flank and the rivers just below the woods. But defeat on it might have meant absolute extermination, since to the rear there was a narrow defile in the angle between the Dronne and the Isle, the only exits being a single bridge and dangerous fords across one or other of the two rivers, difficult for infantry and almost impossible for baggage and guns. In short, the King had resolved to fight in a bottle-necked position, somewhat like that in which Simon de Montfort was caught at Evesham three hundred years back.

Aware that Joyeuse's army was only a march away, Henry sent out long before dawn 200 light horse to observe the road as far as Pointures, half-way to La Roche Chalais, and at an interval behind them 80 men-at-arms. This precaution saved him from being surprised, for Joyeuse had resolved to march at midnight, with the intention of catching his enemy unprepared. His light horse, the same Albanians who had fled from Coutras on the previous afternoon, were supported by 400 gendarmes, and ran in on the Huguenot outriders, with whom they got engaged in a long bickering action, gradually driving them back towards Coutras with much arquebus fire, which warned King Henry of the approach of the Royalists, and gave him time to deploy. Joyeuse's army, in one long column, starting at midnight, had some ten miles to cover along the high road before it got into touch with the enemy, and had then to wheel from line of march into line of battle. Daylight had long arrived before any general collision began.

Coutras was an unwalled village of one long street, with at its south end, commanding the main ford, a castle built by Lautrec—the unlucky general of the old Italian wars.[1] This strong place (long vanished now) had to its right large enclosures, a ' warren,' and a ' park.' Along the outer ditch of the warren Henry placed his right-wing infantry, perfectly safe against cavalry, though its plantations were only low brushwood, save at the most projecting northern end, where there was a well-grown thicket. There were four foot-regiments, about 2000 men,[2] along this front, who could only be dislodged by superior infantry. To the left of the ' warren,' reaching as far as the high road and a little beyond it, were the cavalry, drawn up in four bodies. The right-hand one consisted of the troops who had been skirmishing before dawn, strengthened

[1] See pp. 15, 27, 144, 175.

[2] Apparently Salignac, Bellezonce, Valetaux, and Montgomery.

by Gascon and Poitevin horse under La Tremouille and Turenne, perhaps 400 in all. Next them came three units with which the King intended to deliver his main blow—two columns of 300 cuirassiers apiece, drawn up six deep with a front of fifty—Condé had the right, Henry himself the left corps. To the left of them again was Soissons with his 200 Northern horse, also in column. Between each cavalry block there was a detachment of arquebusiers ' enfants perdus ' five deep, the front rank kneeling, who were under strict orders not to fire till the enemy should come within twenty yards of them. For the King intended to receive the enemy's charge, and only to strike himself when they should be at the shortest possible distance which allowed of getting his own men under way. The ' enfants perdus ' were warned that if they started firing before the crucial moment, they would certainly be ridden down.

On the right hand of Soissons's cavalry block there was a mound, on which Henry placed his modest provision of artillery, three guns worked by his gunner-specialist, Clermont d'Amboise. These came late upon the field, having been on the other side of the Dronne at night, guarded by three foot-regiments,[1] and with them got up not a moment too soon—the enemy being already developing his line of battle. The infantry went into the ' warren ' to reinforce the troops already there—it is rather surprising that they were not brought up on the left of Soissons's horse and the guns, where the King's extreme flank was only guarded by the ' enfants perdus ' on the left of Soissons and 300 arquebusiers detached from the regiments in the ' warren.' [2]

Joyeuse was very late in deploying, all his army having to be drawn out from the high road, where it had been marching in the night through woods. It was only after passing the Bois de la Gelleterie that it could extend itself. The Duke, like the Huguenots, drew up his army with all the infantry on the wings, and all the cavalry in the open ground by the high road in the centre. On the left were two very strong foot-regiments, Picardie and Tiercelin, who would be opposed to the King's infantry in the ' warren ' : they counted 1000 pikes and 1800 arquebuses, about as many men as the four Huguenot corps in front of them, each unit being double the size of one of the

[1] Regiments of Charbonnière, La Borie, and Neuvi.

[2] D'Aubigné expresses surprise that Henry did not bring up more infantry on this flank—it would have taken too long to move them round the rear ? (i. p. 52).

small hostile regiments. In the centre was the cavalry—on the left were the troops which had formed the vanguard during the march, Mercurio's Albanians, several cornets of light horse, and 500 lances under Montigny: these stood opposite La Tremouille's and Turenne's men. On the right, *en haye*, was a long line composed of all the rest of the gendarme companies, not less than 1200 lances. 'All the front rank was formed of counts, marquises, barons, and great lords—' la cavalerie la plus couverte de clinquent et d'orfevérie qui ait esté reunie en France.' Joyeuse's 'cornette blanche' of his personal company, with his banner, was in the centre. This thin line covered the front of Condé's, the King's, and Soissons's array. On the extreme right, beyond the cavalry, were two masses of infantry—the regiments of Cluseau and Verduisant, the latter composed of a number of detached companies formed into a provisional unit—they are reckoned at 2500 men. On their outer flank were some mounted arquebusiers. This wing was on wooded ground with a brook called the Pallard—which falls into the Dronne—making a marsh beside it.

It was not till nine o'clock in the morning that the clash began—so long had it taken for Joyeuse's army to deploy into linear formation. It will be remembered that all, horse and foot, had been on the march since midnight; they must have been weary—while Henry's army had slept comfortably in Coutras, save the outriders who had been bickering along the high road before dawn, and the guard of the artillery.[1]

At nine o'clock,[2] the enemy having completed his formation, but not having yet begun to move forward, the King ordered Clermont d'Amboise to take some long shots at them from his little battery on the mound. The first ball, as we are told by an eye-witness, knocked over the white 'cornet' of Joyeuse's squadron; the succeeding ones were turned against the flank of one of the infantry regiments, in which they made long lanes, killing eighteen or twenty men in a line. The Duke's two guns replied, but had been placed in such low ground that they were ineffective—Lavardin, Joyeuse's 'Maréchal de Camp,' advised his chief to charge at once, as the artillery-fire was doing too much damage, and the whole royalist line moved forward simultaneously.

[1] Sully this day was detached in care of the artillery, and speaks of the fatigue of hauling the guns through the fords of the Dronne—he was soaked through to his middle. *Oeconomies*, xxiii. p. 394.

[2] So d'Aubigné, i. pp. 14, 52.

It must have been at this moment that King Henry bade his chaplain Damours utter a short prayer, after which all the old Huguenots in the army sang in chorus their well-known battle-psalm—verses 24 and 25 of the 118th Psalm :

> ' La voici l'heureuse journée
>   Que Dieu a faite à plein désir ;
> Pour nous soit joye demenée
>   Et prenons en elle plaisir.
> O Dieu Eternel, je te prie,
>   Je te prie ton Roi maintiens
> O Dieu, je te prie et reprie
>   Sauve ton Roi et l'entretiens ! '

For this we have the first-hand evidence of an eye-witness.[1] Less well authenticated is the story that the King cried to his kinsmen, Condé and Soissons, who were standing by him at the moment when the enemy moved, '*Cousins, vous êtes du sang de Bourbon, et vive Dieu ! je vous montrerai que je suis vôtre aîné.*' To which Condé replied, '*Et nous, nous montrerons que vous avez de bons cadets,*' and rode off to place himself at the head of his squadron.

The Catholic infantry on the east end of the battlefield got engaged with the Huguenot infantry in the 'warren,' with no great profit to either side. But at the first clash of the horse, Lavardin, commanding Montigny's gendarmes, the light horse and Mercurio's Albanians, drove in first La Tremouille's and then Turenne's somewhat smaller force, and turned them to such rout that pursuers and pursued went in a confused mass along the corner of the 'warren' and as far as the first houses of Coutras, well behind the King's centre. There was some bad fighting here, says d'Aubigné; Tremouille and Turenne with a handful of others—including eighteen Scottish volunteers—rallied and joined Condé's corps. Some of the fugitives swam the river Dronne and fled as far as Pons. At the extreme other end of the line there was very different resolution shown—the detached arquebusiers who had been sent to 'amuse' Cluseau's infantry, charged into the wood and threw themselves among the Catholic pikes, their captains crying, 'Il faut mourir dans le bataillon !' So surprised were the enemy that their front wavered, and they had not cast off these desperate men till the battle had been settled elsewhere.[2]

[1] D'Aubigné, who was present as one of Henry's squires. The other tale only comes from Père Mathieu, p. 533.

[2] D'Aubigné, i. p. 53.

But the decisive action was in the centre, as Henry of Navarre had intended. Joyeuse with his 1200 cavaliers, charging in line, started to gallop so early that they fell into a most

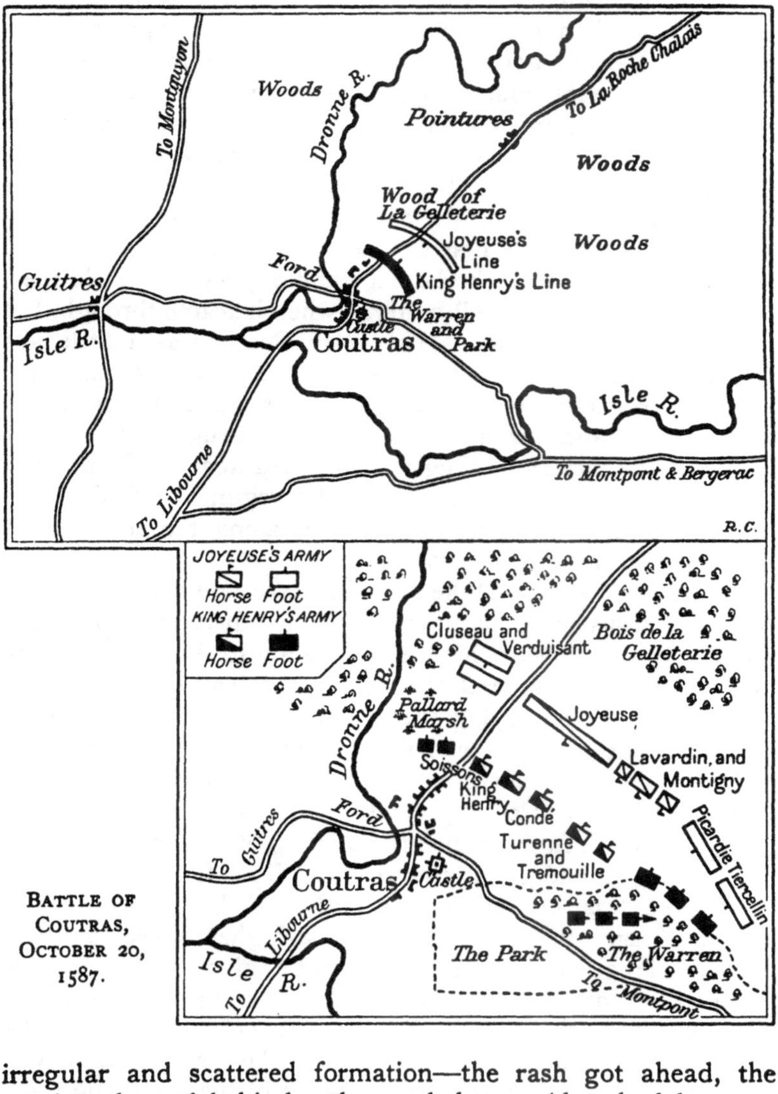

BATTLE OF COUTRAS, OCTOBER 20, 1587.

irregular and scattered formation—the rash got ahead, the cautious lagged behind: the weak horses (they had been on their legs since the past midnight) could not keep up with the rest. There were gaps in the thin line before they got near the enemy: when they were only twenty yards off, the 'enfants perdus' placed between each of Henry's corps blazed into them—not a shot could miss—and the six-deep blocks of the Huguenot squadrons charged the two-deep Royalists at a trot. As La Noue had prophesied, the column went right through the line at every point, in spite of the long lances and the headlong pace at which Joyeuse had come on. The taffety pennons of the gorgeous cavaliers were in a moment tossing in every direction, and in five minutes the battle was settled—the line having dissolved into fugitive individuals.

The King had one of the great days of his life; riding well ahead, he pistoled the first cavalier that he met, and dragged down Chateaurenard, who bore a standard, crying, 'Yield thee, Philistine!': he was well battered when he came through, but in the best of humours. His cousin Condé was a little less lucky: pursuing a gentleman named St. Luc, he was unhorsed —but the other, seeing the day well lost, handed his sword over to the Prince, and surrendered to him as he struggled to his feet. Some of the victorious Huguenots turned against the flank of the royalist infantry, and cut up the regiment of Picardie, which was already frontally engaged, along the edge of the 'warren,' with the King's foot.

The whole of the Duke's army, horse and foot, broke up when the disaster in the centre was seen—there was no reserve to rally upon. Joyeuse himself was riding to the rear, when he was overtaken by several Huguenot captains—he surrendered to the first comer, crying that he was worth 100,000 *écus* of ransom; but a third, whose company had been massacred by the Duke at St. Herai in the last campaign, blew out his brains while he was casting down his sword. There was much slaughter of the fugitives before the King of Navarre succeeded in checking it, and giving the word of 'Good Quarter.'

The contrast between the casualties of the two armies was extraordinary: no authority names a higher figure than 200 for the Huguenot losses; d'Aubigné gives the incredible estimate of only five gentlemen and a score or so of rank and file. On the other hand, the Catholic noblesse was cruelly handled—more than 400 were dead, a larger total than was seen on their side in any previous battle of the wars, and there were nearly 3000 others cut down before the word for 'Quarter' was passed round. Among the dead were Joyeuse, his brother the Count of St. Sauveur, three other counts, two commanders of foot regiments, and many bearers of great names, such as de Brézé, Neury, Rochefort, etc. The not inconsiderable number of prisoners included the Marquis of Piennes, the Count of Montsoreau, St. Luc, titular Royalist Governor of Saintonge, and many captains of gendarmerie,[1] including Montigny, who had beaten the Gascon horse at the commencement of the battle.

After this, as King Henry remarked, no one would ever be able to say again that the Huguenots could not win a battle. It was still open for the cynic to remark that they did not know how to utilize a victory—for the results were practically nil. Both Sully and d'Aubigné, who were with the army, bear witness to the profound dissatisfaction that was felt when no plan of campaign was pursued. Henry, as we have already seen, had been intending to march for Bergerac, and to gather his allies in the south, with the ultimate idea of going up to join the German army of invasion, which was known to have started from the Rhine in September under Dohna, the nominee of the Elector Palatine, and the Duke of Bouillon. Nothing of the kind happened—Condé, it is said, wanted to return into Poitou, and clear that region of royalist garrisons, before going up to the Loire: he opposed the plan for turning south. Turenne and the Gascons wanted to look after Matignon at Bordeaux, and to besiege Sarlat and other places, as well as to recruit their regiments, which were woefully under strength. The King was apparently unable to make up his mind to exert his authority, and—incredible as it may seem—let the army break up, and went off to Béarn, nominally to present his cousin Soissons to his sister Catherine—there was a project of a match between them. But actually he went to visit his mistress the Countess of Grammont, at whose feet he laid down 22 captured royalist standards. His confidant Sully expresses the disgust which he felt—'*la vanité de presenter à cette dame ces enseignes et autres dépouilles de l'ennemi firent que tous les fruicts esperez d'une grande et signalée victoire s'en allerent en vent et fumée.*' Even Soissons's projected marriage did not come off; the King came to mistrust him, and sent him packing!

This is, as we shall see, not the only occasion on which Henry

[1] D'Aubigné says that among them were Santrai and Sansac, the most miserly and the most impecunious officers of rank in the army. Sansac introduced himself and his colleague to the King as two personages from the Parables—Dives and the Prodigal Son. Henry burst out laughing, and let *both* off their ransoms, which vexed Sansac extremely.

showed an astounding inconsequence—he was a fine cavalry general, but frequently showed complete lack of strategical instinct. While he dallied in Béarn, the large German auxiliary army, very badly managed by its chiefs, straggled through Champagne as far as the Loire in September and October, much harassed by Guise and his Leaguers, turned back when they found the King of France with an army of his own in front of them, and finally broke up, some retiring on Alsace, others across the Franche Comté on Geneva, where Bouillon died of a broken heart. Only a wreck got back to Germany in December. All the credit of the campaign went to Guise—the royal army had done nothing.

CHAPTER VIII

THE BATTLE OF ARQUES (SEPTEMBER 21, 1589)

THE strange three-cornered contest between the Huguenots, the King with his slowly dwindling faction of courtiers, officials, and *Politiques*, and the zealous Catholic Leaguers, continued into the year 1588—all parties watching with extreme interest the King of Spain's great venture with the 'Invincible Armada' against Elizabeth of England. Its success would have meant absolute ruin for the Huguenots, no less than for the insurgents of the Netherlands. Henry III thought that it might lead to the extension of Spanish influence throughout France, with the aid of the Leaguers. But long before the Armada came to its disastrous end, and removed this particular source of fear, he had fallen into the hands of his own overweening subjects. In strict disobedience to the royal orders, the Duke of Guise entered Paris on May 9; the city rose in his favour on the 'Day of the Barricades' and overpowered such of the royal guards as tried to make any resistance (May 11). The King fled by night, tried for a few weeks to rally some sort of a party, but failing, capitulated on July 19, and signed the 'Edict of Union,' by which he gave over control to the League, consented to the removal of Epernon and his other favourites from power, and pledged himself never to make any truce or peace with the 'Heretics' till they should be destroyed. Meanwhile the Armada came into the Channel, fled battered through the North Sea, and perished on the rocks of the Hebrides and Ireland. It was probably the news of its disaster which emboldened the King to carry out the most disgusting act of his miserable life. While the States General was sitting at Blois, he lured Guise to pay him a visit in his chamber in the Castle, and had him slain under circumstances of abominable treachery and cowardice by his ruffians of the 'Quarante Cinq' guard (December 23). Two days after, Guise's brother the Cardinal—fourth of his name to wear the scarlet robe—was hacked to death with halberds in a dungeon.

Henry vainly thought that he had killed the League. 'Morte la bête, mort le venin,' he observed, and told his mother that he was King of France once more. The Queen-Dowager, a better judge, cried out in despair that 'he would soon be King of Nothing,' took to her bed, and died within the fortnight (January 5, 1589). Two of the chief authors of the St. Bartholomew massacre had ended suddenly: the third, the King himself, was to join them in the other world within seven months.

The League, so far from being dead, was stung into wild fury by the murders at Blois, and town after town adhered to it when the news went round. The wretched King, finding few to support him save his immediate entourage and his Swiss Guard, was forced to throw himself under the protection of Henry of Navarre and the Huguenots, whom he had sworn a few months back to exterminate. On August 2, 1589, he was assassinated at St. Cloud by the fanatic monk Jacques Clement, who came out from Paris with the deliberate intention of revenging Guise. Before he died he had just time to designate Henry of Navarre as his only legal successor.

The death of this last miserable scion of the House of Valois changed the whole aspect of affairs in France. A very large proportion of the Catholic nobles acknowledged the Navarrese as king, despite of his heresy, though a certain number went over to the League, and others, adhering to neither side, tried to make themselves independent in their own fiefs or governorships. The position of Henry was much complicated by the adhesion to his cause of such old Catholic chiefs as the Dukes of Longueville and Montpensier, and the Marshals Biron and d'Aumont, who had commanded against him in the field so often. They all had to have their loyalty confirmed by gifts and governorships, and kept continually urging him to change his religion, in order to spoil the propaganda of the League. His army, which had hitherto been essentially Huguenot, and composed of enthusiasts, was swamped by these new recruits; and the favour which was shown to them damped the ardour of the old Protestant zealots, of whom many were well aware that their leader's ambition was more certain than his religious convictions. The fighting quality of his troops was for the future not quite what it had been. That bloodthirsty ruffian the Baron des Adrets, who had done astonishing things as a Protestant in the earlier wars, but had shifted over to the

Catholic side, was once, in his old age, asked by an intelligent inquirer why he had accomplished marvels in his first capacity, and little or nothing in his second. His very lucid reply was that he had once been in command of fanatics, but later only of hirelings.[1] And this, to a certain extent, was now the position of Henry IV, for though his old Huguenot friends stuck gallantly to his cause, half his armies were for the future composed of men who had adhered to him from interest only.

He kept the Catholic nobles in hand purely by his personal influence, his geniality, and his reckless courage: they followed him as a fine soldier and a jovial comrade, not from any spiritual enthusiasm. And many of them were found acting with slackness, or even intriguing with the Leaguers, when their loyalty was not being stimulated by sufficiently convincing douceurs. His position would have been even worse if it had not chanced that his old uncle, the Cardinal of Bourbon, whom the Leaguers had proclaimed King by the name of Charles X, happened to be a captive in his hands. Henry III had imprisoned him in the castle of Fontenay after he had murdered the Guises at Blois, and his successor had him in safe custody. The cause of the League was singularly weakened by the fact that its nominal head was a prisoner in the power of his heretic nephew. To strike coins and issue edicts in his name was obviously absurd, when he was inaccessible, in a very comfortable dungeon. Before he died (May 1590) he acknowledged the King of Navarre to be King of France also—leaving the party which had used him as a tool in a hopeless constitutional quandary.

Henry IV, therefore, at the moment of his assumption of the royal title, was forced to work with a combination of somewhat disheartened Huguenot followers, and of interested and rather untrustworthy *Politiques*. That he kept them together, and won his way to complete success after his rather ignominious 'conversion,' is a testimony to his abundant tact and geniality as also to his convenient lack of conscientious scruples. But at the beginning of his reign he was in a very difficult strategical condition—the ill-compacted army with which he and his deceased cousin had been demonstrating against Paris—they could not be said to have been besieging the great city—dwindled rapidly, as successive magnates drifted off, each to secure his own interests in his own region. Henry, left with little more than 8000 men, retired into Normandy, partly hoping

[1] This curious story comes from Brantôme.

to win Rouen, its all-important centre, partly because he was expecting succours from England, which Queen Elizabeth had reluctantly promised, since she understood that the cause of the League was the cause of her old enemy Philip of Spain. He did not retire to the South, where the Huguenot cause was strongest, because he was aware that such a move would lose him the whole of the North, where many places were still held in his name.

On August 28 the King encamped at Darnetal, three miles outside Rouen, with an eye not only on that town, but still more on Dieppe, which was held for him by a 'Politique' governor. It was to this port that Elizabeth had promised to send her succours—Havre was in the hands of the Leaguers, and could not be used.

On September 1, the Duke of Mayenne, Guise's brother and successor as the head of the Catholic party, marched out from Paris with a much larger army than Henry had expected—he had been joined not only by all the Leaguers of the north but by a considerable contingent from Flanders sent by the Duke of Parma, and a large body of Lorrainers under the eldest son of Duke Charles.[1] The whole is said to have amounted to 4000 horse and over 20,000 foot—half of the infantry were Catholic Swiss and landsknechts. The whole force was about three times that which King Henry had in hand at Darnetal. Marching by Mantes and Vernon down the Seine, with the object of driving the King away from Rouen, Mayenne was surprised to find that Henry had moved his army not westward, but down to the coast at Dieppe. He followed, hoping to crush his much inferior enemy against the sea, or at least to drive him into Dieppe—a strongly fortified place—and to besiege him there. Henry's only way of escape would be to take ship for England, a course which would have ruined his cause in northern France.

But the Navarrese was intending to fight, not to allow himself to be besieged or evicted from the kingdom. Hence the extraordinary battle of Arques, one of the most notable cases of the defence of a defile to be found in military history. Though

[1] This Prince Henry, Marquis of Pons, was one of those who had an eye on the French throne. He was the son of Claude, *eldest* daughter of Henry II, while Philip of Spain put in the claim of his daughter, Isabella Clara Eugenia. child of Elizabeth, the *younger* sister of Claude. Catherine de Medici is said to have thought of this grandson as the best substitute for the hated King of Navarre. But he had no general support.

he had made a careful survey of the ground, it must be confessed that he was taking fearful risks, for Mayenne was in such overwhelming strength that it would have been quite possible for him to have shut up the royal army, with its back to the sea, without any general action.

Dieppe lies at the mouth of the Bethune River, whose tidal estuary formed its harbour. For four or five miles upstream there are salt marshes inaccessible even at low tide, through which the Bethune meandered in several courses—now for the most part dry. But in 1589 there were only three ways into the place—the first a road by the coast from Eu to the fishing suburb of Le Pollet, on the east side of the harbour, and thence by a bridge into the town. The second, also on the east bank of the Bethune River, is a high road crossing the marshes opposite the village of Arques, about four miles south of Dieppe. It is blocked by the castle of Arques, on the west bank, an old stronghold modernized by Francis I, and fitted with bastions and artillery platforms commanding the road. This is a perfect defile, as the road descending from the east bank is constricted by high ground about the village of Martinéglise on the right, and a steep thickly wooded hill on the left, the Forêt d'Arques. It crosses the course of the Eaulne, a small tributary of the Bethune, which falls into the marshes of the larger stream below the village of Arques, which rises high on the other side of the depression. The road passes the river-marshes at the bottom of the dip, on made ground not more than 400 yards broad; at the narrowest point of the defile was a chapel, St. Lazare, more commonly called La Maladerie—perhaps a lepers' refuge originally, if one may judge from the name. The dry ground broadens out a little where the Bethune takes its main course, immediately below the village and castle of Arques.

The third way into Dieppe was a high road coming from the south, well to the west of the Bethune valley, and arriving at the town across high ground by Forges and Longueville. If Mayenne were to take it, he left the King the opportunity of getting out of his way by leaving Dieppe by the side of Le Pollet and retreating north-westward, where several towns were in royalist hands—Eu, Neufchatel, etc.—while if either of the other routes were taken the enemy could only go off westward, into the blunt angle of the coast at the end of which lay Havre—in the Leaguers' hands. This may have been the reason why the Duke marched up the right bank of the Bethune, and proceeded to demonstrate against Dieppe from the east. On the 13th of September he appeared in front of Le Pollet, hoping to find that suburb weakly fortified—as indeed it had been till the last few days. If he could capture it, artillery placed in its front would make the harbour of Dieppe unsafe, and force all shipping to run out into the Channel. Henry had been as well aware of this as his adversary, and ever since he had heard of Mayenne's approach from Paris, had been setting his engineers to work, to substitute a formidable line of entrenchments for the slight original wall of the suburb. Not only the garrison, the crews of many ships, and the civil population had been employed, but several regiments from the army, when it had once moved up from Darnetal. They had excavated a deep trench, and palisaded the whole front, whose central projecting point, an old stone mill, had been earthed up into a bastion, and furnished with six pieces of artillery.

When Mayenne had inspected the new defences of La Pollet, and had discovered that the whole royalist army had moved up to Dieppe, and could furnish an ample supply of defenders for the lines, he halted for three days in front of them (September 13–16). We can hardly believe that (as some writers allege) he was in hopes that the King would come out and offer him battle, since he must have been well aware of the immense disproportion between their armies. But on September 17 he made up his mind to force the Arques defile, and marched off to take up his headquarters in the village of Martinéglise, which looks down on it from the east bank of the Bethune. Here he halted for three days more—an inexplicable delay. He cannot but have been aware that the King could make a corresponding move, and transfer his main-body from Dieppe to Arques, where he had before kept only a strong detachment: there are only four miles of good road between the places. It would appear that the Duke was prepared to use brute force of numbers, and did not realize the strength of the defile, nor that the King had taken careful precautions to block it.

While Le Pollet was being refortified at one end of his line, Henry had not neglected the Arques ground. His cavalry reconnaissances had explored the woods and hills beyond it, and having discovered the narrowest points at the bottom of the dip, where the river passes between the heights, he had drawn two successive lines of defence across them. At the Chapel of St. Lazare he had dug a broad trench and earthwork reaching from the thick wood on the east side of the defile to the edge of the marsh on the other hand: there were only two gaps left, through which cavalry might pass—one on the high road, the other at the edge of the marsh. The chapel was barricaded and four guns were placed on made ground, formed

BATTLE OF ARQUES, SEPTEMBER 21, 1589.

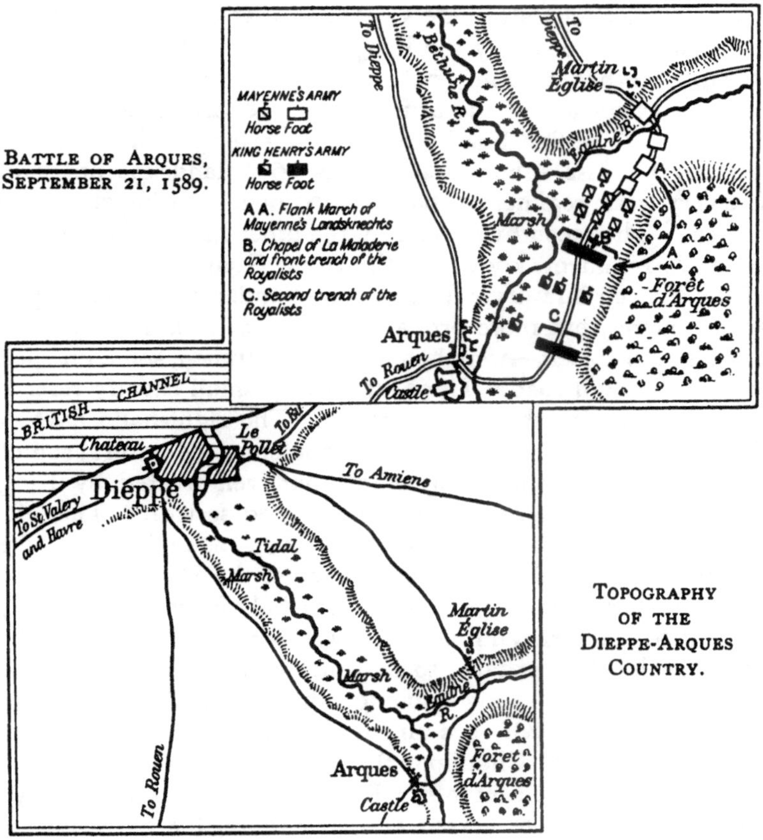

TOPOGRAPHY OF THE DIEPPE-ARQUES COUNTRY.

by earth from the ditch, overlooking the parapet. Five hundred yards back there was a second trench, drawn from the broad marsh to the foot of the wooded hill. The road here was sunken, and bordered by a high hedge forming good cover for arquebusiers. Behind this second trench the ground was commanded by guns placed in the castle of Arques, which could not, however, make themselves felt so far as the position of the first trench at the Chapel of St. Lazare.

On the night of September 20, Mayenne began to file his army down from the heights of Martinéglise to the high road, intending to sweep all before him by numbers, and forgetting (like Xerxes at Thermopylae) that on a narrow front only a limited number of combatants can be engaged. There was some skirmishing on the preceding afternoon, beyond the river, when Henry's outriders detected that the enemy was on the move. Accordingly the King manned his positions before dawn. He had certainly not more than 8000 of all arms: a certain proportion of the army had to be left in Dieppe and Le Pollet. The first line by the 'Maladerie' was held by half the royal Swiss—about 1200 pikes—600 landsknechts, and five very weak regiments—hardly 1200 men—of French arquebusiers. The second was held by the other half of the Swiss—Galatti's regiment, known as the Swiss Guards—and the rest of the French infantry. The cavalry—under 800 according to some sources, but more probably somewhat nearer 1000 [1]—all cuirassed pistoleers—were behind the first trench, some on each side of the high road. There was a very small cavalry reserve placed level with the second trench. The right-hand cavalry corps was under the Bastard of Augoulême, the natural son of Charles IX, quite a young officer but well trusted by King Henry, to whom he did at this time good service.[2] The left-hand cavalry, on the side next the marsh, was under the Duke of Montpensier, who had fought against the Huguenots in so many earlier campaigns. The Marshal Biron, an equally ancient foe of Henry in old days, was with the infantry of the first trench. It is notable to find that there were so many old Catholics with the King—among others Damville, one of the Montmorencies. Of prominent Huguenots we only find present Francis of Chatillon, the son and heir of Coligny, in command of French infantry; Rosny (later the famous Sully), with his company of horse; and Arambure, the old leader of Henry's *chevaux légers*. La Noue and many others were absent, campaigning in other quarters; but of course many staunch old Protestants were leading 'enseignes' and 'cornets.'

Mayenne had at the head of his column of attack masses

[1] Sully will only allow for some 700 horse, d'Aubigné gives the same figure. Davila, on the other side, speaks of 1400. Allowing for the national tendencies of partisans, I should guess at 1000 as a fair estimate.

[2] Sometimes called the Prior of Malta, and at others the Count of Auvergne—one has to remember that the three names only mean one person. He was brother to Madame de Verneuil, the most imperious of Henry's many mistresses.

of horse, commanded by his brother d'Aumâle and the Duke of Nemours, constricted on to the narrow front of the road, but he had a plan for turning the King's position before letting the cavalry loose. The right of the foremost royalist trench was, as has been already mentioned, carried up to the foot of a steep hill covered with closely grown timber, which might have been considered an impassable obstacle. Mayenne, however, detached into it a foot-regiment, the landsknechts of Collalto, with orders to push through it as best they could, and get behind the end of Henry's trench. Two French foot-regiments were to follow. The thickets concealed but disordered the landsknechts, whose leading parties arrived at the destined point scattered and in great disarray. Whether from deliberate treachery, or from fear of being attacked themselves by troops in good order, the first groups of the Germans raised their hands, hoisted their caps on their pikes, and cried that they were Protestants who would not fight their friends, and were intending to desert the League. The garrison of the trench, completely deceived, cheered and held out their hands to them. But when a large mass of the supposed deserters had gathered at the trench, they suddenly formed up, and charged in upon the Swiss companies who held the end of the line. Taken completely by surprise, the King's men fell back in disorder, abandoning the works, which were taken from the rear, and the landsknechts then pushed along the line as far as the Chapel of the Maladerie, which fell into their hands.

Henry himself had been visiting his front line at the moment, and was carried away by the back-rush of his Swiss, all falling back in panic towards the second line and the cavalry. It was then, apparently, that he uttered his well-known cry, 'Are there not fifty gentlemen of France who will come and die with their King?' The appeal was not in vain, for his cavalry to the right of the road came forward, and the Swiss in the second line of trench stood quite firm.

When the companies of horse at the head of Mayenne's main column saw the Maladerie and its trenches fall into the hands of the treacherous landsknechts, they charged down the two gaps which existed in the line, the one by the high road in the centre of the trench, the other on the edge of the marshes to the left. The narrow fronts of both were at once fiercely assailed by the royalist horse, and there set in a long and desperate cavalry mêlée, which lasted for an hour.

Sully, whose company was on the right front, next the wood, has left a very graphic account of these charges and counter-charges. 'When the upper trench was attacked by the landsknechts, who pretended to surrender and then seized the works, the King was in no small trouble. Of the fighting where I was not present in person I do not speak. But below the road we were attacked by 800 or 900 horse in three squadrons, whom we charged and held up at the twist of the road, though we were not more than 150 in all. Then four more squadrons came on, who drove us back at more than a foot-pace as far as a little mound, where we found about 150 of our own horse under the Count of Auvergne, on whom we rallied, and all charged together, driving the enemy off—our pistols in their backs—as far as the same turn of the road; then we could see some 3000 or so of their cavalry waiting for us. They came on, and beat us back as far as a point level with the Chapel, and were only checked by arquebus fire from our infantry. But the landsknechts who had taken the upper trench drove our foot away, and we had to retire to the front of our Swiss battalion (at the second trench) who stopped our pursuers. Here my horse was shot, and M. de Maignan brought me another. The enemy then tried to turn our left, by sending 500 horse through the low ground, to get round our Swiss, but they rode into a quagmire, where the chargers sank up to their girths in the marsh, and most of them had to dismount and struggle off on their legs, leaving their horses completely enbogged.

'But things were in a most uncertain condition, we were mostly tired out, pistols empty and swords blunted, and more fit to fly than to fight. I went to seek for the King, to say that we could hardly defend ourselves. I found him at the lowest point of the road, and cried that he must send us fresh squadrons. "My friend," he replied, "I have no fresh squadrons." But he turned to M. le Grand and said, "Scrape up what you can of our least exhausted men, and follow Rosny along the road; when you get back to our line, then cry, 'Courage, gentlemen, here are the fresh troops that the King has sent us.'" We advanced all together as far as the mound that I have spoken of, and saw before us the whole army of the League, horse and foot, coming on in order, as if to roll over us. No one thought that there was any hope, when suddenly the fog, which had been lying thick all the morning, lifted, and the cannon in the castle of Arques opened on the masses: we saw the balls make four neat gaps in the front of their horse and foot; three or four more similar discharges had a marvellous effect. We saw them give back, to get out of artillery range, pass the turn in the road, and finally retreat towards their old quarters.'[1]

Sully, as he says himself, only gives his personal experiences of what happened on the royalist right, so that it is necessary to supplement his narrative by some account of what happened on the left. Here Montpensier and Damville had spent themselves in many charges against the much superior numbers of d'Aumâle's squadrons, and were driven back as far as the second trench. But the pursuing cavalry were brought to a complete stand by the pikes of Galatti's Swiss, and much harassed by a flank fire from the regiments of arquebusiers, whom Francis of Chatillon had brought up to line the hedgerows of the sunken road. Their attempt to turn the left flank of the Swiss by the fatal move into the marsh has already been mentioned in Sully's narrative. The Leaguers' cavalry was in much disorder, when the cannon of Arques Castle opened upon them, as has been told already. They recoiled as far as, and beyond, the first trench and the chapel, whereupon the King advanced with his Swiss—both the rallied and the reserve—in a column, with the French arquebusiers on their flanks, and stormed the chapel and the earthwork from the rear. The impetus for their charge was given by the Marshal Biron, who with some scores of other gentlemen dismounted, and placed themselves in the front ranks. The landsknechts at the trench, seeing themselves left behind by their cavalry, made a very poor resistance, and fled up the high road. The Royalists halted and reoccupied their original position—the enemy withdrew—though only a third of their infantry and half of their cavalry had ever been engaged. Henry, of course, had put in every man that he possessed.

The casualties were not very heavy—600 it is said on Mayenne's side, not much over 200 among the Royalists—but of the latter there were thirty loyal gentlemen who had sacrificed themselves in the cavalry charges. On Henry's side there fell the Count of Roussy, and two distinguished officers named Bacheville and Montard; of the Leaguers the Count of Sagone

[1] I have, of course, turned into the first person Sully's narrative, so oddly given in the second person, as taken down by his secretaries.

was killed—shot in personal combat, it is said, by the Bastard of Angoulême—and the Baron of St. André. The Count of Belin, Mayenne's *Maréchal-de-Camp*, and the colonel of the Lorraine foot-regiment were taken prisoners.

Thus ended a desperate fight in a defile—the whole front was only 400 yards broad—in which it was proved that superior numbers are useless if they cannot be deployed. The King jocosely remarked that he had been told that M. de Mayenne was a good general, but for the future was prepared to deny it.

This was not quite the end of the campaign of Arques. On September 23 the Duke abandoned his camp, and made a long detour of twenty miles inland, crossing the Bethune high up, and appearing on the other side of Dieppe on September 26. This opened the lines of retreat eastward to the King, but he did not take the opportunity to move off, and drew his army under cover of the fortifications of Dieppe and Arques. He had just received a convoy from across the Channel with money sent by Elizabeth—£22,000 in gold—and a Scottish regiment of foot, and was expecting on October 2 the 4000 men promised by the Queen. After a vain attempt to storm the castle of Arques, Mayenne decamped on October 6, having heard that the English contingent—five regiments under Peregrine Bertie, the 'brave Lord Willoughby'—and Sir Roger Williams, had actually arrived. But a more patent cause for his retreat was that news had come that the Duke of Longueville, the Marshal d'Aumont, and the old La Noue had collected a considerable force behind him, from the Huguenots and 'Politiques' of Picardy and Champagne, and were lying at Gisors, across his natural line of communication with Paris. With his troops much discouraged, and terribly weakened by desertion, the Duke marched on Amiens, where he was to pick up some reinforcements promised him from Parma's army of Flanders. He had no conception that Paris would be in any danger, considering the modest force of the King's army at Dieppe.

But this move on the part of the Duke emboldened King Henry to try one of his most daring—not to say rash—adventures. He would make a dash at Paris in the absence of the army of the League. It would be a matter of sudden surprise, and accurate timing, for Mayenne would be actually nearer to the capital city than the royalist army when the move should begin, and he would certainly be informed of it ere long. If he could reach Paris with his 20,000 men, the game would be up.

On October 19, Henry marched out from Dieppe with his original force and the English contingent, and having been joined on the way by Longueville's and D'Aumont's levies, he appeared before the walls of Paris on October 31 with an army of some 15,000 men—some say with 20,000. On the following day he delivered a desperate assault on the slightly fortified outer line of the faubourgs south of the Seine—St. Germain, St. Jacques, and St. Michel. The city had few regular troops within it, and the defence fell almost entirely on the hands of the urban militia. The Parisians were as fanatically devoted to the cause of the League as ever, and turned out in force, barricading every street. But the Royalists, dashing ahead with three heavy columns of infantry at three points, easily broke into the faubourgs. They were completely checked, however, at the old city wall of Philip Augustus, which protected the original nucleus of transpontine Paris. Vain attempts were made to break in the gates with petards, and the old La Noue failed in a desperate venture to wade round the end of the wall at the Tour de Nesle, where he wrongly thought that the Seine was shallow enough to let him get behind the defences. Miscalculating the depth, he was nearly drowned. After this the Royalists turned to promiscuous plunder of the suburbs—Sully allows (somewhat shamefacedly) that he got over 2000 gold *écus* as his share. The King, however, succeeded in preventing any massacres, rapes, or arson—a thing somewhat surprising when we consider how much of his army was composed of Huguenots, who had the memory of St. Bartholomew in their minds.

The attempt to surprise the city having failed, Henry could do no more. On November 1, Mayenne's cavalry entered the city in the afternoon, his infantry began to arrive in the following night, and by November 2 there were 20,000 Leaguers within the walls, and the King was foiled in his adventure. He withdrew his army to the high ground by Montrouge, south of the city, and there offered battle: but as Mayenne refused to come out, the Royalists retreated by the road to Etampes, and the campaign came to an end. It is doubtful whether this blow at Paris could be justified: supposing that the Royalists had forced their way on October 31 into the network of narrow streets, bristling with barricades, which the Parisians were prepared to hold, it seems that, even if they had won the quarters south of the Seine, they could hardly have crossed the river into the heart of the city. Mayenne's army would have been upon them next day, while they would have been entangled in the wilderness of houses, and undoubtedly would have dispersed to plunder. No sixteenth-century army could resist loot, and Paris was the most tempting place to sack in all Europe. The attempt to take the city by surprise was just the kind of risky venture which appealed to the King's temperament, but was fundamentally unsound.

CHAPTER IX

BATTLE OF IVRY (March 14, 1590)

THE battle of Arques, and King Henry's unsuccessful raid on Paris in the month that followed his victory, had settled nothing. The whole of France was still a prey to local civil war in every province, where petty campaigns were being fought out, without any general effect on the future of the realm. Henry IV had a substitute-capital at Tours, and held cities and whole districts all over the west and south of France. But the League was in possession of Paris, Lyons, Rouen, Amiens, Marseilles, and most of the other towns of first-rate importance, where the writ of its titular King Charles X was still supposed to run, though he himself was a prisoner in the hands of his nephew. Many towns changed hands during the winter of 1589–90, but the only solid gain to either side was that, by a series of short sieges, Henry made himself master of the greater part of Normandy; only Rouen and Havre remained firm to the cause of the League; Alençon, Falaise, Bayeux, Honfleur, Evreux, Lisieux were reduced; Caen was already in the hands of the Royalists. The King was besieging Dreux in March, when Mayenne moved out from Paris against him, determined to stay the progress of the enemy in this direction. Though he had received a considerable contingent of Walloon and Spanish troops from the Duke of Parma, commanded by Philip of Egmont, the unworthy son of the victim of 1569, a fanatical Catholic, Mayenne's army was not quite so numerous as it had been at Arques—apparently about 4000 horse and something over 15,000 foot. Many of his usual helpers were prosecuting minor campaigns of their own in the eastern provinces. King Henry was also weaker than he had been in the preceding autumn; the English contingent had gone home, and many troops were absorbed in garrisoning the newly conquered towns of Normandy, or had not returned from winter quarters in their own native districts. He is credited with 3000 horse and 8000 or 9000 foot.[1] The disproportion between the armies was not so great but that Henry was eager to seek a pitched battle in the open. He had not, as at Arques, to look for a defensive position, or to cover himself with trenches.

Both sides were eager for a decision, but there was some manœuvring before the clash came. The King raised the siege of Dreux on hearing of the Duke's approach on March 12, and on the next day deployed his army on the plain of St. André, between Nonancourt and Ivry, intending, it is said, to fight the enemy when he should attempt to pass the Eure. But Mayenne was already across the river, and his cavalry scouts met those of the King and engaged in a lively skirmish on the 13th—obviously a general action was imminent next day—the royalist headquarters were at St. André, those of the Leaguers in front of Ivry.

There followed on the 14th a deliberate pitched battle in open ground, such as had not been seen since Moncontour, for neither side had any advantage in position, or cover on its flanks. Henry drew up his army with six bodies of horse in line, each flanked by infantry, and with some arquebusiers as 'enfants perdus' thrown out in front. There was a small cavalry reserve—only 150 horse it is said, with two foot regiments, one on each flank of it, placed behind the centre. This was too insignificant a body to be called a second line—it was under the Marshal Biron. Five guns, the modest artillery force of the army, were placed in the centre. King Henry's intention, obvious from his arrangement of his troops, was to smash the enemy's left with a chosen body of his best horse and foot, for while the other cavalry corps in the front line consisted of only some 300 troopers apiece, his own, the fifth from the left, was more than double as strong, and was flanked by his redoubtable Swiss infantry. Not only was his personal guard there, but 'the first rank was almost entirely composed of princes, great lords, and captains.' The Huguenots were in

[1] D'Aubigné, always a 'minimizer' where his own party is concerned, gives Henry only 2000 horse and 6500 foot, attributing to Mayenne some 4000 horse. Davila, on the other side, always prone to larger figures, gives the army of the League as 4500 horse, including 1500 Spanish and Walloon lances and horse-arquebusiers, and nearly 20,000 foot, while he ascribes to King Henry 3000 horse and 8000 foot—less than half the Catholic total. I fancy that he is about right for the royalist army. De Thou gives imperfect details, not vouchsafing figures for some of the King's horse, *e.g.* the German reiters under Schomberg, but his totals will add up to a good deal over 2000 mounted men.

greater force than at Arques. La Tremouille had brought up the pick of the western Protestants, and La Noue, Duplessis-Mornay, and others who had not been present at the battle of 1589 were now with the host, along with Rosny, Montigny, and other old friends. Henry had desired the minister Damours, who had given the battle-prayer at Coutras, to repeat his blessing before certain regiments who had desired it.[1] But, as d'Aubigné tells, he did not 'use the language of Canaan' himself, but, riding at the head of his chosen squadrons, cried, 'Comrades, God is with us; there are his enemies and ours; here is your King—have at them! If you miss my pennon, rally round my white plume; there lies the way to victory and honour!' For, as at Coutras, he was set on leading the main attack in person, whatever the risk.

The array of the royalist army was as follows: The extreme left was formed by 300 horsemen under the Marshal d'Aumont, with a French foot-regiment on each flank. The second corps of horse, under the Duke of Montpensier, had on its left flank 400 landsknechts, and on its right a Swiss regiment of pikes and half a French regiment of arquebusiers. The third corps was composed of two bodies of light horse—400 in all—under the Bastard of Angoulême and the Sieur de Givry—they were somewhat advanced before the general line of the rest of the front. Near them were the royalist guns, directed by an officer named de Guiche. The fourth corps was under the Baron de Biron, son of the Marshal; its cavalry—all pistoleers—were not so strong as some of the other units, but it had 800 picked infantry. Next came, fifth in the line, in the right centre, the King's chosen squadrons, with three French foot-regiments, and his Swiss—four regiments strong—on the right and left. The sixth, or extreme right wing corps, was composed of Dietrich Schomberg's regiment of reiters flanked by French infantry—they touched the village of St. André. Behind the centre, as has been already mentioned, was the Marshal de Biron's small reserve. There was a thin line of skirmishers, 'enfants perdus,' thrown out in front of the battle line.

Mayenne's army, which was a little longer in its front, owing to its superior numbers, was apparently hoping to outflank the Royalists at both ends; it is described as moving somewhat in the form of pinchers or nut-crackers. But, of

[1] Not, apparently, before the whole army, but 'aux bataillons qui l'avaient desirés.'

32

course, in such a design everything depends on whether the crackers are harder than the nut, or the nut than the crackers. In this case the nut was iron, and the crackers only wood. The general disposition of the army of the League was not unlike that of the Royalists—blocks of cavalry flanked by regiments of infantry; there was no proper second line or general reserve. The Duke had organized his forces, on the old principle, into a 'bataille' and an 'avant-garde,' the former under the Duke of Nemours, the latter (which was the weaker) under Charles of Aumâle, his nephew. But he himself took up his position in the left centre of the line, with an independent body of horse, which did not belong either to the 'bataille' or the vanguard, and was composed of the picked squadron of the Catholic noblesse, with the 'cornet of Lorraine,' white with small black crosses, waving above it. It numbered about 700 'lances'—literally so, for while the cuirassed pistoleer was becoming the typical cavalryman of the period, and Henry of Navarre used no other horsemen, there were still in the Leaguers' army many gentlemen who clung to the use of the old heavy lance. And Egmont's Spanish contingent, lent by Parma, was also equipped in the same fashion. Royalist observers noted the heavy clumps of lances in the enemy's line as contrasting sharply with their own front, where not a lance was to be seen.

The Catholic army had been set in array by Jean de Tavannes, Mayenne's 'maistre de camp,' the son of the old Marshal. He was much criticized after the battle for having placed the units too close to each other, so that they had not sufficient space to manœuvre—a thing specially important for the reiter contingent, who were still practising their old (and dangerous) trick of the 'caracole,' which we have already had to mention in so many places. On the extreme right were light horse, then heavy cavalry under Nemours backed by Swiss[1] and landsknecht infantry. After these came Egmont's Walloon lances, or the greater part of them,[2] with French infantry in support. The small provision of guns which accompanied the army was between Egmont's and Nemours' front. After the Walloon horse came the reiter contingent, under

[1] The Swiss were divided, Pfeiffer's regiment on the right centre, Berling's in the left centre.

[2] According to De Thou, Egmont had 400 troopers here, while 300 were more to the left, near the Duke of Mayenne's cavalry.

Eric of Brunswick and a Count of East Friesland. Behind them were other Swiss companies and more French infantry. Next in the line came Mayenne's own companies of lances—700 or more—with 400 Spanish horse-arquebusiers on their immediate left—Aumâle, forming the extreme left wing, had in addition to his own French horse, the Spanish (or rather Walloon) infantry whom Egmont had brought with him, and more French foot-regiments. The very confusing and contradictory details of the deployment of the Leaguers given by De Thou and Davila are not much helped by d'Aubignè's narrative—he frankly admits that 'je voudrais bien vous pouvoir dire les chefs des bataillons—la confusion des memoirs m'empêche d'asseurer,' and then proceeds to speak of dispositions which will not fit in with any other account—*e.g.* that Nemours and Aumâle were with Mayenne in the left centre. The only way to disentangle the confusion is to note which Catholic corps fought with which royalist corps—for of King Henry's line of battle we have a clear and comprehensible description.

The armies for some time approached each other in a deliberate fashion, each waiting for the other to sound the charge first: Henry, we are told, kept shifting his line somewhat to the left, to gain some advantage of sun and wind. At last they were within cannon-shot, and the King ordered de Guiche to try some shooting at the troops opposite him—which were Egmont's Walloon lancers and the reiters on their left: they got two salvos in before the Duke's guns began to reply [1]—against Montpensier's horse. We are told that the Leaguers' artillery shot so wildly that, before the cavalry clash came, they only killed one old Norman gentleman of seventy years of age in Montpensier's front rank,[2] while the royalist cannon-balls made long lanes in the Walloon squadrons and the reiters.

As usual in such battles, the artillery was not allowed much time to play its part, as the cavalry charged at once, to avoid further casualties. On the extreme left of the royalist line the Marshal d'Aumont broke the Leaguers' light horse—though fighting with only 300 men against 500, and forced them right back on to a wood which lay behind the enemy's rear. He disregarded the arquebus fire of the hostile infantry as he drove

[1] De Thou, says *nine* salvos, which seems improbable—Davila only two.

[2] A certain M. de Longaunai, who was 'ictu tormenti initio praelii discerptus.' De Thou, xcviii. p. 849.

past their flank, and took little harm from it. Montpensier's corps closed with that of Nemours, and was for some time engaged in an indecisive mêlée : the Duke himself had a horse killed under him, and his troopers—who must have been much outnumbered—had difficulty in holding their own. Farther up the line Egmont with his Walloons charged furiously against the royalist light horse of the Bastard of Angoulême (' The Grand Prior ') and Givry, and beat them completely. He then turned against de Guiche's line of guns, and cut up the cannoneers and their escort of arquebusiers. But the Walloons were by this time in disorder, and were brought to a stand by being charged in flank by the Baron de Biron's 300 pistoleers, the next unit in the royalist line. The beaten light horse rallied on the infantry supports behind the guns, and resumed the fight. It ended in a complete wreck of the Walloon squadrons ; their commander Egmont was killed by a pistol-ball in the head, his standard was taken, and only a remnant of his men escaped, and took refuge behind the foot-regiments in the Leaguers' right centre.[1]

In the left centre of the Royalists, Mayenne had intended to attack the King's heavy corps with his reiters on one flank—the right—and his Spanish horse-arquebusiers on the other, while striking at Henry's front with his own companies of lancers. Both these flank diversions failed : the reiters, starting out to do their ' caracole,' were first disordered by the last volley of de Guiche's guns, and then, when nearing the royalist line, shaken by an unexpected volley from a line of royalist ' enfants perdus,' arquebusiers, whom they had failed to notice. They discharged their pistols, swerved, and all went to the rear—their leader, Eric of Brunswick, having been killed—without completing their ' caracole.' But, passing in disorder, many of them ran against the left flank of Mayenne's own column of lancers, because the space in the line, from which they had emerged in order, was not broad enough to take them when flying pell-mell. Mayenne had to halt to let the reiters get by,

[1] Davila says that d'Aumont charged the Walloons on their right flank, just when Biron charged them on the left. I cannot see how this can have been the case, for the whole contest between Nemours and Montpensier must have been going on between the sphere of Egmont's charge and that of d'Aumont's operations on the extreme left of the royalist line. D'Aubigné has a slip here also : he says that the *Marshal* Biron came up against the Walloons' flank—but it was not the Marshal, but his son the Baron. For d'Aubigné says that this Biron was wounded—which the Baron was, but not his father.

BATTLE OF IVRY, MARCH 14, 1590.

N.B.—The size of the units is not accurately expressed by this dimension. Those of the Leaguers were individually larger than those of the Royalists.

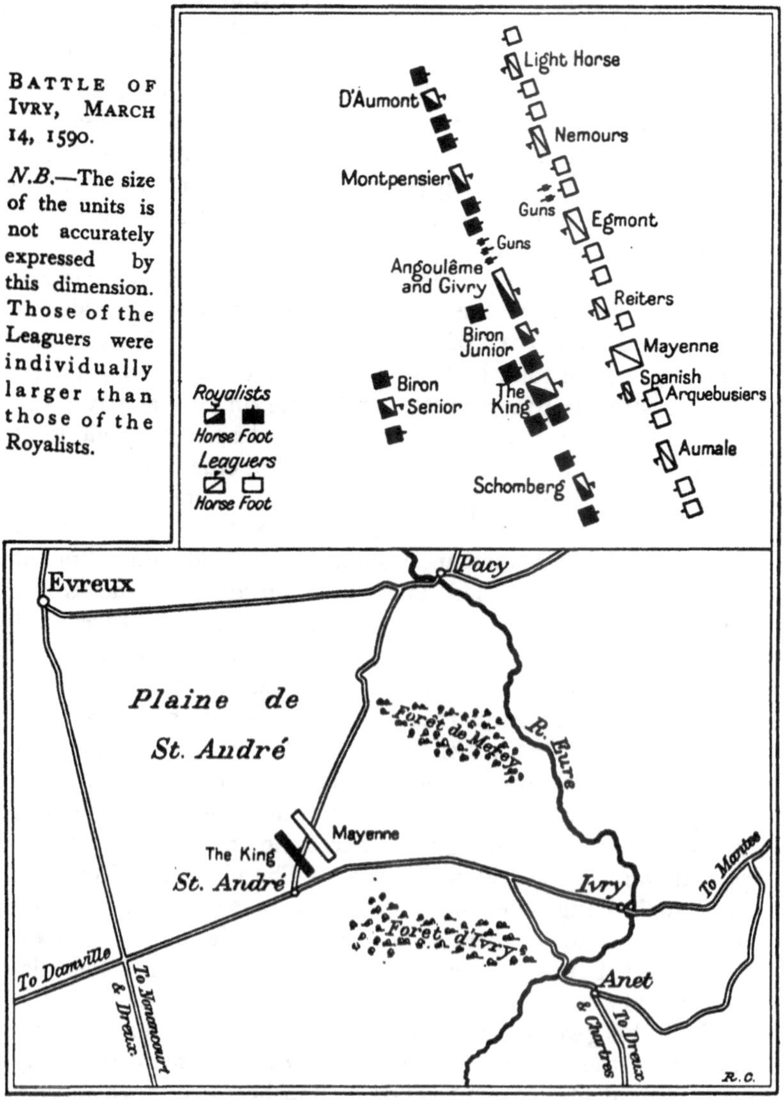

BATTLE OF IVRY, GENERAL TOPOGRAPHY.

and lost the impetus necessary for his charge. On his other flank the Spanish horse-arquebusiers behaved much better, getting within 25 yards of the Royalists, and emptying a volley into the charging column. This volley chanced to kill the young Sieur de Rhodes, who was bearing the King's white pennon: he and it were lost to sight on the ground. Hence it came to pass that, as Henry had promised, his own white plume became the directing point of his squadrons, for want of his ensign. We are told that for some minutes many of his followers feared that he was slain.

But the column of picked men which the King led was not in the least checked by the Spanish volley, rode down the arquebusiers, and then crashed into Mayenne's main body. The Leaguers, we are told, had not got their pace up, owing to the collision with the flying reiters, and came on very slowly; the result of this was that their front rank, despite of their long lances, was broken by the much greater impetus of the Huguenot charge, while the rear ranks, brought to a standstill by the block in front, could not employ their lances at all. The weapon was useless for a horseman brought to a stand, and jammed against comrades pushed back upon him from the front. Wherefore the Leaguers almost all cast down their lances and drew their swords—we are told that they were not equipped with the two pistols of the normal cuirassier of that day.

There followed a quarter of an hour of desperate sword-play and pistol-play, which ended in the complete disruption of Mayenne's column of horse. The King cut his way right through it, and came out on the other side, followed by his best fighters in a rending wedge. Whereupon the Leaguers broke up, and went to the rear in disorder. The victorious Royalists found their King safe and sound, if with somewhat battered plumes, and most of them swept on in pursuit of the fugitives. The actual battle-spot was strewn with wounded or dismounted men, and horses kicking in their last agonies. Sully, who had charged in the King's column, has a curious reminiscence from the moment of victory—a victory which he hardly understood at the moment. For he had been wounded in four places and had received a crushing blow on his head, which left him dizzy. When he came to himself, he found that the battle had passed on, but that seven (!) Catholic gentlemen, mostly wounded and with disabled horses, were asking him to receive their surrender—as the King had won the day. When he explained that he was hardly to be considered a combatant in his present condition, three of them, whose horses could still walk, went off. The other four stuck to him for protection, and one of them, the Sieur de Sigogne, handed him over Mayenne's battle-standard with the crosses of Lorraine, of which he had been the bearer.[1] This was certainly a tame method of capturing the enemy's most cherished ensign. Sully had some difficulty in retaining it, when scattered royalist horsemen came drifting back from the pursuit—a good many people would have liked to present it to King Henry.[2]

Of what went on when the extreme right of the royalist army, the corps of Dietrich of Schomberg, met the extreme left of the Leaguer army under Charles of Aumâle we have very few details—only indeed the interesting note that Schomberg's reiters did *not* practise the 'caracole,' familiar to their class, but charged headlong in a column. The mêlée according to Davila was hard and perilous, and was only brought to an end by 'Biron' intervening on the side of the Royalists; he says the *Baron* de Biron, but must surely mean the *Marshal* de Biron, with the reserve corps.[3] For the Baron was engaged far down the line with Egmont's Walloons, and the whole of the combat between the King's and Mayenne's corps was taking place between the ground where Aumâle and Schomberg were engaged, and the ground by the guns, far to the left, where the Baron fought. Anyhow, the Leaguer cavalry was finally broken, and Aumâle with a remnant was driven off to join Mayenne's equally maltreated horsemen. Schomberg was killed—one of the few officers of note who perished on the royalist side—but his wing was undoubtedly victorious.

All the various corps of the Leaguer cavalry having been routed, and gone to the rear, where some of them tried to rally along the edge of the wood behind their left wing, the infantry was left in line but without support. We hear nothing of its conduct during the cavalry clash save that where it was opposed to royalist foot regiments there was nothing but 'arquebuscades,' and no 'push of pike.'[4] The complete and simultaneous flight

[1] Sully, *Oeconomies Royales*, p. 441.

[2] Every one will remember the lines from Macaulay's 'Ivry' which tell

'How our good Maximilian has the white cornet ta'en,
The cornet white with crosses black, the flag of false Lorraine';

but few readers, probably, know *how* he got it—a matter of disillusion.

[3] Davila, xi. 66.

[4] D'Aubigné, iii. p. 233.

of Mayenne's mounted corps having left the Leaguer foot-soldiery in a desperate condition, most of them broke and went to the rear in disorder. Only the Swiss stood their ground, levelled their pikes, and then offered to surrender if quarter were given. Sully got a glimpse of them as he went to the rear wounded; they and the King's Swiss were facing each other in good order, both holding back and not firing a shot.[1] No doubt the negotiations for surrender were already begun. All around them was chaos and miscellaneous cutting up of fugitives.

The Marshal Biron promised the Swiss quarter, whereupon they threw down their pikes and gave up their ensigns. The King sent the word round that the French foot might also be granted terms, but that there was 'no pardon for the foreigners.' This meant that the landsknechts[2] and Walloon infantry were massacred in large numbers as they fled. The routed Leaguers dissolved into two lines of flight; Mayenne galloped due east to Ivry, where he ordered the bridge to be broken—the result was that great numbers of the fugitives, who had made less speed than himself, were cut down or captured as they tried to pass the Eure. Aumâle, Nemours, and Tavannes, making south, escaped in the direction of Chartres. King Henry, leaving the broken bridge of Ivry alone, passed the Eure at Anet, a few miles to the south, and halted the pursuit at villages on the road to Mantes, through which Mayenne had already made his escape.

The casualties in this essentially cavalry battle, when the infantry made so little show on either side, were fairly heavy. The Leaguers are said to have lost in killed about 800 horse and 3000 foot—the latter largely foreigners massacred in the pursuit or at the fords of Ivry. The Swiss and other prisoners numbered several thousands, and Mayenne's infantry was practically annihilated—there were 80 standards of foot-companies captured, besides sixteen 'cornets' of horse-units. The chief persons slain on the Duke's side were the Count of Egmont, Eric of Brunswick, and the Sieurs de Chataignerie and d'Arconat; but the prisoners were much more numerous, including a Count of East Friesland, who had been second in command of the reiters, and a very large number of captains

[1] *Oeconomies Royales*, 442.

[2] There was a special grudge against the landsknechts for their treacherous behaviour at Arques. See above, p. 489.

and colonels. On the royalist side the casualties were estimated at 500 killed, of whom twenty were gentlemen of some distinction. Among them were Schomberg, who had commanded on the extreme right, the Marquis of Nesle, Clermont, the captain of the King's Guard, de Rhodes, his standard-bearer, and de Crenai, the lieutenant of the Duke of Montpensier. There were many wounded, including the Baron de Biron, Maximilian de Rosny (quoted above), the Counts of Choisy and Lude, and d'O, the old favourite of Henry III.

The field-force of the League was destroyed. Paris was only 35 miles away and had now no relieving army to succour it. The dismayed citizens expected to see the Béarnais before their gates in three days. But he sat down at Mantes for over a fortnight, making no use of his victory, and then marched off to besiege Sens, on the borders of Champagne. Truly he was the most inconsequent and the most un-Napoleonic of generals.

CHAPTER X

HENRY IV AND ALEXANDER OF PARMA (1590–92)

AFTER the comparatively unimportant victory of Arques, Henry IV, as we have seen, had at least tried a raid on Paris, captured its southern faubourgs, and attempted to escalade its walls. He had been forced to withdraw, because the League had still a great field army, which came promptly to save the city. After Ivry that army no longer existed, and the Duke of Mayenne had been forced to fly northward, to beg more succour from the Spaniards in the Netherlands. But Henry made no second attempt to take Paris by a *coup de main*, as his enemies expected. The only immediate result of his triumph at Ivry was that Mantes opened its gates to him; and there he stopped from March 15 to March 28, though Paris was only 35 miles away, destitute of a garrison and stricken with panic. His own best friends found his conduct inexplicable. It is true that the roads were bad, the troops murmuring about pay in arrears, the treasury empty, and that some of Henry's Catholic supporters were rather dismayed at the prospect of a heretic monarch completely victorious and master of the whole realm. 'These people,' says Sully, 'could not support the idea of the domination and prosperity of a Huguenot king, and felt more disgusted at the result of the signal victory that he had won than did the very enemies whom he had beaten.' But Henry, when he chose, could surmount all obstacles, and make his men march, as he had shown on many an occasion. This time he did not do so; not merely the scandalmongers but serious historians ascribe his inexplicable delay at Mantes to the same cause that had led him to disappear after Coutras—the attractions of a lady of the neighbourhood, whom he had added to his long list of mistresses. Be this as it may—he did not march on Paris, and when he did move it was to make a long excursion to the lands of the Upper Seine, where he took Corbeil, Melun,

Provins, Montereau, and then, turning south, failed to take Sens. It was only on May 1, six weeks after Ivry, that he moved back toward Paris.

By this time the Leaguers had been granted leisure to recover themselves. For some days after the news of the defeat of March 14 there was talk in Paris of negotiations for a surrender. Mayenne himself was so unpopular that he did not dare to enter the city. There were no regular troops in the place, very little artillery, and (as it was reported) a dangerous lack of food-stores. Some insincere pourparlers for a truce, or even a general peace to be ratified by a meeting of the States General, were started, but broke down. The less resolute Leaguers were hoping to persuade the 'Politiques' in the King's camp to press their master to abjure—in which case they would accept him. But the more resolute were not ready to go even that far. They did not want Henry to be their master on any conditions—each having his own fears or his own ambitions. However, Henry's failure to appear before Paris at the moment, when the morale of his enemies was low, brought all ideas of a surrender to a close. The Duke of Nemours was appointed governor, some regular troops were collected from the wrecks of the army destroyed at Ivry, provisions were rushed in from all available quarters,[1] and the urban militia was reorganized. Fanatical preachers stirred up the citizens to desperate perseverance, promising help from heaven, and, what was more practical, an army of succour to be sent from Parma in the Netherlands: the Spanish ambassador promised the League that his master would never desert the cause of the Church.

When King Henry at last approached Paris he tried no assaults, but a blockade at some distance from the walls. But the periphery of the city was enormous, especially when outlying positions like St. Denis and Vincennes were being held for the League, and the blockading-line had to be a very thin one. The royalist army had been strengthened by many reinforcements from the South, where things were looking well for the King's cause, but was still of very moderate strength. Twenty thousand men had to hold a circuit of nearly thirty miles, with their lateral communications rendered very difficult by the tiresome disposition of the rivers—the Marne and the Upper Seine dividing the blockading-line above the city, and

[1] It is said that some of the King's officers, notably the Sieur de Givry, winked at the passing of food-convoys which they might have intercepted in April

the two 'hairpin bends' of the Lower Seine below the city complicating matters still more. Henry established his headquarters at a discreet distance outside the north-east side of the enceinte, holding the rising ground of Montfaucon and Montmartre as advanced positions. The Marshal d'Aumont was placed at St. Cloud, with a detachment holding the important bridge of Poissy: but he was cut off from the King by the Leaguer garrison of St. Denis. A third corps, placed on the heights south of the city, watched the transpontine quarters from the bridge of Charenton westward. A separate detachment had to look after the isolated Catholic fortress of Vincennes. To the rear of the line of investment, garrisons had been placed at the bridges of the Upper Seine and Marne—Lagny, Corbeil, Melun, etc., to stop river-traffic, by which in normal times provisions came most copiously to Paris.

The blockading army was so thinly spread that sorties could often pierce it for a moment; but the force available for a sortie—cavalry being almost entirely lacking to the Parisians—could always be driven back after a time by the concentration of the besiegers from the neighbouring sections of their line. The situation of Paris for the four months May to August 1590 was not unlike that of the four months October–January 1870–71, when the Prussian army of William I was repeating the experiment of Henry of Navarre. Sorties always won a little ground, but always failed to break the ring of blockade. The defence in 1870 had the advantage of a circle of strong outlying forts; on the other hand, the besieger possessed modern artillery, which could bombard the greater part of the city. But both the Navarrese and the Prussians were mainly set on reducing Paris by famine: Henry failed, because an army of relief did finally appear, and forced him by manœuvre to evacuate his lines and raise the siege, after the city had been reduced to the extremity of starvation. William of Prussia was successful, because all attempts to intervene from the outside were beaten off by his covering force in the direction of Orleans. But the parallel is interesting and instructive.

Henry IV, no doubt, had miscalculated the amount of food available in the city: he thought that it would be exhausted in some six weeks. He had also misjudged the spirit of the defenders, who showed an astounding resolution—the celebrated parade of the regiment of 1300 friars in arms was something more than the grotesque incident sneered at by cynical observers.

It testified to deadly earnestness—as a matter of fact it was armed ecclesiastics who actually repelled an attempt at escalade on September 9, at the very end of the siege. The tenacity of the defence did not slacken even when all rations had ceased to be issued, when the populace were living first on horseflesh and weeds, and then on cats, dogs, and grass. Thousands died of malnutrition, or absolute starvation, and cases of cannibalism are said to have occurred. There was added spiritual depression, for the help promised to the besieged city from the Spanish army in Flanders seemed never about to materialize: it was delayed for four months. The Duke of Parma had his own troubles in the Netherlands, where he was being hard pressed by Maurice of Nassau, and protested against his master's orders to draw off his main force into France. At first he would only lend a detachment of 3000 or 4000 men to Mayenne, who had rallied the wrecks of the old Leaguers in Picardy. This was not enough, and the small force gathered at Laon was held in check there, by a covering detachment which the King scraped together from his troops in his lines, and led in person to drive off the Duke (June 15).

The fact, however, that Paris showed no signs of capitulating, and that armies of relief would sooner or later be formed in greater strength, impelled the King to more active measures. St. Denis fell on July 9: on July 24 he organized a general attack on all the faubourgs, both north and south of the Seine, with ten columns of infantry headed by his best officers. They were nearly all successful, and the besieged were driven in to the line of the old fortifications of Philip Augustus. This enabled the King to shorten immensely the line of investment, and deprived the garrison of the meagre resources of food which they had hitherto been drawing from suburban kitchen-gardens. Nevertheless, Paris held out, and the demonstrations of starving crowds were fiercely repressed by the authorities.

On August 1, Parma, much against his will, had become convinced that Paris could not be saved unless he brought the main body of the army of the Netherlands to aid the League. Protesting to King Philip that he took the risk unwillingly, and only in obedience to orders, he began to collect his best troops from their garrisons, leaving a minimum force to contain Maurice of Nassau, and a fortnight later crossed the frontier with 14,000 veterans, the pick of the old Spanish army. On August 23 he was joined at Meaux by Mayenne, who brought every man that the Leaguers could raise in northern France, 10,000 foot and 2000 horse, including a strong contingent from Lorraine.

The army with which the King was at this moment blockading Paris was by this time not unequal in numbers to the united strength of Parma and Mayenne He had received of late considerable reinforcements from the South. But, obviously, if he continued to maintain the line of posts all around the city, he could only take away, to drive off the army of succour, a much inferior force. While if he concentrated his army and marched in full strength to face Parma, the roads into Paris would lie open on every side, and the city could re-provision itself. But if he should succeed in inflicting a complete defeat on the enemy within a few days, he might resume the siege before any very great amount of food had entered the walls.

On August 30, Henry resolved that he must not be surprised in his lines by the enemy, but must concentrate, and endeavour to force on a battle. He drew in all his detachments from the south and west sides of Paris, and took post at Chelles, twelve miles on the road to Meaux, with 7000 horse and nearly 20,000 foot, determined to fight Parma at the first favourable opportunity. This chance he was never to obtain, for in the space of seven days, which followed that on which the two armies found themselves in presence, Parma asked to be attacked, but had so placed himself that an attack must fail. Henry had no longer to deal with enemies like Joyeuse or Mayenne, who came forward to fight, but with a general of the ancient Roman type, who covered himself with entrenchments against which cavalry could be of no use. When the King moved, Parma was found to have moved also, and in a single night his laborious veterans had thrown up new earthworks. For five days the armies stood opposite each other, and Henry dared not attack. On September 6 Parma commenced some demonstrations, which made the King hopeful that he might move out of his lines. But these were only intended to distract him—the Spanish army remained safely entrenched, while a detachment passed the Marne and surprised from the rear the fortified bridge and village of Lagny, the main passage of that river which was in the hands of the Royalists. This unblocked the Marne, and opened free passage from north to south for the Leaguers; it also facilitated the resumption of river traffic.

Meanwhile Paris had been re-victualling itself, though with some difficulty, for the surrounding regions had been eaten bare by the four months' sojourn of the besieging army. The game was up, since Parma would not fight save under impossible conditions. On September 9 the King retreated from his position at Chelles, and marched for Bondy and St. Denis. But as a last desperate expedient he detached Francis de Chatillon, the son of Coligny, with a corps of infantry, who were to try to get into Paris by escalade. The point chosen was the one where the garrison was least likely to be on guard—the extreme south of the transpontine half of the city, between the gates of St. Jacques and St. Marcel. But the Leaguers were not taken by surprise—the one ladder that was placed against the walls was thrown down by a picquet of armed Jesuits, and the ramparts soon swarmed with resolute defenders. After this failure Henry retreated north-westward, with a much depressed army, in which Catholics and Huguenots were quarrelling bitterly, and complaining of four months wasted in sight of Paris without any proper attempt to tackle the situation. The main fault had been committed at the very commencement, when Henry lay quiescent at Mantes for a fortnight after his victory at Ivry, and Paris had been in a state of complete panic. Having failed in the great enterprise, Henry dismissed many of his troops to their own regions, as winter was drawing nigh. With a much diminished force he escorted Parma very cautiously to the frontier. For the Duke had refused to undertake the task of ending the war for the benefit of the League, and retired to his own viceroyalty, handing over the conduct of affairs in France to Mayenne. All that he would concede was that he left two Spanish regiments to serve as a permanent garrison for Paris—a boon not desired, as the Leaguers were very suspicious of the designs of their inscrutable ally.

This was not unnatural, as a new political situation had come into existence during the siege of Paris. Charles X, the old Cardinal Bourbon, had died in his prison at Fontenay on May 20, and the League was left without a king—even an invisible one. It was said that before his death he had acknowledged the rights of his nephew of Navarre. The dynastic situation was now a puzzling one—if the Salic law prevailed as the essential rule, Henry of Navarre was undoubtedly king—if he had fallen in battle the next Protestant heir was his little cousin, Henry II of Condé, a child aged three, son of the Condé who had been serving with the Huguenot armies down to his death in 1588. But if a Catholic must be sought for, the nearest available person seemed to be another ecclesiastic, Charles, Cardinal of Vendome, fourth son of the first Prince of Condé, and nephew of 'Charles X.' No one seems to have taken his pretensions very seriously, but he urged them for what they were worth. But if the Salic law was a political fiction of the fourteenth century, as many legists urged, and if the nearest relative of the last generally acknowledged King, Henry III, had to be looked for, without any regard for male descent, then another problem arose. Henry III had three sisters: the eldest was Claude, Duchess of Lorraine, now deceased; she had left a son, Henry, Marquis of Pons and heir of Lorraine; his father, Duke Charles, was still alive—and lived indeed till 1608. The Marquis was an obvious claimant, and Catherine de Medici, his grandmother, had thought of him as a substitute for Henry of Navarre. But the Lorraines were unpopular in France, their younger branch, the House of Guise, had much more influence with the Leaguers.

The second sister of Henry III, Elizabeth, was one of the unfortunate wives of Philip of Spain—she died very young, but left two daughters, of whom the elder, Isabella Clara Eugenia, was considered by her father to be a very fit occupant for the French throne. The third sister of Henry III, Marguerite, was the disgraced but not divorced wife of Henry of Navarre; she had no issue. But the Duke of Mayenne, head of the League, and practically regent for the last two years of those parts of France which were still held by the Catholic faction, had also to be taken into consideration. He would undoubtedly have been willing to make a grasp at the crown if he had considered it possible; but, rightly judging that such ambitions were hopeless, he had some notion of putting forward the head of his house, the young Duke of Guise, son of the man murdered at Blois, as a suitable husband for the Princess Isabella Clara Eugenia—by which he would bring in the Spanish interest to assist the pretensions of his own family, and place a Guise as King-consort on the throne. Unfortunately the young Guise was a prisoner at Tours in 1590, and only got loose in 1591. This project was in the end crushed by Philip II, who thought his daughter worthy of a more lofty alliance.

The net result of this genealogical tangle was that, during the last four years of the war, the Leaguers had no generally recognized pretender to the crown at their disposition. In spite of the old national hatred for Spain there were many who thought that the only hopeful course would be to submit to the exigencies of Philip II, a most indispensable ally. On the other hand, there was a majority which hated the idea of Spanish domination, and were looking round for any other possible expedient. Some favoured the claims of one or other of the possible pretenders whose names have just been cited. But there was an immense proportion of the whole who were not so bitterly hostile to Henry of Navarre but that they would accept him, if only he would 'abjure' and turn Catholic. Hence frequent intrigues with the Catholics in the King's army, the 'Politiques,' who were assured that their master would be acknowledged all round the realm, if only he would make his submission to Rome. For two years more Henry wavered between the two policies, that of endeavouring to win the throne by his own sword without making concession, and that of sacrificing his creed, or what was more real to him, his pride and self-respect, by rejoining the Roman communion, of which he had enjoyed four years of bitter experience during his ignominious court-captivity after the St. Bartholomew (1572–76). 'Politiques' and half-hearted Leaguers alike kept imploring him to be 'instructed,' as the phrase ran. He had had enough experience of such 'instruction' when Charles IX offered him in 1572 the choice between '*Messe, Mort, or Bastille*,' with the result of four years' compulsory orthodoxy. While some of his Catholic followers, not without his knowledge, kept up pourparlers with the anti-Spanish faction of the Leaguers, with secret talk about abjuration, Henry persevered with his endeavour to win by the sword.

It was the inconclusive result of the next two campaigns which finally drove him to swallow the bitter draught. In April 1591 he took Chartres, and during the summer his supporters in many of the provinces achieved considerable successes, but nothing decisive happened, till in November he laid siege to Rouen, far the most important centre of the Leaguers in northern France with the sole exception of Paris. He had got considerable assistance from England: Queen Elizabeth, though she distrusted Henry, had a strong conviction that the Spanish party in France must not triumph. She sent her favourite Essex with 4000 men to help at the siege, and the Dutch lent a considerable contingent also. A large force of Germans, the usual reiters and landsknechts, had also come up. The siege of Rouen lasted for five winter months (November 1591–March 1592); just as the place began to show signs of yielding, the phenomenon of the summer of 1590 was repeated. Stirred up by his master, the Duke of Parma got in motion to relieve Rouen, just as he had relieved Paris in the preceding autumn. He did not intend to fight a general action, but to bring such a force to face the King that Henry would be driven once more to the inevitable alternative of either massing his whole army, to drive away the relieving host, or of maintaining the siege, and taking the chance of being attacked while his troops were strung out in a long line of circumvallation. In the first case, Rouen would be relieved automatically, as Paris had been; in the second, the King would be attacked while his army was dispersed in its trenches on both sides of the Seine. When Parma was known to be on the move, Henry tried to cut matters short, delivering violent assaults on St. Catherine's Mount and other outer defences of the city. But the besieged commander, M. de Villars, was one of the best officers whom the League possessed; he made a splendid defence, and there was no prospect of bringing the leaguer to an end by battery and assault.

At last Parma was found to be drawing near; his progress was slow, for he waited to pick up a new auxiliary force from Italy, which had been subsidized by the Pope, and he intended to have every man with him whom Mayenne could raise from northern France. The Duke was late in coming, for there was trouble in Paris, where he had been forced to put down a rising of the extremist party, and to hang four of the leaders of the celebrated 'Council of Sixteen' who had long controlled the fanatical city mob (December 4, 1591). January had begun, and bitter weather was prevalent when the Spanish and Leaguers united appeared on the Somme. Henry did not on this occasion take the course which had been forced upon him in 1590 at Paris, but resolved to leave all his infantry and guns in the trenches before Rouen, with the least efficient of his horse, in charge of the Marshal Biron. He himself started off with a purely cavalry force, 3000 reiters, 3000 French of his best units, and 1000 mounted arquebusiers,[1] with the

[1] These are Sully's figures; he himself was told by the King not to bring his whole company of gendarmerie, but only the best mounted and most trustworthy, some twenty out of the whole fifty of the company. I presume that the same orders were given to other commanders of units.

intention of hanging on to Parma's flanks, worrying him at defiles, and generally delaying him in such a fashion that he should not be able to approach Rouen. He, like the Duke, had no intention to get involved in a pitched battle.

This was an interesting tactical experiment. Could a very large cavalry force incommode a complete army under a competent general, to such an extent that it could be prevented from carrying out its destined purpose? It was in a way a trial of wits—and Parma's cautious methods proved too much for King Henry's enterprising adventure. The Duke had with him about 6000 cavalry, of which 2000 were French lances commanded by Mayenne and the young Duke of Guise, who had just escaped from his prison at Tours. There were also 700 horse from Lorraine; the rest were Parma's own. Of foot there were about 23,000, of whom 4000 were of Mayenne's Leaguer infantry, 2000 Swiss sent by the Pope, and the rest old Spanish Walloon, German, and Italian regiments of the army of Flanders—eleven veteran corps in all. As far as total numbers went there was no great difference between Parma's expeditionary force and the army with which the King had been beleaguering Rouen; if Henry had considerably more mounted men than the Duke; yet his infantry was inferior by some thousands, and included many weak south-French corps short of pikemen,[1] and 4000 or so of very raw English foot. It was decidedly not so good in quality as that of Parma, whose French Leaguer-infantry were his only weak material. But as Henry had left all his foot, and some 2000 horse, under Biron, in the lines round Rouen, he was trying the problem of foiling an army of nearly 30,000 men by the use of a rapidly moving cavalry force. He was hoping to catch the Catholic army strung out upon the march, or moving in separate detachments, one of which could be molested or delayed in the passage of defiles or rivers.[2]

The campaign of manœuvres which followed was interesting, if most difficult to interpret. Parma was cautious in the extreme; he kept all his troops closely united, marching in a sort of hollow parallelogram or oblong square, the infantry having the heavy cavalry in their middle, while the light cavalry

[1] D'Aubigné, iii. p. 257, says that thirty-two such regiments were so 'harassés et dissipés' that they did not make up much over 4000 men.

[2] All this is explained very well by Davila, xii. p. 363. But he puts the total of Henry's flying force too low, at 1000 reiters, 2000 mounted arquebsiers, 2000 French horse ,and 500 *chevaux légers*. Sully, as quoted above, gives 7000 in all.

was pushed out far both on the flanks and in front, to keep a vigilant look-out for the approach of the enemy. This formation made the march extraordinarily slow. Though Parma had concentrated his army at Amiens on January 16, it was not till February 3 that he got in touch with the King on the Bresle, the little river which separates Picardy from Normandy. Henry had advanced to Aumâle, the last town in Normandy, where he left the Dukes of Nevers and Longueville with the bulk of his force, and pushed on himself with not over 1000 picked men to drive in the Spanish light horse, and discover Parma's dispositions. He crossed the Bresle, and a league beyond got a view of the enemy marching forward in an orderly parallelogram, with no straggling or division into separate columns; so he resolved that he could not dare to meddle with them. He retired that afternoon on to his main body at Aumâle, but hoped that he might get his chance next day when the Spaniards had to cross the Bresle.

Waiting till the Duke's army was drawing near Aumâle, he went forward again with his chosen squadrons, and drove in the hostile cavalry screen—Leaguer horse—on to the neighbourhood of the front side of Parma's square of infantry. He himself led the charge, and was enjoying the pursuit, when he found himself beset on both flanks by the Spanish light cavalry and mounted arquebusiers, who had ridden forward unobserved from both right and left, and were threatening to cut him off from the fords of the river, and his main body of his corps drawn up beside Aumâle. It was necessary to turn at once, and it soon became evident that there was considerable danger of the whole force being annihilated. With his customary rash courage, Henry took charge of the rearguard, and charged several times with a small company to hold back the enemy. In the midst of the skirmish an arquebus ball pierced the pommel of his saddle, and wounded him in the loins, though not severely. He was, however, disabled for the moment, and had to be brought off by trusty friends, barely able to sit his horse. The whole corps were hustled back with great loss—some sixty gentlemen fell, as it is said, and complete destruction was only prevented by Nevers presenting himself with a strong detachment in front of the pursuers. Parma refused to let his main body of cavalry loose against the disordered Royalists, being suspicious that there might be a whole army hidden on the hillsides about the town of Aumâle. Critics held that if the

Duke had struck hard, the King would have been killed or taken before Nevers could have brought up his reserves. Parma justified his caution by remarking that he had supposed that he was fighting a general—not a captain of irregular horse. Who could have guessed that the King would be leading a reconnaissance in person, or that he would have come up to Aumâle with not a single infantry corps with him ?

The Duke of Nevers covered the retreat of the Royalists to Neufchatel ; there was some fighting in the streets of Aumâle, but the whole loss on the King's side was little more than that suffered in the hasty retreat from across the river. The Spaniards plundered the town effectively, brought up their artillery and baggage across the Bresle, and started on their further advance with Parma's wonted caution. King Henry, after his wound had been dressed, was carried to Dieppe, where he was only ten days out of action, and was able to ride again by February 17. Most of the cavalry accompanied him to Dieppe and Arques : it seems surprising that Nevers did not take them to join Biron in the lines before Rouen ; but apparently he was acting on the orders of the King, who was never too much disabled to retain his command.

Parma advanced to Neufchatel, the only fortified place that lay between him and Rouen, and laid siege to it, much to the surprise of many of his officers, who had expected him to march straight on, to cut up Biron and the blockading force which still lay before Rouen. Neufchatel, an old-fashioned fortress, only held out for four days against bombardment, and the governor de Givry, though a tried veteran, surrendered on terms upon February 12. Parma tarried over a week in the place, and then recommenced his march towards Rouen 'with the army always in order of battle, never moving unless the weather was favourable, and all the ground in front well reconnoitred, and halting each afternoon in time to allow of his camping-ground being entrenched.' [1] On February 26 he was at Bellencomble, preparing for another slow move, when he received news that the siege of Rouen was already practically raised. The governor, Villars, knowing that the King and the cavalry were still absent, made a violent sally against Biron's lines, with every disposable man of the garrison, on the night of the 24th. He effected a complete surprise, overran the front trenches, swept along sideways for a mile, and then, falling

[1] Davila, xii. p. 379.

upon the main munitions-depot of the besiegers, blew up their powder-magazine and captured a great many pieces of artillery. Biron, coming up with his reserves too late, was wounded and repulsed: he retired on to his fortified camp at Darnetal, where he rallied his main body. Meanwhile Villars filled up all the trenches in front of the east side of the city, and dislodged all the outlying posts of the besiegers. Biron remained in observation before the place, but it could no longer be said to be surrounded or even blockaded. The Royalists lost 800 men in this affair, including two colonels and fourteen captains, and were thoroughly demoralized.

Now follows one of the perplexing points of this very incoherent campaign. Parma did not march to Rouen to make an end of Biron's shattered force, but only sent into the place a regiment of Walloon infantry and a convoy of food, and arranged for a regular supply of provisions to be sent upstream by boats from Havre and Harfleur, both of which posts were in the hands of the Leaguers. Well-informed chroniclers of the time, both Davila and De Thou, declare that Mayenne dissuaded Parma from entering Rouen for purely political reasons, fearing that it would become a Spanish fortress, and that the second most important city of northern France would pass out of the control of the League. He even, it is said, threatened to withdraw his contingent from the allied army if the Spaniards entered the place.

This was to throw away an immense opportunity, for the Royalists were in a very bad case at the time. The moral effects of the combat of Aumâle and of Villars' great sally had been widespread. The King's troops were deserting by thousands, even many of the noblesse were slinking home, on the excuse that they, their retainers, and their horses, were wearied out with a winter campaign which had ended unhappily. Parma, apparently much disgusted at the recalcitrance of the Leaguers, contented himself with saying that he had raised the siege of Rouen, as he had promised, and cantoned part of his army at Neufchatel, and then—what is most inexplicable—went himself with a very large detachment back into Picardy, to besiege Rue, at the mouth of the Somme, the only town behind him which was held by the Royalists. It would seem that he considered that he had fulfilled his contract, and imagined that the King would pick up his infantry and retire beyond the Seine into central Normandy, which was entirely in his hands After he should have gone off, the whole business would fall into the responsibility of Mayenne, and the Spanish army might return to the Netherlands, where it was much needed, since Maurice of Nassau was giving trouble.[1]

But King Henry did nothing of the kind. He left Biron, in a very perilous position, in front of Rouen, and set himself to molest Parma's cantonments, and to beset the roads between Rouen and Picardy, along which lay the enemy's line of communications. Dieppe served him as a base of operations. Already, before Biron's disaster he had been prowling about the Spanish line of march, and had succeeded on one occasion in cutting up a detachment of Leaguer horse, under the Duke of Aumâle, which was too far from the main body.[2] Learning at last that Parma had vanished, and that Biron was still hanging on before Rouen, he took the obvious course, and went to join him at Darnetal with all his cavalry, which by this time was not over 5000 strong, owing to desertion. He then resumed the half-suspended siege—a thing which had become possible owing to the intervention of a foreign succour. A Dutch fleet with a landing force of some 3000 men on board, under Philip of Nassau, was lying in the mouth of the Seine, and, passing by the hostile fortresses of Havre and Harfleur, had cast anchor off Quillebœuf, where the troops landed. This cut the water-communications of Rouen, on which Mayenne and Parma had been depending for the revictualling of the place, which had hardly begun. The Dutch vessels ran up the river as far as the city, and tried to bombard it from the water-side, but were very soon driven off by shore-batteries, which Villars threw up beyond the walls. But the landing force joined the King's army (March 27).

The Royalists then proceeded to open up their old trenches, and to re-establish their line of posts around the city, bringing up artillery from Pont-de-l'Arche and other places. Villars was seriously disturbed, as his food-supply was now cut off by the Dutch, and his old magazines were almost exhausted. But the King was in no better case, his French infantry having dwindled to a very modest force, and showing signs of demoralization

[1] See below, p. 569.

[2] It was in this skirmish that Chicot, the famous jester of Henry III, unhorsed and took prisoner the Count of Chaligny, one of the house of Lorraine. Chicot, better known to us from Dumas's novels, was a gentleman by birth and a good jouster. He died of his wounds a few days later. See Davila, xii. p. 378, and De Thou, cii. p. 118.

when siege-work was once more begun. Villars sent emissaries to Parma and Mayenne, warning them that if he was not succoured before April 20 he would be in danger of starvation and surrender. For once Parma showed that he could move rapidly when he chose, and, collecting as many of his scattered troops as could be assembled in a hurry, hastened from Rue to the neighbourhood of Rouen in six marches, cutting a long corner by crossing the Somme estuary at the fords of Blanche-Tache, where Edward III had outpaced his enemy, Philip VI, nearly 250 years before, just before the battle of Crécy. He brought with him only 5000 horse and 12,000 foot, his Leaguer-Auxiliaries not having come up in full strength. But the King, after having tried in vain to hold back the approaching Spaniards by cavalry skirmishes, thought himself not strong enough to fight. He sent off his artillery and baggage to Pont de l'Arche, and evacuated his lines of circumvallation, retiring with his whole force to Boos and Goui, nine miles south of Rouen. Parma and Mayenne entered the city in triumph on April 21, and bestowed well-merited praises on Villars for his six-months' defence.

There followed another of those perplexing incoherencies which make this campaign of 1592 so difficult to understand. The Spanish officers, we are told, were in favour of taking the offensive against the King, and forcing him either to fight or to retreat into central Normandy, across the Seine. Mayenne and his followers protested against the risk of a battle—knowing that they were not in full force at the moment—and urged Parma to clear out the King's garrisons from the Pays de Caux, to drive away the Dutch fleet from Quillebœuf, and to open up secure communications with Havre and Harfleur. To do this it was first necessary to take the royalist town of Caudebec, the only important place which blocked the road to Havre; it was also worth having because it contained the central magazines from which the King had been feeding his army during the siege of Rouen. Mayenne assured Parma that the royalist army was for the moment unlikely to interfere or give trouble, owing to its demoralization.

In this view the Leaguer-Duke was hopelessly wrong. For King Henry was furious at having been forced away from Rouen, and was set on resuming the offensive at the earliest possible moment. The moment that Parma's departure from Rue had been announced to him, he had been sending orders all over the provinces of his obedience, to dispatch every possible man to join the depleted field-army. Reinforcements soon began to flock in—the largest was a corps with which the Duke of Montpensier had been vainly besieging Avranches all through the spring: the garrisons of the royalist towns of north France were cut down to a minimum, and a great proportion of the noblesse of the West, who had shirked homeward in March, returned to their colours.[1] In eight days Henry had nearly doubled his marching strength, and was at the head of 8000 horse and 18,000 foot, a force perceptibly greater than that under Parma's orders. He then proceeded to press round Rouen to the west, and to place himself between the Spanish army and its line of communication with Picardy and Flanders, intending to shut it up in the peninsula of Caux, between the Seine estuary and the ocean. If Parma chose to fight, the King was ready; if he should retreat on to Havre and Harfleur, he would presently be starved, since those fortresses could not feed a large army: the sea was in possession of the King's allies, the Dutch and English, and no exit from Havre would be possible.

Parma had marched on Caudebec, and took it in three days (April 24–27), the place having only mediaeval walls with no glacis or bastions. An attempt of the Dutch fleet from Quillebœuf to hinder the siege, by fire from the water-side, was driven off with no difficulty by batteries placed on the shore. But a chance shot from the ramparts of Caudebec before its surrender altered the whole course of the campaign—Parma, while making observations in one of his batteries, was hit by a musket-ball in his right arm: he made light of the wound at first, but the mangling hands of the surgeons who extracted the ball brought on such pain and exhaustion that he fell into a fever, and had to take to his camp-bed. The command of the army fell for a moment into the hands of Mayenne, that of the Spanish troops to Ranuccio of Parma, the Duke's son and heir, who had accompanied his father as a commander of light horse. Mayenne, after having fed the troops for three days on the magazines captured at Caudebec, moved away from the river to Yvetot, the road-centre of the Pays de Caux, from which there were good lines of communication both to Rouen and to Havre. But he had hardly arrived there when he heard that the King was marching against him from the east, with

[1] For a list of the reinforcements, see Davila, xiii. p. 13.

somewhat superior forces, and was threatening to cut him off from Rouen. Following Parma's custom, the Leaguers fortified their position in front of Yvetot, but the intrenchments were not large enough to cover all the cavalry (April 30).

There was heavy skirmishing in front of the lines on the 1st and 2nd of May, without any decisive result, and for more than a week the armies lay opposite each other. The King made several infantry attacks on the enemy's trenches, in which he used his Dutch and English contingents as his most trustworthy units. But failing to break in by frontal attacks, he began to extend his right wing so as to get round Yvetot on its northern side, and finally succeeded in outflanking the light cavalry of the League, who was in charge of this part of the enemy's front, and in driving them in upon the town of Yvetot with considerable loss. Mayenne, nevertheless, held on in his position for some days more, till the army began to suffer severely from privations, having eaten up not only the magazine of Caudebec but all that could be swept up from the surrounding district. Things were looking so gloomy for the League that Parma finally roused himself from his sick-bed, though he had his arm in a sling, and could barely sit a horse, and resumed command. To retreat on to Havre would be ruinous, to endeavour to force his way to Rouen by the main road would involve a battle against superior forces. He therefore devised an ingenious and unexpected scheme for getting out of an unpleasant position.

While holding on to Yvetot for the moment, he sent back detachments to Caudebec, to throw up entrenchments before that place, and at the same time dispatched by water officers to Rouen, who were to order Villars to ship down to Caudebec all disposable boats, and also all the pontoons and landing-stages of the port of Rouen. They could easily slip down when the tide served. It was his intention to bridge the broad tidal Seine, and to transfer his whole army to the other bank opposite Caudebec—a move which the King could not possibly foresee, since every one judged the Lower Seine impassable.

On the night of May 17, Parma marched his whole army out of the lines of Yvetot, and brought it back to the newly planned positions around Caudebec. This was done with complete secrecy, the enemy having been kept occupied by demonstrations against his outposts by cavalry demonstrations conducted by Ranuccio Farnese. When the King had driven them off, and pushed his troops forward, he found the enemy's lines wholly unoccupied. Parma had retired on to Caudebec—a move not unwelcome to his adversary, who judged that he could then pin him against the Seine, and starve him out in a position from which there was no exit. He cast a line of posts all round Caudebec, and broke the roads leading out of it. The

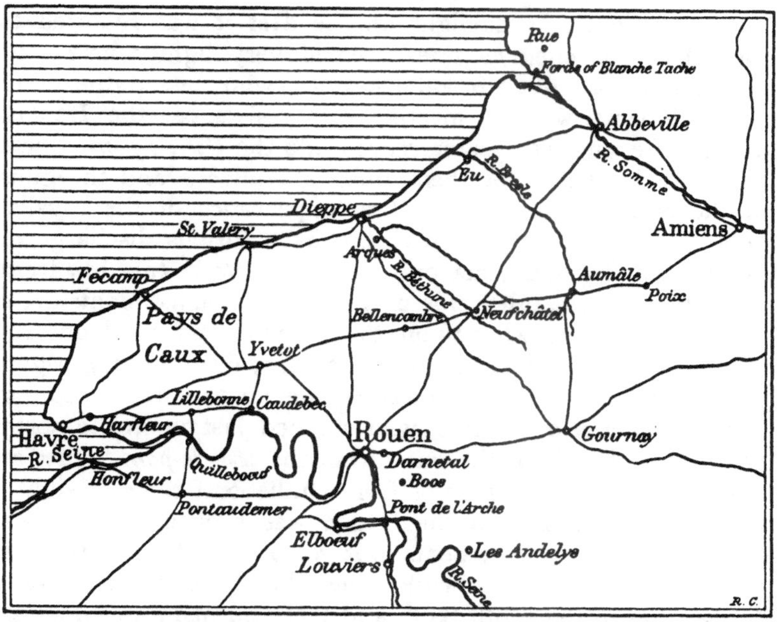

CAMPAIGN OF HENRY IV AND PARMA IN NORMANDY, JANUARY–MAY 1592.

enemy was found entrenched—Parma had already taken his precautions; he had also passed a Walloon regiment across the estuary in skiffs, with engineers who built a redoubt with three emplacements for guns at the spot opposite Caudebec where he intended to lay his bridge. On the near side he had thrown up a *tête-de-pont*, also furnished with embrasures, opposite the redoubt on the other side. On the night of the 20th the pontoons and boats from Rouen were brought into line, and clamped together with ropes and planks. The matter had been so accurately planned that before dawn the heavy cavalry and the French infantry had reached the south bank of the Seine, also the heavy baggage and the guns which were to arm the battery ; as daylight came on the Swiss, Spanish, and Walloon infantry were crossing. The external lines round Caudebec were only held by a couple of thousand Italian foot, and a few hundred light horse. The *tête-de-pont* at the north end of the bridge was manned by Bussy's Walloon regiment and now contained four guns.

All this passed unobserved by the Royalists—the King had been busy throwing lines of contravallation around Parma's outer position. At daylight on the 21st it was reported to him that the enemy's lines seemed very lightly held ; he accordingly brought up some cavalry, who passed the outer entrenchments without meeting any resistance—the Italians and the League light horse retired into the *tête-de-pont*, where they found waiting for them Bussy's Walloons and the guns in the embrasures. At the first appearance of the royalist horse before this work it was met by a furious discharge of musketry and gunfire, not only from the *tête-de-pont* but from the battery on the other side of the river. The assailants drew back in disorder, whereupon the Italians and Walloons retreated across the bridge, just as the royalist infantry began to appear, and to open musketry fire on the the retreating troops. Ranuccio Farnese, who stopped last upon the north bank, brought off all the guns from the *tête-de-pont*. Just at this moment the Dutch ships from Quillebœuf came in sight ; but before they arrived the bridge was broken by Parma's engineers. Some of the boats in it got back to Rouen, but the pontoons, and the woodwork which had joined them, were dragged on to the south bank and there burnt in an immense bonfire. The affair ended by a cannonade across the water between the royalist artillery, which came up too late, and the battery on the south bank near the end of the broken bridge. By this neither party took any harm.

This was a brilliant testimony to the Spanish engineers, who had been so careful and accurate in their calculations that the bridge was clamped together, and capable of bearing baggage-waggons and guns, within a few hours after dark. The month being May, the nights were short, and the width to be spanned was 300 yards of a tidal river. Parma, since the siege of Antwerp, had the greatest confidence in his engineers ; no other general, and no technical staff, could have conceived or attempted such a *tour de force* as to take 15,000 men across the Lower Seine in a single night. The idea that such a thing was possible had never entered into the head of King Henry and his lieutenants.

The neatness of the conception, and the success with which the outer lines were held with a minimum garrison, while the army was absconding, may remind the modern reader of the British evacuation of the Gallipoli lines in 1915, which was a similar triumph of fine calculation and rapid movement under cover of the darkness.

Having his army safely across the Seine, Parma, abandoning his usual slow and formal order of march, went off at a great pace, being determined not to lose the start of several days which he had gained on his enemy. He reached St. Cloud in the almost incredibly short space of five days: he did not turn into Paris, though he threw 1500 Walloon infantry into the city, to strengthen the already existing Spanish garrison, and went back to the Netherlands by way of Champagne. He was much wanted in his viceroyalty, since Maurice of Nassau had been seizing many outlying fortresses in his absence. This was the great Duke's last campaign—he never got over the effects of his wound at Yvetot, though he paid a long visit to the baths of Spa in the hope of recovering his wonted health. He died, a broken man, at Arras on December 2, 1592, and with him expired the last chance of the League in its contest with Henry of Navarre.

N.B.—The dates of the operations about Caudebec are a little difficult to settle. I have followed those given by Davila and De Thou, who are very detailed at this period.

CHAPTER XI

TRIUMPH OF HENRY OF NAVARRE (1593–98)

THERE is little to chronicle of purely military interest in the French Wars of Religion after the curious and incoherent Norman campaign of 1592 had come to an end, and Parma had retired to die in the Netherlands. The events of the succeeding years depended more on politics than on strategy, and the determining factor in the end of the war was King Henry's long-delayed 'abjuration' and reception—for the second time—into the Roman Catholic Church. After that the cause of the League was lost, while the much-disappointed Huguenots had to be contented with good terms of toleration, and to abandon their dream of setting a king of their own creed upon the throne of France.

Military operations there were in plenty in 1593–96, but they were neither decisive nor interesting. The civil wars of the south and west went on as before between extemporized armies, with curious vicissitudes of fortune, in which Lesdiguiéres, the Huguenot commander in Dauphiny, was the only general to achieve a brilliant reputation, for his exploits against the Leaguers and their ally, the Duke of Savoy. But these were of no decisive importance—the real key to the problem was the King's acceptance of Catholicism in 1593, after which his enemies submitted one by one, each making such terms for himself as he could obtain.

A few words, however, must be devoted to the purely military side of the matter. After Parma and the Spaniards had so unexpectedly disappeared from his front on May 20, by crossing the Seine, King Henry found himself faced by a difficult problem: should he pursue the Duke, marching upstream past Rouen and crossing the Seine at Pont de l'Arche, the first bridge in his possession? But this meant an enormous detour, and the Spaniard would have many days' start; what Parma's ultimate intentions were, and how fast he would be moving, no one could guess. Or should the army resume the twice-interrupted siege of Rouen, into which it was soon known that Mayenne had thrown himself with many of the troops of the League,[1] who had not followed Parma eastward? But the memories of the two sieges of Rouen were hateful to the soldiery, and the region around it was a desert, after six months of devastation by two large armies. A third serious attempt to molest Paris hardly came into the region of possibilities, mainly owing to the state of the troops. The force which had opposed Parma had been scraped together from many quarters for an emergency, outlying detachments having been called in, and garrisons cut down to danger-point. Moreover, the important Dutch and English contingents—unpaid as usual—had only been lent for a term of months and were craving to depart. Henry could not keep the army which he had collected together, for a continuous summer campaign. After sending back Montpensier's corps to take up again the defence of western Normandy, where the Leaguer Mercœur had just inflicted a serious defeat on the local Royalists (May 23) at Craon,[2] and filling up his numerous depleted garrisons—Dieppe, Arques, Pont de l'Arche, Pontaudemer, Caen, Mantes, etc.—and permitting the departure of the Dutch and English regiments, he had only the nucleus of an army left. His depleted French infantry companies had to be consolidated into new regiments, or told off into the garrisons. Money was—as usual—lacking, and the old royal Swiss units, inherited from Henry III, were clamouring for long arrears of pay, and were with difficulty kept to their colours.

The King in the end resolved to abandon any idea of a serious campaign, and, allowing his army to break up, went off with a 'camp volant,' as it was called, a mere flying force, on a rather inexplicable expedition. With only 3000 horse—French noblesse and reiters—and 6000 foot, of whom the Swiss, pacified by part-payment, were the most important item, he marched past Paris into Champagne—about the last quarter in which he might be expected to appear. Apparently his original object was to succour Épernay, the only town which he possessed in that region: but it surrendered early in July before he could arrive. Whereupon he set himself to recover the

[1] He had fallen sick, and was not fit for field service. Parma left him the Papal Swiss as a parting present. But this corps had dwindled to less than 2000 men, and was disbanded for mutiny. Davila, xiii. p. 38.

[2] Montpensier died June 2, before he could take up the task of undoing the results of the battle of Craon (May 23), where his son had been defeated by Mercœur.

place (July 20), and lay before it for some weeks. During the siege the old Marshal Biron, who had been acting for the last year as the King's second-in-command, was killed by a chance cannon-ball (July 27). He was a good soldier, but arrogant and somewhat insubordinate: his death was probably more of a relief than a loss to King Henry. Épernay fell (August 8) and Provins soon after (September 2), the local forces of the League failing to make any head against the Royalists, though the expeditionary force was so small. But, after failing before Meaux, the King went round Paris on its northern side, and returned to his old posts on the Lower Seine. Matters had not been going too well for him in that region, not only had the loss of the battle of Craon (May 23) enabled the Leaguers of Brittany to advance into Maine and western Normandy, but the Duke of Mayenne had come out of Rouen and captured Pontaudemer, one of the most important royalist centres of organization.

When things settled down for the winter neither party had much reason to congratulate itself on the events of 1592. The King had failed before Rouen, and had been completely foiled by Parma: his condition was no better than it had been after Ivry—indeed perhaps a little less encouraging, since the considerable succours lent him by Holland and Queen Elizabeth had failed to give any decisive result. On the other hand, the Leaguers were only too conscious that they had been delivered from ruin by the intervention of Parma. Without Spanish help Paris and Rouen might have fallen: but Spanish help threatened to bring about Spanish domination in France. The envoys of Philip II were continually pressing for the calling of a States General, which should elect his daughter, Isabella Clara Eugenia, Queen of France, and he long refused to divulge the husband that he intended to give her: a prince of the Hapsburg line was spoken of. This scheme irritated most of the Leaguer notables—all indeed save the young Duke of Guise, who had some hopes of imposing himself as a bridegroom on the Infanta. There was an extreme Catholic party, which was prepared to accept a Spanish Queen out of desperate hatred for Henry of Navarre. But incessantly the greater magnates were intriguing with the 'Politique' supporters of the King, and intimating that they would join his cause, and throw over the Spaniards, if only Henry would 'abjure.' They always stipulated for enormous rewards for themselves, and demanded that no Huguenot should ever hold office or magistrature, so that their overtures were rejected.

On the other hand, the King was getting more and more convinced that he would never win the crown of France by the sword of his Huguenot co-religionists, and more and more suspicious of his 'Politique' adherents, who after urging 'conversion' upon him for three years were growing restive, and showing signs of treasonable intentions. This was the time when the 'Tiers Parti' was being mooted—the project for the election of some sort of a King of the house of Bourbon other than Henry himself—the Cardinal of Vendôme or another—whom both the moderate Leaguers and the disappointed 'Politiques' in the King's camp might acknowledge. The former were only too anxious to throw up the Spanish connection; the latter, after so many campaigns, had come to look on the triumph of an unconverted Huguenot King, however legitimate, as not only impossible but absolutely pernicious.

There was an obvious way out of the difficulty—the moment that Henry could bring himself to 'abjure,' the idea of a 'Tiers Parti' would vanish, and great masses of the Leaguers would come over to his side, from their hatred for the Spanish domination. It is true that this would bring about bitter disappointment among the Huguenots, even if they were promised the most ample securities and their proper share in the administration of the realm. But the Huguenots had become the smaller part of the King's army, and many of their old chiefs, La Noue, Chatillon (the son of Coligny), and the old Duke of Bouillon were dead. Their existing leaders were mostly politicians rather than enthusiasts, such were Turenne (now Duke of Bouillon by marrying the old Duke's heiress) and Maximilian of Rosny, the famous Sully of later years, who came early to the conclusion that the King could win complete success only by allowing himself to be 'converted.' Lesdiguières, the most brilliant general in the King's service, was such a lukewarm Protestant that he sold his creed for the grand title of Constable of France long years after, and served against his old co-religionists in the troubles that followed his master's death. It is true that there were plenty of zealots who exhorted the King to persevere at all costs, and told him that it was better to win a quarter of France, and to be sure of salvation, than to win the whole and lose his soul. But even among Protestant ministers there were some who doubted whether peace for a ruined realm, combined with toleration, was not worth the sacrifice. Morlas told the King that the Church of Rome was after all a Christian Church, though a corrupt one. It could not be denied that there were many good people among the Catholics, of whose salvation one could not doubt. The Church might be reformed instead of bring broken up—an old cry that had ceased to have much weight since the Council of Trent. One courtly preacher reminded Henry of the prophet Elisha's words to Naaman, who was granted permission to bow in the House of Rimmon—a much more pagan thing than to listen to a Mass.[1]

Matters were brought to a head early in 1593 by the summoning of a States General at Paris by the League, for the purpose of electing a new sovereign, the throne being, in their view, vacant since the death of 'Charles X.' The assembly was wholly unrepresentative, many provinces having no proper delegates present, and was insignificant in numbers, not more than 130 persons having presented themselves. But there was obvious danger to Henry IV in its appearance, perhaps more from the chance of a new French prince being chosen than from that of a Spanish nomination. For a majority of the Leaguers hated Philip II more than they hated Henry of Navarre. Still, the King of Spain was very insistent in bringing forward the claims of his daughter; he had a party of zealots who were prepared to support him, and even those Leaguers who feared him most were afraid that their cause would go down without the continued support of the 'Catholic King.' Parma had saved them twice, and though he was dead his master still lived, and was ready to promise even more military assistance than he was really able to produce.

The States General held their first formal meeting on February 16, 1593, already somewhat influenced by a proclamation issued by the King on January 29 that he was ready to 'receive instruction,' *i.e.* to consider making his submission to the Catholic Church. This declaration somewhat shook the theory that he was an obstinate relapsed heretic, and many members of the States General began at once to doubt the expediency either of electing a new king or queen, who would be only the tool of the extreme party, or more probably the

[1] See d'Aubigné's most interesting chap. xxii. of book iii., the narrative of a firm Protestant who despised compromises, and bitterly censured the complaisant ministers.

mere nominee of the detested Philip II. The Spanish ambassador Feria made matters worse for his master, by declaring that the Infanta, if elected, would marry a Hapsburg prince. The States General could not bring themselves to face this prospect, and took refuge in dilatory constitutional discussion: the most eloquent speakers found a camouflage for their real sentiments by professing an incredible reverence for the Salic Law—sacred, unchangeable, established by divine ordinance, a thing to be regarded with religious awe. On June 23 the majority voted that the election of a foreign king or queen would be null and void, as contrary to the fundamental laws of the realm. In July the States began to disperse, after adjourning further decision as to the choice of a sovereign, though the Duke of Feria, yielding too late (July 20), had at last conceded that his master might give a French prince his daughter's hand. By August 10 the most of the members had gone home.

For matters had already been settled without their sanction. On May 15, King Henry had formally announced that he was prepared to 'receive instruction'—not very necessary since he was already perfectly versed in the details of the Catholic faith, after four years of confirmity to it after St. Bartholomew. He met at Mantes a committee of Catholic bishops and doctors of his own faction, and after some discussion declared himself convinced by their arguments, and took his way to St. Denis, where he had a garrison at the very gates of Paris, on July 12, giving notice that his public profession of orthodoxy was prepared. He had taken great care not to conclude any bargain with the great number of magnates who had given him assurance that they would adhere to him after his conversion, because he intended to make them come to him for terms, not to go to them and bind himself by promises beforehand. On July 25 he presented himself before the closed door of the Basilica of St. Denis, knocked for entry, and was received by the Archbishop of Bourges and seven other bishops, before whom he promised to live and die in the Catholic faith, whereupon he was received into communion, and a High Mass was celebrated. On the day before he had parted with tears from his Protestant chaplains, and assured his Huguenot captains that no harm should ever come to 'those of the religion.' His inner feelings may perhaps be judged from the phrase which he wrote to his most favoured mistress, Gabrielle d'Estrées, on

July 23: 'Apres-demain je fais le saut perilleux'—*faire le saut périlleux* being colloquial French for turning a somersault!

Only a man of genuine religious convictions could have resisted the temptation. Knowing the feelings of the large majority of the moderate Catholics, and being informed day by day of what was going on in Paris, Henry could see certain triumph before him, if he could bring himself to 'abjure.' He was no enthusiast, and he had a real commiseration for the state of the people of the distracted realm, who were calling for peace with ever-growing unanimity. Even in Paris public opinion was turning in his favour, and only the pressure of the Guisard faction and the large Spanish garrison prevented a popular declaration in his favour, now that he was once more a Catholic.

The history of the next two years is that of the gradual defection from the League of the cities and magnates who had formed its strength. That it was not at first quite so rapid as might have been expected was due to the hesitation of Pope Clement VIII to offend Spain by giving his official absolution and blessing to the repentant heretic. It was only on August 5, 1595, that this was done, and long before that date three-fourths of France was in Henry's hands—Paris itself had opened its gates on March 21, 1594. The details of the adhesions of the magnates of the League to the royalist cause were singularly sordid and unedifying. Each haggled for a governorship or a pension. Henry had the upper hand, since his cause was in the ascendant, but he thought it worth while to be liberal when dealing with men of importance. 'Enfin vos sujets vous ont rendu votre royaume,' said a courtier. 'Dites plutôt vendu,' replied the King, and was fully justified in his epigram. Mayenne, almost the last to hold out, was bought in 1596 with the governorship of Burgundy and 500,000 gold *écus*. Mercour, who had maintained himself in Brittany till 1597 with Spanish help, was brought to terms a few months later.

There remained only the war with Spain. When the Leaguers had been purchased, and there were many of them (even Mayenne!) serving in the royal army, it might have been expected that the Spanish Netherlands would have been conquered, and partitioned between France and her Dutch allies. But the successors of Alexander of Parma made a much better fight than might have been expected. They even took the offensive, and captured the under-garrisoned Calais after a siege of a very few days (April 9, 1596), using exactly the same tactics that Guise had employed against the English in 1558. The surprise by stratagem of the great city of Amiens (March 11, 1597) was an even more portentous disaster. Henry recovered Amiens after a long siege (September 17, 1597), but made no head in the project of invading the Netherlands. His strategy does not shine in these years: as to his tactics, he made the last display of his incorrigible addiction to fighting in the front line in cavalry actions at the combat of Fontaine-Française (June 9, 1595), where he got entangled with 300 horse in the vanguard of a whole army of Spaniards, and was only saved by the tardy arrival of his reserves. If he had got killed, what would have happened? He was still without heirs—linked to the 'Reine Margot,' his impossible wife, from whom he was only freed by a papal divorce in 1599. The fate of France, if a new succession-question had arisen, would have been problematical. But Henry could seldom restrain himself when minor cavalry tactics were attracting him.

Philip II, on his death-bed, and with all his schemes foiled, offered peace, and on May 2, 1598, the long wars came to an end at the treaty of Vervins, which left frontiers practically unaltered on the north and east. Only a few days before, Henry had signed an equally important engagement, the famous Edict of Nantes (April 13, 1598), which gave his Huguenot subjects not only toleration but a very privileged position in the State. He could then sit down to reorganize his distressful realm—but his ingenious schemes of domestic policy do not concern us. It is more necessary to remember that his allies in the war, Queen Elizabeth and the Dutch States-General, regarded the Peace of Vervins as a betrayal. It left them with the Spanish war on their hands, deprived of an ally on whom both had lavished a good deal of money and innumerable auxiliary regiments, which had stiffened the royalist armies in many a critical moment of the war. But Henry, no doubt, had first to think of the future of France: the Protestant Crusade against Spain no longer interested him.

BOOK VI

THE REVOLT OF THE NETHERLANDS AND THE DUTCH WAR OF INDEPENDENCE

1568-1609

CHAPTER I

GENERAL ASPECTS OF THE STRUGGLE

AS has already been observed, two long records of wars very different in character and in strategy, but both having as their central fact the clash of religious antagonisms—the Reformation and the Counter-Reformation—fill up the last forty years of the sixteenth century. We have dealt with the Wars of Religion in France; it remains to consider the almost unbroken contest between the King of Spain and his discontented subjects in the Netherlands, which contrasts so strongly with the French troubles, because no long intervals of insincere truces divide the continuity of what was practically one long struggle. The two wars sometimes became for a time blended in a single conflict. Huguenot auxiliaries came to support the insurgent Netherlanders, while William of Orange led an army into Picardy and Champagne in 1568, and a Dutch fleet gave Henry of Navarre useful help on the Seine in 1592. On the other hand, Philip of Spain often lent troops to Guise and Mayenne, and his great general, Alexander of Parma—as we have seen—saved the League from an early demise.

Yet it would be a complete mistake to regard the fighting of the whole period either as a purely religious contest, or as a continuation of the old rivalry between Hapsburg and Valois, which had inspired all the wars of the first half of the sixteenth century. There were indeed French leaders, like the Admiral Coligny, who wished to distract their countrymen from domestic strife by urging a new attack on the Hapsburg power. But their propaganda failed, because French Catholics disliked their Huguenot compatriots more than they did Philip of Spain, the champion of the old faith. Nor did Netherland intriguers profit by electing the wretched Duke of Alençon as their sovereign—he could not commit France to his enterprise of personal ambition. On the other hand, in the last days of the French wars, when Alexander of Parma seemed to be the only possible saviour of the cause of the League, the majority of French Catholics shrank from throwing themselves into the arms of Spain, and choosing Philip's daughter as their queen. National prejudice was stronger than religious fanaticism not only among the 'Politiques,' but among all who were not extremists. The two motives were balanced in perplexing fashion, till Henry of Navarre found the obvious solution, to the discomfiture of all desperate enthusiasts—Huguenots no less than Leaguers.

It is true, nevertheless, that in most countries of Europe there was going on simultaneously for half a century and more the strife between the votaries of the old and the new confessions. There was a contest of parties in each country, leagued with parties of the same faith in other countries. English, French, Scottish, and Netherland Protestants regarded themselves as brethren—'Gentlemen of the Religion' as the phrase rang. In Germany, Protestant princes sent their reiters to serve in the armies of Coligny or William the Silent not purely for financial reasons, while their Catholic neighbours contributed regiments of landsknechts to the armies of Guise or Alexander of Parma. The extraordinary thing in Germany was that this perpetual contribution of auxiliaries to rival religious camps abroad did not lead the two parties to a general war at home—the logical termination of the phenomenon was deferred for two generations, till the outbreak of the Thirty Years' War in 1618, when matters had settled down in France and the Netherlands, and, long before that, in England and Scotland. There had been symptoms which might have portended a general explosion at a much earlier date—*e.g.* the adventure of Gebhard Trüchsess, the Archbishop of Cologne, who tried to turn his vast dominion into a Protestant state by force, and drew in upon himself all his Catholic neighbours. But that danger passed off (1583). For thirty years the Emperor's Palatine and Nassau vassals were conducting reiters by the thousand to the armies of the Huguenots or the Netherland rebels, while Ysenbergs and Rhinegraves and Schaumbergs and Mansfeldts [1] were commanding German units in the army of Alva, Guise, or Don John of Austria. Probably the Turkish danger, insistent till the end of the sixteenth century, was the main

[1] The much-ramified Mansfeldt house supplied generals to both sides—Peter Ernest the Catholic was the most notable, but Wolfrad, who served opposite him at Moncontour, was quite a competent soldier.

cause of the delay of the general outbreak of religious civil war in Germany. But the Emperor Maximilian II must take his share of the credit: his moderating influence was long felt.

THE NORTHERN NETHERLANDS, 1568–1609.

The war in the Netherlands did not at first assume the character of a purely religious struggle, as it started with a widespread opposition to the breaches of old constitutional practice in the governance of the many duchies and countries of the Burgundian inheritance, put in practice by Philip II. The Regent Margaret of Parma provoked discontent, acting purely as the King's agent, but the inexorable Duke of Alva chastised with scorpions, when his predecessor had been using only the whips which her half-brother put into her hand. The King himself was the inspiring spirit, raging against what he considered sedition, and executing faithful Catholics like Egmont—the victor of Gravelines—and the Admiral Horn, till he drove all ranks and classes of the Netherlands into open rebellion. Religion was for a time but a secondary cause of revolt; in many regions the Protestants were but a minority. For a time it looked as if the whole thirteen Provinces of the Low Countries might coalesce into a new state; this seemed possible after Alva's failure before Alkmaar (1573) and that of his successor Requescens before Leyden (1574). At the 'Pacification of Ghent' (1576) the southern provinces joined with the northern, to build up a league founded on toleration in religious matters and independence from the Spanish yoke.

Unfortunately the Union fell to pieces, partly because the crown of Spain was in 1576–78 represented no longer by a tyrant like Alva, but by the affable and romantic Don John of Austria, the bastard brother of the King, still more because, when it came to fighting, the Spanish troops were infinitely more efficient than the heterogeneous levies which the 'Estates' put into the field. But the most powerful efficient cause of disaster was that, after a short period of nominal union, the Catholics found that they could not co-operate happily with the fanatics, iconoclasts, and demagogues who were far too numerous among their Protestant fellows. Image-breaking, church-sacking, and mob violence estranged many from the cause of independence, when once the immediate terror of the rule of such a tyrant as Alva had been removed, and the plausible Don John promised reasonable terms of administration. Hence a gradual drift back of many Catholics who had been constitutionalists in 1576, to submission to the royal authority. The rout of Gemblours (January 31, 1578) proved the military incapacity of the army of the States. The idiotic attempt to win help from France by electing the worthless Duke of Alençon as sovereign of the Netherlands ended in deserved disaster (1581) and the breach between north and south had begun to take shape in the 'Union of Utrecht' (1579) by which Holland, Zealand, Utrecht, Guelderland, and Friesland formed themselves into a separate league, which, while not repudiating the authority of the States General, proceeded to carry out an independent policy of its own, and to persevere in rebellion against Philip II, whatever might happen in the Walloon provinces. This was really the beginning of the Dutch Republic. It only remained to be seen where boundaries were to be drawn.

That the northern half of the old Burgundian heritage was destined to become a Protestant republic, and the southern half to remain a Spanish province down to the beginning of the eighteenth century, was to a large extent due to geographical causes. Belgium (to use a modern term) was for the most part a ' cavalry country,' fit for great battles in the open, from Gemblours to Waterloo. ' The Cockpit of Europe ' is a region that can be lost and won in the field. Not so Holland (to use again a modern word), which is for the greater part of its extent so cut up by arms of the sea, rivers, canals, and marshes, that it is hard to find within its borders room to set a large army in formal array. The fights that it has seen have been narrow-fronted scuffles along the tops of dykes, or desperate attempts to cross difficult water-barriers. Its normal military history has been a long record of sieges of water-girt fortresses, rather than of general actions. The impossibility of deploying large armies for strategical manœuvre in it was discovered alike by the *tercios* of Philip II and the much-irritated battalions of Frederick, Duke of York, in 1799.

There is only one natural phenomenon which can make Holland a country easy to be overrun—a long-continued frost, which can freeze its broad rivers, and make its marshes solid ground fit to bear horse, foot, and artillery. In all modern history there has only been one occasion on which the whole land was subdued in three weeks by an enemy, and that was during the terrible frost of December–January 1794–95, when Rhine and Lech and Yssel became high roads, and the vile marshes of the Biesbosch and the Bommeler Waard looked like moors powdered with snow, save that the shrivelled vegetation rising from them was reeds and sedge, not furze or heather. There was probably never such a long and bitter frost as that of mid-winter 1794–95, but there were many hard seasons in the years of the long war of the King of Spain and the Dutch.[1]

[1] A curious incident occurred in the winter of 1586. A Spanish detachment, blockaded by Dutch ships in the Isle of Bommel, escaped over the ice, when the enemy had to sail off, for fear of being ' frozen in.'

Indeed we have plenty of notes of ice that blocked rivers and canals and complicated tactical plans. But that such chances had no determining effect on the main issue of operations came from the almost universal custom of suspending military movement during the winter, and dispersing armies into comfortable quarters. Gonsalvo de Cordova's victory of the Garigliano (December 29)—told in an earlier chapter [1]—stands almost alone along with Dreux (December 19, 1562), and Gemblours (January 31, 1570) as examples of a mid-winter general action.

The leisurely chiefs of the sixteenth century seldom contemplated the massing of troops for a blow at that season, since the possibility of delivering it depended on the weather, and no one could calculate upon continued stability of atmospheric conditions. Maurice of Nassau at Tournhout (1597) made a most exceptional January stroke. If a dry frost seemed to have set in, yet a thaw followed by inundations might supervene, and wreck a plan that had been set going. The wars of the French Revolutionary and the Napoleonic period differed from those of earlier centuries in that campaigns often continued in the worst season of the year. Beside the conquest of Holland that has been mentioned in 1795, it is only necessary to quote Rivoli (January 14–15), Hohenlinden (December 3), Austerlitz (December 2), Pultusk (December 26), Eylau (February 6), Corunna (January 16), The Beresina (November 25), the Nive (December 10), La Rothière (February 1), and Montmirail (February 11), mostly fought in the snow in weeks when a general of the sixteenth or seventeenth century seldom dreamed of concentrating an army. The nearest approach to such a time of the year seeing a general action in the older annals of the Netherlands wars was on January 31, the date of the rout of Gemblours in 1578.

The strife between the King of Spain and his rebellious subjects was normally a war of sieges—nothing can be more surprising than the extremely small number of general actions that can be catalogued. And of those fights that did happen few save Gemblours and Nieuport (1600), the last of all of them, were battles of a decisive sort fought between the main armies of the two combatants. Indeed it is hard to call any of the others a general action—Heiligerlee, Jemmingen, Mookerheyde, Warnsfeld, Steenbergen, even Tournhout were encounters

[1] See above, p. 120–27.

between comparatively small forces. At Gemblours indeed large numbers were present on either side, but it was a rout, in which the main body of the victorious army was never called upon to strike in. The whole war was mainly one of manœuvres and sieges—on many occasions the dislike for committing everything to the hazard of a decisive battle, which we have noted in the later days of the Great Wars of Italy,[1] is found in the Netherlands. It was marked in the case of the Duke of Alva in 1568, when, even after the preliminary success of the combat on the Gheet,[2] he refused to tackle Orange's main army, and preferred to manœuvre it out of Brabant. Not less clear is William of Orange's reluctance to attack Alva's position covering the siege of Mons in 1572—as his captains urged him to do—and equally so Alva's decision not to press hard on Orange when he had committed himself to retreat.

In the very end of the war, Maurice of Nassau refused to risk a general action even to save the heroically-defended Ostend (1604). As long as the rebels could make any head in Belgium, there were opportunities for stricken fields—seldom utilized.[3] In the middle decades of the war, when the Spaniards had possession of all the south save a few coast-towns in Flanders, it may be said that there was little room for large-scale battles among the canals and marshes of Holland proper. In its last decades, when Maurice of Nassau had begun to press southward, there was more chance of open fighting—and he did justify his tactics and the improved discipline of his army by two victories in the open—the small affair of Tournhout (1597) and the very considerable pitched battle of Nieuport (1600), with which our survey of this long period of war will end. But, as we have already said, he would not fight to save Ostend, though he had got as far as Sluis in 1604.

In the sixteenth century the Low Countries were more thickly strewn with fortified places than most regions of Europe —a testimony at once to their wealth and to a long history of dynastic wars, among the various counts and dukes whose dominions had finally been gathered together by the ambitious House of Burgundy, and then passed on by marriage to the all-devouring Hapsburgs. There were not many open towns (like The Hague) in the whole of the Netherlands. Even small

[1] See above, p. 216–20.

[2] Close to Marlborough's battlefield of Ramillies.

[3] Cf. the way in which Don John, even after Gemblours, refused to tackle at Rynemants the States army under De La Noue in July 1578.

places were habitually fortified, though, when our period commences, they had not been brought up to the level of the ideals of the Renaissance engineer. A glance round the military drawings of the sixteenth century, which are numerous, points out at once a striking peculiarity of the Netherland fortresses. The large majority of them are surrounded by elaborate water defences. Only a few of them, like the high-lying citadel of Namur, Luxemburg, or Bouillon or Vianden, are of the type common in the rest of Europe, where steepness of situation gives the main strength, and a dry ditch or a precipice has to serve, because of the obvious fact that it is impossible to make water run uphill. These strongholds, of course, are to be found mainly in the Ardennes, where the conditions are hilly, almost mountainous. In Flanders or Holland such situations are not to be found in the marshy flats. A wet ditch was, of course, in every way preferable for the defender, since it made mining impossible on so much of the front as it covered, and could only be cancelled by draining off the water, which was impracticable if the supply came from a large river or the sea. If the water was stagnant or fed by small cuts only, it was possible to make approaches across it, by casting in rubbish, gabions, or fascines, or even by completely battering down the wall behind, so that it might fall into the ditch. But an attack in such fashion would have to be made at limited points, and when such points had been designated, the defender could often make work upon them impossible by flank fire, or dominating fire from other parts of the walls. Bergen-op-Zoom and Flushing were proverbially hard to tackle from their complete water-defences, which were fed by tidal sea-water, which the enemy could never drain away. But dozens of less formidable places had long stretches of their fronts covered by marshes or river-channels, or canals, and were only approachable on limited sections of their enceintes, where (of course) specially heavy fortifications would be piled on. Such was the case at Ostend, Ypres, Ghent, Antwerp, Maestricht, Bruges, Haarlem, Leyden, Dordrecht, Groningen, Zutphen, Sluys, and countless other places of less importance.

At the commencement of the War of Independence most towns had only fifteenth-century walls behind their water-defences. But as the years wore on, scientific fortification developed rapidly, and by 1600 all sorts of hornworks, demi-lunes, redans, and ravelins were applied to the trace of the

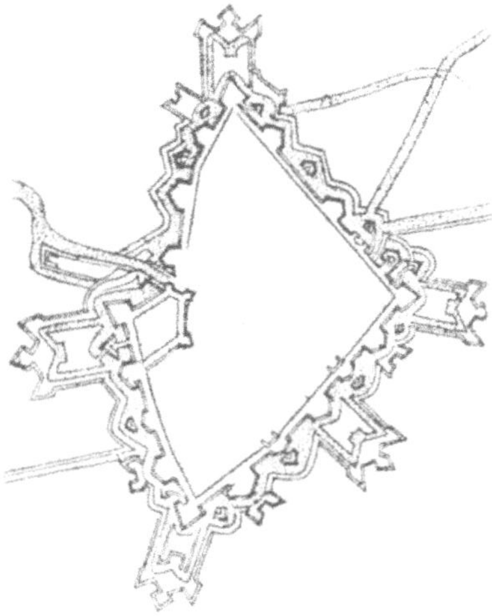

BREDA

BERGEN-OP-ZOOM

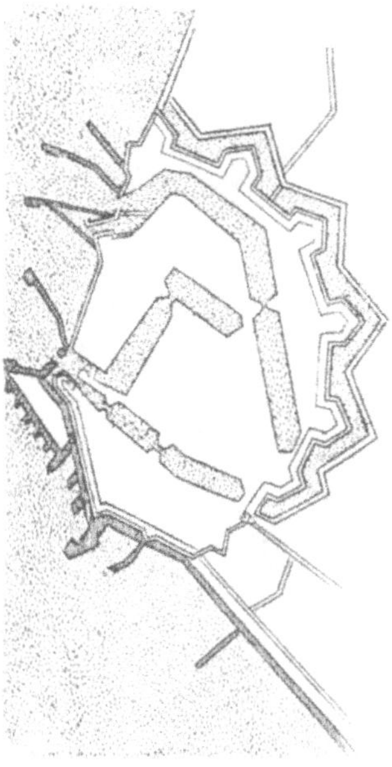

FLUSHING

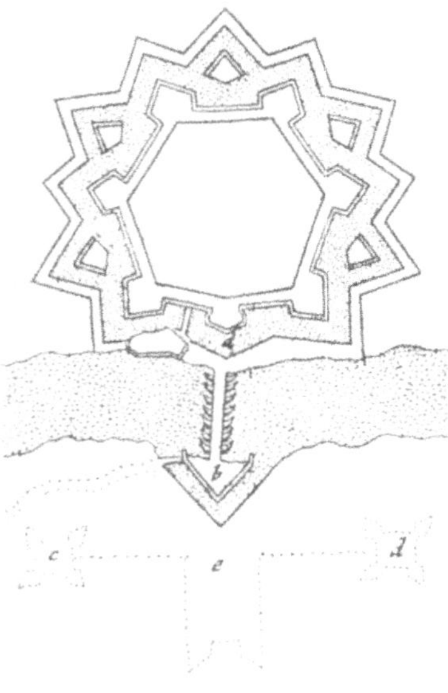

STEVENSWEERT

FOUR DUTCH FORTRESSES REBUILT IN THE MODERN STYLE

From Fournier's 'Traité des Fortifications'

simple original works. Maurice of Nassau was a great constructor, as he was also a great destroyer, of fortifications. In a country of water-defences the military engineer ended by beating the artillerist. When the besieger was forced to attack on limited fronts, as the only accessible points, the besieged heaped line after line of defence behind them—even building new works during the actual progress of a siege. Hence the habitual length of late sixteenth-century and early seventeenth-century leaguers, and the fact that an honestly defended town normally succumbed from starvation, and not by sapping or by storm. In some cases a place fell by treachery—as witness the disgraceful surrenders of Deventer by the English renegade Stanley, of Gelders by the Scot Patten, and of Gravelines and Grave by the Netherlanders Lamotte and Hemart; but this the military engineer could not guard against. In other cases stratagem and surprise were successful against a place that could never have been taken in any reasonable time by orthodox methods—as in the capture of Breda in 1590 by Maurice of Nassau's men hidden in turf-barges arriving at the town-quay. A similar attempt by Martin Schenk at Nijmegen (1589) had failed, because only the leading barges came to hand, the rest having been carried downstream by the current of the Waal. The same condottiere had captured Warrel, by means of men passed into the gates concealed in salt-carts, in the preceding year.

But putting aside such exceptional happenings as treachery or surprise, the normal siege protracted itself for untold months. When a fortress could be revictualled and reinforced by sea, it could hold out practically for ever. The classic instance is the incredibly lengthy defence of Ostend by the Dutch (1601–1604), which beats the record of Sebastopol. But the fourteen months' resistance of Antwerp to the Duke of Parma (1586) and the successful eleven months' endurance of Leyden (October 21, 1373, to October 3, 1574) are almost as well known. In all these cases it was the water-defences which were effective. Ostend, attacked perforce on a narrow front, threw up barrier after barrier to the besieger. Antwerp only fell because Parma, by a marvellous feat of engineering, blocked the broad Scheldt by a fortified bridge, which cut off succour from the sea. At Leyden the besieger had to depart because careful cutting of dykes flooded him out of his camps and trenches, and gave him only the choice between drowning or departing.

There are many points of interest to be recorded in these wars, though they are so barren in the matter of general actions in the field. After the fall of Antwerp and Ghent, and the complete loss of Belgium, save some shore-fortresses in Flanders, by the Netherlanders, the game looked very desperate for them—more especially when the enemy had outlying possessions quite in their rear, such as Groningen and the district round it. The armies of the States on land had experienced an almost unbroken series of defeats. There was no doubt that the Spanish infantry was more efficient than the heterogeneous levies which were sent against it. The troops which William of Orange and his successors took into the field were mostly composed of foreign auxiliaries, English, Scottish, and German infantry, German and Huguenot cavalry. All the best blood of the Dutch themselves seemed to have turned to the naval career; from the first their sea-captains and sailors had a marked ascendancy over the Spaniards. They were permanently in command of the sea, a fact which hampered all the operations of Alva, Requescens, John of Austria, Parma, and their successors. If Philip II could have concentrated a superior fleet in Dutch waters, further resistance would have been impossible. But this he never could do—isolated squadrons were destroyed; the great 1588 venture of the Armada got wrecked upon the fact that the Dutch squadron continued to blockade Parma's army, while the English were dealing with the high-seas fleet that was to transform the whole face of affairs—its threat was as dangerous to Holland as to England.

None can fail to recognize in the early years of the war the incontestable superiority of the Spanish troops in the field, their wonderful *esprit de corps*, and their many desperate feats of valour. Nothing for steadfast courage could surpass the march of Mondragon's men through the waters of the 'Drowned Land' to relieve Tergoes in 1572. To wade at low tide across six miles of water breast-deep, with the knowledge that the ford would be flooded when the tide came in, and that any unforeseen delay meant certain death, was a marvellous achievement for a body of 3000 men to undertake. Wellington's celebrated passage of the Bidassoa estuary in 1813 was child's play in comparison—for there the distance was not much over a mile, and there were only two short channels where the water was up to the thigh at low tide. Add fanaticism to perfect tactical skill and complete self-confidence, and one asks oneself how such troops could be beaten by the levies of the Dutch. But these veterans had their fatal foibles—the most obvious was a deeply rooted tendency to mutiny. When their pay ran short they went on strike, sent off their officers, and chose an *eletto* or temporary chief to negotiate with their general. It is true that the remark was made at the time that Spanish armies always mutinied after a battle, and Dutch armies before it. The Spaniard, having secured military predominance by a victory, turned to blackmailing his commander, when there was no longer any danger from the enemy. But these more or less orderly strikes paralysed a campaign, and sometimes lasted for months rather than weeks, so that all the results of victory were lost.

The regular supply of money, which alone could have kept these stubborn veterans in perpetual harness, often ran short. Not only had Alva's tyranny drained the Netherlands of money, but King Philip was inclined to be penurious, and grudged his subsidies. When he did prepare treasure-ships, there was a chance that the gold might be intercepted on the way (as in 1569) by the English government, or captured by the Dutch squadrons which swarmed in the Channel and the North Sea.

There was also to be reckoned with the King's inscrutable mind—suspicious and perverse. He viewed his successive generals with distrust, and often interfered to change their plans of campaign. Twice when Alexander of Parma seemed to have got the upper hand, he was distracted from his enterprise against Holland—once in 1588 when he was made to concentrate all available troops at Dunkirk to aid the Armada in the invasion of England; the second time in 1590, when he was directed to take his main field army into France to support the failing cause of the Catholics of the 'League.' Both of these distractions were invaluable to the Dutch.

The Protestant field armies which met with so many disasters, from Jemmingen downward, in the earlier years of the war, contained a comparatively small proportion of national troops—a certain number of the nobles served as horse, along with the unlucky princes of Nassau—of whom Adolphus, Louis, and Henry all died in cavalry charges. But the main part of the squadrons were always reiters levied in Germany with Netherland money—they were mercenaries and not Protestant enthusiasts, often mutinied, deserted still more often, and seldom put in a solid day's fighting. French auxiliaries, brought in sometimes by Huguenot chiefs, were better fighters but hopelessly disorderly and undisciplined. Leicester came in 1585 with 1000 English horse, which soon dwindled to 500, but the only good record that we have of them is the desperate onset at Warnsfeld—which was a sort of Balaclava affair, since they charged headlong and were never supported by their main body, so that they had to retire after a useless display of valour. The cavalry of the Dutch Estates only began to achieve respectability in the later years of the war, under the discipline of Maurice of Nassau. But Holland, as we have observed before, is not a cavalry country. In this later period we find the Dutch horse, mostly 'pistoleers,' armed with cuirass and helm, without lances, and relying (like other European cavalry of the time) on the 'caracole' and the successive discharges of their firearms.[1] There were also light horse, some armed with the arquebus, who were used for skirmishing and scouting rather than for the charge. The great day of the Dutch cavalry was at Turnhout, where they drove the Spanish horse completely off the field, so that the abandoned squares of Spanish infantry, beset by horse and foot at once, were finally broken and annihilated.

The infantry of the States' Army was also preponderantly foreign, though not in quite such a high proportion as the cavalry. There were some foot-regiments raised in Friesland, Zealand, and other provinces, of which we occasionally hear. For town-guarding the local burgher militia was pretty efficient—and often behaved better than the mercenaries—it was a matter of fighting for hearth and home, when a place was beset by Spanish troops, to whom rape and massacre came as easily as plunder—more especially when (as was often the case) they had not received any pay from their government for many months. But burgher-guards could not be drawn off for operations in the open, being essentially a sedentary force, and of the field armies the majority of the corps were foreign. Auxiliary companies of volunteers, not raised by the state, but going out under condottiere captains, appear from Scotland and England quite early in the wars. Practically the only

[1] There was a curious example of the fatal results of pistol and 'caracole' practice at Mookerhyde in 1574, when Louis of Nassau's reiters, having driven off the main body of the Spanish cavalry, were all recharging their weapons, and getting back into regular order, when they were suddenly charged by the Spanish reserve of lancers—demi-lances, I suppose—and ridden down by the long weapons before they could reload or rally. But Turnhout was won by 'pistoleer' cavalry.

resistance in the infantry disaster at Gemblours is said to have been made by some Scottish companies (1578). Long before that, in 1572, Morgan had taken over the first regiment of English volunteers which served in the Protestant cause, and went into action at Flushing. This was the first of many corps which were raised in England in the name of the States, and served under Gilbert, Norris, and Roger Williams. They must be carefully distinguished from the authorized government levies which came over with Leicester in 1585—and they were apparently far better material, being volunteers, not pressed men, and ere long practical soldiers. Moreover, they were more regularly paid and fed by the States than were Leicester's regiments by the penurious Queen Elizabeth—whose abominable neglect of her soldiers has been spoken of elsewhere.[1] That the English troops on the Dutch pay-rolls often showed the faults of the mercenary—habits of plunder and of going on strike—is true, but they did not suffer like the Queen's regiments from ignorance of discipline, feckless inability to provide for themselves, and apathetic discontent. It was an immense relief, to the men and officers alike, when the Dutch in 1594 took over the remnants of the old army of Leicester into their service and pay.

Along with the English companies in the army of the States, we invariably find, in somewhat smaller numbers, Scottish and French companies—the latter stray remnants of the Huguenot levies which fared so ill in their native civil wars, and took refuge abroad in considerable numbers, when the 'Gentlemen of the Religion' had to expatriate themselves. All these elements in a Dutch army were rather more trustworthy than the Germans, who were the most numerous but the least reliable of all the foreign auxiliaries—being, as it seems, even more of professional mercenaries than the English Scots or French—all of whom had some tincture of Protestant fanaticism to eke out their love of adventure or of profit. The considerable forces which William of Orange and his successors used to collect on the Rhine for service in the Low Countries, were quite as prone to plunder, and even more habitual deserters than the other foreigners—the moment that pay ran out they melted away from their colours. This was perhaps natural, as they had not enlisted from enthusiasm, and were conveniently close to their fatherland if they wanted to make off. Sometimes they even went over to the enemy, in whose ranks

[1] See above, pp. 373–4.

they knew that thousands of their compatriots were always serving. It must be confessed that similar callousness to military honour was not always unknown among the English and Scots—as witness the disgraceful cases of Stanley, Roland Yorke, Patton, and Sempill.

As a fighting power on land the Dutch always suffered severely from the want of a real national army, whose place was indifferently taken by so many aliens, whose primary interest was not the preservation of the independence of the United Provinces. Maurice of Nassau and Frederic Henry his brother, in the later and more successful days of Dutch military effort, fared better, because they had instituted a more severe discipline, had succeeded in fostering a wholesome *esprit de corps* among their regiments, and always paid their troops on the nail. Their men were long-service mercenaries, not bands lightly raised and frequently disbanded, like the unsatisfactory levies of their father, William the Silent. In the seventeenth century both the English and the Scottish brigades in the Dutch service were highly esteemed, and young men of adventurous spirit used often to take service as officers in them for a time, including several notable figures of the English Civil Wars of Charles I—Essex, Leven, and Skippon among them..

That the Dutch were not finally worn down in the black years that followed the death of William the Silent, and Parma's capture of Ghent and the long-beleaguered Antwerp, is a wonderful testimony to their toughness in defence. The help that they got from abroad was always disappointing—the Alençon experiment for drawing in French aid had ended disgracefully in the 'French Fury' of 1583—the Leicester experiment, intended to commit Queen Elizabeth to open instead of furtive support of the cause of Independence, had brought over a raw and unpaid army and a governor who fostered civil strife (1585–87). The English alliance, which ought to have proved decisive, only staved off disaster, without bringing victory. The best help, indeed, which England gave was non-governmental—Morgan, Francis and Horace Vere and their companions were real assets in the struggle with Spain. On the whole, the Netherlands secured their liberty by their own indomitable perseverance during the most depressing years of the war, when fortress after fortress was lost by force or by treachery. But of course it must be conceded that if Philip II had not been distracted by foreign problems, and had been able to concentrate all his strength against Holland, without being forced to think of Queen Elizabeth, Henry of Navarre, or the Ocean and the Indies, he must have crushed the handful of gallant rebels who held out so persistently.

Perhaps the most potent distracting cause of all which operated for the salvation of the Netherlands was Philip's belated attempt to turn his attention on to France, just at the moment when, the scheme of the invasion of England having failed, he had at his disposition the great army which Parma had collected at Dunkirk and Nieuport.[1] It might have been launched against Holland in 1589—he sent it instead into France, to help the cause of the League against the heretics. Sacrificing his chance in the Low Countries, Philip bade Parma lead his army away from the Scheldt to the Somme. For the next three years the Spaniards had only a 'minimum garrison' in Belgium, while the flower of the field army and its great captain were campaigning in France. The danger of a serious invasion of Holland passed away, and ere long the Dutch were able to undertake that resumption of the offensive, detailed in a later chapter, which had so long been impossible.

[1] See page 509 above.

CHAPTER II

THE CAMPAIGN OF 1568

1. Heiligerlee (May 23, 1568) and Jemmingen (July 21, 1568), William of Orange in Brabant (October–November 1568)

CAMPAIGNS in the open field were, as we have already shown, rare in the history of the Netherland wars. The first and most ambitious of these was the concentric scheme for a combined attack on various points of the Netherlands, which the Protestant leaders framed in the spring of 1568, with the object of distracting Alva's attention from the main invasion, which was to be led by William of Orange himself—it was a complete failure. The raiders into Artois and Guelders were beaten and scattered. Only one enterprise, in a very remote corner of the Netherlands, had a temporary success. Louis of Nassau, the brother of Orange, collected surreptitiously a small mercenary force of German reiters and landsknechts, with which he entered the province of Groningen, with the intention of raising Friesland in revolt. He seized the little port of Dam on the Dollart [1] and issued an appeal to the Protestant party in the north-east. The region was practically ungarrisoned, and he collected several thousand irregular levies before the Duke of Aremberg, royal stadtholder in Overyssol and Guelders, was able to march against him. The Duke, determined to crush the insurrection before it had come to a head, arrived at Groningen on May 22, with a modest force, the *tercio* of Sardinia, four companies [2] more of Spanish and German infantry, six guns, and some cavalry squadrons from Hainault and Artois—about 2000 foot and 400 horse.[3] He could have been joined by the 24th by the Count of Meghen, who was bringing up from

[1] Now quite superseded by Deltzyl as the local port.

[2] 'Vanderas'=banderas, 'ensigns,' *i.e.* the companies of a ten company *tercio*, perhaps 200 pikemen and musketeers apiece.

[3] The ordinance of 1584 cut the company down to 100.

Arnheim four more companies of infantry and Martinengo's three squadrons of light horse,[1] but did not wait for the arrival of this trifling reinforcement. We are told that his officers urged him to strike in at once, on a motley band of insurgents, who would not stand against a bold attack, and not to linger till Meghen appeared—the latter had been delayed by a mutiny of his ill-paid troops and was several days behind time.

Louis of Nassau having a very raw and miscellaneous army gathered round a nucleus of mercenaries, had wisely resolved to fight a defensive battle. He had chosen one of the few pieces of high ground in a flat country, on which lay the Premonastrensian abbey of Heiligerlee. There were woods at the back and on one flank, but in front an expanse of cultivated fields and peat-bogs, separated from each other—as is always the case in Holland—by ditches full of stagnant water, not by hedges. The strength of the position was that these ditches were practically invisible to any one approaching from the plain in front, by the high road leading to the abbey. Louis placed the bulk of his arquebusiers low down on the slope, the main body of his infantry in a clump of pikes above them. A detachment or reserve was concealed in woods on his left flank. His handful of horse, no more than 300, led by his brother, Adolphus of Nassau, was on the high road in the centre. The whole force was somewhat larger than Aremberg's in mere numbers, but of most doubtful quality.

The Duke, searching for the rebels, found them offering battle, when his vanguard emerged from the woods which flanked the high road through the plain. He is said to have had some knowledge of the ground, and to have been aware that there were obstacles before him. But his troops were eager to attack, and his second-in-command Braccamonte, colonel of the Sardinian *tercio*, declared that the Dutch would not stand. Whereupon Aremberg set his artillery to work, first against the line of rebel arquebusiers, who at once gave way uphill, and then against the clump of pikes above, which was not entirely out of range. He then, judging the enemy to be shaken, ordered a general frontal attack by the infantry.

This miscarried hopelessly—the line broke at once, for only the troops on the high road had firm going—those on each side of them floundered into peat bogs, or had to struggle through

[1] So Strada: Motley gives Martinengo's squadrons to Aremberg—I know not on what authority.

deep muddy ditches. Only a fraction ever got to the foot of the hill, at one or two points and in great disorder. The rebel 'battle' of pikes at once charged them downhill, and drove them back into the peat-bogs, or along the road. Aremberg was trying to get into order the remnants of the defeated force, when they were attacked in front by the Dutch cavalry under Adolphus of Nassau, and in flank by the detached reserve of Louis's infantry, which had come round in the woods on the left by a circuit, and appeared almost in the rear of the disordered Spaniards. The Duke tried to save the day by charging in person at the head of his few squadrons of horse—his only intact unit. The narrators of the fight agree that he and Adolphus of Nassau actually met in personal combat, both—and this is a curious point—using the pistol and not the sword. The Duke shot the Prince through the heart; the Prince gave the Duke a mortal wound in the side. Aremberg was helped to the rear by some troopers, but was overtaken by a Protestant captain named Haultain, who clove his head to the eyes by a sword-stroke. By this time the whole Spanish force had melted away, Colonel Braccamonte heading the rout. Whether 500 or 1500 of his army perished with Aremberg is disputed between Spanish and Dutch authorities. The loss of the insurgents was insignificant, save for the death of Adolphus of Nassau, the first of many princes of his house to perish in these wars.[1]

This fight, which was to remain for many years the only victory in the field of a Dutch army, was destined to prove absolutely fruitless. Louis set himself to besiege Groningen, into which the belated Meghen had thrown his own detachment and the wrecks of Aremberg's force: it held out, and no general insurrection followed the victory, though many raw recruits came in. The German mercenaries mutinied for want of pay, and the cause of freedom was by no means made popular by the wholesale requisitions which Louis made in the countryside, both for money and for contributions of all sorts. The rumour that Alva was on the march with an overwhelming force no doubt contributed to damp enthusiasm. At any rate, in the two months which followed Heiligerlee, Louis of Nassau failed to revolutionize the north.

[1] Strada's vivid diagram-picture of Heiligerlee seems to give the topography and manœuvres very correctly, though I should doubt if this artist had ever been in the province of Groningen.

The great Viceroy did not appear till July—a month after he had terrified Brabant by the judicial murders of Egmont and Horn, and set the south, as he supposed, in a proper state of terror. Meanwhile the enterprises of the insurgents in Artois and Guelders had both come to a disastrous end. On July 10 Alva mobilized his field army at Deventer, and started off to deal with Louis of Nassau and his miscellaneous host. He brought with him three complete old Spanish *tercios*, and a number of detached German, Italian, and Walloon companies of infantry, as well as three regiments of cavalry and one of dragoons—'arquebusiers on horseback.' The whole amounted to 12,000 foot and 3000 horse.

On July 14, Alva, marching hard, reached the neighbourhood of Groningen, whose siege was immediately raised by the insurgents. Louis of Nassau, though his total numbers at this time may have been hardly inferior to those of the Royalists, was determined to risk nothing in the open, and to try once more, as at Heiligerlee, the fortune of a defensive battle in a carefully chosen position. With great unwisdom he selected ground very strong in itself, but with the fatal defect of having no proper line of retreat behind it. The position of Jemmingen—on the German frontier—was covered by the broad tidal river Ems on one flank and the estuary of the Dollart on the other; it was fortified from water to water, but there was no road in its rear, only the village of Jemmingen, with a ferry across the Ems, but no bridge. Defeat must mean destruction, and it is difficult to conceive how any general could dare to place himself in such a *cul de sac*, even with the most solid of armies. And that of Louis was eminently unsatisfactory both in organization and in morale.[1]

Alva made a careful reconnaissance of the position, and judged it too strong for storming by a mere frontal onset. He saw the insurgents busily engaged in making it yet more formidable, by cutting the drains and sluices which kept the fields in front of it dry. His idea was to tempt the Dutch out of their lines by weak partial attacks, which they might repulse and then be tempted to come out in pursuit of a presumably defeated adversary—precisely the tactics employed by William the Conqueror in the central episode of the battle of

[1] Motley suggests that, in disgust at the behaviour of his troops, he placed them in a position where flinching would mean certain death—not realizing that when panic sets in a mob ceases to reason, and only runs. He argues from Hoofd, v. pp. 175–6.

Hastings.[1] Accordingly, he kept back the main force of his army, hiding it as far as possible behind dykes and sandhills, and sent out detachments only to open the fight.

The Dutch front consisted of two lines of entrenchments—ditch and parapet—with the gap where the road from Groningen to Jemmingen passes through them covered by a ravelin on each side, garnished with artillery. The main body of the infantry was drawn up behind the entrenchments—the small force of cavalry, useless under the circumstances, in the rear. Alva commenced operations by sending out a squadron of cavalry and 1500 arquebusiers to drive off the men who were at work cutting the sluices in front of the enemy's line. This was done with no great difficulty, but Louis of Nassau then sent out detachments from his line to recover the sluices; these, too, were driven back, and the Spanish vanguard pressed up to the entrenchments, and bickered in front of their left side for three hours at long range without closing. Louis of Nassau was puzzled by the weakness of the attack—and sent out scouts in boats on the Ems upstream, to see whether there were Spanish reinforcements in rear of the demonstration against his left flank. Alva had hidden his main body so well that the report was brought back to Louis that no serious force was visible. Whereupon the Count resolved to drive away the demonstrating force by a vigorous sally from his lines, and come out with half his infantry, flags flying and drums beating, to clear the ground in front of the lines.

The Dutch had not advanced a quarter of a mile when unexpected Spanish troops cropped up from their hiding-places, and closed in on him from front and flank. Seeing the trap too late, Louis ordered a general retreat to the trenches, but this degenerated into a panic, and the pursuing Spaniards crossed the earthworks mixed with the flying enemy. The demonstrating detachment, which had been fighting all day, actually stormed the ravelins and battery on the high road, suffering but a single discharge from the guns posted there. It is said that Louis of Nassau fired the last gun himself, the gunners having all flinched.

The lines having been pierced in several places, the insurgent army made no further stand, but fled blindly to the rear, where only a few of them found boats in which to cross the Ems, whose tidal water was going down at the moment.

[1] Cf. *Art of War*, i. p. 162.

Thousands were cut down; many, who had thrown away arms and armour, tried to swim the river, under fire from the Spaniards on the bank. Some were drowned, some shot 'like wild-fowl' in the water. At least half the Dutch army was destroyed, only horsemen and strong swimmers got over to the German side. It is said that the burghers of Emden, at the mouth of the river, learnt of the rout by seeing thousands of the broad felt hats of the rebel infantry bobbing past their walls on the falling tide. Louis of Nassau swam the river successfully, as did many others. But the army was destroyed—6000 or 7000 men perished, and at little cost to the Spaniards, whose whole loss consisted in seventy or eighty men in the demonstrating force which had opened the fight.

It is difficult to see what the unlucky Count should have done, when he learnt that Alva was marching against him in force. He was aware of the worthlessness of his army, of which all save the German mercenaries were raw recruits, and which had been displaying symptoms of mutiny for some weeks. He had no fortress into which to retire, and he had been cut off from the port of Dam, where his original force had landed. But to fight with bad troops in a position from which there was no retreat, with an unfordable river behind him, was surely insane. The net result was that all heart was taken out of the Protestant party in the whole region; the province of Groningen was destined to be the last corner of the north in which the Spaniards kept a hold. Nearly a generation elapsed before Maurice of Nassau finished the clearing of the land in 1594.

The disaster of Jemmingen was not destined to be the last military event of the unlucky year 1568. When all the other efforts of the Protestants had failed, William of Orange himself started on what had always been intended to be the main-stroke. He had spent all his own money, and all that he could beg or borrow, in raising a really formidable army of mercenary troops in Germany—over 20,000 foot and nearly 9000 horse, as we are told. Trespassing on the lands of minor German States, especially those of the Archbishops of Cologne and Trèves—he crossed the Rhine, and forded the Meuse at Stochem on October 5. He had published a proclamation of defiance against King Philip, and called on all Netherlanders to rise against a tyrant, who had violated all the constitutional privileges of the Netherlands. But the results of the earlier campaigning had cowed the whole of the malcontent party, and he was joined by none save very fanatical reformers, and by a broken band of French Huguenots, the relics of the force which had moved earlier in the year.

When William moved into Brabant he found Alva facing him with an army perceptibly smaller than his own, so much smaller indeed that the Duke refused to commit himself to a general action, but hung about the invaders' flank, always threatening to attack, but never allowing himself to be involved in a decisive fight. The campaign of manœuvres lasted for a month, with marches and countermarches on the borders of Brabant and the bishopric of Liège, around St. Trond, Tongern, and Tirlemont, without any result. The Prince had changed his camping-ground, it is said, twenty-nine times, and the Duke as many. The only serious encounter was when Alva successfully fell upon the rearguard of Orange as it was passing the river Gheet, near Jodoigne,[1] and cut it to pieces before it could be succoured on October 20. A few days later the weather turned to such desperate autumn rain, and the Prince's German mercenaries began to desert in such appalling numbers, that he gave up the game. There had been no signs in Brabant of any general insurrection in his favour. Having failed to induce the Bishop of Liège to allow his army free passage back to Germany, Orange had to turn south towards the borders of France, and finally trespassed over them, closely pursued by Alva. Getting formal protests against his presence on French soil from the ministers of Charles IX, he finally passed eastward and disbanded his army at Strasburg in November.

So ended in complete disaster the first military effort of the Netherland malcontents. It must be confessed that the result shed no lustre on the old-established military reputation of the Princes of Orange—so many of whom had already fallen in the old wars of Italy, and so many more were destined to die on Dutch soil. The general plan of the campaign miscarried hopelessly, partly from bad timing, partly from unwise reliance on the use of mercenary troops and raw levies, partly from the Duke of Alva's display of strategical talent. William of Orange was the most steadfast of men—but, unlike his sons Maurice and Frederick Henry, he was no great general.

[1] Not far from the more famous battlefields of Neerwinden and Ramillies.

It was to be four years later that the second and more formidable revolt of the Netherlands was caused by the maladministration of Alva, whose wholesale persecution of heretics combined with his financial exactions at last forced the whole country into desperate insurrection.

CHAPTER III

MOOKERHEYDE (April 14, 1574)

FOUR more years of the intolerable tyranny of Alva goaded the Netherlanders in 1572 into an insurrection much more formidable than the abortive business of 1568. Following on the reckless exploit of the 'Sea-Beggars' of De la Marck, who seized Brill on April 1,[1] there were a series of popular risings spread over the whole of the provinces from the northern point of Holland to Mons in Hainault. A large proportion of the cities adhered to the cause of independence, and William of Orange once more brought a mercenary army from Germany across the Meuse, and marched across Brabant and Flanders, with the object of raising the siege of Mons, where his brother Louis was being beleaguered by Alva's main field force. In this enterprise he was as completely foiled as in his earlier invasion of 1568; unable to force Alva out of the formidable lines of contravallation which he had drawn round Mons, the Prince had finally to retreat with a dwindling army—his Germans always melted away when pay was a few weeks overdue; it was not with such hirelings that the Netherlands were to be delivered. The ever-unlucky Louis of Nassau having been forced to capitulate at Mons (September 19)—he was fortunate to escape with his life—the formidable viceroy was able to turn his attention to the task of dealing with the insurgent cities of the north—whither Orange had betaken himself in person when his army had broken up.

The campaign of 1572–3 resolved itself into a series of sieges of the rebel towns. Mechlin, Zütphen, and Naarden were successively recaptured—in every case with horrible atrocities to the townsfolk, for Alva carried out a logical theory that all rebels were guilty of high treason and destined to the gallows.

[1] In All Souls Library we have a marvellous series of contemporary sketches by Walter Morgan, one of Orange's English followers, who presented to Lord Burleigh a couple of dozen of his drawings, illustrating all campaigns from Orange's invasion of Brabant down to the end of 1573, including the sieges of Mons and Haarlem, and much naval fighting.

His advance stood still for some time before the desperately defended Haarlem, which held out from December 1572 to July 12, 1573. It was at last starved into surrender at discretion, and Alva put to death 2300 prisoners in cold blood. The first check to his progress was at the important town of Alkmaar in north Holland, which held out successfully from the 21st of August to the 6th of October 1573, till the Spanish army was compelled to draw off by the threat of the cutting of all the dykes on which their camps lay.

This was the last event of importance in Alva's viceroyalty —in the following winter he was recalled to Spain, at his own request, not because King Philip disapproved of his policy or his atrocities. Indeed the King a few years later entrusted him with the greatest military adventure of his reign, the conquest of Portugal (1580), which the Duke accomplished with his usual skill.

In 1574 Luis de Requescens, late Governor of Milan, had to take up the task of completing Alva's unfinished enterprise, the reduction of the large block of rebel cities in Holland and Zealand. The most important point at the moment was Leyden, which had already been invested in March 1574. It was with the hope of relieving this great city by external operations in the rear of the Spaniards that William of Orange planned the third and last of his fruitless diversions with a mercenary force raised in Germany. This time it was to end not in mere manœuvring and disappointment, but in a complete and bloody disaster.

Once more Orange entrusted the leading part in the strife to his ever unlucky brother, Louis of Nassau. One can but wonder that he employed once more a commander whose name was always associated with failure, at the head of the same sort of troops who had disgraced themselves on previous occasions. But it may be pleaded that the only external reinforcements to be got were these same German mercenaries. With 2000 reiters and 6000 foot, Count Louis, accompanied by his younger brother Henry, and by Christopher of the Palatinate, the son of the Elector, started to march along the Meuse, intending to join Orange, who had collected a moderate field army at Bommel, in the marshes between Meuse and Waal. When united, they were to relieve Leyden.

It was obvious to Requescens that he must prevent this junction. Accordingly he collected a covering force under his lieutenant, Sancho de Avila, and sent it out to fend off Count Louis's approach. The latter, after a vain attempt to storm Maestricht, reached the Abbey of Mook on the north bank of the Meuse, just inside the frontier of Guelders, on April 13. Here he found Sancho de Avila facing him and offering battle in a defensive position, with the Meuse on his right and woods on his left. The Spaniards, as Louis had discovered, were inferior to him in numbers—not over 4000 foot and 1000 horse, and despite of the well-known superiority in fighting value of the royalist veterans, he committed himself to a general action. It was known that reinforcements were on the way to Avila and were to arrive in detachments on the succeeding days. Indeed a thousand Walloon foot and horse came on the field only just before the battle began.

Louis intended to engage the Spaniards all along their front, and to smash in their left wing with his very superior force of cavalry, which was all concentrated on his right, at the end of the line farthest from the river. He strengthened his left wing by a trench reaching to the water, and held by ten companies of infantry. Next them was a great 'battle' of his remaining foot, while the cavalry in three successive lines was formed on the right; its front line was led by Henry of Nassau, its reserve line by the Palatine prince (April 14).

The fortune of the day was settled by the disgraceful inefficiency of Count Louis's cavalry. When along two-thirds of the line the Spanish and the insurgent infantry were engaged in a fierce but indecisive combat of musketry, he launched his reiters against the hostile left. He drove in the front line of Avila's horse, which was composed of mounted arquebusiers, and then came, in some disorder, upon their second line of men-at-arms, flanked by skirmishers in the wood which covered the extremity of the Spanish line. We are told that his reiters—all pistoleers—were reloading their weapons and reforming their squadrons, when the Spanish men-at-arms charged in upon them with the lance in close order. The lance prevailed over the unloaded pistol—and the horse of the rebel front line broke, and fell back in disorder on the second, which also failed to stand. Christopher of the Palatinate brought up the reserve—to see it flinch also. In this disgraceful way, fighting against half their own numbers—the German horse lost the battle, for when they fled, their infantry, with its flank now exposed, retreated in disorder, hotly pursued by the Spanish foot.

Count Louis, with his brother Henry and Christopher of the Palatinate at his side, made a last desperate charge at the head of a small band of horsemen, to cover the retreat of the infantry, but all three were ridden down and slain. Strangely enough their bodies were never identified—at first it was supposed that they had escaped, later (when they failed to appear) it was concluded that, having been stripped and perhaps much trampled on by horses, they had been thrown into the common grave of the rebels unrecognized. There was a story that Louis had swam the Meuse, badly wounded, and had been set upon and murdered by marauders, but the tale lacks authority, though it is perpetuated in a corner of Strada's magnificent diagrammatic picture of the battle.

More than half of the beaten army perished, many driven into the Meuse and drowned as they strove to swim over. The loss of the victors was insignificant—perhaps 200 at the most. Considering the details of the battle it is hardly necessary to emphasize the fact—obvious in many other fights—that pistol tactics and the 'caracole'[1] were a snare for cavalry—especially for cavalry operating against troops of their own sort who were ready to charge home with lance and sword. Yet it was not till fifty years had passed by that Gustavus Adolphus impressed upon the military mind the fact that the strength of cavalry lies in impact, and that by squibbing away with pistols or carbines the trooper loses the all-important power of rapid motion, in which his strength resides. La Noue had preached this sermon to deaf ears forty years before.

Oddly enough, the disaster of Mookerheyde had no decisive effect on the general course of the war in the Netherlands. Though Orange was unable to relieve Leyden at the head of a victorious army, he did so by the desperate expedient of cutting the dykes, and flooding all the low country about the course of the 'Old Rhine.' Leyden was relieved by a fleet, not by a land force, when the Spaniards were drowned out of their camps and trenches, and Admiral Boissot's vessels moored themselves before the quays of the starving city (October 3, 1574).

[1] See pp. 86–7 and 469.

CHAPTER IV

GEMBLOURS (JANUARY 31, 1578)

MUCH water had flowed along the dykes of Holland before another pitched battle was seen, which repeated the lessons of Jemmingen and Mookerheyde as to the splendid efficiency of the veteran Spanish soldiery, and the worthlessness of the mercenary levies or raw volunteers which the insurgents of the Netherlands put into the field against them. Independence was won, in the end, by the obstinate resistance of burgher-companies defending the walls of their own cities—though almost helpless outside them—combined with the command of the waters won by the successes of the Sea-Beggars of 1572, and never lost.

When the next general action on a large scale was fought in the Netherlands the political situation had been completely changed since 1574. After years of hesitation the southern provinces had at last thrown in their lot with the northern, and Don John of Austria, King Philip's bastard brother, had found, when he came to take up the viceroyalty of the Netherlands, that Luxemburg, in the extreme south, was the only fortress safe in his hands. Into the complicated political negotiations between the Viceroy and the States General, in which it was vainly hoped to find some *modus vivendi* between the theoretical sovereignty of the King and the practical independence of his discontented subjects, we need not enter. By the end of 1577 it was clear that no pacification was possible, and that arms must settle the controversy. The States General declared Don John deposed from his viceroyalty, and elected in his stead the Archduke Matthias, the young brother of the Emperor Rudolf. They had collected a considerable army, which included not only native regiments but French, English, and Scottish volunteers, besides the inevitable German reiters and pikemen. Their commander-in-chief was not Orange—who would have represented only the Protestant and extremist element, and had deliberately stood aside to avoid offending the Catholic sentiment of the greater part of the States General—but the Sieur de Goignies, an undistinguished veteran who had served in the old wars of Charles V. The army of the States was reckoned at some 20,000 men; that which John of Austria had concentrated at Namur was of about the same strength, but of very different quality.

Gemblours (January 31, 1578) was one of a type of battles not altogether unknown, yet not very frequent, in which an army, endeavouring to withdraw from a position which is becoming rather hazardous, puts out a covering force to shield its retreat, but sees that covering force attacked and routed, and is then exposed, while in columns of route along the road, to a general assault before it can form up.

The army of the States had been for some time drawn up on strong ground, covering Brussels and inviting an attack. But on receiving the perfectly correct information that Don John of Austria, then lying at Namur, had received some heavy reinforcements, the cautious and unenterprising de Goignies, ventured to draw back to a better position farther up the Brussels road. The infantry had marched off, while the bulk of the cavalry, forming a strong rearguard, was formed across the high road, ready to move on when the foot had got well upon their way. The Spanish officers in command of the line of cavalry posts which was watching the enemy, saw the opportunity of molesting him while he was out of battle array, and collecting their squadrons attacked the Netherlanders' cavalry rearguard. They were brought to a stand on the line of a marshy brook, and were making no headway, though they held the enemy well engaged. At this moment the young Duke Alexander of Parma, the hero of so many late campaigns, without the leave or knowledge of his uncle, the commander-in-chief, came up with a small following, took command, and dashing through the marsh, rode round the extremity of the line of the Netherlanders, and charged them in flank and rear. The enemy broke, and galloped off, some of them right up the high road, along which their own infantry were retreating in column, throwing the rear battalions into confusion. The Spaniards, who were following hard behind the fugitives, got in among the infantry, and made havoc of them, falling on the flank of each corps in succession as they came parallel with it. De Goignies and the Netherland horse got away in disorder with no great loss, but the whole of the rear units of the foot were cut off, and massacred or forced to surrender in detail, having never had the chance of forming up in array. Some Scottish companies are said to have been the only troops which made any serious resistance. Several thousands of the foot were cut down, but some quarter was given after the first flush of excitement had passed over. The Spaniards allowed many French and German prisoners to march off, after they had been well plundered, to the bishopric of Liège—theoretically neutral ground—while some of the Flemings and Walloons were forced to swear allegiance to Philip II, and warned that if caught again in arms they would be hanged. A few of them—individuals of some note—are said to have been marched back to Namur and drowned there. Some chroniclers exaggerate the general massacre. The cavalry and the remnants of the Netherland foot scattered, and abandoned the countryside to the victorious Spaniards. Don John gave Alexander of Parma a public rebuke for acting without orders, but then embraced him and congratulated him on having a quick military eye. The battle had been gained with ridiculously small loss [1]—only a few score of troopers—the Spanish infantry had hardly been engaged, and only came up in time to slaughter or take over prisoners. The army of the States had lost at least 5000 or 6000 men.

A fairly close reproduction of the main feature of Gemblours may be quoted from the Peninsular War—at the battle of Alba de Tormes (November 29, 1809). Del Parque's Spanish army, in full retreat, was passing the bridge of Alba, with all its cavalry and two divisions of infantry left to contain the pursuing French, who were known to be some way off. Kellerman with his dragoons, arriving long before the French infantry, saw his opportunity of attacking an army in full retreat, of which half had already passed the Tormes. He scattered the Spanish cavalry rearguard, and then rode headlong into the infantry which was still on the wrong side of the bridge, breaking battalion after battalion, and making a vast number of prisoners.[2] A less close parallel to Gemblours may be found in the battle of the Garigliano (December 29, 1498), where the main damage to the

[1] Some of the chroniclers put it at a dozen or twenty men. One (Bentivoglio) says that the Spaniards had *no* casualties.

[2] See my *Peninsular War*, iii. pp. 100–1.

scattered French army was made during its retreat, by relentless cavalry pursuit. But there had been some heavy fighting with all arms, before the Marquis of Saluzzo gave orders for a 'sauve qui peut' retreat into the fortress of Gaeta.[1]

[1] See above, pp. 124–28

CHAPTER V

THE WORK OF MAURICE OF NASSAU (1590–1600)

THE turn of the tide in the long struggle between Spaniard and Hollander may, as has already been pointed out, be placed in 1589, when Philip II, after the failure of the 'Invincible Armada,' distracted the army which Alexander of Parma had collected at Dunkirk toward France, where it was to be employed for three years, with a 'minimum garrison' alone left in Belgium.

It was undoubtedly this distraction which gave his opportunity to the great soldier whose operations form the most part of the chronicle of the next twenty years, Maurice of Nassau, the heir of William the Silent. The father had been, in essence, more of a politician than a soldier; the son's major activities were all military; his domestic political record is less happy—but with this we are not concerned.

Maurice figures in the history of the Art of War both as an innovator in organization and tactics, and as a master of siegecraft. In the former sphere his changes consisted mainly in the cutting down of the size of units. Companies were reduced from 150 to 115, and afterwards to only 80 rank and file, with an equal proportion of pikemen and musketeers. This, of course, resulted in the corresponding reduction in the strength of regiments. With such units a regiment of ten or twelve companies shrank to what we should now call battalion-size. The depth of array was decreased in the ranks, and the pikemen were trained to stand with at least three-foot intervals. The musketeers were drawn up on the flank of the pikes, still firing (as of old) in successive ranks, and then filing to the rear to reload. The very strongest drill and training was enforced to prevent disorder, or falling back beyond the front of the line of pikes. When a clash and real close fighting ensued—'push of pike' in the language of the time—the musketeers were supposed to align themselves level with the fifth or sixth rank of pikemen, and to continue their rolling fire from this distance. This order could not, of course, be observed in the case of a cavalry charge coming down upon the whole front of the line of battle, when the musketeers were directed to take shelter under the lowered pikes of the flank files of their comrades—a rather hazardous matter, as one would suppose. The main points to notice in Maurice's tactics are the desire to have many small units rather than the very large ones that prevailed for the Spanish *tercio* or the French 'legion,' and his tendency to extend the front rather than to keep to the very deep columnar array of earlier days. His most elaborate and thorough drill-ordinances (partly the work of a cousin specialist) could be better practised with corps of moderate size. To manœuvre a body of 1000 men in battle is far easier than to move one of 3000 or 4000 pikes and muskets massed in one vast block.

Maurice, however, is better remembered for his siege-craft than for his tactical experiments. His record starts precisely with the moment when Philip II had moved off the best of his Netherland troops to the assistance of the French League. While Alexander of Parma was relieving Paris and campaigning in Normandy, the chronicle of Maurice's successful sieges starts on a modest scale. He was in the end to lop off all the outlying fortresses outside the modern Belgium which were still in Spanish hands. The long record begins with his capture of the very important town of Breda—the northernmost of the enemy's strongholds on the Brabant frontier—by combined surprise and stratagem in 1590. It must be remembered that at the moment the Spanish holding stretched far north beyond Meuse and Rhine, up to the Dollart; east of the Zuyder Zee the Dutch held only isolated fortresses in Overyssel and Guelders. It was only in Holland, Zealand, and Utrecht that they had a solid block of territory.

Breda was captured—as has been already mentioned [1]—by surprise and not by regular siege operations. The rest of Maurice's conquests were made by more regular methods. Starting in June–July 1591, an important campaign broke the links between Nijmegen, the Spanish fortress lowest on the Rhine, and Groningen, the chief town and centre of the north-eastern holding. Its great feature was the astonishingly swift capture of the two great towns of Zutphen and Deventer, both of which in former days had stood months of sieges little better

[1] See above, p. 515.

than blockades. By an immense concentration of heavy artillery on small sections of the enceinte, followed by the creation of practicable breaches, he compelled the surrender of Zutphen in seven and of Deventer in eleven days (June–July 1591). In each case he made the yielding easy to the governor, by promising the full honours of war, and free departure with bag and baggage to the garrison. Verdugo, the Spanish stadtholder in Groningen, sent desperate appeals for help to Alexander of Parma, who, though the army that he could collect was small, owing to detachments left in France, drew off Prince Maurice from his northern campaign by an offensive move, moving up to Nijmegen and the Waal, and threatening the province of Utrecht. This brought the Prince down to Arnhem and the opposite bank of the Waal, when Parma, after some ineffective skirmishes, drew off to France, contented to have relieved Groningem by a demonstration. But instead of returning to deal with Verdugo, Maurice shipped his army in barges, and moving across the country by canals and waterways, suddenly appeared in front of Hulst, in Flanders, at the other end of the Spanish line, only twelve miles from Antwerp, which he took by battery after a siege of only five days (September 14), before any Spanish force could be concentrated for its relief. Parma was on his way to France, the garrison of Antwerp was not large enough to furnish a field force.

Having thus drawn the attention of the Spaniards to Flanders and the West, and caused them to concentrate in that direction, Maurice again shipped his little expeditionary army on barges, and brought it by cross-waters to Nijmegen, now the one important place left to the Spaniards in Gelderland, and took it in no more than six days (October 21), again by force of artillery. In none of the later sieges of this most adventurous campaign had the Prince anything to fear from an army of relief. The reason was plain—King Philip had sent so many men to the French frontier, that his remaining force in the Netherlands was hardly more than enough to find full garrisons for the numerous fortresses still in his hands. There was little over to make up an army of operation, unless some of the towns were stripped of their complements down to danger-point. For the burgher-guard could never be trusted—the King's Catholic subjects were in such a state of misery and discontent that even the cause of religion failed to inspire them to help in the defence of their own walls. At Deventer the

Spanish second-in-command had been seen endeavouring to drive the citizens to the defence of the great breach with a cudgel—and failing to move them. It must be remembered that Maurice of Nassau kept his troops in singularly good order—not only did they kill no one in a captured town, but they did not plunder. And for good reason—he executed marauders; at the capture of Hulst in 1591, he shot in front of his company the first man found robbing a woman; at Delfzyl he hung another, whose meagre booty had been a burgher's hat. Wherefore the citizens in a Catholic town had not the reasons for self-defence that were obvious to the citizens of any Protestant town with a Spanish army before its gates.

In November 1591, official figures sent to King Philip with regard to his armies in the north showed that the force in France counted 23,000 infantry and nearly 5000 horse, of all nations—Spaniards, Walloons, Italians, and Germans—while the force left in the Netherlands was a nominal 29,000 men, but with large numbers wanting from sickness and desertion—probably also from false returns made by fraudulent captains, who were numerous.[1] Obviously, then, just half—and that the better half—of the troops who had been available to Parma in 1589 were distracted to co-operate in the civil wars of France. When Maurice of Nassau moved about with an army of 9000 or 10,000 men, it was almost impossible to collect a force capable of meeting him with superior or even equal numbers. For every fortress from Sluys to Luxemburg and from Groningen to Namur was absorbing its hundreds—sometimes, as at Antwerp, its thousands—of necessary garrison. Philip's insane interference in France had changed the balance of military power in the Netherlands—for the first time since the States General's army went to pieces at Gemblours in 1578—and the offensive had passed to the Dutch, whose army, very modest in numbers, was in the hands of a young general whose talent was only equalled by his prudence. Maurice once quoted the dictum of the old Roman general, who said that there were only two cases in which battle should be accepted—the first, when you have an obvious superiority, moral and numerical, over the enemy; the second, when it cannot be avoided without a greater risk than is incurred by fighting. He was much given to field-entrenchment, when the enemy was in force

[1] See the figures from the Simancas archives printed by Motley in detail in *United Netherlands*, iii. p. 138.

opposite him—and this led to his being seldom attacked. But even more was he devoted to elaborate spade-work when he was engaged in sieges—long trenches of approach, and heavy strong-points at the head of the parallels, when they got near a hostile enceinte, and well-covered batteries for the guns when they had been brought up. Nor did he despise mines—much of the soil of Holland is absolutely water-logged, so that it is impossible to dig down many feet without being flooded out; but, when dry ground was to be found, he used mining freely and generally with good effect, though at the siege of Steenwijk a mine, pushed just short of its intended object, once blew up the head of a storming column which was waiting to rush in at a breach that should have been caused!

Maurice's very successful campaign of 1591 which gave him Zutphen, Deventer, Hulst, and Nijmegen was followed by some almost equally brilliant successes in 1592, while Parma was still tied up in his elaborate series of operations in France. In May, he laid siege to Steenwijk, one of the few fortresses still in Spanish hands in the area south of Groningen. It made a very good defence, and only yielded after a siege of thirty-four days. After twenty days of battery with incomplete effect, Maurice finished off the game by mining, which was successful, the governor surrendering when two sections of his enceinte had been blown in, and the besiegers had established themselves on the lip of the breaches (July 3). After securing Steenwijk the Dutch army moved on to Coevorden, the last strong place left to the stadtholder Verdugo in front of his main position at Groningen. This was a strong marsh-fortress, where mining was impossible, and took much battering; the siege, which began on July 26, dragged on till September 12. The place was considered so important that the aged Count, Peter-Ernest of Mansfeldt, who had been left in command in the Netherlands while Parma was absent in France, sent Verdugo some reinforcements by a circuitous eastern route, up the frontier of Germany. When they came to hand the Spanish stadtholder made a desperate attempt to beat up Prince Maurice's siege-lines outside Coevorden. He found them carefully entrenched, and was repulsed with loss (September 7). Whereupon the garrison, being disappointed of succour, and offered easy terms of surrender, hoisted the white flag on September 12. Nothing of importance was now left to the Spaniards in the north save their central stronghold at Groningen.

Alexander of Parma, worn out by his last French campaign, in which he had been severely wounded, died at Arras as winter set in, while drawing back his army from his dangerous operations in Normandy (December 3, 1592). The command of the Spanish armies fell on the senior general in the north, the octogenarian Peter of Mansfeldt, who was past work, and generally disobeyed by his subordinate generals. This gave great opportunities for the continuance of his successful series of sieges to Maurice of Nassau. In the spring of 1593, he sat down before Gertruidenberg, the central fortress of the Spanish line of towns which protected Belgium—leaving the side-issue in Groningen for a time, as its conquest seemed inevitable, now that it was completely cut off from the rest of the Spanish possessions, since Coevorden and Steenwijk had been lost. Gertruidenberg was the nearest of the hostile strongholds to the heart of Holland, and the only one left on the lower Meuse since Breda fell. But it was exceptionally hard to tackle, owing to its protection of marshes between the Meuse and the Donge. Maurice entrenched himself before it in elaborate lines of circumvallation, facing outward as well as inward, as he knew that the enemy would certainly try to raise the siege of such an important strategical centre. The old Mansfeldt did his best to disturb the siege operations, but, owing to the absence of so many Spanish troops in France, he was never able to gather a force sufficient to attack the Prince's entrenched camp, and hovered at a distance from it, sending, so we are told, vain challenges to the young stadtholder to come out and fight—as Hannibal is said to have done to Fabius Cunctator in the Punic War. At last, a lodgment having been made in the outer defences, and the governor killed, the garrison surrendered, on being offered the attractive terms of free departure with bag and baggage (June 25, 1593).

While Maurice was dealing with Gertruidenberg, Verdugo, the Spanish governor in Groningen, finding no field army left in front of him, had made two desperate attempts to recover Coevorden—but his forces were too small, and on each occasion (September 1593 and May 1594) he was forced to return to his stronghold. In the next campaigning season Maurice made an end of the Spanish holding in this remote north-eastern corner of the Netherlands. Verdugo was now cut off from all touch with his friends south of the Rhine and Meuse, and Mansfeldt could spare him no assistance, not only because the best part of the King's army was absent in France, but because the remaining regiments—left long unpaid—indulged in a prolonged mutiny which lasted a year, and only came back to allegiance when cash appeared in December 1594. Maurice laid siege to Groningen on May 20, and took it by capitulation, after much battering and mining, on July 24. The whole province of Groningen was therefore reorganized as a full member of the Union of Utrecht, and William Louis of Nassau, the cousin of Maurice, became its first stadtholder. The 'Seven United Provinces' were really constituted as a permanent political state by this conquest, and their boundaries varied comparatively little for the future.

By this time there was no real chance of winning the Catholic counties and duchies of the south into voluntary union with the Protestant north—all that remained was the chance of conquest, and for this the Dutch power was not strong enough, even when wielded by such a capable general as Maurice of Nassau; and even though the Spaniards were still distracted by the French wars, which lasted down to the Peace of Vervins (1598). Having once thrust the enemy behind the line of the Lower Rhine and the Waal, and obtained a precarious lodgment on the coast of Flanders opposite Zealand, Maurice made little further progress. He had won a defensible frontier, and when marsh and watercourses ended, and the Belgian dry land was before him, the advance slackened. He even in his later campaigns lost his one base of operation against Flanders—the long-defended but too isolated fortress of Ostend. Even his best fought general action, Nieuport (1600), did not secure for him that power to retain this attractive outpost, which might have meant so much, if there had been any reasonable chance of conquering the whole of Flanders.

After 1598, when Henry IV made peace with Spain, the Dutch had lost their chance of expansion southwards. The French campaigns in Artois and Picardy had not sufficed to distract the whole of the Spanish power in the Low Countries. Indeed they had on the whole been unsuccessful with unpleasant surprises at Calais and Amiens,[1] so that the joint pressure from north and south had not resulted in the result that might have been reasonably expected. Maurice won in 1597 his first victory in the open field against a Spanish force at Tourn-

[1] See above p. 533.

hout,[1] but it did not give him Brabant, but only some small outlying places on the Guelders border. This little fight will be dealt with in the next chapter, as an example of the helplessness of sixteenth-century infantry, when its attendant cavalry had been driven completely off the ground. Queen Elizabeth's help to the United Provinces continued—as always— to be parsimonious and ineffective. Now that the danger of a Spanish reconquest of Holland had ceased, English interests were much more concerned with the dominion of the seas, and the hunt after American treasure-convoys. And the Queen seemed quite as likely to make a separate peace with Spain as was her ally the King of France.

Hence after France had come to a bargain with Spain at the Treaty of Vervins, and there was an end to the distraction of Spanish forces southward, so that all resources could be applied to the protection of the King's surviving dominions in the Belgian lands, there was no longer a chance that all the Low Countries might be united into a single state under the Nassau stadtholders. The war lingered on for many years more (1598–1609), but with no decisive results. Philip II on his death-bed had resigned himself to the idea that Holland was unconquerable, but left Belgium (if we may use the term) to his daughter Isabella Clara Eugenia and her husband the Austrian Archduke Albert, as a new state, nominally independent of Spain, though the Spanish army was not removed. He apparently thought that the Seven United Provinces might consent to acknowledge Albert and Isabella as their sovereigns, and might conduct their own affairs, while conceding a nominal allegiance to his daughter and her spouse. The only thing on which he was immovable was a demand that there should be complete provision for the establishment of the Catholic religion—toleration but not dominance for the Protestants. With their experience of the last forty years in their minds, the Dutch very wisely refused any recognition of the sovereignty of the 'Archdukes,' and the war went on till 1609 with small definite profit to either party.

If only Albert and Isabella had been blessed with offspring, there was a fair chance that a new Belgian state might have

[1] The end of the Tournhout campaign was a sally of Maurice into the east, where he took Grol, Bredevort, Entschede, and Oldenzaal, as well as Meurs and Rheinberg, which did not actually belong to the Spaniards, but were in their hands at the moment (August–October 1597), war having spread into the duchy of Cleves and its neighbourhood.

grown up, facing the new Dutch state across the Meuse and the Scheldt. The south Netherlands were now homogeneous in their religion—the Calvinists had all been expelled or exterminated—they were determined not to give up their separate identity, and to be ruled by the heretics of the north. And Albert and Isabella—the patrons of Rubens—were no fanatics, but reasonable and judicious persons set on doing their best for their heritage. Belgium, ruined though she was by the long wars, was capable of recovery; all that was needed was peace and quiet governance. Unfortunately there was no issue to the marriage, and when Albert and Isabella died, their dominions fell back to the natural heir—Philip IV of Spain—for Philip III had died some years before his sister. The unhappy 'Spanish Netherlands' came once more into existence, and the old political and strategical problems cropped up once more—but their further fortunes lie outside the limits of this volume.

It was not till many years had elapsed, and Maurice had been succeeded by his brother Frederic Henry, that the final settlement of the very anomalous frontier between the two halves of the Low Countries was reached. It was a singularly unnatural one—leaving the Dutch in possession of a narrow strip of the Flemish coast, just sufficient to muzzle the mouth of the Scheldt, and to deprive Antwerp of its natural waterway to the sea. Axel, Hulst, and Sluys were nothing in themselves, but they blocked the river of Antwerp. The far more important Ostend had been won—but only to be lost. On the other flank the Dutch had secured (1632) a great isolated fortress on the middle Meuse—Maestricht—untenable in a serious war, because of its advanced position, while in the centre the conquests of Frederic Henry of Nassau had given to the United Provinces Hertogenbosch and a strip of north Brabant south of the Meuse. It would have been wiser to draw the line at the old frontier between the county of Holland and the duchy of Brabant. The outlying annexation was more of a snare than a security. Meanwhile, though Ghent or Brussels never became Dutch, the United Provinces won an empire of a fragile sort upon the high seas; their fleets planted colonies on every coast from Brazil to Java—some ephemeral, some destined to endure—they sought the 'North-East Passage' to no profit, but made the Atlantic, Indian, and Pacific Oceans their own for many a decade. Commerce and colonization, however, do not concern this book—which deals with land war alone.

One thing the obstinacy of the Spanish kings and the tried valour of their veteran *tercios* had accomplished—and certainly to the conscience of Philip II it would have justified all his own turpitudes, and all the atrocities of which his armies and his inquisitors had been guilty—Belgium had been reconquered for the Catholic faith. At the beginning of the struggle it had looked as if there might arise a Protestant state reaching from the Texel to the Ardennes. By the truce of 1609 it had been settled that the Low Countries should be permanently divided, and that a small militant naval state might flourish in the north, but should never amalgamate with itself the peoples of the wealthy plains of the south. In 1814–15 Great Britain tried to undo the work of 1609, and to unite them. But the revolution of 1830 showed the logical consequences of the atrocious perseverance of Philip II two hundred and forty years before.

CHAPTER VI

TOURNHOUT (JANUARY 24, 1597)

AS we have said, no general actions on a large scale are found in the Netherlands after Gemblours — only a record of sieges. It seems hardly necessary to give details as to minor fights at Rynemants (1578) or Steenbergen (1582) or the desperately ill-managed cavalry affair at Warnsfeld, near Zutphen, where Sir Philip Sidney fell (1586). But the small-scale battle of Tournhout is worthy of attention as a complete ' pendant ' to Gemblours—an example of a cavalry victory over an enemy whose horse had fled, and whose infantry was caught on the march, devoid of any help from the mounted arm. Reversing the positions of Spaniards and Netherlanders, the story and the moral of Gemblours are repeated with great precision.

The Spanish viceroys, though their attention at this period was so largely distracted to the side of France, habitually kept a small covering army in front of Antwerp, to protect their holding in north Brabant. There was always a chance of a flanking attack from Bergen-op-Zoom and Zealand. In the winter of 1596–97 the Archduke Albert had left at Tournhout, a very central position, a force under his Master of the Ordnance, De Rie, Count of Varas, an officer from Franche Comté. It consisted of one Spanish *tercio*, that named Naples,[1] under the Marquis of Trevico; two Walloon regiments, La Barlette and Hachicourt; one German regiment, that of the Count of Sulz; and two cavalry units, one of heavy and one of light horse. They made seven ' cornets ' or troops in all—there were three cornets of Spanish lances, under Guzman, Alonzo Dragon, and Juan de Cordova, and four of light cavalry, *herreruelos* as they were now called, since the name of ' stradiot ' was being forgotten. These were under an old officer named Nicolas Basta, and included his own men and some Germans under a

[1] Not Italian in composition, but Spanish—descending from Charles V's Spanish garrison of Naples.

captain with the euphonious name of Grobbendonk. The whole army is stated by Spanish sources to have been not much over 5000 strong. The *tercio* since its reduction by the ordinance of 1584[1] was supposed to have fifteen companies of 100 rank and file, and the German and Walloon corps had a somewhat higher establishment—ten companies of 180 or 200 of all ranks. The cavalry was probably very weak in numbers—Alva had reduced the 'cornet' of men-at-arms to 50 lances—while the companies of light horse had theoretically 100 troopers. The whole seven units cannot have amounted to more than 550 men in all—and probably had much less at midwinter. Similarly, the infantry, which might have had 7500 pikes and muskets, was very probably only 5000 strong in reality—allowing for winter sickness and desertion from badly paid units. Varas had no supports nearer than Antwerp, where there was at the time only a 'minimum garrison' owing to the French War.

Maurice of Nassau, not (as some conceived at the time) because he thought that Varas might have designs on Breda or Gertruidenberg, but because he was aware that the hostile force was in an isolated position, and would not look for trouble at midwinter, resolved to make a sudden raid upon it.

He collected with great secrecy from the various garrisons of Holland fifty companies of foot, and sixteen 'cornets' of horse at Gertruidenberg on January 22. Nearly a third of the force was English, including the three horse-units of Horace Vere, Parker, and Sidney, and the foot-regiments of Francis Vere and Dockwray. There were also eight companies of Scots under an officer named Murray. The whole amounted to over 800 horse—all 'pistoleers'—and some 6000 foot. It is more than twenty-five miles from Gertruidenberg to Tournhout, but Maurice contrived to make his men cover twenty-two of them in one day's forced march, though the weather was a tiresome mixture of frost and thaw, and the road was abominable. On the night of the 23rd his van was at Ravels, only a league from Tournhout, where he hoped to surprise the enemy in his cantonments, for it was an open town, or rather (as contemporaries said) 'the largest village in Brabant.'

But Varas, warned just in time, and hearing exaggerated reports of the strength of the enemy, had evacuated Tournhout in the night, and started off in the dark, intending to cover himself behind the walls of Herenthals, ten or eleven miles to the

[1] See the Conde de Clonard's *History of the Spanish Army*, vol. iv. pp. 269–70.

south, which was a fortified place. He first started off his baggage with an infantry escort, and then followed with his whole army. Progress was slow, for the road was in some places covered with deep slush. At seven in the morning, as daylight came; he was crossing the moor or heath of Thiel, land somewhat higher than the half-inundated countryside. There was a sort of defile at each end, where the road rose from, and then descended into, marshy flats, with ditches on each side of it.

Maurice had rather expected that Varas would stand to fight upon the moor, which offered a position not too broad for his force deployed. But when his first troops came upon the ground—they consisted of some 30 horse, with 300 arquebusiers under Francis Vere, who had found great difficulty in keeping up with the mounted men—the whole Spanish army was descried in a long column of route marching on Herenthals as fast as it could go. At the head of the column was all the cavalry except two 'cornets,' then followed the German regiment of Sulz, and the two Walloon regiments of Hachicourt and La Barlette. The Spanish *tercio* was bringing up the rear. Two 'cornets' of light horse were riding parallel with the right flank of the infantry. The whole force was obviously hurrying, somewhat disordered after a march in the dark, and not particularly anxious for a fight.

Maurice reached the north end of the moor soon after Vere's detachment, and the rest of his cavalry came on behind him. The infantry, save the 300 arquebusiers in the van, were straggling behind along the marshy road, with their four guns dragging in the rear. They would obviously not be on the moor for an hour or more. By that time the Spaniards would be across the open ground, and retreating along the Herenthals road covered by its ditches. Hereupon the Prince took a rather hazardous decision, following an inspiration which only victory could justify. He resolved to attack with his cavalry alone. Vere says in his *Commentaries* that he suggested the move. Maurice ordered the Counts of Hohenlohe and Solms to take four 'cornets' and gallop along the western side of the heath, parallel with the enemy's line of march, and to try to reach its end before the Spanish van had passed the open ground and got to the downhill defile, where it would be practically unassailable. At the same time Vere with three other 'cornets' was to attack the rear of the Spaniards, which was already being vexed and delayed by the 300 musketeers, who had been dogging the steps of the last regiment for some time. Maurice told off three more 'cornets' to support Hohenlohe, if he should prove successful in his first move, and three others to support Vere. He retained two as a central reserve.

Complete success crowned Hohenlohe's dangerous manœuvre. The Spanish cavalry made no proper attempt to prevent the Dutch from reaching the south end of the heath. Some of them were already entering the downward road to Herenthals, and only pushed on. Only part of the rear turned back, and it was broken at the first clash.

It is said that the lancers could not face the fully armoured pistoleers, and were ridden down, and that Grobbendonk's Germans never retraced their steps or came back to help, but rode on, covering the baggage which was at the head of the line of march. Alonzo Dragon's 'cornet' of heavy horse was cut to pieces, and lost its standard, while trying to make head against the flight. The survivors galloped hard down the defile of the narrow road to Herenthals, and caught up the Germans.

Hohenlohe, the moment that he had got his troopers into some order, was reinforced by the three squadrons which had been told off as his reserve, and then charged the German regiment of Sulz, which was at the head of the Spanish marching column. Its musketeers fired only a scattering volley, and fell back against the pikemen—as was usual in the tactics of the day. But the Dutch horse came in upon the disordered front of pikes with a general discharge of their pistols, and broke a hole in the great clump. Whereupon the Germans, instead of closing up, threw down their weapons, and dispersed across the heath in complete rout. They were obviously demoralized by the sight of the desertion of their cavalry, or they would have made a better stand, even when their musketeers were driven in upon the pikes.

Meanwhile Maurice, at the north end of the heath, had seen Hohenlohe's attack launched, and perhaps also the commencement of disorder in the head of the enemy's infantry column—the precise moment at which he directed the other wing of the Dutch cavalry to move is not quite certain. But at any rate he ordered Francis Vere's 300 arquebusiers to press in upon the rear of the Neapolitan *tercio*, and at the same time threw in against its flank first three cornets of horse under Vere, and then three others, which had been set aside as his supports, keeping only two in reserve. Whether it was that a simultaneous attack by horse and arquebusiers was demoralizing to troops hurrying away along a road, or whether the Spaniards had already noted the disaster at the head of the column, we cannot say. But instead of facing outward and levelling pikes to the flank, the men of the *tercio* flinched before Vere's charge and broke, scattering across the moor, every man for himself, as Sulz's Germans had already done.

The horsemen of Vere from one end and Hohenlohe and Solms from the other then fell upon the two Walloon regiments in the centre of Varas's line of march, and the last Dutch reserve of two 'cornets' was thrown in to help—it is said by Parker, the English captain, without the Prince's orders. They found little resistance, and rode the Walloons down. Two thousand of the unlucky host were slain, and 500 prisoners made. Thirty-eight infantry standards were taken, out of the forty which the beaten army had carried. Varas had fallen by a pistol-ball early in the fight, and there had been no general command on the Spanish side during the greater part of the battle. A curious incident ended the day: some of the more resolute of the Spanish horse, finding that they were not pursued by anything save a few stragglers, turned their horses and came up the lane on to the battlefield when all was over. Vere had to collect scattered men to turn them back.[1]

The Dutch infantry struggled up on to the moor only when all the fighting was over. None of them had been engaged save the 300 arquebusiers of Francis Vere in the van. This, therefore, was a victory entirely of cavalry over a mixed force, and the cavalry were only 800 strong. The victor's losses had been insignificant—as at Gemblours; only two captains, Donck and Cabelliau, had fallen, with (it is said) no more than ten troopers—an incredible number—at least fifty or a hundred might be thought a more likely estimate.

The military moral drawn by contemporary commentators was all in favour of the 'pistoleers'—they could face lancers with advantage, and blow a hole by their volleys in an unsteady infantry regiment. A more sound criticism would be that an

[1] Both Vere and the Dutch narrators say that Maurice was for a moment in personal danger at the end of the battle, being left alone with his staff on the moor when Parker had drawn off the whole reserve. He was beset by this troop of returning Spanish horse, from which he was only saved by Vere, according to his own account, or, as the Dutch say, by a Captain Marcellus Bax, who came up with a few troopers from the rout of the Spanish infantry. Compare the story of the personal danger of Charles VIII at Fornovo from flying Italians (p. 112 above).

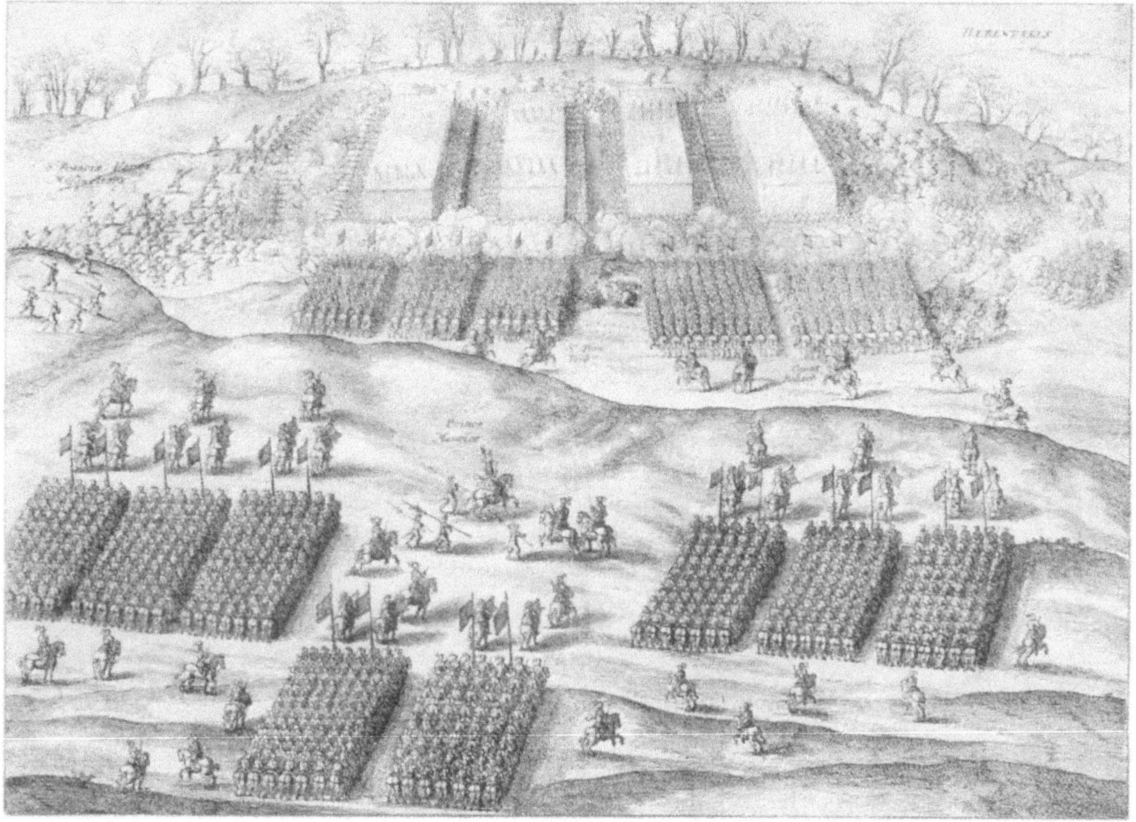

THE ACTION NEAR TOURNHOUT, JAN. 24, 1597

From Vere's 'Commentaries'

army caught in column of march, when every unit is only thinking of getting away, is very helpless. Varas might have faced about, and formed line of battle at the north end of the moor, when he saw that he was caught up. For the Dutch would then have had to debouch from a narrow road through wet ground, and could only have attacked on a small front. There would have been no room for them to manœuvre, or to outflank the enemy, and infantry fire would have told on any leading squadron emerging from the road at the head of the approaching column. Transferring the idea to modern warfare, we may remember how Smith-Dorrien turned to bay at Le Cateau, instead of hurrying on weary men in a continuous retreat. He lost some men and guns, but saved his army. And so, probably, might Varas have done.

NOTE ON TOURNHOUT

There is some difficulty in harmonizing Vere's interesting account of the fight with the Dutch sources, especially as to the exact moment, relative to each other, at which the left and the right wings of the Prince's horse charged the Spanish rear and front. I suspect, in spite of Vere's evidence, that Hohenlohe got in first. The charming picture from his *Commentaries* which illustrates the fight is not quite in accordance with the real chronology, as we know that the two Walloon regiments in the Spanish centre were tackled only after the regiments of Sulz and 'Naples' had been broken, not at the same time. Note that Hohenlohe's men are in two blocks of two 'cornets' each, while Vere's horse are in three single-cornet blocks. The Spanish cavalry is all in rout, flying down the high-road. Vere's artist calls Hohenlohe 'Count Holloch,' the usual English conception of his name.

The other *ordre de bataille* picture in Commelin shows that Hohenlohe's charging line was composed of the four 'cornets' of Rysoor, Marwen, Arozier (Heraugière), and Packet, who are called 'the horse from Breda.' Vere's original three squadrons were those of Marcellus Bax, Dubois, and Donck, which are styled 'the horse from Bergen-op-Zoom.' The troops which finally supported Hohenlohe were the English 'cornets' of Sidney, Vere (under his lieutenant), and Parker; those which followed up Vere's charge were the 'cornets' of Kloet, Edmont, and Riboven.

CHAPTER VII

NIEUPORT (JULY 2, 1600)

THIS is a most interesting fight from the tactical point of view. Though the most brilliant victory achieved by Maurice of Orange during his long military career, it was by no means creditable to him as a strategist, and though he won a complete success, its results were negligible—the town which he had set out to besiege did not fall into his hands, and the campaign was a failure.

In the summer of 1600 the Dutch had a solid hold upon the greater part of the Flemish coast, and Ostend, its most important harbour, was in their hands, as were most of the smaller places, Terneuse, Axel, Hulst, etc., up to the mouth of the Scheldt. The Flemish inland, Bruges, Ghent, and all the rest, was still unconquered, but the States General had come to a conclusion that the most promising means for bringing the war to an end was to seize the two remaining ports by which the enemy could communicate with Spain and the sea. Nieuport and Dunkirk were both nests of corsairs, whose depredations were most pernicious to Dutch ocean-going commerce, which was expanding every day. To extend the narrow slip of occupied coast-territory right up to the French frontier at Calais was no doubt a tempting scheme, and certainly the loss of all touch with the sea would have been an almost fatal blow to the sovereigns of the Belgian lands, Albert and Isabella. They could not have maintained themselves without the constant help from Spain which was their chief reliance—though Philip III was a much more feeble helper to his sister and her husband than his formidable father had been to Spanish viceroys in the century that was just expiring.

The scheme was strategically unsound, as the retention of a string of harbour towns while the inland was in the enemy's hands would be difficult to manage, unless the inland were conquered also. If ambition soared high, and the expulsion of the Spaniards from all the Low Countries were contemplated, the only sound policy would have been to overrun all West Flanders, and then to deal with Nieuport and Dunkirk, which would have been left isolated, and must have fallen, if the enemy had been driven far away from the sea. If the power of the Archduke and his wife had sunk so low as the States General calculated, the first blow should have been delivered at Ghent and Bruges, not at the ports. But the whole naval interest in Holland was clamouring for the destruction of the corsairs of Nieuport and Dunkirk, and it was resolved to ship the best field-force which the States could put together across the Scheldt estuary, and to strike at the two surviving Spanish harbours. Prince Maurice discouraged the adventure, and so did Francis Vere, his formidable English subordinate. Its success, it was urged, depended on the hypothesis that the Archduke could not collect an army competent to face the invading force. But though his finances were known to be in a desperate condition, and though half the Spanish troops were at the moment in mutiny, Maurice held (quite rightly) that in a moment of extreme danger the rebel regiments would come to heel, and that every fortress in Belgium would be stripped of its garrison for a final effort to save the threatened ports. If beaten, the Archduke could still fall back on Brussels and Namur—but if the dreadful chance of a Dutch defeat should occur, there would be no retreat open to the vanquished save into Ostend—there was no land-route available. And from Ostend the troops—or such of them as survived—would have to be brought away by water. Meanwhile all the Dutch frontier on the lower Meuse would be open to attack by the Archduke, if he had put the States' field-army out of action. Even by thinning out the garrisons of Guelders, Overyssel, and Groningen, the army of invasion could not be raised to more than some 12,000 foot and 1800 horse, and this was too small a force for the enterprise. If it met with disaster, no other army could be collected.

Nevertheless, the States General, led by Olden-Barneveldt, Maurice's lifelong foe, decided on the Flemish enterprise, and the Prince, with a protest, consented to undertake it. He was determined not to give up the command of the army, and to see the expedition entrusted to other hands. In June the whole force was collected in Zealand, and on the 21st–22nd it crossed the Scheldt estuary in a multitude of vessels and barges of all sorts, and landed in successive detachments at

Sas van Ghent, one of the string of Flemish coast-towns in Dutch hands. From thence it started to march via Assenede, Eecloo, Jabbeke, and Oudenburg to Ostend, which was to be the base of operations. The movement passed close to Bruges, which shut its gates, and fired fruitlessly on Maurice's scouts. On the 27th serious operations commenced, by the clearing out of the Spanish posts, forts, and sconces, which had been placed round Ostend to restrain the raids of its garrison. These, including Oudenburg, the most important of them, were taken with ease; the Spanish force in Flanders at the moment was only 2000 strong, and retreated into Sluis. The most important capture was Fort Albert, a large redoubt recently erected by the Archduke on the coast only a couple of miles from Ostend, to block the road along the seaside from that fortress to Nieuport. On the 30th of June Maurice, after reinforcing the standing garrison of Ostend with half a regiment of foot [1] and four cornets of horse, started for Nieuport—he had at first intended to go somewhat inland, and capture the fort of Niewendam, which lies south of Nieuport in the marshes of the Yser, but found the way in the marshes impracticable for wheeled traffic and guns, and on the following day came before the place by way of the strand, passing on his left Middelkerke and Lombardzyde—familiar names in the Great War of 1914–18.

Nieuport, an ancient walled town, stands on the south side of the tidal estuary of the Yser River. It lay a mile and a half from the sea, and its harbour was only to be used at high tide; at low tide it is but a series of mud-flats, with the small river meandering through them. So shallow is the estuary, when the water is falling, that it can be forded easily, even by men on foot, for a long stretch quite out of range of the guns of Nieuport. When Maurice arrived in front of the place on July 1, he sent two-thirds of his force—horse, foot, guns, and baggage, across the fords, to blockade Nieuport from the west, while the rearward division under his cousin Ernest of Nassau remained on the east side, watching the place from the other bank of the Yser. A great number of ships from Ostend ran into the estuary and moored themselves there—afloat at high tide, sitting on the mud at low tide; they brought provisions and munitions in plenty. Maurice was on the night of July 1 making preparations for a regular siege—he ordered a bridge

[1] From Ghistelles' regiment. Piron's regiment was about to join.

to be thrown across the estuary, to keep the two sections of his army in touch at high water, and began clearing emplacements for artillery.

Late at night, when commander and army were settling down into their appointed places, startling and ominous news came to hand. The Archduke was close at hand with a field force of unknown strength, and had late on that afternoon recaptured Oudenburg and the other old Spanish sconces and redoubts inland from Ostend. Their garrisons had been mostly massacred after capitulation. Ostend and Fort Albert were in full sight of the hostile army, and sent messages of alarm. Next morning the enemy might cross the Yperle stream at Leffinghem and cut down to the sea, interposing himself between Ostend and the army.

The States General had been entirely deceived as to the improbability of interference in its plans on the part of the Archduke. They had reckoned that the vicious and long-continued mutiny of a great part of the Spanish troops would make it impossible for an army of any value to be collected for the relief of Nieuport. So thinking, they had failed to reckon on two underlying features in the mentality of the mutineers—religious fanaticism and *esprit de corps*. These old bands were sincerely pious in their own strange way, as many gruesome and grotesque incidents of the war suffice to prove. And they were also very proud of their military reputation, and resented the idea of seeing the Belgian lands fall into the hands of the old adversary.

When the Archduke made a desperate appeal to the 'Eletto' and his council at Diest, playing on both these sentiments, as was natural, it met with general approval. The mutineers sent word that they would join him at once, on three conditions—they should serve under their own standards and with the officers whom they themselves had chosen, they should be given the honourable place in the vanguard when a battle was forthcoming, and the discharge of their outstanding pay should be the Archduke's first obligation after victory. Albert naturally accepted the bargain with joy, and was joined in a few days by 600 horse and 800 foot from Diest, to which other detachments kept adding themselves from day to day, till he had the equivalent of two full foot-regiments as well as the horse. It is a curious note of the spirit of the times to find that the mutineers brought with them as their corps-standard an immense blue damask banner, bearing on it a representation of the Virgin Mary standing in a glory, with a star and a moon on each side, inscribed *Ave Gratia plena*, *Stella Maris*, and *Pulcher ut Luna*.[1] Presumably it was church-plunder, but its choice as a rallying-point was a religious demonstration. The Archduke had as his 'vaward' the whole body of the recent mutineers, serving under their own officers; they came from many infantry corps, both Spanish and Walloon,[2] which were embodied in two provisional regiments of foot, making about 2000 men; there were also 600 horse. The command of this very unruly but formidable body was entrusted to Francisco de Mendoza, the Admiral of Aragon, who was popular with them—probably on account of his lavish permission of plunder when he was in command on the Rhine, during the campaign of the preceding year.

The main-battle, which marched behind the 'vaward' during the approach to the battle, was in charge of the Archduke himself, it contained the flower of the army, three old Spanish and one old Italian *tercios*, commanded by Alonzo de Avila, Gasparo Sapena (this was the Italian corps), Luis de Villar, and Jeronimo Monroy. The Archduke had one 'cornet' of the lancers of his guard as escort: and five more were attached to the 'main-battle.'

The 'rearward' consisted of the two Walloon regiments of La Barlotte and Bucquoy, with the Irish regiment of Bostock, who inherited from Stanley the corps that had originally been raised by the latter from the deserting garrison of Deventer, and had been kept in existence by absorbing refugees from Ireland, who drifted over after the suppression of successive rebellions by Queen Elizabeth's troops.[3] Six cornets of horse

[1] There is a detailed description of this great banner, a very complicated piece of stitchery, in Commelin's narrative of the battle, p. 209.

[2] Certainly from regiments Barlaimont, Barbançon, Ponce de Leon, Warambon, Hachicourt, Van den Bergh—from all of which individual officers are found among the prisoners in the Dutch official list, though the regiments as units were not at the battle. They look like Walloons mostly.

[3] It is curious to find in the lists of members of Bostock's Irish regiment, which are preserved in the Dutch prisoners' roll, hardly any old Irish names, with the O' or Mac, but almost exclusively names of English or Anglo-Irish provenance such as Dyer, Green, Flood, Conway, Jarvis, Piers, Carr, Nally, Martin, Herbert, Roberts, Guise, Rogers, Harcourt, Tolleman, Maynard, Ferris, Geraldine, Doyle. I imagine that many of them may have been English Catholic refugees rather than Anglo-Irish. Only one is named Patrick and one Dennis. Possibly Guy Fawkes, who had been serving with the Spaniards in the 'Low Countries,' may have been in the regiment.

were attached to the 'rearward' battle. The artillery consisted of six 'demi-cannons' and two smaller pieces. The total appears to have been about 90 companies of foot whose strength must have varied from an ideal 120 to an actual 100. Probably

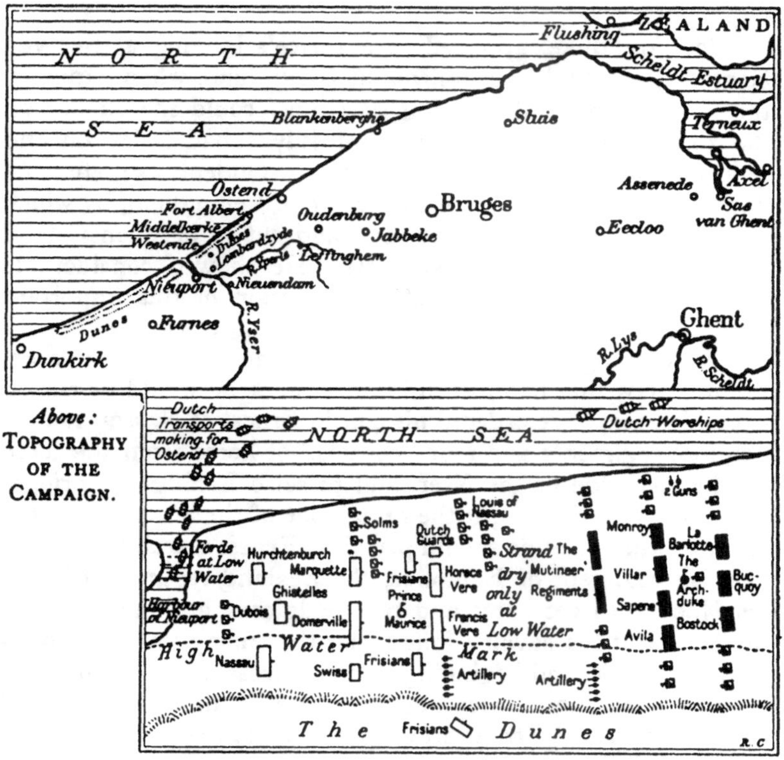

***Above:* Topography of the Campaign.**

***Below:* Position of the Armies at about one o'clock, before the Spaniards moved up to the Dunes. 'Ordre de Bataille.'**

(This arrangement is only diagramatic, as the armies were much further apart at the moment. Nieuport, July 2, 1600.)

10,000 would be something like the real figure.[1] The horse, remembering that the 'cornets' of men-at-arms ('lances') were much smaller than those of 'pistoleers' or light horse

[1] Maurice in his letter to Louis of Nassau estimates the enemy at 11,000 foot and 1500 horse. Ernest of Nassau, in a letter of July 4, allows for 9000 foot only—perhaps arguing from 90 companies being present. [*Nassau Archives*, vol. ii. (2nd series) 16–20.] The Admiral of Aragon declared that seven light horse cornets in his charge had precisely 614 men in the field.

'herreruelos,' may have run to some 1500 in all.[1] The total of the army was very nearly the same as that of Maurice's troops: indeed the Dutch would have had a perceptible preponderance in numbers, but for a disaster on the morning of the 2nd, which ended in the destruction of two of their foot-regiments.

The Archduke, apprised of Maurice's landing at Sas van Gent on June 22, had mobilized at Ghent, and, when joined by the 'mutineers' and outlying garrisons, could march out on the 29th, and on the afternoon of July 1 had fallen on and reduced the small Dutch outposts at Oudenburg, etc. At dawn on July 2 he made for the bridge of Leffinghem, by which he could cross the Yperlé stream, and cut in between Ostend and the Dutch army at Nieuport. He showed considerable activity and skill, for he was in possession of this important bridge at a very early hour, without meeting resistance.

On learning of the unexpected approach of the Spaniards, Maurice saw that he must fight at once, or he would be cut off from his base. He ordered his cousin Ernest of Nassau to take two foot-regiments and four 'cornets' of horse from the force on the east side of the Yser, and to seize Leffinghem bridge, while he himself brought the remainder of the army across the estuary, when the tide should permit. Ernest marched with the Scottish regiment of Edmonds, the Zealand regiment of Van der Noot, with 400 horse and two fieldpieces, and made for the bridge. But his cavalry found that the enemy was already in possession of it, and beginning to pour across it, the 'mutineer' regiments leading. Intending to fight a detaining action, and to retire at leisure, Ernest drew up his small force behind a ditch and road facing the bridge, but the enemy, who was already across in strength, charged right home, and pierced his centre, whereupon the two regiments broke up in panic. The Scots were almost all cut to pieces; seven captains were slain,[2] the Zealanders fared a little better, some of their companies escaping into Fort Albert on the seashore. But over 1000 of Ernest's men were destroyed, and those who got off were either dispersed

[1] We can trace most of the cavalry units—'cornets' of Captains Avilliano, Guevara, Tacon, Sangre, Verdugo, Braccamonte, Belgioso, Visconti, Taillis, Salm, Le Roy, Forest, Goddard. Spaniards, Walloons, and Italians are all represented. There are said to have been six cornets of pistoleers (cuirassiers), four of light horse, and nine of 'lances.' See Meteren, and cf. the prisoner-lists in Commelin.

[2] Their names were Barclay, Kilpatrick, Andrew Murray, Nisbet, Strachan, Stuart, and Mitchell.

or thrust into Ostend and Fort Albert, where their commander and the cavalry also took refuge.[1] Not a man of this detaining force escaped towards Nieuport, save two fugitives, whom Maurice ordered to hold their tongues, and sent them on board a ship, lest the rumour of the disaster should discourage the army.

The Archduke had now thrust himself between Maurice and Ostend, and could force his adversary to fight with his face towards his only base, and a fortified town behind him, or else to ship off his men piecemeal on the numerous transports in the estuary-harbour, where at this hour (7 a.m.) the tide had begun to fall—it would be some time before it would rise again to float off such vessels as were stuck on the mud. If the Prince chose to take this alternative, he would certainly lose his rear-guard, perhaps half his army. But Maurice had no purpose but to fight, and that at once. As soon as the water was low enough—about 8 a.m.—he sent his vanguard across the river by the fords—the projected bridge was not yet ready. The main-battle followed, and the rearward passed last, after staying for a time in front of Nieuport, lest the garrison should make a sally to destroy the stranded ships. These were ordered to put to sea so soon as the tide should serve.

The countryside for the nine miles between Nieuport and Ostend is of a peculiar sort—there is a beach along the sea, so broad at low tide that a whole army could be drawn up upon it, but shrinking at high tide to no more than fifty or a hundred yards of firm sand. Above this are the 'Dunes,' sandhills of considerable height, with a very irregular surface of dips and ridges. They were 700 yards broad in some places, only 350 in others ; for the most part they were composed of loose sand, very tiresome footing, but in patches they were overgrown with low scrub, apparently furze-bushes.[2] On the farther side of the dunes was a country road from Ostend, passing through the villages of Westende and Lambardzyde—it was a broad green track ; beyond it were the cultivated fields of the villages, much cut up by irrigation ditches, and then the small Yperlé stream, which falls into the Yser near Nieuport. There is a gap of

[1] Ernest describes himself, in a letter to his father, as finding himself at the end of the day in safety, with no one about him save the Scottish Colonel Edmonds, and one captain and 180 men of the Zealand regiment. But, of course, many others got off. The cavalry behaved particularly ill, and was thoroughly demoralized.

[2] Lewis of Nassau, in his descriptive letter, calls them 'the low plants which prick a bit' ('qui piquent un peu')—I presume that this means furze.

some 400 yards in the long series of dunes, where the Yser cuts through, but the sandhills recommence on the farther side, and continue as far as Dunkirk along the coast.

Maurice drew up his army, when it first passed the fords, on the firm beach of low tide, but as we shall see, the rising water in the course of the afternoon compelled both him and the Archduke to move up the greater part of their forces into the dunes, where the hardest of the fighting took place on the unstable sand of their ridges.

The 'vaward' this day was under Francis Vere, and contained nearly the whole of the English contingent in Maurice's army. It was composed of the two foot-regiments of the Veres—that of Francis with thirteen companies, that of his brother Horace with eleven: to these there were added a very strong Frisian regiment of 17 companies under a Colonel Hertinga, and the two companies of Maurice's own foot-guards. Altogether the 'vaward' had 41 companies of foot—quite 4000 men, and the best in the army: it was far stronger than either the 'main-battle' or the 'rearward.' The cavalry attached were also picked troops—the 'cornets' of which Prince Maurice, his cousin Louis, and his brother Frederick Henry were the titular captains, Walraven's and Marcellus and Paul Bax's cuirassiers, with three cornets of light horse—9 units in all out of the 17 cavalry corps of the army—or more than half the total of mounted men. The whole were under Count Louis, who was 'Lieutenant-General of the Horse.' Six field guns were attached to the 'vaward,' served, as we learn, by sailors.

Vere deployed the whole of his troops on the shore—which was still broad and dry—in order to cover the passage of the rest of the units over the fords behind him—it was necessary that they should have good time to cross, since the tide would be rising after some hours, and also good space to get themselves in battle order. The whole body of the cavalry under Count Louis was somewhat in advance, with the light horse out in front, to watch for the approach of the enemy. The six guns had been planted on the dry sand, at the foot of the dunes: six companies of Frisians had been sent up into the sandhills to flank the artillery.

It was presently obvious that the Archduke also was advancing along the shore, and not taking the 'green way' by Westende along the other side of the dunes inland. His vaward, all composed of cavalry, could be seen approaching on a

The Battle at Nieuport, July 2, 1600
The moment shown is that of Count Louis' second cavalry charge
From Vere's 'Commentaries'

broad front, occupying the whole space between the sandhills and the sea—some ten ‘ cornets ’ of all sorts were visible. The infantry and the rest of the horse were not yet in sight. Albert had not hurried after the skirmish at Leffinghem bridge—partly because he allowed his troops, who had started before dawn on their march, time to breakfast; partly because he was holding a conference with his captains—some were for striking at once, while the men were cheered by the small victory at dawn; others urged that if they entrenched themselves across the way to Ostend, Maurice would have to take the offensive, and to force his way through, on a narrow front, against troops well covered by earthworks on ground where his cavalry—which the Spaniards dreaded after the experience of Tournhout—could not act. After some discussion the officers of the mutineer regiments out-argued the more cautious—they had the vanguard, and their men were mad for a fight. Some one raised the old Spanish slogan, ‘ Cuanto mas Moros, tanto mas ganancias ’—‘ The more the Infidels, the more the profit,’ and the army moved on in battle order along the strand, which at twelve o'clock was beginning to shrink in breadth (for the tide had turned), but was still very wide.

Vere's vanguard was already in position at the other end of the wet sand, the cavalry in advance, the infantry in formation: the Dutch main-battle had crossed the fords, but the rearward was still crossing. It was clear, however, that Maurice intended to fight. The battle opened with some cavalry bickering—Count Louis sent out his light horse, and the Spanish *herrueleros* pricked out to meet them. The Count had broached to Vere a plan of a false retreat by his skirmishing line, which might induce the Spanish horse to charge, in which case they would come under the fire of the guns, and of the Frisian infantry on the dunes above them, and could be attacked at advantage by the Dutch horse. Vere disagreed, and ordered the Count to bring back his men, and to fall in on the left of the deployed infantry, close to the water's edge. Only the skirmishers remained out in front, and gave way by order; they were hotly pursued by the cornets of the ‘ mutineer ’ cavalry, which suddenly came under the fire of the Dutch guns on the high beach, whose existence they had not suspected, and swerved off in disorder.

There was then a pause—caused by two things. The first was that the Archduke discovered that the strand was rapidly shrinking and the tide coming in; the second and more important consideration came in later, it was that several Dutch ships of war, which had been lying off the coast to escort Maurice's transports, came as far inshore as they could, and began to send long shots into the flank of the Spanish army. At 1.30 the Vice-Admiral of Zealand and another ship put some balls into the Archduke's main-battle. Albert ordered two of his cannon to be brought to the water's edge, to reply—but the water was coming in fast, and more Dutch men-of-war were appearing.[1] Whereupon the Archduke resolved that he must get out of this dangerous flank fire. Had he even ever heard of Gravelines (1558)[2]—possibly, but hardly of the parallel case of Pinkie? At any rate, at about 2.30 the Dutch saw that the enemy was leaving the shore, and turning up into the difficult dunes. Only five guns and some infantry companies in support remained on the beach, facing the Dutch battery: the rest climbed slowly up the sandhills where they were least tiresome: the whole of the cavalry save one 'cornet' crossed them, and arrayed itself on the broad green road by Middelkerke, the infantry drew up on the dunes in its orginal order, and commenced to march slowly westward on their slippery footing. The whole operation took nearly two hours.

This compelled a corresponding change in the order of the Dutch army, and there was time to carry it out, since the enemy had such bad going on his chosen ground. Vere had already six companies of Frisians on the dunes—he now shoved up to join them almost the whole of his force, leaving only five Frisian and two English companies in support of the guns. He had found a dip in the dunes at right angles to the sea, which made a sort of position, and arranged his infantry across it; the cavalry of Louis of Nassau went about and passed the sandhills as best it could, drawing up on the green road somewhere near Westende, continuing the line of the foot on the dunes.

Vere has left us in his interesting *Commentaries* a very elaborate, but not always very clear, account of his dispositions. Projecting on his left front was a high isolated sandhill with a flat top, on which he placed 250 picked men, English and some of the Dutch foot-guards. Some way to their right was a still higher sandhill, up to which were dragged two guns from the battery on the shore. They commanded the 'green road'

[1] See Commelin, p. 195.

[2] See above, pp. 279, 363, for flank fire from the sea on an army upon the shore, and its results.

below them on their right. Before the guns, on lower ground, were 500 Frisian musketeers, who were also well placed for firing at any enemy on the road. The rest of his troops, English and Dutch, were drawn up in support, in small detachments, ready to reinforce the front line. Count Louis's cavalry covered the right.

Meanwhile Prince Maurice brought up the 'main-battle' and the 'rearward' and ranged them behind Vere's line, the infantry on the dunes, mainly to their southern side, the cavalry on the 'green road' in support of the cavalry of the 'vaward.' Only three 'cornets' of horse, belonging to the 'main-battle,' were on the sands, behind the battery.

The 'main-battle' consisted of the strong Huguenot French regiment of Domerville in the centre, flanked on the right by a very small Swiss battalion of only 400 men, and on the left by the Walloon regiment of Marquette—the most doubtful element in the whole army, for it was composed entirely of deserters from the Spanish colours. They showed no slackness, however, in the battle—probably remembering the fact that every one of them was fighting, so to speak, with a rope round his neck. The cavalry attached to this corps six 'cornets,' was commanded by Count Solms, and included two English units, those of Sir Edward Cecil and Captain Pembroke, who was in charge of the 'cornet' of which Francis Vere was the honorary commander. It is curious to note that there was not a single Dutchman in the main-battle, save in the cavalry: all the foot was French, Walloon, or Swiss.

The 'rearward' consisted of the German regiment of Ernest of Nassau—his lieutenant-colonel, Huysmann, was in command—and the two Dutch regiments of Hurchtenburch and Ghistelles: the latter was incomplete, six of its companies having been dropped in the defences outside Ostend. The whole made 26 companies. There was with the 'rearward' a small provision of cavalry only, under a colonel named Dubois, two 'cornets' of Dutch and one of English, commanded by a Captain Hamerton. Prince Maurice, along with his young brother Frederick Henry, who was seeing his first fight at the age of sixteen, took up his position in front of the 'main battle,' being urged to do so by his staff, who were most averse to his risking his person in the firing line.

The Archduke's infantry, advancing in its three lines along the top of the dunes, slowly amid shifting sand, presently came upon Vere's troops ready to receive them. The Spanish cavalry had transferred itself on to the 'green road' on the left of the army. On discovering the Dutch position, the 'mutineer regiments' proceeded to the attack of the advanced sandhill, on which Vere had placed his forlorn hope. It was assailed first by five hundred arquebusiers in a mass, and, when they were beaten, by pikemen and arquebusiers mixed, drawn from the regiments behind. But the ground was advantageous, and no progress was made. This was followed by a more general advance against the whole of Vere's front, accompanied by a forward movement of the Spanish cavalry in the low ground inland. Apparently the Archduke put in his whole 'main-battle,' the three Spanish and one Italian *tercios*. His front-line cavalry were completely routed by Louis of Nassau's horse, much helped by the cross-fire from the two guns on the high sandhill, and Vere's Frisian musketeers. They fled back in disorder, hotly pursued, and took shelter behind the infantry of their own rearguard. Many scattered, some right into the morasses towards Nieuport, and were not seen again. Spanish sources say that 'mutineer' horse did very badly. Louis of Nassau, pressing too far, right along the side of the enemy's army, got into considerable personal danger among the cavalry of the Archduke's second line, and had to be rescued by a small separate charge, made by one of the 'cornets' of Maurice's second line of cavalry. It took a long time for him to get his victorious squadrons into order again on their old ground.

Meanwhile the battle on the dunes grew complicated and confused. The 'mutineer' foot regiments opposite Vere being now exhausted, the Archduke put in, to relieve them, his four old *tercios*. Monroy's and Villar's regiments continued the attack on the left of the Dutch line, Avila's and Sapena's delivered their thrust on its right, south of the two-gun battery on the high sandhill. To meet these last, Maurice brought up the Huguenot regiment of Domerville and the Swiss and Walloons from his 'main-battle.' But he sent no aid to Vere on the other or northern flank, though it was repeatedly requested by the English commander. The battle did not stand still on the dunes, but flickered to and fro in attacks and counter-attacks, with no definite result. Seeing matters still undecided, Louis of Nassau asked for his cousin's leave to try another cavalry charge on the 'green road,' against the enemies' left flank. It was delivered, but was not so successful as his first attack had been; he drove in the Spanish cavalry of the Archduke's main-battle, but was checked by the cross-fire of the infantry of the hostile rearguard, and had to draw back in disorder to his original position, with squadrons exhausted and much thinned.

At last Albert resolved to throw in his third line, the Walloon and Irish regiments of Bucquoy, La Barlotte, and Bostock. This thrust looked for a moment as if it were to be decisive on the dunes, for Vere's part of the Dutch front line at last gave ground; he had used up all his reserves, and had received no reinforcements. In his Memoirs he is content to put the blame on the messengers who had been sent to ask for them—but it is clear that he thought that Maurice had failed him at the crisis. His English companies, and those of the Frisians who were with him on the dunes, were driven off the heights, and had to fall back in disorder to behind the battery on the shore, which was still exchanging a slow fire with the Spanish battery opposite to it. The enemy pursued,[1] but as Vere remarks, tardily and in great disorder, like tired men. He gives, as a proof of their slowness, the fact he himself, bringing up the rear, was badly hurt and thrown from his horse—it was his third wound that day—but that Thomas Heigham and Sir Robert Drury had time to drag him from underneath his charger, and to carry him off before any of the Spaniards got near him.

Francis Vere's officers succeeded in rallying behind the battery a few hundred of his disordered troops, and the guns, though almost at their last provision of balls, kept up their fire. But the decision of the day came from the Princes reserve of cavalry, still intact, the 'cornets' of Pembroke, Balen, and Sir Edward Cecil, which charged right into the pursuing mass of disordered Spanish infantry on the beach, and turned them into utter rout, so that they ran up on to the dunes, falling in upon and turning into confusion their supports. It is said that there were very few Spaniards standing round the ensigns—the large majority had drifted forward and mixed themselves with the fighting line. When Vere's rallied companies under his brother Horace and his sergeant-major Henry Sutton pressed forward, they met little opposition. But this was largely because Prince Maurice, on seeing the success of the

[1] These Spaniards, I take it, must have been Monroy's and Villar's men, already long engaged.

cavalry on the beach, ordered the whole of Count Louis's horse to charge once more, though nearly all its units were exhausted, and at the same time ordered the infantry on the downs—mainly his ' rearward ' regiments—to make a general advance.

Then the enemy, who had fought so staunchly hitherto, suddenly collapsed. The Archduke Albert, in his down-hearted account of the battle, attributes the final disaster entirely to his cavalry, who were thoroughly demoralized already by the previous engagements, and now fled past his flank without making any attempt to stand. The infantry on the dunes, seeing the horsemen's flight, broke up also, and retired in a disorderly mass, pikemen and musketeers all mixed, and the whole crowd made for Leffinghem bridge, their natural line of retreat. The Dutch cavalry rode through and through them, slaying many, but taking prisoners also at their good pleasure.

All accounts agree that if the troops in Ostend had shown any initiative they might have come out and seized the bridge, or blocked it before the retreating enemy, in which case nearly the whole of the Archduke's army must have been captured. But apparently the news of the victory had not reached them, and they were cowed by the memory of the morning's disaster, when the Scottish and Zealand regiments had fared so ill.[1] The members of the States General then present in Ostend had the responsibility.

Nearly half the Archduke's army was destroyed—the fugitive cavalry got off more easily, but the weary and exhausted infantry, who had been on their feet from long before dawn, and had endured three hours' hand-to-hand fighting in the late afternoon, could make no pace, and were slaughtered wholesale. The least estimate of the Spanish loss is 2500 men—the more usual over 4000, which may not be far out, as we have lists of the officers killed or captured, which must represent a fifty-per-cent. proportion of those present. The Archduke himself was wounded—of the Colonels, Bostock was killed, Sapena mortally wounded, Villar taken prisoner, as was also the Admiral of Aragon, who had led the vanguard. Avalos, Bucquoy, and La Barlotte were all wounded—erroneous Dutch reports put the last two in the list of the killed ; Schoonhoven, who commanded the artillery, was captured, as was the

[1] See Commelin, p. 200, and other commentators.

Count of Salm, one of the cavalry commanders. Thirty-three captains were killed, twenty-nine taken prisoners; how many subaltern officers fell we cannot say, but over seventy were in the list of captives, along with a number of the Archduke's staff and retinue, 'gentlemen unattached,' 'reformados,' sergeants and commissaries, which made a total of 203 prisoners, scheduled in the Dutch report as worthy of mention. The lists show that the four old *tercios* and the 'mutineers' must have been pretty well exterminated—the Walloon regiments in the rearward apparently suffered less, as did the cavalry. There were 105 standards of horse and foot taken; there cannot have been more than 120 on the field; most conspicuous was the great blue religious banner of the 'mutineers.'

The Dutch loss was at least 2000, if the casualties in the fight at dawn at Leffinghem bridge are reckoned in. Beside the ten Scottish and Zealand captains of foot there slain, thirteen fell in the main-battle, of whom no less than six were English,[1] three belonged to the French Huguenot regiment of Domerville, two were Frisian, one each from the Walloon regiment of Marquette and the Dutch regiment of Ghistelles. In addition, three captains of horse-units fell—Couteler, Bernardt, and the Englishman Hamerton. It is clear from the casualty list that Maurice's three third-line or 'rearward' regiments (Ghistelles, Nassau, Hurchtenburch) can have been little engaged, while the two English regiments in the 'vaward' lost 600 men out of their 24 companies.

This was in Francis Vere's judgment a very abnormal battle—usually, he remarks in his *Commentaries*, it is the cavalry which win the day—here it was the infantry which, by their long endurance, enabled the tired cavalry to make the final decision. The Spaniards were used up by the evening, so that a comparatively small body of horse was able to turn the tide at the last critical moment. Vere was, of course, looking at the matter from his own side of the battle—otherwise he might perhaps have noted firstly that the Spaniards failed to win from the manifest inferiority of their cavalry, which at every clash was ridden down by the Dutch cuirassiers, and secondly that Maurice at the end of the battle had his third-line infantry much more intact than the Archduke, whose 'rearward' had made its thrust to break down the Dutch line before Maurice's

[1] Tyrrell, Duxbury, Pirton, Yaxley, Honeywood, and C. Drury (brother of the R. Drury who saved Vere's life).

reserve was engaged. It is true that the Prince got this advantage largely by sacrificing Vere and the English and Frisian 'vaward,' who fought not only the Spanish 'vaward' —the 'Mutineers'—but at least two *tercios* of the Spanish main-battle, without receiving any reinforcements. Hence their disproportionate loss, compared with that of the rest of the army.

The Archduke had shown much vigour and dash in bringing on the action before the enemy could expect him, by seizing Leffinghem Bridge, and cutting off the Dutch from their base at Ostend.[1] Possibly he also hoped to catch Maurice with part of his army still on the wrong side of the difficult fords of the Yser. The main weak point of his policy was the neglect to note that by the afternoon his infantry, who had been on the march before dawn, and under arms for twelve hours, would be absolutely exhausted by dusk. It has often been remarked in other centuries that commanders and staff-officers, all mounted, fail habitually to allow for the fatigue of the pedestrian infantry. The second point was that no allowance was made for fire from the sea, which made a battle on the strand too dangerous—for it is clear that the Archduke originally intended to advance on the beach, while the tide was still low, and only turned up into the dunes when he saw the enemy in good order, and the ships beginning to intervene with their flank fire. At the same time the move into the dunes could be defended, on the ground that the Archduke was aware that his cavalry was somewhat inferior in numbers and in quality to that of the enemy, so that getting the decision on ground where horses could not go would be advantageous. His strength and confidence lay in his veteran infantry, while Maurice's foot-regiments were much more heterogeneous, and were reckoned to include some elements of doubtful value. It will be remembered that of the nine corps which formed the Dutch infantry two were English, one French, one Walloon, one German, one Swiss, and only three native Dutch. Sir Francis Vere criticizes the Archduke for not pressing on more vigorously at the moment when the much-tried English and Frisian infantry gave ground, and fell back on to the beach and the battery. Probably he failed to make sufficient allowances for

[1] One may perhaps recall the three humorous paradoxes of Henry IV: 'Three things are true, but no one believes them—namely, that I have become a good Catholic, that the Queen of England died a virgin, and that Archduke Albert is a very good general.'

the fact that the Spanish infantry was by this time not only physically exhausted, but in hopeless disorder, pikemen and musketeers mixed up, and regimental organization lost. His lieutenant, Ogle, noted that the enemy's standards, their centres for rallying, had few men around them, and that the whole army was a mob at the last critical moment; commands could not be transmitted or obeyed, and an orderly advance was impossible. The tired infantry were content to see their opponents apparently giving way in rout, while the twice-beaten cavalry had been completely demoralized, as their subsequent conduct showed.

While the Archduke's strategy had been good down to the morning of the battle, that of Prince Maurice had been—by no fault of his own—most unhappy. He had accepted the orders of the States General to undertake a plan of campaign in which he had no confidence, grounded mainly on the hypothesis that the Archduke could not raise a competent field army, owing to the long mutinies and the exhaustion of his exchequer. When this hypothesis was proved to be wholly erroneous, he was landed in a desperate position, and had to extricate himself from it by hard fighting. Vere's main criticism on him is that he ought never to have sent out the small force of Count Ernest to hold back the enemy, but should either have marched with his whole army, or else have taken up a position on the strand and dunes and invited an attack. In that case he would have fought with 2500 more men in his ranks, and have had an appreciable advantage in numbers over the Spaniards. On the other hand, Count Ernest remarks, in his letter describing his doings, that he had at any rate delayed the Archduke for some hours at Leffinghem Bridge, and so had given his cousin time to bring his whole army across the Yser in good time. Probably the Count's real error was in drawing up his men and offering battle, against overpowering numbers and in a weak position. He could not have foreseen the panic which set in, and made an end of his whole force, but the venture was —at the best—very rash.

Maurice was so entirely disgusted at the folly that had sent him against the Flemish ports, while the Flemish inland was unconquered, that he refused to go on with the plan, and after strengthening Ostend brought back his prisoners and his trophies to Holland in July. He started no campaign in the autumn, and next year his operations were confined to

Brabant and the Rhineland. Hence came the extraordinary fact that in 1601 the Archduke was able first to resume the blockade of Ostend and then to lay regular siege to it. Maurice refused to land in Flanders for the assistance of Ostend, and hoped to relieve it rather by bringing pressure to bear on the Archduke's eastern frontier. The interest of the next period of the war becomes centred on the incredibly long siege-operations conducted around that water-girt fortress, which held out for three years and seventy-seven days; its last governor, Marquette, only surrendered when every outer bulwark had been lost, and after he had received permission from Prince Maurice to haul down his flag, since relief by land was impossible—though the Dutch were so close at hand as Sluys—while the sending in of reinforcements from the sea only protracted the agony. The States had lost 30,000 men, it is said, in defending an unprofitable outwork—it was small compensation that the Archduke had seen more than double that number of his troops perish by fever and cold in the trenches, or by shot and pike-thrust in a hundred fruitless assaults.

Long before Ostend surrendered on September 20, 1604, both parties had become convinced that Spain would never conquer Holland, but that Holland would never succeed in conquering Flanders or Brabant. That the war had gone on so long, and was to continue yet a few years more, was mainly due to the extraordinary genius of that apparently self-taught captain, Ambrogio Spinola, perhaps the greatest of all the Spanish generals—even Alexander Farnese did not surpass him—who saved the cause of Albert and Isabella and of Catholic Belgium, and showed himself the equal, or even the superior, of Maurice of Nassau in the field. A long truce came at last in 1607 and a definite settlement in 1609, and here we may end the chronicle of the military history of the establishment of Dutch Independence—though the truce of 1607 was to prove but a temporary landmark in the long tale of the 'Wars in the Low Countries'—endemic through the seventeenth as well as the later sixteenth centuries.

NOTE.—There are many and valuable sources to be consulted for the battle of Nieuport. Much that cannot be found in Francis Vere's interesting though self-centred commentaries, nor in the series of contemporary letters from Prince Maurice

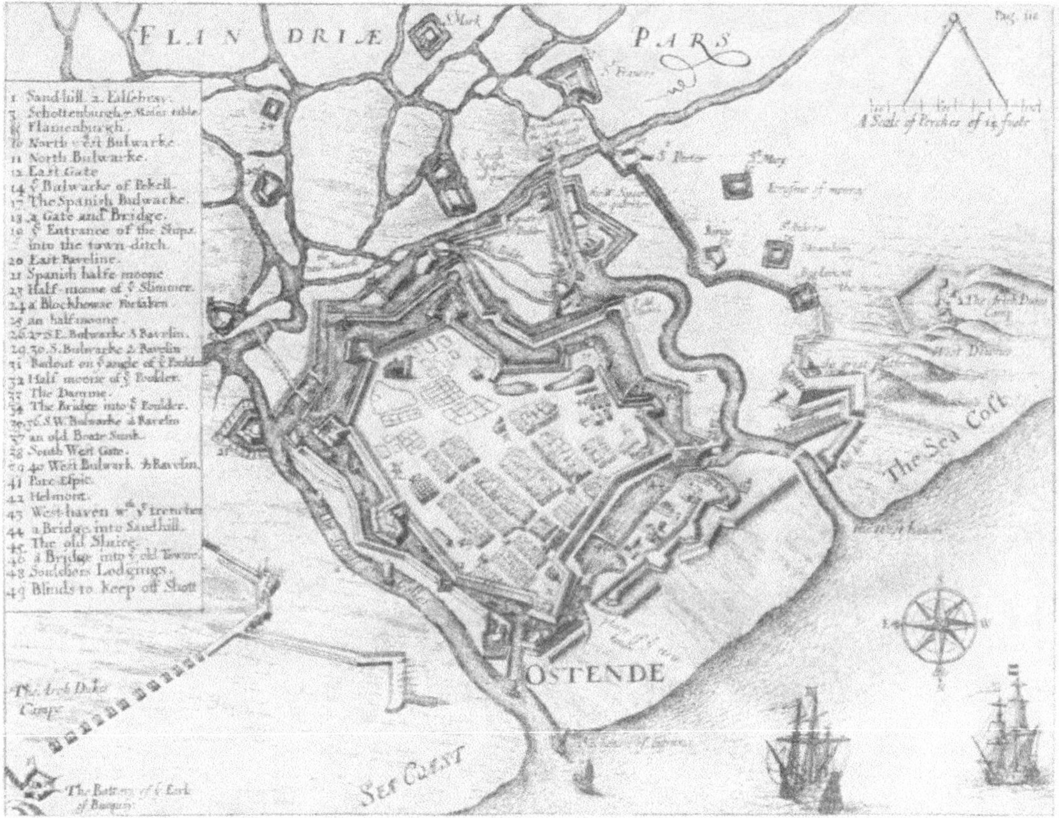

Ostend beleaguered, 1601–1604
From Vere's 'Commentaries'

and his cousins in the Archives of the House of Nassau (ed. Groen van Prinsteren, vol. ii.), may be gleaned from other documents and narratives. Something—though not so much as might have been hoped—may be got from Spanish documents. I found the orders of battle, the plans, and the lists of prisoners in Commelin's *Wilhelm en Mauritz van Nassau'* (Amsterdam, 1651) quite invaluable—though English names are much murdered therein—as indeed are Italian and French names also. There is useful stuff to be got out of Meteren and Bor. On some of the minor Dutch sources I have not been able to lay my hands. The English sources, outside Vere and the notes by Ogle and other captains, published along with Vere's own narrative in the *Commentaries*, are rather disappointing. On the controversy in the nineteenth century between Motley and Clements Markham, I can only say that the former seems to be too hard on Francis Vere, and the latter too insistent on the gallant veteran's infallibility and judicial fairness of mind. The charming picture of the battle, which I insert, comes from Vere's *Commentaries*. It deserves close study, showing every corps accurately, as well as the much disputed sandhill and the two-gun battery on the Dunes.

BOOK VII

THE TURKISH ATTACK ON CHRISTENDOM (1520–1606)

CHAPTER I

PRELIMINARY: THE PERSIAN WARS OF SELIM II (1514–15)—BATTLE OF TCHALDIRAN, AUG. 23, 1514

FROM most points of view the later years of the fifteenth century—the age of Louis XI, Richard III, and Rodrigo Borgia—are a most depressing period; all the inspirations of the Middle Ages—the Holy Roman Empire, the moral supremacy of the Papacy, the Crusading fervour—had passed away. And in this unhappy time there appeared to be in preparation an assault on Christendom at large, such as had not been seen for three centuries. It was an attack from the rear.

In the year 1480 Sultan Mahomet II, having crushed his eastern enemies and imposed a disastrous peace on the Republic of Venice, threw an army ashore on the heel of Italy, which sacked Otranto with circumstances of unusual atrocity, and established itself there. The Sultan announced that he was about to sail next year, with his main fleet and his Janissaries, to continue the invasion. Rumours flew round Italy that he had promised his veteran bands the sack of Rome. This would, indeed, have been a hazardous contract to carry out, for even the jealous and distracted states of Italy must have rallied to defend the Holy City. But it cannot be denied that the opportunity was one such as seldom occurred. Louis XI was making his grab at Flanders and the rest of the Burgundian heritage. Matthias of Hungary was just commencing his assault on the Emperor Frederic III, which was ultimately to put him in possession of Vienna and the Austrian borderlands. Ferdinand and Isabella in Spain were on the eve of opening their final attack on the Moorish kingdom of Granada. England and Scotland were at war—as so often before. A thrust at Rome by the Turk might have compelled all the grasping princes of Europe to put aside their ambitions for a defence of the heart of Christendom. But was this certain?

Fortunately for all parties concerned in the West, Mahomet died of a sudden apoplectic fit on May 3, 1481, just as he was preparing to start for a regular invasion of Italy, after making a preliminary clearance of the Knights of Rhodes, the outpost of the 'Franks' in the Aegean. Rhodes was saved for forty years by the Sultan's sudden death. Italy was never to see a serious lodgment made in its defences—though it was to suffer a transient scare by a Turkish landing in Apulia in 1537, and was to endure many a pirate raid from the Moslem corsairs in the sixteenth century. But these were not invasions but raids.

Chaos set in at Constantinople for the moment, when Mahomet died, and his great armament dispersed. So it came that King Ferrante of Naples was able to besiege and expel the intrusive garrison of Otranto. Credit is due to Ferdinand of Aragon, in that he spared some galleys to assist his cousin of Naples in getting rid of this hornets' nest. As King of Sicily he would have had to strain his resources to the uttermost, and to leave Granada alone, if the Sultan's army had come to Apulia in force.

Of the nine sultans of the House of Othman who reigned between 1360 and 1560, all save one were men of war from their youth up; by a strange—indeed providential—chance Bajazet II, the successor of a formidable father, was the one exception, at the moment when the invasion of Italy seemed the most probable of events. His reign of thirty-one years (1481–1512) covered the period of the great French enterprises in Italy, and the Ottoman power took no advantage whatever of the turmoil across the Adriatic. Bajazet was a strange freak in the bloodthirsty family tree—a poet and a philosopher who loved the arts of peace and hated the camp. In his long reign he seldom headed an army, and never committed himself to a pitched battle. His wars were generally forced upon him by the unruliness of his frontier governors, or by the attacks of his enemies. His reign commenced with civil strife—the rebellion of an ambitious brother—and ended with a forced abdication at the behest of an unscrupulous son. His armies were repeatedly beaten by those of the Mameluke Sultans of Egypt, and by the Hungarians; with both he was forced into unfavourable terms of peace. The only time at which it seemed likely that he might be drawn into the whirlpool of the great wars of Italy was when he engaged in a feeble and inconclusive strife with the Republic of Venice in 1498–1503, which, after much fighting, ended in his capture of the outlying Venetian ports in Greece—Coron, Modon, and Lepanto, but in his surrendering the Ionian island of Cephalonia, of which he had been in possession. As Venice still held Crete, and had recently acquired Cyprus, her power in the Eastern seas was not seriously damaged. It is astounding to find that when the republic was in the most serious danger at home, during the League of Cambray (1508), when the French and Imperial banners were seen on the Lagoons, Bajazet should have refrained from delivering the ' stab in the back ' which would have enabled him to get possession of Crete, the Ionian Islands, and the rest of the Venetian outlying possessions.

But so it was—and what is still more surprising, when Bajazet had been deposed in 1512, and the Sultanate had passed to his bloodthirsty and ambitious son, Selim II, Christendom was to be granted another eight years of respite, before the almost inevitable blow from the East was delivered. The Great Italian Wars were still in progress—Novara and Marignano were fought during Selim's reign—and the opportunity which had seemed so obvious in 1481 was almost as clear in 1512. All the powers of Europe were at war, apparently oblivious of the fact that there was any danger on the side of the Danube and the Adriatic: the thirty years of Bajazet's feeble reign had lulled them into security against a peril that had seemed very real in his father's time. But Selim was what would have been called in 1916 an ' Eastern Front,' not a ' Western Front ' man. He was obsessed with the opportunities that stood before him in Asia, and turned his attention away from Europe. From 1512 to 1519 he was busy in Armenia, Mesopotamia, Syria, and Egypt, and in seven years he doubled the territorial extent of the Ottoman Empire by a series of ruthless campaigns. But precisely in those seven years there was being built up behind him in Europe a power such as had never been seen since the days of Charlemagne—the Hapsburg empire of Charles V, which proved sufficient to turn back the great Ottoman assault, when it came at last in the days of Selim's son, Soliman the Magnificent. In 1512, when Selim went off to the East, Maximilian and Ferdinand of Aragon were both alive and intriguing against each other: their grandson and heir, Charles, was a boy of twelve, living under tutors in the Netherlands. When Selim in 1519 came back from Cairo, all the accumulated heritages of Spain, Austria, and Burgundy had fallen into the hands of a single ruler, and that ruler had just been elected emperor. Charles had reached the age of twenty when he was crowned at Aix-la-Chapelle, and had already shown in Spain that he was a personality of some importance. His realms were sometimes troublesome, but their joint strength was enormous—there was a formidable power waiting behind the buffer-state of Hungary, when the Turkish attack, so long postponed, was ultimately delivered.

Meanwhile Selim's eastern campaigns deserve careful study, as examples of the way in which an Ottoman army was marshalled and employed. When he had put down the rebellion of his brother Ahmed, and made an end of him and several other junior relatives, the Sultan found himself able to contemplate his two Eastern neighbours, the old Mameluke ruler of Syria and Egypt, and the new Shah of Persia, the founder of the Sefavi dynasty. Both of their states were in a condition to tempt the invader. The Mamelukes, though their horsemen had frequently beaten the armies of Bajazet II, were an alien aristocracy, ruling by the sword over millions of discontented fellahîn. Their numbers were limited, and their short-lived rulers generally came to unhappy ends at the hands of conspirators. There was no dynasty, but only a series of successful military adventurers at Cairo, surrounded by ambitious war-lords, each with an eye on the throne, when its occupant should have come to a natural or an unnatural end. On the other flank, the new Persian empire had only just come into existence under Shah Ismail, one of those sudden conquerors, like Tamerlane or Baber, who flash up from time to time in Oriental history. He represented a recrudescence of Persian national feeling against alien Turcoman masters, and of Shiah religious fanaticism against Sunni orthodoxy. There was little to the observer's eye to show that his rising would found a permanent and formidable dynasty. But, as has often been the case in Mohammedan countries, the religious propaganda was being exploited by the leader of the movement—the Shias outside his newly acquired dominions were on the move against their masters, hoping for a general overthrow of established governments. To annoy Sultan Selim there was a dangerous rising of these sectaries in the mountains of southern Asia Minor.

Probably the Ottoman Sultan regarded Shah Ismail as at once the most provocative and the least firmly established of his neighbours. At any rate, after an exchange of diplomatic documents mainly filled with theological abuse, and a systematic massacre of several thousand Shias resident inside his dominions,

Sultan Selim marched against the Persian. His objective was Tabriz, the Shah's newly chosen capital, and he had therefore to penetrate the Armenian mountains, which lie between the valley of the Upper Euphrates and that of the Aras, by a most high-lying and inhospitable route, but the only one available. His army was mobilized at Sivas, the most eastern city in his realm, on June 26, 1514, and was there counted as 140,000 men of all arms, regulars and irregulars—no doubt an exaggeration of the usual Oriental sort. Crossing the passes of the Anti-Taurus, the Turks descended into the valley of the Upper Euphrates by Erzingian and Erzeroum. The whole countryside was found wasted; the Shah had sent out a large force of cavalry, which destroyed the crops and burned the towns and villages. Selim's further progress would have been impossible if he had not taken the precaution of sending a convoy of corn-ships to Trebizond, whose contents were forwarded on mule-back and camel-back, over the hills on to the deserted Erzingian, in time to stop an incipient mutiny of the ever-unruly Janissaries.

After this Selim pushed on through a devastated country by eight marches to Erzerum, and once again by sixteen more, to the entrance of the plain of Tchaldiran on the Aras, having crossed with infinite trouble the watershed between the Euphrates and the upland plateau of Armenia. The army suffered more and more as the days drew on; many horses and baggage camels died in the blazing summer sun, and the ranks of the irregulars were terribly thinned. It was known, however, that the Shah had mobilized all his available forces at Khoi, seven marches in front of Tabriz, and intended to fight; he had held back in order to allow the invader to grow famished, while his own men and horses were well fed in a friendly country.

On descending into the plain of Tchaldiran, on August 23, Selim found the enemy waiting for him, but was allowed to draw up his host at leisure—the Shah intended to break it by a tempestuous general assault, not to start the game by skirmishing. He had won his numerous victories of the last ten years by desperate cavalry charges against enemies whose bands lacked discipline and order, and were, like his own, composed entirely of horsemen without guns or infantry. The Turkish army was of a very different composition, with a nucleus of trained infantry in the Janissaries, and a considerable power of field artillery, which had been dragged over the mountains with infinite trouble.

What were the forces engaged we cannot tell with any approach to accuracy; Italian consuls writing from Constantinople to Venice or Genoa give enormous figures; Turkish historians are always hyperbolical. We need not believe that Selim had 120,000 men of all sorts, or that Shah Ismail owned 80,000 cavalry—reduced by one authority to the much more credible total of 30,000. But the armies were undoubtedly large. The Persian host was composed of bands of those horse-archers and lancers whose likenesses are given in so many sixteenth-century manuscripts. They had the plumed helmet with camail hanging to cover the checks, the chain mail or banded-plate cuirasses, now to be seen in a hundred European museums, with greaves and brassards, and they rode on lightly barded horses. They were trained to rapid movement, and were famous for their archery, being wont, like the Tartars of old, to discharge a formidable arrow flight before charging in with the lance. The Shah practised the familiar tactics of the East, an enveloping horn-shaped attack, intended to outflank the enemy, and get in behind the extremes of his line, before any central assault was delivered.

Selim, on the other hand, was fighting in the more close array which his ancestors had learnt in European battles, in two lines, with a central nucleus of infantry and guns in the second. The first line was a screen of irregular horse and infantry; in the second, the feudal troops, the Timariots of Asia and Europe, formed the right and left wings respectively, while in the centre was the Sultan with his bodyguard of Spahis and his mass of Janissary foot-archers mixed with whom, as we learn, rather to our surprise, there were arquebusiers.[1] These last had their flanks covered by carts and chains (like the Hussites), and guns in their front roped together in batteries, wheel to wheel, with small intervals.

Shah Ismail had his horsemen divided into two corps, each intended to outflank one of the Turkish wings—the centre held back, the two 'horns' pushed forward. He himself led the right corps, Oustadjioglu, the Governor of Diarbekir, the left. The first charge swept away the Turkish screen of light

[1] Apparently the Turks were already taking up the use of the smaller firearms, like the Spaniards and the Bohemians, while the French and English were behind the times. Note the arquebusier in the annexed illustration.

horse and irregular foot without much difficulty, but then serious fighting began—the Shah himself broke the European timariot horse, and flung them back upon the flank of the Janissaries with terrible loss—the 'Beglerbeg of Roumelia,' *i.e.* the commander-in-chief of the European corps, was killed, and along with him the 'sandjak-beys' of Sofia, Silistria, and Prizrend — divisional generals (so to speak) of the Roumelian Timariots. The victorious Persians, however, failed to break into the Janissaries, entrenched behind their carts and camels, and Sultan Selim's horse-guards held their own. On the other wing the Persian left-hand corps came to desperate but unsuccessful fighting with the Asiatic horse of the Turkish army, and, being shoved gradually to their right, came under the gun-fire of the enemy's batteries in the centre. They are said to have cut up some Turkish irregular infantry, and to have got among the guns at one point. But the Sultan's Asiatic horse had the advantage over them, drove them off, and finally were able to close in upon the Shah's own cavalry, which was still engaged with the Janissaries in the Turkish centre. At this moment Ismail was wounded in the arm by a cannon-ball (or, by some accounts, by an arquebus-ball—some of the Janissaries being already possessed of firearms),[1] and when he was carried off the field, all his troops drew away in disorder and galloped to the rear. The Turkish cavalry pursued slowly—we are told that their horses were half-starved and out of condition—but the Persians made no halt and fled along the road to Khoi.

The Shah's camp fell into the hands of the victors, who got there not only great spoil in the way of valuable goods, but, what was more important, ample supplies of provisions. The wounded Shah halted for a moment at Tabriz, and bade the governor to make no attempt to hold the city, but to send off the transportable royal magazines eastward, and to bid the chief inhabitants migrate with their more valuable possessions. Hence when Selim entered the place, after nine short marches from the battlefield, he met with no resistance, and not so much plunder as might have been hoped, though it is said that he found some elephants—trophies of Eastern victories of the Shah's former campaigns. But the spoils of the camp had been more valuable than those of the city (September 5, 1514). A

[1] As in the illustration from the metrical life of Shah Ismail in the British Museum where the Janissary arquebusier appears in the forefront.

war-contribution was levied on such of the inhabitants as had not fled—there was no massacre.

The Persian losses at Tchaldiran are said to have been much smaller than those of the victorious Turks, one of whose wings had been cut to pieces, while the irregulars, horse and foot, had been absolutely massacred. However, an immense number of the Shah's best officers had fallen. Oustadjioglu, the second in command, was dead, along with the Vizier Abdulbahi, and the governors of Bagdad, Meshed, Hamadan, Khorassan, and Moghan. But a beaten cavalry army disperses and gets off easily; beaten infantry (as in the Italian and Huguenot wars) is marked for slaughter.

Selim tarried only a week in Tabriz; he wished to establish himself there and to spend the winter in Armenia. But September had arrived; the feudal horse were resolved not to risk the rigours of highland snows, but to get home, and the Janissaries went into open mutiny. So the Sultan was forced to depart, taking with him some plunder of royal upholstery and precious stuffs, and a thousand artisans skilled in metal work and embroidery. The army went home by a different route from that by which it had advanced, in order to find undevastated country; going by Erivan and Kars, Selim dismissed the feudal cavalry when he had passed his own old border (October 23), and established himself for the winter at Amasia in Pontus with his Janissaries, cantoning the Spahis in Angora. They did not reach their resting-place till the middle of November, after nearly seven weeks' marching from Tabriz, so great were the difficulties of the road.

The battle of Tchaldiran had left in the hands of Sultan Selim the whole valley of the Upper Euphrates from Erzingian to Erzeroum, but all the remoter regions which he had crossed in Armenia and Adherbaijan were immediately reoccupied by the Persians, and the Shah returned to Tabriz. In 1515 the Sultan made no attempt to repeat his adventurous march against the enemy's capital, but set himself to extend his power in the lands of the Middle Euphrates and the Kurdish mountains, where political boundaries were at the moment very uncertain. The state of Sulkadir, held by a prince who ostensibly considered himself as a vassal of the Mameluke Sultan, was overrun in May and June by an army commanded by Selim himself, its last lord was slain in battle, and his towns of Albistan, Malatia, Marash, and Kharput taken and garrisoned. This was a challenge to the Mameluke, who paid little attention to it, since his authority had really never extended into these mountains. Selim was recalled to Constantinople in June by an immense conflagration, which destroyed great part of the city, and called for important reconstruction, especially on the side of the naval arsenals. But his generals pushed on from the conquered Sulkadir into the Kurdish highlands, where some of the chiefs had done homage to Shah Ismail, while others were maintaining an age-long but very precarious independence. The anti-Persian party offered its not very trustworthy allegiance to the Ottoman Sultan, and put Bitlis and Diarbekir into his hands. There followed much confused fighting with the Shah's troops and the section of the Kurds who adhered to them. After a pitched battle in front of Mardin, won by Khosrew Pacha, the 'beglerbeg' of Anatolia, the Persians had to withdraw eastward, and all the western side of Kurdistan passed, at least nominally, under the supremacy of the Ottoman power, as far as Nisibis, Miarfarkin (the site of a very old emirate), and Djezirat-ibn-Omar on the Tigris. The lower valleys of the Euphrates and Tigris remained in the hands of the Persians, the boundary being drawn (as in the old days of the Roman and Sassanian empires) at the desert of central Mesopotamia. The Bagdad country was not to pass into the hands of the Turks till another generation had passed, and Selim's greater son, Soliman the Magnificent, extended his boundaries all down Irak to the sea.

NOTE ON TCHALDIRAN

Details of the battle of Tchaldiran are not wanting, but are most untrustworthy both from the Turkish and the Persian side, and still more so from the dispatches written by Italian consuls from Constantinople and elsewhere. The Persian chroniclers, praising Shah Ismail officially, hardly account for the loss of the battle; the Turkish chroniclers disguise untoward incidents. The one thing certain is that both sides allow that firearms settled the event of the day; but it is difficult to discover how exactly the Turkish guns were placed, or what precisely was the artificial cover under which the Janissaries fought. Persian pictures of the battle exist—emphasizing, of course, the brilliance of Shah Ismail's first charge. In the illustration given, it will be noticed that the Persian horse all wear tufted helmets and cuirasses, and are mostly armed with the bow. A Janissary with an arquebus is evident at the foot of the picture. The MS. from which the scene comes is a metrical encomium by a court poet, drawn up just before Shah Ismail's death, and full of illustrations of his victories. Tchaldiran appears among them!

THE BATTLE OF TCHALDIRAN, AUG. 23, 1514

A Persian version of the battle. Shah Ismail (in centre) kills the Beglerbeg of Roumelia: in front, a janissary with an arquebus shoots a Persian horse-archer. This is the first representation of firearms used by the Turks

From a contemporary Persian MS. in British Museum

CHAPTER II

THE TURKISH CONQUEST OF SYRIA AND EGYPT—BATTLES OF DABIK, AUGUST 24, 1516, AND RIDANIEH, JANUARY 12, 1517

WHILE Selim spent the winter of 1515–16 and the following spring in and about Constantinople, building new naval arsenals, and punishing the Janissaries by a series of executions for their repeated mutinies during the Persian campaign, the Christian powers were a little uneasy as to his future intentions. But he renewed the standing truce with Hungary, in spite of some border raids on both sides which had taken place in 1515, and wrote friendly letters to the Doge of Venice. His whole attention was still fixed on the East, where he knew that he had defied the Mameluke Sultan by annexing the Emirate of Sulkadir, and extending his borders down the Euphrates. Kansou-Ghori was reputed to have sent troops to the northern borders of Syria, and was certainly giving ostentatious shelter to Aladdin, the son of Selim's rebel brother Ahmed, who had taken refuge at Cairo. In the spring of 1516 the grand-vizier, Sinan Pacha, was ordered to mobilize the feudal cavalry of Anatolia at Caesarea in Cappadocia, and then to move them to Marash on the Euphrates, opposite the frontier of the Mamelukes. This showed a clear intention of war, also of the fact that Selim did not propose to attack his neighbour on the Cilician side, through the passes of Taurus, where his father's armies had suffered so many disasters. He was about to outflank Cilicia, by coming down on Syria by the inland route, Marash–Aintab–Aleppo. In June he sent a defiant message, equivalent to a declaration of war, to Kansou-Ghori and started out from Constantinople with his Janissaries and Spahis, with the European feudal horse coming on in his rear.

The Mameluke Sultan, not unwarned, had come up to Aleppo with his best troops, in time to receive there the ambassadors of Selim, to whom he returned pacific messages. But Selim, who was set on open war, treated the bearer of the Egyptian king's letters with contumely, sending him back, with his beard shaved, on a mangy ass, and with a night-cap on his head—indeed he barely refrained from ordering him to be beheaded (August 9, 1916).

Kansou-Ghori, who had reigned for the term (almost unprecedented among Mameluke rulers) of sixteen years, was now aged eighty, and troubled with hernia—a most unhappy commander for the field. But being surrounded by war-lords who were eagerly waiting his decease, he distrusted every one, and headed his own army, for he rightly suspected his lieutenants. His suspicions were often right, for Khair Bey, the Governor of Aleppo, was in secret communication with the Turkish camp, and had sent the Sultan useful topographical information as to the roads of Syria.

The Mameluke army consisted, as of yore, in masses of fully armed cavalry, more lancers than bowmen; they resembled the Persians, with whom Selim had been dealing in 1514, in all respects, save that they wore not the plumed helmet with camail, but immense turbans of many folds, sometimes with steel chains twisted inside them, such as may be seen in Gentile Bellini's famous pictures of the Egyptian court. This head-dress was as good a protection from sword-cuts as any helmet. Mail-shirts were universally employed. The Mamelukes were all trained men-at-arms from their youth up, some were of the old military families which had served the 'Bahri' sultans in the fourteenth century, more of newly-imported Circassian blood, whom the later 'Burji' sultans had sedulously brought in. All were theoretically royal slaves, and many had actually been purchased originally from the Caucasus. There was an inferior caste of this soldiery, the 'Korsans,' who were auxiliaries of mixed nationalities, including all sorts of Asiatic adventurers and even Abyssinians. The Mamelukes individually were much superior as fighting men to the ordinary Turkish horsemen, and better armed.[1] But their numbers were limited, and local Arab levies from Syria and Egypt, whom their Sultans often brought into the field, were inferior in quality and untrustworthy in the matter of loyalty. There was no infantry whatever, and no provision of guns—which were

[1] Oddly enough, Napoleon in 1799 made the same remark about the Mamelukes of his day and the French cavalry of the line. Five Mamelukes, he said, would in close combat beat six or seven dragoons. But, as they lacked drill and order, a French squadron of horse would discomfit 500 Mamelukes.

considered an encumbrance to a lightly moving cavalry army. A Mameluke bey, after the disasters of Dabik and Ridanieh, told the tale of how a Moorish specialist had tried to get Sultan Malik Ashraf, the predecessor of Kansou-Ghori, to purchase cannon and balls on a large scale from the Venetians, who already sold much armour and weapons at Alexandria. But the old Sultan replied that Mahomet himself had described the sword, lance, and bow as the only proper weapons for the True Believers, and that guns were the tools of cowards. To which the Moor replied, 'Those who live long enough will see this kingdom destroyed by the cannon-ball.'[1] The story, however apochryphal, is a true representation of Mameluke psychology.

As was natural in what was really an elective military despotism, where a father very seldom succeeded in leaving his throne to a son, rivalry between eager claimants for the succession was the weak point of the system. Each prominent bey tried to secure the largest following among the minor members of this queer aristocracy: each intrigued against possible competitors, trying to prejudice the reigning sultan against them. When the actual occupant of the throne died or was murdered, civil strife ensued, and several heads might fall before any bey established himself for a few years of power. Since the 'Burji' Mameluke régime began in 1392, there had been twenty-five sultans proclaimed at Cairo! Kansou-Ghori had risen to power in 1501 on the murder of a predecessor who had reigned only three months. He had two sons, the children of his old age, neither of them more than twenty years of age: but a much more likely successor was his nephew, Touman Bey, a great man-at-arms, whom he had made one of his chief ministers, in the hope of keeping the succession in his family at least.

Selim having outflanked the usual lines of defence of the Mameluke state, Taurus and Amanus, had presented himself in the plain of northern Syria, and was met at Merj-Dabik, ten miles north of Aleppo, by the old Sultan's army, in open ground with no local advantage to either side, on August 24, 1516.[2] The Turkish force is said—and the figures for once seem possible —to have consisted of 8000 Janissaries, of whom a considerable number were armed with arquebuses, 3000 'Spahis' of the

[1] This tale is repeated by two Turkish chroniclers of Selim's reign—the words are ascribed to the captive Kourt Bey.

[2] Some chroniclers observed that it was exactly two years since Tchaldiran; but they were out by one day.

royal horse-guard, 15,000 feudal cavalry, European and Asiatic, and 50 guns, with the usual addition of rather worthless irregulars, *Akindjis* and *Azabs*, who always followed a sultan on the chance of plunder. Kansou-Ghori is said to have had with him 13,000 chosen Mamelukes, all tried men-at-arms in mail, and a certain number of Arab horse furnished by the Syrian tribal and local chiefs, of very uncertain fidelity. As at Tchaldiran, the Turkish army was drawn up on a purely defensive scheme, awaiting the attack that an enemy whose force lay in shock-cavalry would probably deliver. And in the usual line of battle the Asiatic horse took the right wing, the European horse the left, the Janissaries were massed in the centre, with in front of them a line of guns, their flanks protected by carts roped together, and the Sultan's Spahis in reserve.

The Mameluke was indeed intending to take the offensive—he had the highest confidence in the fighting value of his men—more in this case than some of them deserved. He had formed his army in four corps—a right wing under Khair Bey, governor of Aleppo, a left wing under Sibei Bey, governor of Damascus—these were both intended to outflank the Turkish wings and charge them diagonally. Much drawn back was a central corps under Berdi-Ghazali, which was not to move till the wing-attacks were developed, and behind this Kansou-Ghori held his own personal guards as general reserve.

The attack was imperfectly delivered. Sibei Bey succeeded in turning the flank of the Asiatic feudal horse, and drove them in with loss upon the side of the Janissaries. But Khair Bey, though he wheeled round the other Turkish wing, did not attack the European feudal horse with vigour, but skirmished with them, while part of his men got into Selim's tents, and cut up the camp followers. This holding back, we are told, was the result of deliberate treachery. Meanwhile the midmost Mameluke corps, seeing the day undecided, rode in at last on the Turkish centre, and received the discharge of Selim's artillery, whereupon they broke, and came back in disorder on the Sultan's reserve, which they carried away with them in rout.[1] The aged Kansou-Ghori was thrown from his horse, and trampled to death according to one account; another story says that he died of an apoplectic fit brought on by rage

[1] There is an elaborate story in some of the chronicles that the *Djelbans*, or Mamelukes of the second class, were indignant at seeing the *Korsans*, or Mamelukes of the third class, preferred to them in the order of battle, and took such offence that they left the field in unjustifiable haste.

and dismay.[1] Seeing the centre broken, the Mameluke wings gave up the fight, and rode off, taking little harm—according to one account they had lost only 1000 men in all. Along with the Sultan there fell Aladdin, Selim's fugitive nephew, and the commander-in-chief of the Mameluke guards—but few other persons of importance. The Turks had serious casualties among the Asiatic feudal horse only.

The leaderless Mameluke army made no attempt to hold the field in Syria. The traitor Khair Bey surrendered Aleppo, and openly joined the Turks; the other emirs, after a hasty retreat to Damascus, failed to agree on the election of a new Sultan, and led their troops back to Cairo, leaving Syria to shift for itself. Selim marched up the Orontes valley, receiving the surrender of city after city without opposition, and then established himself at Damascus. The tribal chiefs of Syria and the Druses of the eastern mountains all came in to offer their allegiance. The Mameluke régime had everywhere been detested. Selim remained at Damascus, reorganizing the administration of the conquered lands, till December 16, but sent on ahead of him a detachment under his vizier, Sinan Pacha, to deal with Gaza, the only town which showed any intention of resistance.

Meanwhile his officers were collecting camels from all quarters, for use during the passage of the desert between Syria and Egypt, where he knew that the question of water and supplies would be difficult.

The general assembly of the Mameluke Beys at Cairo ended by electing the late Sultan's nephew, Touman Bey, as his successor, but only after much quarrelling. There were many competitors, and some who dare not present themselves as candidates tried to nominate Kansou-Ghori's son Mohammed—a boy whose accession would have meant anarchy. Touman's election provoked much jealousy and even—as was to be seen later—treason. But he was obviously the best choice under the circumstances. The moment that he was enthroned he sent a strong corps forward to Gaza, to keep the Turks from establishing a base on the edge of the desert. This force fought a lively engagement with Selim's grand vizier, who had been

[1] There are various stories about Kansou's end: one is that after his death his followers cut off his head, in order that his corpse might not be recognized. Another is that he was beheaded by a Turkish soldier, who found him dead, and brought his trophy to Sultan Selim, who expressed disgust, and nearly had the man cut down on the spot. He gave Kansou's body honourable burial.

sent to the same place for precisely that purpose. The two detachments met at Younis Khan, near Gaza, and the Mamelukes were once more beaten, entirely (we are told) by force of artillery (October 28, 1516). The day ended with a general massacre of the inhabitants of Gaza, who had joined their old masters.

Sultan Touman had expected that he would have the winter before him to organize the defence of Egypt, and that the difficulties of the desert-route from Gaza to Pelusium would compel the Turks to make long preparations for the move. But Selim had determined to strike at once, before the enemy should have had time to make his preparations, and to cross the desert before he was expected. Leaving Damascus on December 16, he made a flying visit to Jerusalem, to worship in the Mosque of Omar, and to Hebron to admire the tombs of the Patriarchs, but was at Gaza toward the last days of the year, where he found that the Grand Vizier had made a very successful gathering of camels and food for the expedition. He resolved to start at once—and discouraged delay by beheading in summary fashion his second vizier, Hussein Pasha, for making gloomy reports about the dangers of the route and the ferocity of the Bedouin.

The passage of the desert by the old caravan route along the coast from Gaza and El-Arish was made in only eight days of marching—we are told that it proved unexpectedly easy, because heavy rain filled all the wadys and wells, and the camels' water-bags were never exhausted. The only military inconvenience came from desultory attacks by the Bedouin of Sinai, who cut off a few stragglers, but were never a real danger to columns marching well closed up. No Mamelukes were seen, and the Sultan reached Salahieh, the first town in Egypt, on January 16, practically unopposed. From thence he started to march directly on Cairo, leaving the marshes and canals of the Delta on his right hand, and keeping to the dry and open ground.

Sultan Touman had taken to heart the character of the recent victories of the Turks, who had won by fighting defensive battles, and allowing their enemies to assail them by cavalry charges. He had resolved to adopt new tactics, to entrench himself behind fortified lines, and to force the enemy to attack. With this object he had thrown up earthworks five or six miles long, across the great pilgrimage-road which goes out of Cairo to the north-east at the village of Matarieh, with the great Nile-canal on his left and the hills of the Jebel Mokattam at some distance to his rear, near the ruins of the ancient Heliopolis. He had scraped together all the artillery that he could collect, ship guns hastily purchased from the Venetians at Alexandria, and some others brought from the walls of Alexandria, Rosetta, Damietta, and other fortified places—for the Mamelukes, though they had despised field artillery, had not ignored its use for seacoast walled strongholds. These guns were placed in redoubts, commanding the high road, and certain gaps left in the lines, behind which divisions of Mameluke cavalry were placed, ready to charge an enemy who should break through the entrenchments. Touman is also said to have laid traps out in front of the lines—deep holes dug in accessible paths, covered with reeds, and then disguised with a light coating of earth—in short, *trous de loup*, in the style of Robert Bruce's 'pottes' at Bannockburn.

But, of course, such a position is only valuable if the adversary is obliging enough to make the desired frontal attack. Selim and his army, marching from Salahieh via Belbeis, worried all the way by light attacks of Arab raiders, reached El-Kankah in front of the lines on January 21, 1517. Here, according to the chroniclers, he was warned of the formidable nature of the works opposite him by traitors in Touman's camp, and advised to turn them by a flank march to his left hand, beyond the end of the trenches. But it seems hardly necessary to presuppose such advice, since any general who has been operating by the method of taking up ground and allowing himself to be attacked, would naturally think twice before assaulting fortifications with cavalry, and would prefer to manœuvre, so as to compel the enemy to come out against him.

At any rate this is what Selim did. On the night of the 21st his whole army turned to the left, marched a distance of several miles round the eastern end of the lines, and was discovered in order of battle on the morning of the 22nd by the village of Ridanieh, on the edge of the desert in open ground. Touman, whose Arab scouts must have been negligent, was thus forced to make a sudden 'front-to-flank' manœuvre, and to draw up in a hasty line facing to the east instead of north-east, and with his carefully prepared earthworks in his rear, instead of in his front. We are told that he made desperate endeavours to drag some of his artillery out of the redoubts into the new line which he was forming, but that the guns were so heavy, being destitute of wheels, and intended to be fired from platforms, that only part of them ever got into action, and those not at points where they were required for the best service.

The battle commenced with a cannonade, in which the Turkish artillery had from the first such a mastery over the few and badly placed Mameluke guns,[1] that Touman, despite of his intentions, was forced to assume the offensive, since his cavalry was suffering serious casualties, and could not stand still to be shot at. He detached his numerous, but not very effective, Arab bands to get round the flanks and rear of the Turkish line, and do what they could to distract the enemy, but made the frontal attack with his own Mameluke squadrons. They were in three corps, he himself in the centre, with the right wing led by Berdi-Ghazali Bey, and the left by the Minister of War (if one may use the term for an Emir-el-Kebir), Helmis Bey.

Sultan Selim's army was drawn up in the usual fashion—Asiatic feudal cavalry on the right wing, European on the left: in the centre, behind the guns, the Janissaries and the Spahis of the horse guard. But, by way of exception, there was a reserve under the Grand-Vizier, Sinan Pasha, consisting of picked men drawn from both wings of the horse and 500 Janissaries. The Mameluke right wing broke the European feudal cavalry, who are said to have been distracted by the Arabs threatening their rear, and then dealt with the reserve, under the Grand-Vizier, who came up to stop the gap. Sinan was killed, and his troops much cut up—but this did not settle the fate of the day. For on the other flank Mustafa Pasha, with the Asiatic feudal cavalry, beat most effectively and chased off the Mameluke left wing—the Emir El-Kebir was mortally wounded.

[1] The Turkish artillery is said to have been managed by an Italian renegade, one Giacomo of Reggio, a specialist who 'overcome by great rewards revolted into the Mahometan superstition.' I cannot refrain from quoting Knolles (i. p. 366), though he is so late an authority, for the passage where he gives such a detailed and vivid description of 'how all the great Egyptian ordnance, as the case required, was turned the contrary way, a thing hastily and disorderly done. For the concourse of people for turning and removing the great artillery, most troubled the ordering of the rest. For many of these were huge iron pieces of great weight, made fast to stocks of wood by rings and iron clasps, after the old and rude manner of ship-ordnance, which for their exceeding and ponderous weight could not be removed save by the strength of many horses and many men, with levers and rollers put under them. And such as were on carriages were drawn through the camp, with great clamour of disordered and hasty people, some drawing, some thrusting them, with tumultuous stir, wonderously troubling the array.'

In the centre there was much hard fighting. Touman charged home with his bodyguard, despite of the gunfire; it is said that he made for the Sultan's standard, determined to settle the day by disposing of his rival—but though his riders cut in deep among the Janissaries, they failed either to reach the Sultan or to break the line. After a desperate mêlée he was forced to give the signal for retreat, being beset in flank and rear by the victorious wing of the Ottoman army—why his own victorious wing did not counterbalance this disadvantage does not appear.

The fighting had been very bloody—a fifth of the Turkish army had fallen, and a third of the Mamelukes.[1] The tents, stores, and artillery of the vanquished fell into the hands of the victors. Sultan Selim finished the day by killing his prisoners, including the wounded Emir-el-Kebir. Touman did not fall back at once into Cairo, which was not regularly fortified, but rallied his broken host beyond its southern suburbs. Selim, on the other hand, only sent into the city a detachment, to take formal possession of it. Noting this, Touman, on January 29, a week after the battle, made a surprise-attack on his capital, and cut to pieces the Turkish garrison. He then barricaded the streets, armed a miscellaneous multitude, including 8000 Soudanese slaves and such of the civil population as would consent to fight, and stood on the defensive. The Turks in four days of desperate street-warfare succeeded in recapturing Cairo; a general massacre followed—the chroniclers say that 50,000 corpses were seen in the streets (February 4). One is reminded of Bonaparte's similar experiences in the 'Cairo Insurrection' of 1799.

Touman escaped with a few thousand followers beyond the Nile to Gizeh and the neighbourhood of the Pyramids, where he rallied outlying detachments—among them the garrison of Alexandria. But many of the Mamelukes and most of the Arabs had given up the game, and surrendered on terms to the Sultan—among them Berdi-Ghazali, who had commanded the right wing at Ridanieh.[2] After much fighting and one

[1] As an example of the hopeless incapacity of Oriental historians to deal with figures, one chronicler says that Touman only had 13,000 men in line, another that 25,000 Mamelukes were left on the field!

[2] Some Turkish chroniclers say that Berdi-Ghazali, who had been a friend of Khair Bey, was scheming to desert his Sultan even before the battle of Ridanieh, and that Touman had been considering his removal from command on suspicion. This seems doubtful when we remember that Berdi-Ghazali's wing at the battle

final defeat the unfortunate Touman fled almost alone into the marshes of the Delta, where he was betrayed by an Arab chief and sent in bonds to Cairo (March 26). Selim at first affected to treat him with magnanimity, expressing a desire to show him as a captive in Constantinople. But on the advice, it is said, of Mameluke renegades, who declared that Egypt would never be quiet while Touman lived, he changed his resolve, and caused this hard-fighting prince to be hanged publicly at the Souweila gate of Cairo (April 13, 1517). So ended the Mameluke dynasty.

This being a military history, it is not necessary to give any account of Selim's reorganization of Egypt and Syria, provinces which brought to the Ottoman empire much wealth but little fighting strength, for the new régime was as alien and unpopular as that which it replaced. It may be worth while, however, to point out that the conquest of Egypt had one important effect in Europe. It was a fatal blow to Venice, who had hitherto kept on the most friendly terms with the Mameluke Sultans,[1] and by paying them very handsome transit-dues had profited by a monopoly of the Indian trade, which used to come up the Red Sea to Suez, and then pass overland to Alexandria, the great central market of the eastern Mediterranean. This route was now blocked, and the Indian merchandise had to take the longer but safe sea-route round the Cape of Good Hope, which Vasco da Gama had opened up only twenty-five years before. Alexandria dwindled down to the decayed city of ruins which Bonaparte discovered in 1798, when next a European fleet came in force to the mouths of the Nile.

routed the Turks opposed to him. There was no question of hanging back, after the fashion of Khair Bey at Dabik.

[1] For Mameluke costume the intelligent observer should study the picture of Gentile Bellini in the Louvre. Having accompanied a Venetian embassy to the court of Malik-el-Ashraf, the painter had the opportunity of studying and reproducing a formal ceremony, where the old Sultan, in an immense white turban with ostrich-plumes, surrounded by Circassian guards in fur bonnets, listens to the oration of the Venetian, who stands before in a flowing red silk gown. An even larger, but unhistorical, picture of Gentile's in the Brera gallery, representing St. Mark preaching at Alexandria, is also full of Mameluke magnates and ladies in queer yashmaks.

CHAPTER III

SULTAN SOLIMAN'S FIRST INVASION OF HUNGARY (1521–22)

SELIM the Terrible returned to Constantinople in August 1518, leaving behind him the newly organized provinces of the East, where he had tarried for nearly eighteen months after the battle of Ridanieh. His reappearance in Europe caused no small perturbation : after his astounding victories in Armenia, Syria, and Egypt he seemed capable of any enterprise. Pope Leo X sent cardinals to France, England, Spain, and Germany, to implore their sovereigns to patch up their disputes and contemplate the possibility of a westward push by the Sultan, corresponding to his recent eastward push. But Selim made no signs of such intentions—he treated with courtesy Spanish and Hungarian ambassadors who came to compliment him, and even received a Venetian envoy with placidity and without threats. It seems that in 1518–19 he was mainly engrossed in the suppression of certain fanatical outbreaks in Asia, where there had flared up one of those Mahdi-insurrections familiar to all students of Oriental history. There were troubles both in Anatolia and Syria, which were only suppressed by large armies and general massacres of the sectaries. The only ominous thing in European eyes was a continual development of the naval arsenals of Constantinople—it was reported, and truly, that this meant an attack on the Knights of Rhodes—but it might mean much more.

The fact, however, was that the Sultan was stricken with cancer—a fact which may account for his frequent and fiendish outbreaks of cruelty, which cost the lives of several viziers. He died, after a long and painful illness, on September 22, 1520, aged only fifty-four, leaving all his neighbours to ponder on the qualities of his only son and successor, Soliman, now a young man of twenty-five, who had been governing a province in Asia Minor for the last few years, in the character of a very obedient son, never provoking his father's irascible temper. No one had fathomed his character—he might possibly be as incompetent an heir to his formidable father as Bajazet II had been to Mahomet II, whose exploits had foreshadowed those of Selim.

Soliman, as events were soon to prove, was no weakling, but the most talented member of his house, which had produced before him many conquerors, but no prince with capacities as great for peace as for war. His record is mainly remembered for his conquests, but he was equally notable as an organizer and a legislator—he fixed the form of the Turkish Empire. Its long survival after his death was in a great measure the result of his work, which it took many generations of decadent heirs to undo. But the main interest of his life for students of European history is that he delivered the last great attack on Christendom from the East. It failed, but only after a protracted struggle with his great contemporary the Emperor Charles V, which ranged not only over the lands of the Danube, but over the whole basin of the Mediterranean from Algiers and Nice to the Aegean. Soliman could not help having Asiatic troubles—his father had left him an unsettled problem, with regard to the new Persian Empire, which had grown up behind the Ottoman state. But his main interest lay in Europe, and he died urging on the last of his many campaigns against Christendom, in his camp on Christian soil, after forty years of battering at the defences of the West.

The first military operations of Soliman were directed to the breaking down of the block of important fortresses by which entry into Hungary was denied to an invader from the south. The Turks had long been in possession of the line of the lower Danube from Semendria (Smederevo) to its mouth; and Wallachia, on the farther bank, had been for some time a tributary principality, though not infrequently it rebelled when it could get Hungarian assistance. At this moment such aid was far from forthcoming, since Hungary, as we shall see, was in a perilous state. But west of Semendria, which was the Turkish headquarters on the Danube, the Christian power had two strong fortresses on the south side of the river-boundary: the more important was the high-lying Belgrade, at the junction of the Danube and Save, which had endured and beaten off repeated assaults of the Sultans, while across the water the secondary stronghold of Semlin, covered Belgrade from the rear. Forty miles farther west Shabatz, on the south bank of the Save, was the second formidable stronghold which blocked the way into the plains of Hungary: from thence the boundary turned south into the mountainous region between Bosnia and Croatia. Here the Hungarians held three important towns, Jaicze (one of the old capitals of the extinct Bosnian kingdom), Banjaluka, and Knin, close to the Adriatic and the Venetian possessions in Dalmatia.

It was obvious that a decisive blow against Hungary would have to be made in the centre, opposite Belgrade, not on either flank, since operations against Croatia on the one side, or against Transylvania (after crossing Wallachia), on the other, would be aimed at secondary objectives, and would get into very difficult mountain country, where Turkish operations had often failed before. For in the fifteenth century the raids of the earlier Sultans had often penetrated far in both directions,[1] but had always ended in retreat and sometimes in disaster. The front door of Hungary lies at the point where the Danube cuts its way between the Carpathians and the Bosnian Mountains. Belgrade was a sort of 'barbican' or *tête du pont*, defensive in one aspect, but a sally port into the Balkan peninsula in another. It was from thence that Louis the Great and John Huniades in earlier days had made their advances into the Serbian lands.

The kingdom of Hungary ought, from historical precedents, to have been well able to defend itself. It was over a century since the Turks had first appeared upon the Danube, and they had failed invariably in attempts to make a lodgment beyond it. Christendom had come to look upon the Magyar realm as a perfect buffer-state against the Moslem, and so it had been in the days of Hunjadi Janos and Mathias Corvinus. Moreover, in 1521 Hungary had been for thirty years joined with the powerful and warlike kingdom of Bohemia, ever since, in 1490, Ladislas of Bohemia had been elected to wear the crown of St. Stephen. On the face of things the strength of the defensive line along the Danube ought to have been much greater than in the fifteenth century.

Unfortunately reality did not at this moment correspond to appearance. Hungary and Bohemia were by no means

[1] Three or four times Turkish raiding bands had cut right across Croatia, and had actually been seen in Venetian territory on the side of Friuli and Udine. But they had never made any lodgment in Croatia, though they had so often molested it.

friendly to each other—the chance that they had a common king did not guarantee common action. And the two sovereigns of this period had been unhappy experiments—Ladislas was a *fainéant*, who had let slip the authority which his predecessors in both kingdoms had maintained by their personal ascendancy. He was the slave of his nobles, because in Hungary he had disbanded the famous 'black bands' of mercenaries, who had been the guarantee of the dominance of Mathias Corvinus, while he was not in Bohemia the head of a hard-fighting religious sect, as the last king, George Podiebrad, had been. In both his kingdoms he was an alien, being a stranger from Poland. His feudal diets had stripped him of the estates and revenues that the crown used to enjoy, and then expected him to carry on the business of the realm. Faction-fights between leagues of nobles had become endemic, and, perhaps in consequence of this anarchy, there had been in 1514 a fierce rising of serfs—a Hungarian *Jacquerie*—under a sort of local Magyar Wat Tyler, George Dosza, whose atrocities during his short reign of power are less well remembered than his horrid punishment with the red-hot crown after his defeat.

King Ladislas, as a legacy to his turbulent realms, left his throne to a son of only ten years of age, when he died in 1516. Minorities in feudal kingdoms were always times of danger, and Hungary and Bohemia after twenty years of such a king as Ladislas were in an exceptionally demoralized condition. Diets in an advanced stage of 'Parliamentarianism,' *i.e.* of endless wrangling between factions, are proverbially unable to govern. And in Hungary there was the special curse of a 'nationalist' party, which purported to yearn for a king of native Hungarian blood, instead of an alien boy—Bohemian or Polish. This party had as its head John Zapolya, a Transylvanian magnate of many ambitions and few scruples, who had schemed to get control of the realm by marrying the young King's eldest sister. She was, however, wedded over his head to Ferdinand of Hapsburg, the younger brother of the Emperor Charles V. This disappointment drove him to ways of secret treason, which ultimately ended in his betraying his country, and becoming the tool of the Turk—but this catastrophe was several years away, in 1521.

When Sultan Soliman opened his campaign against Hungary the King was fifteen years of age, and wholly without any influence. The Councils of regency in his two kingdoms were out of touch with each other—indeed fundamentally hostile and jealous. Each had to deal with factious minorities, which blocked all effective action. It is to this circumstance that we must attribute the surprising fact that, when the Turkish invasion materialized, there was no effective attempt to raise a common army to oppose the invader, and that Soliman was able to reduce the great fortresses of the Danube and Save, without having to fight a general action against the army of succour which ought to have marched against him. He only had to deal with garrisons of the towns which he beset—their governors clamoured in vain for relief from Buda or Prague.

The Sultan started out from Constantinople in person on May 10, but before he reached the front the local commanders on the Danube had been ordered to form the sieges both of Shabatz and of Belgrade, while a detachment of irregular horse was sent to the Carpathians to distract the attention of the Hungarians toward Transylvania. Soliman himself took the route to Shabatz, sending the Grand-Vizier to strengthen the force before Belgrade, which was making little progress. Shabatz, though under-garrisoned, made a fine defence, and repulsed with loss several attempts to storm its battered walls. On July 8, however, it was carried by assault, no help or reinforcements having been received from outside.

Soliman then proceeded to cast a long and solid bridge across the Save, by which his whole army passed into Syrmia, and appeared on the bank opposite Belgrade, which had hitherto only been beset on the south side of the Danube, and had maintained its communications with Hungary, though the governors had received little or no aid during the two first months of the siege, or rather blockade. The Sultan then proceeded to capture Semlin, which faces Belgrade across the water, and to seize Salankemen, Mitrovitza, Ruma, and other Syrmian towns between the Save and the Drave. The Hungarian line of defence was thrown back to the last-named river, and the fortresses of Peterwardein and Essek, forty miles north of Belgrade.

That great fortress was therefore completely cut off from all succour, which could only have been given by an army strong enough to cross the Drave and challenge a general action. Nothing of the kind appeared, only roving parties led by local bans, which occasionally cut up Turkish raiding expeditions. Soliman therefore could settle down unmolested to encircle

Belgrade on both sides of the water. The actual siege-work had, of course, to be done on the south bank, where alone the town was accessible, but useful subsidiary help was given by planting batteries on the islands of the Danube, and bombarding the water side of the fortifications.

The wholly inadequate garrison, part of which was composed of Serbian auxiliaries, made a not discreditable defence, considering that it received no help from outside. On August 9 the lower town had to be evacuated, as its walls had been completely ruined by the fire from the batteries on the islands. There remained the high-lying citadel 'the Palladium of Hungary,' which had defied so many Turkish armies. Soliman worked both by battery and by mines, and made several practicable breaches: on three occasions, the 16th, 26th, and 27th of August, general assaults were delivered under the Sultan's own eye. All failed with great loss: Soliman's diary (fortunately preserved) acknowledges heavy casualties including many high officers. On the 27th August, however, a great corner tower crumbled down from the effect of mines, and though the assault over its ruins failed, the commanders in the place held that the situation had become impossible. There were three of them—a bad arrangement—two Hungarians, Bathy and Morgai, and a French mercenary captain named Blaise d'Obé. They were not on good terms with each other, but might have continued to resist a little longer but for the treason of two officers, Hedervary and Toerok, who raised something like a mutiny, in which the Serbian auxiliaries joined. On August 29 the citadel surrendered, on the liberal terms of a free departure for the garrison—Soliman did not want to waste more lives. The capitulation was kept in the main—a few of the soldiers are said to have been massacred by Turkish rioters, but the officers and the majority were certainly allowed to retire to Peterwardein, while the Serbians were sent to Constantinople. For this the Sultan's diary vouches. On the day of the surrender there were only 400 men fit to fight left in the citadel—so the surrender bore no shame with it, though Hungarian writers seem to think that they ought to have perished in a hopeless defence. The real shame lies on the Hungarian government, which in the three months since the Turks appeared on the frontier had done nothing whatever to save their garrison.

Soliman returned to Constantinople in October, after having

spent some weeks in repairing the fortifications of Shabatz and Belgrade, and finding competent garrisons for them. Every one in Europe expected that the assault on Hungary would be continued in the next year, now that its gates had been broken open—but this was not to be the course of events. Five winters were to elapse before it came.

CHAPTER IV

THE SIEGE OF RHODES (1522)

IN 1522 Soliman turned his attention to the design which his father had conceived in his last year, of using the fleet which had been building in the repaired arsenals of the Golden Horn, for the establishment of Turkish naval supremacy in the Aegean and the Levant. Selim had added Egypt to the Ottoman empire, but the direct sea-route to Egypt was blocked by the line of Christian fortresses drawn across the straight line—in Crete, Rhodes, and Cyprus. He was not only at peace with Venice, the owner of Crete and Cyprus, but well assured that the Republic, in evil state since it had barely survived the perils of the League of Cambray, would stand almost any amount of provocation before breaking with him. There remained the Knights Hospitallers of Rhodes—now without any friends who were likely to help them actively, the Emperor being completely engrossed in his new war with France. This was the year of the battle of Bicocca, and the expulsion of the French from Lombardy; all Europe was aflame.

The Knights of Rhodes were an astounding anomaly—the last surviving trace of the old Frankish supremacy in Eastern waters, which had seemed likely to disappear entirely when Venice had lost the heart to defend it. They owed their survival to the stubborn resistance with which they had beaten off Mahomet II forty years back. Bajazet and Selim had never tried to reverse the disaster of 1480, though the latter had the idea in his brain at the moment of his death. The only thing which counted in the Knights' domain was the single city of Rhodes—they had castles on some of the smaller islands around, but these were of no importance.

But Rhodes itself was at this time probably the best fortified place in the world—three successive grand-masters since 1480, Aubusson, Amboise, and Carretto, living in constant fear of new attacks, had rebuilt the whole of the defences of the city in the most modern style, with full reference to the development of artillery. All the spare cash of the order, from all its commanderies round Europe, had gone into stone and mortar. Rhodes was no longer a mediaeval fortress, but had defence heaped on defence in the most ingenious fashion. Instead of a curtain-wall and a ditch in the old style, it possessed a bastioned front, in which each bastion had ample provision for guns. The more exposed sections had a *fausse braye*, or second lower outside wall, beyond the ditch of the inner wall, with a ditch of its own in front, and then a *glacis* with scarp and counterscarp, which would have to be crossed before even the *fausse braye* was reached. The bastions, especially those which covered the corner angles of the place, and those which guarded gates, were of immense solidity, not very lofty in the old style, but deep and with defence behind defence. The only side of the city not heavily entrenched was the north-eastern, where the port lay; but here the enceinte wall needed not to be complicated, for the port could only be approached by a narrow opening between two strong towers each at the end of a spit—Tower of Naillac and Tower of St. Angelo—between which a massive bronze chain was stretched in time of war: both were heavily gunned.

The care of different sections of the walls was entrusted to the several national 'tongues' of the knights, of which there were now eight, 'Spain' having recently been divided (1462) into the 'Tongues' of Castile and Aragon; while by a survival of political ideas which had been real when the order first came into existence in the twelfth century, the French 'tongue' had parallel with it 'Provence' and 'Auvergne.' The existing units therefore, some much more numerous than others, were France, Germany, Italy, England, Castile, Aragon, Auvergne, Provence. Navarre was treated as a subsection of Aragon, Portugal as one of Castile. This was, of course, quite reasonable, as all these small sub-units had very few members. There were, when the siege commenced, about 700 knights under arms in the city, about 400 more were absent in their manors all over Europe, but were, of course, liable to recall at the time of crisis. Their auxiliary troops of all sorts—'turcopoles' or light horse, marine soldiers from the galleys, western pikemen and arquebusiers, and local mercenaries of Cretan and Greek origin, with some armed burghers—made up 6000 in all.

The northern side of the fortification, from the harbour to the Grand-Master's palace, which formed the north-western corner of the fortress, consisted of the front of the ‘ tongue ’ of France ; it had a strong projecting spur bastion, the ‘ Tower of St. Peter.’ The west front belonged to the German ‘ tongue ’ as far as the great bastion of St. George, then to the ‘ tongue ’ of Auvergne as far as the ‘ Tower of Spain,’ where the Aragonese were in charge as far as the gate and bastion of St. Mary, the south-west angle of the fortress. This section is often spoken of as the ‘ Spanish posts.’

The southern front from the bastion of St. Mary to the gate and bastion of St. John was in charge of the English knights, from whom, starting at the St. John's gate, the ‘ tongue of Provence ’ had its post, holding the other half of the southern ramparts as far as a bastion named from the Grand Master Carretto, which formed the south-eastern angle of the place. The short front facing east from this bastion to the edge of the port belonged to the Italian ‘ tongue,’ whose charge ended at the ‘ gate of the Mills ’ at the edge of the water. From this point, along the inside of the harbour, as far as the commencement of the French front, there was only a simple mediaeval wall, as this section assigned to the ‘ tongue ’ of Castile, was inaccessible by land.

The four main gates, that near the Grand Master's palace in the north-west angle, that of St. Mary at the south-west angle, and those of St. John and Carretto in the south front, were most formidable works, with outer ditches and barbicans, and dominating inner platforms bearing artillery. And the bastions which did not cover gates were almost as strong, with ditches of extra depth. Even the ‘ curtains ’ between the projecting bastions had received special attention at the more exposed points. The Spanish and English posts had each a *fausse braye* outside them, which would have to be demolished before the main wall and its ditch could be reached.

If adequately garrisoned, and supported from time to time by sea-borne reinforcements and munitions, Rhodes might have held out for an interminable siege—as Ostend did eighty years later. But the weak point was that though there was a garrison adequate at the commencement of operations, it was absolutely destitute of succours from without, so that each knight or mercenary killed could not be replaced. If the enemy was ready to sacrifice lives recklessly, there would come a time when the defender, even if he lost only one man for each ten that he killed, would be unable to man the walls adequately. The same was the case with munitions—there was ample powder and balls for many months' use, but they could come to an end some day. The one hope for the gallant Grand Master was that the enemy might suffer such crushing losses in the summer and autumn months of the siege, without making any perceptible lodgment in the defences, that he might lose heart and retire, before winter storms made trench-life intolerable. Meanwhile every possible string was being pulled in the west—the Pope and the Emperor received heart-rending appeals—but it was not likely that either of them could do much. To plead for help from Venice was obviously useless. Meanwhile Villiers de l'Isle Adam,[1] arranged for the concentration in the city of the thinly-spread peasantry of the island, and for the removal of all food-stuffs within the walls.

On the 18th of June the Ottoman fleet sailed from Constantinople, and on the 16th the Sultan with the land-army started on the long march across the western edge of Asia Minor, over the hills and plains of Lydia and Caria, to the long bay of Marmarice, just opposite Rhodes,[2] on whose shores he established his base of operations. The fleet landed 10,000 men on the island on St. John's Day (June 25), but found the countryside bare; the landing force met no opposition, and was able to establish convenient places for the disembarkation of the main army, which did not reach Marmarice till three weeks later. Meanwhile the Turkish engineers looked out sites for batteries opposite all the three exposed sides of the city. Not a shot, however, was fired till the Sultan came across the water on July 28, the day on which the siege actually began. Soliman fixed his pavilion on the slope of the hill of San Stephano, high above the sea, facing the west side of the defences of Rhodes from a distance of 2000 yards. The camps of the various divisions of his army encompassed the city in a semi-circle from water to water. Detachments of galleys kept continually cruising outside the harbour, well out of gun-shot, so that the investment was complete. This did not prevent a little 'blockade running' by individual vessels at night, but no succour of any importance could ever get into the city by water.

[1] Philip Villiers de l'Isle Adam was only recently elected, and had a dangerous passage from France in 1521 owing to Turkish pirates, who nearly caught him on the way. His 'runner-up' at the late election had been Thomas Docwray, an English knight, commander of the horse of the Order.

[2] This was the ample bay in which Sir Ralph Abercromby's army was mobilized for the invasion of Egypt in 1800. It could easily afford harbourage for 300 ships.

The Turkish besieging force was distributed into five groups or corps—the chroniclers guess them at 50,000 men apiece, no doubt a vast over-estimate, and a large proportion of the host did not consist of fighting men. There were some tens of thousands of pioneers, mostly Christians, swept in by conscription from mining districts in the Balkan Peninsula—Bosnians, Wallachs, and Bulgarians, doomed to the dismal task of unprotected trench-building under fire. The disposition of the Sultan's army was as follows: on the left wing facing the French front and the angle of the German post was Ayaz Pasha, governor-general (*beglerbeg*) of Roumelia; in the left centre facing the ports of Auvergne and Aragon was the third vizier, Ahmed Pasha. The right centre was under the charge of the Commander-in-chief (*Seraskier*), Mustafa Pasha, opposite the English line. To the right of this corps was Kassim Bey, governor-general of Anatolia, in front of the 'posts of Provence.' Finally, the Grand Vizier Piri Pasha, with the naval landing forces, ended the line on the sea's edge, facing the Italian front, from the Carretto gate to the water-side of the open sea. The main body of the Janissaries was encamped behind Ahmed Pasha's lines, the Sultan with the remainder, the household troops (Silladar horse, etc.) behind the Seraskier's position.

When the Turkish engineers had marked out the places for their 'first parallel' and the batteries which were to be placed in it, the besieged saw that the main attack was to come against the western and southern fronts, for the northern front, from the Grand Master's palace to the Naillac tower, was never seriously threatened. This was the highest part of the defences and the most deeply ditched. The real attack was directed against the line from the Bastion of St. George—at the end of the German post—to the Carretto bastion in the Italian post, *i.e.* embraced the whole of the fronts of the 'tongues' of Auvergne, Aragon, England, Provence, and Italy. But the most desperate endeavours were made to break the curtain of the Spanish, English, and Italian fronts, between the great bastions and gate-houses, which were practically impregnable. Both the Spanish and the English fronts had exterior *fausse brayes* ('propugnacula') and ditches, but nevertheless they were the selected points for the most serious attack, though it involved the crossing of two instead of one ditch. But batteries were erected opposite all the western and southern fronts, perhaps in the hope of breaking in at points which might be found unexpectedly weak.

The forming of the 'first parallel,' and the bringing into it of the battering cannon, proved a most difficult and expensive business. The engineers of the defence—their chief was a talented Brescian officer named Gabriel Martinengo—had measured all the distances outside the walls, and knew precisely the proper artillery range, so that the moment that trench work began the ground was swept by accurate fire. Moreover, the soil was in many places stony, so that it proved very hard to dig up. The unfortunate pioneers sent out to commence the works were killed by hundreds before they could cover themselves. And several times the garrison made sudden sallies and cut up the trench-guards, before they could be succoured from the camps behind. The Turks, however, showed a stolid persistence in spite of losses, and batteries came into existence. There were two facing the French and German fronts, largely armed with mortars, which threw large hollow shot into the city, but did not much affect the walls. Fourteen batteries of three guns each pounded the curtains of Spain and England, and seventeen worked against the fronts of Provence and Italy. Opposite the Italian front and the angle between the Auvergne and Spanish fronts the enemy built two high 'cavaliers' or 'mounts' to equal the level of the curtain, making them very solid not only with earth, but with hewn stone from quarries in the rear. When the trench line was completed, sap-heads were thrown out from it towards the ditches of the city, from which it was intended that assaults should be delivered, when the artillery should have broken down the walls into the ditches. But it turned out that the walls were so solid that for a long time little crumbling took place, and finally the sap-heads were pushed up to the front, covered with mantlets, and stones and earth cast forward from them into the ditches to level them up.

Some of the Turkish artillery was of very large calibre; one of the narrators of the siege specifies six brass cannon firing stones of four palms in circumference, fifteen iron pieces with stones six palms in circumference, ten very large 'bombards' casting balls of nine or ten palms, and two enormous ones whose balls were of no less than twelve palms. There were twelve bronze mortars dug down into the earth and firing hollow shot in a high trajectory into the city. Smaller guns, 'sakers,' and suchlike were innumerable, and there was a great provision of ' wall pieces '—*espingarderie*, especially on top of the two ' mounts.'

Sultan Soliman's invaluable diary acknowledges that on July 29, the first day on which the besieging batteries opened fire, the trenches and batteries of Kassim Bey, opposite the ' front of Provence,' were swept flat, and that on July 30 the sap-heads were all battered down. It was only on the 31st that effective fire could be opened from several fronts, and that the reply of the defenders slackened because of the retaliation. The whole month of August was spent in the tedious and expensive task of pushing the sap-heads towards the ditches, which were only reached on the English, Italian, and Spanish fronts upon the 18th of that month, nor were the outer ditches filled in certain places before the 28th. The walls stood well: they were much pounded, but while the battlements and the exposed guns on them were damaged, there were no proper breaches as yet. The garrison had begun to suffer perceptible losses among the gunners manning the wall artillery, but the shelling of the town by the mortars did not prove very effective —one witness says that only ten persons (mostly civilians) were killed by bombs in a whole month. But on August 10 the bell-tower of the cathedral of St. John came down, which the Knights had been employing as a useful observation post. On the 19th, since the batteries opposite the Italian front were giving much trouble, a relieving sortie was made against them, and while the Christians were in possession of the ground, the trenches were filled up and many guns disabled. It was ten days before Piri Pasha was able to recommence battering from his reconstructed lines.

The real danger to the place only began when the Turks, having filled up long stretches of the outer ditch, and built covered trench-approaches to the edge of the ramparts, started numerous mines, to burrow under the foundations of the walls. The besieged replied by counter-mining, and discovered many, but not all, of these hidden dangers, out of which they smoked the miners by blasts of pitch and naphtha, and then blew down the roofs with gunpowder. The number of Turkish mines from first to last is given at 54! Gabriel Martinengo is said to have discovered a method of detecting hostile excavations by means of drum-heads fixed down to the soil, which reverberated the stroke of the pick, even when it was still some way off. On September 4 the first successful mine blew down part of the *fausse braye* on the English front, and a storm was attempted but failed. Though the Turks made a lodgment, they were evicted by the heavy flanking fire which the besieged concentrated against the exposed point, and by a counter-attack made by the Grand Master in person. Six days later, however, Mustafa Pasha, who was in charge here, exploded two more mines, which brought down so much stone that a large gap was caused, and a much more serious and prolonged assault was made, which ended in the same way.

The garrison had to deplore the loss of some distinguished

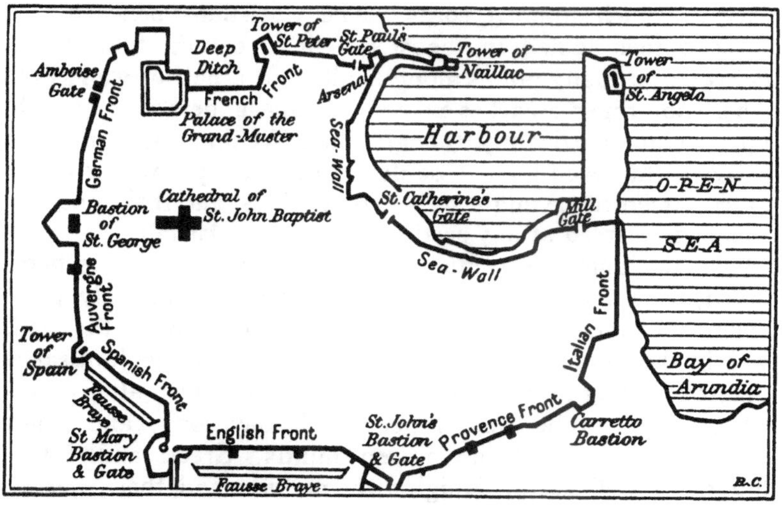

RHODES IN 1522.

officers—among them Henry Mansell, an English knight, the standard-bearer of the Grand Master. The assailants are said to have lost 2000 men, largely from flanking fire on the tail of the column, when the head was jammed in the top of the breach. Nevertheless Mustafa repeated on September 13 his attempt to break in at this same point of the English front; the head of his storming column reached the lip of the breach and planted standards there, but was driven out once more by a counter-attack. After this the Sultan countermanded any further assaults for some days, and directed his officers to concentrate on battering-work and mining, not only on the English front, but all along the western and southern sides of the city, from the German as far as the Italian posts. Ahmed Pasha, opposite the Spanish line, was particularly successful in mining-work, and Piri Pasha also succeeded in bringing down some of the stones of the Italian section.

The net result of this ten days' work looked so promising that the Sultan gave orders on the 23rd for a general assault all along the front. From noon till dusk the criers went round all the camps proclaiming that there would be a storm on the next morning, 'the stones and the ground only for the Padischa, the lives and the property of the infidels will be for the victorious army.' Volunteers were forthcoming from the troops on the unassailed sections of the siege-lines, to reinforce Mustafa Pasha's much thinned corps. The Christians believed that 100,000 men joined in the storm—it is certain that it was delivered at many points along a mile and a half of wall; even where no proper breaches existed, scaling-ladders were brought up. Soliman evidently thought, like Wellington at Badajoz in 1812, that even if the assaults on the most promising points failed, an entrance might be found at others, and if the enceinte was once pierced at any section, the place must fall.

To the immense credit of the garrison the whole scheme failed, with frightful loss to the assailants. The only point where the stormers made a momentary lodgment in the walls was in the Spanish curtain, where the Aga of the Janissaries planted his flag in triumph. But it was cast down after a long struggle. Assaults were repeated all day long, and failed without exception. The citizen militia, Greek and Latin, fought as well as the Knights themselves, and there was no flinching anywhere. The Turks at last drew off, with immense losses, reported (no doubt with exaggeration) to have amounted to 15,000 in all. But since the fighting had come to hand-to-hand strife on many fronts, the garrison had also suffered very heavily, so much so that the down-hearted began to reflect that, after a few more such victories, there would be no defenders left. Sir John Buck, the Prior of the English 'tongue,' was one of those who fell in this assault.

There had been some hope that such a check might make Soliman abandon the siege; but his prestige was dependent on victory, and he persevered. As a sign that he was ready to go on for an indefinite time, he ordered his pavilion to be replaced by a solid stone building, and he sent orders eastward to bring in all the Janissary garrisons from Syria and Mesopotamia to replace his casualties: it was obvious that they could not come up for many weeks. He was (very unjustly as it would seem) discontented with his officers—deposed Ayas Pasha, in command on the left attack, from his command,[1] and ordered Ahmed Pasha to supersede Mustafa Pasha in the conduct of the central attacks on the place—the former was removed from his position as Seraskier (Commander-in-chief) and sent away to govern Egypt, where the Viceroy was just dead. The commander of the fleet was also disgraced for inadequate co-operation with the land forces.[2]

The spirits of the garrison had been much raised by the completeness of the check inflicted on the Turks, and, since October was drawing near, there was some hope that the Sultan might raise the siege before winter came. It was soon discovered that this was not his intention : he sent away his fleet from the somewhat dangerous shore-harbourage where it had been lying since July, to take refuge in the very safe bay of Marmarice on the opposite Asiatic coast : this meant that no departure of the army was intended, and indeed the besiegers began to rear solid buildings on the deserted site of ' Old Rhodes ' for winter cover. No more general assaults were risked, but the original system of filling up the ditches in all directions and continuing mining operations was resumed. Several of the explosions were not unsuccessful—on October 12 the English front was so shaken that the Sultan risked a partial assault—the Janissaries got into the *fausse braye*, but were expelled after their Aga Balibeg had been wounded. A fortnight later similar promising mine-explosions were made on the French and Provençal fronts, followed by assaults repulsed with the same complete resolution. After this there was three weeks of freedom from assaults, but the battering and mining continued, and in many places the walls began to crumble. At the end of November the Spanish and Italian fronts both looked so damaged that Soliman once more ordered a storm—which took place in drenching rain—it failed like the rest, though Ahmed Pasha, opposite the Spanish front, and Kassim Pasha (who had replaced Mustafa) did their best to press on a soldiery profoundly discouraged by so many previous disasters.

But the Knights, though they had held their battered walls for five months of incessant bombardment, were now growing hopeless. Winter had come, and no help from the West had

[1] But Ayas was pardoned and restored after a few days.

[2] To confuse students, this was another Mustafa—surnamed Yailak.

arrived—nor was there any prospect of it. The strength of the garrison was down to half its original force, and could no longer suffice to keep all the fronts properly guarded. When there was an assault, reinforcements had to be borrowed from points which were not actually attacked, which were left practically unmanned. But the worst danger of all was the dearth of gunpowder : the ample stores which had existed in July were used up earlier than had been calculated in incessant countermining, and by November the powder-mills by the harbour, which were worked day and night, barely supplied the consumption of the batteries for each twenty-four hours. It was possible that after some prolonged assaults there might be no more powder forthcoming. The breaches, of late, could only be mended by pulling down houses, and even churches.

The growing despair was marked by the appearance of a lively outburst of spy-mania. It was started, according to Fontanus, the most detailed chronicler of the defence, by a Spanish pilgrim-woman who passed for a seeress, and went round whispering that there was treachery in high places. The idea spread, and led to all sorts of riots and molestation of suspected persons. Fontanus gives curious details of attacks on obviously innocent persons, merchants and citizens. The French cook of the Archbishop was nearly massacred for being seen waving a flag in bravado from a church steeple during one of the storms. But the most sinister incident was the arrest of a Portuguese knight named Andres Amaral, an elderly man and the Chancellor of the Order. He was a factious and unpopular person, who played the part of a consistent 'wet blanket'—had he not said that L'Isle Adam would be the last Grand Master of Rhodes, and criticized the details of the defence ? No one could say that he had shirked his share of fighting, but one day his servant was accused of shooting arrows, with scrolls wound round them, into the Turkish trenches, no doubt conveying useful information to the besiegers. He was arrested and tried for communicating with the enemy, and convicted on very weak evidence—mostly that of his servant put under torture. He might reasonably have been cashiered for persistent '*defaitisme*,' but was actually beheaded for treason (November 8).[1] The only certain case of spying was that of a Jewish doctor, who apparently did communicate with the enemy, as Soliman's diary mentions something of the sort : but the Turks really

[1] Jacques de Bourbon makes no doubt of his treachery—but the evidence is poor.

got their information from renegades and escaped slaves, who from time to time swam out to sea, or dropped by night from the walls, not from ' those in high place.'

When the garrison had reached this stage of despondency and suspicion, they were somewhat surprised to find that the besieger was willing to open negotiations. Soliman did not wish to waste the whole winter before Rhodes: his army was tired out, and the pioneers, we are told, could not well use their spades and picks with frost-bitten fingers. He knew that the Knights were in an evil case, and resolved to offer them the most favourable terms, in return for the mere evacuation of the place. First unofficially (December 1) and then officially he made his proposition known. The garrison might depart, and take with them all the citizens who wished to emigrate, all transportable goods, and even their guns and stores. Any of the Greek population who should elect to stay behind were to be guaranteed possession of their houses and property, and should be exempt from taxation for five years. Their churches were not to be desecrated. A reasonable term of days was to be allowed for the embarkation of the garrison, and Turkish hostages were offered, in order to guarantee the fulfilment of the Sultan's pledge. The Knights' minor forts on other islands were, of course, to be surrendered.

The Grand Master, knowing his own weakness, thought the terms worth discussing, and sent a French knight and a burgess of Rhodes out to the Sultan's camp, to receive the exact details of the Sultan's offer. When these were formally brought back in writing, he called his council together. With the exception of a few desperate fanatics, who advocated fighting to the end as the duty of Christian martyrs, the assembly voted for acceptance of the terms, which were better than could have been expected. The matter was mainly determined on the report of the chief-engineer Gabriel Martinengo, whose valour and competence had been recognized by all during the whole siege. He stated that upon his honour and conscience he held the place untenable: the numbers of the garrison were so reduced that it was no longer possible to move guns in a hurry, or to patch up a new breach. Long fronts were practically unguarded. At dusk on any day there was only twelve hours' expenditure of powder in stock, and the mills could not work up any more, from shortness both of hands and of material. A two days' assault would mean inevitable ruin. A deputation from the Greek citizens, who had hitherto behaved admirably, besought the Master, by the mouth of one of their priests, to remember that if he and the Knights resolved to die fighting, they were giving over 10,000 Christians to massacre, and their wives and children to rape and slavery. This was obviously true, and the alternative was a most honourable surrender, and a free departure both for the garrison and for such citizens as wished to emigrate.

A three days' truce was arranged for the settlement of details (December 13), but the pact was likely to come to nothing, owing to an insane act of a French knight, who, seeing a large group of Turks coming closer to the walls than he thought proper, fired two cannon-shot into them and killed several. In reply, Ahmed Pasha cut off the right hands and noses of three soldiers of the garrison who were caught outside the walls, and sent them back into the city. Firing recommenced all along the line, and on December 17–18, the Turks made their last assault, this time on the Spanish front. On the first day they were repulsed, but on the second day, making their fourteenth assault according to the reckoning of the Knights, they actually broke through, and seized a group of houses within the wall—200 feet long by 150 broad.[1] Their further progress was stopped by the defenders blocking up the narrow side streets on which these houses fronted, but the enemy had at last succeeded in making a lodgment in the town. The Grand Master then renewed the negotiations with the Sultan, and found to his surprise that Soliman was prepared to stand by his former offers. Apparently he viewed with no pleasure the prospect of incessant street-fighting, which would have been necessary if the strife had gone on, and his own troops were worn out.

On December 21 the capitulation was formally ratified, and hostages exchanged on both sides : Soliman even conceded that his troops should leave the trenches and retire for a mile from the city walls. The evacuation commenced at once, and continued till January 1, when the Grand Master and his staff sailed as the last detachment. Only 180 Knights and 1500 rank and file—mercenaries and Latin armed citizens—survived to embark; most of the Greek townsfolk chose to remain behind, and to trust the word of the Sultan as to their treatment. He endeavoured to keep it, but before the evacuation was complete,

[1] Data given carefully by Jacques de Bourbon.

unlicensed bands of Turks, without weapons, slunk into the city and committed minor outrages, pillaging houses, beating civilians, and wrecking the great church of St. John. Soliman was apparently not responsible. He behaved with great courtesy, receiving L'Isle Adam on December 26 at a state ceremony in his camp, at which he invested him with a robe of honour. But (what is much more surprising) he paid a return visit to the Grand Master in his palace on December 29, to which he came without any escort, accompanied only by Ahmed Pasha—who acted as interpreter—and one other attendant. This showed a splendid confidence in the chivalry of L'Isle Adam, which was not misplaced. The riots in the town are said to have been caused by the arrival in the Turkish camp of the Janissary corps from the East, which came too late for glory, and was quite out of hand, and in a state of mutinous discontent after a long and fatiguing winter march.

So ended the last outpost of militant Christendom in the Aegean. The Sultan left Rhodes on January 6, and was back again at Constantinople by the end of the month—there to receive from the Venetian Senate congratulations on his great military exploit! The absolute selfishness of the Republic was degrading; there had been a great concentration of warships in Cretan harbours during the autumn, but this was a mere precautionary measure—there was no intention of helping Rhodes. When L'Isle Adam and the wrecks of his garrison reached Crete, they were passed on to Messina, after having been allowed a week or two to repair their ships, which had been battered by a storm in the Carpathian Sea. At Messina the Grand Master found 2000 mercenaries collected by the funds of the Order, and by the leave of the Emperor Charles, but they had never dared to sail, having no convoy. The Pope had gathered a small sum of money, with which a few companies of adventurers had been raised, but he ended by sending this trifling force to join the Emperor's army in Lombardy. Again transport was the difficulty.

If Rhodes had been succoured from time to time it might apparently have held out for an indefinite period, and there was a great Venetian war-fleet in Crete. But without help the city was doomed from the first. The Knights were granted by the Emperor the lonely isle of Malta for a new residence, where they made their name as famous as it had been at Rhodes, and had the satisfaction in 1566 of beating off the last great attempt of the Ottomans to establish a naval domination in the Central Mediterranean. Of this more in its proper place.

NOTE ON THE SIEGE OF RHODES

The long rhetorical Fontanus is useful, but I have rejected details in him which clash with the Bastard of Bourbon and the diary of Sultan Soliman. Sir Nicholas Roberts's letter to the Earl of Surrey is tantalizingly short. He gives as casualties of the garrison, 703 Knights and 6000 others—but this was apparently the exact total of the whole of the defenders. Turkish losses must have been enormous, but there is no certainty in estimates of them by the Christian writers—to whom 100,000 is a minimum. The plan of Rhodes which I have inserted is reduced from the beautiful official Italian Survey. I have got much topographical help from the account of Rhodes in L. V. Bertarelli's *Possedimenti Italiani*, Milan, 1929.

Obsidio Rhodie.

RHODES BESIEGED

From a contemporary German wood-cut

CHAPTER V

THE CAMPAIGN OF MOHACS (1526)

SULTAN SOLIMAN returned victorious from Rhodes to Constantinople in January 1525. All Europe expected that his eviction of the last Christian colony in the Aegean would be followed by a new attack on Hungary, whose frontier was bare since the loss of Belgrade and Shabatz in 1521. This was not to be: three years more were to pass before the horse-tail standards were to be seen beyond the Danube. The main cause of this delay was a series of dangerous revolts in Egypt, where the old Mameluke faction found an unexpected leader in Ahmed Pasha, the officer who had so much distinguished himself at the siege of Rhodes. Even after his suppression (1524) trouble continued, and while Egyptian affairs were still unsettled there was a dangerous mutiny of the Janissaries in Constantinople itself, which was only suppressed by the execution of their commander and other leaders. It was not till the winter of 1525–26 that the Sultan found his hands free, and began to arrange for a new attack on Hungary.

Meanwhile five years had passed since the fall of Belgrade, without any definite change of the situation on the Danube. The Hungarian 'Bans' of the border were strong enough to wage minor campaigns with their opposite neighbours the Turkish 'Sandjak beys' of Belgrade, Semendria, and Seraijevo, so long as the fighting remained local and the Sultan was occupied elsewhere. The Turks made some destructive raids into Croatia and Slavonia, but failed to win the Hungarian outlying fortresses at Jaicze and Banjaluka. Two of their raiding forces were completely destroyed, one in front of Jaicze by Count Christopher Frangipani, the other in Syrmia by Paul Tomori, the bellicose Archbishop of Calocza, who seems to have taken over complete control of the central frontier. On the other hand, John Zapolya, the ambitious voivode of Transylvania, failed in several attempts to expel the Turks from Wallachia, and to set up there a prince of Hungarophil sympathies. Such border warfare might have gone on indefinitely, if the Sultan and the general levy of the Ottoman Empire had not come upon the scene—a contingency for which the Hungarians had made no provision, though it was certain to happen sooner or later. No ' Wardens of the Marches ' could stop a field army of 100,000 men.

The state of Hungary had not improved since 1521. Indeed its frontier was now on the Drave, and not on the Save, and Peterwardein, the one fortress south of the Danube still held, was a very poor substitute for Belgrade. Syrmia, the region between Save and Danube, was a sort of no-man's-land, ravaged almost bare by continual Turkish incursions. There were only a very few fortified places on the south banks of the Danube and Drave still held : the one physical advantage was that both those rivers flowed among bogs and marshes over which there were a limited amount of crossing-places for an army. Still it was obvious that a serious attack on Hungary must be delivered on this central front: it would not come through the Carpathians on the east or the Bosnian-Croatian mountains on the west. An advance on either of these flanks would be over regions in which a very large army could not live, and would not strike at the heart of the realm, but only at secondary objectives.

King Louis of Hungary had now reached the age of twenty, but was as far from exerting any control over his subjects as when he was only fifteen in 1521. He is described by chroniclers, who were affected by his sad end, as a good-tempered lad, given to hunting, tilting, and pageants, anything but secretive or ambitious. Like his father Ladislas, who had earned the nickname of ' That must do then,' he was accustomed to be ' put upon ' by dictatorial ministers and unruly barons, and took things as they came. He had no personal influence or directive power. The fact that he had been married to Mary of Hapsburg, the sister of Charles V and Archduke Ferdinand, gave him no influence over his subjects, who hated the House of Austria. This was more especially the case in his kingdom of Bohemia, where the anti-German feeling still ruled; while in Hungary there were many nobles who had favoured the pretensions of John Zapolya to the crown, and sneered at the rule of a frivolous boy. In both states the crisis of 1526 came upon a faction-ridden nobility, hopelessly unconscious of the desperate need that was coming.

That Soliman would be on the move some day was obvious—more especially after he had bargained for the neutrality of Poland by a treaty in 1525, and was cajoling Venice into her usual neglect of Christian intents in the East by favourable commercial offers. Every one at Constantinople knew that trouble was at hand—the orders for mobilizing the distant contingents of Asia Minor were issued in the winter, in order that they might be able to serve in the spring, and the accumulation of stores at the Turkish fortresses along the Danube could not be disguised. On March 20 the energetic Archbishop of Calocza, Paul Tomori, came to the court, with detailed information as to the hostile preparations that were being made all along the frontier. His anxious pleadings only led to the summoning of the Diet on St. George's Day, April 23; till it met nothing was done. It so chanced that on that same morning Sultan Soliman had marched out of Constantinople with his Janissaries and household cavalry, with the knowledge that the Asiatic feudal host was on its way behind him, and that the European contingents were ready to be picked up as he passed through their districts. It is a long march from Constantinople to the Danube, and the vanguard under the Grand-Vizier Ibrahim only reached Semendria on June 28, and crossed the Danube to Semlin on July 4. The Sultan with the main body was some days behind, and only reached Belgrade on July 9.

The Hungarians had therefore some forty-five days between the meeting of the Diet and the commencement of hostilities, in which to make their preparations to resist a first-rate invasion known to be led by the Sultan in person. The time was wasted lamentably; after much discussion it was voted that the whole force of the realm should mobilize at Tolna, on the Danube, fifty miles south of Buda, on July 2, and urgent messages were sent to Prague, to urge the Bohemian Diet to order all Bohemian and Moravian contingents to join up at the same place and date.

Meanwhile the Exchequer was nearly empty; but money was to be procured by utilizing a contribution which had been gathered for the Pope for urgent national necessity, and it was voted that a large levy should be made on the treasures of the Church. The clergy were sufficiently convinced of the danger to make the grant. With these moneys it was arranged that mercenary foot should be raised—the arm in which Hungary, essentially a cavalry state, was most deficient. But to get together a sufficient number of mercenary troops—landsknechts and such-like—in forty days was an impossibility. When the campaign began there had only been collected 4000 foot from the west, presumably Germans and Italians, under a condottiere named Hannibal of Cyprus, and 1500 Poles led by an officer named Leonard Gnomski, who had the reputation of knowing all about the fortification of camps by waggon-laagers in the old Hussite style. The King of Poland would do nothing for his nephew, holding himself bound by his recent treaty of neutrality with the Sultan. The mercenary foot did not appear at the front before August had arrived, and can we wonder at it, since they had to be found and hired, and brought from a distance? Ambassadors were sent to the Pope and all the princes of Europe, even to the King of France, who was at this moment secretly intriguing with the Sultan. But no real help could be expected save from the Archduke Ferdinand, who was next to the danger—indeed some Turkish raids had of late penetrated into his duchy of Carniola. The appeal came much too late. One most unhappy plan was for a moment in favour—it was suggested that John Zapolya, with all the forces of Transylvania, should cross the Carpathians and invade Wallachia, where he might count on getting some support from the local boyars, and distract the attention of the Turks by appearing on their flank and rear. This only led to the diversion of all the levies of eastern Hungary; for Zapolya, who liked to play his own game, mobilized a considerable corps, and moved it so far eastward that it never arrived at the real seat of war. Some say that he might have countermarched in time, when he received orders to drop the project and join the main army. He pleaded that contradictory orders had sent him too far afield, and that time failed him in the critical days of the campaign. But there was a general impression that disloyal particularism and selfish jealousy kept him away from the muster on the Danube.

On July 12 the Grand-Vizier and the Roumeliot contingents of the Turkish army appeared before Peterwardein; the Sultan was a little behind, at Semlin, where the feudal horse of Bosnia and Herzegovina joined him. The campaign had begun; in front of the Turks there was at the moment no one but Archbishop Tomori with about 3000 horse and foot, the levies of his own diocese. He threw the infantry into Peterwardein, and retired with the cavalry to the other side of the Danube.

It will be remembered that proclamation had been made to all the nobles of Hungary to be present at Tolna on July 2—ten days before the siege of Peterwardein began. But not a soul was there on the appointed morning; 'all said that they were to meet the King, and they had not heard that the King was yet there.' [1] Much disheartened, Louis collected his personal following and those of his courtiers and ministers and of the Archbishop of Gran, and came out of Buda on July 23 with a miserable muster of 3000 men. He found no one at Erd, and waited several days before moving farther down the Danube to Paks, where some feudal contingents began to drop in, but the army reached Tolna on August 6 still under 5000 strong. Urgent messages were sent to the Voivode of Transylvania to come in at once, but had no effect.

Meanwhile Peterwardein fell after a siege of fifteen days on July 27. The town, which had but feeble walls, was breached and stormed on the 15th. But the garrison, which had retired into the citadel, made a fine defence, and repulsed with loss two vigorous assaults. In the second the Sultan's diary records that 650 Janissaries and other soldiers were killed. The Turks then turned to mining, as at Rhodes, and on the 27th a corner tower came down, and the place was stormed over its ruins. Five hundred of the garrison were massacred, 300 survivors sold as slaves. All this time Archbishop Tomori was watching the siege from the other bank of the Danube, quite unable to give any help with his 2000 light horse. The King and his skeleton army had not yet arrived at Tolna when the citadel fell.

After putting a garrison into Peterwardein the Sultan did not cross the Danube into the waste lands of 'Little Cumania' or Jazygia, where he would have been off the main road to Buda, but marched along the south bank of the river, towards its junction with the Drave, along whose line he rightly expected that the resistance of the Hungarians would begin. He made for Essek, the first large town on the Drave before it falls into the Danube. On his way he captured the small town of Cerevics without meeting resistance on July 28; and Illok, a larger place, by capitulation after a two days' siege on August 8. Continuous and heavy rains delayed the progress of the Turkish army after this, and it was not till the 14th that the Sultan reached the junction of the Danube and Drave, not far from Essek. Behind the broad Drave he expected to find the

[1] Brotheric, p. 1182.

Hungarians on the farther bank, prepared to resist his further progress—but nothing hostile was to be seen across the water.

Meanwhile, as we have seen, the King had reached Tolna on August 2, and at last contingents from Western and Central Hungary began to arrive in his camp. The bishops of Warasdin and Raab came in with their retainers, also George Zapolya, Count of Szepusz—the brother of John, the still-missing Voivode of Transylvania—and other magnates, also the mercenary foot, both western and Polish, but nothing from Croatia, still less from Bohemia or Moravia. Archbishop Tomori was still on the other side of the Danube, watching from the farther bank the progress of the Turks westward, and moving parallel with them; he kept sending regular notice of their movement toward the Drave, but did not yet come across the Danube to join the royal forces.

At this moment, about August 8, there occurred the most deplorable incident, one of several, which marked the indiscipline of the Hungarian nobility. The King proposed to detach his prime minister (if so he may be called), the Count Palatine Stephen Bathori, with a large portion of the army, to occupy Essek and to defend the lower course of the Drave. But the greater number of the magnates who were told off to follow the Palatine refused to march, 'pleading their privilege and baronial duty to serve under the King's standard alone.'[1] Louis was not going to Essek, but only a deputy. The King is said to have burst out in angry declamation: 'The mutineers pretended to be devoted to his person, but were really factious cowards.' But the plan was given up, though the Palatine was anxious to start, in spite of a violent attack of the gout which made it hard for him to mount a horse. The Drave was left unguarded, but the whole army made three marches to the front, and reached Mohacs, thirty miles farther along the Danube, but still thirty more from the all-important line of the Drave. When Mohacs was reached, contentious councils of war became the order of the day. Ignoring the rights of the King and the Palatine, a majority of the magnates voted for two joint commanders, Archbishop Tomori and George Zapolya of Szepusz. This was apparently a deliberate insult to what was supposed to be the court party, engineered by the Zapolya faction. The Archbishop was invited to come across the Danube and join the main army, which he did,

[1] Brotheric, p. 1185.

bringing with him a force which had now swelled to 6000 men. They made a separate camp, some little distance from that of the King.

Meanwhile Sultan Soliman had found the banks of the Drave below the Essek clear of an enemy, and was able to construct a long bridge of pontoons at his leisure; its building took no less than four days—August 16–19. On the 20th a body of light horse passed, and scoured the countryside, bringing in a few prisoners. On the 21st the Grand Vizier crossed with the Roumeliot and Bosnian horse, on the 22nd Soliman himself with the Janissaries, on the 23rd the Anatolian levies and the baggage. In these seven days no sign was seen of any Hungarian troops—the magnates were occupied in wrangling at Mohacs. On the 24th the whole army started off to seek the enemy, whose position had been ascertained by cavalry exploring parties.

The news of the bridge-building on the Drave must have reached the King's camp long before the work was completed, but no effort was made either to hinder it, or to fall on the first hostile corps which crossed, while the rest of the Turkish army was on the other side of the broad water. Wrangling continued; one faction maintained that it would be better to give back till the Croatians, the Voivode of Transylvania, and the Bohemians should come up. The former were now not far off—but John Zapolya was only marching on Szegdin a hundred miles away, and the head of the tardy Bohemian contingents was no nearer than Pressburg on their own frontier. It would be better, said some, to let the whole Hungarian plain from Essek to Buda be ravaged with fire and sword than to fight with an incomplete army—not half of what could be collected. On the other hand, the more bellicose party, headed by Archbishop Tomori, urged that it would be scandalous to allow half a Christian realm to be devastated, when a very considerable army was now collected. For between the 24th of August and the 29th, the day of the battle, some powerful reinforcements came in; the King's artillery train arrived by boat from Buda, with no less than eighty cannon, mostly small. The Ban of Croatia brought up 3000 horse and nearly as many foot, the Bishop of Agram 700 light horse, while 2000 foot with a reputation for excellent archery came from the lands of the Upper Drave and the bishopric of Fünfkirchen. On the 29th there were collected 12,000 horse and 13,000 foot—the former mostly lightly equipped 'Hussars,' but including quite 4000 of the noblesse in full armour. Of the foot 5500 were veteran foreign mercenaries, the rest local levies, mostly from mountain regions, not inexperienced in border warfare. Archbishop Tomori encouraged his partisans by saying that great part of a Turkish host was composed of grooms and camp-followers, and half-armed irregular infantry (Azabs); he did not believe that there were 70,000 real fighting men in Soliman's vast host. None of his cavalry was heavily armed, the Timariots being only light horse, his gunners were Italian and German renegades. There were many other renegades in the Sultan's ranks, whose hearts were not with their master.

The resolution to stand and fight at Mohacs was passed in the council of war, military pride, and the shame felt at the idea of a retreat without a battle, prevailing over the vaticination of some doubters. One of the latter, Francis, bishop of Varasdin, is said to have whispered to the King that the Pope had better make arrangements for the canonization of 20,000 Hungarian Christian martyrs, if there was going to be a battle. He was not far out in his prophecy, and he himself was one of those who fell. But the majority of the Hungarian leaders were carried away by the enthusiasm of the Archbishop, who had a fine fighting reputation behind him, and knew Turkish armies well. But why had the line of the Drave not been held if there was to be a clash?

After burning Essek, which was on the south bank of the Drave and undefended by the Hungarians, the Turkish army took four days in advancing from the bridge to the vicinity of Mohacs; all the time it was in touch with parties of Tomori's light horse, who were observing its approach; there was continual rain and a number of sumpter-beasts are recorded to have perished. On the evening of the 28th, Soliman reached a line of low hills running east and west, from which the plain of Mohacs and the fortified Christian camps beside the town were visible. Collision must obviously come on the next day.

Soliman had been warned of the tempestuous fury of the charge of Hungarian horse, and was aware that he was faced by thousands of heavy-armed knights, as well as by the 'hussars' of the general cavalry levy. He resolved to fight in a much deeper order than his father had adopted when he faced the Mamelukes at Dabik and Ridanieh. Instead of putting the Roumeliot and Anatolian feudal horse on the wings with the guns, Janissaries, and guard-cavalry in the centre, he formed up his army in a triple line: the Roumeliot horse were in front, the Anatolian horse in support of them, the Janissaries flanked by the Spahis and Silladar squadrons in reserve, with (as it seems) all the very large provision of guns. For the Hungarian accounts speak of no hostile cannon in the first episodes of the battle, and make them prominent and decisive at the end, when the front lines of the enemy had been thrust back. Besides scouting parties of irregulars spread before the front of the battle, we are told that Soliman had detached on his left wing a large body of these 'Akindjis' with 4000 Bosnian horse, at a distance of a mile or so from the main order of battle, with orders to come over the hills independently, and outflank the Hungarian army at all costs, when the fighting should grow hot. As to the total numbers of the Sultan's army it is impossible to give any definite estimate, as the Christian writers raise it to 300,000 men in order to give an impression of hopeless odds against the Hungarians; while the Turkish narrators (mostly very inaccurate) also give enormous numbers to Soliman, but at the same time incredible multitudes to the enemy —whose force of 25,000 men of all arms is ascertainable with fair accuracy. On the whole, I should not be surprised if Archbishop Tomori's guess that the Turks put about 70,000 fighting men in line was pretty near the truth. This would allow for 35,000 feudal horse (Roumeliot, Bosnian, and Anatolian), 15,000 for the Janissaries and the regular cavalry, and 20,000 more for the very numerous irregulars, volunteer horse, and foot. But the estimate may be too high all round.

George Zapolya and Archbishop Tomori were aware, from the reports of their exploring parties, that the enemy was lying invisible on the back slope of the low hills which bound the plain of Mohacs to the south and south-west. Since it had been determined that battle should be accepted, there were two alternatives—to make it defensive or offensive. We are told that Leonard Gnomski, the Polish condottiere, pleaded for the use of the old Hussite tactics, to cover the flanks and rear of the army with the massed waggons of the train [1] and to allow the enemy to attack against a front garnished with artillery and missile-bearing infantry. When he should have charged and have been beaten off, the whole army

[1] 'Curribus quibus rex maxime abundabat undique cingentur, ut intra septa et munimenta certamen fieret.' Brotheric, 1190.

should sally out in a counter-attack and would (if old experience went for anything) achieve a complete victory, like the Hussites of old.

To this the objection was that if the royal troops took a purely sedentary position, and waited to be attacked, the enemy would not commit himself to a serious frontal assault,

THE CAMPAIGN OF MOHACS, JULY–SEPTEMBER 1526.

but would manœuvre round the flanks, and cut off the army from Mohacs, its camps and stores, and its communications with the rear. It would be surrounded by very superior numbers, and forced to try to break out, or it would starve. Thus the advantages of the defensive would be lost. Gnomski's proposal was not negatived, apparently, but put into consideration late on the evening of the 28th. On the afternoon of the 29th, when the Turks came over the hills, nothing had been done, and the actual order of battle adopted was to make a very long front, in order to prevent the outflanking to which a short front would be exposed.[1] But this failed to guard against the detached Turkish force far to the northern side, which was no part of the Sultan's main line of battle. The Christian army was finally drawn out on the morning of the 29th—the day of St. John the Baptist's decollation—in two long lines. The front line was formed of nearly all the infantry, ranged in dense columns, with the greater part of the feudal horse of the lay barons drawn up in the broad intervals between each block of infantry.[2] The right of the line was under the command of Francis Bathiani, the Ban of Croatia, and included all the Croatian levies. The left was led by the Ban of Temesvar, Peter Perenni, who had won reputation in frontier fights with the Turks; he was the brother of the Bishop of Warasdin. The guns were placed somewhere in the centre.[3]

The second line was really double, though it is spoken of as a single unit (*acies stativa*) by the chroniclers. The first consisted of three squadrons of feudal horse; behind these rode the King and his selected guard, 1000 fully equipped knights in full armour, with the royal standard above them. Louis's personal safety was entrusted to three gentlemen of approved valour, who brought with them a provision of swift horses for his use, in case of any untoward happenings. On the King's flanks were the levies of the bishops—to the right those of the Archbishop of Gran, and the Bishops of Agram, Warasdin, Fünfkirchen, Nitria, and Bosnia; to the left those of the Bishops of Raab and Vacz, to whom were joined some lay barons and their retinues. Two small bodies of infantry were arranged as flank guard to the *acies stativa*, and outside these there were some light horse.

The front line may have consisted of 10,000 foot and 6000 horse, the double rear line of 3000 foot and 6000 horse. All were native Hungarians save the 5500 mercenary infantry,

[1] 'Quanto latius extendi posset, quo illud est potissimum spectatum ne ab hoste circumveniremur.' *Ibid.*

[2] 'Pretensis peditum cohortibus in oblongam seriem, intemixtis in idoneo loco equitibus.'

[3] Brotheric's odd phrase is that 'tormenta belli statim post principia posita erant,' but as he says (p. 1192) that the action opened with a general discharge of the Hungarian cannon, they cannot have been *behind* the first front, but must have been *in* it somewhere.

and a small party of Bohemians and Moravians, mostly of the retinue of Stephen, Count of Schlik, who were the only help that came from Louis's second realm. They were placed in the array of the feudal horse in the front of the *acies stativa*. The two generals-in-chief, Archbishop Tomori and George Zapolya, are said to have taken no fixed post, but to have moved up and down the front line giving orders 'nullo certo loco, sed ubi res postulasset.' The Palatine, Stephen Bathori, seems to have done the same sort of Marshal's work for the main-battle.

The army was drawn up a mile inland from the town of Mohacs and the camps outside it, and half a mile from the nearest marshy bank of the Danube, which had overflowed from its usual course on account of the recent rains, and had blocked the exits into itself of several brooks and ravines. The ground was absolutely flat and featureless, admirable for cavalry manœuvres, trending up very gently toward the low hills behind which the Turks were known to be lying. The only thing that caught the eye was the village and church tower of Feldvar on the southern down-slope of the ridge.

Soliman was intending to assume the offensive, but it was a very long time before he could get his large army into battle-order; meanwhile he kept behind the hills, invisible to the Hungarians. The latter were drawn up to their fighting disposition, waiting to be attacked, early in the morning. But the enemy came not, though the noise of moving multitudes could be indistinctly heard. The whole of the hours before noon were spent in waiting; the cavalry dismounted and took off their helmets, for the day was warm, and armour is heavy wear. Those who had food took hasty mouthfuls. It was not till three hours after noon that the whole crest of the hills in front became bright with lance heads, and presently the dense masses of Turkish horse began to roll slowly down the slope in excellent order, line after line. Archbishop Tomori is said to have caught some glimpse of the Sultan's turning corps, far to the north, and to have proposed to make a detachment against it from the rear-line or main-battle. But the barons to whom he spoke refused to break their order,[1] whereupon he himself with his personal escort rode out in this direction, but, seeing the

[1] He happened to urge Gaspar Raskay to turn off to the flank, one of the barons to whom the personal charge of the King had been committed. Raskay said that he was responsible for Louis's safety and could not leave his post. Brotheric, p. 1192.

Turkish flanking party making no further advance, returned to the main body where a decision was obviously imminent.

When the enemy had come down the slope and got within cannon-shot of the Hungarian front, the whole line of guns opened on the Roumeliot squadron, with no very great effect—as was usual in sixteenth-century battles. But it was otherwise with the cavalry, who charged most efficiently when the Turks closed, and broke their front in many places. The whole Hungarian line advanced, pushing the enemy before them, and finally thrust the Roumeliot horse against their Anatolian supports. It seems likely that such a general charge must have caused breaks and gaps in the pursuing force, for infantry cannot keep level with horse charging at full speed. Brotheric, our best informant, hints that the complete retirement of the Turks may have been due to a desire to draw the Hungarians under the fire of the row of cannon in the third line of their host.[1] But at any rate the Sultan's feudal horse were thrust far back, and lost heavily in casualties. Andrew Bathori rode back from the front to the King, and reported that the whole Turkish army was giving way, and that it was time for the Hungarian 'main-battle' to come on and complete the victory. Louis gave the order for a general advance, and the second line passed over the ground where the Hungarian guns were still standing, and up the slope in front of them, where Christian and Moslem dead and wounded were lying in great numbers—many more Turks than Hungarians, as was noted.

When the 'main-battle' had come into the rear of the wild *mêlée* which was before them, it became clear that the game was far from won. For the troops of the right flank (the wing under the Ban of Croatia) were not only making no further progress, but giving way.[2] This seems to have been the result of the coming in to the fight of the Turkish detachment which Soliman had thrown out far to his left, with orders to outflank the enemy. The horse of the second line are said to have thrown back their right squadrons *en potence* to relieve the pressure, but to no effect.[3] But the main body joined with the disordered

[1] 'Hostis cedere incipit, sive nostrorum impressione coactus, sive ut nos ad tormentorum loca pertraheret.' *Ibid.* p. 1192. One Turkish chronicler (Petschawi) also says that the retreat of the feudal horse was deliberate, in order to get the Hungarians scattered and mixed up with the yielding squadrons.

[2] 'Inclinare cornu dextrum coepit, et plerique ex eo cornu fugam capessere.' *Ibid.*

[3] This seems implied by the statement of Petschawi and of Soliman's Journal that the Hungarians made a new front to meet the flank attack.

troops of the front line in an effort to break through the Turkish centre. They certainly thrust aside the Anatolian as well as the Roumeliot feudal horse, and got as far as the Janissaries and the line of guns in the third Turkish line. Here they failed to break in, save at a very few points, where individuals are said to have come to handstrokes with the Sultan's guard. But great numbers were mown down by salvos of the chained artillery: 'They were now fighting not on the open ground but only ten paces from the guns, which filled the air with smoke, and scared the horses.' Part of the left wing, we are told, gave way first and retired to marshy ground near the Danube, where it tried to reform for a second charge. But while it was reforming, the centre gave way, 'unable to endure the point-blank discharges any longer,' or the suffocating smoke, and the whole army went to the rear in the greatest disorder, passing through the line of the Hungarian guns, and by the flank of the camp, where it was seen that some Turks had already got in, and were slaughtering the camp-followers. There was no attempt at a rally, and (what is more extraordinary) no general pursuit by the enemy, for the Sultan bade his trumpets sound the order for every man to muster by his standard, and to resume his original position.[1]

About half the Hungarian army perished in the disaster—of the foot only 3000 out of 13,000 got away, and every one of their leaders save Hannibal of Cyprus was left on the field. Broken foot are, of course, the natural prey of victorious cavalry. Among the horse the heavy-armed noblesse fared worse than the light 'hussars'; not only did they charge in more fiercely, but their barded horses were exhausted, and failed to carry them out of the battle when the rout began. The fate of the King is veiled in obscurity: some say that he fought as far as the guns, and only retreated when the general rout began. Others allege that the gentlemen charged with the care of his personal safety took him away early, when the first signs of failure were seen.[2] At any rate, he disappeared: his body was not found next day, when the Turks searched the field. But nearly two months later, when the Danube floods

[1] 'Hostes, visa nostrorum fuga, vel fraudem subesse rati, vel pugna fatigati, diu se intra aciem suam continuerunt, et nequaquam nostros sunt insequti.' Brotheric, p. 1193. The Turkish accounts bear this out, see v. Hammer, v. p. 85. The drums and trumpets were sounding in the Sultan's camp till midnight, and it was only at dawn that Soliman went over the battlefield.

[2] Brotheric gives both stories.

had subsided, his corpse in full armour and the body of his horse were found at the bottom of a ravine or gully, a mile behind the battlefield, which had been full of water and mud on August 29. Apparently he had tried to charge through it, and was embogged and smothered. Near him were the corpses of Stephen Azel, Castellan of Presburg, and of Louis's Polish warden of the gate, Andrew Trepka, who had shared his fate.

The slaughter among the Hungarian magnates had been very great ; not only did the two 'duces belli,' George Zapolya and Archbishop Tomori, meet their deaths, but most of the great officers of the Crown, the Chancellor,[1] the Justiciar, the Lord Chamberlain, the high-butler, and the two wardens of the gate ('janitores regii '). But the clerical leaders fared worst of all—there fell not only Archbishop Tomori, but his colleague Ladislas Zalkan, Archbishop of Gran, and the Bishops of Varasdin, Raab, Canatha, Fünfkirchen, and Bosnia—indeed only three of the episcopal combatants appear to have got off the field. The high nobility had a long list of casualties, including the Croatian Count Frangipani and the Bohemian Count Stephen of Schlik ; 500 of the 'potiores regni' are said to have fallen, nearly all the old noble names can be recognized among them, with four or five thousand other horsemen, 'gentle and simple.' A number of prisoners were made, but the Sultan had them all decapitated next day, as his ancestor Bajazet had done with the French nobles after the battle of Nicopolis in 1396. Only five were spared and held to ransom (as John the Fearless and Boucicault had been at Nicopolis). The heads were made into a ghastly pile, and that of Archbishop Tomori was shown on a pike—according to the disgusting habits of the East.

Of the Hungarian leaders the Palatine, the Bishops of Agram and Nitria, and the Ban of Croatia escaped from the field, with a good many of the light horse, who are said to have behaved very badly in plundering the countryside on the way home.[2] No attempt was made to rally the wrecks of the army. On the day of the battle the first contingents of the Bohemian kingdom had only reached Stühlweissenburg and Raab, a few miles inside the Hungarian frontier. It is clear that if the King

[1] Who was the Archbishop of Gran. Brotheric, the historian of the campaign, acted as his deputy-chancellor.

[2] 'Levis armaturae milites, qui integri superfuere proelio, latrocinio conversi et praedae.' Flaminius, p. 62.

had retired to join them, as some of his advisers had recommended, the junction would have taken place so far to the rear that the whole Danube plain, almost as far as Buda, would have been abandoned to the enemy. Nothing can be more disgraceful to the Bohemian authorities than the fact that, after four months of warning, no one save the Count of Schlik and a very few others had reached the front. Local factions and a general dislike for the Hungarian connection were responsible for this unpardonable slackness. The widowed Queen Mary of Hapsburg fled to the fortress of Pressburg, the last place within the Hungarian frontier, and finally took refuge with her brother the Archduke at Vienna.

Soliman, after three days' halt on the battlefield, advanced unopposed as far as Buda, which he found entirely undefended. On the way his army burned the cathedral city of Fünfkirchen and many smaller places, and swept the whole countryside for plunder and captives. The men were mostly slaughtered, the women and children carried off for the slave-market at Constantinople. On entering Buda (September 10) the Sultan ordered that the place should be spared—but nevertheless it was almost entirely burned, save the royal palace. From this Soliman carried off the treasures and famous library of Mathias Corvinus, also (a thing which shocked some strict Mohammedans) three large bronze statues from Italy—Hercules, Diana, and Apollo—which had been the pride of King Mathias. He had them set up in the Atmeidan of Constantinople opposite the famous Delphic serpents. A later generation of Turks destroyed them, as offensive heathen idols. The irregular horse swept the whole countryside as far as Lake Balaton and the gates of Visegrad, Gran, and Komorn. If the Sultan had cared to press the campaign he might have had these fortresses without much trouble, for the castellans were fled or dead, and the walls were only manned by an irregular gathering of peasants and monks under plebeian captains.[1]

But Soliman had resolved not to annex Hungary, but to make it into a tributary principality, like Wallachia, and in this design he was aided by the abominable ambition of John Zapolya—*pestis Hungariae* as Leonclavius rightly styles him—whose emissaries came to the Sultan at Buda, and offered their master's services to him. The Turkish army left the royal city, crossed the Danube to Pesth by a bridge of boats on September

[1] Brotheric, p. 1195.

23–25, and retired southward across the thinly peopled plains between the Danube and the Theiss. Meanwhile, Zapolya came west with the Transylvanian levies which had been missing on the field of Mohacs, occupied Buda, and had himself crowned at Stühlweissenburg on November 10 by the Bishop of Nitria, one of the few Hungarian prelates who had escaped with his life from the disaster of August 29. Soliman had left garrisons only in Peterwardein and the castles on the Save—the rest of the wasted land was left unoccupied.

But Zapolya was only the head of a party. The obvious claimant for the crown of Louis II was his sister Anne, the wife of Archduke Ferdinand,[1] and the Palatine Stephen Bathori summoned a diet to meet at Pressburg, which elected Anne and her husband as King and Queen of Hungary. This was all the more reasonable because Ferdinand and Anne had already been acknowledged as sovereigns of Bohemia by a diet called at Prague (October 26). For the defence of Christendom it was necessary that Hungary and Bohemia should remain united, and the Austrian archduke was the powerful neighbour who stood next in the line of defence to both the kingdoms of his late brother-in-law. The Hapsburgs were loved neither in Bohemia nor in Hungary—but anything was better than the prospect of subjection to the infidel Turk.

Nevertheless the double election of Zapolya and Ferdinand meant nearly two centuries of disunion for Hungary, and for Europe deadly perils when the Sultan should resume his attack. The buffer-state on the Danube, which had seemed so strong in the days of John Hunniades and Mathias Corvinus, had been destroyed. And one of the two new kings was a traitor to Christendom. There was to be a Hungarian contingent in the Turkish army which besieged Vienna in 1529.

N.B.—Nearly all the accounts of this campaign on the Christian side are borrowed from Stephen Brotheric, *e.g.* Cuspinianus and the rest. There are a few notes of value to be got from Johannes Camerarius and Flaminius and Giovio. For the Turkish narratives, full translations are not to be got.

[1] This had been settled by marriage treaties in 1519, as well as by earlier negotiations in 1491 between the Emperor Maximilian and Ladislas of Hungary. But could treaties bind an elective monarchy?

CHAPTER VI

THE SIEGE OF VIENNA (SEPTEMBER–OCTOBER 1529)

THE Emperor Charles, his brother at Vienna, and indeed all Europe, expected that the Sultan would return to the Danube in 1527 to complete the work of Mohacs. But this was not to be: already while he was deep in Hungary, Soliman had begun to receive very disquieting news from Asia Minor. It was probably the absence in Europe of the whole feudal cavalry of the East which encouraged sedition. Two separate insurrections broke out in the summer of 1526. In August the Turcoman tribes in Cilicia and Karamania rose: they massacred tax collectors and local magistrates, and were only put down after some months by the governors of Adana and Diarbekir. But a more serious outbreak came in the following winter—there was one of those sudden explosions of religious fanaticism which have so often occurred in Moslem countries—the rising of the Soudanese Mahdi in our own time was the most recent and not the least dangerous. A sheikh named Kalendaroglu raised a horde of dervishes in central Asia Minor, defeated and slew the Beglerbeg of Karamania (who had just returned from Hungary), and swept all the regions as far as the Euphrates. The Grand Vizier had to be detached from Constantinople with a force of Janissaries to aid local forces, and Kalendaroglu was only subdued and executed, after much fighting, on June 22, 1527.

This diversion made it impossible for Soliman to undertake serious operations in Europe during the year. But his governors on the Danube front utilized the complete demoralization into which the kingdom of Hungary had fallen, by capturing its last strongholds inside the Bosnian frontier—Jaicze and Banjaluka—which had been held as advanced positions ever since the time of Mathias Corvinus, and they subdued all the smaller places up to the borders of Croatia. Neither of the two rival Kings of Hungary could send any aid to the Croatian Ban.

Ferdinand and John Zapolya were indeed engaged in a lively civil war during 1527; the Austrian got the mastery, and with the aid of Bohemian levies took Raab, Komorn, Gran, Stühlweissenburg, and Buda, and chased the Transylvanian eastward, till after a defeat at Tokay he was driven as far on the frontier of Poland, and sent despairing appeals for help to the Sultan, to whom he had already offered his homage in the preceding autumn. Soliman promised assistance, and sent away from Constantinople with disdain ambassadors whom Ferdinand had dispatched to him proposing a peace, and claiming as his right the old boundaries of the Hungarian kingdom. Why the attack on Austria was not delivered in 1528, but only in the following spring, is not easy to discover. It is true that troubles in Asia Minor had not entirely ceased—there was a recrudescence of sedition in Cilicia early in the year, though it was put down with no great difficulty. But probably the Sultan and his advisers were aware that in challenging the Archduke-King they were also challenging his greater brother the Emperor, and knew that military preparations must be made on a scale such as had never before been required, for an enterprise which meant not only the reconquest of Hungary, but a deliberate thrust at the heart of Christendom. Nor was it of less importance that Soliman expected that his secret alliance with the King of France, who had broken the Treaty of Madrid and sent an army into Italy once more, must be given time to work, before the Emperor was assailed from the East. It is true that Francis in the end left his ally in the lurch, and concluded the unexpected peace of Cambray with the Emperor in July 1529, just as Soliman was reaching the Danube at Belgrade, but this could not have been foreseen in 1528.

The Sultan left Constantinople on May 10, 1529, and reached Essek on the Drave on the 6th of August with an army whose numbers are estimated by the chroniclers, with their usual vagueness, at 120,000 men and 20,000 camels. Up to this point he had found no resistance, and on the 18th he was met on the ill-omened field of Mohacs by John Zapolya, who had brought with him a considerable following from Transylvania—6000 horse—and did homage, proceeding to present to his patron the time-honoured crown of St. Stephen, which had fallen into his hands, in order that it might be restored to him with due formality. With Zapolya clearing the way in front of him by promises of security to the inhabitants of the countryside,[1] the Sultan came before the walls of Buda on September 3. This was the first place that offered resistance; Ferdinand had thrown into it a German garrison, or it might have opened its gates to Zapolya, as Fünfkirchen and Stühlweissenburg had already done. On September 8 the citadel capitulated on terms, which were not kept, for most of the Germans were massacred by the Janissaries as they left the gates, in spite of the Sultan's pledged word. The palace was formally handed over to Zapolya, but a Turkish garrison was left in his capital. On September 10 the Grand Vizier and the vanguard resumed the march on Vienna, the Sultan and the main body following a few days later. Much dismay was caused in Austria by the fact that most of the Hungarian towns on the south bank of the Danube, along which the Turkish advance came, opened their gates to the first summons—pretending to acknowledge the rights of John Zapolya as king, though really they were betrayed by mere panic. The most notable defection was that of the Archbishop of Gran—the successor of the primate who fell at Mohacs three years before. The Turks were kept from spoiling the countryside as long as they were in Hungarian territory, ostensibly friendly ground; it was only after they had crossed the Leitha and got into Ferdinand's archduchy that the normal atrocities began. The only Hungarian town which stood firmly to defend itself was Pressburg—the very last fortress in the realm. The Sultan had provided himself with a flotilla on the Danube, which brought up much of his artillery and munitions. It was cannonaded from Pressburg as it passed, but got through without serious damage.

Ferdinand had received at least three months' warning of the approach of the enemy, and had not only gathered his own Austrian and Bohemian forces, but appealed to his brother Charles, and to the Diet of the Empire, then in session at Speier. The Emperor was entangled in his French war, which did not come to an end till July, and could spare little help till it was too late. But he sent his brother 700 Spanish foot under Luis de Avalos from the army of Italy, whose arquebusiers are said to have done specially good service during the siege of Vienna. The Diet made considerable promises, and chose Frederic, the Elector Palatine, as general; but they were slow in moving. Fourteen 'ensigns' of foot

[1] See Melchior Seiler's *Viennæ Obsessæ Historia*, p. 1208.

under Philip of the Palatinate, nephew to the Elector, had reached the front by the end of August, and a small force of cavalry. But the main body was late, and the Elector himself only reached Krems, on the wrong side of the Danube, with seven more 'ensigns' just too late to enter the city. Other contingents were still more dilatory, and accumulated in the rear while the siege was in progress. Ferdinand's own forces told off for the defence of Vienna were mainly from the arch-duchies—very few Bohemians were forthcoming. But the troops selected for the garrison were mainly veteran units, under experienced captains. There were about 500 Austrian, Styrian, and Carinthian horse, and over twenty 'ensigns' of foot from the same lands, with four 'ensigns' only from Bohemia. The whole force is said to have amounted to 16,000 foot and 500 or 600 horse, to which must be added the burgher-guard of Vienna, who were, however, mostly told off to internal activities—the prevention of fire, pioneering, and transport work, though they were occasionally put into action. This was a very considerable garrison for a city of no great size, for the Vienna of those days extended only as far as the famous 'Ring' which now makes a green belt around the ancient nucleus of the original city.

Vienna was surrounded in 1529 by a mediaeval wall not more than six feet thick in many places, and with very few of the bastions, useful as artillery emplacements, which were the main feature of Renaissance military architecture. On the eastern side there lay under the walls the course of the Wiener Bach, a small river which acted as a wet ditch, and was only crossed by one bridge, at the Stubenthor gate. Similarly, the northern front lay along a broad branch of the Danube, a by-pass from the main stream, which was a protection to about a third of the circumference of the city. Over this there was both land communication by a bridge, and also water communication with Bohemia, until the Turks blocked the waterway during the siege and broke the bridge. The accessible fronts, along which the attack was certain to come, were the western and southern, where from the point at which the city wall turned in from the Wiener Bach (about the modern Stadt Park), and lost its outer wet-ditch, as far as the tower by the Jewry, not far from the Danube bank, there was a simple mediaeval wall. This went along the modern Schottenring, Burgring, and Kärntherring; it was pierced with three fortified gates, the

Schottenthor,[1] Burgthor, and Kärntnerthor, and had one good bastion of recent construction at the last-named gate, which was the most exposed point of all. There was only a dry ditch outside the wall, no water protection as on the eastern and northern sides of the city. The old citadel of the Margraves, a much simpler affair than the modern complicated 'Hofburg' with its annexes, lay immediately behind the Burgthor with the church of the Austin Friars beside it. Close inside the Kärntnerthor and bastion was the nunnery of Saint Clara. It was against the limited space between the Burgthor and the south-eastern angle of the city wall that the whole of the energy of the Turks was directed. Opposite the remainder of the land front—the Schottenring, etc.—they cast up trenches and erected some batteries, but made no serious attempt to break in.

The Palatine Count Philip was in nominal charge of the defence, but was practically guided by the lieutenants, Nicholas Count of Salm, Archduke Ferdinand's best captain—a veteran of seventy years of age—and William von Rogendorf, the Marshal of Austria. The silence of the chroniclers as to any personal activities of Count Philip, and their expressions of enthusiasm as to the conduct of Salm and Rogendorf is sufficiently enlightening. But the details of the defence speak for themselves; every precaution was taken—all houses outside the walls were levelled to the ground; those just inside the city with roofs of wooden shingles or thatch were stripped bare, to guard against red-hot shot or fire-arrows. An inner earthen retrenchment was thrown up within the south-eastern corner of the city, capable of being defended if the main wall was pierced on either side of the Carinthian gate, where the greatest peril was expected, and the bastion beside it was stocked with the best of the guns which the garrison possessed. When the manning of the different fronts of the enceinte had been provided for, separate reserves of picked troops were assigned to each sector. The bulk of the burgher-guard were told off into fire-brigades and munition trains, only two companies being put under arms. Provision was made for breaking the bridge over the Danube-branch in case the Turkish flotilla should land men to occupy the farther bank. The Archduke had some armed launches on this water, but as it was hopeless

[1] So called from the monastery of Scoto-Irish monks, which lay inside the wall—one of the earliest buildings of the town, founded in the eleventh century.

to send them against immensely superior numbers, they were dismantled and sunk in their dockyard.

Speaking generally, the imperial foot companies had charge of the east front, from the bridge and the Rothenthurm Tower as far as the Stubenthor; the Bohemian companies were placed on the north front, from the bridge as far as the tower of the Jewry, along the wall above the Danube-branch. Austrian and Styrian companies had the posts of greatest danger, on the western front from the Carinthian gate to the Schottenthor. The reserves included the whole of the cavalry, who served dismounted save when they were sent out on sallies, the 700 Spaniards, two picked companies of the burgher-guard, several Austrian ' ensigns,' and a body of unattached volunteers, all of noble birth, who had come in unassigned to any regular corps.[1]

On September 23 the garrison got into touch with the advancing enemy; cavalry exploring down the Simmering-Schwechat road being beset by overpowering numbers of Turks, and driven back to the Stubenthor with some small loss in dead and prisoners. While the Sultan's main body continued to come up, and to spread itself progressively along the southern and south-western side of the city, a vast swarm of irregular light horse went ahead, and began to devastate the countryside, all along the Danube and southward also into Styria. All Lower Austria as far as the Enns was ravaged with fire and sword, with the usual massacres of unarmed peasantry and capture of women and boys for slavery.

It took three days (September 23–26) for the Turkish army to develop itself around Vienna; the Sultan encamped with his guard-cavalry by the Church of St. Mark in front of Schwechat. Ibrahim Pasha, the Grand Vizier, with the Roumeliot feudal horse, lay in front of the Sultan, from the bank of the Danube as far as the hill called the Wiener Berg, the highest ground on the south side of the city. On his left was John Zapolya's Hungarian contingent under Peter Pereny and Paul Verdenz. The artillery park, which was reputed to hold 300 pieces large and small, was fixed below the Wiener Berg under the guard of 12,000 Janissaries; the Asiatic and Bosnian feudal horse were posted on the western side, on the lower slopes of the wooded hills of the Wiener Wald. Finally,

[1] Nobiles qui nullo duci addicti proprio aere militaverunt,' as they are called in the *Obsidionis Historia*.

on September 27, the flotilla, which had followed the main branch of the Danube, came down into the smaller Vienna branch from the north, and cut the line of communication with Bohemia, landing several thousand men at Nussdorf.

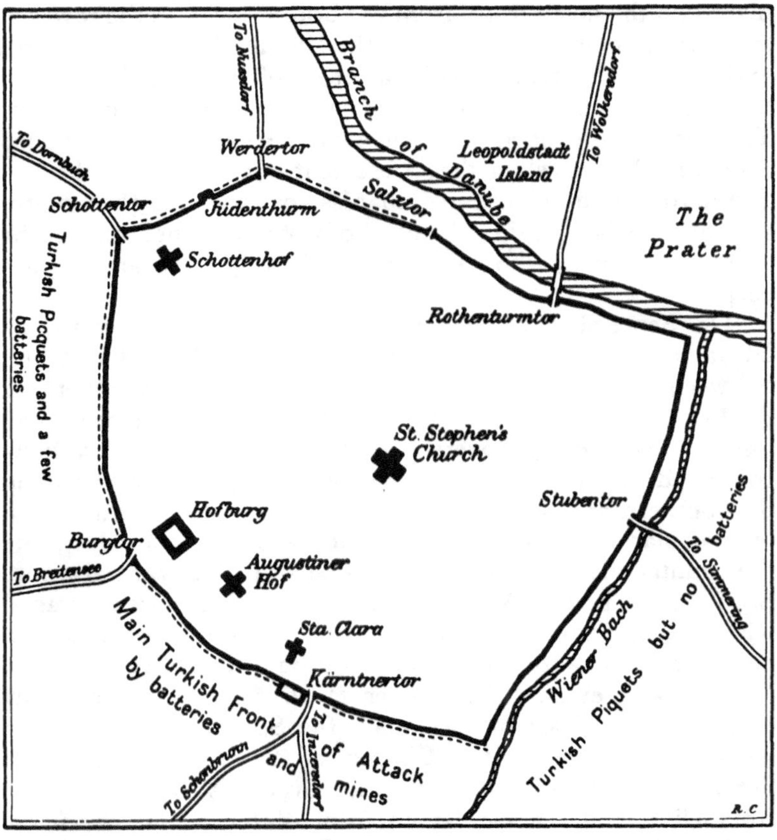

VIENNA IN 1529.

Before the construction of the new bastioned front, which swept away all the old towers existing at the time of the Turkish siege; their position cannot, therefore, be accurately given.

They finished the destruction of the bridge, which the garrison had already begun, by removing two arches of the wooden structure when the approach of the Turkish boats was discovered. Thus Vienna was completely cut off from that touch with Krems and Bohemia, across the bridge and the causeways in the islands of the Danube, which still existed during the three first days of the siege.

After surveying the enceinte of Vienna the Turkish engineers came to the very same conclusion at which the commanders in the city had already arrived—namely, that the best front for attack would be the half-mile on each side of the Carinthian gate. They left unmolested both the eastern walls covered by the Wiener Bach, and (what surprised the Austrians more) the long stretch from the Jews' Tower to the Burgthor on the north-west side of the defences. These fronts were only observed by strong pickets thrown out in front of the camps. But cover was at once sought opposite the south-western walls, trenches being thrown up, and emplacements for artillery sought for, in the ruins and cellars of the houses which had been destroyed by the garrison before the siege began. A system of approaches by zigzag trenches and parallels was sketched out the moment that the investment had begun, and arquebusiers and bowmen were pushed out into the sap-heads to molest the defenders on the walls. It took some days to get the guns forward into the prepared battery-places.

Meanwhile the commanders of the garrison showed that they did not contemplate a mere passive defence. On September 29 a sortie in force, with no less than 2500 foot, was made from the bridge at the Stubenthor, which swept all along the front of the inchoate line of approaches, cut up the workers and the trench guards, and returned with little loss. Ibrahim Pasha, the Grand Vizier, who had been surveying the advanced trenches, narrowly escaped capture. On October 2 a sortie in smaller force was made, in which the Spanish foot took part ; it did some damage, but suffered some loss.[1]

Serious battering by artillery on each side of the Carinthian gate commenced on October 1, and continued regularly with no very decisive results, though battlements were knocked off, and some guns disabled. Meanwhile the Turks from the front trenches kept up a continual fire of arquebuses, and discharge of arrows against the ramparts. But what the commander of the garrison most feared was that mining was going on—suspicious accumulations of earth were seen behind the trenches, and the addiction of the Turks to this form of attack was well known. A Christian refugee, who escaped into the city on

[1] 'Pluresque hostium oppresserunt, ac spolia tulerunt, ipsi tamen suae temeritatis poenas dedere.' The Sultan's diary mentions this.

43

October 1, gave useful information as to the locality of the mining, which was at first all directed against the Carinthian gate. Counter-mining was begun at once, and very soon some Turkish mine-heads were detected and blown in, but others were suspected to be in construction. At dawn on October 6 a very heavy sortie was prepared, with the object of clearing the whole ground behind the enemy's front trenches. Eight thousand men sallied out from the Salzthor, on the Danube side, and passing all along the west front of the city, and sweeping away the Turkish picquets, fell on the approaches opposite the Burgthor and the gate of Carinthia. Immense damage was done, but the retreat of the sallying force into the Burgthor cost many casualties—the troops got jammed in the narrow gate of re-entry, and the rear companies fell into panic and disorder, and were badly cut up. Both sides lost heavily: the Sultan's diary—always reasonably honest—confesses that the Turks lost 600 men, including one Roumeliot officer of rank. The mining, however, was renewed—on October 8 the Sultan's diary records that the whole army was kept under arms all night for fear of another sortie. On the 9th two mines were exploded near the nunnery of St. Clara beside the Carinthian gate, and an assault followed: but the besieged had prepared beams and stakes, with which the gaps were at once blocked, and the enemy failed to get in: the Sultan's diary records hard fighting in which the Pasha of Semendria took a good part—but the failure was complete, and the loss to the stormers heavy.

Soliman, however, was still set on the same game—hoping in vain that the garrison might be exhausted, and unable to resist more assaults, which he was able to deliver with fresh troops owing to his immensely superior numbers. Bombardment went on all day upon October 10, and on the following day another mine was exploded, which made a gap in the wall near the nunnery of St. Clara: it was not very wide, and an assault on it by Albanian contingents [1] was repulsed with loss. On October 12 a council of war in the tent of the Grand Vizier was held, at which many officers urged that the season was growing cold, that food could no longer be got from the ravaged countryside, and that, since no practical progress had been made, the siege had better be raised. It was, however, resolved that a final effort should be made before the order for retreat

[1] Soliman's diary says that the breach was so narrow that only the Beys of Janina and Avlona were sent against it, and failed to break in.

was given. On that same day two successful mines to the east of the Carinthian gate were fired, and shook down two larger sections of the wall than any which had yet been injured, especially one near the extreme south-east corner of the city, where the Carinthian front joined the Stubenthor front. The besieged blocked the gaps with barrels of wet earth during the night, and transferred their best reserve-troops to the threatened points. The expected assault was duly delivered; the Sultan promised a distribution of 1000 aspers a head to the Janissaries, if Vienna should fall. The first man over the walls should receive a timar fief worth 30,000 aspers. Soliman himself visited the trenches and inspected the breaches. But the final attack proved less formidable than had been expected; the Turks had been disheartened by their repeated failures and heavy losses during the earlier assaults, and the troops flinched: the besiegers saw pashas and beys belabouring them with staves and maces to turn them back towards the breaches, but all in vain. 'It was no worse to die by the hands of their officers than by the bullets of the long arquebuses of the Germans and Spaniards.' At three o'clock on the afternoon of the 14th the last feeble assault flickered out, and the Sultan gave orders for the retreat which had already been discussed at the council of war on the 12th.

At midnight the besieged saw the huts and tents of the Turks burst out into flames, much to their satisfaction; but their joy was tempered by the sound of screams and lamentation from the camp—the enemy massacred that night all his prisoners from the Austrian countryside, save young boys and women, who were reckoned worth preserving for the slave-market of Constantinople. A sortie of cavalry from the city came upon thousands of corpses. Meanwhile Ibrahim the Grand Vizier covered with his Roumeliot horse the general retreat, and particularly the transport of the heavy guns to the Danube flotilla, which took up the whole day of October 15. On the 16th the Sultan was well on his way towards Buda, the Grand Vizier always bringing up the rear, and being forced repeatedly to drive back parties of Austrian horse, who followed him persistently, cutting up all stragglers, and getting some booty from foundered transport. The retreat was disastrous, for many of the Turks were half starved when it began, and dropped by the way, while on the 17th snow began to fall, many days earlier than is usual in the Danube valley. The roads became so impracticable that on October 21, as the Sultan's diary records, it became necessary to abandon the transport—the cars were piled together and burned. Horses and camels perished by the thousand. Moreover, many vessels of the flotilla, carrying artillery, were sunk by the guns of Pressburg, as they ran past that fortress.

It was poor comfort that on October 27 King John Zapolya came out of Buda to repeat his homage, and (so the Sultan's diary records) to compliment his master on the 'success' of his campaign. The so-called success consisted in the fact that Soliman had replaced his vassal in the Hungarian capital, had gained possession of several of the Danube fortresses, and had ravaged Lower Austria with fire and sword. It was calculated, however, that he had lost as many fighting men before he got home as he had slaughtered Austrian peasants. And the moral effect on his army had been disastrous—every one knew that it had marched to take Vienna, and to break into Germany, and it had failed completely before a very courageous and scientific defence. The regular infantry of Spain and Germany were the equals or superiors of the Janissaries—it was no longer with hordes of disorganized feudal cavalry—as at Mohacs—that the Turk had to deal, but with trained veteran troops, and skilled artillerists.[1]

This was certainly the most perilous moment for Europe in all the long strife between the House of Hapsburg and the Ottomans. If Vienna had fallen, the Sultan had intended to make it his winter quarters and base of operations for a continued assault on Germany. He started with the knowledge that Charles V was tied up in his French war, and he could not have foreseen the fact that Francis I would make peace with the Emperor in July, just as the Turkish army was reaching Hungary. It was too late in the year for Charles to transfer his energies to the Danube, and his brother had to face the attack alone—but whether Vienna fell or no, the next campaigning season would have seen the full force of the House of Austria turned to the defence of the Empire, and the German Pro-

[1] It must be confessed that the Imperial landsknechts, who had behaved so well during the siege, fell into the riotous habits of their class when it was over. On October 23 they burst out into a sedition, demanding a triple 'battle gratuity,' threatening their officers, and proposing to sack the town! It was with great difficulty that the Count Palatine Frederic pacified them, by promising two gratuities when the Archduke and the Emperor could find the money. See the last paragraphs of the *Obsidionis Historia*, p. 1218.

testants were sufficiently scared to lend their help against the Turk for reasons of self-preservation. It is curious to find that the ever false and selfish King of France quite unintentionally worked for the defence of Christendom on this one occasion. Before he died, he did his best to cancel any credit that might have come to him for signing the Peace of Cambray, by leaguing himself again with the Infidel at moments which he judged advantageous for a stab in the back at his enemy the Emperor.

CHAPTER VII

SOLIMAN'S SECOND INVASION OF AUSTRIA (1532)—SIEGE OF GÜNS—THE PERSIAN WAR (1533–35)

SOLIMAN and his household troops only reached Constantinople on December 16, 1529; the feudal horse had been dismissed and permitted to make their way home on November 11, after the retreating army had passed the Danube at Belgrade. As the weather was abominable—snow and rain alternating—the losses during the last month of the campaign must have been very heavy, during the passage through the defiles of the Balkans. There was no prospect of being able to reassemble an army such as that which had failed before Vienna during the oncoming spring of 1530, and the Sultan let the summer pass by, while useless negotiations were being carried on with ambassadors whom the Archduke Ferdinand sent to Constantinople. As the Austrians suggested that Soliman should recognize their master as King of Hungary, while the Sultan's vassal Zapolya was actually in possession of its capital and three-quarters of its counties, there was not much to be hoped for from such debates. Meanwhile the war continued, and the Archduke, seeing that he was not being attacked, took the offensive. His army recaptured Gran and several other places on the Hungarian border, and pushed as far as Buda, which it failed to reduce, owing rather to the desperate resistance of the Turkish garrison than to any help given it by Zapolya. When Ferdinand's troops had retired, the Pashas of Bosnia and Serbia ravaged Carniola and the parts of northern Hungary which were not in Zapolya's hands—but all this was inconclusive fighting. In 1531 the Sultan took in hand preparations for a second serious attack on the Empire, which was delivered in 1532, with forces perceptibly larger and better organized than those which had appeared before Vienna in 1529—this time there was none of the expectation that Charles V would be distracted by a French war, as Soliman had imagined during his earlier venture. That his brother had been proclaimed 'King of the Romans' in 1531 made little difference—it gave Ferdinand no real authority over the discontented Protestant princes of Germany.

The great campaign of 1532, however, was not to lead to any decisive result. The Emperor had succeeded in drawing, for the benefit of his German subjects, such a picture of the dangers of a Turkish invasion, that the Diet voted him liberal supplies, and Luther himself exhorted the Protestant princes to lend honest assistance for such a crisis. A very large army was collected at Vienna, and Charles himself took the command, and contributed Spanish, Italian, and Netherland contingents. Probably it was the knowledge that such heavy opposition was awaiting him on the Danube, where memories of old disasters were rife, that induced the Sultan to turn his attack against a quarter where he was not expected, though success in that direction could hardly lead to anything very decisive. Possibly he had not considered the disadvantages that would fall to an army consisting mainly of cavalry—both feudal and irregular —in a mountainous country full of castles and fortified towns. At any rate, on leaving Belgrade upon June 25 and Shabatz on July 10, he crossed the Drave at Essek, and then, instead of marching by the usual high road parallel with the Danube, which leads to Buda, and ultimately to Vienna, he turned off westward into the narrow strip of Hungarian territory which was still in the hands of King Ferdinand, all along the borders of the Austrian province of Styria. He took without much difficulty the Croatian town of Belovar (July 26) and the Hungarian town of Koermend (August 3), with many smaller places in the valleys of the Drave and the Mur; but from Koermend, instead of marching due north by Oedenburg, a road which would ultimately have taken him toward Vienna and the Emperor's army, he turned west toward the Styrian mountains, and on August 9 laid siege to the small fortress of Güns, the last town in Hungarian territory.

Soliman sent meanwhile a vast horde of his irregular cavalry—15,000 at the least—to demonstrate against Vienna, and ravage Lower Austria. Some of the raiders got as far as the Enns and the town of Steyr; but almost the whole body was destroyed in September, as it attempted to make its way back across the defiles of the Wiener Wald to rejoin the Sultan. Austrian troops coming out from Vienna had occupied all the passages, while others closed in from the rear, and in a series of engagements in the hills Kassim Bey, the leader of the raid, was killed, and a few days later his successor, Osman Bey. Only a very few of the 'Akindjis' ever rejoined their master.

The Sultan meanwhile lay for no less than twenty days (August 9–28) before the trifling fortress of Güns, without making any headway. The place was small—the garrison consisted of no more than 700 regular troops—but the approaches were difficult, and the governor (a Croatian captain named Nicolas Jurischitz) was an officer of great resolution and resource. Mine after mine was directed against the one accessible side of the place, but most of them were discovered and blown in by the garrison, and though a 'mount' was set up opposite the weakest side of the defence, and some breaches were finally made, twelve separate assaults were beaten back from them—the Sultan's diary mentions four only. Continual rain filled the Turkish trenches. After a final and almost successful assault had been repulsed, on the 28th a most curious bargain was arrived at between Jurischitz and the Grand Vizier Ibrahim. In order to 'save the face' of his master, Ibrahim proposed that the siege should be raised, and the army march off, if the governor would make a formal 'paper' surrender, and deliver over his keys, so that the Sultan might be nominally in possession of the place. Strange as it may appear, hostages were exchanged, and the farce carried out. The Turks then departed, leaving Güns unconquered.

After this, Soliman, instead of marching on Vienna, turned off by cross-roads into the sub-Alpine valleys of Styria, in what was practically a retreat. His diary records painful marches through a mountainous country, during which upland villages were ravaged and two or three small castles taken. But on arriving on September 11 before the walls of Gratz, the capital of the Styrian duchy, the Sultan made no attempt to assault it, and fording the Mur before its walls marched down that river, wasting its valley, till he reached Marburg, which was also left unmolested. Here the Turks halted for four days, raiding the country far and wide, while a bridge was being constructed over the Drave (September 16–19). They then moved off downstream, doing mischief all the way, but dropping much baggage and many stragglers, till they reached on October 9 the border of their own territory, and on October 13 camped opposite Belgrade. During the last days of September the Croatian borders were cruelly harried, and Possega and

Rassina, the last towns in this direction which were still in the hands of King Ferdinand, were taken and pillaged. By November 21 the Sultan was back at Constantinople.

This was a most inconclusive end to a great enterprise. It seems clear that Soliman feared to measure himself against the Emperor and the formidable army collected at Vienna. In his state documents, published after his return, he twisted the situation into a boast that he had marched far into the Empire, and that Charles had never dared to come out to meet him. But if he had wanted a pitched battle, he had only needed to march from Koermend by the Vienna road, and not to turn off into the Styrian Alps. He did an immense amount of damage to Ferdinand's dominions, but did not occupy or annex any of the many small towns which he had sacked. But for the loss of life and property from his widespread raiding, his enemy's position was unchanged. And he had paid for the slaves and plunder that he had carried off by the loss of many thousand fighting men and a perceptible diminution of his prestige. For if the expedition of 1529 was difficult to present as victorious, that of 1532 would only so be styled by official court panegyrists.

This was the last occasion for some time on which the Austrian borders were to see the Turkish standards. Soliman was depressed by the failure of his enterprise, and was beginning to be distracted by news from the Far East. Troubles had broken out upon the Persian border, for the governor of Bitlis had revolted and handed over his city and its province to the Shah. At the same time the Persian governor of Bagdad, who had been meditating treachery, and had opened up secret negotiations with Constantinople, had been detected by his master and put to death. Actual fighting had begun, and the news of a defeat of a Turkish force by the Persians reached Soliman as he was marching on Güns. Obviously it would be unwise to fight on two fronts at once, and the Sultan resolved to turn against the old Persian enemy, and to drop offensive action on the European side. Some additional pressure on his policy may have been brought to bear by the fact that Andrea Doria, the great admiral of Charles V, had appeared in Greek waters, captured Coron, the most formidable stronghold in the Peloponnesus, and thrown a Spanish garrison into the place. He also destroyed the Turkish fortresses at the entrance of the Gulf of Corinth.

At any rate, the summer of 1533 saw an unexpected treaty of peace with King Ferdinand. The Sultan refused to abandon the claim of his vassal John Zapolya to be King of Hungary, but conceded that all the parts of that kingdom which were actually in Ferdinand's hands—still a third of the whole—should be left to him. An odd piece of diplomatic make-believe, which recalled what had taken place at Güns, was devised to give Soliman prestige in the face of his own subjects. Ferdinand professed to surrender the metropolitan city of Gran, the most notable Hungarian place which he held, and keys purporting to be those of its gates were handed over to the Grand Vizier. But Gran was not really evacuated, and the Vizier gave back the keys—real or false—to Ferdinand's ambassadors, Cornelius Schepper and Jerome Jurischitz—the brother of the defender of Güns (July 1533). The treaty brought solace for the moment to the Austrian borders, but was somewhat humiliating to the King of the Romans, since Zapolya was left with Buda and the greater part of Hungary in his possession, and Ferdinand's claim to be the one rightful wearer of the crown of St. Stephen was ignored. Moreover, the invasion of Germany might recommence at any moment when the Sultan should be free from Eastern troubles, and the House of Hapsburg involved once more in a French war. As a matter of fact hostilities broke out again on the Hungarian frontier in 1537.

It is surprising to find that while Ferdinand's ambassadors succeeded in patching up a peace with the Sultan, the envoys of Charles V failed to do so ; nevertheless his brother withdrew from the general war, leaving the Emperor still involved in it, a circumstance that was to have immense European importance during the next ten years.

Meanwhile, the Turkish land-attack on Central Europe was suspended for four years, during which Soliman was engrossed with the details of a tedious and expensive but, on the whole, successful struggle with Shah Thamasp. War in the Armenian and Kurdish mountains was no less destructive to an army mainly composed of cavalry, than war in the Styrian Alps. The first Turkish attack ended in the capture of Tabriz, the capital of the Sefavi dynasty in those days (July 1534). Taking the field himself in the succeeding autumn the Sultan conducted his army through the worst of the rough country between Tabriz, Hamadan, and Bagdad—an unwise march, in which he lost most of his transport and horses and many thousands of men. But the Persian governor of Bagdad fled at his approach, and the half-starved army was able to go into comfortable winter quarters (December 1534–March 1535). Meanwhile, Shah Thamasp turned back, recovered Tabriz after beating the force which Soliman had left to hold it (January 1, 1536), and occupied Van and other places on the old frontier. The Turkish army had to retrace its weary steps to repeat the operations of the previous autumn: it took three months to march from Bagdad to Tabriz (April 1–June 30, 1535), from which the Shah once more retired without fighting. Van was recovered without any trouble, but a circular march into the interior of Persia in pursuit of the Shah had no success—Thamasp very wisely refused to fight, and retired ever eastward when pressed. Soliman returned to Tabriz in August, followed at a distance by the elusive Shah, whose outriders worried his rearguard continually. The whole of Adherbaijan had been so wasted by the two recent campaigns, that Soliman resolved that it would be hopeless to repeat the mistake of holding on to Tabriz for another winter. He sacked the place, burned the Shah's palace, and retired via Bitlis and the Upper Euphrates to his own dominions, finally reaching Aleppo on November 24. The rearguard, which covered the march of the army at a long distance, was badly cut up by Persians on October 13 in the mountains west of Bitlis. Soliman resolved to drop campaigning in Persia in person, and returned to Constantinople in January 1536.

The conquest of Bagdad was a solid achievement, and added much prestige to the Turkish Empire—though Mesopotamia was by now a ruined land, showing little trace of the glories of the Caliphate. Bagdad was held almost without a break down to 1917—though it fell back for a moment into the hands of the Persians for a few years in the seventeenth century. But the campaigns in Armenia and Adherbaijan were fruitless, and were but the sample of many equally unprofitable raids by Turkish armies during the next two centuries. The mountain barrier between the valleys of the Euphrates and the Aras was too broad and too desolate for the invader—far from his base, and operating through the roughest possible country. Soliman had involved himself in a tiresome struggle, which was to last in an intermittent fashion for twenty years. The first peace between Turkey and Persia came only in 1555, and then left Tabriz and Erivan to the Shah, though Van and Erzeroum remained Turkish.

Soliman celebrated his return to his capital by putting to death his great Vizier Ibrahim, the trusted companion and adviser of his earlier years (March 1536). The reason seems to have been that Ibrahim presumed too much—it had become of late a case of 'Ego et rex meus,' as can easily be gathered from many of the reports of European ambassadors with whom the self-confident Vizier had to deal. But possibly there were concurrent causes in the Sultan's discovery of intrigues by which Ibrahim had brought about the death of his rival, Iskender Chelebi, the high treasurer, and in Soliman's disgust at the expensive and inconclusive character of the Persian War, of which Ibrahim had been the main advocate. He never enjoyed again the services of such a capable minister: one is reminded of the contemporary tragedy of Thomas Cromwell and Henry VIII. But Cromwell at least had a formal trial, and did not merely disappear in the Palace.

CHAPTER VIII

HUNGARY AND AFRICA

BUDA, TUNIS, AND ALGIERS (1533–47)—COMMENCEMENT OF THE TURKISH NAVAL OFFENSIVE

DOWN to the year 1532 the Turkish attack on Christendom had been almost entirely on the land side—the naval expedition of Mahomet II against Apulia had been an exceptional phenomenon without any sequel. But from 1534 onward the struggle had an important maritime aspect, due to the sudden expansion of the Ottoman power along the north coast of Africa, combined with the repeated alliances of the French king with the Sultan. When the French and the Turkish fleets were combined, the predominance of Charles V in the Mediterranean was threatened, as it had never been before. This unholy league—which shocked the minds of the better sort of Frenchmen—brought Mohammedan fleets into regions where they had never been seen since the Saracen pirate-kings of the tenth century subdued Sicily, and made lodgments not only at Bari and on the Campanian headlands, but even at Fraxinet on the coast of Provence. There was a serious danger that this unhappy phenomenon might be repeated in the middle years of the sixteenth century. If it failed to materialize, the cause was the utter fickleness of Francis I, who having attracted his Moslem allies into the Gulf of Lyons and the Tyrrhenian Sea, twice made a sudden peace with the Emperor (Truce of Nice, 1538, and Peace of Crêpy, 1544), not because the conscience of the 'Most Christian King' was troubled at the alliance with the infidel, but because his affairs were going badly on his eastern frontier.

The war of Charles V with Sultan Soliman—a war which continued after they were both dead—was therefore at its most perilous crisis whenever the French entered into the game. At other times it showed no signs of coming to a definite decision; the parties were well balanced, and Persian wars hampered

Soliman in much the same way that German religious complications hampered Charles. In 1566, when Soliman died, the land frontier of the Hapsburgs and the Ottoman hardly differed from what it had been in 1532. Only a few more miles of Hungarian borderland had been lost.

The extension of the Turkish attack on to the sea front started in 1533. Before that date the north coast of Africa from the Moluchath River to the Syrtes had been since 1263 in the hands of two native dynasties, the Beni Hafs and the Beni Zian, who, like most Moslem dynasties, were falling into decay after they had passed their second century of existence. They had been of singularly little importance in Mediterranean history, had never shown any signs of becoming a serious naval power, and had standing commercial treaties with Venice, who enjoyed the monopoly of their trade. In the end of the fifteenth century the very decadent Beni Zian house still held Telemsan and Algiers; the Beni Hafs, Tunis and Tripoli. When the urge to transmarine conquest descended on the Christian princes of the Iberian Peninsula, and Portugal conquered Ceuta and Tangier, the Spaniards commenced similar operations against the dominions of the two African families. Ferdinand of Aragon siezed Melilla (1493) and Oran and Marsalquivir (1509), and the regent, Cardinal Cisneros, who bore authority during the minority of Charles V, was a strong supporter of the policy of African adventure. If it had not been for the discovery of America, which turned Spanish energies in another direction, it is possible that Spain might have built up an African Empire. There was no power in the two decadent Moslem houses to prevent it. In 1510 Pedro Navarro, the great engineer of King Ferdinand, even captured Tripoli, the third most important town in Barbary, which Charles V gave to the Knights of St. John, who held it till 1551.

Another element, however, emerged to stop Spanish advance. The weakness of the native dynasties had led to all manner of rebellion and civil strife among their subjects. Among the adventurers who cropped up in the chaos were a group of three Turkish brothers, pirates by profession, Ishaak, Ourouj, and Khaireddin, who entered the service of Selim, the prince who then ruled at Algiers. After having made himself indispensable to this unfortunate ruler as commander of his army and navy, Ourouj murdered him (1518) and usurped his throne. Engaged in simultaneous warfare both with the Spaniards at Oran and Marsalquivir, and with the Beni Zian survivors at Telemsan, both Ourouj and Ishaak fell in battle. Their surviving brother Khaireddin, feeling his position very precarious, sent to offer his homage to Sultan Selim, who had just conquered Egypt, and it was formally accepted (1519), though at the moment it meant little to either party concerned. But the career of Khaireddin gradually grew more brilliant; he collected a very large pirate fleet, manned by adventurers of all sorts—Moors, Turks, and many Christian renegades—with which he swept the Barbary coast, achieved several successes over the Spaniards, captured some of their smaller posts, and even carried out raids on Sicily and the coast of Andalusia. This he did nominally as the servant of Sultan Soliman, with whom he always kept up formal relations, duly sending reports of all his successes to Constantinople.

In 1533, before he departed on his Persian Expedition, Soliman—still involved in war with Charles V, though he had made peace with King Ferdinand—resolved to utilize Khaireddin as a valuable auxiliary against the Emperor. He sent for him to Constantinople, received him with great honour, gave him the command of the whole Turkish fleet, and placed at his disposition a considerable contingent of Janissaries. Soliman was in the wilds of Armenia when Khaireddin, with his own and the Turkish vessels, carried out a great enterprise (1534). Descending on Tunis he expelled Mulei-Hassan, the twenty-second Sultan of the Beni Hafs dynasty, and occupied his capital and the all-important harbour-fortress of Goletta at the mouth of the Tunis lagoons. Mulei-Hassan fled to the Emperor Charles, the hereditary enemy of his race, and craved his aid, pointing out—what was true enough—that if the whole Barbary coast became Turkish territory, the Spanish naval domination in the Mediterranean was endangered. He offered, if restored to his throne, to hold it as a vassal of Spain, and to put several of his harbours in charge of Spanish garrisons. Mulei-Hassan was a very depraved despot—he had murdered over a score of his brothers, and was notorious for his vices—but he retained a following among his late subjects, most of whom looked upon Khaireddin as a foreigner, a mere adventurer, and hated to see the old Moorish kingdom a prey to the Turks. He guaranteed that he could raise many tribes against the usurper, if he were landed in Tunisia and given a Spanish auxiliary force.

The Emperor, being for the moment at peace with France, and having no need to think of the Danube frontier, resolved to accept Mulei-Hassan's offer, and to make the capture of Tunis not a mere affair of restoring the exiled Sultan, but a first-class attack on Africa, where his affairs had not been going well of late, during the times of Khaireddin's expansion. He was particularly irritated by the fact that the Barbary pirates had just surprised and sacked Reggio, and even ran up the Neapolitan coast to harry Fondi, a place three miles from the sea.[1] They had become a perpetual nuisance, and now they were to be backed by the power of the Sultan, which would make them a danger rather than a mere distraction. Accordingly Charles collected at Barcelona a formidable expedition—62 fighting galleys, besides some broadside ships of war—a new feature in Mediterranean naval tactics—with 150 transports, carrying a landing force of nearly 25,000 men. There were Portuguese and Genoese contingents, and the naval side of the affair was entrusted to Andrea Doria, the experienced Genoese captain who habitually served the Emperor as High Admiral.

This was altogether too strong a power for Khaireddin to meet in the open: he retired with his fleet into the Gulf of Tunis, and threw his best fighting men into the fortress of Goletta, which commanded the entrance of the land-locked water. Charles battered and stormed Goletta, beat Khaireddin's land forces outside Tunis, and was received into the city by the townsfolk, who acclaimed Mulei Hassan as their lawful lord. The whole of the pirate fleet was burned, and their admiral sought refuge in Algiers with a mere remnant of his followers. Unfortunately the Emperor's troops sacked Tunis after it had opened its gates (July 21, 1535)—a bad start for the restoration of the exiled Sultan. Nevertheless, Mulei Hassan resumed his position, and carried out most of his promises: not only Goletta but Bona was placed in the hands of Spanish garrisons. More than 9000—some say 15,000 or 20,000—Christian slaves, the victims of Khaireddin's recent raids, were delivered. This was a success of importance for the Emperor, for Mulei Hassan succeeded in maintaining himself at Tunis for eight years, till he was dethroned and blinded by

[1] From whence, according to a popular Italian tale, Julia Gonzaga, the most beautiful lady in Italy, only escaped in her night-shift on a bare-backed horse. Rumour held that Khaireddin had sworn to get her for his harem.

a rebellious son. The Turks did not finally recover Tunis till 1574 from the last of the Beni Hafs.

The Emperor left Tunis victorious in the autumn of 1535. Next year it would have been in his power to finish the expulsion of the Turkish corsairs from Algiers and the other ports still in their hands, if he had not been forced to turn his attention elsewhere. In March 1536 Francis of France delivered his sudden and unprovoked attack on the Duke of Savoy, whose dominions he intended to cross on the way to Milan, where the last heir of the house of Sforza was just dead. Renewing for the tenth time the French claim to Milan, Francis defied the Emperor. He had already secured the help of the Turks—his ambassador Laforêt having signed a secret treaty of alliance with the Grand Vizier Ibrahim, shortly before the fall of that great minister (February 1536), and he had even sent emissaries to Algiers to solicit the help of Khaireddin.

Hence the attention of the Emperor was drawn off to Europe, and the affairs of Africa had to shift for themselves. His invasion of Provence was a complete failure (June–September 1536). Meanwhile Khaireddin showed that he was not yet disposed of—in the spring he made a raid from Algiers with the wrecks of his corsair-fleet, and devastated Minorca, capturing Port Mahon, and carrying off thousands of slaves. He then, at the Sultan's orders, left Algiers in the hands of his lieutenants and came to Constantinople with all his disposable galleys. The naval war had taken a new development, and Soliman had determined to concentrate every vessel that he could collect in Greek waters. For 1537 was destined to be one of the most critical years in the struggle of East and West, the last in which a serious invasion of Italy was threatened. Fortunately for the Emperor Charles, Venice, whose government had for some thirty years observed a complete neutrality, considering commercial advantages all important, and the danger to Christendom a secondary consideration, was now drawn into the turmoil. There had been of late many altercations between the Porte and the Republic—as was natural when the naval struggle between Charles and Soliman had been raging across all the lines of Venetian commerce. The worst had arisen from the molestation by Venetian galleys of a Turkish ambassador and his vessels—taken for pirates according to their captor's excuse. The ambassador was promptly released, but the insult provided a *casus belli*, though the Doge made all

44

manner of excuses—war being the last thing that Venice desired.

But in May 1537 the Sultan marched with a very large army to Avlona on the Albanian coast opposite Corfu, while his fleet at the same time rounded Cape Matapan, and came up to the opening of the Gulf of Venice. Before the command of the sea had been properly secured, a large raiding force was thrown across the Straits of Otranto into the heel of Italy, and for the first time since 1480 a serious invasion of the kingdom of Naples began. Eight thousand irregular horse under Lutfi Pasha and an Italian renegade, Troilo Pignatelli, swept the countryside for ten days, and Ugento and Castro were taken and sacked. Otranto and Brindisi were holding out—when Soliman suddenly withdrew the raiding force,[1] because its communication with its base had been threatened, and nothing was heard of an expected French invasion of Italy from the north. Both Imperial squadrons under Andrea Doria and Venetian squadrons were converging on the straits—wherefore Soliman settled down to the siege of Corfu, and threw up the Italian expedition. From August 18 to September 6 the great Venetian fortress was fiercely assailed, but to no effect, and seeing that there was considerable danger that the Venetian-Imperial fleet might cut off the troops landed on the island, Soliman abandoned the venture, withdrew the besieging army to the mainland, and on September 19 started back for Constantinople—much disappointed with the campaign.

He had, however, his consolations. King Ferdinand dutifully following his brother's orders, had broken the peace of 1533,[2] and collected an army in Styria and Croatia, with which he struck not, as usual, at Buda and the Danube towns, but at Essek, the important fortress over whose bridge on the Drave the road from Belgrade to Buda passed. Advancing down the course of the Drave, this very considerable force under three veteran generals, John Katzianer, Ludwig Lodron, and the Hungarian, Paul Bakicz, laid siege to Essek. But they failed to take it, and being beset on every side by Turkish horse coming in from all quarters, and falling short of provisions, they commenced a retreat in the last days of November. This was a

[1] Lutfi Pasha's expedition started on July 23, and was back at Avlona by August 6, according to the Sultan's diary.

[2] Which, it must be confessed, had been kept very badly by the Turkish pashas in Bosnia and Syrmia—and by their Hungarian neighbours.

disastrous business—snow came on, the army straggled, many soldiers deserted, and in a series of running fights near Valpo, the whole force was destroyed (December 2, 1537). Nearly 20,000 Austrians, Hungarians, and Bohemians are said to have perished: the slaughter was almost as great as at Mohacs. Lodron and Bakicz died, Katzianer, the commander-in-chief, fled early, and saved himself for the moment almost alone. He was accused of cowardice and imprisoned by King Ferdinand. Escaping from his dungeon he fled to the Turks, but was soon afterwards slain by a Hungarian noble, Nicolas Zriny, whom he was urging to join him in his treachery. An ignominious end for a soldier who had greatly distinguished himself at the siege of Vienna eight years before! The rout of Valpo almost atoned for the failure at Corfu in the Sultan's eyes.

It might have been expected that in the next spring Soliman would have delivered a strong attack on King Ferdinand, whose field army had just been exterminated, or—as an alternative—that he might have sent his naval forces to join the French and sweep the Italian waters. As a matter of fact he took neither of these enterprises in hand. He himself collected his Roumeliot levies and took his household troops with him, to deal with a revolted vassal in the north-east. Peter Raresh, Prince of Moldavia, had denied his tribute and made alliance with King Ferdinand. Calling in the Crim Tartars to his aid, the Sultan ravaged Moldavia, burnt Jassy, chased Prince Peter into exile, and set up another ruler in his stead (July–September 1538). Meanwhile Khaireddin's fleet attacked and captured all the outlying Venetian possessions or feudal dependencies in the Aegean, a dozen islands or more—Aegina, Tenos, Sciathus, Carpathus, Scyros, Scriphos, Andros, and others. He raided, but did not attempt to subdue, the great island of Crete, the centre of Venetian power. Endeavours by a land force to take the two surviving mainland fortresses of the Republic in the Morea, the rocky Nauplia and Malvasia, had no success. The garrisons of both made a fine defence.

It was only in the autumn of 1538 that the Turkish fleet took a serious operation in hand. The combined squadrons of the Emperor and Venice were dominating the western waters in the summer, under the command of their usual chief, Andrea Doria. In August Doria appeared before Prevesa, the fortress at the end of a long peninsula which watches the narrow mouth of the great Acarnanian Gulf of Arta. This was a place of strategical importance, the nearest Turkish naval base to Corfu. Khaireddin resolved that it must not be allowed to fall, and brought all his available galleys round Cape Matapan into the western sea. Doria, though his total strength was decidedly superior to the enemy, rather unaccountably refused to commit himself to a general action. He allowed Khaireddin to establish himself at Prevesa, and made no attempt to dislodge him. This enabled the great corsair to come out and to present himself in the open sea. To the surprise and disgust of his captains, more especially of the Venetians, Doria refused to accept the challenge, and held off when the Turks advanced against him. Only some trivial fighting ensued, in which the Christians lost seven galleys. As they had some 200 in line, the affair can hardly be called a battle (September 29, 1538). The most reasonable explanation of the conduct of the veteran Genoese admiral is that he considered that while he had a 'fleet in being' the Turk could make no further attempts to invade Italy. The Venetians, who had clamoured for a fight, were profoundly disgusted. So, we are told, were the Turkish captains, who had expected a pitched battle. There was a general impression in Venice that Doria was acting under dilatory orders from the Emperor, who seems at the time to have been engaged in a secret intrigue with Khaireddin, to whom he had offered the kingship of Algiers if he would desert the Sultan, and become a Spanish vassal, like Mulei Hassan of Tunis.

But a much more likely solution is that Andrea Doria was determined to keep an intact fleet, and to risk nothing, while Khaireddin could not wish to attack a superior force with vessels many of which were newly launched and manned by inexperienced crews. Nor is there any proof that he ever thought of deserting the Moslem cause. An incident in this inconclusive encounter, which might have provoked much hard thinking among naval specialists, but which passed at the time without much comment, was a combat between a single Venetian 'galleon,' *i.e.* a broadside-gunned ship, working by sail alone, and a large number of galleys. The vessel of the Venetian captain, Alessandro Bondulmieri, becalmed for want of wind at some distance from the main fleet, which was using its oars, was beset in succession by three or four groups of galleys, but drove them off again and again by his superior all-round gunfire, taking no damage from ramming.[1] Though

[1] This fine fight is described at great length by Guazzo, pp. 269–72.

much battered and partly dismasted, he got off in triumph. Yet in spite of this suggestive incident, the galley continued to be the master of the Mediterranean for many years. And when Italian naval architects tried to out-class it, they only built 'galeasses'—larger galleys with more sail-power and better cover for the rowers, but still unfurnished with broadside guns.

The end of the campaign of Prevesa was a justification of Doria's dislike of taking risks—though it came by no act of his own. The combined squadrons, after the action, drew up the Gulf of Venice, and based themselves on the fortress and deep fiord of Cattaro. Khaireddin, following them at a distance, was caught by a sudden north-westerly tempest, and driven against the almost-harbourless coast of central Albania. Some seventy of his galleys were lost or disabled, and he was obliged to run into Avlona with the mere wrecks of a fleet. He was put out of action for the rest of the year 1538 and the following winter, and returned in person to Constantinople to petition Soliman for means to reconstruct a fighting force.

Meanwhile Doria, free from any damage on the water, laid siege to Castel Nuovo, close to Cattaro on the north side of the fiord of the 'Bocche,' which was the most important shore-fortress in the Adriatic that was in Turkish hands. After battering it from the sea side, he stormed it with his landing force (October 24) and threw in a large Spanish garrison, after which the fleet dispersed for the winter, the Admiral being aware that there could be no danger on the side of the sea for many a month. The Venetians complained bitterly that the captured Castel Nuovo was not handed over to them. The Turkish governors of Bosnia and Albania made a dash at Castel Nuovo at midwinter, from the land side, but were driven off with loss. It was not till the following midsummer that the place was in any danger, when the Turks had at last reinforced and reconstructed their battered fleet, and appeared before its walls on both the land and the sea fronts, with an army of 60,000 men and 120 galleys. The siege lasted for four weeks (July 13, 1539–August 10) and cost Khaireddin, who had resumed command, no less than 8000 casualties, for the Spanish garrison made a most gallant defence. After this it should have been the turn of the neighbouring Venetian stronghold of Cattaro, but an unexpected end was given to the war in the Adriatic by the conclusion of a truce between the Republic and the Sultan.

Disgusted at the caution of Doria on the day of the battle of Prevesa, and much hurt by the news of the loss of all their insular possessions in the Aegean, the Venetian Senate had sent an ambassador to Constantinople to sue for peace as early as April 1539, before the siege of Castel Nuovo had begun. They were ready to break with the Emperor, and to save their surviving possessions, with small regard to the general welfare of Christendom. Soliman was anxious to get the Venetian fleet out of the way, and to deal with the Emperor alone; he therefore offered terms which, after much haggling, the Senate made up their minds to accept. A truce was first concluded, which brought to an end the operations in the Adriatic, and then, after innumerable proposals and counter-proposals lasting till 1540, a permanent peace came, by which the Venetians gave up not only the lost Aegean islands, but Nauplia and Malvasia on the Greek mainland, which had maintained a gallant resistance against the Turkish land forces ever since the war began. A very moderate war-indemnity was paid to the Sultan, but Venice kept Crete and Corfu, Zante and Cyprus, which was the main advantage secured—if the naval war had gone on they might have been lost. For nearly thirty years she kept up a rather ignominious neutrality.

It may be asked why the Turkish offensive, which had started in such a threatening fashion with the invasion of Apulia in July 1537, had ended so tamely in the fruitless siege of Corfu, in the following autumn, and the equally fruitless naval success at Prevesa in September 1538. The answer is simple—Soliman had counted on the distraction of the Emperor's forces by the French War, and the raid on Italy had been made in the hope that Francis I was overrunning the duchy of Milan at the same moment. But with his usual fickle inconsequence the 'Most Christian King' backed out of his alliance, and on June 18, 1538, three months before the battle of Prevesa, had signed with the Emperor the so-called 'Truce of Nice,' which was scheduled to last for ten years—but actually was in operation only for five. In the last half of 1538, therefore, and during the whole of 1539, Soliman was fighting the Emperor and the Venetians without any help from his Western ally. And he was only too glad in 1540 to allow Venice to pass out of the war on comparatively moderate terms of peace.

From 1539 onward the war resumed the simple aspect of a struggle between the Sultan and the two Hapsburg brothers, and so things went on till 1543, when King Francis delivered one more of his 'stabs in the back' against Christendom—this time with more prospect of success. In 1540 the contest on the land frontier showed signs of assuming a new aspect; King John Zapolya died on July 20, leaving no heir but an infant a few weeks old. His personal popularity with his faction had been the main reason for the adherence of half of Hungary to the Turkish cause. There were many who felt the ignominy of the situation, and though Zapolya's widow, Isabella of Poland, got her son crowned in his cradle, the 'particularist' party showed signs of breaking up. When King Ferdinand came forward with an army, many Hungarian nobles joined him, and many towns, Waitzen, Visegrad, Stühlweissenburg, Fünfkirchen, and others, opened their gates with little or no resistance. Queen Isabella and her faction held out in the royal city of Buda, to which Ferdinand's generals laid siege, and sent desperate appeals to Constantinople for help from the infant King's suzerain.

If Buda had fallen, Ferdinand might possibly have mastered the whole of Hungary, for even in Transylvania, the old stronghold of the Zapolyas, some magnates and towns adhered to him. But the siege of Buda failed, and in 1541 Soliman came to the Danube with a great army, and put an end to the Austrian successes. He took possession of Buda, relegated the infant John Sigismund Zapolya to his father's principality, which he was to hold as a tributary vassal, and divided the rest of the realm into pashalics annexed to the Turkish empire, declaring to Ferdinand's ambassadors, who came to plead with him for peace, that he had now annexed Hungary by pure right of conquest, and had no concern with legalities or ancient treaties. The scission of the 'lands of the Crown of St. Stephen' into three parts was to last for 145 years. It might have been expected that Soliman would have pushed forward from Buda to recover all the cities which had submitted to King Ferdinand, perhaps even to threaten Vienna. But this he did not do, but made his way back to Constantinople when autumn had arrived, postponing further operations to another year.

It would have been well for the Sultan's chief adversary if he, too, had refused to undertake an important campaign in autumn weather. Charles V had considered that all Soliman's energies would be absorbed in the reconquest of Hungary, and that his fleet was not yet strong enough to dispute the command of the Mediterranean. Accordingly he had made up his mind to destroy the remaining Turkish outpost in Africa at Algiers —Tunis was now in the hands of his vassal Mulei Hassan, Tripoli was still held by the Knights of Malta. He collected a great expeditionary force, naval and military, in the ports of Italy and Spain, and gave its component parts a rendezvous in the Balearic harbour of Port Mahon. It was now September, and Soliman was on his way back to Constantinople just as Charles was starting out for Algiers. The Emperor was warned against undertaking naval operations after the autumn equinox alike by his admiral Andrea Doria, by his chief general Del Vasto, the Viceroy of Milan, as also by the Pope—whose knowledge of seamanship was perhaps less valuable than that of the other two advisers. But refusing to listen to gloomy prophecies, he set sail when all his fleet had been collected, and came to land on October 20 in the bay under Cape Matafous, twelve miles (as the crow flies) east of Algiers, where there were long beaches protected by the cape from east and north, but not from westerly winds.

The expeditionary force was very powerful, 1100 horse and over 20,000 foot, of whom 7000 were Spanish, 6000 German, 6000 Italian veterans, and 3000 volunteers of all sorts, not to speak of a small contingent sent by the Knights of Malta. The disembarkation took three days, and when the troops were on shore, and the stores and munitions were beginning to be unloaded, the army marched on Algiers. Khaireddin was absent at Constantinople with his fleet, and his stronghold was in charge of his deputy Hassan, who had a very moderate garrison—under 1000 Janissaries and 2000 or 3000 Moors. The alarm had been sent inland, and clouds of Moorish horse were beginning to gather in the hills above the shore, but there was no force that could have faced the Imperialist army in the field. It looked as if Charles might be able to deal with Algiers as he had dealt with Tunis six years before.

Though the measurement across the water from Cape Matafous to Algiers was only some twelve miles, the army had to go round the head of the bay, which doubled the distance, and to cross three small rivers before coming to the city. It was harassed on the way, but not much incommoded, by the Moorish horse from the inland, who hung about the line of march. On the third day Algiers was encircled on all sides, while a detachment of galleys blockaded the port. On the same night disaster came ; a tempest of storm and rain swept down from the west, which smote and levelled the tents of the army, but—what was infinitely worse—caught the transport fleet, which was lying on a lee shore while munitions and food were being hastily disembarked. This was one of the great shipwrecks of history; in three days of storm, for the wind continued to blow all through the 24th, 25th, and 26th of October, fourteen war-galleys were sunk or driven ashore, and no less than 130 transports, small and great, which were laden with the food and munitions not yet landed. The whole beach from Cape Matafous to near Algiers was strewn with broken timber and barrels and sacks of spoiled provisions. The Moors came down from the hills to cut up the shipwrecked crews, and pick up what plunder they might find, and had to be driven off by large detachments of infantry.

The troops in the trenches before Algiers suffered much from the tempest—it was said that at the height of the storm men could not even keep their feet, but were blown down, like their tents. They had only two days' food with them, and this was largely spoiled. But the most vexatious circumstance was that the besieged delivered two sorties in force during a lull in the storm, and raided the front held by the Italian contingents. The arquebusiers were badly cut up, 'their match and powder being so wet that they could not use their pieces,' and when reserves came up and chased off the assailants, an even worse mischance occurred. In pushing the flying enemy right up to the gates, the Italians came under concentrated artillery-fire from the walls, and were so badly mauled that they gave way *en masse*, and were only rallied by the Emperor himself, who brought up German companies to save them. Charles is said to have conducted the counterstroke himself, and to have shown splendid personal courage and power of leadership. But more than 3000 men are reported to have been lost in this mismanaged affair.

On considering his situation the Emperor came to the conclusion that the expedition must be given up, for there was hardly any food left, and all the munition reserves had been lost in the wrecked ships. The army lay for one day more before Algiers, living on horse-flesh alone, for the Emperor had to order all the transport animals to be slaughtered to give the starving troops a meal. On the 27th October the whole marched off to the original embarkation place, where there were enough of ships left to take off a considerably diminished army. The retreat was accompanied by much loss of life—the troops straggled, and all who fell behind were cut off by the Moors. On the 31st the disorganized and disheartened men were crammed on board the surviving ships, and the flotilla, still much battered by fresh storms, which caused more casualties, made for friendly ports. The wrecks of the expedition came ashore in November at many different harbours, from Trapani in Sicily to Cartagena in Spain. At least a third of the force had perished, and of those who got home many died after landing from the privations that they had suffered.[1]

Such was the dismal result of undertaking a naval campaign after the usual Mediterranean storms of autumn had begun. Oddly enough, the same tempests which had ruined the Emperor's expedition had at least saved him from any danger from the Turkish navy. Khaireddin had been ordered by the Sultan to sail for the help of Algiers, the moment that the news of the departure of the Imperial fleet for Africa came to hand. But he never got into Algerian waters, having wisely preferred to go into harbour when the winds began to blow, and to leave his destined task to them! Galleys were not suitable for campaigns in wintry weather, as all practised seamen knew.

The year 1542 started with gloomy prospects for the Emperor, for the news of his disaster had flown round Europe and heartened up his enemies. They gave special joy to Francis I, who was always waiting his opportunity, and had now a specious *casus belli* and excuse for repudiating the Truce of Nice. Two envoys whom he was sending to Constantinople, Rincon and Fregoso, had been murdered in their boat on the Po by brigands, who were said (not improbably with truth) to have been in the pay of Del Vasto, the Viceroy of Milan (July 1541). It was suspected that the Viceroy was anxious to have their dispatches at any cost. And it is certain that they were published after the outbreak of the subsequent war, and contained proof that King Francis was opening up a new alliance with Soliman. But the fate of the envoys might not have had any decisive political result but for the disaster that happened at Algiers, three months after their murders. It was the news of the Emperor's lamentable failure that

[1] By far the best narrative of this unhappy campaign is that of Villegagnon, the Knight of Malta, an eye-witness who was wounded in the sally of October 25.

enabled the French ambassador at Constantinople, Antoine Paulin,[1] to negotiate an offensive and defensive alliance with the Sultan, to take effect in the next year. While Soliman was to invade the dominion of Ferdinand in force, his fleet under Khaireddin was to go into western waters and join the French for a general attack on the Emperor's Italian possessions—it was only with some difficulty that Soliman consented to risk his galleys so far afield, but they sailed in April 1543.

Meanwhile in the intervening year 1542, King Ferdinand, seeing that no Turkish invasion was yet started, delivered an attack on the enemy's borders in Hungary, having the assistance of a large contingent of German troops voted by the Diet for his aid, and commanded by Joachim, the Elector of Brandenburg. This campaign, made with a very powerful but ill-managed army, was most fruitless—ending in a protracted siege of Pesth which had no success, and in bitter strife between the King's German and Hungarian generals.[2] The army broke up in October, having accomplished practically nothing. It was wanted in the next year—and was not forthcoming.

For 1543 was to be, in a way, the central point of the strife between the Hapsburgs and the Ottoman power, since in this summer the Franco-Turkish alliance was actually working. Soliman came up the Danube in April, having made vast preparations and sent an auxiliary flotilla up the river. Even before he reached the front himself, his local governors began to reduce the fortresses near the Drave, which covered the borders of Croatia, and when he arrived himself he took easily most of the towns which had gone over to the Austrians in the previous year—Fünfkirchen, and others—and then pushed on to besiege Gran, the metropolitan city of the Hungarian church, which had hitherto formed the advanced point of Ferdinand's frontier. It fell after a short resistance of only fourteen days, after which the Sultan turned against Stühlweissenburg (Alba Regia), the city where Hungarian kings were wont to be crowned and buried, and took it after fourteen days of battery (September 4, 1543). In all the captured places he turned the cathedrals into mosques, as a sign that Hungary was now

[1] Better known by his later title of Baron de la Garde, and for his atrocious murders among the Vaudois of the Alps in 1548.

[2] Frederick imprisoned for life Peter Pereni, the most important magnate who had come over from the side of the Zapolyas, on accusation of treachery during the campaign.

Turkish soil, and not a vassal kingdom. King Ferdinand showed no army in the field, he merely massed some 10,000 or 12,000 men at Vienna, and allowed all his fortresses to fall unhelped. The Sultan turned back in September, but after his departure his generals extended the occupation far east of the Danube, and by the spring of 1544 only a narrow strip of Hungary remained unconquered, from Erlau in the north by Pressburg, Szigeth, and Warasdin to the Croatian border. It looked as if one more push would take the Turks to Vienna.

Meanwhile, in this perilous year 1543, Khaireddin, as arranged, had taken his fleet to join the French. He passed unopposed through the straits between Italy and Sicily, stormed Reggio by a sudden attack, and after terrifying the Pope's subjects at Civita Vecchia, whom he nevertheless left unmolested,[1] struck across to Marseilles, and there with 148 galleys joined the French fleet, which was found to be a much smaller affair —only 22 galleys and 18 broadside ships,[2] under the Count of Enghien. Andrea Doria with an Imperialist squadron, hopelessly outnumbered, had retired into the harbour of Genoa. It remained to be seen what would be done; there was an appreciable waste of time, at a crisis when the combined fleet was absolutely master of the sea. But finally a detached Turkish squadron went to devastate the coast of Catalonia, and sacked Rosas and Palamos, while the main body beset Nice, the only remaining possession at the moment of the Emperor's unlucky ally, Charles of Savoy. The town was stormed and taken, but the citadel made a long resistance, till the besiegers, on a false rumour that the Viceroy of Milan was at hand with a relieving army, went on board again and retired to Toulon. Before re-embarking, the Turks sacked and burned the town of Nice, despite of remonstrances by the French commander, who saw the odium that the destruction of a Christian town would bring upon them. Khaireddin had wished to strike at Genoa, but the French objected to this obvious move; Paulin had opened secret negotiations with the party opposed to Andrea Doria, and had hopes of drawing the Genoese out of the Imperial alliance, if they were left unmolested. Nothing came of this, though the anti-Imperialist party actually existed, as was proved by the almost successful conspiracy of the Fieschi three

[1] The French officers with him showing that the Pope was really a friend of France, and had better be left unmolested for political reasons.

[2] Even less, according to discontented Turkish narrators.

years later. The Turkish fleet wintered in Provence, their Admiral grievously discontented that nothing more had been done, when so great a naval force had been accumulated.

He contemplated greater enterprises for the spring of 1544, but finding the French loth to take the sea, went off himself on the most dreadful raid ever launched against Italy; after sweeping Elba, the heritage of the Appiani, he went down the Tuscan coast, capturing the Isle of Giglio and the little ports of Telamon and Porto Ercole. After this he made for the Bay of Naples, devastated and enslaved the islanders of Ischia and Procida, and threw a landing force on shore at Pozzuoli, almost at the gates of Naples. But when the Viceroy came out against him with a considerable army, he went on board again, and, turning seaward, fell on the Lipari Islands, from which he swept away almost the whole population. What more damage he might have done to the Emperor's coasts can be but guessed —for the French alliance suddenly fell through. Francis I, though his army of Italy had won the battle of Cerissoles in April, was so hard pressed by the Emperor and the King of England on his northern frontiers that he signed the hasty and unexpected Peace of Crêpy in September. Khaireddin, deprived of his base in France, went back to Constantinople with 20,000 slaves and incalculable plunder before winter began. This was the last voyage of the great corsair; he never went to sea again, and died in his palace on the Golden Horn on July 4, 1546, leaving the memory of 'Barbarossa'—originally the nickname of his brother Ouroudj—as a terror to generations of Italian coast-folk.

The sudden collapse of the French alliance was undoubtedly the cause of the cessation of Sultan Soliman's main scheme for the invasion of Central Europe. Before the year 1544 was over there was a truce concluded between the generals of King Ferdinand and the Turkish Pasha at Buda. Ferdinand sent ambassadors to Constantinople to sue for terms—the Sultan did not refuse negotiations, but haggled—a general truce for a year was concluded in 1545, which was prolonged into 1546, and finally, on June 13, 1547, a definitive peace was secured by the Dutch ambassador Veltwijk, who represented both Ferdinand and Charles. Territorially it left matters in the *status quo* of the first truce: Gran, Visegrad, Stühlweissenburg, and all the conquered Hungarian regions remained with the Sultan. John Sigismund Zapolya was recognized as Prince of Transylvania under Turkish suzerainty. The humiliating part of the agreement for Austrian pride was that Ferdinand consented to pay a tribute of 30,000 ducats a year for the narrow strip of Hungary left to him. It was disguised under the pretence that it represented rents due in the non-ceded districts to Hungarian magnates who held fiefs therein, but were now Turkish subjects. But this was merely a salve for Ferdinand's *amour-propre* : the Turks called it tribute, and such it was in reality.

So ended the third great eastern crisis that fell into the days of the Emperor Charles : there was yet one more to come, but it was never so dangerous as that of 1543. The Sultan was growing old, and did not talk in his later years of 'reaching out a hand to Vienna,' as he and his Ministers had done in 1531.

CHAPTER IX

THE TURKISH OFFENSIVE BY SEA (1546–65)—SIEGE OF MALTA (MAY–SEPTEMBER 1565)

THE Peace which had been practically settled in 1546, though it was only finally signed in 1547 at the end of a long truce, left all the three sovereigns concerned free to engage in other enterprises. For Charles V it meant that he was at liberty to turn against the Protestant malcontents of Germany; the two campaigns of 1546 and 1547 ended with his complete victory at Mühlberg on April 24, 1547, and the capture or surrender of the rebellious princes. Ferdinand had been involved in the same matter, for his Bohemian subjects had been concerned in the Protestant rising and had taken the field against him. He too was victorious. Sultan Soliman was able to take in hand a long-projected Persian War, which he had been obliged to postpone so long as he was engaged with the two Hapsburgs. Francis of France, who had put in his assent to the treaty, though he was not at the moment at war with the Emperor, turned at leisure to the burning of Lutherans at home, and the massacre of the Vaudois on his Alpine frontier. He died before 1547 ran out, as did his old rival, Henry VIII of England. The unwonted condition of peace on all the European frontiers of the great powers was destined to last only till 1551.

The results of Soliman's second Persian War compared very poorly with those of his first. There was no spectacular conquest like that of Bagdad in 1535, but only some fruitless riding to and fro in the mountains of Armenia and Kurdistan. Van was captured once more, Tabriz occupied for the third time, but Shah Thamasp retreated ever eastward when pressed, and came back to occupy his devastated border when the Turk had gone into winter quarters. Soliman was so set on the adventure that he did not go home to Constantinople as usual in the autumn of 1548, but wintered at Aleppo. His second campaign was as fruitless as the first, and only gave him some obscure fortresses on the borders of Georgia and Armenia. He returned to the Golden Horn in December 1549 decidedly dissatisfied, though he published the usual circular bulletin of victory. The Persian War, however, was by no means ended.

Meanwhile, though there was official peace on the Danube, there had been much unofficial raiding on the part of both Austrian and Turkish local commanders. And things were much the same on the Barbary coast, where Hassan and Torghoud, the successors of Khaireddin in the pirate nest at Algiers, were always bickering both with the Christian garrisons at Oran, Bugia, Goletta, and Tripoli, and with the Beni Hafs at Tunis, where Hamid, the undutiful son of Mulei Hassan, was still ruling —more or less as vassal of Spain. General trouble, however, did not break out again till 1551, and it is rather characteristic of the changed character of the situation that King Ferdinand and not the Sultan was the aggressor. The principality of Transylvania was nominally held as a Turkish fief by John Sigismund Zapolya, now a boy of eleven. His ambitious and unscrupulous clerical minister, George Utschenitz—better known by his mother's name as George Martinuzzi—conspired with King Ferdinand to hand over Transylvania to the Austrians, in defiance of the Sultan. He partly succeeded in his plan—which was to include the transference of his young master to the Silesian duchy of Oppeln, in compensation for his loss—and Austrian troops were introduced into many of the fortresses of Transylvania and the Banat, before the Sultan was aware of the bargain. Hence came a confused war, mainly on Transylvanian ground, involving the sieges of many towns by the Turks, some successful, some unsuccessful. The story becomes hopelessly complicated when Martinuzzi—discontented with the Cardinal's hat, which was almost the only thing that he had got out of his bargain—began to indulge in intrigues against Ferdinand—he wished to be Prince of Transylvania himself. Discovering these plots, the King's general, Castaldo, had the Cardinal assassinated—not without his master's knowledge [1] (December 18, 1551). Consequently Martinuzzi's partisans went over to the Turks. Long campaigning in 1551–2–3 led to the loss by the Austrians of all the fortresses that they had occupied in Transylvania, and even of some of those which

[1] The story somewhat recalls the murder of Wallenstein at the behest of Ferdinand III, eighty years later, under somewhat similar conditions.

they had been holding ever since 1526—the Turks were only stopped before the walls of Erlau in the north and Szigeth in the west, which stood firm.

Meanwhile the Sultan made no personal appearance on the Danube—he was distracted by Eastern affairs. The Persian Shah, who was as useful to the Hapsburgs as the King of France was to the Sultan, had replied to the Turkish invasion of Adherbaijan in 1549 by a sudden raid into the valley of the Upper Euphrates in 1552; he beat the governors of the frontier provinces and advanced as far as Erzeroum. Soliman thought himself forced to take the field in person in 1553 against the Persians, leaving the affairs of Hungary to his pashas. He was far off in the East all through the years 1553–54, to the immense relief of King Ferdinand and his generals. And thus it turned out that a new intervention of the French in the West, when Henry II took up the old policy of his father, Francis I, and renewed the league with Soliman, had not all the effect that might have been expected. While the French were seizing Metz, Toul, and Verdun (April–May 1552), Sultan Soliman was preparing his great army for a counterstroke in Armenia, not for the Danube. And when in 1554 Henry was winning his very dubious victory of Renty, his ally was somewhere between Sivas and Erzeroum. The Persian War only came to an end by the Treaty of Amasia in May 1555, which left frontiers mainly unaltered, though with some territorial gain to the Turk. It is easy, therefore, to understand why matters stood more or less balanced on the Hungarian frontier, when the Sultan was not present to bring the main force of his realm to the help of his local governors; while on the other hand King Ferdinand got no help from the Emperor, who was tangled up in his French war.

The only part of the world where there was some actual co-operation between the Turks and the French was the basin of the Central Mediterranean, where their two fleets (as in 1544) were sufficiently strong to hold the Imperial squadron in check. Hence came the loss of Tripoli by the Knights of St. John in 1551, who made a fine defence, but got no help from Italy. The small number of them who were alive at the capitulation were sent home on the French galleys of Gabriel d'Arramont, who was present at this siege. Alike in 1552 and in the two following years, Turkish squadrons under Torghoud and Sali Reis, the successors of Khaireddin in the management of corsair fleets, did much harm to the coasts of Italy and the Balearic Isles, and in 1553 co-operated with the French galleys of Paulin de la Garde (the murderer of the Vaudois) in a great raid on Corsica, when Bastia was captured from the Genoese. The walls were made over to the French—the garrison allotted to the Turks and sold as slaves. In 1555 Sali Reis, the Governor of Algiers, took from the Spaniards the considerable harbour of Bugia, half-way between Algiers and Tunis, which they had been holding for many years. But all this warfare, though most annoying to the Emperor Charles—whose admirals, even the great Doria, were never able to prevent raids on some points of the vast length of coast-line which they had to guard—had no decisive influence on the general course of the struggle in the West.

Though Soliman got back from his third Persian expedition in the autumn of 1555, he did not lead in person an assault on the Austrian borders on the Danube in 1556 or 1557 or 1558—those three years crowded with events in the West which saw the end of the great struggle between the Valois and the Hapsburgs. Even when the last French invasion of southern Italy took place in 1557, it got no Turkish help, save by incidental corsair raids on the Tuscan and Neapolitan coasts. It is to 1557 that belongs the extraordinary appeal of Cardinal Caraffa, the busy nephew of Pope Paul IV, that bitter enemy of the Hapsburgs, which besought the Sultan to leave Hungary alone, and to turn against Spain all his available force for an invasion of the Two Sicilies. This can only be called high treason against Christendom in the name of its nominal head! Soliman styled himself Caliph, and was champion of aggressive Mohammedanism against Europe; his troops had been seen in Apulia twenty years back—the Caraffas wanted to see them there again!

But the great Sultan's personal vigour was on the wane, though he was yet destined to lead one more assault on Christendom, when it was too late, and to die on Christian ground. In the years after 1555 he was beginning to suffer under the curse that descends on all Eastern monarchies when an autocrat grows old, and has several ambitious heirs waiting for his inheritance. As early as 1553, while he was in Asia engrossed in his Persian war, he had put to death his eldest surviving son Mustafa, on accusations of projected rebellion, invented apparently by harem intrigue. The mother of his youngest sons, the celebrated 'Roxolana,'[1] and his Grand Vizier Rustum conspired to rouse suspicions that were apparently groundless. The murder of a popular prince almost roused a general revolt of the Janissaries, and actually had the curious effect of producing the rebellion of the 'false Mustafa'—a sort of Perkin Warbeck or 'Demetrius Ivanovitch,' which gave Soliman much trouble, till his would-be son was caught and drowned (1555). Mustafa's next brother Jehangir having died—of melancholy at his brother's fate some said, from more sinister causes said others—Roxolana's two sons, Selim and Bajazet, remained to wrangle for their father's succession; there was actual war between them while Soliman was yet alive. This internal friction, which came to a head in 1558–59[2] with the flight of Bajazet to Persia, was the main reason why the last struggle between Hapsburg and Valois was carried to its end without any serious Turkish intervention. The year 1557 saw the battle of St. Quentin and the retreat of the French army from central Italy; it might have seen the complete discomfiture of Henry II if the Hapsburg power had been relentlessly used. But Philip II, the unworthy successor of the great Emperor Charles, was (as always) dilatory in action, and in 1558 the last French defeat at Gravelines had been already offset by Guise's capture of Calais. Hence the Peace of Cateau Cambrésis (March 1559) was but a compromise, not a complete triumph for the House of Hapsburg. But it took France out of the general European war, and the futile death of Henry II in a tournament a few months later, marked the end of the long struggle that had started in 1494. For the future the Sultan could count no longer on French help, however fickle and intermittent, for his attempt to break into Central Europe.

The relapse of France into forty years of religious civil wars left Sultan Soliman without allies, but also in conflict with two enemies whose co-operation was insufficient—Philip II

[1] 'Khourrem Sultana' was a Russian, say most authors, but her origin is disputed. She had been one of the ordinary concubines of the Seraglio till the Sultan formally married her and gave her precedence. Her aim ever after was to get rid of his elder children by other consorts.

[2] Beaten in battle in Asia Minor in 1559, Bajazet fled to Persia, when the old Shah Thamasp most villainously betrayed a guest whom he had received with honour, handing him over to official murderers sent from Constantinople, who strangled him and his sons on Persian soil (1561). In return he received 400,000 gold ducats and the heads of some Persian refugees in Turkey whose destruction he desired. A pleasant bargain!

in the Mediterranean, the new Emperor Ferdinand I on the Danube. The uncle and nephew were so far from working honestly together that Ferdinand made a separate peace with the Porte in 1562, which left the Hungarian frontier practically unchanged from the delimitation of 1547, and saw John Sigismund Zapolya once more installed in Transylvania as the vassal of the Turk.

Philip II, on the other hand, left engaged in a naval war for the domination of the Mediterranean, fared badly in the contest. His first effort, after he was quit of the French war, was to resume the invasion of Africa which his father had carried out in 1535 and 1541. The corsair-pashas who represented the Sultan in African waters were in possession of Tripoli and Algiers, but not of Tunisia, where the Beni Hafs still reigned as Spanish vassals. An attempt by them to establish themselves at Mahadia, half-way down the Tunisian coast, had been foiled by Andrea Doria in 1550, when the pirate colony was for the moment exterminated. But in 1558 Torghoud, the most famous of Barbarossa's heirs, made himself a new base of aggression on the Isle of Jerba, at the southern end of Tunisia. Against it Philip sent a considerable armament, led by Gian Andrea Doria, nephew of the famous Andrea, and the Viceroy of Sicily; it contained contingents sent by the Pope, the Duke of Florence, and the Knights of Malta. Doria captured the town of Jerba (March 7, 1560), and threw in a Spanish garrison, but was surprised, while his fleet was at anchor, by an immense armament sent from Constantinople; a great number of Christian galleys were destroyed, the rest escaped to Malta. But no less than 5000 veteran troops were stranded in Jerba, where the Turks besieged them for three months and starved them out. Some of the most noted Spanish generals were taken prisoners to Constantinople, among them Requescens (afterwards Viceroy of the Netherlands), Alvaro de Saude, and Sancho de Leyva, all soldiers of great repute. In the next year, 1561, the Turks were for a time masters of the narrow seas between Italy and Africa, and executed most furious raids on Sicily, before Philip could assemble another fleet; when he did, it was wrecked in a hurricane, and twenty-five galleys perished with the admiral Juan de Mendoza (1562).

So debilitated appeared the Spanish naval power that in 1564 Soliman devised a plan of campaign which would have made him master of the central Mediterranean. A great fleet under Mustafa Pasha, one of the Viziers, sailed from Constantinople on April 1, 1565, and was to be joined by Torghoud and the African corsair-squadrons. The objective was Malta, the rocky and deserted island which the energy of the Knights of St. John had turned into a formidable naval base for every Christian fleet that operated in the narrows between Sicily and Africa. Though far from becoming as yet the first-rate fortress that later generations knew, the old 'Borgo' and galley-harbour of Malta, with its narrow entrance, were the home of a small but active squadron which joined every Spanish expedition that went to Africa, and was in close touch with the viceroys of the Two Sicilies. Here fleets mustered, or took refuge after an abortive venture. Twice Malta had beaten off casual attacks by the Turkish admirals, but it had never before been assailed seriously by a large force.

Its fortifications were ever growing, as the Knights realized the increasing danger from the eastern front. The old 'Borgo' of Malta had been but a small place on a projecting headland in the harbour, with the Castle of St. Angelo at its point, and the galley-harbour on its left side. There was an outlying fort, St. Elmo, covering the harbour mouth. But the last two Grand Masters, La Sangle (1553–57) and the famous Jean de la Valette, who took over charge eight years before the Turkish invasion came, had been great builders. La Sangle had fortified the sister-promontory which faces the old city on the left side of the harbour, and built on it the Castle of St. Michael. Named after him, 'Sanglea,' this quarter became an integral part of the fortress. La Valette had made good bastioned cover on the short land fronts of both the 'Borgo' and the Sanglea, and had greatly strengthened St. Angelo, the central point of defence, and cut it off from the 'Borgo' by a deep ditch. He had much desired to occupy the opposite heights, where the modern city of Valetta stands on its steep many-staired eminence, but the plan was too large for his resources, and all that was done was to rebuild the old fort of St. Elmo at the extreme seaward end of this ridge, which was then known as Monte Schebarras. St. Elmo both commanded the entrance of the outer harbour at its narrowest point, and also had a flanking fire on the mouth of the other bay, Marsa Muscetto, which lies north-eastward, and offered a convenient landing-place for hostile forces so long as it was left unprotected. La Valette made St. Elmo a first-rate work, a star fort after the best and most modern fashions, with scarp and counter-scarp, ravelins, and a 'cavalier' inside. But it was very remote from the rest of the defences, and could not only be attacked from the land side, but battered from across the water if the enemy took possession of neighbouring headlands.

La Valette had ample notice of the approaching storm, and his masons had been working to the last moment, all through the winter and early spring months, when on May 18, 1565, the expected Turkish armament came in sight. The Vizier, Mustafa Pasha, was in command, Piale the 'capitan-pasha' led the fleet, and after it had arrived at Malta the great corsair Torghoud came in with the Tripolitan squadron. There were 130 galleys and 50 transports, and other non-military vessels. The landing force comprised 4500 Janissaries, 7500 Asiatic 'Spahis'—serving, of course, dismounted—13,000 infantry, and 1500 irregular horse from the eastern provinces, with 5000 from the Roumelian sanjaks—altogether 30,000 troops of varying quality. To meet them La Valette had within the walls a very considerable garrison—about 500 Knights of the Order, 1300 of their regular mercenary troops, 1000 marines disembarked from the galleys, and 4000 armed Maltese levies, and many volunteers of all nations, who had drifted into Malta on the news of the approaching storm, from the love of adventure, or the wish to work off past delinquencies by good service for the Cross against the Infidel. This made up about 9000 men of all sorts. Reinforcements were expected, for all the Knights absent in their 'commanderies' all over Europe had been ordered to rejoin. There was a considerable accumulation of them at Messina soon after the siege began, waiting to join the fleet which Garcia de Toledo, the Viceroy of Sicily, had promised to collect for the relief of La Valette's garrison.

The Turks landed in bays on the west coast of the island, at a considerable distance from the fortress, mainly at the bay of Marsa Sirocco, where the fleet remained, and marched overland to the eastern front of the harbour, where they pitched their camp on Monte Schebarras, opposite St. Elmo. It was noted with some surprise by the Grand Master that they did not detach any force to deal with the old inland town—Citta Vecchia, or Citta Notabile as it was called—on the south side of the island, where most of the inhabitants had taken shelter under the protection of ancient walls and a small garrison. All their efforts were directed against the maritime fortress and its harbour. The Vizier concentrated his first efforts against the isolated fort of St. Elmo, by whose capture he hoped both to close the mouth of the Great Harbour on one side, and to make it possible for his fleet to enter the very convenient bay of Marsa Muscetto on the other. As long as the guns of St. Elmo commanded the entry of this bay, Turkish ships could not get into it.

The siege of Malta, therefore, falls into two parts; during the first—May 20 to June 24—the whole of the efforts of the Vizier were directed against St. Elmo. Only when it fell, after a most gallant resistance, did he bring his fleet into the Marsa Muscetto, and set to work against the main stronghold of the Knights, the twin promontories of the Borgo and the Sanglea, with their Castles of St. Angelo and St. Michael, on each side of the galley-harbour. The second series of operations lasted from June 24 to September 7, when the Vizier raised the siege and went off with the wreck of his army.

The defence of St. Elmo for a full month, when the Turkish trenches were pushed right up into the ditch, and an overpowering force of artillery had opened several practicable breaches, was a very gallant achievement. The Knights repulsed every assault, and built up the breaches almost as fast as they were made. For many days the Grand Master was able to keep up communication with the garrison by boat across the harbour, and to pour in reinforcements and munitions. But this convenience came to an end when the corsair Torghoud, newly arrived from Tripoli with his squadron, showed the Vizier how to place guns on the shore opposite the Castle of St. Angelo, which made it impossible for boats to cross the harbour mouth even by night. He at the same time erected a new battery at the rocky point on the other side of the Marsa Muscetto harbour-mouth, which completely enfiladed the right flank of the wall of St. Elmo and reduced it to a ruin.[1] The garrison, as long as it was being reinforced, beat off assault after assault, even after the Turks had blown in the counterscarp and made a secure lodgment in the ditch. But each successful repulse of the stormers cost lives to the defenders, which they could not afford, when their communication with

[1] This headland, still called Torghoud's Point, was afterwards guarded by Fort Tigné, but in 1565 it was unoccupied.

the other side of the harbour had been cut off.[1] A very violent assault on June 23 was repelled, but the garrison was then reduced to a handful, and on the following day the Turks got in at points where there was practically no defence for sheer lack of men, and St. Elmo fell. The survivors were massacred and their bodies sent floating into the harbour, with red crosses

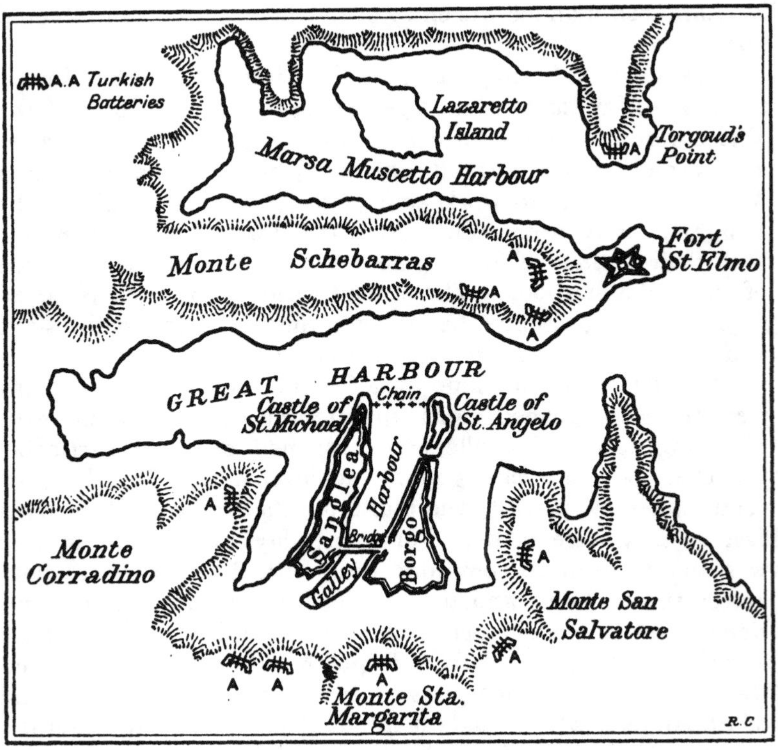

SIEGE OF MALTA, 1565.

cut on their mutilated breasts. In the month-long operations against this fort the garrison had lost 130 Knights and over 1200 of their rank and file. The besiegers are said to have lost 8000 men, among them the corsair-pasha Torghoud, who was mortally wounded in the trenches while he was directing one of the later assaults (June 16) and died on the day of the final

[1] La Valette made an attempt to bring off the garrison on the night of the 23rd, but the Turkish battery opposite St. Angelo opened on the boats at once, and those not sunk had to turn back.

storm. Piale, the commander of the fleet, was hurt by splinters from a ricochetting shot, but not so as to disable him.

Surveying the ruined fort, Mustafa is said to have asked himself, 'If the child has cost me so many casualties, how much of the army will remain when we have done with the parent?' But he was now at least free to commence the major operation, and started by bringing the whole fleet round to the convenient harbour of Marsa Muscetto from its distant roadstead at the Marsa Sirocco. The greater part of the army moved round the head of the great harbour, and camped itself on the slopes of the Margarita, Corradino, and Salvatore heights, looking down on the land-fronts of the Borgo and the Sanglea, against which operations could now commence. But several of the batteries on Monte Schebarras, which had been used against St. Elmo, were turned sideways to batter the Castle of St. Angelo from across the water. Trenches were begun opposite the new objectives in the last days of June, but it was not till July 15 that serious attacks began. Batteries were built not only opposite the land-fronts, but on outlying headlands on the flanks, which would give an enfilading fire against parts of the enceinte walls which could not be reached by guns in the main attack.

Just before the second series of operations began, La Valette received the only reinforcement which reached him during the siege. Weary of waiting for the viceroy of Sicily, who refused to stir from Messina before he had collected every available ship, eighty Knights of Malta in their own galleys with some 600 mercenary foot, ran across the narrow seas, and landed on the south side of the island, not far from the Citta Vecchia. Having learned of the situation, they made a hazardous march on a night of dense fog (a rare thing in Malta in summer) and got through the Turkish lines, by some special providence as they imagined (June 29). La Valette was glad of the reinforcement, though somewhat depressed by the reported tardiness of the Viceroy, whose best promise was that he might arrive with a respectable force sometime in July—a pledge which he did not keep.

From July 15 down to September 7, for more than seven weeks, the inner defences of Malta were exposed to a concentric fire from all sides, since they were raked not only from the frontal attacks but from the batteries on Monte Schebarras across the water, and from those on the headlands east and west of the Borgo and the Sanglea. There was no place of absolute safety from fire save in the casemates of the Castle of St. Angelo—whose upper works, however, were much battered. But bombardment does not capture a fortress properly defended, storming has to follow, and this was attempted by the Turks on a dozen different days, when they thought that the walls were sufficiently shaken. The main attacks were, of necessity, made on the two short land-fronts of the Borgo and the Sanglea, both very well-built stretches of masonry, with ditches and bastions in the approved modern style, well furnished with artillery. Though the besiegers drove their trenches right up to the ditches, and made lodgments close in to the battered bastions, they never succeeded in breaking in, so scientific was the artillery defence, and so firm the line of pikes and arquebuses that fronted the storming party at a new breach.

Early in July, Hassan, the Pasha of Algiers, turned up with a rather belated corsair contingent. The son of Khaireddin was an arrogant and self-confident person, who expressed surprise that the Vizier had stayed so long before St. Elmo, and boasted that he could storm the Sanglea defences with his Algerines alone. Mustafa lent him several thousand of his own men in addition, and bade him justify his promise. Hassan's notion was that he would combine an assault on the land-front of the Sanglea defence with a boat-attack on weak points of its water-front. He had a great number of small boats dragged by the labour of his galley-slaves across the shoulder of Monte Schebarras, and launched in the head-waters of the harbour above the town. The double attack was duly made: Hassan's own attempt on the land-front was turned back decisively, with a vast loss of his Algerines. The water-attack was an equal failure—it was found that the weak point in the wall of the Sanglea had been covered by a boom of masts and beams chained together, which the Turkish boats could not cross. In desperation, the leader of the flotilla, a renegade named Candelissa, thrust his boats against the rocky point of the spur, beyond the boom, and made a violent attempt to escalade the extreme end of the work. His men scrambled ashore, and some of them actually got a footing in a breach, but were driven out and mostly exterminated, as their boats had nearly all been sunk by cross-fire from the Castle of St. Angelo. Half of Hassan's Algerines perished in one attack or the other.

This was the only attempt made by the Turks at an assault

from the water. They turned to their usual methods, mining as well as battery, and built a high ' mount ' opposite the Borgo front, from which a tiresome fire was kept up day by day. Many breaches were made, but all attempts to storm them failed—especially bloody repulses were suffered on July 28 and August 1 on the Sanglea front, where the besiegers ran a temporary causeway across the ditch, which had to be cast down. On the 7th of August an attack in great force was made on the Borgo front, which lasted for five hours before the last stormers were thrust down. Between the 18th to the 23rd, four similar assaults were made, after a heavy bombardment had brought down considerable sections of the wall: on one occasion the Turks made a lodgment in the Borgo ditch, and stockaded themselves in it, from which they had to be expelled by a costly sally. By this time Mustafa's army was terribly reduced—he is said to have had only 14,000 fighting men left, and 10,000 sick and wounded in his camps. He had heard that the Viceroy of Sicily had now a large fleet in hand, and might arrive at any moment, while forty of his own galleys in the Marsa Muscetto harbour could not take the sea, because their crews had been used up in the land operations. But he resolved to make one final assault, which was accordingly delivered on September 1, but without the dash or confidence shown in earlier attempts, and repelled as usual. After this he made up his mind that the enterprise must be abandoned. The tardy appearance of the Spanish fleet only hastened his departure.

Garcia de Toledo had set sail from Messina on August 20 with 70 galleys and a landing force of 10,000 men, and after a short stay at Syracuse made for Malta, but was caught by an east wind off Cape Passaro, and driven 200 miles toward the Straits of Gibraltar. The ships were much scattered and only reassembled once more at Trapani on September 3, from which they at last made their much-delayed crossing to Malta, and anchored off Gozo on September 5. On the 6th the army disembarked and marched to Citta Vecchia, while the fleet rounded the south end of the island and appeared in the open sea, only three miles from the harbour mouth, and exchanged salutes with the wearied garrison.

The Turks then hurried to their ships and fled, abandoning 24 great guns in their batteries, burning masses of stores and munitions, and many disabled and unmanned vessels in the port of Marsa Muscetto. Why the Spanish fleet did not attack them as they moved off is difficult to understand—but the Viceroy showed as great a timidity in these days as he showed torpidity during the last three months. So cautious was he that Mustafa, on a false rumour that the army of relief was only 3000 strong, disembarked his troops in St. Paul's Bay, on the north side of the island, and only drew back after a skirmish which showed that the Spaniards were in great force. Over a thousand Turks were driven into the sea as they made a hasty re-embarkation, after which Mustafa made off as best he could, with a shattered armament, and returned in disgrace to Constantinople.

It is said that the total Turkish loss during the expedition amounted to 24,000 men, including no doubt the dead among the galley crews. The gallant defenders of Malta, who could count up their casualties with greater accuracy, made up a list of 240 knights who had fallen—no less than 130 of them in the operation at St. Elmo—and nearly 5000 of the rank and file and volunteers. This must have represented a 50 per cent. loss, as the original garrison was only 9000 strong, and the one reinforcement received on July 29 was under 1000 of all ranks. But we must suspect exaggeration when we read that, after deducting sick and wounded, La Valette had only 600 men on the walls on the day that the Turks departed. The most notable officers lost were two Spaniards, Medrano and Eguaras, who had been in command at St. Elmo; Robles, 'maestre de campo' and commanding in La Sanglea; Parisot de la Valette, the Grand Master's nephew; and Federigo de Toledo, son of the tardy Viceroy of Sicily, who was serving as a volunteer. Of 'priors,' 'commanders,' and 'bailiffs' there is a considerable casualty list.

Though naval warfare was to continue for several years more, this was the last attempt made by the Turks to get control of the 'narrows' of the Mediterranean, by occupying Malta and making it their central base. Their subsequent operations only aimed at clearing the Spanish power out of the African mainland, and establishing a superiority in the eastern half of the Mediterranean, if they could not master the western half. Their corsair raids remained frequent and vexatious, but no attempt appeared at making a lodgment on European soil. The all-important chance of French co-operation had ceased to operate. Looking forward a few years, it may be well to notice that the Spaniards were never cleared out of Oran,

Melilla, Marsalquivir, and some other headland-fortresses, but their vassals the Beni Hafs were for the second time expelled from Tunis, when in 1569 Ouroudj Ali, Pasha of Algiers (better known to Christians as Ochiali) defeated and drove away Hamid, the last Sultan but one, and occupied his capital. But the third and final expulsion of the Beni Hafs only came in 1574, as will be told in a later chapter.[1]

Yet when Tunis, Tripoli, and Algiers were all once more Turkish corsair-haunts, under pirate viceroys, there was never another resolute attempt made on Malta. The raids were a nuisance, but never a serious danger to the Spanish naval control of the central Mediterranean.

Meanwhile Malta became more than ever the centre of restraint on the pirates. The gallant Grand Master La Valette lived long enough to double the enceinte of his fortress by founding on Monte Schebarras the city which bears his name as Valetta, and to which he transferred the headquarters of the Knighthood (1571). Later generations took into the circuit the isthmus between Valetta and the mainland—the Floriana fortifications (1635–38), and later still threw round the head of the old galley-harbour the 'Cottonera' outer enceinte—named after Grand Master Nicolas Cottoner (1670). But all these complicated outworks never had to repel a Turkish attack. Their strength remained untested till, in the days of the degeneracy of the Order of St. John, they were surrendered without any defence to the piratical enterprise of Napoleon Bonaparte (1798).

N.B.—There are very good detailed accounts of the siege of Malta—unfortunately the dates in many of them do not tally exactly with those in others. I have taken as correct those which can be found in La Valette's dispatches to the Viceroy of Sicily, the Pope, and to Hohnstein, the Prior of Germany. La Valette fixes the fall of St. Elmo on June 24, the first appearance of the Turk on May 18, and the raising of the siege on September 7, though the festival instituted in commemoration of it has always been kept on September 8. Von Hammer's dates, taken from Turkish sources, are often inaccurate.

[1] See p. 739 below.

CHAPTER X

SZIGETH, CYPRUS, AND LEPANTO—THE LAST TURKISH OFFENSIVE (1564–73)

THE Emperor Ferdinand had died in 1564, but the peace concluded with him in 1562 was continued after the accession of his son Maximilian II, although endemic border strife on the Turkish-Austrian frontier in Hungary never entirely ceased: both parties were undoubtedly guilty of raids and trespasses disavowed by their sovereigns. However, it is somewhat surprising that so early as January 1566 open war broke out again—though the Sultan was seventy-two and failing in health. He was by now much dominated by self-confident ministers, who presumed on his growing senility: the family tragedies of the preceding ten years seem to have broken his spirit. It was probably the desire to restore the prestige of the Turkish power, which had suffered such a humiliating blow at Malta in the preceding autumn, that led the ambitious vizier, Mohamed Sokolli, to induce his master to imprison the Austrian envoys at Constantinople, and to declare war (January 1566); Soliman announced that he would, as on previous occasions, head his own army for an invasion of the Emperor's dominions—this was the thirteenth time that he took the field, counting from his first successful expedition against Belgrade in 1521.[1] He was no longer able to use a horse, being riddled with gout, but travelled sometimes in a carriage, sometimes in a closed litter. Forced to proceed slowly, he only reached Belgrade after forty-nine days of march. On June 27 he received at Semlin the homage of John Sigismund Zapolya, and promised to make him king of all Hungary. His plan of campaign, as we are told, had originally been to cross the Danube and march on Erlau, but he turned aside westward, on receiving news that one of his flank detachments had been

[1] The expeditions were those to Belgrade (1521), Rhodes (1522), Mohacs (1526), Vienna (1529), Güns (1532), Bagdad (1533), Corfu (1537), Moldavia (1538), Buda (1541), Gran (1543), Tabriz (1548), and Erivan (1552).

cut to pieces at Siklos, near the Drave, by Count Nicholas Zriny, commander for the Emperor of the Hungarian borders on the side of Croatia, and governor of Szigeth, the most important fortress in that region. As this swerve from the line of his original march took him away from the army which the Emperor had collected in northern Hungary, and from the road to Vienna, in order to deal with a sudden minor provocation, it is hard to explain save by mere irritable senility. That Soliman was in a state of nervous irascibility was shown by an incident of August 3. When he was two marches from Szigeth, the Pasha in charge at Buda, Mohammedbeg Arslan, came into his camp to report, with some dismay, that he had been beaten by the Emperor's forces, and had lost the towns of Vesprizm and Tata. Though he was a soldier of approved courage and loyalty, Soliman had him strangled in the royal tent without any semblance of a trial.

On August 5 the Turkish army—at least 100,000 strong, came before the walls of Szigeth, where Zriny had shut himself up with a garrison of 2300 men. This was a strong place, not because it lay high, but because it was surrounded by a complicated marsh defence, the backwaters and bogs of the river Almas, a tributary of the Drave. Save where the high road entered and left it, there was a broad belt of impracticable ground. Szigeth was composed of three blocks, each separated from the others by low marshy depressions, and joined by bridges only—the old town, with a mediaeval wall, the new town which had grown out from it, and the citadel, a modern building with a keep and bastions. After surveying the objective, the Sultan's engineers reported that it could be battered easily enough from a distance, but that it could only be stormed by building causeways across the marshes—a tedious business. And so it proved—Szigeth held out for a full month—August 5 to September 8, 1566. Zriny evacuated and burnt the new town early in the siege, and the Turks established their main batteries in it, and commenced to build causeways from it toward the old town. These works were accomplished under a continuous and murderous fire from the garrison, but on August 19 the old town was breached and carried by assault. The last stage of the siege, when the well-built and water-girt citadel had to be dealt with, occupied nineteen days, and cost the besiegers a terrible casualty list—three successive assaults by the Janissaries, across improvised causeways, were repulsed with dreadful loss. Among the slain were the Pashas of Cairo and Basrah, the high treasurer, and most of the chief commanders of the Janissaries. The citadel, though terribly battered, was still holding out, when on the night of September 5–6 the Sultan died in his tent—worn out by dysentery, according to some Turkish authorities; of an apoplectic fit, caused by rage at the failure of the last assault, according to others. The Grand Vizier and private secretaries of Soliman concealed his death—partly because the place was bound to fall in a few days, and the army must not be discouraged, partly because news had to be sent in haste to Selim, the only surviving heir to the throne, to bid him repair to Constantinople without delay.[1] Orders and some forged dispatches continued to be sent out from the royal tent for many days, and Soliman was declared to be merely prostrate from repeated attacks of gout.

On September 8, Zriny, unaware of course of the Sultan's death, resolved that he could hold out no longer, his bastions being one mass of débris, and only the keep intact; wherefore he determined to die fighting. He laid a match to his store of powder in the keep, and charged out of the main gate at the head of the small remainder of his garrison. He fell almost immediately, with two bullets in his breast and an arrow in one eye. His men perished with him. The Turks cut off his head and rushed into the citadel, when the magazine in the keep exploded and slew hundreds of them. The Grand Vizier issued a forged bulletin of victory in the Sultan's name, and ordered a lavish distribution of gratuities to the soldiery. He then announced the return of the army to Constantinople—victory having been achieved, and the Master's health remaining unsatisfactory, so that he would be unable to continue the campaign. Soliman's corpse was borne off in a closed litter, and his personal attendants pretended to keep up communication with him. How long this farce was continued is uncertain—surely not for the space of three weeks of which Turkish historians speak—adding that as a matter of precaution the Sultan's personal physician was strangled—lest he might be indiscreet.

So ended not the last Turkish invasion of the Emperor's dominions, for there were to be others, but the last of the series which had once been a real danger to Christendom.

[1] It is said that the courier bearing to Selim the news of Soliman's death, rode from Szigeth to Kutahia in Asia Minor in the incredibly short time of eight days.

The final one of all was to be made so late as 1683—but this was a desperate adventure on the part of a decadent power—quite out of date in face of the Europe of the later seventeenth century. There had been a graver danger during the Thirty Years War; but in that unhappy period the Turks had not been in a position to take advantage of the tribulations of the Emperors Ferdinand II and Ferdinand III. Decadence had set in since the death of the great Soliman—perhaps even before he vanished from the scene—but of this more hereafter.

Meanwhile, the death of Soliman before Szigeth left the Turkish Empire at strife with the Emperor Maximilian on the Danube front, and with his cousin, Philip of Spain, on the Mediterranean front. The two wars had a singularly disconnected character. The general aspect of the struggle between the Sultans and the House of Hapsburg was profoundly affected by the fact that Soliman's successor was as unlike his father as could be well conceived. With the exception of Bajazet II, the great-grandfather of the new ruler, the Sultans of the House of Othman had been fighters—generally successful fighters—men able by their personal ascendancy to control the complex military machine of the Turkish Empire, a machine which was always requiring personal attention to some of its joints. Selim II was anything but a fighter—he was now well on in middle life—in the forty-third year of his age—and had been for the greater part of his existence in the precarious position of a possible heir to the throne with rival brothers. At any time he might have disappeared, as the result of some intrigue among those about his imperious father, for Soliman's autocratic temper was capable of sudden gusts of suspicion.

Mustafa, Selim's elder half-brother, had been put out of the way in 1553 by their father; but his full-brother, Bajazet, had bid fair to win the succession, till his flight to Persia in 1559 and his subsequent murder in 1562, a fate which was shared by his four young sons, all of whom might have become Selim's rivals. There is some reason to suppose that the stress of anxious years of waiting, with a possible bowstring in his mind's eye—turned Selim into a secret drinker—after his accession he became a very obvious drunkard, and goes down in Turkish chronicles as 'Selim the Sot.' He was not, however, one of those drunkards whom liquor stirs up to sudden bursts of frantic energy, but a confirmed tippler of the sedentary sort—the one thing that he could not face was an emergency, and the main resolve of his mind was that he would never take the field at the head of an army, which he felt sure that he was incapable of conducting. His first interview with the soldiery was terrifying; he hurried to Belgrade to meet his father's corpse, only to find the mourning ceremonies troubled by a general strike of the troops, who demanded a donative with insulting cries. Selim took refuge in his tent, and granted it on the most lavish scale, after which he made off for Constantinople, to shut himself up in the Seraglio, sending on his father's body, under a very small escort, to be buried with no better attendance than that of the Chief Eunuch. There was a second sanguinary *émeute* when the Sultan reached home himself, appeased only by further lavish donatives.

Selim would probably have come to an early end but for two things. One was that he had no relatives (thanks to his father's cruelty) who could be set up against him, the only survivors of the royal house being his own six young sons. The second was that he put himself under the protection of his father's Grand Vizier, Mohammed Sokolli, a drastic person who cleared away rival ministers by getting them killed. By his help chaos was averted, and the short reign of the timid and drunken monarch passed by without any supreme disaster. Peace was made on the Hungarian border, after much haggling with the ambassadors of the Emperor Maximilian, without any appreciable loss or gain of territory to either party (February 1568). This left the Grand Vizier free to push on the war with Spain in the Mediterranean without interference, where (as we have already recorded) the Pasha of Algiers had made an end of the Spaniard's vassal-king of Tunis, and driven King Philip's troops into Goletta and their other coast fortresses (1569). The King of Spain was at the moment tied up in two unnecessary wars provoked by himself—the great Morisco rebellion (1568–69) and the outbreak of the insurrection in the Netherlands, where the battles of Heiligerlee and Jemmingen had only recently been fought. He had little attention to spare for the Mediterranean, and the Kapitan-Pasha Piale had been ravaging the coasts of Naples almost unhindered.

But for a revolt in Arabia (1569) and an unlucky attempt to aid the Crim Tartars to seize Astrakhan in the same year,[1]

[1] Recently occupied by Ivan the Terrible, who had driven out the last Tartar Khan. The expedition failed completely, and the end for which it is said to have been sent out, to cut a canal from the Don to the Volga, sounds chimerical.

it is probable that Selim might have been induced to put energy into the Spanish War as early as 1569, after the recovery of Tunis. But apparently regarding King Philip as tied up in other difficulties, and out of the game for the present, Selim committed himself to another enterprise.

Venice had withdrawn from the league against Sultan Soliman so far back as 1540, and had for a quarter of a century allowed the Hapsburgs unaided to keep up the struggle with militant Mohammedanism. Her neutrality was commercially profitable, and she had refused to be moved by echoes of crusading propaganda. But obviously her colonial empire in the Levant was only safe from the Turk so long as the latter was not preponderant on the water, and still feared to see the Venetian navy acting along with that of the King of Spain. Cyprus in particular was strategically indefensible, if the Sultan were master at sea—Crete and the Ionian Isles were not quite such far-flung possessions. Selim was convinced in 1570 that the danger from Spain had become negligible, and that if he were to attack Venice she would get little or no help from outside. It is said that his great minister, Mohammed Sokolli, was not specially attracted by the project, and that the Sultan was for once acting on his own initiative, impelled by other advisers, the admiral Piale and the Vizier Lala Mustafa. It is perhaps unnecessary to lay stress on the oft-repeated story that his favourite beverage was the strong white wine of Cyprus, and that he was determined to be master of the vineyards from whence it came. But the tale appears in every chronicle. It is sufficient to cast a glance at the map to see that Cyprus is an untenable position, unless its owners have command at sea. It is far too close to the ports of Syria and Anatolia, too far from Venice. And Venice, unless helped by other powers, was no longer strong enough to dominate the Levant.

There were always petty subjects of dispute to hand between the Republic and the Porte, which could be vamped up into a *casus belli*, but Sultan Selim reduced the problem to one of religion. 'When a country once held by the Moslems has been conquered by infidels, is it not the duty of a pious prince to recover it for Islam?' was the question which he put to the Mufti Ebnsaoud. And the reply, as expected, was in the affirmative. It may be mentioned that the period when Cyprus had been in Mohammedan hands ran from A.D. 690 to 965.

So the question showed at least an intelligent interest in ancient history, usefully applied to a modern problem.

War having duly been declared on the Venetian republic in the first months of 1570, a Turkish fleet of 116 galleys with many smaller vessels appeared before the Cypriot harbour of Limasol on July 1, under the Admiral Piale Pasha, and put ashore an army of some 50,000 men under Lala Mustafa, including 6000 Janissaries. No hostile squadron was at hand to hinder the disembarkation. The Venetians were occupied in administering crusading propaganda to the Pope, the King of Spain, and the minor Mediterranean powers, though they themselves had been so singularly deaf to such appeals ever since 1540. They were conscious that they could not face the Turkish navy with their own resources, and would not risk all their available galleys in the waters of the extreme East. As to the needs of the first months of the war, Cyprus had a garrison which would have been considerable if it had been an island no larger than Rhodes or Malta, but which could not cover a kingdom 100 miles long by 60 broad—nine or ten thousand men in all, including horsemen. A few hundred stradiots and perhaps 1000 Cypriot cavalry, the survivals of the old Frankish noblesse of the defunct kingdom of the Lusignans, and of infantry 1500 Italian mercenaries, 3000 native Venetian foot, and 3000 or 4000 Cypriot militia, Greek or Frankish by race. Clearly this force could not hope to face 50,000 Turks in the open field, and would have to take refuge in fortresses.

Cyprus is full of old Lusignan castles in inaccessible mountains, but obviously it would have been useless to fritter away the garrison of the island in detachments holding antiquated fastnesses. The governor, Nicolo Dandolo, left the countryside to be ravaged, and concentrated all the troops in the two great cities, Nicosia the capital, in the central plain, and Famagosta, the strongly fortified harbour, which had always formed the naval base of Venetian fleets. Nicosia had quite recently been refortified according to the latest European systems, with a regular front of eleven bastions, bearing 250 guns, and having good scarps and counterscarps. To round off the enceinte and provide a broad glacis, all manner of mediaeval buildings had been scrapped, including many churches, and the monastery-palace in which the Lusignan kings had often resided. The place was quite strong according to all sixteenth-century military engineering, and Famagosta was also a good post, with a front covered in part by the sea, though its fortifications had not been brought up to date like those of Nicosia.

The Turkish conquest of Cyprus consisted of only two central episodes of siege—the open country having been left unprotected, and the mediaeval castles in the hills practically abandoned. These two episodes were the sieges of Nicosia (July 22 to September 9, 1570) and of Famagosta (September 15, 1570, to August 1, 1571). It will be noted that the former was a comparatively short affair of seven weeks—the latter a very long one of eleven months. It is clear that only a concentration at very short notice of all available Christian naval resources in the Mediterranean could have saved Nicosia. But Famagosta ought to have been succoured—eighteen months having elapsed between the outbreak of war and the fall of the great Venetian naval base. It was only in May 1571 that the treaty for a general Crusade for the relief of Cyprus was signed by the Christian powers, and their united fleet only mustered at Messina in September 25, 1571, eight weeks after Famagosta had fallen.

The difficulty which the Venetians found in persuading the other Mediterranean powers to co-operate with them for the defence of Cyprus was wholly due to their own neutrality during the last twenty-five years. They certainly had not deserved well of Christendom when they allowed Charles V and Philip II to bear the whole brunt of the Turkish wars at sea. No enthusiasm could be felt by others, when they suddenly preached a Crusade in the moment of their own danger. And as a matter of fact while they were suing for help at Rome and Madrid, they were making repeated secret attempts to buy off the Sultan by something less than the complete cession of Cyprus.

The summer of 1570 passed by without any attempt of the Venetians to relieve Cyprus. Some vessels were sent down the Adriatic, and bickered along the coasts of Albania and Dalmatia with such squadrons as the Turk had in those waters, after his main fleet had gone to Cyprus. In the autumn news came of the invasion of the island, and a squadron was got together under Zani, which being joined by some 40 Spanish vessels under Gian Andrea Doria went as far as Suda Bay in Crete by September 22. Here news came that Nicosia had fallen, and that the Turkish Admiral Piale had left only a small squadron before Cyprus, and had come to Rhodes, where he had picked up detachments from Constantinople and the Aegean, and was in great force, ready to fight. Therefore Zani gave up the game, holding that the season was too late for a serious naval campaign, which he dared not risk against superior numbers. Famagosta must shift for itself till a larger force had been collected. The Venetian galleys returned to Corfu, and the Spanish contingent to Messina.

Nicosia had fallen much earlier than the Venetians expected. The usual Turkish methods of combined battering and mining failed to prepare the walls for storm during the first weeks of the siege, and two assaults were beaten off. But the garrison, though considerable, proved insufficient to protect all the points on which attacks were made at a third general assault on September 9. Entrance was made at several breaches, the resistance collapsed, and the governor, Dandolo, and the wrecks of his troops were slaughtered in the fortified palace in which they had sought refuge. The whole population, Frank and Greek, was exterminated ; 20,000 persons are said to have been massacred, and only a few thousand boys and women saved to be sent to the slave-market of Constantinople.

Lala Mustafa than overran the whole island, only the port of Famagosta was found ready to defend itself. Baffo, Cyrenia, Limasol, and Larnaka were taken without resistance. On September 18 he appeared before the walls of Famagosta, and sent in, along with a summons to surrender, the head of the unfortunate Dandolo in a basket, as a warning to Marc-Antonio Bragadino, the commandant of the garrison. A proper reply of defiance being made, the Pasha sat down in front of the place, and threw up a battery with which he began to bombard the tower at the angle of the port. He also constructed lines, from which trenches might be directed towards the land-front of the place. But learning of the arrival of the Venetian fleet at Crete, and seeing the ships of his colleague Piale gone off to meet it, he took into consideration the possible arrival of the enemy in strong force, and as October rains had begun, withdrew his army and cantoned it far back, leaving only a blockading force in his lines, and seven galleys to watch the harbour. The siege became a mere blockade for the whole of the winter. Bragadino was able to send out blockade-running vessels, which reached Venice, setting forth his dangerous position and asking for help. It was probably, however, before these appeals reached the Signory that he received a useful reinforcement: twelve galleys under the command of Antonio Quirini, who had been told to risk winter storms, ran into the harbour beating off the blockading squadron, and landed 1600 mercenary Italian infantry, and a large supply of munitions (January 23, 1571).

So things remained for the whole winter, the Turkish fleet having settled down partly at Rhodes, partly at Constantinople, the Venetian partly at Venice, partly at Corfu. Zani, who was thought to have shown cowardice, was removed from command and replaced by Sebastian Venier. Bragadino had five months to prepare for the renewal of the Turkish offensive against his walls, for active operations only began again in April 1571. He made good use of his time, strengthening his walls, clearing out his ditches, and erecting new artillery emplacements on walls which were mostly only provided with mediaeval towers, not with bastions. He had a garrison of 200 stradiots, 2500 Italian infantry, and about the same number of Cypriot militia—not at all too many for the defence of a rather old-fashioned fortress.

During the whole of the winter and spring of 1571 the Venetians were sending appeals all round Europe for help against the Sultan. They even succeeded in getting an envoy to the court of Shah Thamasp of Persia, who obtained little profit by his visit. Pope Pius V was propitious—more so than any other sovereign; but Rome could give little military help—only money and mercenaries: the money was used to hire galleys, on which the mercenaries were embarked. The Italian minor states offered their small contributions—Cosimo of Tuscany (whose coasts had been sadly wasted) offered twelve galleys, the Duke of Savoy three, the Grand Master of Malta three also: the Duke of Urbino, and Alessandro Farnese, heir of Parma, offered personal service, money, and mercenaries, since they had no galleys. But the main encouragement was that Philip II was at last free from his distracting Morisco war, and announced that he could lend not only (as usual) his Sicilian and Neapolitan squadron, but the main body of his fleet. Also that he could spare the services of his distinguished half-brother, Don Juan of Austria, who had just succeeded in putting an end to the Morisco rebellion with much credit to himself. He was only twenty-six years of age, but was already a commanding figure and inspired confidence in all who had personal touch with him. But the last Moriscos had only surrendered in 1570, and it was clear that Don John and his galleys could not be available till summer was far spent. Elsewhere appeals met with disappointment. The Emperor Maximilian and Sigismund of Poland both reported that they had concluded very recent treaties with the Sultan, and were bound by them. The young King of France, who had just got through one of his Huguenot wars (Peace of St. Germain, August 8, 1570) was averse at the moment to doing anything to help Spain—and offered futile ambassadorial mediation. It took the whole summer of 1571 to gather an allied fleet at Messina, and by the time that it was ready to sail (September 22) Famagosta had fallen, and all Cyprus was in the hands of the Turk.

Antonio Bragadino made a splendid defence, and repulsed every assault, from the opening of the regular siege operations on May 16, 1571, down to August 1. But he was oppressed by overwhelming artillery fire, and mine after mine shook his walls. Every breach was stopped and defended, and when the Turks broke open one gate by a series of conflagrations in front of it which smoked the garrison back, and finally brought down the gate, Bragadino had constructed a tenable earthen rampart behind the gap. All through the midsummer months he was expecting news of relief, being well aware that the 'grand alliance' had been formed, and that a fleet able to sweep the Turks off the sea ought to have been collected. But no help came, and on August 1, the stock of powder having been reduced to six barrels, and the garrison worn down to 2000 men, the governor opened up negotiations for surrender with Lala Mustafa Pasha. Then followed a horrible piece of treachery—much like the Cawnpore Massacre of 1857 on a larger scale. The Turk offered suspiciously easy terms—free departure to Crete for the garrison on fourteen ships to be made over to them, and safety of life and goods to the civil population. On August 4, the garrison having been shipped and the walls evacuated, Bragadino went with his chief officers to the Pasha's tent, to hand over the keys of the city. After a few words exchanged, Lala Mustafa burst out into a feigned rage and a flood of abuse, ordered the Venetians to be seized, and beheaded all the officers except Bragadino himself, for whom a more dreadful fate was reserved. His nose and ears were cut off, he was subjected to various tortures for several days, and finally he was flayed alive. At the moment of the first treachery Turkish warships fell on the transports on which the garrison had embarked, and captured them all—the soldiers who resisted were slain, the rest sold as slaves. There can be no doubt about the atrocity of the affair—Mustafa was wild with rage at the number of his best troops who had fallen in the fruitless assaults, and he was ready to promise any terms in order to get the Venetians out of the place. As far as he knew, a great fleet of succour might have arrived before Famagosta fell, as it was known that the Christian armada was assembling in Italian waters, and might sail any day.

The news of the fall of Famagosta percolated to Messina before the whole of the Christian armament was collected, as did also intelligence that the large Turkish fleet, which had been collected in Greek waters, waiting for a start of their enemies to relieve Cyprus, was still lying ahead, and had been ravaging Crete, the Ionian Isles, and the Dalmatian coast by flying squadrons. Such a force having been collected as that which Don John had in hand, it would have been absurd to dismiss it because Famagosta had fallen. The alternatives were to seek out the hostile fleet at all costs, and endeavour to destroy it, or to fall upon some one of the Turkish seaports in Albania and Peloponnesus, from Castro Nuovo to Modon, in the hope that such an attack would bring out the enemy and induce him to fight. The former and simpler plan was adopted on Don John's advice, and on September 25 the fleet sailed for the mouth of the Gulf of Corinth, the last news being that the main Turkish fleet lay at Lepanto, though detached squadrons had been operating elsewhere along the Adriatic coast. The enemy had observing vessels out at sea, and the moment that the start of the Christian fleet was reported, orders went round for the whole Turkish force to concentrate at the jaws of the Corinthian Gulf. And here it was found on September 7 when Don John's armament dropped down from Corfu past Paxos, St. Maura, and Cephalonia to the Echinades or Curzolari Islands, near the mouth of the Achelous. This was narrow water, with the fortified towns of Lepanto on the north, and Patras on the south side of the straits.

Don John formed his armada in a wide crescent, the main line consisting of galleys close ranked, with a reserve squadron half a mile to the rear. The fleet consisted of 108 Venetian galleys, with six great 'galeasses' (of which more hereafter), 81

Spanish galleys, 12 Papal galleys, and about 20 more from the minor allies—Tuscan, Maltese, and Savoyard. It is to be noted that there were no broadside ships in line, and only two apparently in the whole fleet, though the efficiency of such vessels under favourable conditions had been shown as far back as 1540 at the battle of Prevesa. The only exceptional units were the six Venetian 'galeasses,' which were not broadside-gunned ships, but galleys of double size, higher built, with overhead protection for the rowers, and larger and more numerous guns on the bows. They were so solid and strong-beaked that it was calculated that they ought to sink an ordinary galley by ramming. These were placed in pairs in front of the main line, with the idea that their impact would break the enemy's array when the crash came. The galleys of different nationality were not kept together, but mixed, 'that by the equality of danger they should have more call to relieve each other than if each nation served apart.' The idea was the same as that of Villeneuve at Trafalgar, when he alternated his French and Spanish men-of-war.

The right wing consisted of 53 galleys under Gian Andrea Doria, the Genoese captain who was as regular a commander in Spanish fleets as his more famous uncle Andrea had been in the last generation. The left wing, of similar force, was led by the Venetian Agostino Barbarigo. In the centre was Don John himself with 70 galleys; he was in his flagship, on each side of which were those of the Papal Admiral Marc Antonio Colonna, Duke of Palliano, and of the Venetian admiral Sebastian Venier. Half a mile in the rear of all was a reserve of 37 galleys under Alvaro de Bazan, the Spanish admiral who was destined twelve years later to win the battle of Santa Cruz over the French fleet of Strozzi. There was a gap only of the width of three galleys' breadth between the central squadron and the two wings, and the whole array was so long that it extended over nearly five miles of front. The six galeasses advanced well in front of the line, which was otherwise crescent-shaped, the wings a little curving inward. This was to be a battle in the old style, to be settled by ramming and boarding, not by gunfire, for the galley had but a few cannon in its bows. Hence the ships were crowded with land-soldiers, of whom nearly 20,000 had been taken aboard: arquebusiers were specially well represented; there were 400 on Don John's flagship alone—all Sardinians we are told. He had drafted

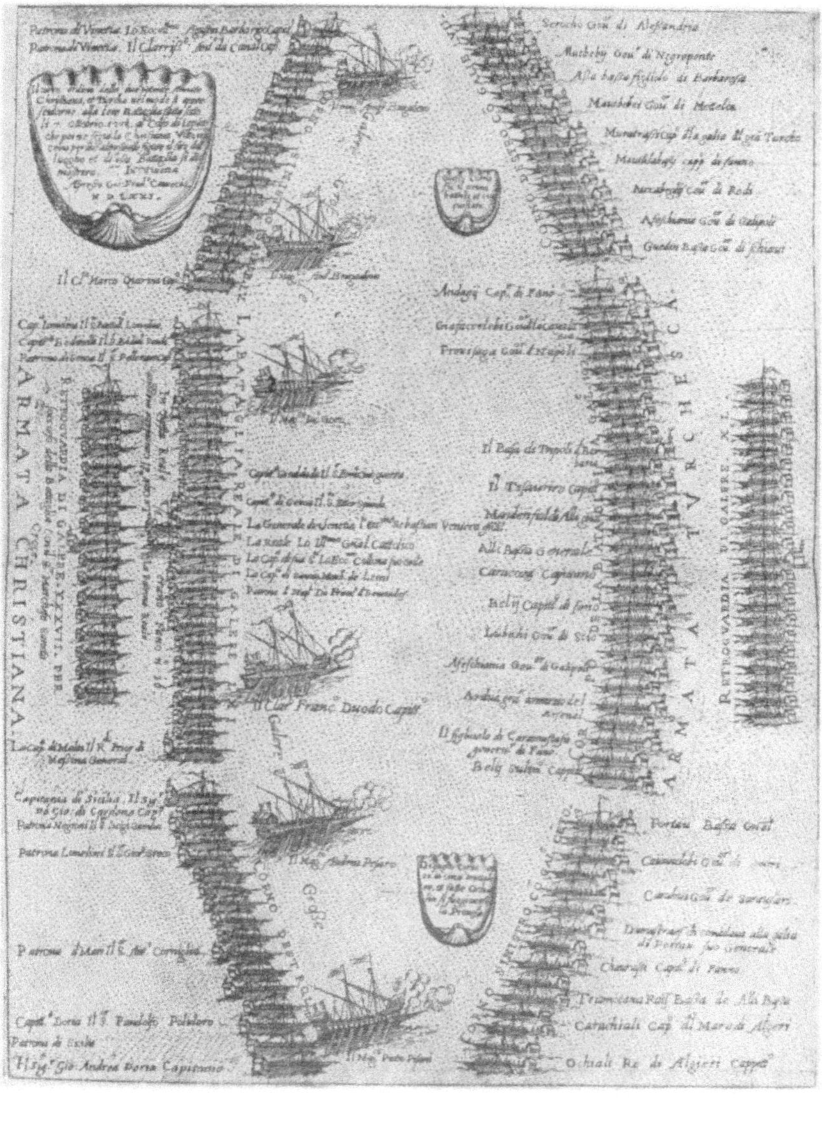

THE BATTLE OF LEPANTO, 1571 (1)
Disposition of the fleets before the battle
From the Venetian engraving by G. B. Camocio

Spanish detachments on to some of the Venetian galleys, considering that the fighting men on many of them were too few. This led to some unseemly national squabbles, the presence of the incomers being resented by the Italian captains.

The wind on the battle day was westerly, favourable to the Christians, and unfavourable to the Turks, who had to row against a breeze which caught their sails. The Ottoman fleet was under Ali Mouezinzade, the 'Kapitan-Pasha,' who had the central position opposite Don John, with the galley of the Vizier Pertev Pasha on one side of him, and that of the Treasurer Mustafa Tchelebi on the other. The whole force was decidedly superior in numbers to that of the Christians, apparently about 270 galleys against Don John's 220; but some of them were rather smaller than the average, and others, recently drawn out from Constantinople, had inexperienced crews. About 16,000 land-service troops were on board, including 6000 Janissaries. Unlike Don John the Kapitan-Pasha had made one of the wings stronger than the other: the right wing, under Mohamed Schaoulak, bey of Negropont, had only 56 galleys, as against 95 on the left, largely Algerines, under Ouloudj Ali, the Pasha of Algiers. There was a reserve of 40 galleys under Murad, son of the celebrated corsair Torghoud, who met his death at Malta. The centre under the 'Kapitan-Pasha' counted about 100 vessels. It is possible, but not certain, that the strength of the left wing was intended to favour a turning movement, which was actually carried out—whether intended or not—against the Christian right or southern horn of Don John's armada. But it is disappointing to the student of the Art of War to find that this battle, the largest naval contest since Actium (B.C. 31) fought in the Mediterranean, was for the most part an unscientific mêlée, not a battle of manœuvre. The decision came not from fleet-tactics, but from the superior management of individual galleys, and the relative fighting efficiency of the land-soldiers on each side. So much of the struggle was mere boarding-work that the Christian marines, wearing more armour—morions and breastplates—and carrying far more firearms than the Turks, had a decisive advantage over enemies many of whom were still armed only with bows and arrows, and who were for the most part unprotected even by mail-shirts.

The impact of the six great galeasses did, as had been expected, break the Turkish line, and throw it into some confusion, but four or five hostile galleys swarmed around each galeass, and there were six group-fights going on before the main bodies crashed together. The galeasses acquitted themselves well, ramming some Turkish vessels successfully, and disabling others by superior gunfire. But they did not settle the day. The decision was really given by close fighting, not by tactics. What it was like may be judged from the chronicle of Don John's own flagship, which chances to have been preserved. He charged for the galley of the Kapitan-Pasha, which was directly opposite him in the line. They met with such violence 'head on' that both their beaks were broken off by the shock, and fell into the sea. The two vessels were jammed together, and the Spaniards entering the Turkish vessel, beat the crew back as far as their main mast, where the first assault ended, and the enemy thrust the boarders back into the bows. Don John kept sending in every man that could be collected, and thrice, it is said, the mêlée flickered back to the main mast. At last Ali Mouezinzade was shot through the head and fell. The Spaniards decapitated him, and made an end of his disheartened crew—no quarter being given. Don John ordered the Pasha's severed head to be thrown into the sea, but some one stuck it on a pike and hoisted it on the poop of the captured vessel.

Similar encounters were going on all along the line where individual galleys got locked together, and others of both sides crashed in—like the *Bucentaure*, the *Victory*, and the *Redoubtable* at Trafalgar. Barbarigo, the Venetian admiral, had five Turks on him at once—he was mortally wounded, and his ship would have been taken had not two other Venetians, Loredano and Malpieri, pushed in and taken off the stress. The two fleets became so mixed, many galleys of each having got through the main line of the other, that Alvaro de Bazan, commanding the Christian reserve squadron, found himself usefully employed in cutting off Turkish galleys which had got through, and were attacking Christian galleys from the rear. This confused fighting with cannonade and attempts to ram, ending generally in the locking together of several ships, went on for a long time—perhaps three hours before decision came. It started on the northern flank, where the somewhat weaker Turkish wing was mostly driven ashore against the coast of Acarnania, and practically exterminated.

On the south flank, nearer the Peloponnesian shore, there was much longer and more dangerous fighting. Ouloudj Ali, the Algerian war lord, engaged with Gian Andrea Doria's squadron. Taking advantage of somewhat superior numbers, he turned his adversary's right wing with a large detachment. Doria threw back his outer ships *en potence*, to avoid being taken entirely in the rear, a manœuvre which separated them somewhat from his main line, and fifteen galleys, Venetian, Maltese, and Tuscan, got cut off, and were hard beset, and in two or three cases captured. One Venetian captain is said to have blown up his own galley with the Turkish boarders who had just mastered it. Doria, with his main body, lying behind one of the great galeasses, kept off, and contented himself with cannonading against Ouloudj Ali's chief force. He did not 'grapple,' like the other Christian admirals had done; for this he was much criticized at the time. But the fight in this direction suddenly slacked off, for Ouloudj Ali noted the rout of the Turkish central squadrons, and seeing many Christian galleys from the main fight swerving off towards him, collected all his available vessels and made off up the Gulf of Lepanto, hugging the southern shore. He escaped with forty of his own vessels, taking with him one captured Venetian galley and the great ensign of St. John, from the Maltese flagship under the Prior of Messina, who had been killed in a lost fight.

These were the only Turkish ships which escaped. Some sixty ran (or were driven) ashore, whose crews scrambled to land, leaving their vessels to be burned by the victors. One hundred and fifty were sunk or captured—apparently forty were sunk, the rest taken, with a horrible massacre of their fighting crews. There were, however, some 300 prisoners made, including the two sons of the Kapitan-Pasha, who had sailed on their father's flagship, and Mahomet Bey of Negropont, who had commanded the Turkish right wing. The dead included, beside the Kapitan-Pasha, the High Treasurer, Mustafa Tchelebi, and the sanjak-beys of Alexandria, Nicopolis, Mitylene, Rhodes, Chios, Bigha, Nauplia, and Angora, with Karaguez, one of the best-known Algerine pirate captains. The Vizier Pertev Pasha escaped from his flagship when it was captured, by throwing himself into a small skiff and getting ashore. Ouloudj Ali got off with the Algerine squadron up the gulf. Fifteen thousand Christian slaves are said to have been liberated from the chained galley-benches of the captured vessels; but thousands more must have perished from the cannonade, and from the sinking of the vessels on which they were confined. No one could ever calculate the total loss of the Turks—wild figures up to 120,000 were given!

Of the Christians we have more trustworthy casualty lists. There were in all fifteen galleys sunk, and 7566 dead. Agostino Barbarigo, the Venetian admiral, was mortally wounded; two of his kinsmen, Giorgio and Andrea, killed, with not less than twenty senators more, bearing the best old names of the Republic—Contarini, Loredano, Paschaligo, Venier, Quirini, etc. Two Spanish captains, the brothers Cardona—one the Admiral of Sicily—fell, also a Caraffa from Naples, and two Orsini from Rome. The Maltese lost the Prior of Messina, their commodore, and Joachim Sport, Commander of Mainz. But undoubtedly the most notable Christian casualty was Miguel Cervantes, who, serving as a volunteer, and in charge of a skiff with twenty men only, had his left hand carried away by a cannon-ball—and was cut off from chivalrous enterprises for the rest of his chequered life. His right hand and his pen remained, also much memory of blighted hopes.

The Turkish fleet was absolutely annihilated; it remained to see what Don John could do, with many battered ships, grave gaps in his crews, and a hundred or more of leaky Turkish prizes. The day was October 7, and no Spaniard could fail to remember what had happened to Don John's imperial father, when he went against Algiers in the stormy month. Cyprus was entirely in the hands of the enemy, Famagosta having fallen before the fleet of relief had even started from Messina. An enterprise against the island, where some 50,000 or 60,000 Turks were in garrison, with the landing forces from the fleet, reduced by battle and wholly destitute of cavalry, would have been absurd. Considering the time of the year, there was nothing to be done save to return in triumph to Italy, and make preparations for next spring. Church bells rang everywhere, the victorious young Admiral was saluted by the Pope as 'a man sent from God whose name was John,' [1] and Vassari's rather schematic picture of the battle was painted on the entrance wall of the Vatican chapel—to be balanced a year later by another picture, that of the murder of Admiral Coligny on the night of St. Bartholomew (1572)—a strange pendant.

The Turkish high-seas fleet having been annihilated, it remained to be seen how the victors could utilize their triumph

[1] John i. 6.

THE BATTLE OF LEPANTO, 1571 (2)

General mêlée of the fleets at the moment of decision

From the Venetian engraving by G. B. Camocio

when the spring should come round. All sorts of possibilities were forthcoming—an expedition to recover Cyprus, a blow at the Turkish corsair-bases in Africa, with special care for Tunis, where Spain's old vassal, the last of the Beni Hafs, was clamouring for his restoration, or a dash at the Dardanelles, where the Turks were reported to be rebuilding in haste the entrance forts, or the more unenterprising and tedious process of recovering all the outlying Venetian ports on the east side of the Adriatic and in the Aegean. The problem was to find a project agreeable both to Venice and the King of Spain, whose views were not likely to be identical; for Philip remembered that the Venetians had been for many years leaving him alone to fight the Turk, and had only come in to the Christian alliance when their own interests were threatened; while the Venetians were thinking of their own lost possessions, and had no particular desire to see the King of Spain omnipotent in the Mediterranean.

There would in any case have been cross-purposes in 1572, but the face of affairs was suddenly changed by the news that the Turks had once more a great fleet on the sea, and that the naval predominance of the allies had to be vindicated once more. This extraordinary fact shows that there was still rallying power in the Ottoman State—due not at all to the bibulous Sultan, but to his dominant Grand Vizier Mohammed Sokolli. He was determined that a fleet must be reconstituted against time and at any cost—his proceedings remind the reader of ancient history of the last effort of Athens in 406 B.C. after the disaster of Mitylene. Ouloudj Ali had brought home 47 galleys—the wrecks of Lepanto—Piale Pasha collected what ships had been left before Cyprus and in the Levant. There were a certain amount of vessels in the docks of the Golden Horn, under repair or unfinished. But all this would not have made up a fleet fit to take the seas against Don John. The extraordinary move made by the Grand Vizier was to order that 180 new galleys were to be laid down on the arsenal-slips and finished against time, and crews for them must be found even by conscripting Mohammedan rowers, since the bulk of the old galley slaves had been lost at Lepanto. The fighting men must be replaced by a great draft from the Janissaries, and by ordering 'Spahis' and 'timariots' to go to sea.

All this was done, and in June Ouloudj Ali—now named Kapitan-Pasha—came out of the Dardanelles, with a fleet of some 160 galleys—it was not a good fleet, the crews being inexperienced, and the new galleys by no means all properly equipped, but it was a formidable force, and its appearance was wholly unexpected by the Christian allies. The Venetians during the spring months had been waiting for the arrival of Don John and the Spaniards, and meanwhile had been using detached squadrons in small enterprises—the recovery of forts in Dalmatia, and raids from Crete to cut communication between Cyprus and Constantinople. Jacopo Foscarini was to be their admiral for this year, the elderly Venier being superseded, mainly because he had quarrelled with Don John. Foscarini lay at Corfu in June, waiting for the Spaniards, who did not appear. And it was ascertained that there were only 50 of King Philip's ships at Messina, for the Genoese Gian Andrea Doria's galleys, were still expected. Don John had orders not to move till the whole were collected. One real reason for the delay of the Spaniards was that King Philip was set on the expedition against Tunis, which was carried out in the following year, but another reason was found in the plea that a war with France was possibly imminent. This was the time, before the St. Bartholomew, when Coligny was urging the half-convinced Charles IX to strike at King Philip, and when the Huguenots of Genlis actually invaded the Netherlands at the moment of the siege of Mons (May–July 1572). Kept back undoubtedly by his half-brother's orders, Don John would only send some twenty vessels to join the Venetians. Tired of waiting, Foscarini sailed in August for the Aegean with about 150 galleys, and the six galeasses which had figured at Lepanto. He found himself opposed about Cerigo and Cape Matapan by Ouloudj Ali's fleet, which was as strong or stronger than his own. This was a sad surprise, for it was not suspected that the Turks would show at sea. Foscarini would not fight without the galeasses, which were slow movers, and the Kapitan Pasha evaded him. At this moment (September was come) a dispatch was brought from Don John, to say that he was now arriving, and that the whole armament should concentrate at Corfu. Accordingly the Venetians retired thither, and met the main Spanish fleet. The combined force, over 200 vessels, then sailed to look for the Turks. But the Kapitan Pasha had betaken himself to the strong harbour of Modon, from which he refused to come out, having a fortified town at one side of his refuge and batteries thrown up on a headland at the other. Don John refused to thrust his fleet into the fortified harbour, and, after blockading its entrance, proceeded to disembark troops to take the enemy from the land side. This enterprise failed before the town of Navarino,[1] which covered Modon on the north at a distance of some six miles, though the expedition was conducted by Alessandro Farnese of Parma, who a few years later was to be acknowledged the best general in the world. He was held off by Turkish troops which came down in strength from the interior of the Peloponnese. This failure determined Don John to break up the fleet, because October was come, and he would not face a winter campaign. Accordingly the Venetians returned to Corfu, the Spanish to Messina, to the great relief of Ouloudj Ali. For the Turkish fleet was quite helpless in Modon harbour —munitions had run out, the hastily equipped galleys were leaky, and the crews had been thinned by an epidemic. It is said that on many vessels there were not 120 men left. As soon as the departure of the Christians was discovered, the dilapidated fleet returned to Constantinople to refit. But by its mere existence it had saved the Sultan from losing the command of the sea—the gains of Lepanto had been thrown away, mainly by the tardiness of King Philip in coming to a resolution in the spring.

NOTE ON THE BATTLE OF LEPANTO

Nothing can give a clearer idea of the battle than the two excellent Venetian engravings by Giovanni Francesco Camocio which are here inserted. They are absolutely contemporary, and represent, the first, the order of the fleets before the clash, and the second the mêlée which followed. The names of most of the commanders of galleys on both sides are given. Note in the second picture one Turkish galley on fire, and several sunk. In the lower right hand corner Ouloudj Ali and his squadron are escaping.

[1] The modern town of Navarino does not represent the old Pylos, which was at the other end of the great bay that is formed by the isle of Sphacteria, so famous to Spartans and Athenians of old. Pylos was at its northern entrance, Navarino at its most southern point. The Turks of 1572 preferred the narrow harbour of Modon to the larger one so famous in 1827.

CHAPTER XI

AFTERMATH OF LEPANTO—END OF THE TURKISH OFFENSIVE (1574–1606)

THE miserable and inconclusive naval campaign of 1572 marks the end of an epoch. No Turkish admiral ever again risked the repetition of a battle of Lepanto, nor was a Turkish fleet ever seen again ready to fight in the waters west of Cape Matapan. A Pasha once observed to a Venetian that by taking Cyprus the Sultan had lopped off a limb from Christendom, while by destroying Mouezzedin Ali's fleet at Lepanto Don John had merely shaved off a beard that would grow again. This was an ingenious and plausible epigram, but as a matter of fact what was lost at Lepanto was the self-confidence and offensive spirit of the Turkish navy. The fleet that was blockaded at Modon had never wanted to fight.

The year 1573 saw an unexpected end to the grand alliance—the Venetians were so disgusted with the behaviour of the Spanish king that in the summer they opened negotiations for peace with the Sultan, who for his part was overjoyed to see them retiring from the war, since it was their contingent which made the allies superior at sea. His ministers had no belief that their improvised fleet would really have been able to face the Christians, if the league held, but were prepared to go on against Spain alone. The Venetians had no hope of recovering Cyprus after their experiences of 1572, and resigned themselves to let it go, in consideration of securing safety for Crete and their other outlying possessions. They agreed to hand back the few places that they had captured on the Adriatic coast,[1] and to pay a war indemnity of 300,000 ducats, to be spread over three years—no very great sum. This was an exact repetition of their conduct in 1540, when they had once before slipped out of an alliance with the Hapsburgs at the sacrifice of some outlying ports—Nauplia, Malvasia, and the islands in

[1] Margarita, Lopot, and Marino. But the Turks kept Dulcigno and Antivari, captured in 1571.

the Cyclades. The treaty of peace was concluded in February, after much haggling over details, and published at Venice on March 15, 1574. It caused intense disgust at Rome, as well as Madrid, since it put an end to what had been considered a sort of crusade; but the Venetians never showed much Christian zeal, but only acute perception of their own dangers when the Turk turned ambitious on the seas. They were now quit of wars with the Sultan for many years; it was far into the seventeenth century before they had to shed Crete, as they had shed Cyprus.

This was not quite the end of the crisis which had started with Lala Mustafa's invasion of Cyprus, and whose central point had been the battle of Lepanto. When he found that he could count on no further help from Venice, King Philip was free to carry out an adventure which had long been in his mind. Don John, with a purely Spanish fleet of 167 galleys, swooped down on Tunis in August 1573, expelled the Algerines, and placed on the throne Mohammed, last of the House of Beni Hafs—not his brother Hamid, whose conduct had been viewed with discontent of late years. The Prince then departed, after having built new outworks to Goletta, the great Spanish haven at the mouth of the Gulf of Tunis, and left there a strong garrison of 4000 Spanish and Italian mercenaries, under Juan Zanoguera, Juan Salazar, and Gabriel Serbelloni. If the vassal kingdom of Tunis had been permanently restored, there would have been something to show as a solid gain from the battle of Lepanto. But when Mohammed had been reigning only thirteen months a great Turkish fleet under Ouloudj Ali and Sinan Pasha—over 200 galleys carrying a landing force of 20,000 men—appeared before Goletta, took it after a stout defence, and made an end of the kingdom of Tunis. The Sultan Mohammed and several Spanish captains were taken in chains to Constantinople (August–September 1574). No help came from Spain before it was too late, apparently as a result of King Philip's habitual procrastination. Don John of Austria sailed from Naples, with a fleet of succour of no great strength, in September, but he was caught in a storm which wrecked many vessels, and was driven back to Sicily. Before he could refit his expedition, the fall of Goletta was reported to him. He had been prepared to help the garrison, but not to reconquer the whole of the Tunisian lands, now that they were lost. He informed his master that he was not strong enough for an invasion of Africa, and the King agreed with him, and cancelled the enterprise.

Thus Algiers, Tunis, and Tripoli, in line once more, became the strongholds of three piratical beys, prepared to support the Sultan so long as it suited them, but more and more inclined to serve their own interests, not those of the Ottoman Empire in general. As from this very date onward the Sultanate fell into decay, the co-operation of the African corsairs became less and less valuable, as each worked for himself, and not for any general scheme of Mediterranean domination. From 1574 onward the Barbary pirates continued to be a perpetual nuisance to the inhabitants of the Spanish seaward provinces, from Sicily to Cadiz. But they are never found again as partners in a great imperial Ottoman venture, as at Prevesa, Malta, or Lepanto. Their quarrels among themselves, their relations with the semi-independent Arab and Kabyle tribes in the Atlas, and an interminable series of wars with the new dynasty of the Sherifs in Morocco, gave them sufficient occupation. Strange as it may seem, the precarious Spanish hold on Oran and other African coast fortresses was never broken, even in the decadent days of Philip III and Charles II. Oran was lost in the eighteenth century, but Melilla and Ceuta remain Spanish till to-day.

But to recur to 1574, and the aftermath of Lepanto. The crisis, which might have meant so much in the history of the Mediterranean, passed away without any startling results. Sultan Selim drank himself to death on the 9th of December 1574, and his successor, Murad III, a mere harem-haunter, was no better than his father. He started his reign drastically enough by murdering all his five brothers, but this display of ruthless energy was not borne out by the remainder of his twenty years' reign. As long as his father's great Vizier, Mohammed Sokolli, lived, the empire held together more or less, but this useful minister was assassinated by a disgraced petitioner early in the reign, and after his death the Sultan was the tool of a succession of corrupt and misguided viziers and court favourites, and the victim of constant mutinies of Janissaries and provincial revolts. The great event of his reign was the resumption of the Persian Wars, which ran on for many years, and left the Danube front comparatively quiet. For a treaty, which was formally only a truce, was arrived at in 1575, and though bickering in Hungary never ceased, it did not swell for many years into regular imperial war.

The strength of the Ottoman Empire had been that it was a 'one-man power'; under sultans like Mohammed II, Selim I, and the great Soliman it was a public danger to all its neighbours—to Central Europe most of all. But under weak sultans, who seldom or never took the field, and were dominated by selfish and corrupt grand viziers, it started on the downward road. Selim II was a sot, Murad III a nymphomaniac, Mohammed III somewhat of a madman, in the next generation Mustafa I actually an idiot. While Philip II was absorbed in his Netherland rebellions and his interference in French religious politics, the ministers of his neighbour in the Mediterranean, Murad III, chose to embroil themselves in a Persian war, which was as fruitless and as exhausting as any Spanish campaign against the Oranges or the Huguenots or Elizabeth of England.

The old Shah Thamasp having died in 1576, there was an outbreak of fratricidal wars and palace assassinations among his sons. Lala Mustafa (the ruffian who conquered Cyprus) persuaded Sultan Murad that the moment had come to destroy the Persian Empire. But the Shias pulled themselves together again under Mohammed Khodabunda, the least unlucky of the sons of Thamasp, and, after some initial successes, the Turks found themselves involved once more in a series of indecisive campaigns in Armenia, Georgia, and Adherbaijan, such as Soliman the Magnificent had known too well. Several times the Turks occupied Tabriz and Erivan: they installed a vassal on the throne of Georgia, and even on one occasion conquered right up to the Caspian Sea—a Tartar army from the Crimea coming round through the Caucasus to assist. But the successes were never permanent, and as many Turkish as Persian armies were exterminated. The struggle lasted for fourteen years (1576–90), till Shah Abbas, who had succeeded his father Mohammed Khodabunda in 1585, thought it worth while to buy peace, by surrendering Tabriz and parts of Adherbaijan, Georgia, and Shirvan—all of which he was destined to recover during the course of his long and glorious reign from Murad's unlucky successors. The ceded lands, in a dreadful state of desolation, were of little profit to the Sultan.

This remote and prolonged struggle relieved the situation for Christian Europe, where there was constant raiding, but no formal war, on the Danube, while Philip II, comparatively reassured as to the situation in the Mediterranean, could turn his attention to the Dutch, the French Huguenots, and Queen Elizabeth's tortuous fluctuations between war and peace, without having to face any serious danger in the south. The Barbary pirates continued to give trouble—it is recorded that 1585 was the first year in which they executed a raid outside the Straits of Gibraltar, but their operations were never on a large scale, and they never combined to expel the Spanish garrisons along the Algerian coast. Indeed Philip got a more secure lodgment in Africa when in 1583 he annexed Portugal, and became master of Ceuta and Tangier. The Portuguese 'Algarve de acquem Mar' was a solid block of territory in Morocco, a much more valuable base than Oran and the other old Spanish fortresses, and Ceuta and Gibraltar, facing each other across the narrow strait, never ceased to be a check on the corsairs, as valuable as was Malta in the central Mediterranean.

Naval operations after 1575 were spasmodic and resultless for many years. The Venetians, being out of the game, and resolved never to enter it again, were content to endure many small vexations from Turkish and Algerian piracy, and contributed a little provocation themselves. Both Turk and Venetian suffered when the Dalmatian 'Uscocks,' freebooters based on the many small harbours of the irregular Adriatic coast, raided outlying coastal villages. And each accused the other of giving secret harbourage to piracy. But the Venetians could point out with greater certainty undoubted outrages by Algerines or 'Marteloses,' the local Turkish seafaring robbers. Meanwhile Venetian 'Bailos' were in high favour at Constantinople, and exercised considerable influence on all commercial matters in the Levant, now that the claims on Cyprus were finally surrendered.

Small-scale expeditions by Spain and her minor allied powers are occasionally recorded, though Philip II was by now much more interested in the Atlantic and the North Sea than in the Mediterranean. The Knights of Malta were never long idle, and (what is more surprising) we hear of descents by Tuscan squadrons on Asia Minor; the 'Knights of St. Stephen' established by Duke Cosimo, as a crusading order, occasionally put in an appearance at unexpected points. Turkish retaliation was mainly by Algerine raids on the coasts of Sardinia, Sicily, and Naples. There were no serious maritime expeditions made by either party, and the one Ottoman descent which requires more than passing notice was an isolated landing at Manfredonia in Apulia, long after the last year of the sixteenth century, when the sack of a considerable harbour city provoked panic for a moment in south Italy. But this was a wholly spasmodic and inconsequent effort by a single Turkish admiral, without any authorization from headquarters. Mediterranean wars had ceased to have any appreciable effect on the struggle between the Ottomans and Christian Europe.

Indeed the battle of Lepanto, though its territorial gains in Africa were lost in 1574, marks the end of the period of Turkish maritime aggression. As to the land attack on Central Europe up the Danube, there was incipient war once more in Hungary from 1590, the last year for which the Emperor (Rudolf had succeeded Maximilian) paid up the annual 30,000 ducats, which the Turks called tribute and the Austrians compensation for alienated feudal estates.[1] Wranglings and frontier raids on both sides in 1591–92 led to a definite commencement of general hostilities in 1593. Murad III was now free from his Persian entanglements, and his bloodthirsty Grand Vizier, Sinan Pasha, turned his imbecile master's attention westward once more. The final provocation had been given by a tremendous defeat inflicted on Hassan, governor of Bosnia, who, raiding in force into Croatia, though war had not yet been formally declared, was exterminated with the whole of his horde in front of Sissek (the old Roman Siscia) by Rudolf's generals (June 20, 1593). On hearing of this, Sinan threw the Emperor's ambassador into prison, and marched against Hungary with the whole of the Sultan's European levies and 12,000 Janissaries. He laid siege to Vesprizm—at that moment the Emperor's most advanced fortress—and took it after a short siege of three days (October 13, 1593), but got no further, as the Janissaries went on strike when he proposed a winter campaign, and insisted on his return to Belgrade (November 1593). After the Turkish army had retired, the Emperor's generals revenged the fall of Vesprizm by capturing Neograd and other frontier places.

In the following year 1594, it looked as if the old grand-scale invasions of the Hapsburg dominions were to be resumed, for the Grand Vizier, discontented with his practical failure in the last autumn, called out not only the European feudatories but all those of Asia, and the Tartars of the Crimea. This summons

[1] See above, p. 702.

gave him an immense army such as had not been seen in Europe since the days of Sultan Soliman—at least 100,000 men as it was reported. It came into the field late, but compelled the Austrians to raise the siege of Gran which the Archduke Matthias, the Emperor's brother, had taken in hand, and to retire beyond the Danube (June 1). Since the Christian field army refused to risk a general engagement, the Grand Vizier was able to capture the important city of Raab, after a siege of twenty days, and some smaller places, but failed before Komorn, which held out till the autumn, when the clamours of the Asiatic timariots and the Tartars compelled him to disband his army, for a winter campaign was a terror to these auxiliaries from far distant regions. The result was a disappointment—the largest Turkish army seen for thirty years on the Danube had only captured a single town of any note, and that one on the extreme frontier. The old days of broad-spreading conquests were over.

Sultan Murad III, that wretched nymphomaniac—he had over 100 children—the tool of his viziers, died in January 1595, a victim apparently to melancholia, and physical exhaustion due to his deplorable private life. His eldest son, Mohammed III, succeeded, and opened his reign by the massacre of the nineteen surviving younger brothers with whom the inordinate uxoriousness of his father had provided him. This was the largest royal hecatomb in Turkish history. Mohammed's short reign of eight years marks another stage in the decadence of the Ottoman power. He was not such a despicable character as his father—he did once take the field at the head of the whole levy of the Empire, and win a victory, the last ever achieved in Europe by a Sultan in person, but he was not the man to reorganize a failing state, or to face a crisis, such as a general revolt of the Janissaries. He sometimes displaced a vizier, while his father had allowed viziers to displace each other, and he was not such an habitual denizen of the harem, though he allowed his mother, the 'Sultana Validé,' to have a considerable share in promotions and dismissals of ministers. He was apparently, if tales are true, occasionally liable to fits of mental derangement—insanity came out strongly in his son Mustafa I.

The crisis which faced Mohammed III at his accession was in one respect of a new sort. The Zapolya princes of Transylvania had been on the whole docile subjects of the Sultans, who had always supported their outworn pretensions to the crown of Hungary, and had found them useful tools against the House of Hapsburg. But the Zapolyas had died out in 1570, and their place had been taken by the three Bathori princes, who were much less trustworthy. Stephen Bathori, without disowning his homage to Constantinople, had got himself elected King of Poland (1575–86), and, after finishing off for the Poles a dangerous war with Ivan the Terrible of Moscow, was believed

HUNGARY AND THE NEIGHBOURING LANDS, 1521–1606.

to be about to attack his suzerain at the head of a Polish army. But he died suddenly in 1586, and it was not till some years later that his nephew and successor, Sigismund Bathori, throwing up the old policy of his predecessors, made proposals to the Emperor Rudolph to surrender his illusory claims on the Hungarian crown, in return for a grant of his actual dominion to his house as a vassal-principality, and of the hand of a Hapsburg princess for himself. This defection of Transylvania uncovered the flank of the Turkish possessions from Buda all down the Danube, and ought to have resulted in the expulsion of Mohammed's governors from all the conquests of 1526. That it failed to do so is rather surprising, as it was accompanied by insurrections in Wallachia and Moldavia, under the two voivodes Michael and Aaron. The former made a clearance of the garrisons in Wallachia, and rolled back the Turkish frontier to the lower Danube. When the Grand Vizier Sinan had assembled a great army at Constantinople, he was forced to go with the larger part of it to Rustchuck and Giurgevo, and though he penetrated as far as Bucharest, and burned the city, he had to retire behind the Danube in October. Only strong reinforcements could be sent to Hungary, not the main Ottoman army. Hence the summer and autumn months of 1595 saw such a recoil of the Turkish frontier on the Hungarian front as had not been seen since the time of Charles V.

Charles of Mansfeldt, who had come over from the Netherlands this year to act as the Emperor's generalissimo, had been joined by considerable detachments of troops lent by the minor princes of Germany, to whom Rudolph had been making urgent appeals for help: there were also Walloons, brought over by Mansfeldt, and Italians in his army. On August 4 he beat the Turkish governor in a pitched battle before the gates of Gran, and captured that ancient metropolis, lost as far back as 1543, and the high-lying fortress of Visegrad soon after. At the same time the Turks gave ground on the border of Croatia also, and the Wallachian voivodes crossed the lower Danube into Bulgaria, and sacked Silistria and Varna. Mansfeldt died of dysentery in the summer, but his successor, the Archduke Matthias, continued to capture small places of some strategical value all through the autumn.

The prospect for the Ottoman Empire looked so black that the Grand Vizier Sinan insisted that the new Sultan must take the field in person for a spring campaign in 1596. Sinan himself died in April, but his successor Ibrahim Pasha was of the same opinion, and Mohammed came out of his palace, not too willingly, to put himself at the head of a general levy of the whole Turkish forces of Europe and Asia. The army marched on Erlau—not up the Danube to Gran or Komorn as might have been expected.[1] This place surrendered earlier than had been judged probable, apparently owing to treachery among foreign mercenaries in the garrison; and the Imperial troops,

[1] Probably the idea was to strike at the junction point of the Austrians and their new Transylvanian ally.

who had been besieging Hatvan all through August, appeared just too late to relieve it (September 21, 1596). The Archduke Maximilian and the Austrian generals had been joined by Sigismund Bathori of Transylvania, and were in great force; hence it came that after much manœuvring round Erlau they made up their minds to risk a pitched battle—a rare thing in these Danubian wars—but they had been encouraged by their success before Gran in the previous year, and resolved to encounter the Sultan, though he had the whole strength of his Empire behind him. Hence came the two-day battle of Kerestes—some twelve miles south-east of Erlau, near the Theiss—the last general engagement between Turk and Christian in all the wars which we have to chronicle (October 24–26, 1596).

The Archduke had under him an army even larger than that which Mansfeldt had led in 1595, and, owing to the presence of the Transylvanians, it was much better provided with cavalry than was usual for an Imperial host. It is said that, by a muster-roll compiled just before the battle, there were 3000 reiters and 4000 foot from the levies voted by the German diet, 5000 Transylvanian horse and the same number of foot under Prince Sigismund, 5000 horse and 5000 infantry from the Austrian archduchies, while the Hungarian Palatine Nicholas Palffy had brought 13,000 men, more of the light-horse 'hussars' than of foot soldiery. In the whole force there were actually more cavalry, heavy and light, than infantry—a thing most rare in Imperial armies. There was also a tremendous provision of artillery—no less than 97 guns, small and great.

Maximilian fortified himself on October 22 behind a marshy depression formed by a brook named the Cincia, a tributary of the Theiss, fordable in most places but forming a decided military obstacle, and asked to be attacked. The position had been occupied only after a long skirmish with a detaining force of Turks, who were driven across the bottom in disorder with the loss of many guns. On the following day the main Turkish army came in view—the Sultan had been drawing together outlying detachments, including a large body of Asiatic feudal horse which had been left at Erlau, and many predatory bands of Tartars. He was determined to fight in full force—though whether (as the Hungarians thought) he had 150,000 men may be doubted. Turkish reports gave the Imperialists 300,000 in all—about seven or eight times their real number! If Mohammed had 100,000 horse and foot at the end of a severe campaign it is more than we dare concede, but the figure is given by the most respectable Turkish authority. On the afternoon of the Sultan's arrival, a large body of Tartar horse tried the passages of the marsh, and were driven off. The army encamped a cannon-shot only on the other side of the depression, directly opposite to the fortified Imperial position. The two camps were but a mile apart.

On the 24th the Turks made a serious attempt to cross the obstacle with masses of cavalry, commanded by Hassan the Pasha of Buda, Cicala the renegade junior Vizier, and the Grand Vizier Ibrahim's Kiaya, or second in command. The Imperialists, in battle order, repulsed the attack with loss: many of the Sultan's horse-guard was killed, also the Beglerbeg of Anatolia. But heavier fighting was expected for the 26th. The Archduke, as we are told, had three 'battles' in his first line: the 'vaward' contained Hungarians under Palffy and the Germans under a count of Schwartzenberg, in the 'main-battle' were the Transylvanians under their prince and the Archduke's own guards, in the 'rearward' Austrians and some Hungarian detachments under Tieffenbach, Marshal of Austria. In charge of the camp there were left as a reserve 6000 men, including Silesian, Lusatian, Austrian, and Hungarian foot, and some Moravian horse-arquebusiers. How infantry and cavalry were arranged in the front line is not properly indicated by any authority—presumably in each 'battle' there were heavy infantry squares of pikes, with cavalry in the broad intervals between them.

The Sultan, according to Turkish accounts, was by no means anxious to make a second assault on the Imperialist position, and even proposed to go to the rear himself and hand over the command to the Grand Vizier Ibrahim, who might then act as he pleased. All his councillors, however, advised him that this would be morally ruinous, and that if the army saw the Padishah depart, and then drew off without fighting, demoralization would set in. So large a force having been collected, and the Tartar scouts having reported that the enemy was much inferior in numbers, a serious general attack must be delivered, with the usual outflanking manœuvres by Tartar and other light horse.

Accordingly, before dawn on October 26, the Turks crossed the brook and marsh without much difficulty, and were found coming forward in a great half-moon formation, with the wings advanced and the centre somewhat 'refused.' The European feudal cavalry had the right—as usual—the Asiatic the left:[1] in the centre were the Janissaries and many guns chained together, also the Sultan's guard-cavalry: he himself was with them, having the sacred standard of the Prophet at his side. There was, as at Mohacs, a detached flanking corps led by the junior Vizier, Cicala, which was kept out of sight far to the left. It does not seem certain that the Janissaries and guns crossed the marsh, or did more than advance to the line of the Cincia brook: it must be remembered that the armies were very close, and that if the centre drew up to the edge of the depression, it was not out of touch to any great extent with the cavalry wings. The best Hungarian account of the battle says that the Turkish line was drawn up 'tam citra vada quam ultra,' which seems to show that part of the enemy (the centre of course) was on the other side of the fords from the wings.

The Imperialist army was perfectly prepared to receive the attack, having been under arms since dawn, and when the clash came the cavalry charged out to meet it. On the right wing, where the first meeting seems to have come, the German and Hungarian horse of Palffy and Schwartzenberg completely defeated the Asiatic horse, who fled in shameful rout, abandoning some guns which they had brought forward with them. They fell into hopeless disorder when recrossing the marshy brook, and could not be checked: nearly the whole of them disappeared from the field. Many are said to have been next heard of at Szolnok—over twenty miles to the rear. Some brought news of a disaster to Buda.

The Archduke had ordered that the Turks were not to be pursued beyond the Cincia, but his troops disobeyed all orders; the leaders were as much to blame as the led. The Palatine Nicolas Palffy came back to beg his commander not to spoil a half-won victory, and Sigismund of Transylvania joined in the prayer. The right wing, horse and foot, crossed the brook in pursuit of the Asiatics, then the centre; a body of Janissaries who were found holding a ruined church at the main passage

[1] According to Naima, the Asiatic cavalry had the right—this would be contrary to all Turkish etiquette, as when a battle was on European soil the Roumeliot cavalry had the place of honour, when on Asiatic soil the Anatolian. There seems no reason why this custom should have been violated, and it seems certain that the Asiatic cavalry, which fled, was opposed to Palffy and Schwartzenberg, who had the Imperialist right: *i.e.* must have been on the left.

of the marsh were rolled over and exterminated. The Christian infantry, 'panting and disordered,' followed the victorious horse at such pace as they could make, and all together burst into the Sultan's camp ; in front of it they captured 50 Turkish guns chained together, but useless when taken in flank, and with their infantry supports driven off. The Sultan, with his horse bodyguard, took up a position behind the abandoned camp, being barely restrained from flight by his staff. It is *not* true that, as the best of the Hungarian chroniclers states, he made off, and did not stop till he reached Szolnok. His tents were plundered, and a number of his eunuchs, mutes, and pages slaughtered.

What was happening meanwhile on the Christian left wing, where Austrian and Hungarian horse were engaged with Roumeliot feudal cavalry and Tartars, is not clear. We are only told that they drove off a flank attack by Tartars, who were trying to circle round them, and came forward in good order—having enemy still in front of them it is not clear whether they crossed the marsh ; probably they did not.

But a sudden change was given in the aspect of the battle by the coming in on to the flank of the Germans and Hungarians, who were plundering the Turkish camp in perfect neglect of all precaution, of the large detached body of Turkish horse under the renegade Cicala, who had been lying out on the Sultan's left, completely hidden from the Imperialists. We are told that the victorious troops were in disgraceful chaos—the horsemen had dismounted in order to scramble along with the foot for valuable pickings—and that the Archduke had barely two squadrons in order—the rest having melted away.[1] The sudden charge into the confused mob of a large body of fresh enemies started a panic, and the whole of the troops of the right and centre of the Imperialist army bolted in disorder back across the marsh and the brook. Maximilian tried to rally them, but in vain ; no one would stand, and the whole horde vanished in disorder, making for Miskolcz, Kashau, and the other towns northward. Only Tieffenbach, who had commanded the left wing, brought off his men without serious loss, as they had never been engaged in the scramble in the Turkish tents ; he is said to have been barely pursued, and to

[1] Istvanfi puts special blame on two Hungarian captains, who were the first to dismount, and to break into the chests of gold in the Turkish war-treasurer's tent, instead of keeping their squadrons together.

have returned to the camp and picked up his impedimenta before going on his way to Miskolcz and Kashau. But all the 97 Austrian guns were abandoned on the field.

The slaughter of the Germans and Hungarians in this rout was naturally very great, as there was a hopeless crush and confusion at the passage of the marsh. The loss was at least 5000 or 6000 horse, and quite as many foot. The casualties included two dukes of Holstein, who had been serving in the contingents sent by the German diet, and many captains from the south German contingents, also a number of members of the great Hungarian families—a nephew of the Palatine Palffy, and Forgachs, Esterhazys, Telegdis, etc., with the Szekler Stephen Lazarus, who had commanded the Transylvanian foot. The Turkish losses must also have been very great; they did not push the pursuit far, and marched back on Belgrade almost immediately.

The doings in the Turkish camp on the day after the battle were extraordinary. The Sultan disgraced the Grand Vizier Ibrahim, and gave his post to Cicala, the hard-fighting renegade whose charge had won the battle. This violent and reckless soldier raged against those who had failed on the previous day—he beheaded the chief of the camp-guard and several other officers, paraded the Pasha of Aleppo, who had led the fugitive wing, in women's clothes on a donkey around the camp, and caused the Sultan to decree that all the Asiatic feudatories who had fled should forfeit their fiefs, and the men of the regular cavalry should be cashiered. The anger of those disgraced and disendowed cavaliers was so bitter that they were found *en masse* at the head of the rebellions in Asia Minor, which broke out a little later. They were styled 'Firaris' or 'Runaways' from the day of Kerestes onward. Cicala also deposed the Khan of the Crim Tartars, and made a general sweep of all his personal enemies. His severities provoked such general discontent that, only a few months later, the Sultan was compelled to dismiss him from office, and to restore the former Grand Vizier Ibrahim to power.

The best proof of the absolute decadence of the Ottoman Empire in the last years of the sixteenth century is that this very notable victory of Kerestes was followed, not by any further territorial gains in Hungary, but by a considerable disaster in the very next year. In 1597 the Sultan and Grand Vizier did not either of them take the field, and the junior

Vizier, Mohammed Satourdji, left in command on the Danube, suffered marked reverses. Despite of the slaughter at Kerestes the Emperor had again two armies in the field by the next summer: one under the Archduke Maximilian took Papa and Totis, while the Prince of Transylvania with the other besieged Temesvar, the most important fortress in the south of Turkish Hungary. Mohammed Satourdji failed with loss in an attempt to storm Waitzen, and was so hard pressed that he opened negotiations for a truce with the Archduke for the winter, and sent home the Asiatic contingents of his army, when he learned that Temesvar at least had not fallen.

In the next year (1598) matters went still worse for the Turks; the Imperial generals, the Austrian Schwartzenberg, and the Hungarian Palffy, took the field early and recaptured Raab, the only place of importance lost of recent years, on March 29, and then took Vesprizm and laid siege to Buda, the centre of Turkish power on the Danube; they pressed it hard but could not take it. On the other hand, Mohammed Satourdji failed hopelessly before Grosswardein, and it was by no merit of his, but only by the approach of winter, that the Austrians had to give up the long siege of Buda (October 1598).

In 1599 the Grand Vizier Ibrahim thought himself forced to take the field, with great reinforcements, and started his campaign by ordering the execution of his predecessor in command, Mohammed Satourdji. He advanced as far as the gates of Gran—but only to recommence the negotiations for a peace which had proved futile in the last year; he did not lay siege to Gran, because he dared not risk a pitched battle with the Austrian covering army, and in November retired to Belgrade.

The next campaign (1600) was not quite so inactive on the part of the Turks; the Grand Vizier got back Papa, by the treachery of some French mercenary officers who sold the place to him, and though he dared not march on Gran, took the fortress of Kanishka after a long siege, in face of an Austrian army under the exiled Guisard duke, Mercœur, who was in command for the Emperor on the Danube in this year (September 1600). But Kanishka was a place of very secondary importance compared with Raab or Gran, and the garrison had secured an honourable retreat by capitulation, not perished in a storm.

Nevertheless, the date 1600 was a very dismal one, despite

of the fall of Kanishka, for the Turkish power, as it saw the rapid development in Asia of troubles which were to grow worse each year. There had been already in 1598 a revolt in Irak, under an adventurer who proclaimed himself the long-expected Mahdi, but this was unimportant compared with the rising in Eastern Asia Minor headed by one Abdulhalim,[1] who had been one of the officers cashiered for cowardice by Cicala after the battle of Kerestes. He gathered round him a nucleus of the unlucky Asiatic Timariots, who had been deprived of their fiefs at the same time, those who were called the ' Firaris ' or ' Runaways ' by the rest of the army. These ruined men were joined by all manner of malcontents, largely Kurdish tribesmen who lived in a state of chronic rebellion in the hills, and professional brigands, and formed an army which beat in the open field the Vizier, Hadji-Ibrahim, who was sent against them (April 1600). Abdulhalim proclaimed the restoration of all banished men, and the abolition of all taxation, a form of propaganda which drew round him forces with which the governors of Asia Minor were unable to cope. He swept all the open country, took several important towns, and though he was beaten in battle in 1601, and died soon after, his brother Deli Hassan (Hassan the Mad) kept up the revolt for two years, and finally came to terms with the Government in 1603, by which his followers were restored to their estates or pardoned, or taken into the Sultan's regular army, while he himself was made governor of Bosnia.

But this did not end rebellion in Asia Minor—fragments of the rebel army, under new chiefs, kept many regions unquiet all through 1604–5–6, and, while they were by no means disposed of, the much greater danger of a Persian War supervened. Shah Abbas, who had kept quiet since the peace of 1590, and had devoted himself to restoring the internal strength of his empire, saw his chance in 1603, and while the Anatolian rebellion was still alive, and the war in Hungary lingering on, fell upon the flank of the Turkish possessions in Armenia. After winning a battle over the local governors at Sofian (September 25), he besieged and captured Tabriz (October 21), the long-lost capital of his father. And in the following spring he took Erivan after a long siege, thus making an end of all the Turkish annexations of 1590.

Just after the disastrous outbreak of the new Persian War

[1] Better known in Turkish chronicles as Karayasidji, the Black Scribe.

became known at Constantinople, Sultan Mohammed III died, in a mood of depression caused by the general aspect of affairs, for besides the Asiatic revolts and the declaration of war by Shah Abbas, he had been suffering from a series of internal mutinies and disputes among the guard-soldiery in Constantinople. The Janissaries and Spahis came to open war in the streets of the capital (January 1603)—a sedition which led to the execution of a Grand Vizier. He was not the first of Mohammed's great ministers executed—another Grand Vizier had perished by the machinations of his successor, and several times the Sultan had to sacrifice his favourites to riotous insurrection. The final blow to his vitality was apparently given by the discovery of a conspiracy, real or alleged, of his eldest son Mahmoud, whom he suddenly put to death in June 1603, along with his mother [1] and a number of important officials. All that was proved against the prince was that he kept soliciting military commands, which was interpreted by the Sultan's favourites as an indication of an intention to seize power by force—and discontented soldiery were numerous and reckless.

On Mohammed's death the succession went to his second son Ahmed, a lad of fourteen—his elder brother having been executed only six months back, probably by the intrigues of those who would gladly see a boy on the throne. A knot of greedy and unscrupulous pashas had to face a disastrous Persian War, the lingering struggle in Hungary, and to cope with the still lively rebellion in Anatolia. It is no wonder that they desired peace in Europe, and this proved more attainable than might have been expected. The Emperor Rudolf's attempt to draw Transylvania into his allegiance by his alliance with Sigismund Bathori had proved a complete failure after the disaster of Kerestes. The anti-Austrian party in Transylvania disowned Sigismund, and elected as prince first Michael of Wallachia (1599), and after his death (1601) Stephen Bockskai, who did homage to the Sultan and was promised by him in return the title of King of Hungary. Particularism and hatred of the House of Hapsburg sufficed to drive the majority of the Transylvanian nobles to prefer vassalage to the Infidel to subjection to Austria. After Bockskai's death they elected Sigismund Rakoczy (1607).

The chance of withdrawing Transylvania from the Turkish

[1] Mahmoud's mother was not the mother of Ahmed the succeeding sultan, but another denizen of the seraglio.

suzerainty having failed, and the campaigns of the last three years having been unlucky, both Gran and Stühlweissenburg having been recovered by the Turks, the Austrians were as tired of the war on the Danube as their adversaries—hence came the Peace of Sitva Torok[1] (November 11, 1606), more interesting from the form which it took than from its contents. Territorially it accepted the *status quo*, leaving the boundary where it actually stood, with Erlau, Gran, Kanishka, and Stühlweissenburg in Turkish hands, and Raab, Komorn, and Waitzen, as the limits of the Emperor's military front, and it acknowledged that the Prince of Transylvania was a vassal of the Porte and not of the Holy Roman Empire. So far as frontiers went there was practically no change since the last abortive pacification in 1575, which had lasted—in spite of much informal bickering—till 1590. But the form of the Treaty was all-important, since it was drawn up as a settlement on equal terms between two sovereign powers. The document of 1575 had been shaped in ignominious language as a grant by the Padishah to a prince styled in contemptuous terms 'the King of Vienna,' and it had involved the annual payment of 30,000 ducats of tribute—which was actually paid by Rudolf II from 1575 to 1590. The tribute now ended, and the next Imperial ambassador who entered Constantinople rode in with trumpeters before him, and the standard of the double-headed eagle waving over his head, a fact which gave great offence to the Turkish populace. The Emperor, it is true, paid over 200,000 thalers on this occasion, but he stipulated that Turkish ambassadors sent to Vienna should always be officials of high rank, and should bring with them suitable gifts to lay before the Imperial throne. It was fortunate for the House of Hapsburg that this pacification took place well before the outbreak of the Thirty Years War, and at a moment when the Ottoman Empire was destined for many years to be the prey of internal rebellions and unlucky Eastern wars, under a series of Sultans who became the slaves of their Janissaries. The descendants of the mighty Soliman were no danger to Christian Europe, and the great Turkish offensive, which had ceased to be a dominating fact since Lepanto, had at last come to a formal end, when the Padishah acknowledged that he was a sovereign like other sovereigns, and not the destined dictator of the world.

[1] A little place on the Sitva River, not far from Komorn.

N.B.—For these long campaigns I have followed in the main the magnificent narrative of Nicolas Istvanfi, 'the Livy of Hungary,' whose account of the years during which he himself was engaged in the wars is most trustworthy. He was writing still, when he was struck down by paralysis at the coronation of the Archduke Matthias as King of Hungary in June 1608. In early years he copies Brotheric very closely. Turkish authorities, available only through Von Hammer, seem to be singularly untrustworthy in dates and often in facts. Venetian dispatches, useful for what happened in the capital, are often singularly inaccurate for outlying regions.

CHAPTER XII

THE TURKISH PERIL TO CHRISTENDOM—ITS CAUSE AND ITS END

SURVEYING the whole course of the Turkish attack on Christendom in the sixteenth century, from the Fall of Belgrade in 1521 to the battle of Kerestes in 1596—which marks the last serious offensive move on the side of the Ottomans—it is not hard to draw general conclusions.

The first and most obvious has been already noted. The Ottoman Empire was essentially a 'one-man power.' All depended on the personality of the Sultan, and it so chanced that for generation after generation from 1300 to 1566 there was a succession of sovereigns of great warlike ability—the only exception was the unfortunate Bajazet II (1481–1512) whose thirty years of weak governance provided a welcome respite to the neighbours on whom his ancestors had been encroaching. Precisely those thirty years saw the growing up of the great Hapsburg power, and the union of the Spanish, Austrian, and Burgundian inheritances into a single unit. When Soliman the Magnificent resumed the assault on Christendom in 1521, he found himself faced by the Emperor Charles V, who from 1519 onward was in possession of such military resources as no foe of the Ottoman had ever before commanded. But for the perpetual stabs in the back which the French kings kept delivering against the Emperor, the Ottoman attack would have been foiled from the first. When the turbulent and disorderly Hungarian monarchy had fallen at Mohacs (1526) the Sultan found himself facing an enemy such as none of his ancestors knew, an Emperor with disciplined armies and large fleets. If it had not been for the incessant French wars, and the Reformation troubles in Germany, Charles V could have dealt drastically with the Ottoman attack—he could certainly have reconquered Hungary, instead of merely saving a frontier-slip of it: perhaps he might even have crossed the Danube and thrust the Turks behind the Balkans, or even beyond the

Straits of Gallipoli. Optimists dreamed of such triumphs—La Noue of the Iron Arm, years later, put on paper his plans for a general advance by reconciled Catholics and Protestants—when an army from every German state with French, Spanish, and Italian contingents, should be at the gates of Constantinople, and a Spanish, French, Genoese, and Venetian combined fleet should force the Dardanelles.

That there was a serious danger of a Turkish irruption into Central Europe during the long reign of Soliman (1519–66) was mainly due to the fact that the power of Charles V was being perpetually distracted by the French interventions. That such a malevolent betrayal of Christendom had no worse result was caused by the inconsequent nature of the policy of Francis I, who generally made a peace or a truce with the Emperor at the moment most inconvenient for his Turkish ally. The most perilous crisis, no doubt, was in 1529, when Soliman was besieging Vienna, while the armies of Charles V, only just set free by the peace of Cambray, had not yet had time to reach the Danube. The length of the siege fortunately drove the Turks into the time of autumn cold and rain, and they departed. By the next spring the forces of the Emperor were available, and the danger was waning.

Two points must be stressed, when we reflect on the limitations of Turkish offensive power, even under the most capable of Sultans. The first was that the most important part of the Ottoman army consisted of the feudal Timariot cavalry, who could not face a winter campaign : if they were kept in the field so late as November their horses perished, and they themselves melted away or mutinied. Hence it was the rule to dismiss them to their homes in October. And they could hardly be brought together in force till May—at least Anatolian horse could not easily be at Belgrade before that month, nor Roumeliot horse at Erzeroum. The campaigning season, if the whole force of the Empire was levied, was therefore very short. Moreover, since all administration centred round the person of the Sultan, so long as he was the real master of his empire, it was inconvenient for him and the Grand Vizier and other ministers, who always accompanied him, to be absent from the Golden Horn for more than six months. Almost invariably Soliman came home for the winter—there seem to be only two exceptions : at Christmas 1534 he was at Bagdad, at Christmas 1548 at Aleppo. He got back in January to Constantinople in 1522, after the fall of Rhodes. But when the Sultan and his guard-troops had departed, and the feudal horse had been sent home, there was nothing left on recently occupied lands but garrisons and local governors, who were naturally unable to continue the offensive, and were liable to be overwhelmed by attacks of the local enemy in the spring, before the main force of the Ottoman Empire could be again collected. This accounts for many counter-offensive blows of the Austrian or Persian, which undid much of the work of the campaign of the preceding summer. Not unfrequently, after an exhausting burst of conquest, the Sultan held off for the succeeding year, and only resumed the offensive after a twelvemonth's interval—probably because the feudal horse had been so tried by one hard year of field operations that it was thought well to give them a rest, two successive campaigns at a vast distance from home being more than they could endure. Hence came a certain intermittance in the Turkish offensive. Soliman's enemies of the House of Hapsburg had, as the century wore on, a considerable advantage in the fact that their armies were not solely feudal, but contained regular regiments of troops in permanent service, who could go into winter quarters, and did not disperse to a thousand distant homes. The Sultan's corresponding nucleus of regular units—the Janissaries and the squadrons of the guard-cavalry—were a comparatively small (though a most important) part of his whole campaigning force. But military tradition in Turkey presupposed that when the Sultan returned to Constantinople his household troops returned with him.

It was this tradition, as it would seem, that led Soliman, before he had been many years on the throne, to increase the numbers of the Janissary corps, so that while the main force still acted as his bodyguard, and kept about his person, he might leave detachments for permanent garrison-duty in places of high importance remote from the capital, such as Buda, Bagdad, or Cairo.[1] He seems to have raised the number of Janissaries from the 12,000 of his father's time to 20,000 by 1530. But at the end of his reign there are indications that, including the outlying companies, it was very much greater—one authority states that there were as many as 48,316, including the guard-cavalry and certain small corps with special duties, like the *Solaks* of the innermost life guard, the *Bostanjis* or palace guards, and also apparently the *Janissari-Oglus*, or

[1] The permanent Janissary garrison of Cairo seems to have been 3000 men.

junior companies, destined to be drafted into the central body on promotion.

By the end of Soliman's time the detachments left in garrison for long periods were becoming localized, and particularist in feeling, with a marked jealousy for the more favoured nucleus which enjoyed the privilege of barrack-life in Constantinople, and occasional donatives. But all Janissaries, wherever quartered, showed a contempt and dislike for other troops, even for the squadrons of the guard-cavalry. It was often noted that if the 'sipahis,' and 'silladars' favoured one possible heir to the throne, the Janissaries would normally show their sympathy for another—his rival. When a detachment of the Janissary corps took the field under the command of some pasha designated as 'Seraskier,' or commander-in-chief, they were difficult troops for their general to handle, as they had always an under-feeling that the Sultan was their only proper leader. Soliman was the first sovereign who regularly sent out large corps of Janissaries without accompanying them himself. Under his wretched successors the exception became the rule—with evil consequences both to the prestige of the sultan, and to the discipline of the soldiery.

Quite early in his reign Soliman had permitted Janissaries to marry, and yet to remain in the ranks. When at its end he began to admit sons of Janissaries into the corps,[1] a marked deterioration in its fanatical pugnacity was noted. These boys had been brought up in family life, and not in the austere and harsh training to which the tribute-children had been subjected. They were, it is said, much less tough and reckless than the old recruits, after they had become a hereditary caste, and not an assembly of disciplined fanatics knowing no relative save their master the Sultan. Their class-consciousness increased as they became a family group, if their military excellence waned. The old Janissary body had grown unruly sometimes, when a weak ruler like Bajazet II was on the throne. By the end of Soliman's time, as he grew old, the unruliness had increased. His three weak successors found themselves faced by a praetorian guard, set on getting donatives on all possible occasions. Only two generations later the Janissaries were found deposing Sultans, because a new Padishah meant a new distribution of sequins on a lavish scale. In the golden year November 1517–November 1618 there were *three* donatives,

[1] Some say that it was only his son Selim who yielded to the petition at his accession.

produced by the murder of Sultan Ahmed, the proclamation and deposition of the idiot Mustafa I, and the final installation of Osman III.

Soliman's son and successor, Selim, was the first Ottoman ruler who never took the field at the head of his official bodyguard, but lurked in the Palace. His heir, Murad III, was even more contemptible. Mohammed III did once appear at the head of his army, which won the battle of Kerestes in 1596, but his personal conduct had not been inspiring. A theoretical commander-in-chief who never went out to war seemed to his Janissaries an absurd contradiction in terms, and they contemned and criticized mere Pashas put in temporary command. Were they not, in their own eyes, a royal bodyguard which ought to have the sovereign at its head, not one of his servants?

It may be added that the unique value which the Janissaries had possessed in the fifteenth century, as being the only trained and disciplined infantry in the world had become relatively less by the end of Soliman's reign. There was still nothing to oppose to them on the Persian front, but on the Western front they had now to face not tribal or feudal levies, like those of the old Christian states of the Balkan Peninsula or of Hungary, but veteran infantry, highly organized and confident, trained in the Hapsburg service in the wars of Italy. The pikes and arquebuses of an old Austrian or Spanish regiment were a formidable combination, and formed corps able to manœuvre with accuracy. The Janissary battalions never had anything like Western drill—and continued to work with arquebus and sabre, not with arquebus and pike. They depended on close order and fire properly controlled—which was very effective against Persian or Hungarian cavalry charges, but was less certain against a combined regiment of pikes and arquebuses, which could answer fire, and also charge in with the long weapon. But as in the usual Turkish array the Janissaries were massed in the reserve, along with the artillery and the Sultan's horseguard, direct clash with Western regular infantry was not the normal opening of a pitched battle—of which there were singularly few in these wars. On the other hand, in siege operations Janissary companies were always put in as the core of a storming force, and it is in assaults that we most frequently hear of them.

Fighting, after the awful example of Mohacs, was more often defensive than offensive on the side of the Christian armies. And when it was offensive it almost invariably took the form of sieges. On the two or three occasions on which Austrian armies entered Hungary in force they seldom met the Turks in the open field, but usually got entangled in siege operations, complicated by the appearance of armies of relief, with which partial engagements took place—generally with unfavourable results—as at Katzianer's disaster before Essek in 1537, Ferdinand's own operations before Buda in 1532 and again in 1540, and Joachim of Brandenburg's invasion of 1542, which failed before Pesth. In none of these campaigns was there a general engagement—several times there were running fights in which the retreating Austrian army got badly mauled by pursuing cavalry.

But the same phenomenon appears when we observe Soliman's offensive operations: he sometimes complained—as in 1532 after Güns—that he had sought the enemy but could not find him. Yet he never seems to have made up his mind to seek a decisive action at all costs. He might have made a fair bid for it if he had marched on Vienna in 1533 or 1540 or 1543, probably also in his last campaign in 1566. But on each occasion he lapsed into siege operations, and made no direct stroke at the point where the enemy was known to have an army mustered. This was specially obvious in 1532, when he deliberately swerved sideways against the insignificant town of Güns, and started homeward when he failed to take it. Though he knew that Ferdinand had a large force concentrated at Vienna in 1540, when he contented himself with raising the siege of Buda, and again in 1543 when he had taken Gran and Stühlweissenburg, he turned back in September in each case.

An army whose main force consists of masses of cavalry cannot live for more than a short time in a hostile and devastated countryside. It must keep on the move into fresh districts, or it must starve. When the Sultan settled down to a long siege, as at Vienna or Güns in 1529 and 1532, he had no system of transport supply behind him, and raids into outlying regions, for the purpose of gathering food, always meant the destruction in detail of many of the foraging parties. In 1532 nearly the whole of the detached cavalry was entrapped and exterminated in the woods and hills of the Wienerwald. The same thing happened sometimes on the Persian front, where rearguards or detachments were cut up in difficult country, if they went too far from the main army.

The plain of Hungary had been won by the single battle of Mohacs because it was a plain, with very few fortified places. But when the Turks had to operate beyond the Hungarian plain, and ran against the Styrian Alps or the Little Carpathians, in a border moreover full of fortified places, the matter ceased to be simple. More especially was it complicated when, as the years went on, the border fortresses of the Hapsburgs had been in many cases rebuilt according to modern systems, and were held by regular troops, not by local levies. Long sieges were the ruin of armies of which the best fighting force consisted of cavalry, with only a comparatively small nucleus of trustworthy infantry composed of the Janissary corps. The Turkish irregular infantry, the 'azabs' and volunteers, seldom carried a breach—the precious Janissary companies had to be put in, and were gradually thinned down, even when they were successful in the end. Dismounted feudal horse were often used in a storm, but disliked foot-service to which they had not been trained, and do not seem to have been very effective—though they were courageous enough when fighting in their own style. It will be remembered that the army which failed before the walls of Malta in 1565 was very largely composed of Asiatic 'timariots,' but the most desperate storming was done by Janissaries and Algerines. The irregular infantry got killed off by hundreds in trench-building, or in filling up the ditches of hostile forts, but seldom achieved a success.

As to the masses of feudal horse, which formed the best part of an Ottoman army, they were formidable for their courage and their fine horsemanship, but they had to operate on favourable ground if they were to be used to good effect, though they were nearly always opposed to an enemy whose cavalry was inferior in numbers. They could not stand before fully equipped men-at-arms in a frontal clash; but they were capable of hanging together, and gradually wearing down an enemy whose first impetus was spent, and who had lost his order. Wherefore, says the acute author of the *De Turcica Militia*, our Christian men-at-arms may be safely used against the Turkish horse if only they take care not to go far from our infantry, and so find themselves surrounded and scattered. 'Even our light horse no less than the heavy can charge with success, if they know how to get back in order to our infantry. The Turkish foot-soldiers, however numerous, can never be equal to the combination of German landsknechts and Spanish and Italian arquebusiers. If their Janissaries, the core and strength of their army, are beat, the Christians may get without difficulty their guns, their camp, and all its plunder. For though vast quantities of their horse may escape, it does not much matter, since they will not stand if the Janissaries are once routed.'

Alas! the old fault of rash pursuit which had ruined so many arrays of Christian men-at-arms at Mansourah, Nicopolis,[1] and other ancient fights, still prevailed in the sixteenth century on too many occasions. 'The Germans were often put to the worse, who mounted on their great heavy horses fit only for set battle could not so easily charge the enemy and pursue him in his flight as could the Turks—with their nimble and light horses, so well accustomed to the manner of flying fight—wheel about to frustrate the first charge of the heavy horsemen, and by and by come upon them with fresh charges, and so oft retire and come again, until they had either wearied or overthrown them. But Hungarians, acquainted with that manner of fight as well as they, and better armed, did easily encounter the Turks and foil them, even though they might be in number more.' [2]

The Hungarian light horse, indeed, ought to have been the salvation of Imperialist armies, since they could fight Turks or Tartars with their own tactics and equal courage. But it was the curse of Hungary that, ever since the unhappy election of John Zapolya to the kingship in 1526, the military force of the realm was divided into two halves, and the particularist party was to be found fighting in the Turkish ranks, not on the side of Christendom. There was no union of the two factions till Sigismund Bathori did homage to the Emperor, and disowned the vassalage to the Sultan in 1595. At the battle of Kerestes in 1596, every Hungarian in the field was fighting on the same side—a new phenomenon. And for this reason the army of the Archduke Maximilian had a larger proportion of cavalry to infantry than had been seen in a Christian host for the seventy years that had passed since Mohacs. Unfortunately on that unhappy field Hungarian indiscipline undid the possibilities of Hungarian union, and Bathori's Transylvanians vied with Palffy's western Hussars

[1] See *Art of War in the Middle Ages*, i. p. 344.

[2] All this apropos of the fighting round Buda, when besieged by the army of Ferdinand in 1540.

in breaking their ranks, and racing for the plunder, before the battle was really won. Disgusted at the failure of the Prince's submission to the Austrian, his followers disowned him, elected new 'voivodes,' and relapsed into the vassalage to the Sultan. Hungary was not to be united till the end of the seventeenth century, and then by force, not by the goodwill of the particularists of Transylvania.

An Imperialist army, however, in the long years of struggle with the Turk between 1526 and the end of the century, did always comprise a certain not inconsiderable contingent of Hungarian light horse, and found it most useful, both for the covering of flanks in open fights, and for cutting off Turkish raiders when the enemy dispersed to plunder.

The ideal disposition of a Hapsburg army, intending to face a Turkish Sultan in pitched battle, chances to have been preserved for the campaign of 1532, when Charles V and his brother Ferdinand were, for once, in company, with a very large force, such as had never before been seen on the Danube, to resist the expected appearances of Soliman for a second attack on Vienna. The Sultan—as we have seen—did not challenge them for a general action, but swerved off to besiege Güns. 'The Christian army lay in a great field before Vienna in this order. In the centre were three very great squadrons of pikemen, each a great way distant from the other, but with a like and equal front, so that all the horsemen divided into two parts, might well be ordered in the great spaces between the three squadrons, for it was not considered convenient to oppose so small a force of horsemen in the open field without the footmen, against so vast numbers of the Turk's horsemen. The right wing of the horse was led by the Emperor himself, and the left by King Ferdinand. Before and behind the squadrons of pikes, saving in the spaces left open for the horsemen, were placed 20,000 nimble arquebusiers, five in a rank, so that when the first rank had discharged, the second, and so the rest, orderly coming on, might deliver their bullets against the enemy. And if they found themselves opposed, they might easily retire among the pikes, standing so close at hand. In front of the arquebusiers were planted the great ordnance, of which the Emperor had such good store that he could therewith have compassed his whole army as with a trench. Only the Hungarians, men well acquainted with the manner of the Turkish fight, lay in the open field in two great wings under the leading of their two valiant captains, Valentine [Turacz] and Paul [Bakicz]. Many noble gentlemen wept for joy on seeing such a goodly army; but Soliman, certainly advised of the Emperor's strength, got him over the river of Mur, and passed away.'

It is characteristic of the inconsequent strategy of these wars that this great combined army of Charles and Ferdinand never took the field for an offensive campaign, and broke up in the autumn—not without some tiresome mutinies. The Sultan did not move against Hungary in the succeeding year, and concluded a stop-gap truce with King Ferdinand, because he was on the eve of his great Persian War. And so a crisis which might have been solved by a great pitched battle, if either party had pressed for one, passed away without a solution. Though the conditions and the character of the armies were so different from those which prevailed in the West, during the later campaigns of Charles V against the French, the results were much the same. Neither party cared to risk great pitched battles—Ceresole (1544) was the only exception in the middle decades of the century. St. Quentin and Gravelines were not occasions on which both parties were wishing for a general action.

There can be no doubt that the quality of the Turkish armies was steadily going down not only after Soliman's death, but even before. How the Janissaries came to deteriorate in morale we have already seen. But an even greater decline was to be noted in the efficiency of the Timariot feudal horse, the largest element in the Turkish military power. This was directly due to the worthlessness of Soliman's successors, who no longer kept a tight control over the army, because they never saw it, and left everything to their ministers. Originally the fief always went to an approved warrior, who had won it as a reward for service in the field, and served in person. But under corrupt viziers fiefs might be bought, by unavowable services or hard cash, by almost any one. Court officials, clerks, secretaries, even the eunuchs of the harem got them. There was still the liability to produce a fighting man for each timar, but the sort of deputies, cheap adventurers, who were sent in by civilian masters, were not efficient soldiers after the type of the old feudal Timariot. Often they were not real soldiers at all, but only hired men on horses, more careful of their lives than of their honour. Under Murad III and Mohammed III it was said that the ladies of the harem dabbled in fief-mongering, especially the old Sultana Validé, who had a great influence over viziers, and foisted on them those who had got her favour by cash or intrigue. There was a slang name for timars which had come under harem influence—they were said to have 'fallen into the work-basket.' By 1600 the feudal levy was of very mixed quality, genuine warriors of the old stock serving along with some very worthless hirelings. A state document of the early seventeenth century complains that the Timariot horse was going down in numbers as well as in military value, because court favourites who had accumulated many fiefs, cheated the state, by producing less horsemen than their acreage of granted land should have furnished. This they could do because their influence was so great that no vizier would dare to call them to account for the shortage. A Grand Vizier who once tried to hold at Adrianople a review of Roumeliot fief-holders, with obligation of personal presence, found himself defeated, because he recognized, among the horsemen who turned up, the domestic servants and hangers-on of other great dignitaries, whom he dared not offend. There was also something like the abuse of the 'passe-volants' who were found in Western armies—useful transitory persons, lent by one captain to another for a review or muster-calling. One man might appear on different occasions as representing several capacities. Military organization was bound to decay when Constantinople had become the centre of a corrupt bureaucracy of pashas, always ousting each other by intrigue or murder from the control of a worthless sovereign. When a Janissary mutiny swept away both Sultan and pashas, it only let loose a new set of leeches upon the empire.

While Soliman was still reigning and Turkish military efficiency was in its prime, an acute observer made a catalogue of the points of superiority of a Turkish over a Christian army. They are worth noting. The first was the Sultan's ruthless power of dealing with his officers, which produced an instant obedience rare in Western hosts—summary strangulation or beheading on the field of great officials was accepted as normal; they were all the Sultan's slaves, however high their position. Western kings could not slay dukes or counts without any semblance of a trial in such fashion. Obedience among the Turks was secured by imminent fear of death. The second point was that Moslems were more fanatical and fatalistic than Christians—God, they thought, settled a man's hour of death, which he could not avoid, and if he died fighting the infidel he was sure of heaven. The third point was that the Turk was more abstemious than the ordinary Western soldier —he could live on less and never got drunk. He was better at starving, and never indulged in alcoholic brawls in the camp. In short, his discipline was better than that of landsknechts, largely because indiscipline meant instant execution. Mutinies did sometimes occur, but they were not, as in Western armies, drunken riots, or strikes because pay was overdue. Fourthly, the Turk had an immense superiority in light horse work—his riders overran a whole countryside, and had carried scouting to perfection—his camp was never surprised by the negligence of his outlying picquets, as Western camps often were. Fifthly, his army marched in better order, without the straggling and trailing to the rear too well known in a Christian host.

Nevertheless, says the author of the *De Turcica Militia*, the Turk ought to be beaten, because his infantry, except the few battalions of Janissaries, is worthless, and cannot stand against the properly marshalled regiment of pikemen and arquebusiers. And his cavalry, all light horse, cannot hold against a charge of men-at-arms or reiters, if only the latter will remember to hold together, not to scatter, and not to go far from their infantry. A Christian general, he holds, ought to seek for a pitched battle, because his troops are individually superior. 'Ne bellum protrahetur, et facultas illico confligendi nobis detur.' The Turk has no training for orderly manœuvres, and is still using in large numbers the out-of-date bow, which cannot compete with the arquebus. Destroy his Janissaries, and the cavalry will go off the field. 'Si statim confligatur verisimile est peditatum nostrum nobis victoriam pariturum.'

The weak and strong points of this screed of criticism are fairly obvious. That Turkish efficiency went down when there was no masterful Sultan with the army, to put to death slack or disobedient officers without a moment's warning, is clear. After Soliman the pashas in a field army quarrelled with each other, and lapsed into factious argument, because there was no supreme control. And the fanatical fury, which had struck the observation of Western onlookers, was much tamed down, when the proportion of hirelings and substitutes in the feudal cavalry had grown large. The discipline which had been so stern under a Sultan's eye deteriorated, when the Sultan was in his seraglio at Constantinople, and his armies were under upstart pashas, who might lose their place by some intrigue of the palace. Indeed the Janissaries became the most undisciplined force in Europe—worse than the Spanish strikers in the Netherlands, for these last had not the opportunity of murdering their sovereigns.

This being so, one asks what was the cause that the Peace of Sitva Torok left the Turkish frontiers in Hungary much on the same line that the earlier truces of 1533 or 1575 had indicated. The main answer must be that the interests of the two branches of the House of Hapsburg being now separated, and the attention of Philip II being drawn off to his Dutch, French, and English wars, the defence of the frontier of Christendom on the Danube became a purely Austrian affair. The second answer is that religious scission made it impossible for the Emperor to count on the whole of the strength of Germany. The resources of Maximilian II and Rudolf II just sufficed to hold off the weakening Turkish attack, and no more. If a second Soliman the Magnificent had arisen, he would have found no Charles V to oppose him on every front by land and sea. The Emperor Rudolf, whose long reign covered the years from 1576 to 1611, though he was neither vicious, cruel, nor uncultured, was a most unsatisfactory warden of the marches against the infidel. He was, as a Catholic of extreme devotion, more interested in checking the advances of Protestantism in his own dominions, and if possible in the whole Empire, than in Turkish wars. He never took the field himself, and was generally at odds with his brothers Matthias and Maximilian, the fighting members of the family. Next to the Jesuits his favourites were the professors of occult arts. He spent his days and nights in casting horoscopes, or seeking for the 'Philosopher's Stone' along with charlatans from every country of Europe—even the English alchemist and visionary Dr. Dee—a most ineffective person—succeeded in engrossing his attention for some time. While he was wrangling with the Protestants of Bohemia, and toying with magic mirrors and alembics, the affairs of Hungary interested him little. At last he became so difficult of access, and so subject to fits of melancholia, that his brothers gradually deprived him of the administration of all his estates, and ended by deposing him (1611).

Murad III and Mohammed III were fortunate as having for their contemporary an Emperor who was, if not so contemptible as themselves, at least as unsatisfactory as a practical administrator in time of war. The most unfortunate thing for Christendom was that Rudolf's militant Catholic policy estranged the states of Germany, and led the Diet to grant him unfrequent and inadequate supplies of men and money, even when the situation on the Turkish frontier looked threatening. Practically the struggle was sustained only by Austrian resources, and he was not blessed with any great soldier to lead his armies. Of his brothers, Matthias made a show of much incompetence in the campaign of 1594, and Maximilian was partly, if not wholly, responsible for the awful disaster of Kerestes in 1596. Of the long list of generals who served Rudolf, including foreigners like Mansfeldt and Mercœur, none has left a reputation behind him. The annals of the wars of 1590–1606 are dreary reading, and their final result was to leave frontiers almost as they had been at the first truce between Soliman and Ferdinand I in 1533. Hungary was still divided, and the only sign of a changed balance of power was that the Sultan, for the first time, recognized the Emperor as an equal. Yet the Turkish offensive had certainly come to an end, and the form of words employed at Sitva Torok recognized the fact in the clearest fashion. The aggressive fury of the Muslim had passed away, and the peril was over, though this was hardly yet recognized in Christendom. The conquest of Cyprus was not the last territorial gain that the Ottoman power was to make, but it was the last that had a serious threat behind it. The Thirty Years War went on uncomplicated by any sudden interruption from the East, and when the last spurt of Ottoman energy was seen in the later seventeenth century, the siege of Vienna in 1683 had no such ominous meaning for Europe as the siege of the same imperial city in 1529. Before another generation had passed away the Turks had been driven beyond the Danube, and all the conquests of Soliman the Magnificent had been lost. The memory of old terrors had passed away, and the Sultan was already beginning to be 'the Sick Man of Europe.'

CHRONOLOGICAL LIST OF BATTLES, COMBATS, AND SIEGES DEALT WITH IN THIS VOLUME.

N.B.—Battle plans, or maps of the country covered by a campaign, or siege-plans, are to be found for the names marked with a star (*).

INDEX

PRINTED BY MORRISON AND GIBB LTD., LONDON AND EDINBURGH

For Product Safety Concerns and Information please contact our EU representative GPSR@taylorandfrancis.com
Taylor & Francis Verlag GmbH, Kaufingerstraße 24, 80331 München, Germany

www.ingramcontent.com/pod-product-compliance
Lightning Source LLC
LaVergne TN
LVHW021000300926
849160LV00007B/209
* 9 7 8 1 1 3 8 5 6 3 4 7 6 *